PRINCIPLES OF HUMAN PHYSIOLOGY

PRINCIPLES OF

Gerard J.

Ronald L.

Nicholas P.

1817

HUMAN PHYSIOLOGY

TORTORA
Bergen Community College

EVANS
California State University, Fresno

ANAGNOSTAKOS
Bergen Community College

HARPER & ROW, PUBLISHERS, New York

Cambridge, Philadelphia, San Francisco,
London, Mexico City, São Paulo, Sydney

Sponsoring Editors: Steven P. Heckel/Claudia Wilson
Project Editor: Holly Detgen
Designer: T. R. Funderburk
Production Manager: Willie Lane
Photo Researcher: Myra Schachne
Compositor: Ruttle, Shaw & Wetherill, Inc.
Printer and Binder: The Murray Printing Company
Art Studio: Vantage Art, Inc.
Cover Art: George Kelvin

PRINCIPLES OF HUMAN PHYSIOLOGY

Library of Congress Cataloging in Publication Data

Tortora, Gerard J.
    Principles of human physiology.

    Includes index.
    1. Human physiology. I. Evans, Ronald L., 1932–
II. Anagnostakos, Nicholas Peter, 1924–
III. Title. [DNLM: 1. Physiology. QT 104 T712p]
QP34.5.T69   612   81-6365
ISBN 0-06-388776-2   AACR2

CONTENTS IN BRIEF

CONTENTS IN DETAIL

PREFACE

"Since I have my body with me wherever I go, I thought it would be appropriate to know something about it." This was one reason given by a student when asked why she was taking a class in human physiology. While there may be more profound and altruistic reasons for studying human physiology, there is perhaps none more obvious.

When students begin learning human physiology, they often feel overwhelmed with the volume of seemingly complicated information present in most textbooks on the subject. Indeed, the human body seems to be a very complex organism. It contains many interacting systems, each integrated with the others and regulated or affected by the others. Physiology is complex and complicated because our understanding is always limited or incomplete. In recent years, investigators have been exploring the human body at cellular and molecular levels. As the results of their investigations are made known, it often happens that the current complex theories and confusing and conflicting ideas are replaced with one simple easy-to-understand model of operation. These breakthroughs are always beautiful in their simplicity and when proposed, one intuitively wants to say, "Why of course, it's so simple and obvious. Why was that so difficult to figure out?"

The human body functions by employing simple control mechanisms to keep the trillions of human cells in harmony with each other. When Watson and Crick proposed the double helix model for DNA in 1953, it explained very beautifully and simply a mechanism for how the genetic machinery operates. The interdigitating filament model proposed by Huxley and Hanson in the early 1960s very simply explained the cellular mechanism of muscle contraction. Earl Sutherland in 1956 discovered the intracellular messenger, cyclic AMP, and in so doing eventually outlined the elegantly simple amplification cascade mechanism to explain how a hormone could regulate a cell's activity.

These examples also indicate the operational level presently used to explain physiological phenomena. Scientists have continually searched for answers to "How does the human body work?" The answers came first from the systemic levels, then the organ levels, then the tissue and cellular levels. At present, research explores the cellular and molecular components of the human body to understand normal and abnormal operations. This research is made possible primarily by the great technological developments of recent years. The electron microscope, ultracentrifuges, radioactive tracers, and human tissue culture techniques are relatively new tools. Of course, the list could be expanded greatly. Computers are used to simulate physiological situations and mathematics and physics are routinely utilized to explain physiological actions.

OBJECTIVES

It is our purpose in this textbook to provide students with a sound basis for understanding human physiology which can then serve as a foundation for future learning. The content of this book is as current as possible and it contains material that is generally accepted by most of the scientific community. In keeping with the concept of simplicity, all explanations, even though described at the molecular level, are expressed in as clear and simple language as possible.

Principles of Human Physiology is designed for use in introductory physiology courses and assumes no previous study of physiology. The text is geared to students in biological, medical, and health-oriented programs. Among those who might use this text are those students aiming for a career as a nurse, medical assistant, physician's assistant, medical laboratory technologist, radiologic technologist, inhalation therapist, mortician, or medical record keeper. But, because of the scope of the text, it is also useful for students in the biological sciences, premedical and predental programs, science technology, liberal arts, rehabilitation therapy, and physical education.

The principal objectives of *Principles of Human Physiology* are twofold. First, we have attempted to emphasize the key concepts needed for a basic understanding of the functions of the human body. In this respect, we have selected only material regarded as essential. Our second objective is to present the material at a reading level average students can handle. To support this objective, we have not avoided difficult concepts. Rather, we have attempted to develop step-by-step, easy-to-comprehend explanations of each important concept.

Two distinct themes dominate *Principles of Human Physiology*. One is homeostasis. Throughout the text, students are shown how dynamic counterbalancing forces interact to maintain homeostasis. The second theme is disorders, the effects of homeostatic imbalances. Accordingly, we present a large number of clinical topics and contrast them with specific normal processes. Both themes are presented within the context of a sufficient amount of anatomy. We feel that this anatomical presentation enables students to better understand normal physiology and disorders.

ORGANIZATION

The text is organized as follows. The first four chapters deal with a given introduction to physiology and a presentation of mathematical, physical, and chemical principles useful for an understanding of physiological concepts covered later in the text. The next six chapters deal with a detailed treatment of the concept of homeostasis and its operation at the cellular level of organization. The next six chapters deal with the importance of the nervous system in coordinating homeostasis through nerve impulses. The importance of the endocrine system in coordinating homeostasis through hormones and the relationship of hormones to reproduction are emphasized in the next four chapters. The final six chapters of the text consider body systems that maintain homeostasis on a day-to-day basis.

FEATURES

Principles of Human Physiology contains a number of special features. Among these are the following:

1. Student Objectives. Each chapter opens with a comprehensive list of student objectives that describe a knowledge or skill students should acquire while studying the chapter.
2. Chapter Summary Outline. This summary, at the end of each chapter, provides a checklist of major topics students should learn. It consolidates the major points learned in the chapter.
3. Review Questions. These help students to synthesize the key points in the chapter and meet the objectives stated at the beginning of the chapter.
4. Exhibits. These summary tables organize data and deemphasize rote learning of concepts covered in the narrative.
5. Homeostatic Imbalances. These sections, grouped at the end of selected chapters, contain disorders that result from homeostatic imbalances. They are clinical applications that illustrate the results of disruptions in homeostasis.
6. High-Interest Topics. We have included many topics of contemporary interest such as cells and aging, cells and cancer, biofeedback, prenatal photography, choosing the sex of offspring, hypertension, hemodialysis, and the Heimlich maneuver.
7. Line Drawings. Illustrations are unusually large so that details are clearly discerned and labeled.
8. Photographs. The photographs amplify the narrative and line drawings. Numerous photographs, photomicrographs, and electron micrographs have been included.
9. Glossary of Terms. Definitions of many of the formal terms used in the text appear here, as a reference aid for students.

ACKNOWLEDGMENTS

Since the inception of this textbook, Harper & Row, Publishers, Inc., has made available to us a number of specialists in their respective disciplines to assist in the preparation of *Principles of Human Physiology*. Those to whom we wish to express our deepest gratitude are listed below.

For reviewing the outline and providing a comparison to other texts: John Buuck, Concordia College; Harry Reasor, Miami-Dade Junior College; and Ronald Reuss, State University College at Buffalo.

For reviewing the manuscript: Paul R. Barnes, San Francisco State University; Gordon Bradshaw, Phoenix College; Martin A. Kramen, University of Texas at San Antonio; Steven H. Mills, Central Missouri State University; Richard Nuccitelli, University of California, Davis; James A. Orr, University of Kansas; Ann Smith, Kansas State University; and Irwin Spear, University of Texas at Austin.

For reviewing the manuscript and providing chapter-by-chapter comments: Philip Dowling, Diablo Valley College; Richard W. Heninger, Brigham Young University; and Richard Tullis, California State University, Hayward.

As the rather lengthy list above indicates, the participation of numerous individuals of diverse talent and expertise is essential to the production of a textbook of this scope and complexity. We, therefore, would like to invite readers of this first edition of *Principles of Human Physiology* to send us their reactions and suggestions to help us in formulating plans for subsequent printings and editions.

GERARD J. TORTORA
Biology Department
Bergen Community College
400 Paramus Road
Paramus, New Jersey 07652

RONALD L. EVANS
Department of Biology
California State University, Fresno
Fresno, California 93740

Chapter 1

INTRODUCTION TO HUMAN PHYSIOLOGY

STUDENT OBJECTIVES

After reading this chapter, you should be able to

- Trace the origin of the human organism
- Explain why humans have been able to survive so long
- Discuss how human beings can control their internal environment
- Describe the basic concepts of the French physiologist Claude Bernard
- Define and explain the meaning of the word homeostasis as introduced by the American physiologist Walter Cannon
- Discuss extracellular fluid and its principal subcompartments
- List the prerequisites necessary for an organism to be in homeostasis
- Explain why homeostasis is a state that results in normal body activities and why the inability to achieve homeostasis leads to disorders
- Identify the effects of stress on homeostasis
- Define physiology and explain its various specialized branches
- Explain humanness by correlating the human body with the human mind
- Define anatomy and physiology
- Determine the relationship between structure and function
- Compare the levels of structural organization that make up the human body
- Define a cell, tissue, organ, system, and organism

THE HUMAN ORGANISM: A CASE STUDY IN ADAPTATION

Most recent fossil evidence indicates that Africa was the cradle of human origins and development. The earliest human record traces back to between 12.5 and 14 million years ago, and the genus *Homo* extends back to about 5 million years ago. The human organism is a unique creature, not only strong and durable but most adaptable. By the power of a superior intellect and questing spirit, humans have pushed themselves into every corner of the earth and beyond. Humans have adapted in extreme environments no matter how inhospitable. They eke out an existence as nomads in the hot, dry, barren deserts of northern Africa. They survive in the low-oxygen atmosphere of the Tibetan mountains and withstand the subfreezing Arctic snows. From their earliest origins, humans have been hunted and killed as they also have hunted and killed in order to adapt and survive. Over hundreds of millenia, humans have survived the onslaught of bacterial wars, viral epidemics, loneliness, fear, and every anxiety imaginable. Humans have survived when other creatures have failed because humans can imagine. They can invent, create, plan ahead, refine, and improve their status among other living things. They have created complex societies and shared labor and talent.

Fundamental to the survival of human beings has been their ability to control their *internal environment* (watery environment surrounding body cells), even when the external environment ranges to extremes. Of the 100 trillion cells that make up the human organism,

not one single cell, nor any process occurring within a cell, can exist by itself. Each cell and probably every molecule that constitutes the human body is located precisely where it functions best. The cells of the human body adapt and survive because they are interdependent and they cooperate to maintain their common environment. The watery environment surrounding the body cells, reminiscent of the ancestral oceans that gave birth to all life, is perpetuated and maintained by the selfsame cells that it nurtures. This internal environment is regulated, within narrow limits, with respect to temperature, composition, concentration, volume, and pH (acid-base balance).

Though these limits beyond which life could not survive were probably apparent to even ancient physicians, the formalization of the fundamental concept of the constancy of the internal environment was announced only over a century ago. In the summer of 1870, the French physiologist Claude Bernard (1813–1878) announced the idea that the living cells that make up an animal (or a plant) are not in contact with the external environment—the air, sea, or earth in which it lives—but rather with an internal environment. "The constancy of the internal environment is the condition that life should be free and independent." This was not meant to say that the animal is isolated from its surroundings or their influence. As Bernard put it, "The higher animal, far from being indifferent to the external world is, on the contrary, in a precise and knowing relationship with it, in such a way that its equilibrium results from a continuous and delicate compensation, established as by the most sensitive of balances." Bernard's statement can be expressed more simply by saying that higher animals possess control systems that adjust the interactions and exchanges with their surroundings. These control systems keep the physical state and chemical composition of the aqueous internal environment constant. Bernard placed before the scientific world a simple concept. This idea still guides medical thinking and scientific research in all considerations of health and disease.

HOMEOSTASIS: A BASIS FOR PHYSIOLOGY

Bernard's concept of the *milieu intérieur,* or internal environment, was championed and elaborated further by the American physiologist Walter Cannon. It was he who introduced the term *homeostasis*. The word comes from two Greek words: *homeo* meaning similar, like, same; and *stasis* meaning to stand or stay. The combined terms indicate "a standing the same," more freely expressed as keeping the status quo. **Homeostasis** is a condition in which an organism maintains a relatively constant internal environment. Cannon, in his research, repeatedly demonstrated many homeostatic processes that were continually monitoring and adjusting body functions to maintain the *milieu intérieur* within the prescribed limits suitable for cellular life. He published a book in 1932 entitled *The Wisdom of the Body*. Even the title emphasizes the fact that control of the internal environment is essential to the whole design and organization of an animal, ensuring that its component parts work in cooperation with one another.

In studying physiology it is necessary for us to discover which activities of the animal are controlled. This discovery requires examination of the *stimulus* (input) and *response* (output) of the control system under study; how the discrepancy between them is detected and corrected; and how accurately the system works. Such an examination is done chiefly by applying the methods of experimental physiology together with those of biophysics and biochemistry.

The body fluid that bathes the body cells is termed the *extracellular fluid.* This fluid compartment can be conveniently divided into two subcompartments: the interstitial fluid and the plasma. *Interstitial* comes from the Latin verb *intersistere,* which means to put between. *Interstitial fluid* is the fluid flowing in the submicroscopic spaces between the cells. *Plasma* is that portion of the extracellular fluid that is confined to the interior of the heart and blood vessels.

Among the substances in extracellular fluid are gases, nutrients, and electrically charged chemical particles called ions—all needed for the maintenance

of life. Extracellular fluid circulates by flowing through the blood vessels, leaking out through the microscopically small spaces between the cells of the vessel walls into the spaces between the tissue cells, and flowing back into blood and lymphatic vessels. Thus, it is in constant flow throughout the body. Essentially, all body cells are surrounded by the same fluid environment, and for this reason, extracellular fluid is often called the internal environment of the body.

An organism is said to be in homeostasis when its internal environment (1) contains the appropriate physiological ranges of gases, nutrients, and ions; (2) has an optimal temperature; and (3) has an optimal pressure for the health of all its cells. Failure to maintain homeostasis results in illness. If the body fluids are not eventually brought back into balance, death occurs.

Homeostasis in all organisms is continually disturbed by *stress,* which is any stimulus that creates an imbalance in the internal environment. The stress may come from the external environment in many forms such as heat, cold, loud noises, or lack of oxygen. The stress may originate within the body as, for example, high blood pressure, pain, tumors, or even unpleasant thoughts. Most muscular exercise creates temporary chemical and temperature imbalances in the internal environment. Occasionally, though, the body is subjected to extreme stresses such as poisoning, overexposure, severe infection, and surgical operations.

Fortunately, the body has many regulating devices that oppose the forces of stress and bring the internal environment back into balance. In fact, high resistance to stress is a striking feature of most organisms. Consider the examples of the people who live in barren deserts where the daytime temperatures easily reach 120°F and of those who dwell in subzero weather. Both maintain an internal body temperature of 98.6°F. Mountain climbers exercise strenuously at high altitudes where oxygen content of the air is low. Once they adjust to the new altitude, they do not suffer from oxygen shortage. The extremes in temperature and in oxygen content of the air are stresses from the external environment, yet the body is able to compensate and remain in homeostasis. The muscular exercise performed by the mountain climber is an example of an internal stress. Walter Cannon noted, for example, that the heat produced by the muscles during strenuous exercise would curdle the body's proteins if the body did not have some way of dissipating heat quickly by sweating. In addition to heat, muscles that are being exercised also produce a great deal of lactic acid. If the body did not have a homeostatic mechanism for reducing the amount of lactic acid, the extracellular fluid would become so acidic that it would destroy the cells and eventually the entire body.

PHYSIOLOGY

Physiology is the science that describes how organisms function and survive in environments that are continually changing. Over the years, scientists have been able to disclose an increasing number of the adaptations that various organisms have developed for surviving these changes. Even the smallest unicellular organism possesses simple but reliable regulating mechanisms that enable it to survive—sometimes only long enough to reproduce. The science of physiology is subdivided into several specialized branches. *Comparative physiology* is the subdivision of physiology that examines and compares the ways in which widely diverse organisms carry out certain similar life-requiring functions. How is energy extracted from the environment by simple microorganisms as compared to the much more complex organisms such as the gigantic redwood trees or huge whales? How are the waste products produced by metabolic processes of such widely diverse organisms eliminated? Each species has developed different ways of solving these and many other problems. Physiology uses the laws and rules, as well as the instruments and techniques, of the more exact and basic sciences of chemistry and physics.

As the study of physiology became more specialized, it narrowed. Scientists could devote their time and interests to only certain types of organisms such as plants or animals or microbes. The fields of *plant physiology, animal physiology,* and *microbial physiology* thus developed. Even these areas are so large that they are subdivided and redivided. Plant physiology can include the physiology of vascular and nonvascular plants or a very specific field such as physiology of photosynthesis. The study of animal physiology is just as diverse and includes the physiology of different groups of animals. Insect physiology, fish physiology, bird physiology, reptile and amphibian physiology, and mammalian physiology are only a few. In the area of microbial physiology, there are subdisciplines that study the physiology of algae, parasites, bacteria, and viruses, even though viruses are probably not considered alive according to the strictest definitions.

Cellular physiology demonstrates that at this level all organisms are more alike than they are different. It explores those basic characteristics common to most living organisms and attempts to explain function in molecular terms.

Mammalian physiology is the field of study that has taught us much about how our own body functions. The study of how organ systems and tissues cooperate to regulate the internal environment of the mammal helps us to understand how these same functions operate in the human body. Since humans are mammals, it would seem likely that we function alike in most respects. Indeed, it becomes very difficult to describe *human physiology* as distinct from mammalian physiology. In many courses of mammalian physiology, the mammal most frequently described is, in fact, the human. *Medical physiology* strives, by its very nature, to center only on the human body in order to correct those internal regulatory mechanisms when they malfunction. It is self-centered, self-serving study. At what point, then, does mammalian physiology become only human physiology and not intrude into medical physiology? Perhaps the key lies in the word *human*. Without attempting to answer the unanswerable question of what it is to be human, we can look at some related fascinating disciplines and glimpse a few ideas.

Being human is more than having a body that scientists can identify and describe in physical and chemical terms. For example, we often use words such as *inhuman* or *inhumane* and *humanity*. It is an attempt to add some unique quality to our beings that no other living mammal possesses. This notion of humanness is expanded by philosophers, sociologists, psychologists, ethnologists, theologians, and many others. Some scientists search for the key to humanness by tracing back through evolutionary history hoping that the past will help us to understand our present humanness. Other scientists search for clues by noting and recording human behavioral responses to changes in the environment.

It is clear that humanness cannot be confined simply to the description of how the complex internal machinery of our bodies operates. The nature or unique quality of being human is not explained by, nor does it lie within the purview of any one discipline. Rather, it is interrelated with many. All the forces of history, genetics, and environment are part of the concept of humanity. Even more amazing is the study of individuals who are forged and tempered by their own history and genetics and then are thrust into coping with the changing environment.

The most fruitful search for a unique human quality explores the major attribute that sets humans apart from other animals: the human mind. The human mind creates and employs symbols to convey generalities and abstract ideas. The hallmark of humanness somehow resides in a complex interplay between the human body and the creations of its mind. This interdependence embodies the notion that anything that impinges upon a human being affects both body and mind simultaneously. For example, all nonhuman organisms are subject to the same environmental changes of light, temperature fluctuations, wind forces, and dry and wet weather as is the human body. The environmental changes affect not only the human body but also have an indirect effect on the human mind. Therefore, the physical response to environmental changes becomes tempered by that indirect effect the changes in environment also have on the mind. Some mind symbol is generated that captures the event in a way that later enables the human being to respond just to the symbol. Thus, later responses to actual changes in the environment become very personalized as they reflect the sum total of all the past experiences.

The further study of the human mind can be pursued in many other interesting courses and the student is encouraged to do so. In this book we shall examine the human body and the internal homeostatic devices that enable the total body to survive.

THE HUMAN BODY

The study of the human body involves many branches of science, each of which contributes to a more comprehensive understanding of the parts of the body and how they work. Two branches that contribute directly to this understanding are anatomy and physiology. **Anatomy** (or **morphology**) refers to the study of *structure* and the relationships among structures. Anatomy is a very broad science and is often a field of study all by itself. In this book only that anatomy required for supporting the description of function is included. Whereas anatomy and its branches deal with structures of the body, **physiology** deals with the *functions* of the body parts. Each structure of the body is custom-modeled to carry out a particular set of functions. For instance, the interior of the nose is lined with hairs that allow the nose to perform the function of filtering dust from inhaled air. Bones are able to function as rigid supports for the body because they are constructed of hard minerals. In a sense, then, the structure of a part determines what functions it will perform. In turn, body functions often influence the size, shape, and health of the structures. For example, glands perform the function of manufacturing chemicals. Some of these chemicals stimulate bones to build up minerals so they become hard and strong. Other chemicals from other glands cause the bones to give up some of their minerals so they do not become too thick or too heavy. Physiology and anatomy are closely interwoven disciplines, and physiology is best studied with a prior knowledge of anatomy.

Levels of Structural Organization

The human body is a complex organism consisting of levels of structural organization that are associated with each other in several ways. The lowest level of structural organization, the *chemical level*, includes all chemical substances essential for maintaining life. All these chemicals are made up of atoms joined together in various ways (Figure 1-1). The chemicals, in turn, are put together to form the next higher level of organization, the *cellular level*. **Cells** are the basic structural and functional units of the organism. Among the many kinds of cells found in your body are muscle, nerve, and blood cells. Figure 1-1 shows several isolated cells from the lining of the stomach. Each of these cells has a different structure, and each performs a different job.

The next higher level of structural organization is the *tissue level*. **Tissues** are groups of cells and their intercellular material that perform specific functions. For example, when the isolated cells shown in Figure 1-1 are joined together, they form a tissue called epithelium, which lines the stomach. Each kind of cell in the tissue has a specific function. Mucous cells produce mucin, a secretion that lubricates food as it passes through the stomach. Parietal cells produce hydrochloric acid in the stomach, and chief cells produce enzymes needed to digest proteins. Other examples of tissues in your body are muscle tissue, bone tissue, and nervous tissue.

In many places in the body, different kinds of tissues are joined together to form an even higher level of organization, the *organ level*. **Organs** consist of two or more tissues that perform a particular function. Examples of organs are the heart, liver, lungs, brain, and stomach. Referring to Figure 1-1, you will see that the stomach is an organ since it consists of two or more kinds of tissues. Three of the tissues that make up the stomach are shown in the figure. The serous tissue layer (also called the serosa) protects the stomach and reduces friction when the stomach moves and rubs against other organs. The muscle tissue layers of the stomach contract to mix food and pass the food on to the next digestive organ. The epithelial tissue layer produces mucus, acid, and enzymes.

The next higher level of structural organization in the body is the *system level*. A **system** consists of an association of organs that have a common function. The digestive system, which functions in the breakdown of food, is composed of the mouth, salivary glands (saliva-producing glands), pharynx (throat),

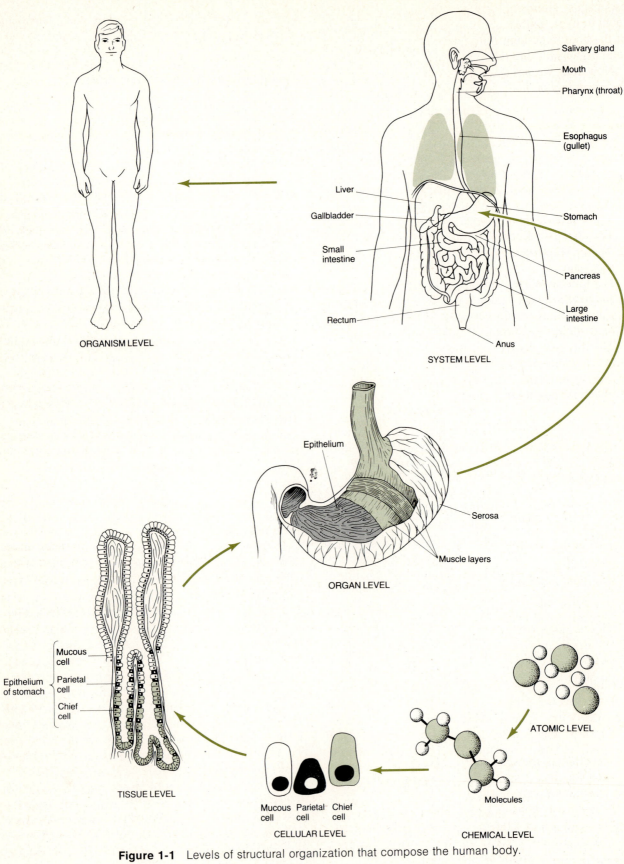

Figure 1-1 Levels of structural organization that compose the human body.

esophagus, stomach, small intestine, large intestine, rectum, liver, gallbladder, and pancreas. All the parts of the body functioning with each other constitute the total **organism** — one living individual.

Now that we have surveyed the general nature of the meaning of homeostasis, its importance to the human body, and the levels of structural organization that compose the human body, we can begin to look at some of the important physical and chemical principles that will help you to understand how your body functions and how it maintains homeostasis.

CHAPTER SUMMARY OUTLINE

THE HUMAN ORGANISM

1. The earliest human record traces back to between 12.5 and 14 million years ago.
2. The genus *Homo* extends back to about 5 million years ago.
3. Humans have been able to survive because of their superior intellect, ability to adapt, and imagination.
4. Fundamental to the survival of human beings is their ability to control their internal environment.
5. The French physiologist Claude Bernard was the first to announce the idea that living cells are not in contact with the external environment, but rather with an internal environment.

HOMEOSTASIS

1. The American physiologist Walter Cannon introduced the term homeostasis..
2. Homeostasis is a condition in which an organism maintains a relatively constant internal environment.
3. Extracellular fluid is the body fluid that bathes the body cells.
4. This fluid is divided into two subcompartments: interstitial fluid and plasma.
5. Stress is any stimulus that creates an imbalance in the internal environment.
6. If a stress acts on the body, homeostatic mechanisms attempt to counteract the effects of the stress and bring the system back to normal.

PHYSIOLOGY

1. Physiology is the science that describes how organisms function and survive in environments that are continually changing.
2. Comparative physiology is one subdivision of physiology which examines and compares the ways in which widely diverse organisms carry out certain similar life-requiring functions.

3. As physiology became more specialized scientists could devote their time and interests to only certain types of organisms such as plants, animals, or microbes. The fields of plant physiology, animal physiology, and microbial physiology thus developed.
4. Cellular physiology explores those basic characteristics common to most living organisms (cells), and attempts to explain function in molecular terms.
5. Mammalian physiology is the field of study that has taught us much about how our own body functions.
6. Medical physiology centers only on the human body in order to correct those internal regulatory mechanisms when they malfunction.
7. The hallmark of humanity resides in a complex interplay between the human body and the creations of its mind.

THE HUMAN BODY

1. The study of the human body involves many branches of science, each of which contributes to a better understanding of the parts of the body and how they work.
2. Anatomy deals with the study of structure while physiology deals with the functions of the body parts.

Levels of Structural Organization

1. The human body consists of levels of structural organization from the chemical level to the organismic level.
2. The chemical level is represented by all the atoms and molecules in the body. The cellular level consists of cells. The tissue level is represented by tissues. The organ level consists of body organs, and the system level is represented by organs that work together to perform a general function.
3. The human organism is a collection of structurally and functionally integrated systems.

REVIEW QUESTIONS

1. How far back can you trace the human? The genus *Homo?*
2. List as many factors as you can that explain the long survival of human beings.
3. What is the relationship of this long survival with human beings' ability to control their internal environment?
4. Describe Claude Bernard's concept of internal environment.
5. Who was responsible for the word homeostasis? Define homeostasis.
6. What is your interpretation of extracellular fluid? What are its two subcompartments?
7. Why is extracellular fluid called the internal environment of the body?
8. What is stress? How is stress related to homeostasis? Give several examples.
9. How is homeostasis related to normal and abnormal conditions in the body?
10. Substantiate this statement: "Homeostasis is a cooperative effort of all body parts."
11. Define physiology. Differentiate between five different types of physiology, explaining each of them.
12. Explain fully the statement: "Being human is more than having a body that scientists can identify and describe in physical and chemical terms."
13. Define anatomy. Give two specific examples that explain the statement: "The structure of a body part determines what function it will perform."
14. Construct a diagram to illustrate the levels of structural organization that characterize the body. Be sure to define each level.

Chapter 2

SCIENTIFIC MEASUREMENTS AND THEIR APPLICATIONS

STUDENT OBJECTIVES

After reading this chapter, you should be able to

- Define measurement
- Distinguish between a fundamental unit of measurement and a derived unit of measurement
- Describe the units of measurement used in the United States system
- Describe the units of measurement used in the metric system
- Compare the units of the United States system with those of the metric system
- Convert decimals and large numbers into exponential form
- Discuss the important features of the apothecary system of measurement
- Define matter and energy
- Contrast the kinds and forms of energy
- Define density and describe how it is measured
- Define specific gravity and explain how it is measured
- Describe the use of a hydrometer for measuring the specific gravity of a liquid
- Define pressure
- Explain the operation of a barometer
- Discuss the importance of pressure to the body
- Define heat
- Describe temperature as the measure of the heat of an object
- Explain the operation of a thermometer
- Compare the calibrations of the Fahrenheit, Celsius, and Kelvin temperature scales
- Convert Celsius degrees to Fahrenheit degrees and convert Fahrenheit degrees to Celsius degrees
- Contrast the use of the calorie and the Calorie as units of heat energy
- Explain the operation of a calorimeter
- Define specific heat
- Relate specific heat to the maintenance of normal body temperature

Whether you realize it or not, you are always involved with measurements. In the course of an ordinary day, you are constantly exposed to measurements involving time, weight, temperature, size, length, and volume. As part of the studies in your science program, you will measure and compare various things. Even when you become a practitioner in your field, you will still be expected to measure an assortment of objects. What, then, is the basis of measurement and why is it so important in your life and your work?

WHAT IS MEASUREMENT?

Whenever you **measure** anything, you are *comparing* it with some kind of standard scale to determine its *magnitude*. In other words, how big is it? Some measurements are made directly by comparing the unknown quantity with the known unit of the same kind. Weighing a patient and taking the reading directly in pounds is an example of a direct measurement. Other measurements, however, are indirect and must be done by calculation. An example of this kind of measurement is counting a person's blood cells in a certain number of squares on a microscope slide and then calculating the total blood count.

Regardless of how a measurement is taken, it always requires two elements—a *number* and a *unit*. In recording the weight of a patient, you would not just say 145. You would have to say 145 pounds, that is, you would have to give both the number (145) and the unit (pounds). When you count white blood cells, you report the measurement as 10,000 (number) white blood cells per cubic millimeter of blood (unit). Think of how many times you apply the number and unit concept when you buy food.

Units of Measurement

The numerous units in use can all be expressed in terms of one of three special units called *fundamental units*. These fundamental units are the units of length, weight, and time. All other units are referred to as *derived units* because they can always be written as some combination of the three fundamental units. Units are grouped into systems of measurement. The two principal systems of measurement commonly used in this country are the United States system and the metric system. Another system, the apothecary system, used by physicians and pharmacists for years, is gradually being replaced by the metric system.

Mass is a term that is perhaps unfamiliar to you. *Mass* is simply the amount of matter that an object contains. The mass of this textbook is the same whether it is measured in your laboratory, under the sea, on top of a mountain, on the moon, or in space where no gravity operates because wherever you take it, it still has the same quantity of matter. *Weight*, on the other hand, is a force and is determined by the pull of gravity on an object. This textbook will not have the same weight on the earth as on the moon because of the differences in gravitational forces. In fact, the gravitational force on the moon is one-sixth that of the gravitational force on the earth. However, weight and mass are often considered synonymous terms because the force of gravity on the surface of the earth is

Exhibit 2-1 METRIC UNITS OF WEIGHT AND SOME UNITED STATES EQUIVALENTS

Metric Unit (Symbol)	Metric Equivalent	English Equivalents
1 kilogram (kg)	1000 g (10^3 g)	2.205 pounds (lb)
1 hectogram (hg)	100 g (10^2 g)	
1 dekagram (dkg)	10 g (10^1 g)	
1 gram (g)	1 g (10^0 g)	1 lb = 453.6 g
		1 ounce (oz) = 28.35 g
1 decigram (dg)	0.1 g (10^{-1} g)	
1 centigram (cg)	0.01 g (10^{-2} g)	
1 milligram (mg)	0.001 g (10^{-3} g)	
1 microgram (μg)	0.000,001 g (10^{-6} g)	
1 nanogram (ng)	0.000,000,001 g (10^{-9} g)	
1 picogram (pg)	0.000,000,000,001 g (10^{-12} g)	

nearly constant. Thus, weight remains nearly the same regardless of where the measurements are taken on the earth. Even though there is a fundamental difference between mass and weight, they will be considered synonymous terms in this textbook. Exhibit 2-1 lists metric units of weight along with some United States equivalents.

United States System

The **United States (U.S.) system** of measurement was introduced to this country when the land was still a British colony. It is still used for everyday household work, in most industry, and in some fields of engineering. The fundamental units in the U.S. system are the foot (length), the pound (weight), and the second (time). The standard of length of the U.S. system is the *yard*. The official standard is the distance between two lines on a bronze bar kept at 62°F in the Office of the Exchequer in London, England. It is believed to have originated as the length of the stride of an average man. The standard of weight of the U.S. system is the *pound*. It has been arbitrarily set as the weight of a bronze bar also kept at the Office of the Exchequer in London. The standard of time in the U.S. system is the second.

The U.S. system lacks *uniform* progression from one unit to another. For example, consider the standard unit of length, the yard (yd). If you measure the length of an object at 2.5 yd and you want to convert to feet, you have to multiply 2.5 yd by 3 ft, since there are 3 ft in a yard. If you want to convert the same length in yards to inches, you have to multiply 2.5 yd by 3 by 12 (or 36), since there are 12 inches in a foot. In other words, in order to convert from one unit of length to another, it is necessary to use *different* numbers each time. As you will see, conversions in the

Exhibit 2-2 UNITED STATES UNITS OF MEASUREMENT

Fundamental or Derived Unit	Units and Equivalents
Length	12 inches = 1 foot (ft) = 0.333 yards (yd) 3 ft = 1 yd 1,760 yd = 1 mile (mi) 5,280 ft = 1 mi
Weight	1 ounce (oz) = 28.35 grams (g); 1 g = 0.0353 oz 1 pound (lb) = 453.6 g = 16 oz; 1 kilogram (kg) = 2.205 lb 1 ton = 2,000 lb = 907 kg
Time	1 second (sec) = 1/86,400 of a mean solar day 1 minute (min) = 60 sec 1 hour (hr) = 60 min = 3,600 sec 1 day = 24 hr = 1,440 min = 86,400 sec
Volume	1 fluid dram (fl dr) = 0.125 fluid ounces (fl oz) 1 fl oz = 8 fl dr = 0.0625 quarts (qt) = 0.008 gallons (gal) 1 qt = 256 fl dr = 32 fl oz = 2 pints (pt) = 0.25 gal. 1 gal = 4 qt = 128 fl oz = 1,024 fl dr

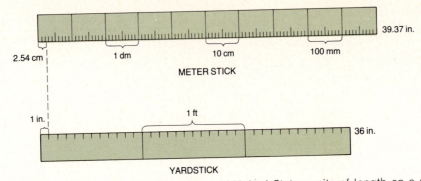

Figure 2-1 Comparison between metric and United States units of length on a meter stick and a yardstick.

metric system are much easier since they are based on progressions of the number 10.

Exhibit 2-2 lists U.S. units of measurement.

Metric System — Fundamental Units

The **metric system** was introduced in France in 1790 and is now used by all major countries of the world except the United States. Scientific observations are almost universally expressed in terms of metric units. The standard of length in the metric system is the *meter (m)*. It was originally defined in 1790 as one ten-millionth of the distance from the North Pole to the Equator. Later, in 1889, it was defined as the distance measured at 0°C between two lines on a bar of platinum iridium kept at the Bureau of Standards in France. Three facsimiles of this bar are also kept at the United States Bureau of Standards in Washington, D.C. The meter is equal to 39.37 inches.

Recall that a major advantage of using the metric system is that each unit is related to the others by some factor of 10. For example, one meter is the same as 10 decimeters (dm), or 100 centimeters (cm), or

1,000 millimeters (mm). If you were working with the U.S. system of length, you would have to say that 1 yd is the same as 3 ft or 36 inches. Refer to Figure 2-1 for an illustration of the differences between metric and U.S. conversions using the metric meter stick and the U.S. yardstick. Exhibit 2-3 lists metric units of length. Read it carefully and be sure that you understand the relationships between the different units of length.

Notice in Exhibit 2-3 that some of the decimal equivalents such as 1 Å = 0.000,000,000,1 m are cumbersome to work with. In a case like this, the decimal can be converted to an *exponential form*, a convenient way to express very small and very large numbers in an abbreviated form. This conserves time and space. We can express such numbers as powers of 10. The exponential notation has the form $M \times 10^n$, where M is the one-digit number to the left of the decimal point and n is a positive or negative value. Two operations are necessary to change numbers into exponential form: (1) Determine M by moving the decimal point so that only one nonzero digit is to the left of it. (2) Determine n by counting the number of places the

Exhibit 2-3 METRIC UNITS OF LENGTH AND SOME UNITED STATES EQUIVALENTS

Metric Unit	Meaning of Prefix	Metric Equivalent	U.S. Equivalents
1 kilometer (km)	kilo = 1,000	1,000 m	3,280.84 ft or 0.62 mi; 1 mi = 1.61 km
1 hectometer (hm)	hecto = 100	100 m	328 ft
1 dekameter (dam)	deka = 10	10 m	32.8 ft
1 meter (m)	Standard unit of length		39.37 inches or 3.28 ft or 1.09 yd
1 decimeter (dm)	deci = 1/10	0.1 m	3.94 inches
1 centimeter (cm)	centi = 1/100	0.01 m	0.394 inch; 1 inch = 2.54 cm
1 millimeter (mm)	milli = 1/1,000	0.001 m = 1/10 cm	0.0394 inch
1 micrometer (μm)	micro = 1/1,000,000	0.000,001 m = 1/10 000 cm	3.94×10^{-5} inch
1 nanometer (nm)	nano = 1/1,000,000,000	0.000,000,001 m = 1/10 000 000 cm	3.94×10^{-8} inch
1 angstrom (Å)		0.000,000,000,1 m = 1/100 000 000 cm	3.94×10^{-9} inch

decimal point was moved. If the decimal point was moved to the left, n is a positive number; if it was moved to the right, n is a negative number. Let us apply these simple rules. Another way of stating 1 Å equals 0.000,000,000,1 m is to state that 1 Å = 1×10^{-10} m. Why? If you determine M by moving the decimal point so that only one nonzero digit is to the left of it, you have

$$0.000,000,000,1.$$

Therefore, $M = 1$. Now, since you have moved the decimal point 10 places to the right, $n = {}^{-10}$. Thus,

$$1 \text{Å} = 1 \times 10^{-10} \text{ m}$$

The wavelength of yellow light is about 0.000,059 cm/second. Convert this number to exponential form. If your answer is 5.9×10^{-5}, you are ready to move on. If not, go back and reread the discussion.

Now let us suppose that we are working with a very large number instead of a decimal. The same rules apply, except that our exponential value will be a positive rather than a negative one. For example, refer to Exhibit 2-3 and notice that 1 km equals 1,000 m. Even though 1,000 is not a very cumbersome number, we can still easily convert it into exponential form. First move the decimal point so that there is only one nonzero digit to the left of it to determine M.

$$1.000.$$

Now, since the decimal has been moved three places to the left, n equals $+3$ or simply 3. Thus,

$$1 \text{ km} = 1 \times 10^3 \text{ m}$$

The speed of light is about 30,000,000,000 cm/second. Convert this number to exponential form. The answer is 3×10^{10} cm.

Review Exhibit 2-3 again to note some common metric and U.S. equivalents. Note also the exponential forms. Do you know what they mean? To find out, convert them back into decimals or large numbers.

The second fundamental unit of the metric system is weight. The standard unit of weight is the *kilogram (kg)*. A kilogram is defined as the weight of a bar containing 90 percent platinum and 10 percent iridium kept at the Bureau of Standards in France.

The standard pound is defined in terms of the standard kilogram by the relation 1 lb = 0.4536 kg. Thus, 1 kg = 2.205 lb; 453.6 g = 1 lb; and 28.35 g = 1 oz.

The third fundamental unit of measurement is common to both the U.S. and metric systems. This unit is time, and the standard of time is the *second*. It is defined as 1/86,400 of a mean solar day. The average length of all days throughout the year is called the mean solar day. There are 86,400 seconds in a mean solar day ($24 \times 60 \times 60$). Thus, a second is 1/86,400 of a mean solar day. Units of time are used in measuring pulse and heart rate, metabolic rate, x-ray exposure, and intervals between medications.

Derived Units

It was stated earlier that derived units can always be written as some combination of the three fundamental units. Consider volume as an example. *Volume,* or capacity, in the U.S. system may be expressed as cubic feet (ft^3), cubic inches ($inch^3$), and cubic yards (yd^3). Notice that the units of volume are derived from the units of length. Volume in the metric system may also be expressed in cubic units of length such as cubic centimeters (cc or cm^3), or in terms of the basic unit of volume, the *liter*. A liter is defined as the volume occupied by 1,000 g of pure water at 4°C. Since 1 cm^3 of water at this temperature weighs 1 g, then 1,000 g of water also occupies a volume of 1,000 cm^3. This means that 1 liter is equal to 1,000 cm^3 and 1 milliliter (ml) is equivalent to 1 cm^3. Because of this relationship, many volume-measuring devices, such as hypodermic needles, may be graduated in either milliliters or cubic centimeters. Exhibit 2-4 lists metric units of volume and some of their U.S. equivalents.

The Apothecary System

In addition to the U.S. and metric systems, there is a system of measurement called the **apothecary system.** This system is still commonly used by physicians in prescribing medications and pharmacists in preparing them. The important units and conversions of this system are presented in Exhibit 2-5.

Exhibit 2-4 METRIC UNITS OF VOLUME AND SOME UNITED STATES EQUIVALENTS

Metric Unit (Symbol)	Metric Equivalent	U.S. Equivalents
1 liter	1,000 ml (1×10^3 ml)	33.81 fl oz or 1.057 qt; 946 ml = 1 qt
1 milliliter (ml)	0.001 liter (1×10^{-3} liter)	0.0338 fl oz; 30 ml = 1 fl oz; 5 ml = 1 teaspoon
1 cubic centimeter (cc or cm^3)	0.999972 ml	0.0338 fl oz

Exhibit 2-5 APOTHECARY SYSTEM OF WEIGHTS AND VOLUMES WITH CONVERSIONS

Apothecary Units	Metric Equivalents
Weight	
60 grains = 1 dram	1 g = 15 grains
8 drams = 1 ounce	4 g = 1 dram
12 ounces = 1 pound	30 g = 1 ounce
	1 kg = 32 ounces
Volume	
60 minims = 1 fluidram	1 ml (or cm^3) = 15 minims
8 fluidrams = 1 fluidounce	4 ml (or cm^3) = 1 fluidram
16 fluidounces = 1 pint	30 ml (or cm^3) = 1 fluidounce
	500 ml (or cm^3) = 1 pint
	1000 ml (or cm^3) = 1 quart

Frequently Used Conversions Based on the Milligram

1000 mg (1 g)	= 15 grains
600 mg (0.6 g)	= 10 grains
300 mg (0.3 g)	= 5 grains
60 mg (0.06 g)	= 1 grain
30 mg (0.03 g)	= 1/2 grain
20 mg (0.02 g)	= 1/3 grain
10 mg (0.01 g)	= 1/6 grain
5 mg (0.005 g)	= 1/12 grain
4 mg (0.004 g)	= 1/15 grain
1 mg (0.001 g)	= 1/60 grain
0.5 mg (0.005 g)	= 1/120 grain
0.1 mg (0.0001 g)	= 1/600 grain

APPLYING UNITS OF MEASUREMENT TO MATTER AND ENERGY

We have seen in the preceding discussion that it is possible to measure length, weight, time, and volume by using various units of measurement. Aside from time, we are really measuring only two things, matter or energy. What, then, are matter and energy?

Definition of Matter

Matter may be defined as anything that occupies space and has mass. It is usually measured by units such as milligrams, grams, ounces, pounds, and tons. Matter exists as a solid, liquid, or gas, and most matter may be changed from one state to another by adding or removing pressure or heat. One example is the conversion of water from a solid (ice) to a liquid to a gas (steam) as temperature increases.

Definition of Energy

Unlike matter, energy does not have mass. Thus, energy does not occupy space. **Energy** is the capacity to do work and is measured by its effects on matter. More energy must be expended to move greater masses of matter.

Types of Energy

One type of energy, called *potential energy*, is stored (inactive) energy. One example of potential energy is a steel ball at rest on a table (Figure 2-2). When the ball is dropped, its potential energy becomes kinetic energy since work is being done. *Kinetic energy*, unlike potential energy, is the energy of a moving object. There is a conversion of potential energy into kinetic energy as the ball falls. Note in Figure 2-2 that as the steel ball falls, the sum of the energies is equal as potential energy is converted into kinetic energy.

Among the forms of energy related to biological systems are electrical, radiant, mechanical, and chemical energy. Each may exist in either the potential or kinetic states.

1. **Electrical energy.** This results from the movement of charges such as electrons or ions along or through a conductor. In living systems, electrons are associated with energy changes and ions are absolutely necessary for the conduction of nerve impulses by neurons.

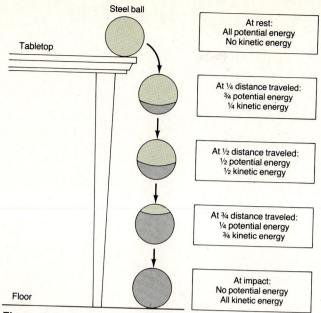

Figure 2-2 Transformation of potential energy into kinetic energy. The ball resting on the edge of the table represents potential energy, or stored energy. If the ball is dropped, potential energy becomes kinetic energy, or energy in action. As the ball falls to the floor, potential energy is converted to kinetic energy and the sum of the energies remains the same.

2. **Radiant energy.** This form of energy travels in waves and includes heat and light. Heat is used by living systems to speed up atomic and molecular movements to accelerate chemical reactions. Light, as you probably know, is the primary source of energy for all organisms. It is required by photosynthetic organisms to produce food, and in the process, light energy is changed into chemical energy.
3. **Mechanical energy.** This form of energy is required to move matter and is illustrated in Figure 2-2. An example of mechanical energy is the movement of bones when muscles contract and pull on them.
4. **Chemical energy.** This is energy stored in molecular bonds. Chemical energy is released when the bonds are broken and stored when the bonds are formed.

Let us now examine some of the ways in which matter and energy may be measured by considering density, specific gravity, pressure, heat and temperature, and specific heat.

Density

It was stated earlier that matter occupies space and has volume. Just from everyday experiences, we recognize that materials have different weights. For ex-

ample, we say that gold or lead is heavy and cork or paper is light. But this has little meaning unless we are talking about *equal volumes* of gold, lead, cork, and paper, since a truckload of cork would be heavier than a handful of gold. In order to get around this problem of varying volumes when comparing two substances, we use a concept called density. The **density** of a substance is its mass per unit volume. This means that when we compare the densities of two substances, we are comparing equal volumes of the substances. In calculating or measuring densities of solids or liquids, the basic unit of mass is the gram and the basic unit of volume is the cubic centimeter. The densities of gases are usually given in grams per liter. The density of a substance is expressed by the equation:

$$D = \frac{m}{V}$$

The density of pure water is 1 g/cm³. It is written as follows:

$$D_{\text{water}} = \frac{m}{V}$$

$$D_{\text{water}} = \frac{1 \text{ g}}{\text{cm}^3}$$

The density of lead is 11.34 g/cm³ and that of cork is 0.24 g/cm³. It can be seen that lead is 11.34 times denser than water and 50 times denser than cork. Water is about four times denser than cork. Examples of the densities of other substances are shown in Figure 2-3. Keep in mind that the density of a sub-

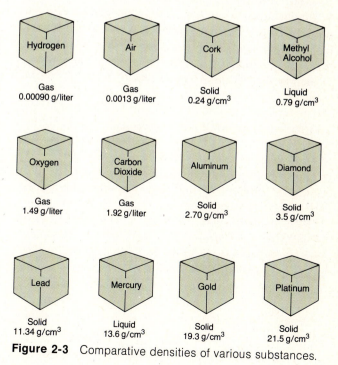

Figure 2-3 Comparative densities of various substances.

stance is always expressed in mass per unit volume. Thus, it is necessary to say that the density of methyl alcohol is 0.79 g/cm³; the density of oxygen is 1.49 g/liter; and the density of lead is 11.34 g/cm³.

Specific Gravity

Another property of matter that we can measure is its specific gravity. **Specific gravity** is defined as the density of a substance relative to the density of water. When we divide the density of a substance by the density of water, the units cancel each other out, so the value for specific gravity is a number only and has no units. Consider the following example for the specific gravity of methyl alcohol:

$$\text{Specific gravity} = \frac{\text{density of a substance}}{\text{density of water}}$$

$$\text{Specific gravity} = \frac{0.79 \text{ g/cm}^3}{1.00 \text{ g/cm}^3}$$

$$\text{Specific gravity} = \frac{0.79}{1.00}$$

$$\text{Specific gravity} = 0.79$$

Essentially, the density and specific gravity of a substance are the same. The only difference is that density has units (mass/volume), whereas specific gravity has none. Thus, the density of methyl alcohol is 0.79 g/cm³ but its specific gravity is simply 0.79. The density of water is the standard reference for solids and liquids; the commonly used standard of reference for gases is air (1.29 g/liter).

Practical use is made of the concept of specific gravity in the medical sciences with an instrument called a *hydrometer* (Figure 2-4). It is a sealed glass cylinder weighted on one end and bearing a graduated scale on the other. When placed in a liquid whose specific gravity is to be measured, the hydrometer floats and the reading is taken directly from the calibrations at the liquid line. Hydrometers with calibrations from 0.500 to 1.000 are generally used for liquids whose specific gravities are lower than that of water (1.000); ones with readings from 1.000 to 2.000 or higher are used for liquids whose specific gravities are higher than that of water.

Hydrometers used for analyzing the specific gravity of urine are called urinometers. It is important to know the specific gravity of urine because abnormal values indicate to the physician that there is an abnormality in a person's body. Normally, the specific gravity of urine is between 1.015 and 1.030. Any significant deviations from the normal range inform the physician of existing or potential problems.

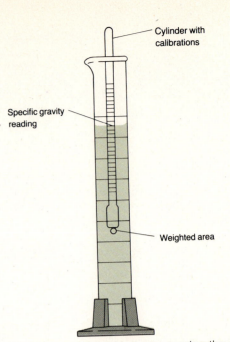

Figure 2-4 Typical hydrometer for measuring the specific gravity of a liquid.

Labels: Cylinder with calibrations; Specific gravity reading; Weighted area

Pressure

How many times have you heard your local weather forecaster say that there is a high or low pressure system in the area? How many times have you been told that your blood pressure is normal, or high, or low?

Pressure, like density, is expressed in units of measurement. **Pressure** is defined as the number of units of force on a unit area. It is expressed by the equation:

$$P = \frac{F \text{ (force)}}{A \text{ (area)}}$$

Let us see what this means. Assume that you have two blocks of stone, each of which weighs 1,000 lb (Figure 2-5). Each block is 40 inches × 20 inches × 10 inches. Block 1 is placed on its end, where the area is 200 inch² (20 inches × 10 inches), and the pressure is as follows:

$$P = \frac{1,000 \text{ lb}}{200 \text{ inch}^2}$$

$$P = 5 \text{ lb/inch}^2$$

Block 2 is lying on its side, where the area is 400 inch² (40 inches × 10 inches), and its pressure is as follows:

$$P = \frac{1,000 \text{ lb}}{400 \text{ inch}^2}$$

$$P = 2.5 \text{ lb/inch}^2$$

The pressure exerted by block 1 is 5 lb/inch², while the pressure exerted by block 2 is 2.5 lb/inch². In

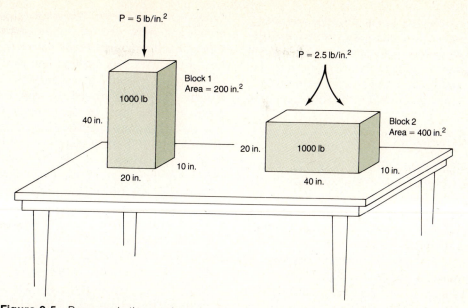

Figure 2-5 Pressure is the number of units of force on a unit area. The pressure of block 1 is greater than the pressure of block 2.

other words, the pressure exerted by block 1 is twice as great as the pressure exerted by block 2. Notice the important concept—pressure is force per unit area. It is not a total force.

The air surrounding the earth's atmosphere is acted on by gravity causing it to exert a pressure on the earth. The gravitational force on the air closest to the earth is greatest, and as a result, air in this region is compressed. This compression increases both the density and pressure of the air, so the density and pressure of air decrease as altitude increases.

Atmospheric (air) pressure is measured by an instrument called a *barometer* (Figure 2-6). The idea for this instrument was first suggested by an Italian physicist in the seventeenth century named Torricelli. A barometer may be constructed from some mercury, a glass tube about a meter in length sealed at one end, and an open dish. The tube is filled with mercury, the open end is closed by holding something over it, and the tube is inverted in the open dish, which is about half-filled with mercury. When the closure is removed, mercury drains out of the tube until the pressure exerted by the column of mercury counterbalances that exerted by the atmosphere. The height of the column of mercury at this counterbalanced state is about 29.92 inches or 760 mm at sea level. It is read as 760 mm of mercury. Standard pressure is the pressure exerted by a column of mercury at exactly 760 mm of pressure per square inch. One application of this standard is in weather forecasting. If the air pressure rises above 760 mm, it will force the mercury

column up. This is called high pressure and indicates fair weather. Conversely, if the air pressure falls below 760 mm, the mercury column will fall. This low pressure indicates that rain or a storm will follow. Another application of the standard is in taking blood pressure. Blood pressure also is measured in millimeters of mercury (mm Hg), and the apparatus used for taking the pressure is a sphygmomanometer. A blood pressure reading of 125 indicates that the pressure is sufficient to raise the level of a mercury column above that of the atmosphere.

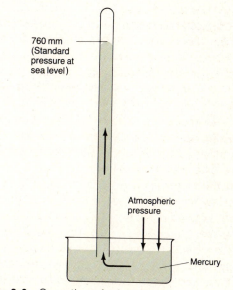

Figure 2-6 Operation of a barometer.

Differences in pressure are responsible for many activities of the body that we take for granted. Breathing, blood flow, and supplying body cells with oxygen are examples. Pressure differences also provide the basis for the operation of syringes, suction drainage, siphons, pumps, autoclaves, and artificial respirators.

Heat and Temperature

In its simplest terms, **heat** is a form of energy, which when put into an object, increases the energy of the object. The effect of heat on an object is to increase its kinetic energy. A warmer object cools as it releases heat, and a cooler object warms as it acquires heat. **Temperature,** on the other hand, is a measure of the energy of an object. Although a number of instruments have been devised for measuring temperature, the most familiar and simplest is a mercury *thermometer.* Basically, it consists of a slender glass tube sealed at one end and having at the other a small bulb containing mercury. As the temperature rises, the mercury increases in volume. When the temperature falls, the mercury column shrinks. Mercury has been chosen because it expands uniformly when heated.

Thermometers presently in use are calibrated differently and thus have different temperature scales. The commonly used temperature scales are Fahrenheit, Celsius (centigrade), and Kelvin (absolute). See Figure 2-7. The original reference points for the

Fahrenheit scale were the temperature of a mixture of salt and snow and the armpit of a human subject. When the thermometer was placed in the salt-snow mixture, a mark was made on the glass at the level of the mercury column and designated 0°F. The thermometer was then placed between the arm and the body, and the level of the column was marked at 100°F. The distance between the two points was then divided into 100 equal degrees. The boiling point of water was marked at 212°F. We now know that normal body temperature on the Fahrenheit scale is 98.6°F and not 100°F, the body temperature of the subject first used to calibrate the scale.

The points of reference for the *Celsius* (C) *scale* were the temperature of an ice-water mixture marked at 0°C and the temperature of boiling water marked at 100°C. Normal body temperature on the Celsius scale is 37°C. How, then, are the Fahrenheit and Celsius scales related? To convert from one scale to another, you must use two simple equations.

Suppose that you wanted to convert 30°C to degrees Fahrenheit. In order to do this, you would use the formula:

$$°F = 9/5 \; (°C) + 32°$$
$$°F = 9/5 \; (30°) + 32°$$
$$°F = 54° + 32°$$
$$°F = 86°$$

Thus,

$$30°C = 86°F$$

What about converting degrees Fahrenheit into degrees Celsius? Suppose this time that you wanted to convert 77°F into degrees Celsius. You would use the formula:

$$°C = 5/9 \; (°F - 32°)$$
$$°C = 5/9 \; (77° - 32°)$$
$$°C = 5/9 \; (45°)$$
$$°C = 25°$$

Thus,

$$77°F = 25°C$$

Additional sample problems of Fahrenheit and Celsius conversions are provided at the end of the chapter.

A third temperature scale, the *Kelvin* or *absolute scale,* is also commonly used. If all the heat were removed from an object, it would then be at its lowest possible temperature (or lowest possible kinetic energy). This temperature, called absolute zero, would be equal to −459.6°F or −273.16°C. If the Celsius scale is renumbered so that absolute zero becomes 0°, we have the Kelvin scale. On the Kelvin scale, the boiling point of water is 373.16°K (or simply 373°K), the freezing point of water is 273.16°F (or simply 273°K), and normal body temperature is 310.16°K (or simply 310°K). Therefore, degrees Kelvin = degrees

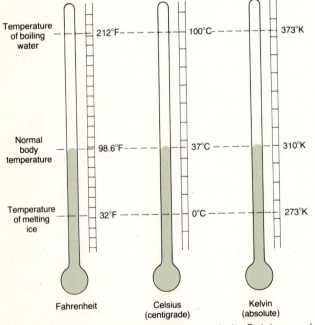

Temperature of boiling water	212°F — 100°C — 373°K	
Normal body temperature	98.6°F — 37°C — 310°K	
Temperature of melting ice	32°F — 0°C — 273°K	
Fahrenheit	Celsius (centigrade)	Kelvin (absolute)

Figure 2-7 Comparison of the Fahrenheit, Celsius, and Kelvin temperature scales. Notice the boiling point, freezing point, and normal body temperatures on each.

Celsius + 273°. Kelvin temperatures are used to measure the behavior of gases.

In scientific work it is important to know the quantity of heat produced by a substance. The amount of heat energy is expressed in units called calories. A **calorie (cal)** is the amount of heat energy required to raise the temperature of 1 g of water 1°C. The calorie is so small a unit that, for many practical purposes, the kilocalorie or Calorie is used. The **kilocalorie** or **Calorie** is equal to 1,000 calories and is defined as the amount of heat required to raise the temperature of 1,000 g of water 1°C. The Calorie is the unit used to determine the caloric value of foods, and as you will see later, to measure rates of body metabolism.

The apparatus used to determine the caloric value of foods is called a *calorimeter* (Figure 2-8). It operates as follows: A weighed sample of a dehydrated food is burned completely in an insulated metal container. The energy released by the burning food is absorbed by the container and then transferred to a known value of water that surrounds the container. The change in temperature of the water is directly related to the number of Calories released by the food. Determination of the caloric value of foods is important because if we know the amount of energy used by the body for various activities, we can adjust our food intake. In this way, we can attempt to control body weight by taking in only enough Calories to sustain our activities.

A final measurement of matter that we shall consider is specific heat, a concept of extreme importance to the normal functioning of the body.

Specific Heat

It is a fairly common observation that different kinds of matter store the same amount of heat but show different changes in temperature. Take water and an aluminum pot as examples. While the water seems to take forever to boil, the aluminum pot quickly reaches a very high temperature. This observation is explained by the concept of specific heat. **Specific heat** is the quantity of heat required to raise the temperature of 1 g of a substance 1°C. In other words, it takes more heat to raise the temperature of water than of an equal quantity of aluminum. Specific heat is recorded in terms of calories per gram per degree Celsius. That is, the specific heat of water is 1.0 cal/g/°C; the specific heat of aluminum is 0.21 cal/g/°C; that of glass is 0.19 cal/g/°C; and that of alcohol is 0.55 cal/g/°C. Notice that of the substances listed, water has the highest specific heat. More heat must be applied to water before its temperature will increase. Your body is mostly water, and the ability of water to absorb large amounts of heat without any significant change in temperature has a lot to do with your body temperature. When it is very hot in your environment, you still maintain a fairly constant body temperature of 98.6°F. If it were not for the high specific heat of water, the water in your body would boil and you would die. Moreover, the water in your body also gives off its heat very slowly. Thus, even if your environmental temperature is freezing, your body water will not freeze. There are other factors that help you

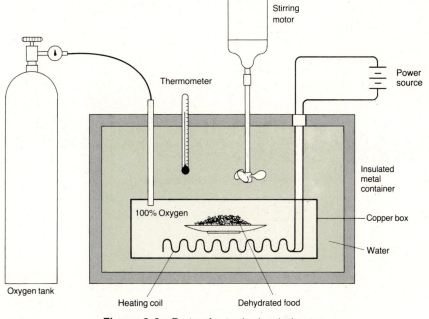

Figure 2-8 Parts of a typical calorimeter.

maintain a constant body temperature, but the high specific heat of water is a very important one.

This chapter has been concerned with the nature and kinds of measurements and their applications to matter and energy. Using much of this information, we shall now explore a few basic physical principles that are important for an understanding of the human body.

CHAPTER SUMMARY OUTLINE

MEASUREMENT

1. Measuring means comparing to a standard to determine magnitude.
2. All measurements involve a number and a unit.
3. The fundamental units are length, mass, and time.
4. Derived units are combinations of the three fundamental units.

United States System

1. The fundamental units are the foot, pound, and second.
2. The major problem with the system is a lack of uniform progression from one unit to another.

Metric System

1. The standard for length is the meter, the standard for weight is the kilogram, and the standard for time is the second.
2. Decimals or large numbers are converted into exponential form to save time and space.
3. Mass is the amount of matter an object contains; weight is a force determined by the pull of gravity.
4. Volume is a derived unit of length; the metric standard is the liter.
5. The apothecary system is used primarily by physicians and pharmacists.

APPLICATIONS OF MEASUREMENT

1. Matter is anything that has mass and occupies space.
2. Energy is the capacity to do work and is of two kinds—potential and kinetic.
3. Forms of energy include chemical, mechanical, radiant, and electrical.

Density

1. Density is mass per unit volume; it is expressed by the equation $D = m/V$.
2. The standard for density is pure water, which is 1 g/cm^3.

Specific Gravity

1. Specific gravity is the density of a substance relative to the density of water.
2. The density of water is the standard for solids and liquids; the density of air is the standard for gases.
3. Specific gravity is measured by a hydrometer.
4. A urine hydrometer is called a urinometer.

Pressure

1. Pressure is the number of units of force on a unit area; it is expressed by the equation $P = F/A$.
2. Air pressure is measured by a barometer.
3. Standard pressure is 760 mm Hg/inch².
4. Weather highs and lows are measured in millimeters of mercury.

Heat and Temperature

1. Heat is a form of energy that increases the kinetic energy of another object.
2. Temperature is a measure of the energy of an object.
3. Temperature is measured by a thermometer.
4. Commonly used temperature scales are Fahrenheit, Celsius, and Kelvin.
5. Degrees Celsius are converted to degrees Fahrenheit by the formula $°F = 9/5(°C) + 32°$.
6. Degrees Fahrenheit are converted to degrees Celsius by the formula $°C = 5/9(°F − 32°)$.
7. Degrees Kelvin = degrees Celsius + 273.
8. Heat energy is expressed in calories or Calories.
9. The caloric value of foods is determined in a calorimeter.

Specific Heat

1. Specific heat is the quantity of heat required to raise the temperature of 1 g of water 1°C.
2. The specific heat of water is 1.0 cal/g/°C.
3. The high specific heat of water is important in maintaining a constant body temperature.

REVIEW QUESTIONS

1. Define measurement. What are the two characteristics of any measurement? Give several examples.
2. Distinguish between a fundamental unit of measurement and a derived unit of measurement.
3. Describe the fundamental units of the United States system of measurement. What is the basic advantage of the United States system of measurement?
4. What is the standard length in the metric system? How was it determined?
5. Define each unit of length in the metric system.
6. Solve the following conversions of length:
   a. If a bacterial cell measures 100 μm in length, how many nanometers is it?
   b. How many meters are in one mile?
   c. If a road sign reads 35 km/hour, what would your speedometer have to read in order for you to obey the sign?
   d. How many millimeters are in one kilometer? In one inch?
   e. A person's arm measures two feet in length. How many centimeters is this?
   f. Convert 0.40 m to millimeters.
   g. How many millimeters are in 5 inches?
   h. If you had to run 295.2 ft, how many meters would you have to run?
   i. Why would you not measure the length of a virus in decimeters?
   j. If the distance to the moon is 239,000 mi, calculate the distance in meters.
7. What is an exponential notation? What two rules must be kept in mind when converting a large decimal or large number into exponential form?
8. Convert the following into exponential form: (a) 0.00018, (b) 1,000,000,000, (c) 186,000, (d) 0.003690.
9. Convert the following exponents into decimals or large numbers: (a) 2.818×10^{-13}, (b) 10^6, (c) 5.9×10^{-5}, (d) 10^{27}.
10. What is the standard of mass in the metric system? How was it determined?
11. Distinguish between mass and weight.
12. Solve the following conversions of mass:
   a. How many milligrams are in 0.4 kg? In one pound?
   b. If a bottle contains 1.42 g of a substance, how many centigrams does it contain?
   c. What is the mass of a cubic foot of water in grams? What does it weigh in pounds?
   d. The indicated dosage of a certain drug is 50 μg. How many milligrams is this?
   e. If your body weight is 110 lb, how many kilograms do you weigh?
   f. How many centigrams are there in one gram?
13. What is the standard of time in the metric system? How was it determined?
14. Why is volume considered a derived unit? Give an example.
15. What is the standard of volume in the metric system? What is the relationship between a cubic centimeter and a milliliter?
16. Solve the following conversions of volume:
   a. If you excrete 1,200 ml of urine in a day, how many liters is this?
   b. How many milliliters are there in 2 liters?
   c. Convert 2 pt to milliliters.
   d. If you remove 15 cm³ of blood from a patient, how many milliliters have you removed?
17. Describe the apothecary system of measurement. Who uses the system?
18. Convert the following apothecary system units:
   a. 30 grains = _____ drams
   b. 500 mg = _____ grains
   c. 4 fluidounces = _____ pint
   d. 5 ml = _____ fluidram
   e. 0.5 grains = _____ mg
   f. 3 ounces = _____ grains
   g. 1 pint = _____ ml
19. Distinguish between matter and energy.
20. Contrast potential and kinetic energy.
21. Compare chemical, mechanical, radiant, and electrical energy.
22. Define density. State the equation for calculating density.
23. What is the difference between density and specific gravity? Give several examples.
24. Relate specific gravity to a hydrometer.
25. Define pressure. State the equation for calculating pressure.
26. Distinguish between pressure and total force. Give an example.
27. How is atmospheric pressure measured? Describe the operation of a barometer.
28. What is the difference between heat and temperature? How is temperature measured?
29. Explain the operation of a mercury thermometer.
30. What are the basic differences between the Fahrenheit, Celsius, and Kelvin temperature scales?

31. What formula would you use to convert degrees Celsius to degrees Fahrenheit? Using the formula, convert the following to degrees Fahrenheit: (a) 0°C, (b) 18°C, (c) 37°C, (d) 60°C, (e) 100°C.

32. What formula would you use to convert degrees Fahrenheit to degrees Celsius? Using the formula, convert the following to degrees Celsius: (a) 32°F, (b) 68°F, (c) 98.6°F, (d) 125°F, (e) 212°F.

33. How would you convert degrees Celsius to degrees Kelvin? Using the formula, convert the following to degrees Kelvin: (a) 0°C, (b) 9°C, (c) 37°C, (d) 43°C, (e) 100°C.

34. What unit is used to express the quantity of heat energy produced by a substance?

35. Distinguish between a calorie and a Calorie.

36. How are caloric values of foods determined by using a calorimeter?

37. Define specific heat. How is it related to the maintenance of normal body temperature?

Chapter 3

INTRODUCTORY PHYSICAL PRINCIPLES

STUDENT OBJECTIVES

After reading this chapter, you should be able to

- Define mechanics
- Describe motion
- Compare speed, velocity, and acceleration
- Explain Newton's three laws of motion
- Define inertia and friction
- Apply Newton's laws of motion to the human body
- Contrast centripetal and centrifugal force and apply the forces to the operation of a laboratory centrifuge
- Define gravitation and apply the concept to sedimentation rate
- Describe center of gravity in terms of body mechanics
- Define a simple machine
- Explain the operation of a lever and classify levers into types
- Identify levers used in clinical settings
- Define an inclined plane and apply its principle to the construction of clinical instruments
- Define a pulley and explain its use in traction
- Describe the kinetic theory of matter and distinguish the states of matter on the basis of the theory
- Explain the characteristics of a gas
- Describe the behavior of a gas in terms of Boyle's law, Charles' law, Dalton's law, and Henry's law
- Apply the gas laws to body functions such as breathing and the exchange of gases between the blood and body cells
- Explain the cause of bends and the operation of a hyperbaric chamber on the basis of Henry's law
- Explain the characteristics of a liquid
- Define liquid pressure and its application to clinical procedures
- Describe Pascal's law and its application to a hydraulic press
- Define Archimedes' principle and explain its application in the use of a hydrometer and in physical therapy
- Compare adhesion and cohesion and note their relationship to surface tension
- Relate surface tension to the action of antiseptics and disinfectants
- Explain the phenomenon of capillarity
- Describe the relationship between liquid flow and pressure
- Define viscosity
- Explain the characteristics of a solid
- Define the elasticity of a solid in terms of Hooke's law
- Compare the effects of stress and strain on solids
- Identify tension, compression, bending, and twisting as types of stresses that act on solids

In this chapter, we shall consider a few basic physical principles that are important for an understanding of how the body works. The body is constantly acted upon by a variety of physical forces, and it functions on the basis of some very simple physical principles. This chapter has been organized into two major areas of study in order to illustrate these principles: mechanics and properties of matter. In later chapters, they will be discussed in more detail as they apply to a specific body function.

MECHANICS

Mechanics is a special branch of physics that deals primarily with the state or motion of objects resulting from the action of applied external forces. Our discussion of mechanics will center around motion and simple machines.

Motion

Motion is produced when an external force acts on a body or object. The motion of blood through veins and arteries is due to contraction of the heart, the movement of your eyes is due to contraction of the muscles attached to the eye, and the motion of urine from the kidneys into the bladder is due, in part, to the force of gravity. No matter where motion occurs, it is always the result of some kind of force acting on an object or body. Let us look at a few concepts that apply to motion.

Speed, Velocity, and Acceleration

Speed is distance traveled in a given period of time. For example, if you take a 400-mi trip from your home and it takes 8 hours, your average speed is 50 mi/hour. Speed may be determined by the following equation:

$$\text{Speed} = \frac{\text{distance traveled}}{\text{elapsed time}}$$

Thus,

$$\text{Speed} = \frac{400 \text{ mi}}{8 \text{ hours}}$$

$$\text{Speed} = 50 \text{ mi/hour}$$

Speed tells us nothing about direction of motion. Since speed has magnitude but no direction, we say that it is a *scalar quantity*. Other examples of scalar quantities are volume, area, mass, and temperature. Another useful piece of information about an object in motion is its velocity. **Velocity** is constant speed in a given direction, usually a straight line. Refer back to the 400-mi trip you took. If you traveled south in a straight line, your velocity would be 50 mi/hour south. Since velocity has both magnitude and direction, it is referred to as a *vector* quantity.

Very few objects move at a constant velocity. Generally, velocity changes with respect to magnitude or direction, or both. Motion in which the velocity changes is termed accelerated motion and the rate at which it changes is called **acceleration.** Acceleration may be represented by the following formula:

$$\text{Acceleration} = \frac{\text{final velocity} - \text{initial velocity}}{\text{elapsed time}}$$

Consider an example of an automobile in motion. Assume that the initial velocity of the automobile is 30 ft/second as it passes a traffic light, that it increases to a final velocity of 50 ft/second as it passes a second traffic light, and that it takes 5 seconds to go from the first to the second traffic light. By substituting these values into the equation, we arrive at the following:

$$\text{Acceleration} = \frac{50 \text{ ft/second} - 30 \text{ ft/second}}{5 \text{ seconds}}$$

$$\text{Acceleration} = \frac{20 \text{ ft/second}}{5 \text{ seconds}}$$

$$\text{Acceleration} = \frac{4 \text{ ft}}{\text{second/second}}$$

The acceleration is 4 ft/second/second, or 4 ft/second². This means that the velocity increases 4 ft/second every second of time. The initial velocity is 30 ft/second. Thus, an increase of 4 ft/second means that at the end of one second the velocity is 34 ft/second; at the end of two seconds, the velocity is 38 ft/second; at the end of three seconds, the velocity is 42 ft/second; and so forth (Figure 3-1).

Laws of Motion

The motion of bodies has been described in terms of speed, velocity, and acceleration. We shall now consider the cause of motion by examining Newton's laws of motion. Sir Isaac Newton published these laws in 1687 in his monumental work called *Principia*.

Newton's **first law of motion** states that a body at rest remains at rest and a body in motion remains in motion at constant speed and in a straight line unless acted upon by an external force. If a slide is placed on a microscope, it will remain there until acted upon by an external force. If you move the slide, you are the external force. If you place a thermometer on a table, it will remain there unless some force acts on it. The tendency for a body to remain in its present state, whether at rest or in motion, is called **inertia**. It is common to all material bodies. Thus, the microscope slide and the thermometer remain at rest because of their inertia.

A body in motion will stay in motion at a uniform speed in a straight line unless an external force is applied. This is a bit harder to understand based on our everyday experiences. For example, if you were to roll a ball on a dirt field, it would go only so far, and then it would stop. If you rolled the same ball on a sidewalk, it would go farther. If you rolled it on ice, it would go farther yet. The only reason that the ball would stop at all is because of **friction.** The external force of friction is less on a sidewalk than on dirt and even less on ice. If friction could be removed, the ball would continue to move. The reason the earth moves around the sun at a uniform speed is that there is very little friction.

Newton's first law of motion, concerning bodies at rest or moving with constant velocity, assumes that no forces are acting to change their state. Newton's **second law of motion,** however, assumes that such a force is acting and describes the resulting change in motion. The law states that a force acting on a body produces an acceleration that is directly proportional to the force and inversely proportional to the mass. In other words, for a given mass, the greater the force, the greater the acceleration produced. Conversely, for a given force, the greater the mass, the less the acceleration. We can illustrate this law in a hospital situation. Suppose a radiologic technologist has to push a 150-lb patient (mass) on a wheeled table or in a wheelchair from the patient's room to the radiation laboratory. The greater the force employed by the technologist in pushing, the greater the acceleration. Suppose that later in the day, the technologist has to push a 325-lb patient. If the technologist uses the same degree of force as for the 150-lb patient, there will be less acceleration.

Newton's **third law of motion** states that for every acting force there is an equal and opposite reacting force. You have seen the application of this law many times. When a baseball strikes a bat, the bat exerts a force on the ball and the ball exerts an equal and opposite force on the bat. When you weigh a patient, the person's mass exerts a force on the scale and the scale exerts an equal and opposite force against the

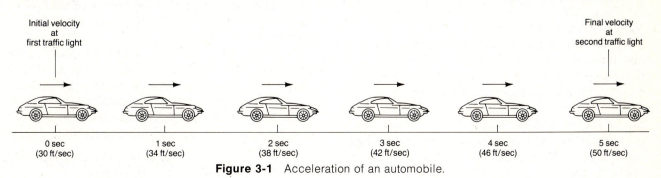

Figure 3-1 Acceleration of an automobile.

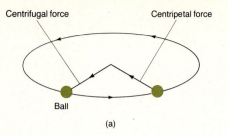

Centrifugal force   Centripetal force

Ball

(a)

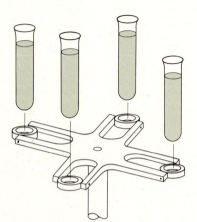

Tubes are placed in the centrifuge

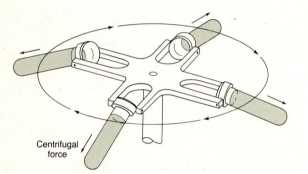

Centrifugal force

Creation of centrifugal force by rotating the tubes

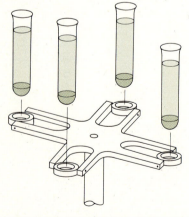

Heavier components fall to the bottoms of the tubes

(b)

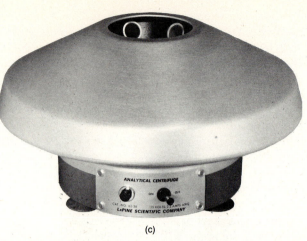

(c)

Figure 3-2 Circular motion. (a) Centripetal and centrifugal force. (b) Principle of a laboratory centrifuge. Centrifugal force concentrates the heavier components to the bottoms of the tubes. (c) Photograph of a typical laboratory centrifuge. (Courtesy of LaPine Scientific Company.)

person. When a rifle is fired, the bullet goes forward (action) and the rifle recoils (equal and opposite reaction). The many rockets that are fired into outer space also work on the same principle. The thrust from the burning of the fuel (action) causes the rocket to move forward (equal and opposite reaction).

Circular Motion

When a ball is whirled at the end of a string, an inward force is exerted by the string on the ball (Figure 3-2a). This force is called **centripetal force,** which means force toward the center. According to Newton's third law of motion, there must also be an equal and opposite force acting outward. This force is called **centrifugal force.** It is the outward force the ball exerts on the string. If the string breaks, there is no longer a centripetal force and no longer a centrifugal force. The ball no longer moves in a circular path but follows a straight-line path.

Circular motion nicely demonstrates an important principle that is utilized by medical science personnel all the time. The principle is that heavier objects are thrown out farther from the center than lighter ones. One application of the principle is the use of a laboratory apparatus called a *centrifuge* (Figure 3-2b and c). Such an apparatus is used to separate substances with different densities. When a centrifuge rotates, a centrifugal force is created. This force concentrates the heavier components in the centrifuge tubes to the bottoms of the tubes. This is a very efficient procedure for separating blood cells (heavier component) from plasma (lighter component). When you shake down a thermometer, you are applying the same principle. You shake the thermometer in a circular path, causing the heavier mercury to fall to the bottom of the thermometer.

Gravitation

Centuries ago, Aristotle taught that heavy bodies fall proportionately faster than light ones. It took nearly 2,000 years before anyone challenged this belief. One of the first challengers was Galileo in 1590. Galileo provided us with the important principle that all bodies, large and small, fall with the same acceleration, assuming that friction is neglected. For example, if a feather and a steel ball are dropped together in a tube in the absence of air, they will reach the bottom of the tube at the same exact time (Figure 3-3a). In other words, in the absence of air friction, all bodies fall with the same acceleration. It has been shown by very careful experiments that a body falls at a constant rate of acceleration at 32 ft/second².

Acceleration due to gravity = 32 ft/second²

This means that at the end of one second a body falls with a velocity of 32 ft/second, at the end of two seconds it falls with a velocity of 64 ft/second, at the end of three seconds with a velocity of 96 ft/second, and so on (Figure 3-3b). Experiments have also shown that the distance a body falls is 16 ft in the first second, 64 ft at the end of two seconds, 144 ft at the end of three seconds, etc. (Figure 3-3b). This distance a body falls is determined by use of the following formula:

s (distance) = ½ × a (acceleration) × t² (time)²

If an object has an acceleration of 32 ft/second², then after falling for 1 second it will have travelled a distance of

$$s = \frac{1}{2} \times 32 \times 1^2$$
$$s = 16$$

Look at Figure 3-3b to see if this answer is correct. If an object has attained a velocity of 128 ft/second at the end of four seconds, how far has it fallen? Look at Figure 3-3b again to check your answer.

Newton explained why objects fall as they do. According to his **law of gravitation,** the centers of any two objects are attracted to each other with a force that is directly proportional to their masses and inversely proportional to the square of their distance apart. In very simple terms, the law can be represented as follows:

F (force of gravity) =
$$\frac{M_1 \text{ (mass of one object)} \times M_2 \text{ (mass of another object)}}{d^2 \text{ (distance squared)}}$$

Suppose that we want to measure the attraction of the earth for a lead ball. Let us assume that the ball weighs one pound at the surface of the earth. In this position the ball is 4,000 mi from the center of the earth because the radius of the earth is 4,000 mi. If the ball is carried 8,000 mi from the center of the earth (twice the distance from the surface), the weight of the ball is decreased to one-fourth of a pound. At a distance of 16,000 mi, the weight is decreased to one-sixteenth of a pound. What we are saying is that the earth and the ball are attracted to each other with a force that is directly proportional to the product of their masses and inversely proportional to the square of the distance between them. If the one-pound ball was carried to a distance of 32,000 mi, how much would it weigh?

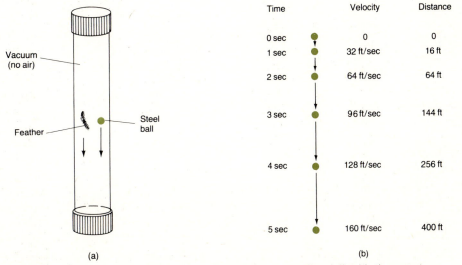

Time	Velocity	Distance
0 sec	0	0
1 sec	32 ft/sec	16 ft
2 sec	64 ft/sec	64 ft
3 sec	96 ft/sec	144 ft
4 sec	128 ft/sec	256 ft
5 sec	160 ft/sec	400 ft

(a) (b)

Figure 3-3 Gravity. (a) If friction is neglected, all bodies fall with the same acceleration. (b) All bodies fall at a constant rate of acceleration of 32 f/second.²

Newton's gravitational law conveniently explains why objects fall to the ground. They fall because of the attraction of the earth (one mass) for the object (the other mass). The law also explains why planets move around the sun. The force of gravity in this case is the centripetal force that pulls the planets inward. Tides are also caused by the pull of the moon on the waters of the earth.

Gravity also has applications in clinical settings. For example, the force of gravity helps liquids to flow in drainage, irrigation, catheters, and intravenous infusions. The principle is also applied in a very important diagnostic test called a *sedimentation rate,* or simply sed rate. When blood stands in a test tube for a period of time, the red cells settle to the bottom, leaving a clear liquid, the plasma, on top. The time it takes for the cells to gravitate to the bottom is important because when there is an infection in the body, the red cells settle faster than when no infection is present. A sed rate is determined by measuring, in millimeters per hour, the rate at which the interface between the red cells and the plasma settles toward the bottom of the sedimentation tube. The Wintrobe-Landsberg method for determining sed rate is explained in Figure 3-4.

We also use gravity to adjust blood pressure. In cases of head injury, the head may be raised to reduce blood pressure and therefore slow down bleeding. Conversely, when a person feels faint, the head should be lowered to increase the pressure of blood in the brain. Gravity causes the blood to flow back to the head. As the oxygen in the blood is passed on to the brain cells, the tendency for fainting is reduced. Can you guess why bedridden patients with circulatory problems often have their feet elevated?

Before leaving our discussion of gravity, we shall say a few words about the center of gravity. The *center of gravity* is the point at which the entire weight of an object may be assumed to be concentrated. The center of gravity of the body in a standing position lies in the pelvic cavity on a line that passes through the area between the feet (Figure 3-5a). The body is said to be stable if a line drawn straight down from the center of gravity passes within its base (area between the feet). For example, if you are carrying a heavy object close to your body, the added weight causes a forward displacement of the center of gravity. However, since the line still falls within the base, you do not topple over (Figure 3-5b). But take the same weight and hold it out at arm's length. Now the center of gravity line is moved so far forward that it extends beyond the base, and you tend to fall forward because your body is unstable (Figure 3-5c).

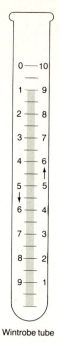

Wintrobe tube

Figure 3-4 Application of gravity to sedimentation rate. When the Wintrobe tube is filled, it contains 10 cm of blood. The numbers 1 through 10 represent centimeters. The sed rate is determined after the tube has been in a rack for one hour, when the red cells have settled and an even line is visible between the clear plasma on top and the red cells below. If the line is at the 1 mark, the red cells have settled 1 cm or 10 mm in one hour, and the sed rate is reported as 10 mm in one hour. The normal sed rate using this method is 0 to 9 mm in one hour for males and 0 to 20 mm in one hour for males and 0 to 20 mm in one hour for females.

You have probably noticed that spreading your feet apart widens the base and thus increases your stability.

Similar reasoning applies to picking up heavy objects from the floor. First you should stoop down close to the object, bending your knees and keeping your back straight. Then, reach for the weight, bring it close to your body, and stand erect. In this way, your stronger leg muscles and not your weaker back muscles are doing the bulk of the work. Take advantage of your center of gravity. It will make your work easier and will protect you from unnecessary injuries.

Simple Machines

A **simple machine** is a device in which mechanical energy is applied at one point and mechanical energy is delivered in a more useful form at another point. A few examples of simple machines are described below.

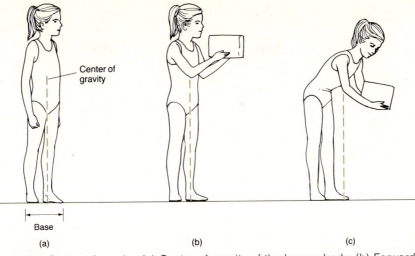

Figure 3-5 Center of gravity. (a) Center of gravity of the human body. (b) Forward displacement of the center of gravity, but stability still maintained. (c) Forward displacement of the center of gravity to such an extent that instability results.

The Lever

A simple **lever** is a rigid bar that moves about an axis called a *fulcrum* or *F*. Any lever consists of three parts—a fulcrum, a resistance, and an effort. The resistance (*R*) is the weight or load to be lifted and the effort (*E*) is the force required to overcome the resistance. Depending on the position of the fulcrum, resistance, and effort, levers are of three types called first-class, second-class, and third-class levers.

In a *first-class lever,* the fulcrum is between the resistance and the effort (Figure 3-6). An everyday example is a seesaw. Notice the relative positions of the fulcrum, resistance, and effort. Many first-class levers may be found in your school in clinical settings, and

your body. Consider a pair of scissors, a hammer, and the muscle action holding your head erect. When you use a pair of scissors, the resistance is an object to be cut, the fulcrum is the screw holding the blades together, and the effort is the force you must apply to cut the object.

Second-class levers are similar to the mechanics of a wheelbarrow (Figure 3-7). In this case, the resistance is between the fulcrum and the effort. The resistance is the weight that you are moving in the wheelbarrow, the effort is the force you must apply to lift and push the wheelbarrow, and the fulcrum is the axle through the wheel. Oxygen-tank carriers work on the same type of lever system. You are employing second-class levers when you use a nutcracker, stand on toptoe,

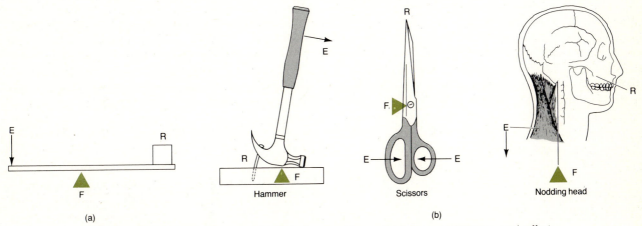

Figure 3-6 First-class lever. (a) Relative positions of fulcrum, resistance, and effort. (b) Examples of first-class levers.

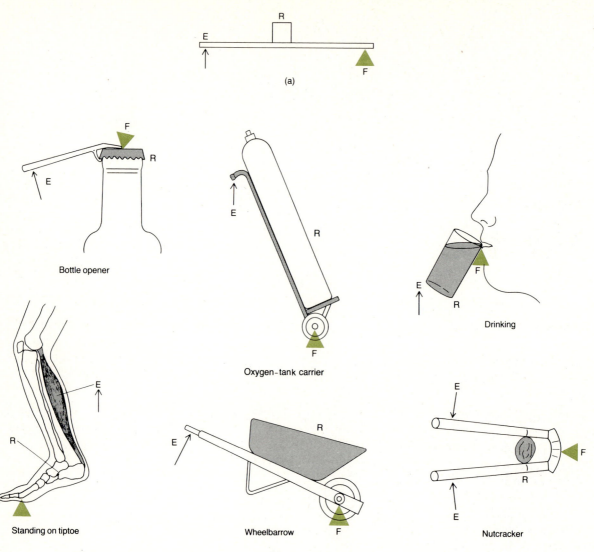

Figure 3-7 Second-class lever. (a) Relative positions of fulcrum, resistance, and effort. (b) Examples of second-class levers.

drink a glass of water, and open the door of a refrigerator or sterilizer.

In *third-class levers*, the effort is between the resistance and the fulcrum (Figure 3-8). An everyday example is lifting a loaded shovel. The resistance is the weight of the loaded shovel to be lifted, the effort is the force you must apply to lift it, and the fulcrum is the handle of the shovel in your hand. An example of a third-class lever in a clinical setting is a pair of forceps.

You will see in a later chapter that in bringing about body movements, bones act as levers, joints function as fulcrums of the bones, and muscular contractions provide the effort to perform work.

The Inclined Plane

An **inclined plane** is a simple machine with which a heavy load may be lifted by applying a relatively

small force. One example is a ramp leading from a lower to a higher level (Figure 3-9). It is much easier to push a patient in a wheelchair up a ramp than to lift the patient the same vertical distance. In fact, the longer the incline—that is, the less steep—the less effort it takes to push the patient. Some inclined planes, such as wedges, are movable. A small force applied at the large end of the wedge causes a greater force to be exerted at the narrow end (point) of the wedge. Examples of the application of this principle are hypodermic needles, suture needles, chisels, and threads of screws.

The Pulley

A **pulley** is a simple machine that consists of a wheel with a grooved rim so that it can carry a rope, cable, or chain. Pulleys are classified as fixed if they are attached to some stationary object or movable if they

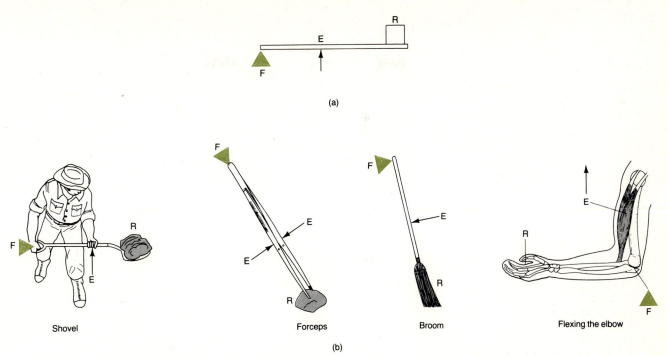

Figure 3-8   Third-class lever. (a) Relative positions of fulcrum, resistance, and effort. (b) Examples of third-class levers.

move with the weight (Figure 3-10). With a fixed pulley, the force required to lift the load is equal to the weight of the load. A common clinical use of the fixed pulley is in traction cases, in which the amount of force applied for traction is equal to the weights applied. A movable pulley, on the other hand, may be used to reduce the amount of effort required. In fact, with a movable pulley, the effort required is only one-half the force of the weight. This means that you can lift a 30-lb weight by applying a force of only 15 lb. However, even though a movable pulley reduces the amount of effort required by one-half, you must pull the rope twice the distance on a movable pulley than on a fixed pulley.

Now that we have studied a few principles of mechanics related to motion and simple machines, we shall examine some very important concepts related to properties of matter.

PROPERTIES OF MATTER

It was mentioned in Chapter 2 that matter exists in three states—gaseous, liquid, and solid. Now we shall examine some of the properties of matter that are essential for an understanding of the structure and function of your body.

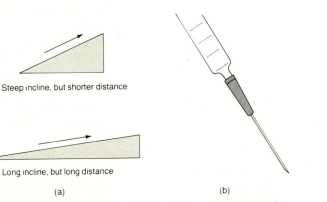

Figure 3-9   Inclined plane. (a) The longer the incline, the less effort it takes to move an object. (b) A wedge at the end of a hypodermic needle.

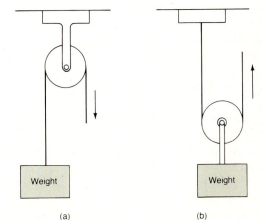
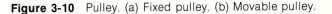

Figure 3-10   Pulley. (a) Fixed pulley. (b) Movable pulley.

The **kinetic theory** helps to explain the properties ɔf gases, liquids, and solids in terms of the forces between the particles of matter and the energy they possess. The basic assumptions of the kinetic theory are as follows:

1. All matter is composed of tiny particles called *molecules*. In gases, the molecules are relatively far apart, in liquids they are closer, and in solids they are closer yet.
2. The molecules of matter are in constant motion. This energy of motion, called *kinetic energy*, depends on temperature.
3. Molecules collide with each other and with the walls of their container. When the molecules collide, there is no loss of energy. Such collisions are said to be elastic.

With these assumptions in mind, let us look at some of the properties of gases, liquids, and solids.

Gases

A **gas** does not have a definite shape or a definite volume. It occupies the entire volume of any container into which it is introduced. Gases exert pressure because the gas molecules are constantly striking the walls of the container. Since the molecules also strike each other and the walls of the container with the same frequency, the same pressure is exerted uniformly over the container walls. Gases are also very compressible. We shall see what this means shortly. Finally, gases expand upon heating. Let us take this information and describe some of the gas laws that greatly affect your body.

Boyle's Law

Boyle's law states that the volume of a gas varies inversely with pressure, assuming that temperature is constant. To see how this works, suppose a gas is introduced into a cylinder with a movable piston and a pressure gauge (Figure 3-11a). Initially the pressure is 1 atmosphere (atm). This pressure is created by the gas molecules striking the walls of the container. Now, if we place weights on the piston to increase the external pressure, the gas molecules are compressed and the gas sample is now concentrated in a smaller volume. The gauge indicates that the pressure doubles when the gas is compressed to one-half its volume (Figure 3-11b). The same number of molecules in half the space produce twice the pressure. Conversely, if the piston is raised to increase the volume, the pressure decreases. Thus, the volume of a gas varies inversely with pressure. This law is applied in

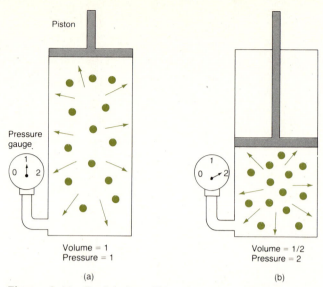

Figure 3-11 Boyle's law. The volume of a gas varies inversely with the pressure. If the volume is decreased to one-fourth, what happens to the pressure?

the operation of a bicycle pump and in blowing up a balloon. Differences in pressure force air into our lungs when we inhale and force the air out of our lungs when we exhale.

Charles' Law

According to **Charles' law,** the volume of a gas is directly proportional to its absolute temperature, assuming that the pressure remains constant. Let us use the cylinder again to demonstrate this law. Assume that the gas in the cylinder exerts an initial pressure of 1 atm (Figure 3-12a) when the piston is halfway down. If the temperature of the gas in the cylinder is raised, the pressure on the walls of the cylinder increases, since the gas molecules are moving faster. As a result, the piston moves upward. As the space within the cylinder increases, the gas molecules have to travel a greater distance to strike the cylinder walls (Figure 3-12b). The number of collisions decreases, so the pressure decreases to 1 atm. The pressures of the two cylinders are constant, but the volume increases in proportion to the temperature increase.

Dalton's Law

According to **Dalton's law,** each gas in a mixture of gases exerts its own pressure as if all the other gases were not present. This pressure is called *partial pressure (p)*. The total pressure of the mixture is calculated by adding all the partial pressures together. Consider a sample of the air we breathe as an example. Atmospheric air is a mixture of several gases—oxygen, carbon dioxide, nitrogen, water vapor, and a

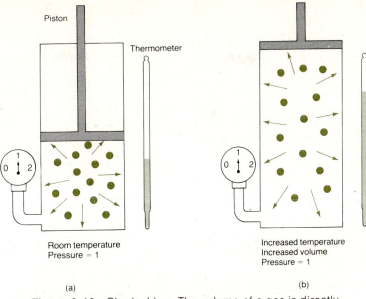

Piston

Thermometer

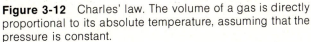

Room temperature
Pressure = 1

(a)

Increased temperature
Increased volume
Pressure = 1

(b)

Figure 3-12 Charles' law. The volume of a gas is directly proportional to its absolute temperature, assuming that the pressure is constant.

number of other gases that appear in trace quantities. Atmospheric pressure, which is 760 mm Hg, is actually the sum of the pressures of all of these gases.

Atmospheric pressure (760 mm. Hg) =
$$pO_2 + pCO_2 + pN_2 + pH_2O$$

We can determine partial pressure by multiplying the percentage of the mixture that the particular gas constitutes by the total pressure of the mixture. For example, to find the partial pressure of oxygen in the atmosphere, we simply multiply the percentage of atmospheric air composed of oxygen (21 percent) by the total atmospheric pressure (760 mm Hg). Thus,

> Atmospheric pO_2 = 21 percent × 760 mm Hg
> Atmospheric pO_2 = 159.60 or 160 mm Hg

Now suppose that we want to calculate the atmospheric pCO_2. Since the percentage of CO_2 in the atmosphere is 0.04, we multiply this figure by 760 mm Hg. Thus,

> Atmospheric pCO_2 = 0.04 percent × 760 mm Hg
> Atmospheric pCO_2 = 0.3040 or 0.3 mm Hg

Can you calculate the pN_2 knowing that the percent of nitrogen in the atmosphere is about 78 percent? When we say that atmospheric pressure is 760 mm Hg, we mean that the total atmospheric pressure is the sum of all the partial pressures of the gases in the atmosphere. You will see later that the partial pressures of gases in your blood are very important in determining the movement of oxygen from your blood to your body cells and the movement of carbon dioxide from your body cells to your blood.

Henry's Law

You have probably noticed many times that a bottle of soda or beer makes a hissing sound when the cap is removed and bubbles rise to the surface for some time afterward. The gas in carbonated beverages is carbon dioxide. Its ability to stay in solution (dissolve in the soda or beer) depends on the partial pressure of the gas. The higher the partial pressure of a gas over a liquid, the more gas will stay in solution. Since the soda or beer is bottled under pressure and capped, the CO_2 remains dissolved as long as the bottle is unopened. However, once you remove the cap, the pressure is released and the gas begins to bubble out. The decrease in pressure lowers the partial pressure of the CO_2, and the gas cannot stay in solution. This common observation is explained by **Henry's law**, which states that the quantity of a gas that will dissolve in a liquid is proportional to the partial pressure of the gas, when the temperature remains constant.

Henry's law also explains a condition called caisson disease or the bends. Normally, even though the air we breathe contains nitrogen, very little of it gets into the blood because it does not dissolve at body temperature. But when a diver breathes air under pressure, the quantity of dissolved nitrogen in the blood increases. If the diver surfaces too quickly, the pressure of the dissolved nitrogen increases to above atmospheric pressure. As a result, nitrogen comes out of solution and tiny bubbles of it may find their way into body tissues. The bubbles may affect the impulses conducted by the nervous system and cause severe pain. The only treatment is to reapply the pressure and then remove it slowly, so that the bubbles do not form. Divers can avoid the bends by breathing a mixture of helium and oxygen instead of air. Helium is only about 40 percent as soluble as nitrogen.

A very important clinical application of Henry's law is the hyperbaric chamber (Figure 3-13). Using pressure to cause more of a gas to dissolve is an effective technique in treating patients who have been infected by what are called anaerobic bacteria. These bacteria, which cause diseases such as tetanus and gangrene, cannot live in the presence of oxygen. A hyperbaric chamber contains oxygen at a pressure of 3 to 4 atm (2,280 to 3,040 mm Hg). The infected body tissues pick up the oxygen and the bacteria are killed. Hyperbaric chambers are also used for treating certain heart disorders and carbon monoxide poisoning.

Liquids

Liquids, unlike gases, occupy a definite volume; they assume the shape of their container. However, a liquid does not expand and completely fill its container like a gas. Liquids have one free surface, and the other surfaces are supported by the container walls. Liquids may also be poured from one container to another. Since the molecules of a liquid are much closer together than those of a gas, a liquid is not as compressible as a gas. Liquids also have higher densities than gases, and if a liquid is left in an open container, it may evaporate.

Liquid Pressure

Pressure has already been defined in Chapter 2 as the number of units of force on a unit area as expressed by the equation $P = F/A$. For liquids, the magnitude of the pressure at any depth is equal to the weight of a column of liquid of unit cross section reaching from that point to the top of the liquid (Figure 3-14). Note in the figure that at a depth of l_1 the pressure, p_1, is determined by the weight of a column of liquid 1 cm² in cross section and l_1 cm high. At a greater depth, l_2, the pressure, p_2, is determined by the weight of a column of liquid 1 cm² in cross section but at a depth of l_2 cm. In other words, the pressure at any given point in a liquid is determined by the weight of the liquid above it. In addition, the pressure at one point is equal to the pressure at any other point at the same level. Liquid pressures are usually measured in pounds per square inch, dynes per square centimeter

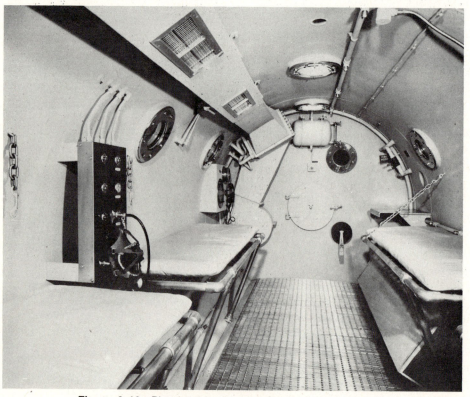

Figure 3-13 Photograph of a hyperbaric chamber. (Courtesy of Harvey Markinson, Sales Manager, Vacudyne Altair, Chicago Heights, Illinois.)

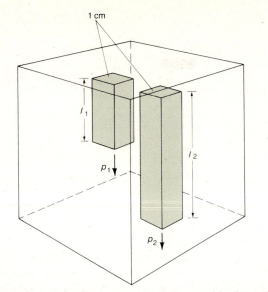

Figure 3-14 Pressure at any depth is equal to the weight of a column of liquid of unit cross section reaching from that point to the top of the liquid. Pressure depends on depth; as depth increases, pressure increases.

(1 dyne (dyn) = 1 g × 1 cm/second²), or newtons per square meter (1 newton (N) = 1 kg × 1 m/second²).

In order to find the weight of a column of liquid, it is necessary to know the density of the liquid. To find density, simply determine the volume and mass of a given substance and divide one by the other. Thus,

$$D = \frac{M}{V}$$

The mass of any column of liquid is determined by $M = DV$. The weight of any mass (M) is determined by multiplying the mass by g (the acceleration due to gravity). Since weight is the total downward force, F, of a liquid column, then $F = Mg$ or $F = DVg$. The volume, V, of a column of liquid is given by the product of h, the height, times A, the area: $V = hA$. If we put all of the relationships together, we find that pressure in a liquid is equal to height of the liquid times the density of the liquid times the acceleration due to gravity. Therefore,

$$P = hDg$$

Liquids exert a pressure on a container because the molecules are attracted to the earth in accordance with Newton's law of gravitation. As we have just seen, the greater the depth of the liquid, the greater the pressure. This has important consequences in procedures such as enemas, irrigations, and intravenous infusions. The higher the containers are placed, the greater the pressure their liquid exerts. Limits of height are observed so that the pressure of the liquid does not become so high as to cause the patient pain from the excessive pressure.

Let us examine a few more characteristics of liquid pressure. The force that a liquid at rest exerts on any surface is perpendicular to the surface, and the pressure acts in all directions. You can demonstrate equal force in all directions by filling with water a hollow steel ball containing several metal tubes leading from the sides and bottom. Then attach a tube to the top of the ball fitted with a handle and a plunger (Figure 3-15). If you push down on the handle, the plunger forces water out of the tubes on the sides and bottom. The height of the water coming out of all of the tubes is the same. This means that the water exerts equal force in all directions.

Because pressure is proportional to the height of a liquid, the pressures at the bottoms of several containers are equal if they all have the same depth of liquid, regardless of the size of the containers. This is illustrated in Figure 3-16. Several vessels of different size containing the same height of water are attached to a pressure meter. Even though the amount of water varies in each vessel, the measured pressure is the same because the height of water is the same in all the vessels.

Pascal's Law

Another characteristic of liquid pressure deals with pressure transmission. This relationship is summarized by **Pascal's law,** which states that an increase

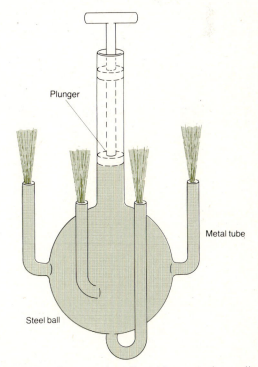

Figure 3-15 Pressure in a liquid is exerted equally in all directions.

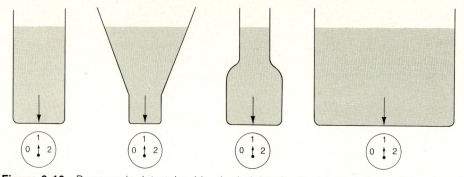

Figure 3-16 Pressure is determined by the height of a liquid, regardless of the size of the vessel.

in pressure applied to a part of a liquid at rest is transmitted undiminished to all parts of the liquid. This law has widespread application. In a hydraulic press system (Figure 3-17), a force is applied on one part of the liquid and a load is moved at another part of the liquid. In the apparatus shown in Figure 3-17, two pistons, one large and one small, are connected by a pipe and filled with a liquid. When pressure is applied to the small piston, the pressure is transmitted undiminished through the liquid to the large piston. Since the large piston has a greater area, a greater force is exerted on it because a greater area has a greater force on it. Thus, the hydraulic press can be used to lift heavy weights with the application of less force. For example, if the larger piston has 10 times the area of the smaller one, a load 10 times that of the applied force can be lifted.

Pascal's law is applied in the hydraulic press used to raise and lower a dentist's chair and an operating table. The chair or table is placed over the large piston and a foot pedal is placed over the smaller piston. Another application of the law is the sphygmomanometer, an instrument used to record blood pressure.

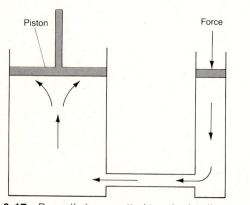

Figure 3-17 Pascal's law applied to a hydraulic press. An increase in pressure applied to a part of a liquid at rest is transmitted undiminished to all other parts of the liquid.

Archimedes' Principle

Liquids exert a buoyant or upward force when objects are placed in them. For example, when you place an object such as a rubber ball in water it apparently loses weight because it is easier to push the ball in water than on land. This observation was recognized by Archimedes (287–212 B.C.), a Greek mathematician and inventor, and is known as **Archimedes' principle.** The principle states that a body immersed in a fluid (liquid or gas) is buoyed up by a force equal to the weight of the fluid displaced. This can be demonstrated by placing a piece of wood in water (Figure 3-18). When the block of wood is first placed in the water, it sinks until the buoyant force of the water becomes great enough to equalize the downward force, which is the weight of the block. The buoyant force is equal to the weight of the volume of the water displaced. If the buoyant force is equal to or greater than the weight of the object, the object will float. If, however, the buoyant force is less than the weight of the object, the object will sink.

A submarine operates on Archimedes' principle. When the vessel submerges, water is permitted to enter its tanks to make the submarine heavier than the weight of the water it displaces. When the vessel surfaces, the tanks are emptied of water to make the submarine lighter than the weight of the water it displaces.

The densities of liquids are measured by the buoyant force they exert on a hydrometer. The hydrometer (see Figure 2-4) is buoyed up by a force equal to the weight of the liquid displaced. When the hydrometer is placed in a liquid of low specific gravity, it sinks deeper to displace a greater volume of liquid so as to produce enough buoyancy to cause it to float. When the hydrometer is placed in a liquid of high specific gravity, it does not sink as deeply because not as much of the liquid is needed to produce a weight equal to that of the hydrometer. Physical therapists also employ Archimedes' principle in the treatment of

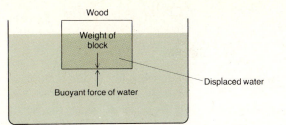

Figure 3-18 Archimedes' principle. A body immersed in a fluid is buoyed up by a force equal to the weight of fluid displaced.

muscular and nervous disorders. Patients with these conditions are placed in water, and because they are buoyed up by the water, they seem to weigh less. Thus, less force is required by therapists to exercise patients' muscles.

Surface Tension

Another very important characteristic of liquids is surface tension. In order to understand what this means, we must first define adhesion and cohesion. **Adhesion** is the molecular force that keeps molecules of different substances together. Adhesive tape sticks to skin because of adhesion. **Cohesion** is the molecular force that keeps molecules of the same substance together. Bones and many other body structures are held together by cohesion.

The differences between adhesion and cohesion may be demonstrated by using the apparatus shown in Figure 3-19. A glass plate supported by one arm of a balance is counterbalanced by weights so that the plate touches the surface of the water. Upon contact, adhesion sets up an attractive force between the water and the plate. If additional weights are added, the plate breaks free from the surface of the water. If the plate is examined, it can be seen that water still clings

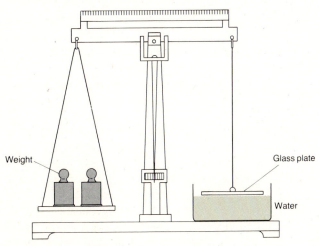

Figure 3-19 Adhesion and cohesion.

to the plate. This means that the break came between the water molecules and not between the water and the glass plate. In other words, adhesion between the water and glass is greater than cohesion between water molecules. In effect, the added weights are a measure of cohesion.

If we take the same apparatus, and substitute mercury for the water, the results are different. In this case, the glass plate pulls away cleanly from the mercury and the added weights are a measure of the adhesive force between the glass and mercury molecules. The cohesion between mercury molecules is greater than the adhesion between mercury and glass.

The cohesion of molecules in liquids produces **surface tension,** a phenomenon in which the surface of a liquid acts at all times as though it has a thin membrane stretched over it. Such a "membrane" is under tension and attempts to contract. It is for this very reason that raindrops and soap bubbles assume a spherical shape. Surface tension is produced because the molecules at the surface of a liquid are attracted more strongly to those next to and below themselves than by air molecules above them. Thus, the surface molecules are pulled inward by the greater downward attraction of the molecules below.

Knowledge of surface tension is important in the medical field. For example, the lower the surface tension of a substance, the more quickly the substance will spread out. It is no coincidence that disinfectants and antiseptics have a lower surface tension than water, the most abundant chemical in the body, and thus spread out more quickly to exert their effects on microbes. Incidentally, soap also lowers the surface tension of water.

Capillarity

Adhesion and cohesion have a lot to do with a phenomenon called capillarity. When a long glass tube is placed in a dish of water, the water rises in the tube until it reaches a certain height above the water level of the dish and then stops (Figure 3-20). The smaller the diameter of the tube, the higher the column will rise. This phenomenon is called **capillarity.** It occurs because the adhesive force for glass and water is greater than the cohesive force of water. Surface tension prevents any of the water from dropping back. The water will continue to rise until the surface tension is equalized by the weight of the liquid in the tube. In cases where the cohesive forces of a liquid are greater than the adhesive forces, the liquid in the tube falls below the level of the liquid in the dish. Mercury is an example of such a liquid.

Since mercury is not adhesive toward glass, it does not stick to the glass and wet it (see Figure 3-19).

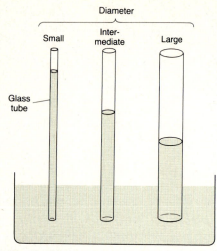

Figure 3-20 Capillarity.

This is one reason that mercury is used in thermometers. If mercury did adhere to the inner glass wall of the thermometer, it would be very difficult to get accurate readings.

Liquid Flow

One of the factors determining the flow of water, oil, or gas through a pipe, or the flow of blood through the vessels of the body, is the resistance of flow offered by the confining walls. Let us examine the apparatus in Figure 3-21 to see what this means. Water in the vertical tank at the left flows through the horizontal tube at the bottom. The pressure of the water flowing through the horizontal tube is measured by five vertical tubes that are equally spaced. The velocity of flow is controlled by the valve at the right. When the valve is closed, the water rises to the same level in each vertical tube, indicating that the pressure in the horizontal tube at the bottom is the same throughout (recall that the pressure in a liquid at rest is the same

at all points of the same height). When the valve is partially opened and a steady flow is attained, the water level falls successively lower in each of the five vertical tubes. This means that there is a successive decrease in pressure along the length of the horizontal tube. In fact, the faster the rate of flow, the greater the drop in pressure along the horizontal tube.

In your body, a similar situation occurs in your blood vessels. As blood flows through your vessels from large arteries to small arteries, to arterioles, to capillaries, to venules, to small veins, to large veins, the pressure successively decreases. By the time blood has returned to the heart, the pressure may be zero, or even negative. We shall examine this principle later when we discuss the circulatory system.

Viscosity

The internal resistance of a liquid to flow is referred to as the **viscosity** of the liquid. Viscosity is caused by the friction between molecules of the liquid as the liquid flows. You can "see" the viscosity of a liquid as it flows. Liquids with lower viscosities, such as alcohol or water, flow or pour more easily than liquids with higher viscosities, such as syrup or motor oil. The more viscous a liquid, the greater the force required to set it in motion and the greater the pressure it exerts once it is moving. The viscosity of blood is 4.5 to 5.5 times greater than that of water. The viscosity of blood is one of the factors involved in maintaining blood pressure since blood resists movement.

Solids

Solids have a definite shape and the shape is independent of their containers. Unlike liquids or gases, solids do not flow under ordinary conditions. All the surfaces of a solid are free surfaces and thus the volume of a solid is independent of its container. The

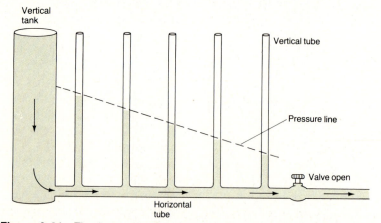

Figure 3-21 The faster the rate of flow, the greater the drop in pressure.

pressures required to decrease the volumes of solids are even greater than those required for liquids. Thus, for all practical purposes, solids are noncompressible.

Elasticity

After a solid substance has been relieved from some kind of external force, it may return to its original size and shape. This property is called **elasticity.** As an example, consider a spring supported at its upper end (Figure 3-22). If a 5-lb weight is placed on its lower end, it stretches 2 inches. If the weight is removed, the spring returns to its original size and shape. This is elasticity. If the 5-lb weight is replaced with a 10-lb weight, the spring stretches 4 inches. Replacement with a 15-lb weight stretches the spring to 6 inches. The amount of stretch is directly proportional to the weight applied. As the weight is doubled or tripled, the amount of stretch is doubled or tripled.

Stress and Strain

When a force of any magnitude is applied to a solid body, the body becomes distorted. The greater the applied force, the greater the distortion. In fact, if too great a force is applied, the body becomes permanently distorted or it breaks. The force that produces distortion is called a **stress** and the distortion produced by the force is referred to as the **strain**. As we have seen in the spring demonstration in Figure 3-22, stress is proportional to strain. This relationship is summarized in **Hooke's law,** which states that the stress within an elastic body is directly proportional to the strain caused by the applied force.

Many kinds of stresses exist. Among the more common types are tension, compression, bending, and twisting. In *tension,* a solid object is stretched as a result of an applied force. In some cases, such as tension weights used in traction, this is beneficial. However, tension may also be injurious to the body. For example, if the elastic limits of tension on ligaments and tendons are exceeded, a sprain may result. In *compression,* a solid object is decreased in length as a result of an applied force. If the elastic limits of

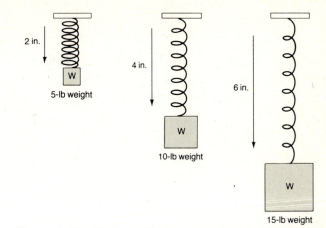

Figure 3-22 Elasticity. The amount of stretch is directly proportional to the weight applied. When the weight is removed, the spring returns to its original size and shape, within limits.

compression on the skull are exceeded, a fracture may result. Under normal circumstances, the long bones of your body, especially those in your legs, are continually subjected to compression from your body weight. But because these bones have a hollow center, they can resist the stress.

Bending involves both tension and compression. Imagine a beam supported at both ends and a force applied at its center to cause it to bend. Now imagine that the beam consists of an upper layer and a lower layer. When the force is applied to the upper layer of the beam, that layer is compressed. At the same time, the lower layer is subjected to tension. If the elastic limits of bending on a bone are exceeded, a fracture called a *greenstick fracture* may result. In this kind of fracture, the end of the bone is not only broken, it is also bent.

Most of us are familiar with *twisting* in the application of a tourniquet, a device used to arrest bleeding. Twisting is also applied to place stress on teeth to change their position during orthodontic work. If the elastic limits of twisting on a bone are exceeded, a fracture called a *spiral fracture* may result.

CHAPTER SUMMARY OUTLINE

MECHANICS

Motion

1. Motion is a branch of physics that deals with the state or motion of bodies resulting from the action of applied external forces.

2. Motion is produced when an external force acts on an object.
3. Speed is distance traveled in a given period of time.
4. Motion in which velocity changes is called acceleration.

5. Newton's laws of motion state that a body at rest remains at rest and a body in motion stays in motion at a constant speed and in a straight line unless acted upon by some external force; a force acting on a body produces an acceleration that is directly proportional to the force and inversely proportional to the mass; and for every action there is an equal and opposite reaction.
6. Motion in a circle involves centripetal and centrifugal force; one application is the centrifuge.
7. All bodies fall at a constant rate of acceleration of 32 ft/second².
8. Newton's law of gravitation states that the centers of any two objects are attracted with a force that is directly proportional to their masses.
9. A clinical application of gravity is sedimentation rate.
10. If the center of gravity of an object is maintained, stability is achieved.

Simple Machines

1. Simple machines are devices that deliver more useful mechanical energy than that applied.
2. Levers consist of a fulcrum, resistance, and effort and are classified into three kinds.
3. In an inclined plane, a heavy load may be lifted by applying a relatively small force.
4. A pulley consists of a wheel that can carry a rope; used in clinical settings in traction.

PROPERTIES OF MATTER

1. The kinetic theory states that all matter is composed of molecules that are in constant motion.
2. Gases do not have definite shape or volume.
3. The behavior of gases is explained by Boyle's law, Charles' law, Dalton's law of partial pressure, and Henry's law.
4. Liquids have a definite volume and shape.
5. The pressure of a liquid is determined by depth.
6. According to Pascal's law, pressure is transmitted undiminished to all parts of a liquid; an application is the hydraulic press.
7. Archimedes' principle states that a body immersed in a fluid is buoyed up by a force equal to the weight of the fluid displaced; the principle is applied in a hydrometer.
8. Surface tension is determined by adhesion and cohesion; antiseptics and disinfectants have a low surface tension.
9. The faster the rate of flow of a liquid, the greater the drop in pressure.
10. The internal resistance of a liquid to flow is called viscosity.
11. All solids have a definite shape.
12. Solids exhibit elasticity, stress and strain, tension, compression, bending, and twisting.

REVIEW QUESTIONS

1. Define mechanics. Why is a knowledge of mechanics important to an understanding of your body?
2. What produces motion? What are the fundamental differences between speed, velocity, and acceleration?
3. If you run 1 mi in 4 minutes, what is your average speed in feet per second? Miles per hour?
4. If a ship travels 2,300 mi from California to Hawaii in 4 hours, what is its average speed in miles per hour? Kilometers per hour?
5. Given the velocity of an object, do you also know its speed? Explain.
6. Distinguish between a scalar and a vector quantity.
7. How is acceleration calculated?
8. Explain Newton's first law of motion. How is inertia related to the law?
9. Compare inertia and friction.
10. Describe Newton's second law of motion. Apply this law to pushing a patient in a wheelchair.
11. What is Newton's third law of motion? How is this law applied to weighing a patient?
12. Contrast centripetal and centrifugal force. How are these forces related to a centrifuge?
13. What is Galileo's contribution to our understanding of gravity?
14. Explain Newton's law of gravitation. Describe a sedimentation rate. What is the importance of gravity in this procedure?
15. What is the center of gravity? Why is it important in lifting objects?
16. Define a simple machine.
17. What is a lever? How are levers classified? Give several examples of the different types of levers.
18. Define an inclined plane. How is the principle important in the construction of hypodermic needles?

19. What is a pulley? How is it used in a clinical setting?
20. Describe the kinetic theory of matter.
21. Explain the characteristics of a gas.
22. Define each of the following, and where possible, make an application to your body or to an everyday situation: Boyle's law, Charles' law, Dalton's law, and Henry's law.
23. Explain the characteristics of a liquid.
24. Define liquid pressure. How is this concept applied to intravenous infusions?
25. How would you demonstrate that liquid pressure acts in all directions?
26. Explain Pascal's law. How is the law applied in the operation of a hydraulic press?
27. What is Archimedes' principle? Apply the principle to the operation of a submarine and a hydrometer. How is the principle employed by physical therapists?
28. Compare adhesion and cohesion. How do they help to explain surface tension?
29. Why is a knowledge of surface tension important in the medical field?
30. Relate adhesion, cohesion, and surface tension to capillarity.
31. Describe the relationship of liquid flow to resistance. Why is this important to the flow of blood through vessels?
32. Define viscosity. Why is the viscosity of blood important?
33. Explain the characteristics of a solid.
34. What is elasticity? Explain it in terms of Hooke's law.
35. Distinguish between stress and strain.
36. What are the essential differences between tension, compression, bending, and twisting?

Chapter 4

INTRODUCTORY CHEMICAL PRINCIPLES

STUDENT OBJECTIVES

After reading this chapter, you should be able to

- Define matter, chemical element, and chemical symbol
- Identify the chemical elements found in the human body
- Diagram the structure of an atom
- Define atomic number, atomic mass, mass number, and Avogadro number
- Explain how atoms combine on the basis of their outer energy level electrons
- Discuss valence as the combining capacity of an atom
- Distinguish between a molecule and a compound
- Define a chemical bond
- Compare ionic and covalent bonding
- Distinguish between synthesis, decomposition, exchange, and reversible chemical reactions
- Describe the mechanism of a chemical reaction according to the collision theory
- Identify the factors that influence chemical reactions
- Contrast exergonic and endergonic chemical reactions
- Classify all matter into pure substances and mixtures
- Identify elements and compounds as pure substances
- Identify solutions, colloidal dispersions, and suspensions as mixtures
- Define the characteristics of a solution
- Explain the process of solvation
- Distinguish between saturated, unsaturated, and supersaturated solutions
- Explain the concentration of a solution in terms of percentage, milligram percent, ratio, parts per million, serial dilution, molarity, and molality
- Define the characteristics of a colloidal dispersion
- Explain Brownian movement and the Tyndall effect as they apply to mixtures
- Describe the characteristics of a suspension
- Define and distinguish between inorganic and organic compounds
- Discuss the functions of water as a solvent, suspending medium, chemical reactant, heat absorber, and lubricant
- List and compare the properties of acids, bases, and salts
- Define pH as the degree of acidity or alkalinity of a solution
- Describe the role of a buffer system as a homeostatic mechanism that maintains the pH of body fluids
- Contrast the structure of deoxyribonucleic acid (DNA) and ribonucleic acid (RNA)
- Define the roles of DNA and RNA in heredity and protein synthesis
- Identify the function and importance of adenosine triphosphate (ATP), cyclic adenosine-3,5-monophosphate (cyclic AMP), and prostaglandins (PG)

In this chapter we shall examine the structure of matter by analyzing the structure of atoms. You will learn something about how atoms react with one another and how matter is organized or classified.

STRUCTURE OF MATTER

Chemical Elements

All living and nonliving things consist of **matter,** which is anything that occupies space and possesses mass. Matter may exist in a solid, liquid, or gaseous state. All forms of matter are made up of a limited number of building units called **chemical elements.** At present, scientists recognize 106 different elements. Elements are designated by letter abbreviations, usually derived from the first or second letter of the Latin or English name for the element. Such letter abbreviations are called *chemical symbols.* Examples of chemical symbols are H (hydrogen), C (carbon), O (oxygen), N (nitrogen), Na (sodium), K (potassium), Fe (iron), and Ca (calcium).

Approximately 24 elements are found in the human organism. Carbon, hydrogen, oxygen, and nitrogen make up about 96 percent of the body's weight. These four elements together with phosphorus and calcium constitute approximately 99 percent of the total body weight. Eighteen other chemical elements, called trace elements, are found in low concentrations and compose the remaining 1 percent.

Atomic Structure

Each element is made up of units of matter called **atoms.** An element is simply a quantity of matter com-posed of atoms all of the same type. A handful of the element carbon, such as pure coal, contains only carbon atoms. A tank of oxygen contains only oxygen atoms. Measurements indicate that the smallest atoms are less than one 250-millionth of an inch in diameter, and the largest atoms are one 50-millionth of an inch in diameter. In other words, if 50 million of the largest atoms were placed end to end, they would measure approximately 1 inch in length.

An atom consists of two basic parts: the nucleus and the electrons (Figure 4-1). The centrally located *nucleus* contains positively charged particles called *protons* (p+) and uncharged particles called *neutrons*

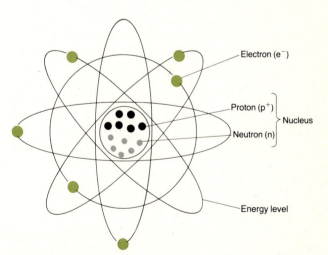

Figure 4-1 Structure of an atom. In this highly simplified version of a carbon atom, note the centrally located nucleus, which contains neutrons and protons. The electrons orbit about the nucleus at varying distances from its center.

(n). Because each proton has one positive charge, the nucleus itself is positively charged. The second basic part of an atom contains the *electrons* (e⁻). These negatively charged particles spin around the nucleus. The number of electrons in an atom always equals the number of protons. Since each electron carries one negative charge, the negatively charged electrons and the positively charged protons balance each other—and the atom is electrically neutral.

What makes the atoms of one element different from those of another? Part of the answer lies in the number of protons. Figure 4-2 shows that the hydrogen atom contains one proton. The helium atom contains two. The carbon atom has six, and so on. Each different kind of atom has a different number of protons in its nucleus. The number of protons (or electrons) in an atom is called the atom's *atomic number*. Therefore we can say that each kind of atom, or element, has a different atomic number (Exhibit 4-1).

In addition to atomic number, atoms have an atomic mass. The actual masses of atoms are very small. An atom of oxygen, for example, has a mass of 2.65×10^{-23} g, while the mass of a hydrogen atom is 1.67×10^{-24} g. Since these very small numbers are difficult to handle, a relative scale of the masses of atoms is used by chemists. One atom has been selected and arbitrarily assigned as mass value. This atom is carbon, and its assigned mass is 12. Thus, an atom of hydrogen which has a mass $\frac{1}{12}$ that of carbon (1.007825), has a relative mass of about 1. An oxygen atom has a mass about $\frac{4}{3}$ that of carbon, so its relative mass is 15.99491 or 16. A magnesium atom has a mass almost double that of carbon (23.98504), so its relative mass is 24. The mass of an atom expressed relative to that of carbon's mass is called the *atomic mass* of an atom. The *mass number* indicates the total number of protons and neutrons in the nucleus (Exhibit 4-1).

The number of atoms in a quantity of an element is its atomic mass in grams. For example, in 12 g of carbon there are 6.02×10^{23} carbon atoms. In 1 g of hydrogen there are 6.02×10^{23} hydrogen atoms. In 16 g of oxygen there are 6.02×10^{23} oxygen atoms. This quantity of atoms, 6.02×10^{23}, is called the **Avogadro number**. We shall use this very important number later in this chapter when we talk about the concentration of a solution. Refer to Exhibit 4-1 again. How many grams of helium, boron, nitrogen, magnesium, chlorine, and calcium contain the Avogadro's number of atoms?

Atoms and Molecules

When atoms combine with other atoms, or break apart from other atoms, the process is called a **chem-**

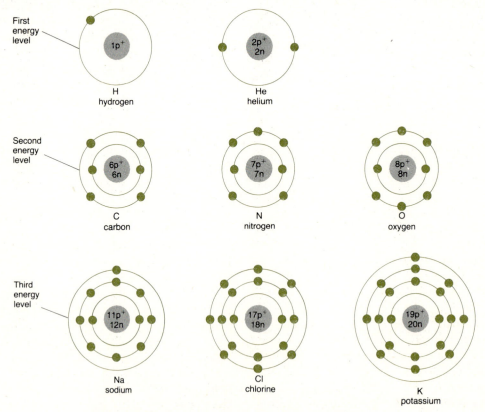

Figure 4-2 Energy levels of some representative atoms.

Exhibit 4-1 ATOMIC NUMBERS, MASS NUMBERS, AND VALENCES OF REPRESENTATIVE ATOMS

Atom and Symbol	Atomic Number	Mass Number	Valence
Hydrogen (H)	1	1	+1
Helium (He)	2	4	0
Lithium (Li)	3	7	+1
Beryllium (Be)	4	9	+2
Boron (B)	5	10	+3
Carbon (C)	6	12	+4
Nitrogen (N)	7	14	+5, −3
Oxygen (O)	8	16	−2
Fluorine (F)	9	19	−1
Neon (Ne)	10	20	0
Sodium (Na)	11	23	+1
Magnesium (Mg)	12	24	+2
Aluminum (Al)	13	27	+3
Silicon (Si)	14	28	+4
Phosphorus (P)	15	31	+5, −3
Sulfur (S)	16	32	−2
Chlorine (Cl)	17	35	−1
Argon (Ar)	18	40	0
Potassium (K)	19	39	+1
Calcium (Ca)	20	40	+2

ical reaction. Chemical reactions are the foundation of all life processes.

The electrons of the atom actively participate in chemical reactions. The electrons spin around the nucleus in orbits, shown in Figure 4-2 as concentric circles lying at different distances from the nucleus. We call these orbits *energy levels*. Each orbit has a maximum number of electrons it can hold. For instance, the orbit nearest the nucleus never holds more than two electrons, no matter what the element. This orbit can be referred to as the first energy level. The second energy level holds a maximum of eight electrons. The third level also can hold a maximum of eight electrons—and if there is a fourth level, the third level can hold a maximum of 18 electrons.

An atom always attempts to stabilize an outermost orbit with the maximum number of electrons it can hold. To do this, the atom may give up an electron, take on an electron, or share an electron with another atom—whichever is easiest. Take a look at the chlorine atom. Its outermost energy level, which happens to be the third level, has seven electrons. However, the third level of an atom can hold a maximum of eight electrons. Chlorine can be visualized as having a shortage of one electron. In fact, chlorine usually does try to pick up an extra electron. Sodium, by contrast, has only one electron in its outer level. This again happens to be the third energy level. It is much easier for sodium to get rid of the one electron in its third energy level than to fill the third level by

taking on seven more electrons. Atoms of a few elements, like helium, have completely filled outer energy levels and do not need to gain or lose electrons. These are called *inert* elements.

If we know how many electrons an atom has in its outer energy level, then we know the number of electrons it must gain, lose, or share in order to fill its outermost energy level. This number is called the **valence** of an atom. The valences of elements that tend to lose electrons are marked + and the valences of elements that tend to gain electrons are marked −. From what we have just said, you can see that the valence of chlorine is −1, that of sodium is +1, and that of helium is 0. Refer again to Exhibit 4-1 and note the valences of the atoms listed.

Atoms with incompletely filled outer energy levels, such as sodium and chlorine, tend to combine with other atoms in a chemical reaction. During the reaction, the atoms can donate, receive, or share electrons and thereby fill their outer energy levels. Atoms that already have filled outer levels generally do not participate in chemical reactions for the simple reason that they do not need to gain or lose electrons. When two or more atoms combine in a chemical reaction, the resulting combination is called a **molecule**. A molecule may contain two atoms of the same kind, as in the hydrogen molecule: H_2. The subscript 2 indicates that there are two hydrogen atoms in the molecule. Molecules may also be formed by the reaction of two or more different kinds of atoms, as in the hydrochloric acid molecule: HCl. Here an atom of hydrogen is attached to an atom of chlorine. A molecule that contains at least two different kinds of atoms is called a **compound**. Hydrochloric acid, which is present in the digestive juices of the stomach, is a compound. A molecule of hydrogen is not.

The atoms in a molecule are held together by forces of attraction called **chemical bonds**. Here we shall consider the two basic types of chemical bonds: ionic bonds and covalent bonds.

Ionic Bonds

Atoms are electrically neutral because the number of positively charged protons equals the number of negatively charged electrons. But when an atom gains or loses electrons, this balance is destroyed. If the atom gains electrons, it acquires an overall negative charge. If the atom loses electrons, it acquires an overall positive charge. Such a negatively or positively charged atom or group of atoms is called an **ion**.

Consider the sodium ion (Figure 4-3a). The sodium atom (Na) has 11 protons and 11 electrons, with one electron in its outer energy level. When sodium gives up the single electron in its outer level, it is left with

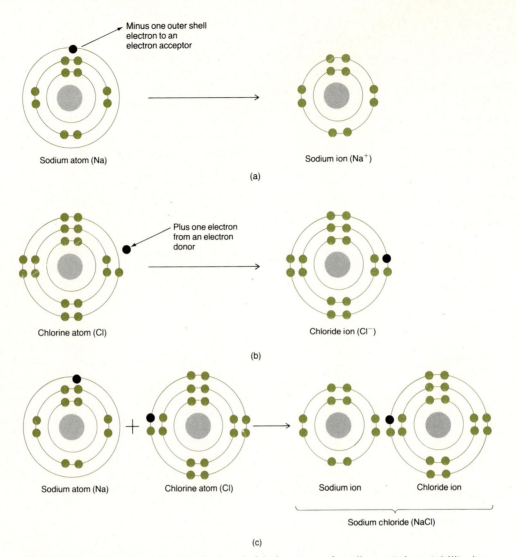

Figure 4-3 Formation of an ionic bond. (a) An atom of sodium attains stability by passing a single electron to an electron acceptor. The loss of this single electron results in the formation of a sodium ion (Na⁺). (b) An atom of chlorine attains stability by accepting a single electron from an electron donor. The gain of this single electron results in the formation of a chloride ion (Cl⁻). (c) When the Na⁺ and the Cl⁻ ions are combined, they are held together by an attraction called an ionic bond, and a molecule of NaCl is formed.

11 protons and only 10 electrons. The atom now has an overall positive charge of one (+ 1). This positively charged sodium atom is called a sodium ion (written Na⁺).

Another example is the formation of the chloride ion (Figure 4-3b). Chlorine has a total of 17 electrons, 7 of them in the outer energy level. Since this energy level can hold 8 electrons, chlorine tends to pick up an electron that has been lost by another atom. By accepting an electron, chlorine acquires a total of 18 electrons. However, it still has only 17 protons in its nucleus. The chloride ion therefore has a negative charge of −1 and is written as Cl⁻.

The positively charged sodium ion (Na⁺) and the negatively charged chloride ion (Cl⁻) attract each other since unlike charges attract one another. The attraction, called an **ionic bond,** holds the two atoms together, and a molecule is formed (Figure 4-3c). The formation of this molecule, sodium chloride (NaCl) or table salt, is one of the most common examples of ionic bonding. Thus an ionic bond is an attraction between an atom that has lost electrons and another that has gained electrons. Generally, atoms whose outer energy level is less than half-filled lose electrons and form positively charged ions called *cations*. Examples of cations are potassium ion (K⁺), calcium ion (Ca²⁺), iron ion (Fe²⁺), and sodium ion (Na⁺). By contrast, atoms whose outer energy level is more than half-filled tend to gain electrons and form negatively charged ions called *anions*. Examples of anions include iodine ion (I⁻), chloride ion (Cl⁻), and sulfur ion (S²⁻).

Notice that an ion is always symbolized by writing the chemical abbreviation followed by the number of positive (+) or negative (−) charges the ion acquires.

Hydrogen is an example of an atom whose outer level is exactly half-filled. The first energy level can hold two electrons, but hydrogen atoms contain only one. Hydrogen may lose its electron and become a positive ion (H⁺). This is precisely what happens when

hydrogen combines with chlorine to form hydrochloric acid (H⁺Cl⁻). However, hydrogen is equally capable of forming another kind of bond: a covalent bond.

Covalent Bonds

The second chemical bond to be considered here is the **covalent bond.** This bond is far more common in organisms than is the ionic bond. When a covalent bond is formed, neither of the combining atoms loses or gains an electron. Instead, the two atoms share their electrons. Look at the hydrogen atom again. One way a hydrogen atom can fill its outer energy

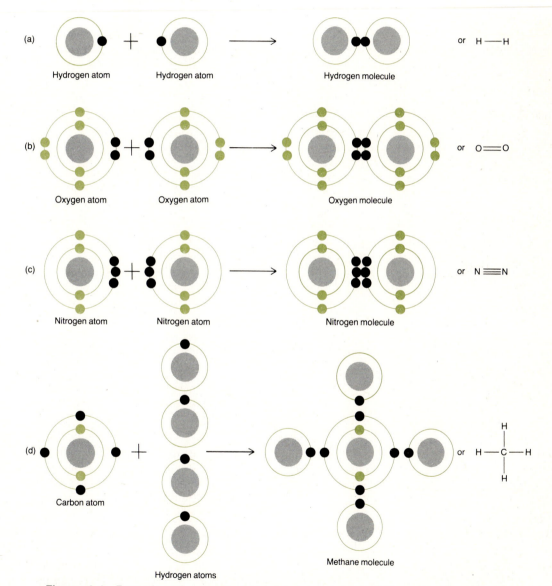

Figure 4-4 Formation of a covalent bond between atoms of the same element and between atoms of different elements. (a) A single covalent bond between two hydrogen atoms. (b) A double covalent bond between two oxygen atoms. (c) A triple covalent bond between two nitrogen atoms. (d) Single covalent bonds between a carbon atom and four hydrogen atoms. The representations on the far right are another way of showing the covalent bonds. Each straight line between atoms is a single covalent bond.

level is to combine with another hydrogen atom to form the molecule H_2 (Figure 4-4a). In the H_2 molecule, the two atoms share a pair of electrons. Each hydrogen atom has its own electron plus one from the other atom. The shared pair actually circles the nuclei of both atoms. Therefore the outer energy levels of both atoms are filled. When only one pair of electrons is shared between atoms, as in the H_2 molecule, a *single covalent bond* is formed. For convenience, a single covalent bond is expressed as a single line between the atoms (H—H). When two pairs of electrons are shared between two atoms, a *double covalent bond* is formed, which is expressed as two parallel lines (=) (Figure 4-4b). A *triple covalent bond*, expressed by three parallel lines (≡), occurs when three pairs of electrons are shared (Figure 4-4c).

Methane (CH_4) is an example of covalent bonding between atoms of different elements (Figure 4-4d). Commonly called swamp gas, methane is responsible for the odor of decaying vegetation. The same principles that apply to covalent bonding between atoms of the same element apply here. The outer energy level of the carbon atom can hold eight electrons but has only four of its own. Each hydrogen atom can hold two electrons but has only one of its own. Consequently, in the methane molecule the carbon atom shares four pairs of electrons. One pair is shared with each hydrogen atom. Each of the four carbon electrons orbits around both the carbon nucleus and a hydrogen nucleus. Each hydrogen electron circles around its own nucleus and the carbon nucleus.

Elements whose outer energy levels are half-filled, such as hydrogen and carbon, form covalent bonds quite easily. In fact, carbon always forms covalent bonds. It never becomes an ion. However, many atoms whose outer energy levels are more than half-filled also form covalent bonds. An example is oxygen. We shall not go into the reasons why some atoms tend to form covalent bonds rather than ionic bonds. You should, however, understand the basic principles involved in bond formation. Chemical reactions are nothing more than the making or breaking of bonds between atoms. And these reactions occur continually in all the cells of your body. As you will see again and again, chemical reactions are the processes by which body structures are built and body functions carried out.

Radioactive Tracers

Atoms of an element, although chemically alike, may have different nuclear mass, so the atomic weight assigned to an element is only a mean. Each of the chemically identical atoms of an element with a particular nuclear mass is an *isotope* of that element. Isotopes are named by a number that indicates their atomic mass, the sum of their neutrons and protons. Certain isotopes called *radioisotopes* are unstable—they "decay" or change their nuclear structure to a more stable configuration. In decaying they emit radiation that can be detected by instruments. These instruments estimate the amount of radioisotope present in a part of the body, or in a sample of material, and form an image of its distribution.

Radioisotopes of iodine were among the first discovered, and their specificity for the thyroid gland made them the cornerstone for the study of thyroid physiology. Nuclear medicine now uses ^{32}P to treat leukemia and ^{59}Fe for the study of red blood cell production. Moreover, short-lived agents such as technetium-99m pertechnetate (^{99m}Tc) have improved the quality of the images and reduced the patient's radiation dose.

Chemical Reactions

In this section, we look at four basic types of chemical reactions. These reactions are simple yet central to life processes. Once you have learned them, you will be able to understand the chemical reactions discussed later.

Synthesis Reactions—Anabolism

When two or more atoms, ions, or molecules combine to form new and larger molecules, the process is called a *synthesis reaction*. The word *synthesis* means to put together, and synthesis reactions involve the *forming of new bonds*. Synthesis reactions can be expressed in the following way:

$$A \quad + \quad B \quad \rightarrow \quad AB$$

| Atom, ion, or molecule A | Atom, ion, or molecule B | Combine to form new molecule AB |

The combining substances, A and B, are called the *reactants;* the substance formed by the combination is the *end product*. The arrow indicates the direction in which the reaction is proceeding. An example of a synthesis reaction is:

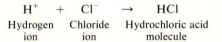

$$H^+ \quad + \quad Cl^- \quad \rightarrow \quad HCl$$

| Hydrogen ion | Chloride ion | Hydrochloric acid molecule |

All the synthesis reactions that occur in your body are collectively called anabolic reactions, or simply *anabolism*. Combining glucose molecules to form glycogen and combining amino acids to form proteins are two examples of anabolism. The importance of anabolism is considered in detail in Chapter 25.

Decomposition Reactions — Catabolism

The reverse of a synthesis reaction is a *decomposition reaction*. The word *decompose* means to break down into smaller parts. In a decomposition reaction, the *bonds are broken*. Large molecules are broken down into smaller molecules, ions, or atoms. A decomposition reaction occurs in this way:

$$AB \rightarrow A + B$$

| Molecule AB breaks down into | Atom, ion, or molecule A | Atom, ion, or molecule B |

Under the proper conditions, methane can decompose into carbon and hydrogen:

$$CH_4 \rightarrow C + 4H$$

| Methane molecule | One carbon atom | Four hydrogen atoms |

The subscript 4 on the left-hand side of the reaction equation indicates that four atoms of hydrogen are bonded to one carbon atom in the methane molecule. The number 4 on the right-hand side of the equation shows that four single hydrogen atoms have been set free.

All the decomposition reactions that occur in your body are collectively called catabolic reactions, or simply *catabolism*. The digestion and oxidation of food molecules are examples of catabolism. The importance of catabolism is considered in detail in Chapter 25.

Exchange Reactions

All chemical reactions are based on synthesis or decomposition processes. In other words, chemical reactions are simply the making and/or breaking of ionic or covalent bonds. Many reactions, such as *exchange reactions*, are partly synthesis and partly decomposition. An exchange reaction works like this:

$$AB + CD \rightarrow AD + BC$$

Here the bonds between A and B and between C and D are broken in a decomposition process. New bonds are then formed between A and D and between B and C in a synthesis process.

Reversible Reactions

When chemical reactions are reversible, the end product can revert to the original combining molecules. A *reversible reaction* is indicated by two arrows:

$$A + B \underset{\text{breaks down to}}{\overset{\text{combines with}}{\rightleftharpoons}} AB$$

Some reversible reactions occur because neither the reactants nor the end products are very stable. Other reactions reverse themselves only under special conditions:

$$A + B \underset{\text{water}}{\overset{\text{heat}}{\rightleftharpoons}} AB$$

Whatever is written above or below the arrows indicates the special condition under which the reaction occurs. In this case, A and B react to produce AB only when heat is applied, and AB breaks down into A and B only when water is added. Figure 4-5 summarizes the basic chemical reactions that can occur.

Mechanism of Chemical Reactions

One theory that offers a widely accepted view of how chemical reactions occur is called the **collision theory.** According to this theory, all molecules, atoms, and ions are in constant motion. A chemical reaction occurs when the interacting substances collide. Whether a chemical reaction results from the collisions depends on the velocity of the colliding particles, their energy, and the chemical configurations of the particles. If the velocity is greater, the probability of eventual collision is greater. If a specific minimum energy requirement is met, then the probability of a chemical reaction increases. If the colliding particles are joined at their surfaces, the probability of a reaction is greater. Enzymes hold colliding particles at their surfaces in living systems. Let us compare a chemical reaction without an enzyme and with an enzyme.

First we assume that the following chemical reaction occurs in the absence of an enzyme: X (reactant) $\rightarrow$ Y (product). In the entire population of reactant molecules, some have little kinetic energy, the majority have average kinetic energy, and a few have high kinetic energy. It is only the high-energy reactant molecules that are capable of entering into a chemical reaction to form a product. The term given to the minimum kinetic energy of collision for a chemical reaction to occur is **activation energy** (Figure 4-6). The frequency of collisions can be increased by increasing temperature (Figure 4-7a). In this way, more reactant molecules can achieve the required activation energy. In addition to heat, other nonenzymatic agents of acceleration are higher concentrations of reactants and increased pressure. They also permit more reactant molecules to attain the necessary activation energy for reaction.

Now assume that the same chemical reaction, X (reactant) $\rightarrow$ Y (product), occurs in the presence of an enzyme (Figure 4-7b). In this case, the enzyme joins with molecules (substrates) that have a wider range of kinetic energies, not only those that are energy rich. As a result, the presence of an enzyme

GENERAL NATURE   SPECIFIC EXAMPLE

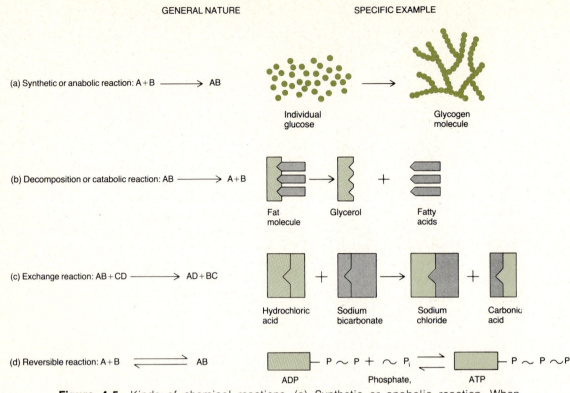

(a) Synthetic or anabolic reaction: A + B ⟶ AB

Individual glucose

Glycogen molecule

(b) Decomposition or catabolic reaction: AB ⟶ A + B

Fat molecule   Glycerol   Fatty acids

(c) Exchange reaction: AB + CD ⟶ AD + BC

Hydrochloric acid   Sodium bicarbonate   Sodium chloride   Carbonic acid

(d) Reversible reaction: A + B ⇌ AB

ADP   Phosphate,   ATP

Figure 4-5 Kinds of chemical reactions. (a) Synthetic or anabolic reaction. When linked together as shown, molecules of glucose form a molecule of glycogen. Glucose is a sugar that is the primary source of energy. Glycogen is a storage form of that sugar found in the liver and skeletal muscles. (b) Decomposition or catabolic reaction. Shown in the example is a molecule of fat breaking down into glycerol and fatty acids. This particular reaction occurs whenever a food containing fat is digested. (c) Exchange reaction. In this exchange, atoms of different molecules are exchanging with each other. Shown is a buffer reaction in which the body eliminates strong acids to help maintain homeostasis. (d) Reversible reaction. The molecule of ATP (adenosine triphosphate) is an important source of stored energy. When the energy is needed, the ATP breaks down into ADP (adenosine diphosphate) and PO_4 (phosphate group), and energy is released in the reaction. The phosphate group is symbolized as P_i. The cells of the body reconstruct ATP by using the energy of foods to attach ADP to PO_4.

lowers activation energy. Thus, more reactant molecules participate in the chemical reaction and are converted to product in the presence of an enzyme. The enzyme is capable of speeding up the reaction without an increase in temperature. This has important consequences to living systems since enzymes are proteins and can be destroyed by high temperatures. Although enzymes will accelerate chemical reactions, they will not make a chemical reaction proceed if it normally does not occur.

Factors Affecting Chemical Reactions

One factor that influences the rate of a chemical reaction is the concentration of the reactants. Within limits, as the concentration of reactants increases, the rate of the reaction increases. This is related to the fact that more molecules of reactant increase the fre-

quency of collision. Another factor is temperature. Once again, within limits, as temperature increases, the rate of reaction increases. This is because a rise in temperature speeds up the velocity of reacting molecules, thus increasing their frequency of collisions. An increase in pressure has usually the same effect as an increase in temperature. Another factor that influences a chemical reaction is the presence of a *catalyst,* a substance that usually speeds up the reaction rate. Enzymes are catalysts, and as noted earlier, they increase reaction rate by lowering the level of the energy of activation.

The Energy of Chemical Reactions

Some form of energy is involved whenever bonds between atoms in molecules are formed or broken during the chemical reactions taking place in the body. *When*

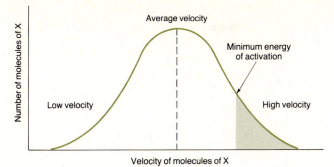

Figure 4-6 Distribution of velocities among a large population of molecules. The bell-shaped curve indicates that some molecules of X have a low velocity, some have a very high velocity, and a major proportion of the population possesses average velocity. In order for a chemical reaction to occur, molecules of X must have sufficient activation energy. That portion with the minimum energy of activation is represented by the colored area. Only this proportion of molecules of X is capable of being converted to molecules of Y.

a chemical bond is formed, energy is required. Conversely, when a bond is broken, energy is released. This means that synthesis reactions need energy in order to occur, whereas decomposition reactions give off energy. The building processes of the body, such as the construction of bones, the growth of hair and nails, and the replacement of injured cells, occur basically through synthesis reactions. The breakdown of foods, on the other hand, occurs through decomposition (hydrolysis) reactions. When foods are broken down, they release energy that can be used by the body for its building processes. Released energy can also be used to warm the body by taking the form of heat energy.

Chemical reactions that release more energy than they absorb are called *exergonic*. The general form of such a reaction is:

$$\text{Reactants} = \text{Products} + \text{Energy}$$

Reactions that release energy from the foods we eat are exergonic. Other reactions in the body are *endergonic*, that is, they absorb more energy than they release. The general form of an endergonic reaction is:

$$\text{Reactants} + \text{Energy} = \text{Products}$$

Reactions in your body that store energy for later use are endergonic.

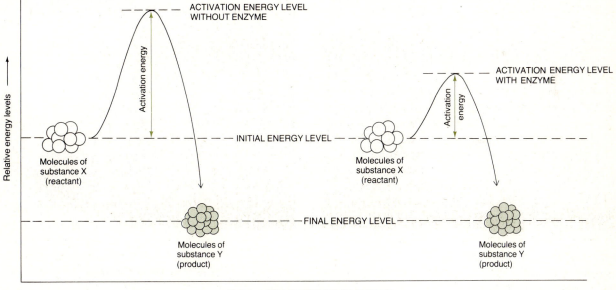

(a)   (b)

Figure 4-7 Energy requirements of a chemical reaction with and without an enzyme. Before the reaction can occur, molecules of X must collide with one another with sufficient energy (activation energy) to form molecules of Y. In reaction (a), the nonenzymatic reaction, the frequency of collisions may be increased by raising the temperature. In reaction (b), an enzyme-catalyzed reaction, the presence of the enzyme lowers the required activation energy. In this way more molecules of X are converted to Y (at a lower temperature) since a larger fraction of molecules of substance X possess the required activation energy for reaction.

ORGANIZATION OF MATTER

We have already noted that matter exists in three main states — solid, liquid, or gas. We have also learned that the atoms are very closely packed in a solid, spread out more in a liquid, and spread out a great deal in a gas. Now we are going to organize or classify all matter — solid, liquid, or gas — into a system that will help us understand how and why matter behaves in certain ways under certain conditions.

The organization of matter into categories, like any other classification scheme, is arbitrary and is based upon certain relationships. For example, if you were asked to organize a stamp collection, you might organize the stamps according to country; or, you might organize them by year of issue; or, you might organize them by value. Your criteria for organization are arbitrary and are based on particular relationships. In fact, even the standards of measurement you studied in Chapter 2 are arbitrary. So it is with our organization of matter.

For purposes of our subsequent discussion, we shall organize matter into two main categories: (1) pure substances and (2) mixtures (Figure 4-8). **Pure substances** are samples of matter in which all parts are identical in composition. Two kinds of pure substances are elements and compounds. In any sample of the element oxygen, all parts (all of its atoms) are identical in composition. The same holds true for elements of mercury and copper. Now consider the three compounds, methane or CH_4 (a gas), water or H_2O (a liquid), and table salt or $NaCl$ (a solid). In any sample of them, all of their parts (all of their molecules) are identical in composition, but obviously different from each other. Notice what we are saying: Elements and compounds are pure substances.

Most matter does not exist in pure form. Rather, it exists with other forms of matter as a mixture or dispersion system. A **mixture** is composed of two or more pure substances but they are *not joined to each other by chemical bonds*. Sand in water is a mixture. The grains of sand do not form chemical bonds with the water molecules and the two can be separated very easily. Because no bonds are formed, the sand and the water retain their individual identities. The fact that no chemical change occurs when a mixture is made is the chief difference between a mixture and a compound.

When particles of one kind of matter are spread through the particles of another to make a mixture, a **dispersion system** is created. Dust in the air, gases of the atmosphere, salt in water, and starch in water are examples of dispersion systems or mixtures. If you examine Figure 4-8 again, you can see that there are three principal kinds of mixtures: solutions, colloidal dispersions, and suspensions.

Solutions

A mixture formed by dissolving one substance in another is referred to as a **solution.** The dissolved substance is called the **solute** and the substance in which it is dissolved is called the **solvent.** Among the kinds of solutions are solids in liquids (sugar in water), liquids in liquids (vinegar — acetic acid in water), and gases in liquids (carbonated beverage — carbon dioxide in water). Other examples are shown in Exhibit 4-2. In any solution, the individual atoms, ions, or molecules of solute are separated by atoms, ions, or molecules of the solvent. Let us see how solutions are formed.

Water is by far the most abundant and important solvent in living systems. Solutions containing water as the solvent are said to be *aqueous solutions*. Consider a solution of salt in water. A crystal of sodium chloride consists of positively charged sodium and negatively charged chloride ions ionically bonded together (Figure 4-9a). When a salt crystal is placed in water, its weak ionic bonds are broken by the positive

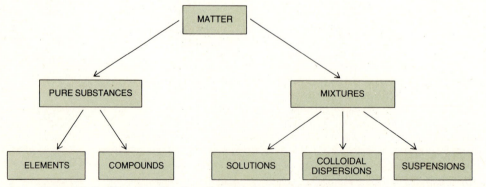

Figure 4-8 The classification of matter into pure substances and mixtures.

Exhibit 4-2 KINDS OF SOLUTIONS

Solute and Solvent	Example
Gas in a liquid	Carbonated beverage such as champagne (carbon dioxide in alcohol)
Liquid in a liquid	Vinegar (acetic acid in water)
Solid in a liquid	Sugar in water
Gas in a gas	Atmospheric air
Liquid in a gas	Humid air
Solid in a gas	Vapor from mothballs in air
Gas in a solid	Alloy of palladium and hydrogen
Liquid in a solid	Dental fillings (mercury in cadmium)
Solid in a solid	Steel (carbon in iron)

and negative charges of the water molecules. The hydrogen portion of a water molecule has a positive charge that attracts the Cl^- ions, and the oxygen portion of the water molecule has a negative charge that attracts the Na^+ ions. This attraction occurs first at the surface of the crystal (Figure 4-9b). As a new surface becomes exposed, the process continues. In time, all of the Na^+ and Cl^- ions of the crystal are pulled away and each is surrounded by several water molecules. This process is called *solvation*. The accumulation of water molecules around each ion shields the oppositely charged ions from making contact and attracting each other. Thus, the ions of the salt are kept apart in the solution.

Experience tells us there is a limit to how much solute a solvent can hold at a given temperature. Just put too much sugar in your coffee and you will see. The *solubility* of a substance is generally expressed as the number of grams of solute that will dissolve in 100 g of solvent at a given temperature. A solution that contains all the solute it can is said to be *saturated*. An *unsaturated* solution contains any smaller amount of solute. A solution that contains more than the amount of solute that can be dissolved is said to be *supersaturated*.

Characteristics of Solutions

The following is a list of some of the more important characteristics of solutions:

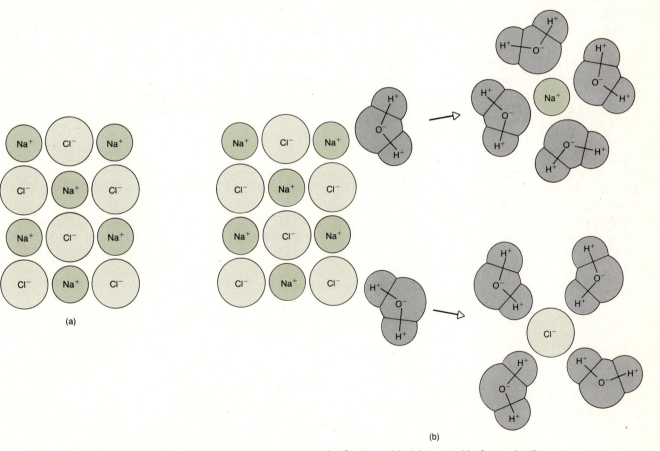

Figure 4-9 The solvating property of water. (a) Sodium chloride crystal before solvation. (b) During solvation, water molecules pull the ions in the crystal away from each other. Once separated, water molecules surround the ions of the crystal. The ions are thus kept apart in solution.

1. Solutions are homogeneous. *Homos* means the same and *genos* means kind. This means that a solution is uniform throughout. A sample taken at the top, middle, or bottom is the same.
2. Solutions are clear, but not necessarily colorless. For example, a solution of Easter egg dye is clear, but it also has a distinctive color.
3. In any solution, the individual particles of solute are always smaller than 10 Å in diameter.
4. The solute in a saturated or unsaturated solution does not settle to the bottom of its vessel upon standing. In a saturated or unsaturated salt solution, the salt stays dispersed among the water molecules; it does not settle.
5. The solute in a solution can pass through a filter. In fact, many solutes also pass through cell membranes. The importance of this fact will be seen later.
6. The components of a solution may be separated from each other by physical means, such as evaporation or distillation. In *evaporation* of a solution, the solvent usually changes to a vapor at a lower temperature than the solute. In the case of our salt solution, the water (solvent) evaporates and the salt (solute) is left behind. In *distillation*, the temperature of the solution is raised gradually until the two components vaporize. They are then conducted to different containers where they are cooled and returned to their liquid states. Distillation is an effective way to separate a liquid from a liquid solution.

Concentration of Solutions

It is common to describe the concentration of a solution as either concentrated or dilute. Concentrated implies that the solute constitutes the greater portion of the solution. Dilute implies that the solute constitutes a lesser portion of the solution. In the medical field, these terms are only relative and are not at all precise. For example, when we talk about the strength of a disinfectant or antibiotic, it is essential to indicate the precise concentration of the solution. It is not enough merely to say that the solution is concentrated or dilute.

Chemists refer to the *concentration* of a solution as the quantity of solute in a given amount of solvent. Among the common ways that we can express concentration are percentage, milligram percentage, ratio, serial dilution, molarity, and molality.

Percentage Solutions. The *percentage concentration* of a solution is usually expressed as the number of grams of solute in 100 ml of solution. Percent means

parts in 100. For example, a 10 percent sugar solution is prepared by placing 10 g of solid sugar in a calibrated container and then adding enough distilled water to make 100 ml of solution (Figure 4-10a). In a percentage solution in which the solute and solvent are a solid and liquid, respectively, there is a weight (10 g) to volume (100 ml) relationship. In a percentage solution in which the solute and solvent are a liquid in a liquid, there is a volume to volume relationship. For example, in order to make a 20 percent alcohol solution, you would place 20 ml of 100 percent pure alcohol in a graduated container and then add enough distilled water to make 100 ml.

The formula that we use to determine the percentage of solute in a solution is as follows:

$$\text{Percentage} = \frac{\text{grams (or milliliters) of solute}}{\text{volume of solution}} \times 100$$

If 16 g of salt is dissolved in 100 ml of water, what is the percentage of the salt solution? Substitute the known values into the formula.

$$\text{Percentage} = \frac{16 \text{ g}}{100 \text{ ml}} \times 100$$

$$\text{Percentage} = \frac{1600}{100}$$

$$\text{Percentage} = 16$$

If 4 g of sugar is placed in 50 ml of water, what is the percentage of the sugar solution?

$$\text{Percentage} = \frac{4 \text{ g}}{50 \text{ ml}} \times 100$$

$$\text{Percentage} = \frac{400}{50}$$

$$\text{Percentage} = 8$$

If 38 ml of 100 percent pure alcohol is added to enough distilled water to make 75 ml of solution, what is the percentage of the alcohol solution?

$$\text{Percentage} = \frac{38 \text{ ml}}{75 \text{ ml}} \times 100$$

$$\text{Percentage} = \frac{3800}{75}$$

$$\text{Percentage} = 50.6$$

We use the factor of 100 in all cases because percent means parts in 100.

Milligram Percentage. The concentrations of many substances are so low that they are more conveniently expressed in milligrams percent (mg percent). For example, many solutes in body fluids such as blood in urine are of low concentration. *Milligrams percentage*

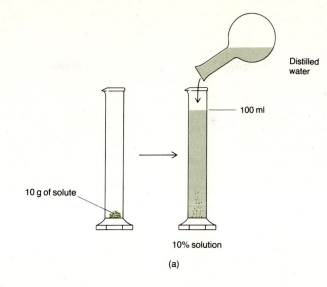

(a)

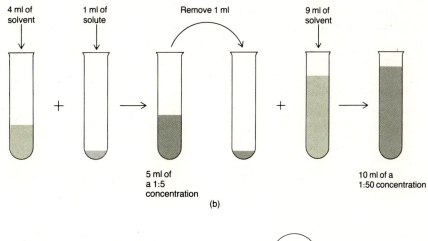

(b)

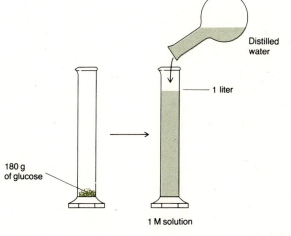

(c)

Figure 4-10 Preparing concentrations of solutions. (a) Preparation of a percentage solution. (b) Serial dilution. (c) Preparation of a molar solution.

is the number of milligrams of solute in 100 ml of body fluid. The concentration of glucose in blood normally ranges between 80 and 100 mg/100 ml of blood, or 80 to 100 mg percent. If we expressed this as a simple percentage concentration, we would have to say 0.080 to 0.10 percent. As you can see, the use of milligram percentage for expressing certain concentrations eliminates an awkward decimal fraction. Here is another example. The concentration of sodium in urine is normally 0.35 mg percent. Express this value as a simple percentage value. Which is easier, percentage or milligram percentage?

Ratio. Many chemical solutions, such as disinfectants and antiseptics, are furnished to institutions in a concentrated form. However, prior to their use, they must be diluted in order to be effective. The concentration of a dilute solution is frequently expressed as the *ratio* between the parts of solute and the parts of solvent. For an antiseptic that has a concentration of 1:1,000, one part of the solute is present in 1,000 parts of the solvent. Some of the concentrations employed in many microbiology laboratory studies are so small that the concentration is expressed in a ratio of *parts per million (ppm)*. The concentration of chlorine in a swimming pool that effectively prevents the spread of disease is 0.5 to 1.0 ppm. Can you convert this value to milligram percentage and simple percentage?

Serial Dilution. Very frequently, laboratory procedures in microbiology and physiology involve the use of several concentrations of the same solution. For example, it might be necessary to prepare dilutions of bacteria of 1:5 and 1:50. This means 1 part bacteria to 4 parts water and 1 part bacteria to 49 parts water, respectively. The obvious way to prepare a 1:50 solution is to combine 1 part bacteria with 49 parts water. However, this may involve making 50 volumes of the solution, an unnecessarily large quantity. This problem can be avoided by diluting the 1:5 solution until it is $\frac{1}{10}$ as strong. Mixing one volume of the 1:5 solution with nine volumes of water will result in a 1:50 solution. This method of preparing dilutions of different concentrations of the same material is called *serial dilution*. (Figure 4-10b.) In some cases, it may be necessary to make serial dilutions of 1:100, 1:1,000, 1:10,000, 1:100,000, 1:1,000,000, and so on.

Molarity. The *molarity* of a solution is the concentration of the solution expressed in terms of the number of moles of solute per liter of solution. A molar solution of any substance is written as $1M$, or M, and contains one mole of solute in the amount of solvent that produces a total volume of one liter. In order to explain this, we must first define a mole. One *mole* of a substance contains the Avogadro number of particles (atoms, molecules, or ions). In other words, a mole is a measure of the number of particles in a solution. One mole of sulfur contains 6.02×10^{23} atoms and weighs 32 g (see Exhibit 4-1). One mole of oxygen contains 6.02×10^{23} atoms and weighs 16 g. The mass in grams of one mole of a substance is the *gram atomic weight* of an element. Thus, the gram atomic weight of sulfur is 32 g, for oxygen it is 16 g, for carbon it is 12 g, for hydrogen it is 1 g, and so on. If we wish to weigh out one mole of carbon atoms, we must take one gram atomic weight of carbon, that is 12 g. How would you make a 1 molar ($1M$) solution of carbon atoms? You would weigh out 1 mole (12 g) and add enough water to make 1 liter of solution. How would you make a $2M$ solution of carbon? You would weigh out 2 moles (24 g) and add enough water to make 1 liter of solution. How about a $\frac{1}{10}$ molar solution ($0.1M$) of carbon? Weigh out 0.1 mole (1.2 g) and add enough water to make 1 liter of solution.

Molecular oxygen is symbolized O_2 because oxygen atoms only exist bonded together; such molecules are said to be diatomic. Since one mole of oxygen (O) contains 16 g of oxygen atoms, two moles of oxygen atoms (O_2) contain 32 g of oxygen atoms. We have simply added the separate atomic weights (16 + 16). The sum is called *molecular weight*. What is the molecular weight of H_2O? It is 2 + 16 or 18. What is the molecular weight of Cl_2? CH_4? NH_3? Actually, we have added more than atomic weights. We have added the masses in grams. So we call it *gram molecular weight*. The gram molecular weight of O_2 is 32 g. The gram molecular weight of a diatomic molecule is one mole of molecules of the element. Thus, 16 g of oxygen atoms, 32 g of oxygen molecules, 1 g of hydrogen atoms, and 2 g of hydrogen molecules all contain the same number of particles.

What about atoms of different elements in a molecule? One mole of H_2O contains 2 moles of H atoms and 1 mole of O atoms. Two moles of H atoms has a mass of 2 g and 1 mole of oxygen atoms has a mass of 16 g. One mole of water, therefore, has a gram molecular weight of 18 g. One mole of methane (CH_4) contains 1 mole of carbon atoms (12 g) and 4 moles of hydrogen atoms (4 g). The gram molecular weight of 1 mole of methane is 16 g. The gram molecular weight of a substance is one mole of molecules of the substance. This means that there are the same number of molecules in 18 g of water and 16 g of methane. Moles of all molecular substances contain the same number of molecules, the Avogadro number, 6.02×10^{23} molecules per mole.

Let us try a final example of preparing a 1M solution of glucose ($C_6H_{12}O_6$). We calculate the gram molecular weight of glucose by adding $72(C_6) + 12(H_{12}) + 96(O_6) = 180$ g. To make a 1M glucose solution, we place 180 g of glucose in a beaker and add enough water to make 1 liter of solution (Figure 4-10c). How would you make a 0.5M and a 2M glucose solution?

Molality. The *molality* of a solution is the number of moles of solute in one kilogram of solvent. The symbol for molality is *m*. A half-molal solution (0.5m) contains 0.5 mole of solute per kilogram of solvent. A 2m solution has two moles of solute in one kilogram of solvent. Molal solutions are important because for a given solvent two solutions of equal molality have the same ratio of solute to solvent molecules. Molality is used to prepare solutions when temperature and pressure changes may otherwise change solvent volumes.

Colloidal Dispersions

A **colloidal dispersion** is a heterogeneous (not uniform) mixture in which the components do not separate. A colloidal dispersion is thus somewhat like a solution because the particles do not settle. It differs from a solution in several important respects. We have already learned that the solute particles in a solution are less than 10 Å. In a colloidal dispersion, the solute particles are larger, ranging in size from 10 Å to 1,000 Å. In addition, although colloidal particles can pass readily through most filters, they cannot pass through membranes. This fact is very important in the operation of artificial kidneys. Like solutions, colloidal dispersions may consist of any combination of the three states of matter. These are listed in Exhibit 4-3. Here, the solute is called the *dispersed phase* and the solvent the *dispersing medium*.

The terms *solute* and *solvent* are not used in reference to a colloidal dispersion because there is no

solvation. In other words, one substance does not dissolve in another. Thus, we use the terms dispersed phase and dispersing medium. Yet the dispersed phase does not settle out. Why? Colloidal particles are in a state of constant although sluggish motion because they are continuously bombarded by the molecules of the suspending medium. This random motion is called *Brownian movement*, and was first observed in 1827 by an English botanist named Robert Brown. He originally observed the movement of pollen grains in water. The continuous, random movement of colloidal particles may be observed through a microscope. Brownian movement is sufficient to overcome the force of gravity in the colloidal particles and thus they do not settle.

Because solutions are clear, light rays pass through a solution without any significant change. However, when light rays pass through a colloidal dispersion, the rays are scattered somewhat and can be seen passing through the dispersion. This light-scattering property of colloidal dispersions is referred to as the *Tyndall effect*. Because particles in a colloidal dispersion are large enough to reflect light, the beam can be seen passing through. This light-scattering property of colloidal dispersions gives such solutions a cloudy character.

Colloidal particles also tend to remain suspended because they have an electrical charge. The charge may be positive or negative, depending on the nature of the dispersed phase. But all the particles in any given dispersing medium have the same charge. Do you remember what this means? Like charges repel each other. This overcomes the tendency of the particles to attract each other, form large masses, and settle out under the influence of gravity.

Colloids have another property called adsorption. *Adsorption* is the ability of colloids to hold and attract substances on their surfaces. This is so because colloids have very large surface areas. One application of this fact is the use of colloidal poison gas masks. The charcoal is able to adsorb gas molecules and remove them from the air. Another application employed clinically is the use of burnt toast that adsorbs swallowed poisons in the stomach before they can get into the blood.

In some colloidal dispersions, the dispersed phase has a tendency to settle out. In such cases, a substance called a *protective colloid* is added which prevents the particles from clumping together and settling out. Protective colloids are frequently used in pharmaceutical preparations to keep the medication suspended in the dispersing medium so that it has maximum effect in the body. A similar mechanism is employed in your body in the digestion and transporta-

Exhibit 4-3 TYPES OF COLLOIDAL DISPERSIONS

Dispersed Phase	Dispersing Medium	Example
Gas	Liquid	Foam, whipped cream
Gas	Solid	Taffy
Liquid	Gas	Mist, fog
Liquid	Liquid	Cream, milk
Liquid	Solid	Butter, cheese
Solid	Gas	Smoke
Solid	Liquid	Gelatin, paints
Solid	Solid	Alloys, pearls, ruby glass

tion of fats. This mechanism operates as follows. An *emulsion* is a colloidal dispersion of a liquid in a liquid. A good example is oil in water. We know from experience that if oil and water are shaken together, the oil drops become dispersed among the water molecules. But if the mixture is allowed to stand, the two separate. The dispersed phase has settled out. How can this be prevented? Soap can be added to the mixture to function as a protective colloid. The soap covers the oil drops with a thin film with a negative charge. As a result, the oil drops repel each other and stay in solution. Such protective colloids that stabilize emulsions are called *emulsifying agents.* The casein in milk is the emulsifying agent in milk. Milk therefore is an emulsion of butterfat in water. Egg yolk is the emulsifying agent in mayonnaise, an emulsion of olive or corn oil in water. Emulsification is very important in our bodies as a means of keeping dietary fats in very small droplets for efficient digestion and transportation to all body cells. The emulsifying agent for fats in our bodies is called bile, a substance made by the liver and stored in the gallbladder.

Suspensions

A **suspension** is a mixture in which the diameter of the dispersed particles is greater than 1,000 Å. Like the particles in a colloidal dispersion, those in a suspension do not undergo solvation; they are insoluble. If we note the effect of light on a suspension, we find that the suspended material blocks the passage of light, as does a colloidal dispersion. Particles in a suspension do not pass through a filter or through a membrane. Suspended particles tend to settle upon standing because of their large size; they are pulled by gravity. Suspensions are also heterogeneous. A good example of a suspension is clay in water. Another example is blood. Here the cells are suspended in plasma.

A few examples of pharmaceutical suspensions are milk of magnesia (magnesia in water); Kaopectate (aluminum hydroxide in water); and Crysticillin (penicillin in water).

PHYSIOLOGICALLY IMPORTANT MOLECULES

Most of the chemicals in the body exist in the form of compounds. Biologists and chemists divide these compounds into two principal classes: inorganic compounds, which usually lack carbon, and organic compounds, which always contain carbon. **Inorganic compounds** are usually small, ionically bonded molecules

that are vital to body functions. They include water, many salts, acids, and bases. **Organic compounds** are held together mostly or entirely by covalent bonds. They tend to be very large molecules and are therefore good building blocks for body structures. Organic compounds present in the body include carbohydrates, lipids, proteins, nucleic acids, ATP, cyclic AMP, and prostaglandins. The chemistry and importance of carbohydrates, lipids, and proteins will be considered in Chapter 25, where metabolic physiology is described.

Inorganic Substances

Water

One of the most important, as well as the most abundant, inorganic substances in the human organism is **water.** In fact, with a few exceptions such as tooth enamel and bone tissue, water is by far the most abundant material in all tissues. About 60 percent of red blood cells, 75 percent of muscle tissue, and 92 percent of blood plasma consist of water. Although there is no specific amount of water that must be present in living matter, the average water content is 65 to 75 percent. The following functions of water explain why it is such a vital compound in living systems:

1. Water is an excellent solvent and suspending medium. The solvating property of water is essential to health and survival. For example, water in the blood forms a solution with some of the oxygen you inhale, allowing the oxygen to be carried to your body cells. Water in the blood also dissolves much of the carbon dioxide that is carried from the cells to the lungs to be exhaled. Furthermore, if the surfaces of the air sacs in your lungs are not moist, oxygen cannot dissolve and therefore cannot move into your blood to be distributed throughout your body. Water, moreover, is the solvent that carries nutrients into your body cells and wastes out.

2. Water participates in many chemical reactions. During digestion, for example, water can be added to large nutrient molecules in order to break them down into smaller molecules. This kind of breakdown is necessary if the body is to utilize the energy in nutrients. Water molecules are also used in synthesis reactions—reactions in which smaller molecules are built into larger ones. Such reactions occur in the production of hormones and enzymes.

3. Water absorbs and releases heat very slowly. In comparison to other substances, water requires a large amount of heat to increase its temperature and a great loss of heat to decrease its temperature.

Thus water maintains a more constant body temperature than other solvents, despite fluctuations in environmental temperature. Water thus helps to maintain a homeostatic body temperature.

4. Water serves as a lubricant in various regions of the body. It is a major part of mucus and other lubricating fluids. Lubrication is especially necessary in the chest and abdomen, where internal organs touch and slide over each other. It is also needed in joints where bones, ligaments, and tendons rub against each other. In the digestive tract, water moistens foods to ensure their smooth passage.

Acids, Bases, and Salts

When molecules of inorganic acids, bases, or salts are dissolved in water in the body cells, they undergo *ionization* or *dissociation*. This is, they break apart into ions. An **acid** may be defined as a substance that dissociates into one or more *hydrogen ions* (H^+) and one or more negative ions or anions. An acid may also be defined as a proton (H^+) donor. A **base,** by contrast, dissociates into one or more *hydroxyl ions* (OH^-) and one or more positive ions or cations. A base may also be viewed as a proton acceptor. Hydroxyl ions, as well as some other negative ions, have a strong attraction for protons. A **salt,** when dissolved in water, dissociates into cations and anions, neither of which is H^+ or OH^- (Figure 4-11).

Many salts are found in the body. Some are more concentrated in cells, whereas others are more concentrated in the body fluids, such as lymph, blood, and the extracellular fluid of tissues. The ions of salts are the source of many essential chemical elements. Chemical analyses reveal that sodium and chloride ions are present in higher concentrations than other ions in extracellular body fluids. Inside the cells, phosphate and potassium ions are more abundant than other ions. Chemical elements such as sodium, phosphorus, potassium, and iodine are present in the body only in chemical combination with other elements or as ions. Their presence as free, un-ionized atoms could be instantly fatal.

Acid–Base Balance

The fluids of your body must maintain a fairly constant balance of acids and bases. In solutions such as those found in body cells or in extracellular fluids, acids dissociate into hydrogen ions (H^+) and anions. Bases, on the other hand, dissociate into hydroxyl ions (OH^-) and cations. The more hydrogen ions that exist in a solution, the more acid the solution. Conversely, the more hydroxyl ions that exist in a solution, the more basic, or alkaline, the solution. The term **pH** is used to describe the degree of *acidity* or *alkalinity (basicity)* of a solution. Biochemical reactions—reactions that occur in living systems—are extremely sensitive to even small changes in the acidity or alkalinity of the environments in which they occur. In fact, H^+ and OH^- ions are involved in practically all biochemical processes, and the functions of cells are modified greatly by any departure from narrow limits of normal H^+ and OH^- concentrations. For this reason, the acids and bases that are constantly formed in the body must be kept in balance.

A solution's acidity or alkalinity is expressed on a *pH scale* that runs from 0 to 14 (Figure 4-12). The pH scale is based on the number of H^+ ions in a solution expressed in chemical units called moles per liter. A pH of 7 means that a solution contains one ten-millionth (0.0000001) of a mole of H^+ ions per liter. The number 0.0000001 is written 10^{-7} in exponential form. To convert this value to pH, the negative exponent (-7) is converted into the positive number 7. Thus a solution with 0.0000001 mole of H^+ ions per liter has a pH of 7. A solution with a concentration of

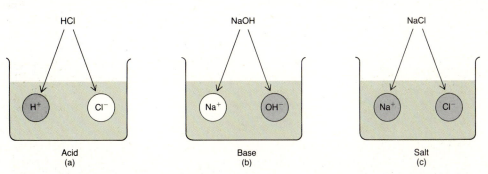

Figure 4-11 Ionization of acids, bases, and salts. (a) When placed in water, hydrochloric acid, HCl, dissociates into H+ and Cl⁻ ions. Acids are proton donors. (b) When the base sodium hydroxide, NaOH, is placed in water, it dissociates into Na+ and OH⁻ ions. Bases are proton acceptors. (c) When table salt, NaCl, is placed in water, it dissociates into positive and negative ions, neither of which is H+ or OH⁻.

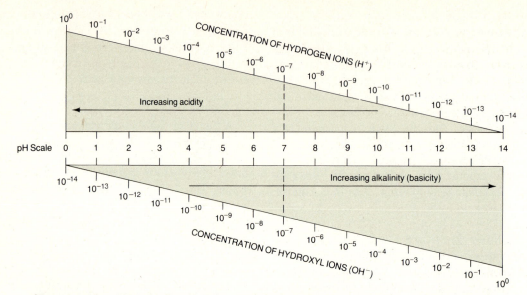

Figure 4-12 pH scale. At pH 7 (neutrality), the concentration of H+ and OH− ions is equal. A pH value below 7 indicates an acid solution; that is, there are more H+ ions than OH− ions. The lower the numerical value of the pH, the more acid the solution is because the H+ ion concentration becomes progressively greater. A pH value above 7 indicates an alkaline (basic) solution—there are more OH− ions that H+ ions. The higher the numerical value of the pH, the more alkaline the solution is because the OH− ion concentration becomes progressively greater. A change in one whole number on the pH scale represents a tenfold change from the previous concentration.

0.0001 (10^{-4}) has a pH of 4, a solution with a concentration of 0.000000001 (10^{-9}) has a pH of 9, and so on.

A solution that is zero on the pH scale has many H+ ions and few OH− ions. A solution that rates 14, by contrast, has very many OH− ions and practically no H+ ions. The midpoint in the scale is 7, where the concentration of H+ and OH− ions is equal. See Exhibit 4-4. Any substance that has a pH of 7, such as pure water, is neutral. Any solution that has more

H+ ions than OH− ions is an *acid solution* and has a pH below 7. If a solution has more OH− ions than H+ ions, it is a *basic* or *alkaline solution* and has a pH above 7. A change of one whole number on the pH scale represents a tenfold change from the previous concentration. That is, a pH of 2 indicates 10 times fewer H+ ions than a pH of 1. A pH of 3 indicates 10 times fewer H+ ions than a pH of 2 and 100 times fewer H+ ions than a pH of 1.

Exhibit 4-4 RELATIONSHIP OF pH SCALE TO RELATIVE CONCENTRATIONS OF HYDROGEN (H+) AND HYDROXYL (OH−) IONS

Concentration of H+ Ions			pH		Concentration of OH− Ions	
1.0	10^0		0		10^{-14}	0.00000000000001
0.1	10^{-1}		1		10^{-13}	0.0000000000001
0.01	10^{-2}		2		10^{-12}	0.000000000001
0.001	10^{-3}		3		10^{-11}	0.00000000001
0.0001	10^{-4}		4		10^{-10}	0.0000000001
0.00001	10^{-5}		5		10^{-9}	0.000000001
0.000001	10^{-6}		6		10^{-8}	0.00000001
0.0000001	10^{-7}	(Neutrality)	7	(Neutrality)	10^{-7}	0.0000001
0.00000001	10^{-8}		8		10^{-6}	0.000001
0.000000001	10^{-9}		9		10^{-5}	0.00001
0.0000000001	10^{-10}		10		10^{-4}	0.0001
0.00000000001	10^{-11}		11		10^{-3}	0.001
0.000000000001	10^{-12}		12		10^{-2}	0.01
0.0000000000001	10^{-13}		13		10^{-1}	0.1
0.00000000000001	10^{-14}		14		10^0	1.0

Buffer Systems

Although the pH of body fluids may differ, the normal limits for the various fluids are generally quite specific and narrow. Exhibit 4-5 shows the pH values for certain body fluids compared with common substances. Even though strong acids and bases are continually taken into the body, the pH levels of these body fluids remain relatively constant. The mechanisms that maintain these homeostatic pH values in the body are called **buffer systems.**

The essential function of a buffer system is to react with strong acids or bases in the body and replace them with weak acids or bases that can change the normal pH values only slightly. Strong acids (or bases) ionize easily and contribute many H^+ (or OH^-) ions to a solution. They therefore change the pH drastically. Weak acids (or bases) do not ionize easily. Therefore they contribute fewer H^+ (or OH^-) ions and have little effect on the pH. The chemicals that change strong acids or bases into weak ones are called **buffers** and they are found in the body's fluids. Of the several buffer systems, only the *carbonic acid–bicarbonate buffer system,* the most important one in extracellular fluid, will be described.

The carbonic acid–bicarbonate buffer system consists of a *pair* of compounds. One of them is a *weak acid,* and the other is a *weak base.* The weak acid of the buffer pair is *carbonic acid* (H_2CO_3); the weak base is *sodium bicarbonate* ($NaHCO_3$). The carbonic acid is a proton (H^+) donor and the bicarbonate ion of the sodium bicarbonate is a proton acceptor. In solution, the members of this buffer pair dissociate as follows:

Acidic component: $$H_2CO_3 \rightleftharpoons H^+ + HCO_3^-$$
Carbonic acid, Hydrogen ion, Bicarbonate ion

Basic component: $$NaHCO_3 \rightleftharpoons Na^+ + HCO_3^-$$
Sodium bicarbonate, Sodium ion, Bicarbonate ion

Each member of the buffer pair has a specific role in helping the body maintain a constant pH. If the body's pH is threatened by the presence of a strong acid, the weak base of the buffer pair goes into operation. If the body's pH is threatened by a strong base, the weak acid goes into play.

Consider the following situation. If a strong acid, such as HCl, is added to extracellular fluid, the weak base of the buffer system goes to work and the following reaction occurs:

$$(1) \quad HCl + NaHCO_3 \rightleftharpoons NaCl + H_2CO_3$$
Hydrochloric acid (strong acid), Sodium bicarbonate (weak base of buffer system), Sodium chloride (salt), Carbonic acid (weak acid)

The chloride ion of HCl and the sodium ion of sodium bicarbonate combine to form NaCl, a substance that has no effect on pH. The hydrogen ion of the HCl could greatly lower pH by making the solution more acid, but this H^+ ion combines with the bicarbonate ion (HCO_3^-) of sodium bicarbonate to form carbonic acid, a weak acid that lowers pH only slightly. In

Exhibit 4-5 NORMAL pH VALUES OF REPRESENTATIVE SUBSTANCES

Substance	pH Value
Gastric juice (digestive juice of the stomach)	1.2–3.0
Lemon juice	2.2–2.4
Grapefruit	3.0
Cider	2.8–3.3
Pineapple juice	3.5
Tomato juice	4.2
Clam chowder	5.7
Urine	5.5–7.5
Saliva	6.5–7.5
Milk	6.6–6.9
Pure (distilled) water	7.0
Blood	7.35–7.45
Semen (fluid containing sperm)	7.35–7.50
Cerebrospinal fluid (fluid associated with nervous system)	7.4
Pancreatic juice (digestive juice of the pancreas)	7.1–8.2
Eggs	7.6–8.0
Bile (liver secretion that aids in fat digestion)	7.6–8.6
Milk of magnesia	10.0–11.0
Limewater	12.3

other words, because of the action of the weak base of the buffer system, the strong acid (HCl) has been replaced by a weak acid and a salt, and the pH remains almost constant.

Now suppose a strong base, such as sodium hydroxide (NaOH), is added to the extracellular fluid. In this instance, the weak acid of the buffer system goes to work and the following reaction takes place:

$$\text{NaOH} + \text{H}_2\text{CO}_3 \rightleftharpoons \text{H}_2\text{O} + \text{NaHCO}_3$$

(2)

| Sodium hydroxide (strong base) | Carbonic acid (weak acid of buffer system) | Water | Sodium bicarbonate (weak base) |

In this reaction, the OH^- ion of sodium hydroxide could greatly raise the pH of the solution by making it more alkaline. However, the OH^- ion combines with an H^+ ion of carbonic acid and forms water, a substance that has no effect on pH. In addition, the Na^+ ion of sodium hydroxide combines with the bicarbonate ion (HCO_3^-) to form sodium bicarbonate, a weak base that has little effect on pH. Thus, because of the action of the buffer system, the strong base is replaced by water and a weak base, and the pH remains almost constant.

Whenever a buffering reaction occurs, the concentration of one member of the buffer pair is increased while the concentration of the other decreases. When a strong acid is buffered, for example, the concentration of carbonic acid is increased but the concentration of sodium bicarbonate is decreased. This happens because carbonic acid is produced and sodium bicarbonate is used up in the acid buffering reaction (see reaction 1). When a strong base is buffered, the concentration of sodium bicarbonate is increased but the concentration of carbonic acid is decreased. This happens because sodium bicarbonate is produced and carbonic acid is used up on the basic buffering reaction (see reaction 2). When the buffered substances, HCl and NaOH in this case, are removed from the body via the lungs or kidneys, the carbonic acid and sodium bicarbonate formed as products of the reactions function again as components of the buffer pair.

Organic Substances

Organic substances are chemical compounds that contain carbon and usually hydrogen and oxygen as well. Carbon has several properties that make it particularly useful to living organisms. For one thing, it can react with one to several hundred other carbon atoms to form large molecules of many different shapes. This means that the body can build many compounds out

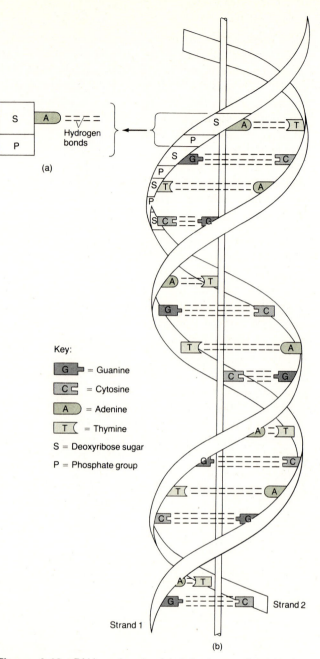

Figure 4-13 DNA molecule. (a) Adenine nucleotide. (b) Portion of an assembled DNA molecule.

Key:
G = Guanine
C = Cytosine
A = Adenine
T = Thymine
S = Deoxyribose sugar
P = Phosphate group

of carbon, hydrogen, and oxygen. Each compound can be especially suited for a particular structure or function. The large size of carbon molecules and the fact that they do not dissolve easily in water make them useful materials for building body structures. Carbon compounds are mostly or entirely held together by covalent bonds, and they tend to decompose easily. This means that organic compounds are also a good source of energy. Ionic compounds are not good energy sources because they form new ionic bonds as soon as the old ones are broken.

Nucleic Acids

Exceedingly large organic molecules containing carbon, hydrogen, oxygen, nitrogen, and phosphorus are **nucleic acids,** compounds first discovered in the nuclei of cells. Whereas the basic structural units of proteins are amino acids, the basic units of nucleic acids are *nucleotides*. Nucleic acids are divided into two principal kinds: **deoxyribonucleic acid (DNA)** and **ribonucleic acid (RNA).**

A molecule of DNA is a chain composed of repeating units called deoxyribonucleotides. Each deoxyribonucleotide consists of three basic parts (Figure 4-13a):

1. It contains one of four possible *nitrogen bases,* which are ring-shaped structures containing atoms of C, H, O, and N. The nitrogen bases found in DNA are named adenine, thymine, cytosine, and guanine.
2. It contains a pentose sugar called *deoxyribose*.
3. It also contains *phosphate groups*.

The deoxyribonucleotides are named according to the nitrogen base that is present. Thus one containing thymine is called a *thymine deoxyribonucleotide*. One containing adenine is called an *adenine deoxyribonucleotide,* and so on.

The chemical composition of the DNA molecule was known before 1900, but it was not until 1953 that a model of the organization of the chemicals was constructed. This model was proposed by J. D. Watson and F. H. C. Crick on the basis of data from many investigations. Figure 4-13b shows the following structural characteristics of the DNA molecule:

1. The molecule consists of two strands with crossbars. The strands twist about each other in the form of a *double helix* so that the shape resembles a twisted ladder.
2. The uprights of the DNA ladder consist of alternating phosphate groups and the deoxyribose portions of the deoxyribonucleotides.
3. The rungs of the ladder contain paired nitrogen bases. As shown, adenine always pairs off with thymine and cytosine always pairs off with guanine.

Cells contain hereditary material called genes, each of which is a segment of a DNA molecule. Our genes determine which traits we inherit, and they control all the activities that take place in our cells throughout a lifetime. When a cell divides, its hereditary information is passed on to the next generation of cells. The passing of information is possible because of DNA's unique structure.

RNA, the second principal kind of nucleic acid, differs from DNA in several respects. RNA is single-stranded; DNA is double-stranded. The sugar in the RNA ribonucleotide is the pentose ribose. And RNA does no contain the nitrogen base thymine. Instead of thymine, RNA has a nitrogen base called uracil. At least three different kinds of RNA have been identified in cells. Each type has a specific role in reacting with DNA to help regulate protein synthesis reactions.

Adenosine Triphosphate (ATP)

A molecule that is indispensable to the life of the cell is **adenosine triphosphate (ATP).** This substance is found universally in living systems and performs the essential function of storing energy for various cellular

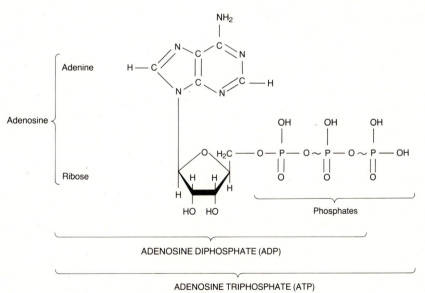

Figure 4-14 Structure of ATP.

activities. Structurally, ATP consists of three phosphate groups and an adenosine unit composed of adenine and the five-carbon sugar ribose (Figure 4-14). ATP is regarded as a high-energy molecule because of the total amount of usable energy it releases when it is broken down by the addition of a water molecule (hydrolysis).

When the terminal phosphate group, symbolized P_i, is hydrolyzed, the reaction liberates a great deal of energy. This energy is used by the cell to perform its basic activities. Removal of the terminal phosphate group leaves a molecule called **adenosine diphosphate (ADP).** This reaction may be represented as follows:

$$ATP \rightleftharpoons ADP + P_i + E$$

| Adenosine triphosphate | Adenosine diphosphate | Phosphate | Energy |

The energy supplied by the catabolism of ATP into ADP is constantly being used by the cell. Since the supply of ATP at any given time is limited, a mechanism exists to replenish it—a phosphate group is added to ADP to manufacture more ATP. Logically, energy is required to manufacture ATP. The reaction may be represented as follows:

$$ADP + P_i + E \rightleftharpoons ATP$$

| Adenosine diphosphate | Phosphate | Energy | Adenosine triphosphate |

The energy required to attach a phosphate group to ADP is supplied by various decomposition reactions taking place in the cell, particularly by the decomposition of glucose. The body does not use the energy from the decomposition of glucose because glucose cannot be stored in cells. Some glucose can be converted to glycogen and stored in the liver, but the rest is decomposed immediately. ATP, by contrast, can be stored in every cell, where it provides potential energy that is not released until needed.

Cyclic AMP

A chemical substance closely related to ATP is **cyclic adenosine-3,5-monophosphate (cyclic AMP).** Essentially it is a molecule of adenosine monophosphate with the phosphate attached to the ribose sugar at two places (Figure 4-15). The attachment forms a ring-shaped structure—thus the name cyclic AMP. Cyclic AMP is formed from ATP by the action of a special enzyme, called *adenyl cyclase,* located in the cell membrane. Although the cyclic AMP was discovered

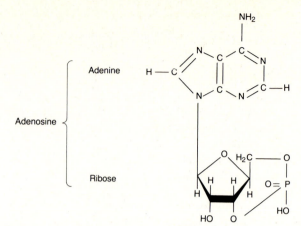

Figure 4-15 Structure of cyclic AMP.

in 1958, only recently has its function in cells become clear. One function is related to the action of hormones, a topic we explore in detail in Chapter 17.

Prostaglandins

Prostaglandins (PG) are a large group of membrane-associated lipids composed of 20 carbon fatty acids containing a cyclopentane ring. (A cyclopentane contains five carbon atoms joined to form a ring.) Prostaglandins were first discovered in prostate gland secretions, but they are now known to be present in many body tissues. Prostaglandins are produced in cell membranes in minute quantities and are rapidly decomposed by catabolic enzymes. One function of prostaglandins is to control the activity of adenyl cyclase and thus the production of cyclic AMP. This function is discussed in Chapter 17.

Although synthesized in very small quantities, prostaglandins are potent substances and exhibit a wide variety of effects on the body. Basically, prostaglandins mimic hormones. Among the effects produced by prostaglandins are:

1. Lowering and raising blood pressure.
2. Stimulating and inhibiting uterine contractions.
3. Causing abortion.
4. Inducing labor.
5. Transmitting nerve impulses.
6. Regulating metabolism.

Investigations are now under way to determine the role of prostaglandins in preventing peptic ulcers, opening bronchial passageways, clearing nasal passages, and inducing menstruation.

CHAPTER SUMMARY OUTLINE

STRUCTURE OF MATTER

1. Matter is anything that occupies space and has mass. It consists of chemical elements represented by symbols.
2. Elements consist of atoms that contain protons and neutrons in a nucleus and electrons that orbit the nucleus.
3. Atomic number is the number of protons; the mass of an atom relative to that of carbon is called atomic mass.
4. The number of atoms in a quantity of an element is its atomic mass in grams and is expressed by the Avogadro number, 6.02×10^{23}.

How Atoms Combine

1. Electrons move about the nucleus in energy levels; each level has a fixed number of electrons that it can hold.
2. Outer energy level electrons enter into chemical reactions.
3. The combining capacity of an element is called its valence.
4. The result of chemical combination of two or more atoms is called a molecule.
5. A molecule that contains at least two different kinds of atoms is called a compound.
6. Atoms are held together by forces of attraction called chemical bonds; the two principal kinds are called ionic and covalent bonds.
7. In ionic bonds, electrons are lost and gained; in covalent bonds, electrons are shared.

Chemical Reactions

1. Four principal kinds of chemical reactions are synthesis, decomposition, exchange, and reversible.
2. The mechanism of chemical reactions is explained by the collision theory.
3. Among the factors influencing chemical reactions are the quantity of reactants and products, the temperature, and the presence of catalysts.
4. In chemical reactions, breaking bonds produces energy and making bonds requires energy.
5. Reactions that release more energy than they absorb are called exergonic; those that absorb more than they release are endergonic.

ORGANIZATION OF MATTER

1. Matter is organized into pure substances (elements and compounds) and mixtures (solutions, colloidal dispersions, and suspensions).
2. In mixtures, the forming substances are not joined by chemical bonds.

Solutions

1. A solution is a mixture formed by dissolving one substance (solute) in another (solvent).
2. Depending on the amount of solute, solutions may be saturated, unsaturated, or supersaturated.
3. Solutions are homogeneous, clear, with solute size smaller than 10 Å in diameter; the solute does not settle; the solute can pass through a filter and membranes; the components of a solution may be separated by evaporation or distillation.
4. The concentration of a solution can be expressed as a percentage, in milligram percent, in parts per million, by serial dilution, by molarity, or by molality.

Colloidal Dispersions

1. Colloidal dispersions are heterogeneous mixtures in which the components do not separate.
2. Particles range from 10 to 1,000 Å in diameter and pass through filters but not membranes.
3. The two components of a colloidal dispersion are the dispersed phase and the dispersing medium.
4. Colloids remain suspended because of Brownian movement. A light ray passed through a colloidal dispersion is scattered. This scattering phenomenon is known as the Tyndall effect.
5. Colloidal particles have a great capacity for adsorption.
6. Protective colloids stabilize colloidal dispersions by preventing clumping.
7. Bile is a protective colloid in the body that keeps fat droplets apart so that they do not clump.

Suspensions

1. A suspension is a mixture in which the dispersed particles are greater than 1,000 Å in diameter.
2. The particles do not undergo solvation, do scatter light, do not pass through filters or membranes, and settle upon standing.

PHYSIOLOGICALLY IMPORTANT MOLECULES

1. Inorganic substances usually lack carbon, contain ionic bonds, resist decomposition, and dissolve readily in water.
2. Organic substances always contain carbon and usually hydrogen and oxygen as well. Most organic substances contain covalent bonds and are insoluble in water.

Inorganic Substances

1. Water is the most abundant substance in the body. It is an excellent solvent and suspending medium, participates in chemical reactions, absorbs and releases heat slowly, and lubricates.
2. Acids, bases, and salts ionize into ions in water. When an acid ionizes, it produces H^+ ions; when a base ionizes, it produces OH^- ions. When a salt ionizes, it produces neither H^+ nor OH^- ions. Cations are positively charged ions; anions are negatively charged ions.
3. The pH of different parts of the body must remain fairly constant for the body to remain healthy. On the pH scale, 7 represents neutrality. Values below 7 indicate acid solutions, and values above 7 indicate alkaline solutions.
4. The pH values of different parts of the body are maintained by buffer systems, which usually consist of a weak acid and a weak base. Buffer systems eliminate excess H^+ ions and excess OH^- ions in order to maintain pH homeostasis.

Organic Substances

1. The body can build many compounds out of carbon, hydrogen, and oxygen. Each compound can be especially suited for a particular structure or function.
2. Organic compounds do not dissolve easily in water. This makes them useful materials for building body structures. Organic compounds are also a good source of energy.
3. DNA and RNA are nucleic acids consisting of nitrogenous bases, a sugar, and phosphate groups. The structure of DNA is a double helix and it functions as the primary chemical in genes. RNA differs slightly in structure and chemical composition from DNA and it functions in protein synthesis reactions.
4. The principal energy-storing molecule in the body is ATP. When its energy is liberated, it is hydrolyzed to ADP and inorganic phosphate. ATP is manufactured from ADP using the energy supplied by the food that is eaten.
5. Cyclic AMP is closely related to ATP and participates in certain hormonal reactions.
6. Prostaglandins are lipids that control the production of cyclic AMP. They mimic the effects of hormones.

REVIEW QUESTIONS

1. Define matter. What is a chemical element? How are chemical elements abbreviated? Give several examples.
2. Describe the chemical elements that are found in the human body.
3. Diagram an atom of hydrogen, of carbon, and of oxygen.
4. Distinguish between atomic number, atomic mass, and mass number.
5. How is atomic mass related to the Avogadro number?
6. What is a chemical reaction?
7. Explain how electrons occupy energy levels.
8. What is the valence of an element?
9. Distinguish between a molecule and a compound.
10. What is a chemical bond? Explain how an ionic and a covalent bond are formed.
11. Diagram each of the following kinds of chemical reactions: synthesis, decomposition, exchange, and reversible.
12. Explain how the collision theory relates to chemical reactions.
13. Define activation energy. How does an enzyme change the activation energy of a chemical reaction?
14. Describe several factors that influence the rate of a chemical reaction. What is a catalyst?
15. Explain the relationship of making and breaking chemical bonds to the energy of a chemical reaction.
16. How do exergonic and endergonic reactions differ?
17. Describe how matter may be organized into pure substances and mixtures.

18. Distinguish the three kinds of mixtures.
19. List and explain the defining characteristics of a solution.
20. Explain how the process of solvation occurs.
21. Compare a saturated, unsaturated, and supersaturated solution.
22. Why are evaporation and distillation effective methods for separating the components of a solution?
23. What is meant by the percentage concentration of a solution?
24. If 32 g of sugar is added to 56 ml of water, what is the percentage concentration of the solution?
25. What is the milligrams percent of a solution? Why is this expression preferred over simple percentage in some cases?
26. How is ratio used to express the concentration of a solution? Explain the use of a ratio by describing the expressions parts per million and serial dilution.
27. What is the molarity of a solution? Define a mole.
28. Distinguish between gram atomic weight and gram molecular weight.
29. Calculate the molecular weight of CCl_4, CO_2, H_2S, $C_{12}H_{22}O_{11}$, $CaCl_2$, H_2SO_4, HCl, HNO_3, $NaOH$, H_2CO_3, and $C_{10}H_{44}O_{22}$.
30. What is the basic difference between a $1M$, $0.5M$, and $2M$ solution of glucose?
31. How would you prepare a $1M$ solution of sucrose, $C_{12}H_{22}O_{11}$?
32. What is meant by the molality of a solution?
33. Define a colloidal dispersion.
34. List the distinguishing properties of a colloidal dispersion.
35. What is meant by a dispersed phase? Dispersing medium?
36. Describe several examples of colloidal dispersions.
37. What is Brownian movement?
38. Describe the Tyndall effect.
39. Why do colloidal particles remain suspended in their dispersing media?
40. Define adsorption. What practical and medical use does it have?
41. What is a protective colloid?
42. Define an emulsion and an emulsifying agent. Explain the importance of bile as an emulsifying agent.
43. Define a suspension.
44. List the distinguishing properties of a suspension.
45. Describe some examples of pharmaceutical suspensions.
46. How do inorganic compounds differ from organic compounds? List and define the principal inorganic and organic compounds that are important to the human body.
47. What are the essential functions of water in the body?
48. Define an acid, a base, and a salt. How does the body acquire some of these substances?
49. What is pH? Why is it important to maintain a relatively constant pH? What is the pH scale? List the normal pH values of some common fluids, biological solutions, and foods.
50. Refer to Exhibit 4-5 and select the two substances whose pH values are closest to neutrality. Is the pH of milk or the pH of cerebrospinal fluid closer to 7? Is the pH of bile or the pH of urine farther from neutrality? If there are 100 OH^- ions at a pH of 8.5, how many OH^- ions are there at a pH of 9.5?
51. What are the components of a buffer system? What is the function of a buffer pair? Diagram and explain how the carbonic acid–sodium bicarbonate buffer system of extracellular fluid maintains a constant pH even in the presence of a strong acid or strong base. How is this an example of homeostasis?
52. Why are the reactions of buffer pairs more important with strong acids and bases than with weak acids and bases?
53. What is a nucleic acid? How do DNA and RNA differ with regard to chemical composition, structure, and function?
54. What is ATP? What is the essential function of ATP in the human body? How is this function accomplished?
55. How is cyclic AMP related to ATP? What is the function of cyclic AMP?
56. Define a prostaglandin. List some physiological effects of prostaglandins.

Chapter 5

HOMEOSTASIS

STUDENT OBJECTIVES

After reading this chapter, you should be able to

- Describe the interrelationships between the circulatory, respiratory, and digestive systems in maintaining homeostasis
- Explain the general characteristics of a nonbiological control system using a tropical fish aquarium as the example
- Describe some control systems in the body
- Compare the role of the nervous and endocrine systems in maintaining homeostasis
- Define a feedback system and explain its role in homeostasis
- Contrast the homeostasis of blood pressure through nervous control and blood sugar level through hormonal control
- Explain why the inability to achieve homeostasis leads to disorders

HOMEOSTATIC MECHANISMS

In the first chapter the idea of homeostasis was introduced and described as the underlying principle of physiology. Recall that every body structure, from the cellular to the system level, contributes in some way to keeping the internal environment within normal limits. The homeostatic function of the cardiovascular system, for example, is to keep the fluids of all parts of the body constantly mixed and moving. When we are at rest, fresh blood is circulated throughout the entire body about once every minute. But when we are active and our muscles need nutrients rapidly, the heart quickens its pace and pumps fresh blood to the organs as much as five times a minute. In this way, the cardiovascular system helps compensate for the stress of increased activity.

The respiratory system, which includes the lungs and air passageways, offers another example of a homeostatic mechanism in the body. This system supports cellular activity by providing oxygen and removing carbon dioxide, a waste product that could be harmful to body cells if its concentration were too high. Cells need more oxygen and produce more carbon dioxide when they are active. Therefore, during periods of activity, the respiratory system must work faster to keep the oxygen in the extracellular fluid from falling below normal limits and to keep excessive amounts of carbon dioxide from accumulating.

The digestive system and related organs help to maintain homeostasis by reducing the consumed food to molecular-sized pieces that can be transported into all body cells. Undigestible food components are eliminated. As circulating blood passes through the organs of digestion, the products of digestion are transported to the body fluids so they can be used as nutrients by the cells. The liver, kidneys, endocrine glands, and other organs help to alter or store the products of digestion. The kidneys help to remove the wastes produced by cells after they have utilized the nutrients.

General Characteristics of Control Systems

A homeostatic control system can be easily understood by first examining a familiar nonbiological control system. Consider an aquarium filled with tropical fish such as you may see in a laboratory or in your home (Figure 5-1). It is essential that the temperature of the water in the aquarium be maintained at 30°C. Since the air temperature of the laboratory or your home is usually 20°C–30°C, it is clear that there will be a continual loss of heat from the water in the aquarium to the air in the room. Therefore, a heater is always installed in the tank to heat the water. However, some control system is also necessary to regulate the amount of heat needed to keep the water at 30°C. A common system consists, first of all, of a special temperature sensing device, immersed in the tank, that conducts a small electric current. The amount of current carried by this sensor increases as the water temperature becomes colder and decreases as the water temperature approaches the desired set point of 30°C. If you examine Figure 5-1, you will see that the current generated by the sensing device is conducted

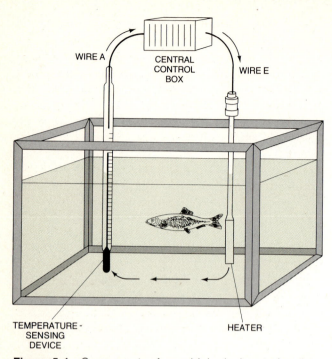

Figure 5-1 Components of a nonbiological control system as illustrated by an aquarium.

through wire A to a central control box. The control box responds to the input signal by generating a larger response current which flows out via wire E in direct proportion to the signal current that entered the central control box via wire A. Wire E carries this response current to a heater in the tank which proceeds to warm the water. As the water temperature in the aquarium rises, the strength of the current produced by the sensing device decreases and the central con-

trol box reduces its output proportionately. Hence, the rate of heating slows and the water temperature gradually reaches 30°C. Actually, there are oscillations in water temperature around the set point of 30°C caused by the alternations in the heating and cooling of the water (Figure 5-2). Overall, the system is operating effectively and is maintaining a fairly constant temperature. In a control system, the set point is determined by the properties of all the component parts, with the central control box usually designed to be a component with that responsibility.

When Bernard, Cannon, and others studied the constancy of the internal environment, they concentrated primarily on water, oxygen, carbon dioxide, mineral salts, and temperature. The homeostatic control systems regulating the exchange of these substances and of heat are well known today. Each has its receptor device and some effector organ concealed within the body. Unless specifically searched for, their actions are seldom obvious. In addition, there are many other control systems, less likely to be noticed but just as important.

Body Control Systems

Control systems regulate the positions and movements of body parts. Muscular responses are responsible for the maintenance of positions and for movements because muscles are attached to parts of the skeleton. Muscle movements are controlled, in great part, by the information received by the skin, eyes, ears, and nose. The rate of blood flow to any particular organ or even to a group of cells is varied by a control system. Blood pressure to assure the adequate flow of blood to various organs is carefully controlled. The rate and depth of breathing are varied by control systems to meet the changing needs of the body as its level of activity fluctuates.

The homeostatic mechanisms of the body, such as those performed by the cardiovascular, respiratory, and digestive systems, are themselves subject to control by the nervous system and the endocrine system. The nervous system regulates homeostasis by detecting when the body departs from its balanced state and by sending messages to the proper organs to counteract the imbalance. For instance, when muscle cells are active, they produce a large amount of carbon dioxide which is picked up by the blood. Certain receptor nerve cells in the brain stem detect the chemical changes of CO_2 occurring in the blood. The cells in the brain stem send a message to the heart to pump blood more quickly to the lungs so that the blood can give up its excess carbon dioxide and take on more oxygen. Simultaneously, the cells in the brain stem

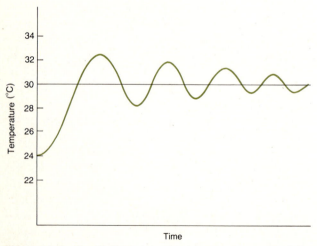

Figure 5-2 Oscillations in water temperature around the set point of 30°C caused by the alternating heating and cooling of the water.

send a message to the lungs to breathe faster so that the excess carbon dioxide can be exhaled and more oxygen can be inhaled.

Homeostasis is also a function of the endocrine system. This system consists of a series of glands that secrete chemical regulators, called *hormones,* into the blood. Whereas nerve impulses coordinate homeostasis rapidly, hormones coordinate homeostasis much more slowly. Both nervous and endocrine controls are directed toward the same end. Two specific examples of homeostasis that can be considered here are the nervous control of blood pressure and the hormonal control of blood sugar level.

HOMEOSTASIS OF BLOOD PRESSURE: NERVOUS CONTROL

Blood pressure is the force exerted by the blood as it presses against and attempts to stretch the walls of the blood vessels, especially arteries. The technique for measuring blood pressure is a standardized procedure and routinely measures the blood pressure in the larger arteries of the body as near to the heart as possible. The brachial artery in the arm is most commonly employed. Pressure is maintained in the arteries primarily by two factors: the rate and strength of the heart beat, and the peripheral resistance encountered by the blood as it flows into the smaller blood vessels.

If some stress, such as a sudden loss of blood (hemorrhage) occurs, the following sequence occurs (Figure 5-3). The sudden loss of blood reduces the total blood volume of the body, causing an abrupt drop in arterial blood pressure. The dropping pressure is detected by special pressure-sensing devices built into the walls of the carotid and aortic arteries. The receptors send impulses to the brain. The brain interprets the message and sends impulses to the arterioles to cause constriction and to the heart to increase its rate.

The continual monitoring of the blood pressure in these strategic locations by the nervous system results in a blood pressure that is maintained at a level adequate to drive blood continually to the important organs of the body. This reflex nervous control of the blood pressure is an example of a feedback system.

A **feedback system** is a device or situation wherein information about the "status of things out there" is continually reported to some central control region. The central control region can then cause appropriate alterations to occur so that the "status of things out there" is always optimal for what is required. As will be seen, a feedback system may be either *negative* or *positive.* In Figure 5-3, the regulation of blood pressure, the input of information to the central control region comes from the pressure receptors, and the appropriate response or output of the system is to increase heart rate and constrict arterioles in an effort to restore the blood pressure to a normal level. Figure 5-3 shows that the system runs in a circle. The pressure-sensing cells are continually reporting the blood pressure to a center in the brain stem, and the

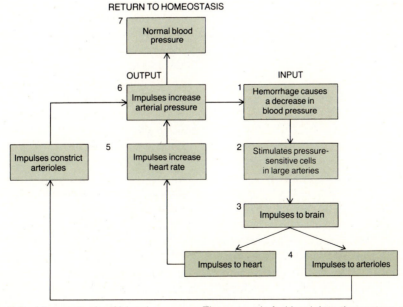

Figure 5-3 Homeostasis of blood pressure. The output is fed back into the system, and the system continues to increase blood pressure until there is a return to normal conditions.

output continues until the signal from the pressure-sensing devices to the center is adequate.

A feedback system, such as the one just described, wherein changes that are straying from a desired value are reversed so that they move toward the desired value are termed **negative feedback systems.** The reaction of the body, the output, counteracts the stress, the input, in order to restore homeostasis. Most of the feedback systems of the body are negative feedback mechanisms. In **positive feedback systems,** the movement away from the desired level is continually accelerated and increased rather than reversed. Hence, these systems tend to be explosive in their rate of response and therefore usually destructive (pathological). Examples of positive feedback systems are few and short-lived in the body.

Referring again to Figure 5-3, you will see that a second negative feedback control is also involved in maintaining a normal blood pressure. Small blood vessels, called arterioles, have muscular walls that can constrict upon receiving an appropriate signal. When the blood pressure decreases, the receptors located in the carotid and aortic arteries send messages to a center in the brain, and the center puts out signals that cause the muscular walls of the blood vessels to constrict. The diameter of the blood vessels thus decreases. As a consequence, the blood flowing through the arterioles is reduced; that is, resistance to flow increases and blood pressure in the large arteries rises. In sum, a sudden drop in blood pressure is brought back to normal by mechanisms that accelerate heart output and decrease the diameter of the arterioles. Conversely, a rise in blood pressure is counteracted by decreased heart activity and an increase in the diameter of the arterioles. The net result, in either case, is a change in blood pressure back to normal.

HOMEOSTASIS OF BLOOD SUGAR LEVEL: HORMONAL CONTROL

A crucial homeostatic system regulated by hormonal activity is well illustrated by the mechanism that controls the blood sugar level. Since every cell in the body relies upon the constant availability of the sugar glucose for its supply of energy, it is therefore important that the blood that flows by every cell carry sufficient amounts of glucose. Because the cells constantly remove glucose from the blood, glucose from some stored source needs to be released and added to the blood. To ensure that the rate of addition to the blood equals the rate of removal, the system monitors and maintains the amount of glucose dissolved in the

blood at a relatively constant level of 90 mg percent (90 mg of glucose dissolved in 100 ml of blood plasma).

This level is maintained primarily by the actions of the two pancreatic hormones, insulin and glucagon. Suppose that you have just eaten a giant-sized chocolate bar on an empty stomach. The sugar from the chocolate bar enters the stomach and small intestine, and the molecules of glucose are rapidly carried through the intestinal wall into the blood. The sudden entrance of this glucose into the blood becomes a stress because it elevates the blood sugar level above normal. Some of this blood passes through the pancreas. The cells of the pancreas that constantly monitor the blood sugar level (receptors) detect this elevated blood sugar and release insulin into the blood (Figure 5-4a). Insulin released into the blood is distributed to all parts of the body and causes two processes to take place. First, it causes most body cells to transport glucose into their interior at an accelerated rate, lowering the blood sugar level. Second, insulin causes the cells of the liver and muscles to accumulate glucose in large amounts and store it in the form of a large compact molecule called *glycogen.* This storage of glucose inside these cells also removes large amounts from the blood, thereby further reducing the blood sugar level. In essence, the action of the hormone insulin lowers the blood sugar concentration until it reaches a more normal level. When it reaches this level, less insulin is released.

The other hormone released from the pancreas, called glucagon, has the opposite effect of insulin. This time suppose that you have not eaten for several hours. Your body cells continue to remove glucose from the blood as they require it for energy; therefore, the blood sugar level steadily decreases. The decrease in blood sugar level is also detected by the cells of the pancreas, and this time they release the hormone glucagon (Figure 5-4b). This hormone causes the glucose stored in the liver cells to be released into the blood. Blood sugar level can thus be increased until it reaches a normal level. Then glucagon release tapers off to a low level.

In summary, the blood sugar level is regulated by two hormones, both produced and released by the pancreas. Insulin lowers the level of blood sugar if the concentration increases above normal, and glucagon raises the level of blood sugar if it has fallen below normal.

In the chapters that follow, each system of the human body can be considered a structural and functional complex that participates cooperatively and in coordination with all the other systems to keep all the body cells alive. Each system operates as a homeo-

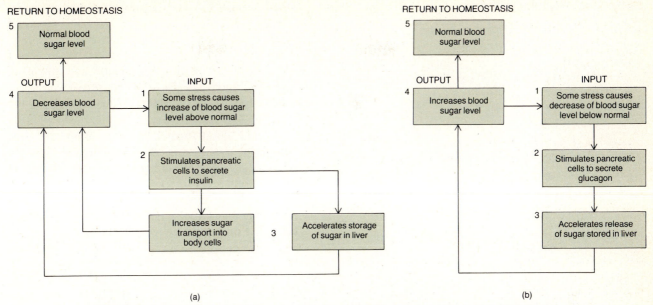

Figure 5-4 Homeostasis of blood sugar level. (a) Mechanism that lowers high blood sugar level. (b) Mechanism that raises low blood sugar level.

static system designed to maintain a specific variable at a preset desirable level. Homeostatic systems operate on the principle of negative feedback.

When a particular system achieves homeostasis, it functions normally. When all control systems are "go" and are functioning normally, the body is in a state of health. Significant disruptions of one or more systems from a homeostatic condition results in abnormal functions or disease. The study of abnormal functions is termed *pathology*.

Pathology describes the origins of the disruptive process and the nature in which the disruptions manifest themselves. Signs and symptoms together with physical and laboratory findings are the tools of the pathologist. *Disease* is characterized by alterations in the normal structure of some microscopic component of the body. Disease is a process that is continually changing. It is not a static state. It may end with restoration to normal, or almost normal, function, or it may terminate with death. Disease can begin suddenly and quickly ravage the entire body, or it may be a subtle, ongoing process such as aging.

In subsequent chapters, the normal physiology of a system will be described first and then certain pathologies related to the system will be examined.

CHAPTER SUMMARY OUTLINE

HOMEOSTATIC MECHANISMS

1. The homeostatic function of the cardiovascular system is to keep the fluids of all parts of the body constantly mixed and moving.
2. The respiratory system offers an example of a homeostatic mechanism in the body by supporting cellular activity. It does this by providing oxygen and removing carbon dioxide, a waste product that could be harmful to body cells if its concentration were too high.
3. The digestive system and related organs help to maintain homeostasis by reducing food to molecular-sized pieces that can be transported into all body cells. Undigestible food components are eliminated.

General Characteristics of Control Systems

1. In a control system, the set point is determined by the properties of all the component parts, and the central control box is usually designed to be the component with that responsibility.

Body Control Systems

1. The homeostatic mechanisms of the body, such as those performed by the cardiovascular, respiratory, and digestive systems, are themselves subject to control by the nervous system and the endocrine system.
2. The nervous system regulates homeostasis by detecting when the body is headed out of its balanced state and by sending messages to the proper organs to counteract the imbalance.
3. The endocrine system coordinates homeostasis through chemical regulators called hormones.

HOMEOSTASIS OF BLOOD PRESSURE: NERVOUS CONTROL

1. Blood pressure is determined by the rate and strength of the heartbeat, the amount of blood, and arterial resistance. A dropping blood pressure rises toward normal because of a negative feedback system of blood pressure regulation.

HOMEOSTASIS OF BLOOD SUGAR LEVEL: HORMONAL CONTROL

1. A normal blood sugar level is maintained by the actions of insulin and glucagon. Insulin causes a rising sugar level to fall back to normal. Glucagon brings a low level up to normal.
2. Significant disruptions of one or more systems from a homeostatic condition results in abnormal functions or disease called pathology.

REVIEW QUESTIONS

1. Compare the homeostatic functions of the cardiovascular, respiratory, and digestive systems.
2. Describe a nonbiological control system.
3. Describe the interrelationships of body systems in maintaining homeostasis.
4. Contrast the nervous system with the endocrine system in controlling homeostasis.
5. What is a feedback system? Contrast negative and positive feedback mechanisms and explain their roles in homeostasis.
6. Discuss briefly how the regulation of blood pressure and blood sugar level are examples of homeostasis.
7. What is pathology? Explain disease.

Chapter 6

CELLULAR STRUCTURE AND FUNCTION

EVOLUTION OF THE CELL

The best scientific estimates date the origin of the solar system around 4.5 to 5 billion years ago (5×10^9 years). Life on the earth is estimated to have arisen around 3.2 to 3.5 billion years ago. Between the time that the solar system formed and life appeared there was a period during which *chemical evolution* is said to have occurred. Chemicals are believed to have evolved into living forms during this time. Most scientists believe that the atmosphere of the primitive earth probably contained the gases methane (CH_4), ammonia (NH_3), water (H_2O), and molecular hydrogen (H_2). During the period of chemical evolution, these gasses interreacted with each other, aided by energy from electrical discharges, heat, and ultraviolet and gamma radiation from the sun. These interreactions produced a large variety of simple carbohydrates, amino acids, and organic bases. These organic molecules then accumulated in continually increasing amounts in the bodies of water covering the earth.

Scientists have verified the likelihood of chemical evolution. From the above simple basic molecules, scientists using simulated primitive earth conditions have produced many of the larger, more complex organic molecules found in living systems. The larger complex molecules such as phospholipids and hydrocarbons have been tested under various conditions to demonstrate a variety of ways in which they could be assembled into cell-like structures called *microspheres*.

The first "cells" that emerged from chemical evolu-

tion were not cells as we know them today but *anaerobic heterotrophic cells* on the threshold of life. By definition, anaerobic heterotrophic cells survive in the absence of oxygen and obtain their energy for growth and replication from preformed simple organic molecules similar to those formed during the period of chemical evolution. The question of when life first appeared can never be answered with certainty because the definition of life itself is debatable. It is generally agreed, however, that a minimum requirement for life is the presence of some type of informational or coding molecule that can direct its own replication. Since both proteins and nucleic acids have informational properties, the question of which type of molecule evolved first is still debated in many scientific arenas.

In the course of this evolutionary process, once a self-replicating system was added to a set of catalytic molecules and a surrounding membrane, the subsequent steps of biological or cellular evolution became easier to describe. The first anaerobic heterotrophic cells carefully selected and consumed from their environment those molecules that increased their chances of survival. Conditions changed, however, when the primordial seas began to be depleted of organic molecules.

This depletion led to the evolution of some early form of *autotrophic cell*—a cell that could convert the carbon dioxide of the atmosphere to a simple sugar. Energy for such conversions was obtained from the oxidation of inorganic compounds such as H_2S to S or H_2 to H_2O (chemosynthetic autotrophs) or by the absorption of light radiation from the sun (photosynthetic autotrophs). When the evolution of photosynthesis took place, oxygen began to accumulate in the atmosphere, changing it from a hydrogen, reducing one to an oxygen, oxidizing one. *Heterotrophic cells,* capable of utilizing the oxygen for a more complete extraction of energy from the simple organic molecules of the oceans, were the next evolutionary products.

Only after these developments, which may have taken over half of the time of the earth's existence, did the next gigantic jump in cell evolution occur. The appearance of *eucaryotic cells,* that is, cells with a membrane-bound nucleus and organelles, was one of the most important steps in biological evolution. These cells, identified by their complex of internal membrane-delimited compartments, are thought to have resulted when a large anaerobic heterotrophic cell engulfed a smaller photosynthetic or aerobic heterotrophic cell. The cell-within-a-cell relationship where both cells are mutually supportive is called *symbiosis.* In this symbiotic relationship, the host cell furnished protection and the anaerobic capabilities, and the parasite furnished either photosynthetic or aerobic capability. Such *(endo)symbionts* may be the precursors of the organelles called the chloroplast and mitochondrion.

Is life as we know it today, and as we have learned about its evolution, the inevitable outcome of the conditions that prevailed when the earth first formed? If so, could similar or even identical conditions exist or have existed on other planets? The term *exobiology* has been coined to refer to the study of life beyond our planet. Much research, funded by the federal government, has been conducted to probe for signs of life on the moon and Mars and other planets. Most biologists around the world concentrate on the study of life forms that presently exist on the earth. Some simple life forms called procaryotic cells have persisted for millions of years relatively unchanged. Procaryotic cells are the simple forms of unicellular organisms that first evolved.

PROCARYOTIC CELLS

Procaryotic cells are much simpler in structure than the complex eucaryotic cells that will be described shortly. Procaryotic cells lack a membrane around the nucleus, and all of the internal vacuolar membranes are also absent. The nucleus lacks a mitotic apparatus and nucleoli, and the genetic information is stored in one single chromosome that is in the form of a circle. Procaryotes, as they are also called, are distributed all over the surface of the earth and share such features as rapid growth and short generation times. They are

used by scientists in many disciplines for research into the genetics and biochemistry of simple systems with the hope of producing clues to the operation of more complex systems. Representative procaryotes are bacteria, blue-green algae, and mycoplasmas. We shall first outline some of the important features of bacteria.

Bacteria

Bacteria comprise the largest group of procaryotes. They are a group of microorganisms that are barely visible with the light microscope. Most bacterial cells are 2 μm to 5 μm in length. Bacteria have been divided into three general groups based upon their overall shapes: the rodlike bacilli (singular = bacillus), the spherical cocci (singular = coccus), and the corkscrew-shaped spirilla (singular = spirillum). These are shown in Figure 6-1a. All bacteria reproduce by a process of division called *binary fission* (Figure 6-1b). During binary fission, a bacterium grows larger and larger, reaches a critical size, then divides in half to form two equal-sized daughter cells. Bacterial divisions can occur as rapidly as every 20 minutes.

When the environment of certain species of bacteria becomes inhospitable, the bacterium forms a virtually impenetrable membrane — a *spore coat* — within its cytoplasm. This coat surrounds the genetic components and a small part of the cytoplasm, and the structure is called an *endospore* or simply a *spore.* The remaining parts of the cell surrounding the spore disintegrate. Spore formation enables the organism to survive in a sort of quiescent state until conditions more favorable for survival are restored. Spores can withstand a temperature of 100°C for several hours in a slightly alkaline solution, whereas the original bacterial cells would be killed almost instantly at this temperature. Exposure to a temperature of about 120°C for just a few minutes destroys most bacterial spores. Because water cannot be heated to such a temperature at a normal atmospheric pressure, medical and laboratory equipment is sterilized effectively by using higher pressures in a device called an autoclave, in which water boils at about 120°C.

Bacterial invasion of the body produces disease in a number of ways. In some cases, the bacteria find the body so favorable an environment that they divide rapidly, producing great populations. These invaders then compete with the body cells for nutrients and oxygen, drain resources away from the host cells, and cause cellular death. Other bacteria produce poisonous chemical substances called *toxins* that interfere with or inhibit the normal cellular processes and thus cause cellular death. One of the most potent

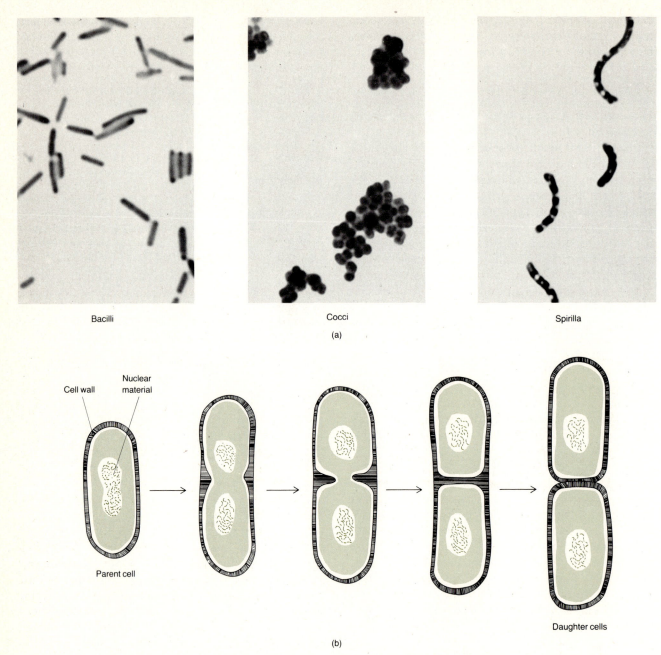

Figure 6-1 Bacteria. (a) Representative groups based on shape. (Courtesy of Carolina Biological Supply Company.) (b) Binary fission.

toxins is produced by the bacterium *Clostridium botulinum*. This toxin is so lethal that as little as 3.5×10^{-7} g can kill a human.

Bacteria that are most familiar to people are those that cause diseases in humans, in domestic animals, and in cultivated crops. Bacterial invasions of the human body may cause typhoid fever, cholera, plague, dysentery, scarlet fever, tuberculosis, diphtheria, and wound infections such as gangrene. In defense of these minute organisms, however, they are generally

more beneficial to humans than harmful. They play vital roles in the production of cheese, buttermilk, vinegar, and other foods. They are indispensable for the proper operation of sewage treatment. Industry uses bacteria profitably to remove hair from hides in the preparation of leather. Bacteria have the prime responsibility for causing the decay of dead organisms and the restoration of carbon and nitrogen to the ecosystems in forms usable by plants. Residing in the human intestinal tract in a symbiotic relationship are beneficial bacteria that produce vitamin K, an essential molecule required by the human body for normal blood clotting.

Blue-Green Algae and Mycoplasmas

Blue-green algae (Figure 6-2) are presently represented by nearly 1,500 species, all of which are photosynthetic and contain photosynthetic pigments—chlorophyll (a green pigment) and phycocyanin (a blue pigment). Some species contain additional colored pigments. Many can live in moist earth and survive long, dry spells by forming resistant resting spores. Some species survive temperatures as high as 70°C and are found growing in hot springs. The multicolored formations in springs and ponds in Yellowstone National Park are colored mainly by various

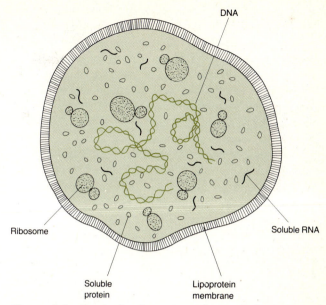

Figure 6-3   Generalized structure of a mycoplasma.

species of blue-green algae containing a wide variety of accessory pigments.

Mycoplasma cells are responsible for one form of human pneumonia and pleuropneumonia, a highly contagious disease of cattle. These cells range in size from 0.1 μm to 9.25 μm and represent the smallest, simplest living systems known to exist (Figure 6-3).

EUCARYOTIC CELLS

The cells of animals, plants, and most microorganisms are similar in their basic structure. The **eucaryotic cells** (a scientific combination of words from the Greek to mean a true or real nucleus) are enclosed by a cell membrane and possess a nucleus enclosed by a nuclear envelope. Specialized structures, termed organelles, are distributed in the cytoplasm between the nuclear envelope and the cell membrane. Eucaryotic cells exist in a variety of shapes and sizes. They range in size from the ostrich egg, which is over 170 mm long, to microscopic cells that are around 10 μm in size.

Membranes divide the eucaryotic cell into organelles that function independently but cooperatively with each other. These internal membrane-bound compartments include the mitochondria, chloroplasts (in plants), lysosomes, peroxisomes, the large enclosed spaces of the endoplasmic reticulum, and Golgi complex.

Membranes serve many purposes. They present physical barriers that separate and maintain internal

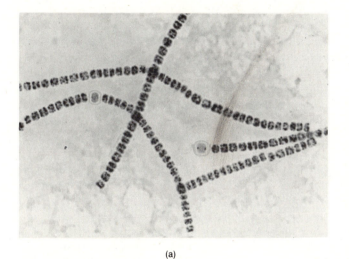

(a)

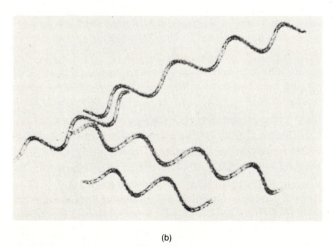

(b)

Figure 6-2   Representative blue-green algae. (a) *Nostoc*. (b) *Arthrospira*. (Courtesy of Carolina Biological Supply Company.)

environments so that metabolic functions occur under optimal conditions. Enzymes are also incorporated into the structure of certain membranes. In the mitochondrion, enzymes are arranged in intricate repeating complexes in and on the mitochondrial membrane in order to ensure the utmost efficiency in energy capture. The sequential delivery of substrates to successive enzymes that are arranged in the membrane is a most remarkable evolutionary feat.

Nucleoli, ribosomes, chromosomes, centrioles, and microtubules are all specialized structures lacking membranes. Other functional units include filaments such as the actin and myosin of muscle cells.

It is easy to obtain the impression from photographs, drawings, and prepared microscope slides that the cell interior is a static environment in which all structures are solidly entrapped. In fact, the cell interior is dynamic with all structures demonstrating active movement and contributing to homeostasis. Organelles (and other functional units) move within the cell, membranes flex, and filaments twist or shorten. Cytoplasm streams from one region of the cell to another, the nucleus rotates, granules migrate along membrane surfaces, and the cell membrane engulfs food particles while microvilli beat rhythmically. Some organelles seem to divide independently, free from nuclear control. Most organelles demonstrate the phenomenon of *turnover*. In the process of turnover, molecules making up a structure are continually replaced by new ones so that, in time, all molecular parts are exchanged for new ones; yet the structure is always present. This is another excellent example of homeostasis.

Membrane Structure

The **cell membrane** or **plasma membrane** encloses every cell. Its presence is readily accepted today, yet its unequivocal existence was proven to everyone's satisfaction only about 30 years ago by the use of the electron microscope. This membrane, between 75 and 100 Å thick, is the primary barrier to molecules trying to enter or leave the cell. It is by way of cell membrane associations that tissues are formed, and the tightness of these associations determines the filtration properties of the tissues as fluids attempt to flow among cells.

Most membranes contain about 40 percent lipid and 60 percent protein, but there is considerable variation. For example, the inner mitochondrial membrane contains between 20 and 25 percent lipid, whereas the myelin membrane that surrounds certain nerve cells contains up to 75 percent lipid. It is now fairly well established that membrane lipids are in a state of flux and are constantly exchanging with other lipids present in the surrounding medium. Nevertheless, membrane lipid composition appears to be fairly constant, varying somewhat with diet, temperature, pH, ionic composition, and metabolic state.

Membrane lipids are mainly of a class of compounds called *phospholipids,* and to a lesser extent include varying amounts of cholesterol. Molecules of phospholipids have some interesting properties. The portion of the molecule called the *polar head* possesses an electrical charge asymmetry that makes that end of the molecule water soluble *(hydrophilic).* See Figure 6-4a. The long, *nonpolar tails* exhibit no such charge distribution and are not soluble in water *(hydrophobic).* Molecules that possess a polar head and a nonpolar hydrocarbon tail are called *amphipathic (amphi* = on both sides; *pathic* = suffering, feeling). Most investigators agree that the matrix of the plasma membrane is a bilayer (double row) of phospholipid molecules oriented with their hydrophilic heads facing outward. Although Figure 6-4a suggests an orderly array of molecules, the phospholipid molecules are somewhat intertwined and randomly disrupted by the molecules of cholesterol. This arrangement creates a lipid matrix or core to the membrane that is fluid and has the consistency of a light oil.

The cholesterol content of biological membranes, however, is highly variable. Even in a given cell, the cholesterol content of various organelle membranes varies greatly. The addition of cholesterol to the phospholipid bilayer results in a more condensed packing of all the molecules. The addition of cholesterol reduces the area occupied by phospholipid molecules, which makes the membrane more impermeable to aqueous solutions.

Much research has been done on identifying the protein components of the membrane by isolating them from the membrane. Such experiments indicate that there are two easily identifiable classes of proteins: (1) those that are readily removed by simple detergentlike compounds and (2) those that are firmly bound and difficult to extract. These two classes of proteins are termed the *peripheral (extrinsic)* and *integral (intrinsic)* proteins respectively.

At present, the Singer-Nicolson model of membrane architecture (Figure 6-4b) provides the most satisfactory basis for explaining membrane functions. Proposers of this model indicate that most of the protein molecules are of the integral type, firmly bound to phospholipids. If proteins are embedded deeply in the phospholipid bilayer they, like phospholipids, must exhibit *amphipathic* properties—the deeper or centrally located portions are hydrophobic and the outer regions of the molecule are hydrophilic. Accordingly, some of the proteins traverse the phos-

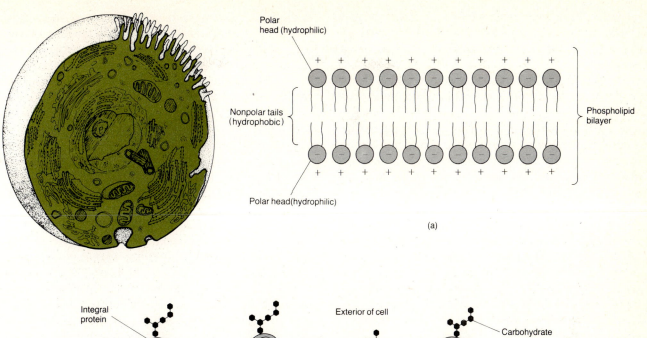

Polar head (hydrophilic)

Nonpolar tails (hydrophobic)

Phospholipid bilayer

Polar head (hydrophilic)

(a)

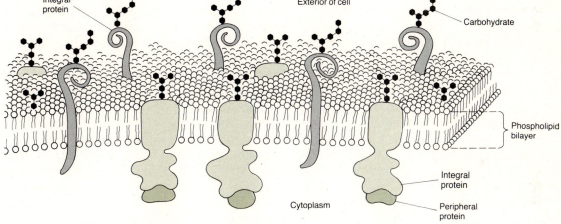

Integral protein

Exterior of cell

Carbohydrate

Phospholipid bilayer

Integral protein

Cytoplasm

Peripheral protein

(b)

Cytoplasm

Vacuolar membrane

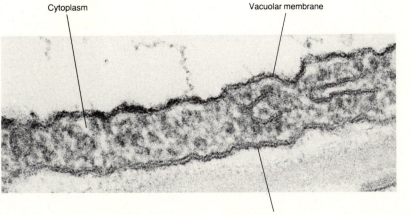

Plasma (cell) membrane

(c)

Figure 6-4 Structure of the plasma membrane. (a) Suggested arrangement of phospholipid molecules forming a phospholipid bilayer in the matrix of a plasma membrane. (b) Singer-Nicolson model of membrane structure. (c) Electron micrograph of the plasma membrane of a red blood cell showing the trilaminar appearance. (Courtesy of Myron C. Ledbetter.).

pholipid bilayer and extend freely on both surfaces. Other protein molecules are embedded only partially yet firmly in the central regions of the phospholipid bilayer.

The peripheral proteins are more superficially located on the surface of the phospholipid bilayer and are weakly bound. The distribution of proteins is random, and indeed the two surfaces of the membranes have distinctly different protein profiles. In addition, the outer surface of the cell membrane has a variety of short-chain sugars also protruding outward from the phospholipid bilayer. The asymmetry of membranes accounts for many of the diverse functions located within the membrane. In addition, the proteins and phospholipids can flow laterally within membranes. Membranes are thus very dynamic structures. When viewed under the electron microscope, the membranes are visible as a uniformly appearing trilaminar (three-layered) structure (Figure 6-4c). The trilaminar appearance is a consequence of the standard preparatory and staining techniques.

More recent techniques called freeze-fracture and freeze-etching have supported the Singer-Nicolson model. These techniques capitalize on the idea that a tissue, when rapidly frozen and fractured, breaks along planes of weakness (Figure 6-5a). This often results in the exposure of the interior of a membrane separating the two phospholipid layers. The exposed surfaces show particles embedded in the membrane (Figure 6-5b). Often, most of the particles remain with one of the separated surfaces while the other surface shows a complementary pattern of pits or depressions. These particles seem to be large enough to extend the width of the membrane (75 Å) and are believed by many investigators to be proteins. Other surface proteins on adjacent cells are known to intermingle readily, perhaps contributing to cellular adhesions.

Cell Membrane Associations: Cell Junctions

The cell membrane and cell surface of mammalian cells can be highly specialized and modified depending on relationships to adjacent cells. Cells are held together to form tissues by what are called *cell junctions*. Four categories of cell junctions are recognized: (1) tight junctions, (2) gap junctions, (3) intermediate junctions, and (4) desmosomes.

Tight Junctions

In the *tight junction* shown in Figure 6-6a, the electron microscope reveals the trilaminar membranes fusing so that the two outermost dark bands form only one single band equal in width to one of the original bands. The appearance of the tight junction, therefore, is of a five-layered complex—three dark and two light. This type of junction is common in the epithelial cells lining the digestive tract. It prevents the movement of molecules across epithelial cells via the intercellular channel. The junction encircles each cell around its free margins and links the cells into a continuous impermeable sheet.

Gap Junctions

In *gap junctions* (Figure 6-6b), the outer bands of the trilaminar-appearing membranes approach each other and leave a small gap between them. This gap distance is 20 Å and is bridged by structures that link the cells together. These structures may have a hollow core through which ions and molecules can pass from one cell to the next. This arrangement would enable adjacent cells to communicate. For example, in heart muscle the cells are linked to each other by such gap junctions, and this presumably enables the cells to contract in an orderly and coordinated manner.

Intermediate Junctions

An *intermediate junction* (Figure 6-6c), like the tight junction, encircles each cell that is joined. The two cell membranes are separated by a larger than normal intercellular space averaging about 200 Å. This space seems to provide a hollow collar around each of the connecting cells. Inside the collar is usually a glycoprotein material that appears as a relatively clear region when viewed with the electron microscope. Electron-dense material is found along the cytoplasmic sides of the membrane participating in this type of junction. Also, apparently anchored in this electron-dense material are *tonofilaments* (see Figure 6-6d), threadlike strands that extend into the microvilli (described later) where they serve as cores for these structures.

Desmosomes

Desmosomes (Figure 6-6d) are spot connections that form firm intercellular attachments between adjacent cells. Viewed with the electron microscope, these regions demonstrate an accumulation of electron-dense material on the cytoplasmic surface of the two neighboring cell membranes. The membranes of the two adjacent cells do not come into direct contact but are separated by an intercellular space of about 250 Å. The electron-dense region of the desmosome is disk-shaped and has embedded in it tonofilaments that enter and make a u-turn back into the cytoplasm. Regions lacking desmosomes permit interstitial fluid to

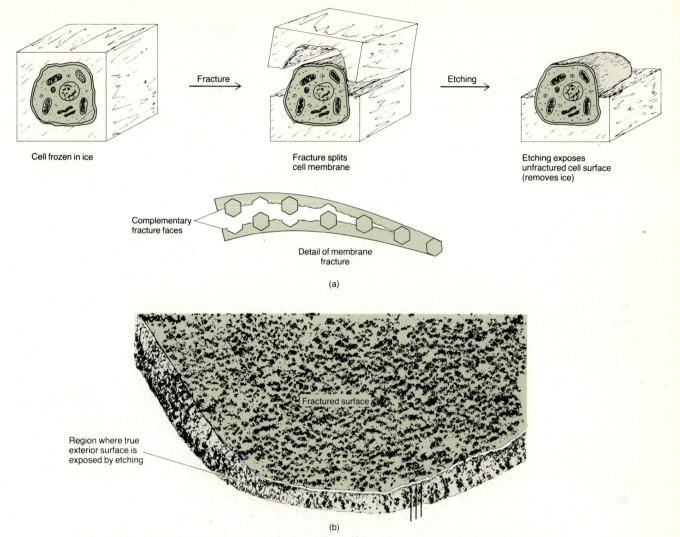

Cell frozen in ice

Fracture

Fracture splits
cell membrane

Etching

Etching exposes
unfractured cell surface
(removes ice)

Complementary
fracture faces

Detail of membrane
fracture

(a)

Fractured surface

Region where true
exterior surface is
exposed by etching

(b)

Figure 6-5 Freeze-fracture and freeze-etching applied to the plasma membrane. (a) A frozen cell is fractured with a sharp blow. Since the fracture line very often runs through the hydrophobic interior of membranes, the two leaflets of the lipid bilayer separate and reveal membrane proteins otherwise buried therein. Etching exposes unfractured membrane. (b) A fractured red blood cell. The fractured surface shows particles believed to be membrane proteins. The edge of the fracture is indicated by arrows. Adjacent to the fracture edge is a thin region where the true exterior surface is exposed by etching.

circulate between cells, an important physiological process.

Cell Membrane Modifications

Electron microscopic studies have shown many membrane modifications. Membranes are arranged in unique patterns for specific purposes in certain cells. For example, membranes of cells lining the small intestine increase their surface area greatly with the presence of thousands of small, fingerlike projections called microvilli (Figure 6-7a). *Microvilli* are cytoplasm-filled extensions or evaginations of the membrane that are devoid of cell organelles except for a varied number of tonofilaments that form a central core. The intestinal epithelium has microvilli that are so straight and uniform in height that this region is often referred to as the *striated border*. The surface of the membrane of these microvilli is covered by a fuzzy coating of intertangled filaments called the *glycocalyx*. These filaments provide the microvilli with a sticky surface that probably aids in the collection of molecule-sized nutrients. A similar arrangement of less orderly and more densely packed microvilli is

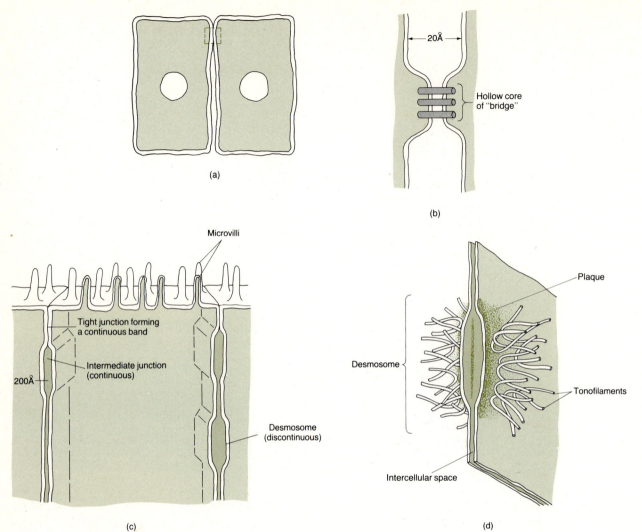

(a)

(b)

(c)

(d)

Figure 6-6 Cell junctions. (a) Tight junction. The trilaminar membranes share a common outer dense region at the junction. (b) Gap junction. (c) Intermediate junction. (d) Desmosome.

visible with the electron microscope on the free surface of kidney tubule cells. This region in the proximal kidney tubule is called the *brush border*.

Another example of a membrane modification is the *myelin sheath* that surrounds and electrically insulates portions of nerve cells (Figure 6-7b). During the embryonic development of the nervous system, special supporting-type cells called *Schwann cells* begin wrapping themselves concentrically around a small portion of the nerve cell. This envelopment causes a thinning out of the cytoplasm between the membranes of the Schwann cell. As the process continues the cytoplasm is ultimately displaced completely. The membranes of the Schwann cell are then brought into intimate contact so the outer layers of the trilaminar membrane fuse together. This membrane wrapping constitutes the myelin sheath of the nerve cell. The

Schwann cell is a specialized cell that donates its entire cell membrane to an insulating function. It is the lipid in the layers of the Schwann cell membrane that imparts the high electrical resistance or insulating property to the myelin.

CYTOPLASM

Just as the whole body has an internal environment, cells also have their own internal environment. This is the **cytoplasm** (see Figure 6-8a). It is here that we find the machinery of the cell. The components of the cytoplasm express the function of the cell and maintain homeostasis. The cytoplasm is a very heterogeneous system. Intracellular membranes subdivide the cytoplasm into numerous compartments, each with its own specific internal matrix. The cytoplasmic

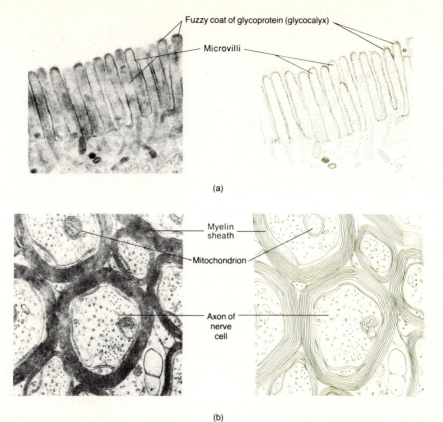

Fuzzy coat of glycoprotein (glycocalyx)

Microvilli

(a)

Myelin sheath

Mitochondrion

Axon of nerve cell

(b)

Figure 6-7 Modified plasma membranes. (a) Microvilli. Electron micrograph of a portion of the small intestine at a magnification of 20,000×, and diagram. (b) Myelin sheath. Electron micrograph of several myelin sheaths in cross section at a magnification of 20,000×, and diagram. (Courtesy of E. B. Sandburn, University of Montreal.)

matrix consists of a gelatinous solution called a colloid that can undergo sol-gel (liquid-jelly) transformations; its viscosity changes. Cytoplasm is responsible for intracellular movement and amoeboid movement, it promotes spindle formation, and it participates in cell division.

In specialized cells, the cytoplasm is the site of fibril synthesis. For example, keratin, a protein fibril component of the outer dead skin layer, is formed by the deep skin cells. Myofibrils are formed by muscle cells. Microtubules are synthesized by all cells and are used for shaping and maintaining the cells. Dissolved in the matrix or in suspension in the cytoplasm are many soluble proteins and enzymes. These proteins and enzymes constitute 20 to 25 percent of the total protein of the cell. Mechanical properties such as elasticity, rigidity, cohesion, and contractility are in large part attributed to the cytoplasmic matrix.

The electron microscope has revealed the existence of fine *microfilaments* within the cytoplasm. Microfilaments have a diameter of 4 to 14 nanometers (nm) and are divided into two types. One type forms a network next to the inner surface of the cell membrane and appears to be inserted into it. Recall the terminal web of interstitial cells. In the second type, the microfilaments are elongated and tend to be laid down in parallel array. Studies suggest such filaments represent contractile systems, which may be involved in cytoplasmic movement and amoeboid movement. Cells also contain *microtubules* either free in the cytoplasmic matrix or forming part of the centriole's cilia and flagella. They are tubules 25 nm in diameter, uniform in size, and remarkably straight, and they extend in length up to several microns. These structures form the cytoskeleton that seems to dictate the shape of the cell and its processes. They also play a key role in the spindle of dividing cells, causing the chromosomes to migrate in opposite directions. Finally, they are responsible for the motile force that causes ciliary beating and flagellar whipping (see Figure 6-14).

CELL ORGANELLES

Despite the myriad of chemical activities occurring simultaneously in the cell, there is little interference of one reaction with another as the cell attempts to maintain homeostasis. This is because the cell has a system of compartments that isolate specialized functions into distinct cytoplasmic structures that are collectively called **organelles**. These structures are specialized portions of the cell that assume various roles in growth, maintenance, repair, and control. An understanding of the structure and function of the organelles will aid you in understanding subsequent discussions of systems in the body. Organelles to be described include the nucleus, the vacuolar systems (endoplasmic reticulum and the Golgi complex), lysosomes, peroxisomes, mitochondria, ribosomes, cilia, centrioles, microtubules, and cell inclusions.

Nucleus

The growth and continual development of every living organism depends on the growth and multiplication of its cells. New individuals develop from a single primordial cell, the *zygote*. The multiplication of this cell and its descendants determines the development and growth of the organism.

Every cell generally has two periods in its life cycle: *interphase* (when it is not dividing) and *division* (when it produces two identical daughter cells). During interphase, the **nucleus** and certain of its structures are usually obvious. A *nuclear envelope* is visible under the electron microscope as two concentric membranes enclosing a *perinuclear space* (Figure 6-8a). *Nuclear pores* are also evident in the envelope where the two concentric membranes seem pinched together. The pores are octagonal in shape and appear to be plugged by a cylinder of protein material that extends through the pore and onto both sides of the pore margins, thus forming a doughnut-shaped ridge encircling the pore. This is the *annulus*, and together with the pore constitutes the *nuclear pore complex* (Figure 6-8b). The nuclear envelope controls the passage of ions and small molecules. It is very likely that certain large molecules such as ribonucleic acid (RNA—see Chapter 7) and enzymes may leave and enter the nucleus via the operation of the nuclear pore complex.

The shape of the nucleus is usually related to the shape of the cell, but sometimes it may be quite irregular. In spherical or cuboidal cells the nucleus is generally isodiametric (spherical). In cylindrical or spindle-shaped cells, it tends to be elliptical. Irregularly shaped nuclei are found in some of the leukocytes (white blood cells) where they appear horseshoe-shaped or multilobed. Almost all cells possess only one nucleus (are mononucleate) but some cells have more than one. Some liver and cartilage cells have two nuclei, and striated muscle fibers (cells) may have a hundred or more.

Inside the nucleus, certain prominent structures can be distinguished by fixing and staining techniques (Figure 6-8a). The *nucleoplasm (nuclear sap)* is a gelatinous fluid seen as the clearer regions of the nucleus when viewed with the electron microscope. It is in this region where *chromatin* (nucleoprotein) is largely dispersed. Chromatin is genetic material, consisting principally of DNA. It is observable when a cell is in interphase. The *nucleolus* appears as a large, dense accumulation of granules. It is now well established that the nucleolus is responsible for the synthesis of the components of the ribosomes.

Vacuolar Systems

The cytoplasm of eucaryotic cells is partitioned by a system of membrane-bound tubules, vesicles, and flattened sacs that are interconnected. This system of membranes subdivides the cytoplasm into two main compartments; one is enclosed within the membranes, the other is excluded (the cytoplasmic matrix).

The term *cytoplasmic vacuolar system* is used to describe this organization of membranes. The two components commonly recognized as subdivisions of the cytoplasmic vacuolar system include the endoplasmic reticulum (ER) and the Golgi complex. These membrane-delimited compartments carry out vital processes among which are the separation or association of certain enzyme systems.

Endoplasmic Reticulum

The **endoplasmic reticulum (ER)** may be further subdivided to include (1) the nuclear envelope, which we have already discussed, (2) the rough or granular ER, so-called because it is studded with small particles called ribosomes, and (3) the smooth or agranular ER, so-called because it is free of ribosomes.

The *rough* or *granular ER* is always characteristic of cells engaged in protein (enzyme) synthesis. In pancreatic cells, the rough ER is so highly developed that stacks of neatly arranged, flattened membranous sacs called *cisternae*, almost completely covered with ribosomes, occupy most of the cell's volume (Figure 6-9). In liver cells, the rough ER is distributed as small groups of cisternae throughout the cell. To illustrate the extent of rough ER in a typical liver cell, it has been calculated that 1 ml of liver tissue contains 11 m² of ER, two-thirds of which is the rough type. The *smooth* or *agranular ER* forms a network continuous with the rough type. The smooth type of ER is a characteristic of cells that secrete steroids, such as hormone-secreting cells. Some interesting research using radioactively tagged precursor molecules suggests that smooth ER is synthesized from the rough type.

Numerous functions related to homeostasis are

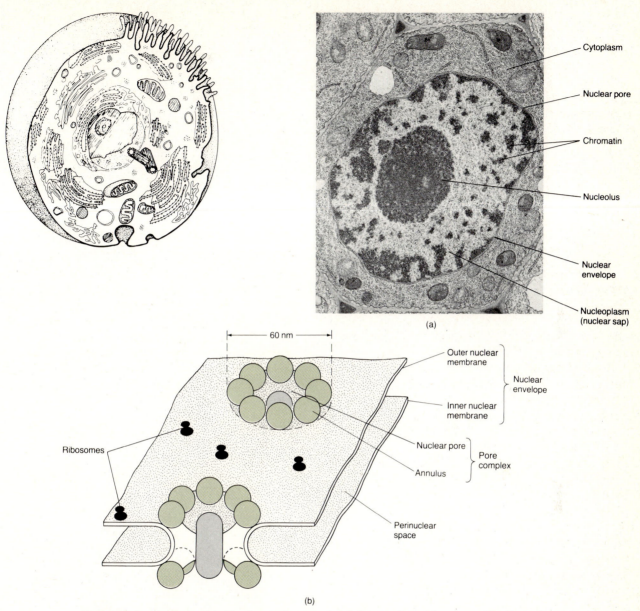

Cytoplasm

Nuclear pore

Chromatin

Nucleolus

Nuclear envelope

Nucleoplasm (nuclear sap)

(a)

60 nm

Outer nuclear membrane

Nuclear envelope

Inner nuclear membrane

Nuclear pore

Pore complex

Annulus

Ribosomes

Perinuclear space

(b)

Figure 6-8 Nucleus. (a) Electron micrograph of the nucleus at a magnification of 25,500×. (Courtesy of Myron C. Ledbetter.) (b) Nuclear pore complex.

attributed to the endoplasmic reticulum. It contributes to the mechanical support and distribution of the cytoplasm. It is involved with the intracellular exchange of materials between the cytoplasmic matrix and its own internal compartment. The membranes also conduct intracellular impulses, such as in muscle cells. Various products are transported from one region of the cell to another by way of the channels of the ER, so the ER can be considered an intracellular circulatory system. Portions of the ER membrane often pinch off and carry packaged material to other regions of the cell, particularly to the cell membrane where the contents of the package are released to the outside of the cell. One of the most important func-

tions of the rough ER is the synthesis of protein for export. Smooth ER is associated with the synthesis of lipid-type molecules and seems to have an important role in inactivating many toxic compounds.

Golgi Complex

The electron microscope makes the **Golgi complex** of the vacuolar system visible (Figure 6-10a). Generally the Golgi complex is located between the nucleus and the cell surface where secretion or absorption takes place. The membrane surfaces are devoid of granules (ribosomes), and the complex is located in a zone of cytoplasmic matrix that also seems to exclude free ribosomes. The Golgi complex consists of

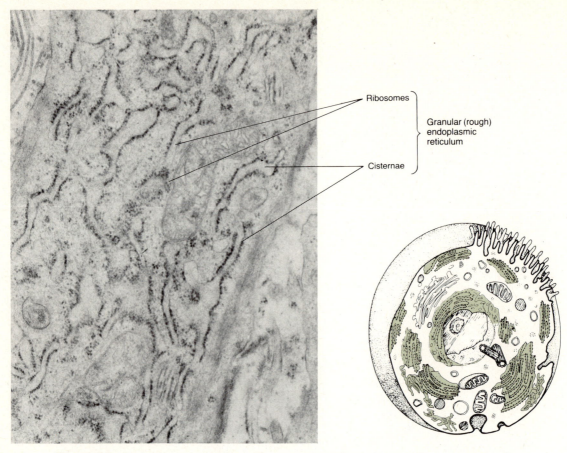

Ribosomes

Granular (rough) endoplasmic reticulum

Cisternae

Figure 6-9 Endoplasmic reticulum. Electron micrograph of rough ER in a pancreatic cell at a magnification of 8,000×. (Courtesy of Rotker, Taurus.)

three membranous parts: (1) flattened sacs called *cisternae*, (2) collections of *tubules* and *vesicles*, and (3) larger *vacuoles* that are filled with granules. The number of cisternae stacked upon each other varies between three and seven. In the stacks, the cisternae are slightly curved and separated from each other by a space of 200 to 300 Å. The membrane seems to be formed continuously at the *proximal* or forming face, that is, the portion of the stack nearest to the nuclear envelope or smooth ER. The other end of the stack, the *distal* or maturing face, is used up in the formation of small membrane-limited vesicles.

Coupling of cellular functions as part of maintaining homeostasis is well illustrated by the Golgi complex and the other compartments of the vacuolar system (Figure 6-10b). Some of the protein products of the ER are modified into glycoproteins as they are transported from the ER and through the channels of the Golgi complex. Enzymes that are present in the Golgi system bond sugar components to the proteins thus forming the glycoprotein. The glycoproteins are packaged into membrane-bound vesicles at the distal face and released to migrate through the cytoplasm. Many

vesicles arrive at the boundary of the cell where the membranes of the vesicle and cell coalesce. The contents of the vesicle are spilled into the extracellular space, and the vesicle membrane is incorporated into the cell membrane. Other vesicles that remain within the cell boundaries are loaded with special digestive enzymes, and through a series of changes, become functioning lysosomes.

Lysosomes

When viewed under the electron microscope, **lysosomes** appear as membrane-enclosed spheres ranging in size from 0.2 to 0.8 μm (Figure 6-11). They were named lysosomes (*lysis* = dissolution; *soma* = body) because of their powerful digestive capability. Potent hydrolytic enzymes are enclosed by the lysosomal membrane and are maintained in an inactive form. The enzymes become active when the membrane is exposed to an acidic environment. Lysosomes demonstrate polymorphism (*poly* = many; *morphos* = forms). *Primary lysosomes* are dense, newly formed particles that are pinched off from the Golgi complex. These

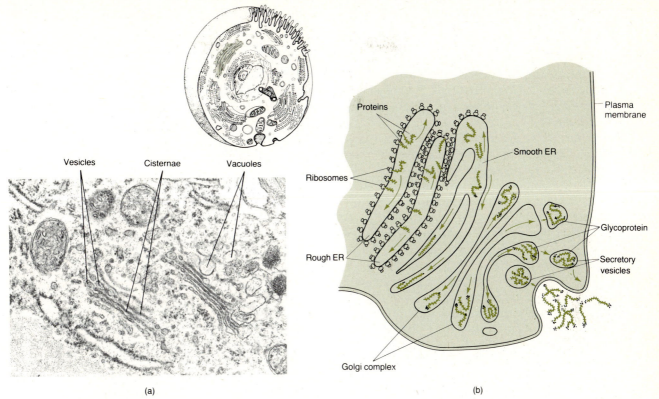

Figure 6-10 Golgi complex. (a) Electron micrograph of two Golgi complexes at a magnification of 46,800×. (Courtesy of Myron C. Ledbetter.) (b) Diagrammatic representation of the synthesis and secretion of glycoproteins by the cellular vacuolar system.

primary lysosomes fuse with vacuoles containing engulfed material and form a *secondary lysosome*. It is at this stage that digestion of the phagocytosed material is broken down by the hydrolytic enzymes. Often, the contents of the secondary lysome are not completely digested. These undigested particles are known as *residual bodies* and are often eliminated from the lysosome and the cell. The elimination or secretion of material from the cell is referred to as *exocytosis*. Other residual bodies consist of a pigment called

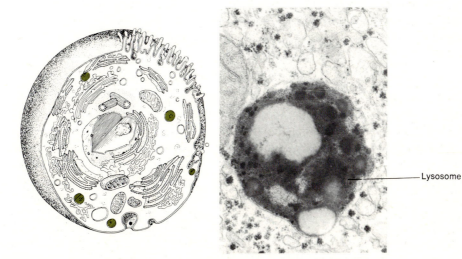

Figure 6-11 Electron micrograph of a lysosome at a magnification of 55,000×. (Courtesy of F. Van Hoof, Université Catholique de Louvain.)

lipofucsin that is a result of wear and tear within the cells of certain structures such as heart muscle, the liver, and nervous tissue. Some investigators feel that accumulation of residual bodies is an important component of the aging process. Pigment inclusions are found in the nerve cells of old animals and may interfere with normal cellular process by clogging up enzymatic machinery.

In some metabolic diseases in which homeostasis is disrupted, the lack of certain lysosomal enzymes may lead to enormous accumulations of cellular products. Severe pathologic disturbances result when glycogen accumulates in lysosomes and is not metabolized for normal cell use.

Partial cellular breakdown occurs when lysosomes engulf and digest other cellular organelles. During starvation, the liver cell has many autophagic (*auto* = self; *phago* = eat) vacuoles containing mitochondria. The cell has a self-devouring mechanism that can degrade its own parts without completely destroying itself. To further support the normal physiology of this process, the pancreatic hormone glucagon is known to stimulate this process.

Electron microscopic and experimental investigations on white blood cells have shown that these cells phagocytose bacteria. Lysosomes within the white blood cells fuse with the phagosome and the bacterium is soon dissolved.

Lysosomes are crucial in the removal of cell parts, whole cells (a sort of suicide process), and even extracellular material, which may be the process underlying bone removal. In bone reshaping, especially during the growth process, a special bone-destroying cell, called an *osteoclast,* actually dissolves bone. Particles of bone may be engulfed and digested by these cells. Bone tissue cultures given excess amounts of vitamin A remove bone from the culture apparently through an activation process involving lysosomes. Animals overfed on vitamin A interestingly develop spontaneous fractures, suggesting greatly increased lysosomal activity. On the other hand, cortisone and hydrocortisone, steroid hormones produced by the adrenal gland, have a stabilizing effect on lysosomal membranes. The steroid hormones are well known for their anti-inflammatory properties, which suggests that they reduce destructive cellular activity by lysosomes.

Peroxisomes

Peroxisomes are ovoid structures that are limited by a single membrane and contain a fine, granular material, most likely condensed protein. Peroxisomes are rich in enzymes such as peroxidase, catalase, and certain other oxidase types; hence the name. Peroxisomes isolated from liver cells contain four enzymes related to hydrogen peroxide (H_2O_2) metabolism. Three of the enzymes produce H_2O_2 and the fourth, named catalase, breaks it down. Hydrogen peroxide is toxic to the cell, and therefore it appears as though catalase plays an important homeostatic role by converting H_2O_2 into H_2O and O_2 as follows:

$$2H_2O_2 \rightarrow 2H_2O + O_2$$

Hydrogen peroxide   Water   Oxygen

Mitochondria

Mitochondria (*mito* = thread; *chondrion* = granule) are present in the cytoplasm of all eucaryotic cells. When viewed with the electron microscope, mitochondria have a variety of shapes and sizes. Often granular, filamentous, or rod-shaped with a relatively constant width of about 0.5 μm, their length varies from 2 μm to 7 μm. The number of mitochondria per cell varies. In a normal liver there are about 1,200 per cell, whereas in oocytes there can be as many as 300,000 per cell.

Figure 6-12a illustrates the structure of a mitochondrion. The *outer membrane,* about 6 nm thick, surrounds the mitochondrion. Within this membrane is the second or *inner membrane.* Appearing as though it is too large to be contained within the outer membrane, the inner membrane is thrown into many folds. These infoldings or *cristae* project into the mitochondrial center *(matrix)* and provide a large surface area for the orderly incorporation of enzymes, coenzymes, and special transporting molecules. These incorporated molecules interact with each other to carry out cellular energy transformation. The energy of the foodstuffs taken into the cell is extracted and converted to the more universally usable form adenosine triphosphate, ATP. In recent years, the mystery of the energy-producing process has been one of the most intensely pursued problems of biological research. Using a special technique called negative staining, electron microscopists discovered that the inner mitochondrial membrane is covered with particles about 8.5 nm in diameter. These so-called *elementary particles,* or F_1 particles or inner membrane spheres (IMS) as they are alternately named, are spaced regularly every 10 nm along the inner surface of the inner membrane (Figure 6-12b). These particles have been shown to be the sites of a special enzyme that produces the ATP molecule.

Mitochondria are distributed throughout the cytoplasm and are located in higher concentrations where the energy requirement is greatest. For example, mitochondria of the kidney tubule are located between the infoldings of the cell membrane on the basal

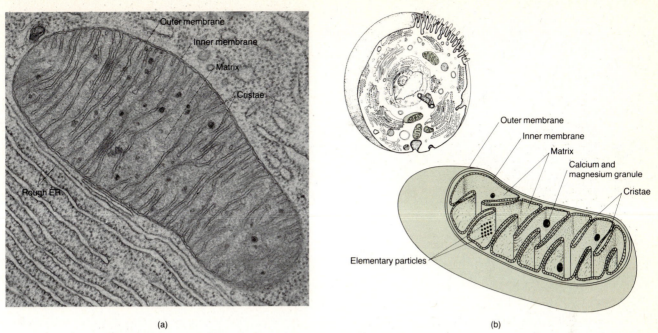

Figure 6-12 Mitochondria. (a) Electron micrograph of a mitochondrion at a magnification of 85,000×. (Courtesy of Keith R. Porter.) (b) Ultrastructure of a mitochondrion showing elementary particles.

surface. Mitochondria behave within the cell with considerable independence. They contain DNA and ribosomes, and probably synthesize some of their own proteins. Furthermore, they divide and duplicate their own DNA. The mitochondrial DNA is a circular strand, and the ribosomes are smaller than those of the cytoplasm of the cell. It has been shown that mitochondria behave much like procaryotic organisms, and this is used as evidence to support the evolutionary origin of eucaryotic cells, as was mentioned early in this chapter.

Ribosomes

Ribosones are very small particles composed of RNA and protein that are found in all cell types. In eucaryotic cells they have a diameter of about 20 nm. Ribosomes in the cytoplasm are either dispersed randomly or attached to the membranes of the endoplasmic reticulum (see Figure 6-9). A few ribosomes are located in the nucleus and mitochondria. Often the ribosomes are linked together in long chains called *polyribosomes* or *polysomes* (Figure 6-13a).

Ribosomes are composed of a large and a small subunit and are attached to the endoplasmic reticulum membrane by the large subunit. The large subunit and small subunit fit together in a way that creates a small groove between them through which slips the messenger RNA molecule, as shown in Figure 6-13b. This intimate relationship permits the ribosome to

read the messages carried by the mRNA. The number of ribosomes strung out in a polyribosome varies and seems to be regulated by the length of the mRNA that is to be read. Each ribosome synthesizes a polypeptide molecule. In this way, many identical protein chains are formed simultaneously. The process of protein synthesis involves many cellular interactions and is discussed in detail in Chapter 7.

Flagella, Cilia, Microtubules, and Centrioles

Some cells of the human body possess hairlike projections used for propelling the entire cell or for moving substances along the surface of fixed cells. The projections contain cytoplasm and are surrounded by the cell membrane. If the projections are few and long in proportion to the size of the cell, they are called **flagella**. An example of a flagellum is the tail of a spermatozoan (sperm cell) (Figure 6-14a). The tail is used for the locomotion of the sperm. If the projections are numerous and short, resembling many hairs, they are called **cilia**. In humans, ciliated cells of the respiratory tract move lubricating fluids over the surface of the tissue and trap foreign particles (Figure 6-14b).

Flagella and cilia are made up of bundles of paired **microtubules,** called **axonema** (Figure 6-14c), that are supported by the cytoplasmic ciliary matrix. These axonema are surrounded by the ciliary or flagellar membrane that is continuous with the cell membrane

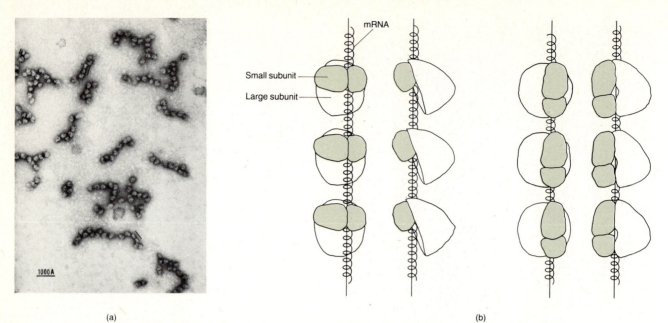

(a)   (b)

Figure 6-13 Ribosomes. (a) Electron micrograph of several polyribosomes. (Courtesy of Dr. David Sabatini, New York University School of Medicine.) (b) Possible relationship between ribosomes and mRNA. The ribosomes are shown in various orientations.

(Figure 6-14c). The tubules of the axoneme are arranged in a characteristic pattern of nine double microtubules arranged in a circular pattern around two single microtubules. The cilium or flagellum is anchored in the cytoplasm of the cell by a structure called a *basal body*, which is about 0.5 μm long and 0.15 μm in diameter.

Centrioles are paired cylindrical structures located near the nucleus. The wall of the centriole has nine clusters of microtubules arranged in a circular pattern (Figure 6-15). Each cluster has three microtubules (rather than two as in the cilium) arranged in a slanted pattern. The microtubules are named A, B, and C, from inner to outer. Centrioles lack the two central single microtubules found in flagella and cilia. Centrioles assume a role in cell division.

Cell Inclusions

Cell inclusions are a large and diverse group of chemical substances. These products are principally organic in nature and may appear or disappear at various times in the life of the cell. *Hemoglobin* lies inside red blood cells. It performs the function of combining with oxygen molecules and carrying the oxygen to body cells. *Melanin* is a pigment stored in the cells of the skin, hair, and eyes. It protects the body by screening out harmful ultraviolet rays from the sun. *Glycogen* is a polysaccharide that is stored in liver and skeletal muscle cells. When the body requires

quick energy, the cell organelles break down the glycogen into glucose and release the glucose. *Lipids*, which are stored in fat cells, may be decomposed when the body runs out of carbohydrates for producing energy. *Zymogen* (*zymo* = ferment; *gennein* = produce) *granules* are obvious in cells that synthesize enzymes for export. The zymogen granules are highly condensed concentrations of inactive enzyme molecules. These granules, upon release from the cell, are activated by constituents outside the cell.

The cytoplasm of the nerve cell body contains *Nissl bodies*, a dark-staining aggregation of flattened stacks of rough ER with numerous ribosomes and polysomes scattered randomly between the individual cisternae. The ribosomes of the nerve cell are important because they continually synthesize important molecules that are constantly being moved down the length of the nerve cell to be used for stimulating adjacent cells. A final example of an inclusion is *mucus*, which is produced by cells that line organs. Its function is to provide lubrication.

EXTRACELLULAR MATERIALS

Substances that lie outside cells are collectively called **extracellular materials.** They include the body fluids, which provide an aqueous medium for dissolving, mixing, and transporting substances. Materials that are produced by the cells and secreted into the extra-

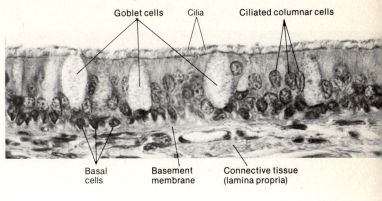

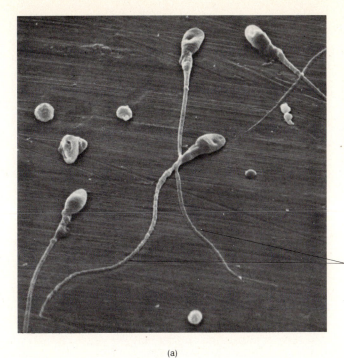

(a)

(b)

Flagella

Microtubules

Flagellar or ciliary membrane

Cell membrane

Ciliary plate

Basal body

Flagellar or ciliary membrane

Microtubules

(c)

Figure 6-14   Flagella and cilia. (a) Scanning electron micrograph of several spermatozoa showing flagella at a magnification of 2,000×. (Courtesy of Fisher Scientific Company and S.T.E.M. Laboratories, Inc., Copyright 1975.) (b) Photomicrograph of the tracheal epithelium showing cilia at a magnification of 640×. (Courtesy of Edward J. Reith, from *Atlas of Descriptive Histology,* by Edward J. Reith and Michael H. Ross, New York: Harper & Row, 1970.) (c) Structure of a flagellum or cilium in cross section and longitudinal section (left) and cross section (right).

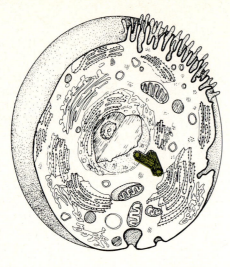

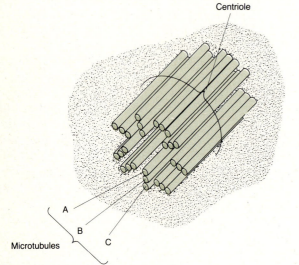

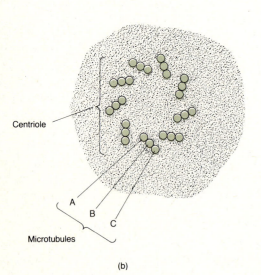

Figure 6-15 Centrioles. (a) Longitudinal section. (b) Cross section.

cellular space, such as acid mucopolysaccharides, collagen, reticular fibers, and elastic fibers are also considered extracellular materials. The secretory products can be combined in different ways to produce an intercellular matrix with special properties. Cartilage and bone are two examples of tissues whose component cells function primarily to secrete extracellular materials.

Acid Mucopolysaccharides

Matrix materials are produced by certain cells and are deposited outside their plasma membranes. The matrix supports the cells, binds them together, and gives strength and elasticity to the tissue. Some of the matrix materials are *amorphous,* which means they have no specific structure. These include hyaluronic acid and chondroitin sulfate, two compounds classified as *acid mucopolysaccharides* because they have acidic properties and are composed of modified sugars that have a mucuslike consistency.

The acid mucopolysaccharides collectively play an essential role in controlling the water content of connective tissue. Acid mucopolysaccharides are very hydrophilic and therefore bind many water molecules. Thus, almost all water in the intercellular spaces of connective tissues is incorporated into the jellylike *sol.* The acid mucopolysaccharides contribute to the process of calcification of cartilage and bone; they help to control and localize the concentrations of nutrients and significant ions; and they are important in wound healing.

Hyaluronic acid is a lubricant in the knee joint, where it has the consistency of a syrup. In the skin, where it functions as a cushion, it has a great ability to take up and hold water and to behave with the consistency of a jelly. Generally, it binds and supports cells and lubricates joints. It is a major component of the vitreous humor of the eye, and therefore it helps to maintain the shape of the eyeball.

Chondroitin sulfate is abundant in the intercellular matrix of cartilage and to a lesser extent in other connective tissues where it imparts an adhesive property. It is also found in the cornea of the eye, umbilical cord, skin, heart valves, and bone.

Collagen

Fibrous materials provide strength and support for tissues. Among these are *collagen* or *collagenous fibers.* Collagen (*kolla* = glue) is the most abundant protein in the human body, making up about one-third of the body's total protein. Collagen is found, for example, in all bones, tendons, ligaments, and the cornea of the eye. Where collagen is present in large amounts, the tissue takes on a shiny white color. Collagen fibers are arranged in parallel bundles in tendons, and this arrangement imparts great strength but no ability to stretch to the tissues. Under the electron microscope, collagen fibrils reveal a characteristic cross-striated appearance. Collagen is assembled outside the fibroblast cell by the orderly arrangement of smaller subunits called *tropocollagen*. The tropocollagen molecules are laid down in a head-to-tail pattern in linear lines. The linear lines are then arranged in parallel array in a distinctive overlapping pattern. Biochemists have determined that each tropocollagen molecule is constructed of three polypeptide chains.

Reticular Fibers

Reticular fibers (*rete* = net) are a much finer version of collagen consisting of collagen fibrils with the additional complexity of a carbohydrate coating. Some investigators feel that the carbohydrate coating prevents the collagenous fibrils from complexing to form the true collagen fiber. Reticular fibers are arranged in certain organs (for example, liver, kidneys, endocrine glands, spleen, lymph nodes, and bone marrow) to form an internal framework or scaffolding that supports the cells of the organ and determines the organ's overall shape.

Elastic Fibers

Elastic fibers are composed of protein chains that are covalently bonded together to make up an elastic two-dimensional sheet. These elastic fibers branch and interlace with each other and form an irregular tangle. When examined while still fresh, elastic fibers have a characteristic yellow color, and for this reason they are sometimes called the yellow fibers. Elastic properties are familiar to all students who have ever stretched a rubber band. In a similar way, elastic fibers yield to pulling forces but return to their original shape when the force stops. Elastin is a very tough, almost indestructible, molecule that is therefore quite long-lived. A recent discovery is that the elastic fibers, as well as the collagen fibers, of the arterial wall are made by the smooth muscle cells, cells that had been thought to be involved only with the contractile process. Elastin is most likely synthesized by the ER of the smooth muscle cells and secreted much like the process for collagen. In addition to their presence in arteries, elastic fibers are found in loose connective tissues, skin, lungs, and the cartilage of the epiglottis and external ear.

Cartilage

Both cartilage and bone represent special examples of tissues in which the cellular material is the major component. The cells that are present in cartilage are called *chondrocytes* (*chondro* = cartilage; *cyte* = cell), and they secrete, in addition to collagen and elastic fibers, an amorphous mucopolysaccharide that contains large amounts of chondroitin sulfate. One type of cartilage is a tough substance that is often located over the articular ends of bone to provide a slick and smooth surface. When lubricated with special fluids, as in the ball-and-socket joints of the shoulder or hip, cartilage provides free and easy movement. Research has shown that the relative proportions of the various components of cartilage vary with age and change during certain disease states.

Bone

The process of bone formation is best described in histology or anatomy texts, and you are referred to them for a detailed description. However, you should be acquainted with the idea that even though the intercellular spaces of bone tissue are made impermeable by the deposition of calcium salts, the bone cells survive. These cells, called *osteocytes* (*osteo* = bone), are kept alive because the bone tissue is permeated with microscopic canals called *canaliculi* through which fluids, nutrients, and wastes diffuse.

The intercellular material of bone has two components. One is organic in nature; the other is inorganic. About 75 percent of the dry weight of bone is inorganic substance and the rest is organic. Of the organic material, almost 90 percent is collagen. The remainder of the organic material is the amorphous ground substance that contains neutral and acid mucopolysaccharides (chondroitin sulfates). These materials are secreted and maintained by the bone cells. The inorganic salts include bicarbonate, citrate, magnesium, potassium, and sodium. The most abundant mineral that is laid down in the matrix is a complex salt of calcium and phosphate called hydroxyapatite, $Ca_{10}(PO_4)_6(OH)_2$. Crystals of this salt complex are needle-shaped slivers that lie along or within the collagen fibers. Though the crystals are not secreted by the bone cells, an enzyme released by the cells promotes the formation of the hydroxyapatite crystals.

Having considered the principal structures and functions of the cell and their relationship to homeostasis, we will now examine cellular control and reproduction. Cellular control, perhaps more than any other concept, is an excellent example of homeostasis at the cellular level.

CHAPTER SUMMARY OUTLINE

EVOLUTION OF THE CELL

1. Life on the earth is estimated to have arisen around 3.2 to 3.5 billion years ago.
2. During chemical evolution various gases interreacted, aided by energy from electrical discharges, heat, and ultraviolet and gamma radiation from the sun.
3. These interreactions produced a large variety of simple carbohydrates, amino acids, and organic bases.
4. From these simple basic molecules scientists have produced many of the larger, more complex organic molecules found in living systems.
5. The first "cells" that emerged from chemical evolution were anaerobic heterotrophic cells. These were followed by autotrophic cells and eucaryotic cells.

PROCARYOTIC CELLS

1. Procaryotic cells lack a membrane around the nucleus, and all internal vacuolar membranes are absent.
2. The nucleus lacks both a mitotic apparatus and nucleoli, and the genetic information is stored in one single chromosome that is in the form of a circle.
3. Examples of procaryotes include bacteria, blue-green algae, and mycoplasmas.

Bacteria

1. Bacteria comprise the largest group of procaryotes, and are divided into three general groups based on their shapes: bacilli, cocci, and spirilla.
2. All bacteria reproduce by a process of division called binary fission.
3. Some bacteria protect themselves from inhospitable environments by producing spores.
4. Bacteria may invade the body, producing disease and sometimes death. Familiar diseases in humans caused by bacteria include typhoid fever, cholera, plague, dysentery, scarlet fever, tuberculosis, diphtheria, and wound infections such as gangrene.
5. Generally however, bacteria are more beneficial to humans than harmful. They play vital roles in the production of cheese, buttermilk, vinegar, and other foods.

6. They are indispensable in sewage treatment, cause the decay of dead organisms and the restoration of carbon and nitrogen, and produce vitamin K in the human intestine.

Blue-Green Algae and Mycoplasmas

1. Blue-green algae are photosynthetic, and contain photosynthetic pigments such as the green chlorophyll and the blue phycocyanin.
2. Mycoplasma cells are responsible for producing disease, and represent the smallest, simplest living systems known to exist.

EUCARYOTIC CELLS

1. Eucaryotic cells are enclosed by a cell membrane and possess a nucleus enclosed by a nuclear envelope.
2. Membranes divide the eucaryotic cell into organelles that function independently of but cooperatively with each other.
3. Organelles are internal membrane-bound compartments that include the mitochondria, lysosomes, peroxisomes, endoplasmic reticulum, and Golgi complex.
4. Nucleoli, ribosomes, chromosomes, centrioles, and microtubules are all specialized structures lacking membranes.
5. Most organelles demonstrate the phenomenon of turnover. In this process, molecules making up any structure are continually replaced by new ones, so that, in time, all molecular parts are exchanged for new ones, yet the structure is always present.

Membrane Structure

1. The cell membrane encloses every cell, and is the primary barrier to molecules that enter or leave the cell.
2. The cell membrane is composed of proteins and lipids.
3. Membrane lipids are mainly phospholipids and to a lesser extent cholesterol. The proteins are classified as peripheral and integral.
4. Most investigators agree that the core of the plasma membrane is a bilayer (double row) of phospholipid molecules somewhat intertwined and randomly disrupted by cholesterol molecules.

5. Most of the protein molecules are of the integral type, firmly bonded to phospholipids.
6. At present, the Singer-Nicolson model of membrane structure fits the current information most satisfactorily. The electron microscope reveals a uniformly trilaminar (three-layered) structure.

Cell Membrane Associations: Cell Junctions

1. The cell membrane and cell surface of mammalian cells can be highly specialized and modified depending on relationships to adjacent cells.
2. Four categories of cell junctions are recognized: tight junctions, gap junctions, intermediate junctions, and desmosomes.

Cell Membrane Modifications

1. The membranes of certain cells are arranged in unique patterns for specific purposes.
2. Microvilli are cytoplasm-filled extensions or evaginations of the membrane that increase the surface area for absorption.
3. The myelin sheath of nerve cells surrounds and electrically insulates them.

CYTOPLASM

1. The cytoplasmic matrix consists of a gelatinous solution called a colloid that can undergo sol-gel (liquid-jelly) transformations; its viscosity changes.
2. Cytoplasm is responsible for intracellular movement and amoeboid movement possibly through microfilaments; it promotes spindle formation; and it participates in cell division.
3. Mechanical properties such as elasticity, rigidity, cohesion, and contractility are attributed to the cytoplasmic matrix.
4. Microtubules in the cytoplasm play a key role in the spindle of dividing cells, and are responsible for the motile force that causes ciliary beating and flagellar whipping.

CELL ORGANELLES

1. Organelles are distinct cytoplasmic structures organized into a system of compartments, each with a specialized function.
2. Organelles include the nucleus, endoplasmic reticulum, Golgi complex, lysosomes, peroxisomes, mitochondria, ribosomes, cilia, centrioles, microtubules, and cell inclusions.

Nucleus

1. The nucleus controls cellular activities including cell division.
2. The nuclear pore complex controls the passage of ions and of small and large molecules into and out of the nucleus.
3. The nucleus contains the nucleolus, which is responsible for the synthesis of the components of the ribosomes.

Vacuolar Systems

1. The term cytoplasmic vacuolar system describes a system of membrane-bound tubules, vesicles, and flattened sacs that are all interconnected.
2. Two subdivisions of this system include the endoplasmic reticulum (ER) and the Golgi complex.
3. The endoplasmic reticulum may be further subdivided to include: (1) the nuclear envelope, (2) the rough or granular ER, and (3) the smooth or agranular ER.
4. Functions of the ER include: protein synthesis, intracellular circulatory system, transportation system, mechanical support and distribution of cytoplasm, and conduction of intracellular impulses.
5. The Golgi complex functions to synthesize glycoproteins.

Lysosomes

1. Lysosomes contain powerful hydrolytic enzymes in an inactive form.
2. Lysosomes within white blood cells can phagocytose bacteria.
3. Lysosomes are crucial in the removal of cell parts, whole cells, and even extracellular material such as bone.

Peroxisomes

1. Peroxisomes are rich in enzymes such as peroxidase and catalase.

Mitochondria

1. Mitochondria have an inner membrane that is thrown into many folds. These folds or cristae provide a large surface area for enzymes, coenzymes, and special transporting molecules.
2. This inner membrane is covered with particles called elementary or F_1 particles or inner membrane spheres that are the sites of a special enzyme that produces the ATP molecule.

3. Mitochondria are located in higher concentrations where the energy requirement is greatest.

Ribosomes

1. Ribosomes are very small particles composed of RNA and protein.
2. Each ribosome synthesizes a polypeptide molecule, so many identical protein chains can be formed simultaneously by a polyribosome.

Flagella, Cilia, Microtubules, and Centrioles

1. Some cells of the human body possess hairlike projections for moving the entire cell or for moving substances along the surface of the cell.
2. If the projections are few and long in proportion to the size of the cell, they are called flagella.
3. If the projections are numerous and short, resembling many hairs, they are called cilia.
4. Flagella and cilia are made up of bundles of paired microtubules called axonema.
5. The wall of the centriole has nine clusters of three microtubules arranged in a circular pattern.

Cell Inclusions

1. Cell inclusions are a large and diverse group of chemical substances principally organic in nature.
2. Hemoglobin is an example of a cell inclusion. It lies inside red blood cells and functions by collecting oxygen molecules and carrying them to body cells.
3. Another inclusion is melanin, a pigment that protects the body by screening out harmful ultraviolet rays from the sun.
4. Other cell inclusions are: glycogen, lipids, zymogen granules, Nissl bodies, and mucus.

EXTRACELLULAR MATERIALS

1. Substances that lie outside cells are called extracellular materials.
2. Materials that are produced by the cells and secreted into the extracellular space are also considered extracellular materials. Some examples follow.

Acid Mucopolysaccharides

1. The acid mucopolysaccharides play an essential role in controlling the water content of connective tissue.
2. They contribute to the process of calcification of cartilage and bone. They help to control and localize the concentrations of nutrients and significant ions.
3. They are important in wound healing. Two examples are hyaluronic acid and chondroitin sulfate.

Collagen

1. Collagen is the most abundant protein in the human body and is found in all bones, tendons, ligaments, and the cornea of the eye. Collagen imparts great strength to a tissue.

Reticular Fibers

1. Reticular fibers are much finer than collagen, and form an internal framework that supports the cells of various organs.

Elastic Fibers

1. Elastic fibers are made up of protein chains that branch and interlace with each other. These fibers yield to pulling forces but return to their original shape when the force stops.

Cartilage

1. Cartilage is a special tissue in which the intercellular material is the major component. Cartilage cells are called chondrocytes, and there are different types of cartilage.

Bone

1. Bone is also a special tissue in which the intercellular material is the major component. Bone cells are called osteocytes, and the intercellular material of bone is 75 percent inorganic material and the rest organic.
2. The most abundant organic material is collagen, while the most abundant inorganic salt is called hydroxyapatite.

REVIEW QUESTIONS

1. Describe what is meant by chemical evolution.
2. What are the main differences between anaerobic heterotrophic and autotrophic cells?
3. Define symbiosis and give one example of this relationship.
4. Define exobiology.
5. Distinguish between procaryotic and eucaryotic cells.
6. Describe some characteristics of bacteria and contrast them with the blue-green algae.
7. What are mycoplasma cells?
8. What are some functions of membranes?
9. Discuss the structure of the cell (plasma) membrane.
10. Distinguish between the four different types of cell junctions, giving functions and examples of each type.
11. How are cell (plasma) membranes modified for various functions?
12. Give examples and the advantages of two such membrane modifications.
13. Discuss the chemical composition and physical nature of cytoplasm. What is its function?
14. What is an organelle? By means of a diagram, indicate the structure and describe the function of the following organelles: nucleus, endoplasmic reticulum (ER), Golgi complex, lysosome, peroxisome, mitochondria, and ribosome.
15. Differentiate among flagella, cilia, and microtubules.
16. Define a cell inclusion. Provide examples and indicate their functions.
17. What is an extracellular material?
18. Describe in detail, with functions, each of the following extracellular materials: acid mucopolysaccharides, collagen, reticular fibers, elastic fibers, cartilage, and bone.

Chapter 7

CELLULAR REPRODUCTION AND CONTROL

STUDENT OBJECTIVES

After reading this chapter, you should be able to

- Discuss the significance of cell division
- Define chromosomes indicating the various types
- Contrast the chemical composition of deoxyribonucleic acid (DNA) and ribonucleic acid (RNA)
- Define the roles of DNA in heredity and protein synthesis
- Explain how the Watson-Crick model of DNA correlates with the process of replication
- Describe the stages and events involved in cell division
- Define cytokinesis
- Explain the genetic code
- Define transcription and translation and contrast the functions of the various kinds of RNA
- Describe the functions of RNA in the biosynthesis of proteins
- Explain mutations
- Discuss the regulation of protein biosynthesis
- Compare the negative and positive control of transcription
- Explain how hormones are able to regulate transcription
- Describe the relationship of aging to cells
- Explain cancer as a homeostatic imbalance of cells

Most of the cell activities mentioned thus far maintain the life of the cell on a day-to-day basis. However, the cells that make up the tissues of the human body undergo a constant turnover (replacement). Cells become damaged or diseased, or grow old, wear out, and die. Thus, there is a need for the continual production of new cells from existing ones. The turnover rate of the cells varies from one tissue to another. Skin cells and cells that line the digestive tract are reproduced rapidly compared to cells of the pancreas or thyroid gland. Moreover, during the rapid growth period of human life, new cells are produced more rapidly than they are destroyed. A notable exception to the turnover process is nervous tissue. The nerve cells of the body carry out such a specialized function that they lose the ability to regenerate or multiply shortly after birth.

Cell division is the process by which cells reproduce themselves. For our purposes, assume that cell division may be one of two kinds. The first kind is the process by which a single parent cell duplicates itself. This process includes two phases: *mitosis* (nuclear duplication), and *cytokinesis* (cytoplasmic duplication). It is by these processes that body cells replace themselves. The second kind of cell division, *meiosis,* is the mechanism that enables the reproduction of an entirely new organism. Meiosis will be discussed in detail in Chapter 18.

Cells without a nucleus cannot divide. In fact, such cells can survive for only a short time. For example, erythrocytes or red blood cells, when released into the circulation from bone marrow where they are produced, are without a nucleus and survive for only about 120 days.

Recall that every cell has two principal periods in its life cycle: *interphase* (nondividing) and *division.*

The two periods of the life cycle are repeated with each new generation of cells. It is in the nucleus that obvious structural changes occur and signal the transition from the interphase part of the cycle to the division phase. The nuclear envelope and nucleolus disappear, and the genetic material that is dispersed as *chromatin* begins to aggregate and form dark-staining structures, the *chromosomes* (*chroma* = color; *soma* = body).

CHROMOSOMES

Chromosomes have been studied for over one hundred years, and consequently, much is now known about their structure and function. These nuclear components have the unique and amazing capabilities of replicating themselves and maintaining their chemical composition and physiological properties intact from one generation to another.

During the early stages of the division phase, when the chromatin condenses to form the chromosomes, it is possible to study and identify each of the chromosomes. The 46 chromosomes from a normal human male are shown in Figure 7-1. Such a display is called a *karyotype,* and the grouping of chromosomes is based on their overall length and the location of the centromere. The *centromere* is the primary constriction that divides the chromosome into two parts.

In practically every cell of the body the chromosomes are paired. One member of each pair is contributed by each of the parents. Each member of the chromosome pair is referred to as a *homologue.* The human species has 46 chromosomes or 23 homologous pairs of chromosomes. In a nucleus beginning to di-

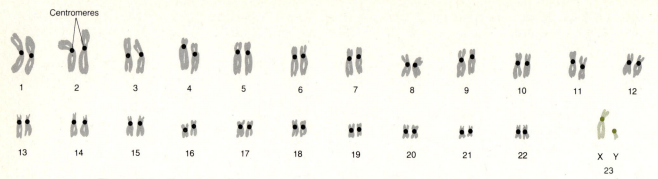

Figure 7-1 Karyotype of human male chromosomes. The chromosomes are arranged according to size and centromere location of each chromosome pair. The X and Y chromosomes designate the male sex chromosomes.

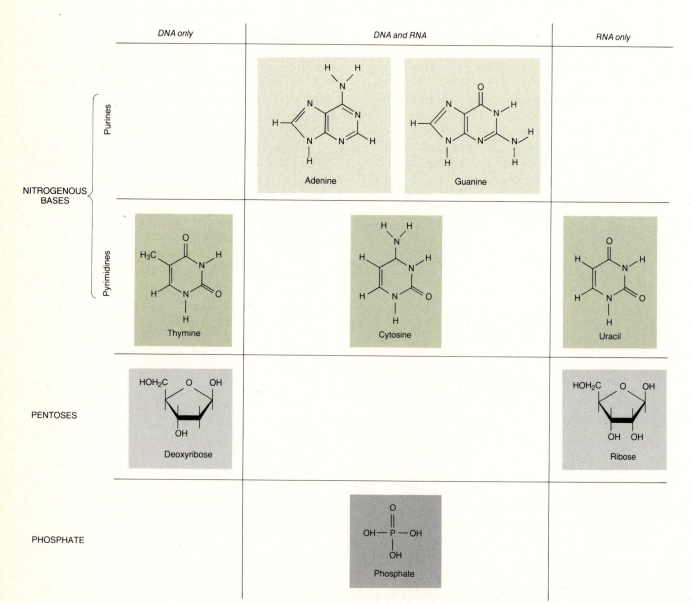

Figure 7-2 The basic parts of DNA and RNA.

vide, the homologues are not always near their partners. In fact, any single chromosome can be located anywhere within the nucleus. Cells in which the chromosomes are present in pairs are referred to as *diploid*. The karyotype is characteristic of an individual species or genus. The chromosomal pattern of a normal human male consists of 22 matching pairs of chromosomes called *autosomes* plus two other chromosomes that are unpaired and are called the X and Y chromosomes. These are the *sex chromosomes* (see Figure 7-1). The presence of one X and one Y chromosome indicates that the sex of the individual is male. A normal human female, by contrast, possesses the same 22 matching pairs of autosomes, but the remaining two chromosomes are a matching pair of X chromosomes. The difference between male and female, therefore, is in the X and Y sex chromosome distribution.

Chemical Composition

Biochemical analyses of chromosomes show that they consist of a molecule of *DNA* (deoxyribonucleic acid)

plus a special type of protein called *histone*. The combination of DNA and the histone is often designated *nucleoprotein*.

DNA is the ultimate repository of all cellular information. Encoded within the DNA molecule are the blueprints for the construction of all the enzyme molecules and other proteins the cell will ever need for its homeostasis. The discovery and continual elucidation of the mechanisms by which this remarkable molecule functions are among the most exciting developments in recent biological history.

DNA is composed of subunits termed deoxyribonucleotides. *Deoxyribonucleotides* share a structure very similar to those of the *ribonucleotides* that are found in RNA. These two groups of nucleotides each contain three basic parts: (1) a *nitrogenous base* that is a derivative of either pyrimidine or purine, (2) a five-carbon sugar called a *pentose*, and (3) a molecule of *phosphoric acid*, also called a *phosphate group*. The basic parts of DNA and RNA are compared in Figure 7-2. The four assembled deoxyribonucleotides found in DNA are shown in Figure 7-3.

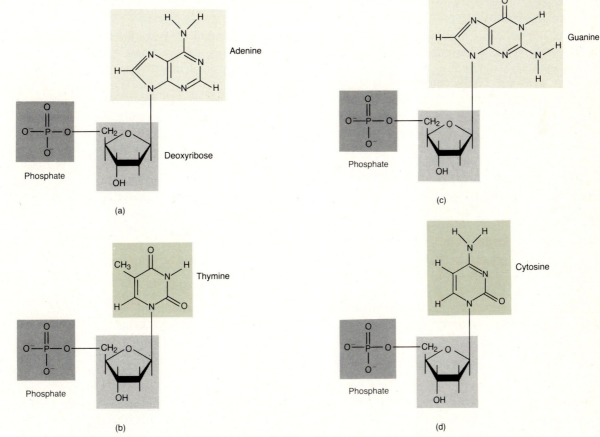

Figure 7-3 Deoxyribonucleotides of DNA. Each deoxyribonucleotide consists of a specific nitrogenous base attached to the deoxyribose sugar, which, in turn, is attached to the phosphate group. (a) Adenine deoxyribonucleotide. (b) Thymine deoxyribonucleotide. (c) Guanine deoxyribonucleotide. (d) Cytosine deoxyribonucleotide.

Structure of DNA

The structure of the DNA molecule was first proposed by James Watson and Francis Crick in 1953. Their proposed model consisted of two long chains of deoxyribonucleotides twisted around each other to form what is termed a *double helix*. Figure 7-4a shows the following structural characteristics of the DNA molecule. The molecule resembles a flexible ladder that is twisted so it takes ten rungs to make one revolution of the spiral. The side rails of the ladder are made of alternating pentose and phosphate groups. The pentose sugar in DNA is called deoxyribose; in RNA, it is called ribose. The flat rungs or steps of the ladder are the result of the nitrogenous bases of each chain meeting and fitting together precisely.

Replication

The key to the Watson-Crick model was the discovery of the precise "fits" of the opposing nitrogenous bases. They discovered that adenine of one chain always and only paired with thymine of the other chain, and cytosine and guanine always and only paired with each other. As a result of this discovery, it is always possible to list the sequence of bases in one chain if the sequence in the other is known. The chains are said to be complementary to each other and are illustrated in Figure 7-4.

The weak hydrogen bonds linking the complementary nitrogenous bases are also shown in Figure 7-4. The two chains are held together by thousands of weak hydrogen bonds. It is important to understand the benefits of such a structure. Though each pair of bases can be separated quite easily (since the individual hydrogen bonds are very weak), the long chains are held together with considerable strength because of the large number of the hydrogen bonds. During the process of cell division, when the chromosomes are replicated, the two chains gradually uncoil and separate starting at one end; each chain is used as a pattern to guide the correct assembly of a new partner chain (Figure 7-4b). When the chains have completely uncoiled and each original chain has a new partner, the new double-stranded chains migrate to opposite ends of the cell.

The replication of the exact sequence of bases in each strand is thus assured. Each new generation of DNA molecules is carefully synthesized in a complementary fashion to fit exactly an original chain. Because each new double helix thus always contains one chain of the old DNA and a new complementary partner chain, this method of replicating the complete DNA molecule is called *semiconservative replication*.

CELL DIVISION

The process of *mitosis* is the sequence of events in which the nuclear material divides equally. Thus, two new sets of chromosomes are formed, and they are separated and packaged into the nuclear envelope of each new cell. As mitosis concludes, the process of *cytokinesis* begins. In this process, the cytoplasm of the parent cell is divided more or less equally between the two newly forming cells. An inward movement of the cell membrane from opposite sides of the cell eventually pinches the cell into two separate cells. The resultant cells are called *daughter cells*.

The function of mitosis and cytokinesis is to replace cells in the body. The process ensures that each new daughter cell has the same number and kind of chromosomes as the original parent cell. After the process is complete, the two daughter cells have the same hereditary material and genetic potential as the parent cell. This kind of cell division results in an increase in the number of body cells. Mitosis and cytokinesis are, therefore, the means by which dead or injured cells are replaced and also the means by which cells are added for body growth. In a 24-hour period, the average human adult loses about 500 million cells from different parts of the body. Obviously, these cells must be replaced in order to maintain homeostasis. Cells that have a short life span, such as the cells of the outer layer of skin, the cornea of the eye, and the digestive tract, are continually being replaced.

The succession of events that takes place during mitosis and cytokinesis is plainly visible under a microscope after the cells have been stained in the laboratory. When a cell reproduces itself, it must replicate its chromosomes so that its heredity may be passed on to succeeding generations of cells. As just described, a chromosome consists of a duplex strand of DNA plus a special basic type of protein termed a histone. The strands of DNA in turn contain *genes*, which are composed of groups of nucleotides strung out along the DNA molecule. Each human chromosome contains about 20,000 genes.

The process of mitosis is the replication of chromosomes and the distribution of the two sets of chromosomes into two separate and equal nuclei. For convenience, biologists break down the process into four stages: prophase, metaphase, anaphase, and telophase. These are arbitrary classifications, as mitosis is actually a continuous process, one stage merging into the next. However, before a cell passes through its divisional stages, it must first pass through a stage called interphase.

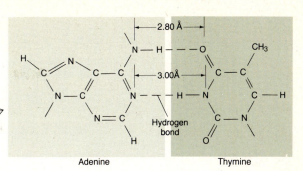

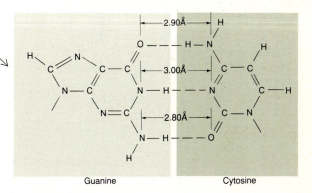

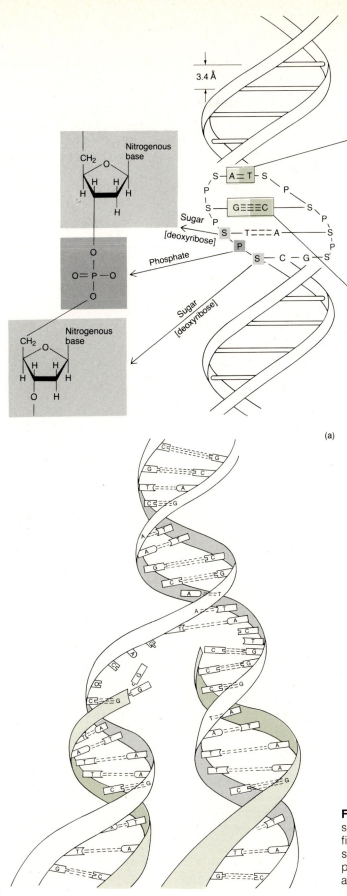

(a)

(b)

Figure 7-4 Structure and replication of DNA. (a) DNA structure. In the center is shown the double-helical configuration. The details of nitrogenous base pairing are shown to the right and the details of the deoxyribose and phosphate bonding are shown to the left. Relative distances are indicated in angstroms. (b) Replication of DNA.

Interphase

One of the principal events of *interphase* is the replication of DNA. During this phase, the chromatin is dispersed throughout the nucleus, and when stained, shows an unequal stain distribution. Some regions stain darker than others (see Figure 6-8a). The variation results from differences in the tightness of DNA coiling.

During interphase, the cell is also synthesizing other kinds of nucleic acids and proteins. It is producing chemicals so that all cellular components can be doubled during division. When examined under a microscope, the cell nucleus has a clearly defined membrane, nucleoli, and chromatin (Figure 7-5a). As interphase progresses, a pair of centrioles appears. The centrioles duplicate and the resulting two pairs of centrioles separate. Once a cell completes its activities during interphase, mitosis begins.

Prophase

During *prophase* (Figure 7-5b), the centrioles move apart and project a series of radiating fibers called *astral rays*. The centrioles move to opposite poles of the cell and become connected by another system of fibers called *spindle fibers*. Together, the centrioles, astral rays, and spindle fibers constitute the *mitotic apparatus*. Simultaneously, the chromatin has been shortening and thickening into chromosomes. The nucleoli have become less distinct, and the nuclear membrane has disappeared. Each prophase "chromosome" is actually composed of a pair of separate structures called *chromatids* (remember the duplication of DNA during interphase). Each chromatid is a complete chromosome made of a double-stranded DNA molecule. Each chromatid is attached by a small spherical body called the *centromere*. During prophase, the chromatid pairs move toward the equatorial plane region or equator of the cell.

Metaphase

During *metaphase* (Figure 7-5c), the second stage of mitosis, the chromatid pairs line up on the equatorial plane of the spindle fibers. The centromere of each chromatid pair attaches itself to a spindle fiber. The lengthwise separation of the chromatids now takes place. Each centromere divides, and the independent chromatids begin moving to opposite poles of the cell.

Anaphase

Anaphase (Figure 7-5d), the third stage of mitosis, is characterized by the division of the centromeres and the movement of complete sets of chromatids (now the new chromosomes) to opposite poles of the cell. During this movement, the centromeres that are attached to the spindle fibers seem to drag the trailing parts of the chromosomes toward opposite poles.

Telophase

Telophase (Figure 7-5e), the final stage of mitosis, consists of a series of events that are approximately opposite those of prophase. By now, two identical sets of chromosomes have reached opposite poles. New nuclear membranes begin to enclose them. The chromosomes start to assume their chromatin form. Finally, nucleoli reappear, and the spindle fibers and astral rays disappear. The centrioles also replicate so that each new cell has two centriole pairs. The formation of two nuclei identical to those in cells of interphase terminates telophase. A mitotic cycle has thus been completed (Figure 7-5f).

Cytokinesis

The division of the cytoplasm, called *cytokinesis*, often begins in late anaphase and terminates at about the same time as telophase. Cytokinesis begins with the formation of a *cleavage furrow* that runs around the cell's equator. The furrow progresses inward, resembling a constricting ring, and cuts completely through the cell to form two separate portions of cytoplasm (Figure 7-5d to f). We shall consider how genetic information is passed from one part of the cell to another.

GENETIC CODE

The information needed by each cell to carry on its life processes is transmitted to each new generation by means of the DNA molecules. This information must be relayed into the cellular cytoplasm to the ribosomes. There, the information is used to direct the synthesis of protein molecules, many of which are enzymes. Enzymes, in turn, determine the functional characteristics of the entire cell.

The transfer of information out of the nucleus to the ribosomes is performed by *ribonucleic acid (RNA)*. RNA molecules are used by all living organisms to carry out protein synthesis. Similar to the DNA molecule already described, RNA is composed of subunits termed ribonucleotides. The ribonucleotides of RNA are constructed by combining a five-carbon sugar called ribose with phosphoric acid (phosphate group), shown in Figure 7-2, and one of the following

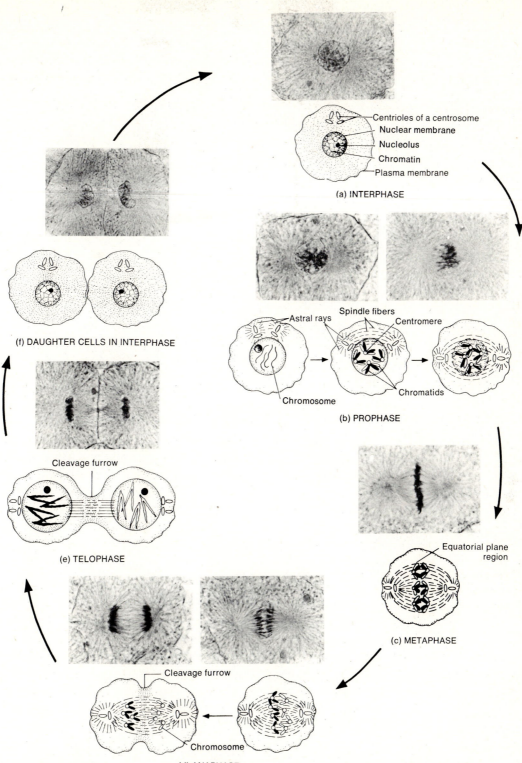

Figure 7-5 Cell division: mitosis and cytokinesis. Photomicrographs and diagrammatic representations of the various stages of cell division in whitefish eggs. Read the sequence starting at (a), and move clockwise until you complete the cycle. (Courtesy of Carolina Biological Supply Company.)

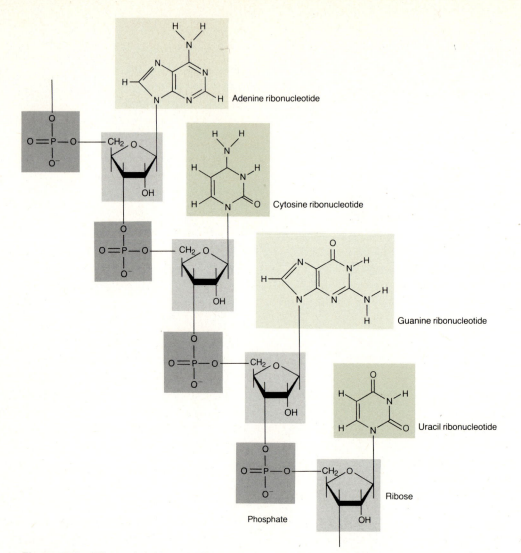

Figure 7-6 Ribonucleotides of RNA. Whereas the pentose in DNA is deoxyribose, the pentose in RNA is ribose. Also, RNA does not contain thymine; instead it contains uracil.

Labels on figure: Adenine ribonucleotide; Cytosine ribonucleotide; Guanine ribonucleotide; Uracil ribonucleotide; Ribose; Phosphate

nitrogenous bases: adenine, uracil, guanine, or cytosine. Thus, RNA consists of four different ribonucleotides (Figure 7-6). RNA is formed by linking the sugar of one ribonucleotide to the phosphate group of the next ribonucleotide. RNA is a single long chain of ribonucleotides that has a backbone of alternating ribose and phosphate and nitrogenous bases projecting outwards. Three types of RNA molecules are employed by all cells in the process of making proteins, and each has its characteristic structure and function. The three types are messenger RNA (*m*RNA), transfer RNA (*t*RNA), and ribosomal RNA (*r*RNA). Let us first examine how RNA is synthesized by the cell.

Transcription

All RNA molecules are created in a manner similar to that used in replicating DNA. A section of a double-stranded DNA molecule uncoils, and one of the exposed strands is copied. The sequence of molecular events occurring during the copying process is termed *transcription*. The strand of RNA is formed by aligning complementary ribonucleotides opposite the nitrogenous bases of the DNA molecule. The same base-pairing rule applies in RNA synthesis as in DNA replication, except that a ribonucleotide containing uracil pairs off with one containing adenine. The nitrogenous base thymine is not used in RNA. Different sections of the DNA molecule are copied, thus producing a variety of different RNA molecules. Though only one strand of the DNA molecule is copied for a given RNA molecule, both strands of the DNA molecule can be used for transcription purposes.

A segment of a DNA molecule that encodes in its nucleotide sequence all the information necessary for the synthesis of a complete polypeptide chain is called a *structural gene*. In some instances, the information necessary for two or more polypeptides can be copied into a single mRNA molecule. Such a messenger molecule has special signals that indicate to the ribosome the end of the first polypeptide and the beginning of the second. For example, a set of enzymes that operates in a particular sequence within the cell can be transcribed onto the same *m*RNA molecule in this manner. Let us now examine the functions of the various kinds of RNA.

Messenger RNA (*m*RNA)

Messenger RNA or simply *m*RNA carries the coded information from the nucleus to the ribosomes. The *m*RNA dictates the exact order in which the amino acids must be assembled to construct the required protein as determined by DNA. The number of different proteins that can be assembled from 20 different kinds of amino acids, from very simple short chain lengths to some with hundreds of amino acids, is astronomical. How does *m*RNA code for the building of any given protein? Recall that the four ribonucleotides present in RNA are available for coding the sequence of amino acids in a protein. The four ribonucleotides, symbolized by A (adenine), U (uracil), G (guanine), and C (cytosine), can be grouped to "spell out" certain "words" in a kind of ribonucleotide language. By combining the symbols in pairs it is possible, for example, to produce 16 different two-letter "words." A two-letter-word language, however, would not provide enough "words" to designate each of the 20 amino acids. But, if the ribonucleotides were used to form three-letter words, then the four ribonucleotides could be grouped to make up 64 different three-letter combinations. Though this provides many more "words" than would seem to be needed, research has proven that the information is carried in this three-letter-word language. Scientists have determined which amino acids are coded for by the 64 possible triplet nucleotide sequences. Figure 7-7 shows the 64 combinations and the corresponding 20 amino acids. This relationship of the triplet ribonucleotides and the amino acids is known as the *genetic code,* and the triplet sequences of ribonucleotides are called *codons.*

The codons in *m*RNA are arranged in linear order along the molecule and are read starting at the first codon at one end, then the next codon, and so on down the length of the *m*RNA molecule.

It was a surprising discovery that the genetic code was universally employed by all living systems. This fact is strong support for the idea that all life on earth emerged from a common cellular line.

An examination of Figure 7-7 also reveals that certain codons are used as "stop" and "start" signals in the process of reading the message. The start or initiation signal is required on the messenger RNA to set the correct beginning point for the first amino acid of the polypeptide to be synthesized. The stop signal on the messenger is required to cause the release of the completed polypeptide from the site of synthesis on the ribosome.

Transfer RNA (*t*RNA)

Transfer RNA (tRNA) is the smallest of the three types of RNA, containing around 85 ribonucleotide molecules. Although it is linear and single-stranded, regions of the molecule base-pair with themselves to form double-stranded portions. The unpaired regions

Second nucleotide

First nucleotide		A or *U*			G or *C*			T or *A*			C or *G*			
A or *U*	AAA AAG AAT AAC	*UUU* *UUC* *UUA* *UUG*	Phenylalanine Leucine	AGA AGG AGT AGC	*UCU* *UCC* *UCA* *UCG*	Serine	ATA ATG ATT ATC	*UAU* *UAC* *UAA* *UAG*	Tyrosine "Stop"	ACA ACG ACT ACC	*UGU* *UGC* *UGA* *UGG*	Cysteine "Stop" Tryptophan	A or *U* G or *C* T or *A* C or *G*
G or *C*	GAA GAG GAT GAG	*CUU* *GUC* *CUA* *CUG*	Leucine	GGA GGG GGT GGC	*CCU* *CCC* *CCA* *CCG*	Proline	GTA GTG GTT GTC	*CAU* *CAC* *CAA* *CAG*	Histidine Glutamine	GCA GCG GCT GCC	*CGU* *CGC* *CGA* *CGG*	Arginine	A or *U* G or *C* T or *A* C or *G*
T or *A*	TAA TAG TAT TAC	*AUU* *AUC* *AUA* *AUG*	Isoleucine Methionine ("Start")	TGA TGG TGT TGC	*ACU* *ACC* *ACA* *ACG*	Threonine	TTA TTG TTT TTC	*AAU* *AAC* *AAA* *AAG*	Asparagine Lysine	TCA TCG TCT TCC	*AGU* *AGC* *AGA* *AGG*	Serine Arginine	A or *U* G or *C* T or *A* C or *G*
C or *G*	CAA CAG CAT CAC	*GUU* *GUC* *GUA* *GUG*	Valine	CGA CGG CGT CGC	*GCU* *GCC* *GCA* *GCG*	Alanine	CTA CTG CTT CTC	*GAU* *GAC* *GAA* *GAG*	Aspartic acid Glutamic acid	CCA CCG CCT CCC	*GGU* *GGC* *GGA* *GGG*	Glycine	A or *U* G or *C* T or *A* C or *G*

Figure 7-7 The genetic code. The DNA codons are shown in black and the complementary RNA codons are in color. The 20 amino acids are specified by 61 of the 64 triplet codons. The remaining three codons are "punctuation marks" that signal the end of a genetic message in a sequence of nucleotides. The codon for methionine is the recognized "start" signal. All polypeptides are synthesized beginning with this amino acid. It is often removed after the polypeptide is completed.

form outlying loops that have special functional roles. Figure 7-8 illustrates the common characteristics of all the tRNA molecules.

All *t*RNA molecules have the same terminal sequence of nucleotides in the protruding *amino acid arm*. It is here that a specific amino acid is attached. In the outlying loop termed the *anticodon arm*, each *t*RNA has a special triplet sequence of ribonucleotides that will enable it to base-pair with one of the codons of the *m*RNA. This triplet nucleotide sequence is termed the *anticodon*. When attached to its specific amino acid, the *t*RNA will travel to the ribosome where the messenger is located and align its anticodon to a codon of the *m*RNA. In this way, the required amino acid is selected and properly oriented for bonding to the growing polypeptide chain. A second outlying loop is the *ribosomal recognition arm*, which is common to all *t*RNA molecules and orients the *t*RNA molecules correctly onto the ribosome. The third outlying loop, the *enzyme recognizing arm*, serves in the recognition process whereby a special activating enzyme recognizes the *t*RNA and then joins it to the correct amino acid.

Translation

The processes of DNA replication and transcription of DNA into RNA sequence have been outlined. A third process necessary for the expression of the genetic information is called *translation*. In this process, the sequence of codons in *m*RNA is employed to assemble amino acids into a protein. Fundamental to this process also is the base-pairing process. The *t*RNA molecule performs the critical step of translating "nucleotide language" into "protein language" by acting as an interpreter between the amino acid and the codon carried on the *m*RNA.

Activating enzymes attach the *t*RNA to the correct amino acid. Every cell contains at least one *t*RNA for each of the 20 amino acids. Some cells have several *t*RNA molecules for certain amino acids. The *t*RNA with its amino acid is then available to function at the ribosome, where the assembly of the protein occurs.

Ribosomal RNA (*r*RNA)

Ribosomal RNA or *r*RNA makes up about 65 percent of the mass of the ribosomal particle. Ribosomal RNA constitutes the largest proportion of the total RNA of the cell, and yet its function in the ribosome is not clear. Some investigators suggest *r*RNA keeps the ribosome functional by orienting a large variety of important enzymes in the correct way. It may also be important in aligning the *m*RNA in the ribosome so that incoming *t*RNA molecules can base-pair conveniently.

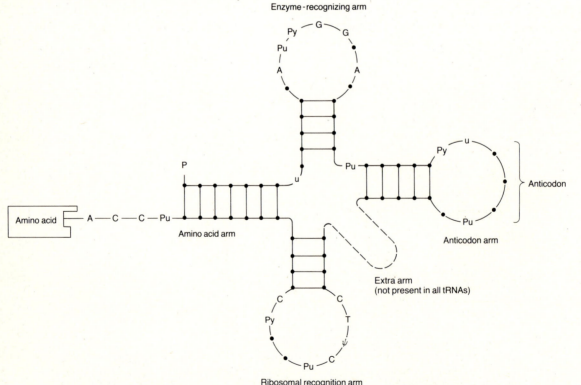

Figure 7-8 Structure of transfer RNA. The black dots represent ribonucleotides and the crosslines represent hydrogen bonding between nitrogenous base pairs.

Transfer RNA and ribosomal RNA molecules are relatively long-lived, and are reused many times. Messenger RNA molecules represent a small percentage of the total cellular RNA, are present in a variety of sizes (dependent on the size of the protein for which they code), and are much shorter-lived than the other two types.

BIOSYNTHESIS OF PROTEINS

Protein biosynthesis, the formation of protein molecules, can be described by first considering the role of *t*RNA. In the cytoplasm of the cell, a series of activating enzymes match the 20 amino acids to their corresponding *t*RNA molecules and bind the amino acid to the amino acid arm of the *t*RNA molecule. For example, the *t*RNA with the anticodon GCU is recognized by one particular activating enzyme and is matched to the amino acid arginine. The process of attaching the *t*RNA's to their amino acids requires the input of energy in the form of ATP (Figure 7-9).

The *m*RNA arriving in the cytoplasm, with its information transcribed from the nucleus, binds to a free-floating small ribosomal subunit. The end of the *m*RNA that has the start codon ends up in the peptidyl site (P-site) of the small ribosomal subunit. The start codon AUG is in place, and a *t*RNA with the correct anticodon, UAC, then base-pairs to the start codon. At this point, the complex of molecules aggregated together make up the *initiation complex.* The initiation complex then combines with a large subunit

ribosomal particle to form a complete functional ribosome (Figure 7-10). Adjacent to the P-site on the small subunit of the ribosome is another site in which the second codon of the *m*RNA is located. This second site is termed the *amino acyl site (A-site).* It is into this A-site that the next *t*RNA arrives and base-pairs anticodon to codon. At this time, both the P-sites and A-sites are plugged with *t*RNA's that are attached to their respective amino acids.

Because the *t*RNA molecules are the same size and are aligned together on the ribosome, the amino acid each *t*RNA carries is positioned so they are easily linked together by an enzyme located in the large subunit (Figure 7-11). When this occurs, the *t*RNA in the P-site relinquishes its hold on its amino acid and slips out of the P-site. The two amino acids are thus bonded together and linked to the *t*RNA still in the A-site. Next, this remaining *t*RNA with its two amino acids attached is translocated to the P-site as the *m*RNA molecule shifts its second codon from the A-site to the P-site. In this manner, the *m*RNA and ribosome have shifted positions relative to one another by the length of one codon. This also introduces the third codon into the A-site. This new codon signals for a specific *t*RNA with its amino acid, which then base-pairs anticodon to codon by plugging itself into the A-site. The two amino acids bonded together on the *t*RNA in the P-site are in position to bond to the new amino acid held by the *t*RNA in the A-site. The *t*RNA in the P-site relinquishes the two amino acids (a dipeptide) and slips out of the P-site. A sec-

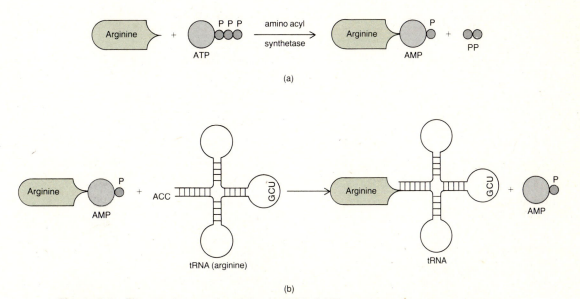

(a)

(b)

Figure 7-9 The attachment of arginine to *t*RNA. (a) The amino acid arginine is activated when the enzyme amino acyl synthetase adds AMP to the arginine. The AMP is obtained from ATP, and when the reaction is complete, pyrophosphate (PP) is left over. (b) The amino acyl synthetase enzyme selects the correct *t*RNA for arginine. The *t*RNA has the anticodon GCU. The enzyme then binds the activated amino acid to the *t*RNA.

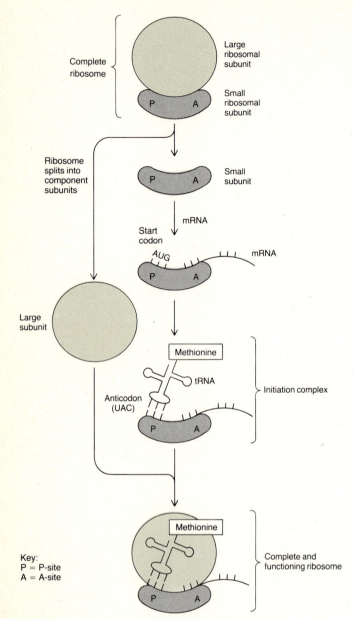

Figure 7-10 Formation of the initiation complex and the functional ribosome. (From *The Physical Basis of Life.* Copyright © 1972 by CRM Books. Reprinted by permission of CRM Books, a Division of Random House, Inc.)

ond codon shift occurs when the *m*RNA and ribosome again move the length of one codon. The process repeats and continues until a stop codon is reached, at which time the completed protein is released from the ribosome and its *t*RNA.

The synthesis of a protein containing 200 amino acids may be accomplished in a fraction of a minute. Several ribosomes may follow each other in linear fashion along a given *m*RNA molecule. Each ribosome synthesizes the identical polypeptide. You will recall that when the ribosomes are strung along the *m*RNA like beads, they form a structure termed a *polyribosome* or *polysome*. When the *m*RNA has been read by all the ribosomes, it is released and usually degraded; the individual ribonucleotides are reused for a new *m*RNA. The two subunits of the ribosome come apart and are reused to form complete functional ribosomes by the process described.

The preceding discussion of protein biosynthesis nicely illustrates the concept of homeostasis under normal conditions. However, there are times when homeostasis is disrupted and a mutation may result.

MUTATIONS

A **mutation** is any change in the nitrogenous base sequence of DNA. In general, it represents a loss or a substitution of one or more of the nitrogenous bases. In the normal sequence of life activity, mutations occur rarely, but the rate of occurrence can be increased by agents such as ultraviolet light, x-rays, and certain chemicals that react or interact with the DNA.

Mutations are usually seen only when they cause some change in physical appearance. Many mutations occur and are not noticed because some base substitutions in DNA do not cause a change in the amino acid sequence. For example, UCU is the codon for the amino acid serine. Any substitution of the third base in the codon to UCC, UCA, UCG, still results in the correct insertion of serine into the protein (see Figure 7-7).

If there is a deletion or an insertion of a nucleotide, a so-called *frame shift mutation* results. Recall that the nucleotides are read three at a time starting from one end of the molecule and the result of the loss or addition would be a message very different from the original. Figure 7-12 shows an original sequence of DNA that might be part of a DNA molecule. This sequence of nucleotides is then transcribed into mRNA. This can easily be determined from the base-pairing rule that A transcribes to U, G to C, T to A, and C to G. These messenger codons are next translated to the amino acid sequence that can be ascertained by reference to Figure 7-7.

If there were a deletion in the original DNA nucleotide sequence so that the fourth nucleotide in the sequence was lost, the new sequence would be that shown on the right-hand side of Figure 7-12. The resultant transcriptions and translations are also shown. The resultant amino acid sequence is very different from the original and in a cell would produce a totally nonfunctional protein. Such an inactive product would produce a loss of catalytic function and greatly affect the normal function of the cell. This is an excellent example of a disruption in homeostasis.

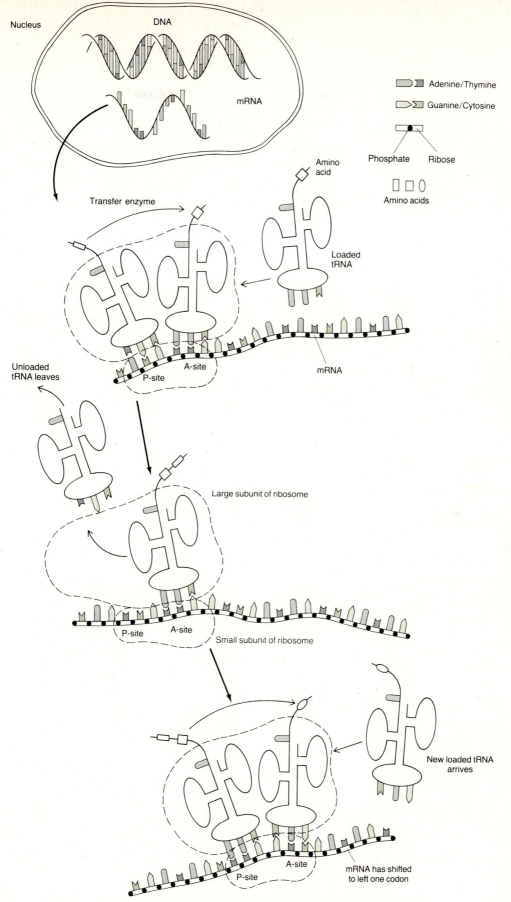

Figure 7-11 Summary of protein biosynthesis.

Nucleus

DNA

mRNA

Adenine/Thymine

Guanine/Cytosine

Phosphate   Ribose

Amino acids

Amino acid

Transfer enzyme

Loaded tRNA

Unloaded tRNA leaves

P-site   A-site   mRNA

Large subunit of ribosome

P-site   A-site   Small subunit of ribosome

New loaded tRNA arrives

P-site   A-site   mRNA has shifted to left one codon

Frame	Original								After deletion of G in second frame						
	1	2	3	4	5	6	7	8	1	2	3	4	5	6	7
DNA sequence	TAC	GAG	CTG	ACC	CGG	CCA	GCT	T	TAC	AGC	TGA	CCC	GGC	CAG	CTT
mRNA sequence	AUG	CUC	GAC	UGG	GCC	GGU	CGA	A	AUG	UCG	ACU	GGG	CCG	GUC	CAA
Amino acid sequence	meth	leu	asp	trypt	ala	gly	arg		meth	ser	threo	gly	pro	val	glut

Figure 7-12 Frame shift mutations. Shown are changes that result in DNA, mRNA, and the consequent amino acid sequence when a single nucleotide is deleted in DNA.

REGULATION OF PROTEIN BIOSYNTHESIS

Enzymes catalyzing major biochemical reactions in a cell may be present in tens of thousands of copies, whereas enzymes catalyzing minor reactions may be present in numbers less than ten. Also, certain enzymes are present only at special times in the lifetime of a cell. Enzymes are programed to appear and disappear as the cell goes through certain phases. The system that controls the expression of the genetic information does so with incredible accuracy and dispatch so that the cell always has the correct number and types of proteins at the right time. This is an excellent example of the maintenance of homeostasis.

Three general categories of enzymes have been described on the basis of how their synthesis is controlled by the cell. *Constitutive enzymes* are those formed at constant rates by the cell regardless of its metabolic state. These enzymes can be thought of as absolutely necessary for every cell. For example, the primitive enzymes that evolved to provide the first cells with energy-extracting capability (the *glycolytic pathway* that breaks down glucose) are present in all cells and in similar amounts regardless of the concentration of glucose in the medium. However, the concentration of many enzymes fluctuates rapidly and considerably in response to changes in substrate concentrations. These are the *inducible enzymes* and *repressible enzymes*.

Consider the example of a cell that depends on the substrate glucose for its source of energy. If glucose is unavailable, and lactose is substituted, within minutes the cell will synthesize inducible enzymes that can break down lactose to simpler molecules, glucose and galactose, which the cell can use for fuel. If the induced cell is supplied with glucose but no lactose, the previously induced new enzymes are degraded and disappear. Enzyme induction thus occurs only when needed for cell survival.

The production of repressible enzymes, on the other hand, ceases when the product of the reaction is available from other sources. Consider the example of a cell which does not have access to the amino acid histidine but must synthesize this amino acid from an ammonium salt and a carbon source. Then if histidine is made available to the cell, the enzymes previously required for making histidine from ammonia and the carbon source are quickly degraded. This effect is called *enzyme repression* and also reflects the principle of cellular economy. When the enzymes are not required, their production ceases.

Inducible enzymes generally catalyze breakdown (catabolic) reactions. Repressible enzymes generally catalyze synthetic (anabolic) reactions.

Negative Control of Transcription

According to the "central dogma" of molecular biology:

$$DNA \xrightarrow{transcription} RNA \xrightarrow{translation} PROTEIN$$

DNA is capable of self-replication, and it can be transcribed to form RNA. RNA in turn is translated to form a protein.

Consider a multicellular organism such as the human body. It develops from repeated mitotic divisions of a single fertilized cell called the *zygote*. As each cell of the human body forms, it retains one complete identical copy of the genetic information present in the original zygote. If each cell possesses the identical information for synthesizing proteins, how then does the developing organism create the variety of specialized cells each with its own specialized functions? In other words, how can a diverse array of cell types with different functions appear when the same set of genetic instructions is used? As the number of cells in the organism increases, the environment of some of the cells changes. This triggers an internal cellular response so that genes previously dormant are selectively uncovered and transcribed. As a result, new proteins appear within the cell; some act as enzymes to catalyze new biochemical reactions, while others act to modify the cell's appearance.

The cell needs a method for such *selective gene*

expression in order to maintain its homeostasis. In fact, the cell exerts control at each step of protein synthesis from DNA to the functioning enzyme. Only recently have some of the methods of control been explained. The regulatory mechanism best understood today is the control of *m*RNA as described by Jacques Monod and Francois Jacob. These scientists proposed that DNA was composed of two different kinds of genes: structural genes and regulatory genes. *Structural genes* code the information for the amino acid sequence of a functional protein. *Regulatory genes* control selective gene expression. The concept is termed the *operon model*. The **operon** is defined as a group of functionally related structural genes located very close together along the DNA molecule, plus a set of regulatory genes that determine whether the structural genes can be transcribed (Figure 7-13). Consider the example used before. A cell deprived of glucose is provided with lactose as its energy source. In the proposed *lactose operon* model, the following sequence of events illustrates the homeostatic control mechanism.

When glucose is present and lactose is not, the region designated the regulator gene (R-gene) on the DNA is transcribed. A special protein, called the *repressor*, diffuses from the ribosomes, where it is formed, to the DNA molecule, where it binds to the region of DNA called the *operator* or the *operator locus*. The protein repressor thus bonded to the operator, which is located at the front end of structural gene sequence z,y,a, prevents the transcription of the structural genes into a *m*RNA molecule. On the other hand, when glucose is absent and lactose (functioning as the *inducer* molecule) is substituted as the energy source, the lactose molecule specifically binds to the repressor molecule to form a repressor-lactose complex, and this prevents the repressor from bonding to the operator. When uncovered, the operator accepts the transcribing enzyme that produces the polygenic *m*RNA coding for the enzymes a,b,c. These enzymes than permit the cell to metabolize lactose. In this example, the lactose functions as an inducer molecule, allowing the synthesis of new proteins.

More recent investigations have identified a *promoter site* of the operon. The promoter is on the opposite side of the operator from the structural genes. Like the operator, the promoter is a recognition site on DNA. Neither operator nor promoter is transcribed, but both are regions where regulating molecules combine. The promoter is the site to which the transcribing enzyme RNA polymerase binds and begins copying the DNA to form *m*RNA. If the operator is blocked with repressor protein, then the enzyme does not have access to the promoter and the

structural genes are not transcribed. If the inducer (in this example, lactose) is added, it removes the repressor from the operator by bonding to it. This makes it possible for the RNA polymerase enzyme to function.

The operon model also explains the repression of enzyme synthesis. Recall the histidine example. The normal situation is for the cell to produce histidine from ammonia and a carbon source. That is, the cell has enzymes in operation to produce this needed molecule. The operon concept proposes, in this instance, that the repressor molecule present is an incomplete one. When the inducer, histidine, is provided for cellular use, it binds to the incomplete repressor and generates a functionally complete repressor molecule. This complete repressor now bonds to the operator locus and prevents further transcription of the nine structural genes that were required to synthesize the histidine. The histidine in this case is called a *corepressor molecule*.

Positive Control of Transcription

The controls of the transcription process just outlined are designed to operate with the use of a repressor molecule. That is, they operate to turn off the system and thus represent a negative control. Recently, some rather surprising experimental results have shown that *positive* or activating molecules exist to turn on or enhance transcription. The molecule *cyclic adenosine-3',5'-monophosphate (cyclic AMP)*, and the cyclic AMP-binding molecule called *cyclic AMP receptor protein (CRP)*, operate as a team to increase the rate of transcription of inducible enzyme systems. Presumably, cyclic AMP binds to CRP, which then interacts with the promoter to stimulate the initiation process to take place and hence speed up transcription.

Regulation of Transcription by Hormones

It has been shown in several laboratories that hormones are also able to function as positive regulators of protein synthesis. Glucocorticoids (a group of hormones produced by the adrenal cortex) and thyroxine (produced by the thyroid gland) bind in a similar way as does the cyclic AMP to specific receptor proteins in the cytoplasm of the cell susceptible to the hormone. After binding to the receptor protein, the receptor-hormone complex finds its way into the cell nucleus, where it binds to the chromatin. An increase in the rate of *m*RNA results, followed shortly thereafter by an increase in enzyme activity. The exact mechanism of the positive regulation is still unknown and requires more research.

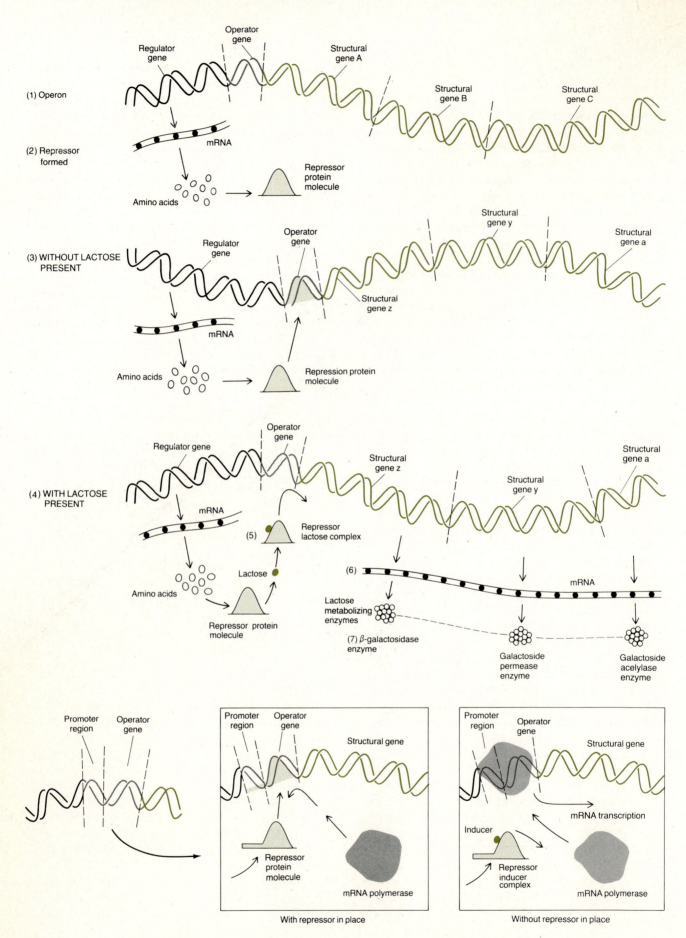

(1) Operon

Regulator gene · Operator gene · Structural gene A · Structural gene B · Structural gene C

(2) Repressor formed

mRNA

Amino acids → Repressor protein molecule

(3) WITHOUT LACTOSE PRESENT

Regulator gene · Operator gene · Structural gene z · Structural gene y · Structural gene a

mRNA

Amino acids → Repression protein molecule

(4) WITH LACTOSE PRESENT

Regulator gene · Operator gene · Structural gene z · Structural gene y · Structural gene a

mRNA

(5) Repressor lactose complex

Lactose

Amino acids → Repressor protein molecule

(6) mRNA

Lactose metabolizing enzymes

(7) β-galactosidase enzyme

Galactoside permease enzyme

Galactoside acelylase enzyme

Promoter region · Operator gene

Promoter region · Operator gene · Structural gene

Repressor protein molecule

mRNA polymerase

With repressor in place

Promoter region · Operator gene · Structural gene

Inducer

mRNA transcription

Repressor inducer complex

mRNA polymerase

Without repressor in place

CELLS AND AGING

Frédéric Verzar, the Swiss dean of gerontologists, once said: "Old age is not an illness; it is a continuation of life with decreasing capacities for adaptation." Only recently has his view of aging as a progressive failure of the body's homeostatic adaptive responses gained wide acceptance. There has been a strong tendency to confuse aging with many diseases frequently associated with it—especially cancer and atherosclerosis. Each, in fact, probably accelerates the other.

The obvious characteristics of aging are well known: graying and loss of hair, loss of teeth, wrinkling of skin, decreased muscle mass, and increased fat deposits. The physiological signs of aging are gradual deterioration in function and capacity to respond to environmental stress. Thus, basic kidney and digestive metabolic rates decrease, as does the ability to maintain a constant internal environment despite changes in temperature, diet, and oxygen supply. These manifestations of aging are related to a decrease in the actual number of cells in the body (100,000 brain cells are lost each day) and to the disordered functioning of the cells that remain.

The extracellular components of tissues also change with age. *Collagen* fibers, responsible for the strength in tendons, increase in number and change in quality with aging. These changes in arterial walls are as much responsible for the loss of elasticity as those in arteriosclerosis. *Elastin,* another constituent of the intercellular matrix, is responsible for the elasticity of blood vessels and skin. It thickens, fragments, and acquires a greater affinity for calcium with age—changes that may be associated with the development of arteriosclerosis.

Several kinds of cells in the body—heart cells, skeletal muscle cells, neurons—are incapable of replacement. Recent experiments have proved that certain other cell types are limited when it comes to cell divisions. Cells grown outside the body divided only a certain number of times and then stopped. The number of divisions correlated with the donor's age. The number of divisions also correlated with the normal life span of the different species from which the cells were obtained—strong evidence for the hypothesis that cessation of mitosis is a normal, genetically programed event.

Just as the factors that limit the life of an individual cell are unknown, so are those that restrict the growth or life of a tissue or organ. At menopause, the ovary ceases to function. Its cells die long before the rest of the female body. Perhaps similar mechanisms determine longevity.

Some investigators have studied aging from the standpoint of immunology. The ability to develop antibodies is said to diminish with age. Senescence, according to researchers, results in the older person's immunological system having a "shotgun," rather than a specific, response to foreign protein. This shotgun response may include an autoimmune reaction that attacks and gradually destroys the individual's own normal tissue and organs.

The following generalizations on aging can be made:

1. Aging is a general process that produces observable changes in structure and function.
2. Aging produces increased vulnerability to environmental stress and disease.
3. Evidence suggests that the life span is about 110 years, but maximum life expectancy is about 85 years.
4. The mechanism behind aging is not known. Improvements in life expectancy may be due to an overall reduction in the number of life-threatening situations or to modification of the aging process.
5. Eventually the aging process may be modified and life span and life expectancy lengthened.

In this chapter, it has been demonstrated that homeostasis, that governing principle essential for survival of the organism, operates at the cellular level. Control systems abound within the cell to assure

---

Figure 7-13   The operon. (1) The operon consists of a regulator gene, an operator gene, and structural genes. (2) In the absence of lactose the regulator gene produces a repressor protein. (3) Without an inducer, such as lactose, the repressor binds onto the operator region. This action prevents the activation of the structural genes, and no unnecessary enzyme synthesis results. (4) In the presence of lactose, repressor molecules are synthesized as above, but lactose acts as an inducer for its own enzymatic breakdown. (5) Lactose and the repressor combine to form an unstable complex that cannot bind to the operator region. (6) The "switch" is now turned on, and the structural genes transcribe *m*RNA. (7) This, in turn, guides the synthesis of the three enzymes (a,b,c) necessary for lactose breakdown. (From *The Physical Basis of Life.* Copyright © 1972 by CRM Books. Reprinted by permission of CRM Books, a Division of Random House, Inc.)

survival not only of the cell but of future generations of similar cells. The search for the mechanisms of control is carried out in laboratories all over the world. Scientists delve continually, not only into the anatomy of the cell, but into the nature of the molecular components of the cells, examining the forces that cause specific interactions to occur. Physiology is moving ever more rapidly into the cellular-molecular sphere to explain the meaning and purpose of phenomena that have been obvious to the naked eye for centuries.

HOMEOSTATIC IMBALANCES OF CELLS

Cancer

Cancer is not a single disease but many. The human body contains more than a 100 different types of cells, each of which can malfunction in its own distinctive way to cause cancer. When cells in some area of the body duplicate unusually quickly, the excess of tissue that develops is called a growth, or *tumor*. Tumors may be cancerous and fatal or quite harmless. A cancerous growth is called a *malignant tumor,* or *malignancy.* A noncancerous growth is called a *benign growth.*

Cells of malignant growths all have one thing in common: They duplicate continuously and often very quickly. The growth continues until the victim dies or until every malignant cell is removed or destroyed. As the cancer grows, it expands and begins to compete with normal tissues for space and nutrients. Eventually, the normal tissues regress and die. The organ functions less and less efficiently until finally it ceases to function altogether. Cancer cells may spread from the original, or *primary,* growth. Cancer of the breast, for instance, has a tendency to spread to the lungs. The spread of cancer to other regions of the body is called *metastasis.* Metastasis occurs when a malignant cell breaks away from the growth, enters the bloodstream, and is carried through the body. Wherever the cell comes to rest, it establishes another tumor: a *secondary* growth. Usually death is caused when a vital organ regresses because of competition with the cancer cells for room and nutrients. Pain develops when the growth impinges on nerves or blocks a passageway so that secretions build up pressure.

Cancer cells multiply without control. Individual cells vary in shape and size, and the orderly orientation of normal cells is replaced by disorganization which may be so extensive that no recognizable structures remain.

At present, cancers are classified by their microscopic appearance and the body site from which they arise. At least 100 different cancers have been identified in this way. If finer details of appearance are taken into consideration, the number can be increased to 200 or more. The name of the cancer is derived from the type of tissue in which it develops. *Sarcoma* is a general term for any cancer arising from connective tissue. *Osteogenic sarcomas* (*osteo* = bone; *genic* = origin), the most frequent type of childhood cancer, destroy normal bone tissue and eventually spread to other areas of the body. *Myelomas* (*myelos* = marrow) are malignant tumors, occurring in middle-aged and older people, that interfere with the blood-cell-producing function of bone marrow and cause anemia. *Chondrosarcoma* is a cancerous growth of the cartilage (*chondro* = cartilage).

Benign tumors are composed of cells that do not metastasize—that is, the growth does not spread to other organs. Removing all or part of a tumor to determine whether it is benign or malignant is called a biopsy. A benign tumor may be removed if it impairs a normal body function or causes disfiguration.

Cancer has been observed in all species of vertebrates. In fact, cellular abnormalities that resemble cancer—such as crown gall of tomatoes—have been observed in plants as well. Cell masses that resemble cancers of higher animals have also been produced and studied in insects.

What triggers a perfectly normal cell to lose control and become abnormal? Scientists are uncertain. First there are environmental agents: substances in the air we breathe, the water we drink, the food we eat. The World Health Organization estimates that these agents—called carcinogens—may be associated with 60 to 90 percent of all human cancer. Examples of carcinogens are the hydrocarbons found in cigarette tar. Ninety percent of all lung cancer patients are smokers. Another environmental factor is radiation. Ultraviolet light from the sun, for example, may cause genetic mutations in exposed skin cells and lead to cancer, especially among light-skinned people.

Viruses are a second cause of cancer, at least in animals. These agents are tiny packages of nucleic acids without life of their own that are capable of infecting cells and converting them to virus producers. Virologists have linked tumor viruses with cancer in many species of birds and mammals, including primates. Since these experiments have not been performed on humans, there is no absolute proof that viruses cause human cancer. Nevertheless, with over 100 separate viruses identified as carcinogens in many species and tissues of animals, it is also probable that at least some cancers in humans are due to virus.

CHAPTER SUMMARY OUTLINE

1. Cell division is the process by which cells reproduce themselves.
2. Mitosis is the process by which a single parent cell duplicates itself.
3. Meiosis is the mechanism that enables the reproduction of an entirely new organism.

CHROMOSOMES

1. Chromosomes have the capabilities of replicating themselves and maintaining their chemical composition and physiological properties intact from one generation to another.
2. The 22 matching pairs of chromosomes in the human are called autosomes.
3. The remaining two chromosomes, which are unpaired, are called the sex chromosomes.

Chemical Composition

1. Biochemical analyses of chromosomes show that they consist of a molecule of DNA plus a special type of protein called histone.
2. Nucleotides each contain three basic parts: (1) a nitrogenous base, (2) a sugar called pentose, and (3) a molecule of phosphoric acid.

Structure of DNA

1. The structure of the DNA molecule was first proposed by James Watson and Francis Crick in 1953.
2. Their proposed model consisted of two long chains of deoxyribonucleotides twisted around each other to form what is called a double helix.

Replication

1. During the process of cell division, when the chromosomes are replicated, the two chains gradually uncoil and separate starting at one end; each chain is used as a pattern to guide the correct assembly of a new partner chain.
2. Replication results in the exact sequence of bases in each strand.

CELL DIVISION

1. The process of mitosis is the sequence of events in which the nuclear material divides.

2. The process of cytokinesis divides the cytoplasm of the parent cell equally between the two newly forming cells.
3. The function of mitosis and cytokinesis is to replace cells in the body. Each new daughter cell has the same number and kind of chromosomes as the original parent cell.
4. The two daughter cells have the same hereditary material and genetic potential as the parent cell.
5. Mitosis is broken down into four stages: prophase, metaphase, anaphase, and telophase.
6. Before a cell passes through its divisional stages, it must first pass through a stage called interphase.

Interphase

1. The principal events of interphase are the replication of DNA and the synthesis of other kinds of nucleic acids and proteins.

Prophase

1. During prophase the centrioles move apart projecting a series of radiating fibers called astral rays, and themselves become connected by fibers called spindle fibers. All of these structures are collectively called the mitotic apparatus.

Metaphase

1. The chromatid pairs line up on the equatorial plane of the spindle fibers, and the lengthwise separation of the chromatids now takes place.

Anaphase

1. Anaphase is characterized by the division of the centromeres and the movement of complete sets of chromatids to opposite poles of the cell.

Telophase

1. Telophase consists of events opposite those of prophase, and two identical sets of chromosomes reach opposite poles.

Cytokinesis

1. Cytokinesis begins in late anaphase and terminates at about the same time as telophase. The cleavage

furrow cuts completely through the cell to form two separate portions of cytoplasm.

GENETIC CODE

1. The information needed by each cell to carry on its life processes is transmitted to each new generation by means of the DNA molecules.
2. The information is relayed to the ribosomes, which direct the synthesis of protein molecules many of which are enzymes. Enzymes, in turn, determine the functional characteristics of the entire cell.

Transcription

1. All RNA molecules are created in a manner similar to that used in replicating DNA. The sequence of molecular events occurring during the copying process is called transcription.

Messenger RNA

1. Messenger RNA carries the coded information from the nucleus to the ribosomes.

Transfer RNA

1. Transfer RNA attaches to a specific amino acid, brings it to the ribosome, and properly orients it for bonding to a growing polypeptide chain.

Translation

1. In the process of translation the sequence of codons in messenger RNA is employed to assemble amino acids into a protein.

Ribosomal RNA

1. Ribosomal RNA makes up about 65 percent of the mass of the ribosomal particle. It is suggested that ribosomal RNA keeps the ribosome functional by orienting a large variety of important enzymes in the correct way.

BIOSYNTHESIS OF PROTEINS

1. In the cytoplasm of the cell a series of activating enzymes match the 20 amino acids to their corresponding transfer RNA molecules and bind the amino acids to the amino acid arm of the transfer molecule.

2. The messenger RNA arriving in the cytoplasm with its information transcribed from the nucleus binds to a free-floating small ribosomal subunit.
3. Both the P-sites and A-sites of a ribosome are occupied by transfer RNA's that are attached to their respective amino acids.
4. The amino acids are positioned so that they are easily linked together by an enzyme. After a codon shift another amino acid can be bonded to the first two amino acids.
5. A second codon shift occurs when the messenger RNA and ribosome again move the length of one codon. The process repeats and continues until a stop codon is reached, at which time the completed protein is released from the ribosome and its transfer RNA.
6. The synthesis of a protein containing 200 amino acids may be accomplished in a fraction of a minute.

MUTATIONS

1. A mutation is any change in the nitrogenous base sequence of DNA. In general, it represents a loss or a substitution of one or more of the nitrogenous bases.
2. Mutations occur rarely, but the rate of occurrence can be increased by agents such as ultraviolet light, x-rays, and certain chemicals that react or interact with the DNA.

REGULATION OF PROTEIN BIOSYNTHESIS

1. Enzymes are programed to appear and disappear as the cell goes through certain phases.
2. The system that controls the expression of the genetic information does so with incredible accuracy and dispatch so that the cell always has the correct number and types of proteins at the right time.
3. Three general categories of enzymes have been described on the basis of how their synthesis is controlled by the cell. They are: constitutive, inducible, and repressible enzymes.

Negative Control of Transcription

1. A special protein, called the repressor, diffuses from the ribosomes, where it is formed, to the DNA molecule, where it binds to the region of DNA called the operator.
2. The repressor thus bonded to the operator pre-

vents the transcription of the structural genes into a messenger RNA molecule.

Positive Control of Transcription

1. Positive or activating molecules exist to turn on or enhance transcription.

Regulation of Transcription by Hormones

1. Hormones are able to function as positive regulators of protein synthesis.

CELLS AND AGING

1. It has been said that, "Old age is not an illness; it is a continuation of life with decreasing capacities for adaptation."
2. Many theories of aging have been proposed, but none successfully answers all the experimental objections.

HOMEOSTATIC IMBALANCES OF CELLS

1. An example of a homeostatic imbalance of cells is cancer, in which cells multiply without control into various sizes and shapes.
2. At present cancers are classified by their microscopic appearance and the body site from which they arise.
3. At least 100 different cancers have been identified in this way.

REVIEW QUESTIONS

1. Define cell division and differentiate between mitosis and meiosis.
2. What is your interpretation of a chromosome?
3. Define each of the following: karyotype, homologue, diploid, autosomes, and sex chromosomes.
4. What are the three basic parts of a deoxyribonucleotide?
5. Who first proposed the structure of the DNA molecule? Describe this molecule in detail.
6. How does DNA replicate itself?
7. What are the main characteristics of each stage of mitosis?
8. Describe the genetic code.
9. Explain fully the terms transcription and translation.
10. Explain the functions of the various kinds of RNA.
11. Describe in detail the biosynthesis of proteins.
12. Define mutation and explain how its rate of occurrence can be increased.
13. How is the biosynthesis of proteins regulated?
14. Differentiate between the three general categories of enzymes found in cells.
15. Explain negative control of transcription and give a specific example.
16. Explain positive control of transcription and give an example.
17. How is transcription regulated by hormones?
18. List the five generalizations on aging that can be made based on current information.
19. Give an example of a homeostatic imbalance of cells.
20. How are the different types of cancer classified?

Chapter 8

THE CELLULAR ENVIRONMENT AND TRANSPORTATION

STUDENT OBJECTIVES

After reading this chapter, you should be able to

- Discuss the importance of water in living organisms
- Define calorie, specific heat, heat of vaporization, and heat of fusion
- Describe the structure of a water molecule
- Explain the distribution and importance of water in the human body
- Discuss how ions in body fluids are expressed, and describe the distribution of ions in the three major fluid compartments
- Correlate the movement of body fluids with osmotic pressure
- Explain how the movement and distribution of water between intracellular and extracellular compartments is governed by their concentrations of osmotically active particles
- Describe how materials are moved across plasma membranes
- Define passive and active processes
- Compare the two passive processes called bulk flow and diffusion
- Contrast the passive processes called osmosis, filtration, and dialysis
- Compare active transport with active and passive mediated transport

This chapter will examine the cellular environment and the various ways that chemical substances are transported within that environment. The **cellular environment** is the environment inside the cell (intracellular) and the environment outside the cell (extracellular). Chemically, the cellular environment consists mostly of water. Dissolved or suspended in the water are an assortment of chemicals. Because of the predominance of water and its importance in maintaining the homeostasis of the cellular environment, the examination of the cellular environment begins with the study of water.

WATER

Life as we know it on earth is not possible without water. Water comprises 60 to 90 percent of the weight of most living organisms. It is the continuous matrix in which all substances present in the organism are dissolved or dispersed. It is a highly reactive substance that enters into many metabolic reactions and is often a product of others. Water is such a common part of our daily experience that its real significance and unusual properties are never fully appreciated.

Physical Properties of Water

Water is the substance that is used to establish the *calorie* as the unit of energy. A *calorie* is the amount of heat required to raise the temperature of 1 g of water from 15°C to 16°C. Water was selected as a measure of the calorie because it is so common and has such a high specific heat. *Specific heat* is the number of calories required to raise the temperature of 1 g of a substance 1°C. Only liquid ammonia has a higher specific heat (1.23 cal/g). The heat of vaporization of water is 540 cal/g, which is higher than any substance that is liquid at room temperature. *Heat of vaporization* is the heat required to convert water at 100°C to steam at 100°C. The next-closer substance is methyl alcohol, which has a heat of vaporization of 263 cal/g. The amount of heat necessary to convert 1 g of ice at 0°C to water at 0°C is known as the *heat of fusion*. The heat of fusion of water is 80 cal/g. The heat of fusion of ammonia exceeds that of water with a value of 108 cal/g. These physical properties of water and its high melting and high boiling points are a result of the strong intermolecular forces in water. It is therefore worth examining the structure of a water molecule in some detail.

Structure of a Water Molecule

A water molecule is shaped like an isosceles triangle (Figure 8-1). The oxygen atom of the molecule exhibits a very strong electronegativity which creates a permanent *dipole property* within the molecule, that is, the oxygen atom has a negative charge and the two hydrogen atoms have slight positive charges. As a consequence of this permanent dipole property, adjacent water molecules orient themselves so that the positive ends of one molecule face the negative end of an adjacent molecule (Figure 8-2). This is what creates the cohesive property of water as evidenced by the physical properties previously mentioned.

When ice melts, the distance between neighboring

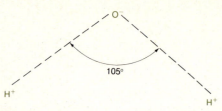

Figure 8-1 Structure of a water molecule. Note the strong electronegative property of oxygen and the positive charges on the hydrogen atoms. This sets up a permanent dipole between the oxygen and hydrogen atoms.

atoms *increases* from 2.76 Å to 2.9 Å. As the atoms move further from each other, the number of nearest neighbors of each oxygen atom *increases* from 4 in the ice state to 4.5 in water at the melting point. The increase of nearest neighbors sets up a denser packing of the water molecules that continues to increase to 4°C. As the temperature increases above 4°C, the higher thermal energy that is absorbed increases the disruption of the orderly crystalline structure of water and pushes the water molecules further apart.

Ice has a density less than that of water, and this fact has had profound influences on the development of life. If ice were heavier than water, ice would form at the bottoms of oceans, lakes, and rivers. If temperatures continued to drop, the ice would gradually expand to reach the surface of the water, thus completely freezing the entire body of water and destroying most of the marine life. In warmer weather, in deep bodies of water, ice could persist at the bottom while very warm water existed at the surface.

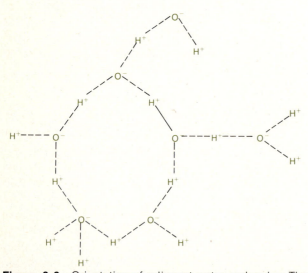

Figure 8-2 Orientation of adjacent water molecules. The water molecules are oriented with the O^- of one water molecule directed toward the H^+ of neighboring water molecules. This intermolecular bonding pattern explains many of the physical properties of water that suggest the strong cohesiveness of water.

However, since ice is less dense than water, it forms first on the surface of water. The ice protects the deeper water from additional cooling, and the deeper water stays liquid. As water is cooled, it becomes denser and therefore heavier and sinks. At 4°C the maximum density is reached, and upon further cooling, it becomes lighter again and rises toward the surface. These upward and downward currents are important because they serve to stir up food for many marine organisms thus enabling them to survive conditions that are otherwise quite inhospitable.

We can now consider the distribution and importance of water in the human body.

Body Water

Water is the most abundant component of the human body ranging from 45 to 75 percent of the body weight. With such a wide range, you might think that the percentage of body water in any individual might vary considerably. Actually, body weight and water content remain constant from day to day in the normal adult in caloric balance. For example, drinking a large volume of water produces a large urine volume, usually within 2 to 3 hours. The range of water percentage with respect to body weight is primarily due to the amount of adipose tissue or fat deposited in the body. As obesity increases, the percentage of body water decreases. In other words, lean bodies have a higher percentage of body water. Fifty-two percent of body weight of young adult females is water, whereas 63 percent of the body weight of young adult males is water. These percentages reflect the observations that females have larger fat deposits.

Exhibit 8-1 shows the distribution of water in various tissues of a lean male weighing 70 kg (154 lb). Most of the water of the human body is contained in skin and muscle; the skeleton and adipose tissue contain the least amount of water.

Body Fluid Compartments

The water contained within the human body can be conveniently subdivided into two major compartments: (1) The *intracellular water compartment* consists of water contained within body cells; (2) the *extracellular water compartment* consists of the water distributed outside the body cells.

Intracellular Fluid. *Intracellular fluid* is the total of all the water contained within the trillions of body cells. Different cell types contain varying amounts of water, and the chemical composition of the fluid inside cells also varies greatly. It is sometimes a misleading notion to use the term *compartment* when in truth we are dealing with countless, small, membrane-enclosed

Exhibit 8-1 DISTRIBUTION OF WATER AND PERCENTAGE OF BODY WEIGHT IN VARIOUS TISSUES OF 70-KG (154-LB) MAN

Tissue	% Water	% Body Weight	Liters of Water per 70 kg
Skin	72	18	9.07
Muscle	75.6	41.7	22.10
Skeleton	22	15.9	2.45
Brain	74.8	2.0	1.05
Liver	68.3	2.3	1.03
Heart	79.2	0.5	0.28
Lungs	79.0	0.7	0.39
Kidneys	82.7	0.4	0.25
Spleen	75.8	0.2	0.10
Blood	83.0	8.0	4.65
Intestine	74.5	1.8	0.94
Adipose tissue	10	±10.0	0.70

Source: From H. Skelton, *Arch. Int. Med.* **40**:140, 1927.

units containing water. These units are the cells. The amount of the total body water contained by the cells ranges from 30 to 40 percent of body weight.

Extracellular Fluid. The *extracellular fluid* compartment includes the water in blood plasma, interstitial fluid, and lymphatic fluid. Its properties will be described later in this chapter in more detail. The plasma flows through the blood vessels rapidly, whereas the interstitial fluid between the cells percolates slowly. The lymphatics collect and return the plasma proteins and fluid that leak out of the capillaries. The flow of lymph in lymphatics is sluggish.

Recall that the interstitial fluid constitutes the internal environment of the body that we talked about with respect to homeostasis. This fluid is constantly losing nutrients and oxygen to the cells and accumulating waste materials and carbon dioxide from the cells. In turn, the interstitial fluid is renewed with fresh nutrients and oxygen from the rapidly circulating plasma while it simultaneously unloads into the plasma certain wastes and carbon dioxide. These exchanges occur because concentration differences exist across short distances.

Some physiologists occasionally include a *transcellular fluid* compartment in their discussions of body water. This compartment is a special subdivision of the extracellular compartment and consists of cerebrospinal fluid around the brain and spinal cord, intraocular fluid in the eyes, pleural fluid around the lungs, peritoneal fluid around internal organs, synovial fluid in joints, and digestive fluids. These volumes are insignificant fractions of the body weight with the possible exception of the digestive juices, which represent between 1 and 3 percent.

Measurement of Body Fluid Compartments

The technique of measuring unknown volumes is based on the *dye dilution principle*. This simple technique is illustrated by the following problem. Assume that we have a container like the one shown in Figure 8-3. In order to measure the volume of water it contains, it is first necessary to carefully weigh an *amount* of some soluble salt or dye. For example, we can use 100 mg of salt sodium chloride (NaCl). The salt is dissolved in the unknown volume, and the contents are thoroughly stirred. Then, a small sample of the solution is removed and its *concentration* is determined. If the concentration is found to be 1.0 mg/ml, then the volume of the water in the container is calculated to be 100 ml. This determination can be made by using the following equation:

$$\text{Volume of compartment (ml)} = \frac{\text{amount of material added to compartment (mg)}}{\text{final concentration of material (mg/ml)}}$$

If we substitute the values used in the sample problem, we arrive at the following:

$$\text{Volume (ml)} = \frac{100 \text{ mg introduced into unknown volume}}{1 \text{ mg/ml of final concentration}}$$
$$= 100 \text{ mg} \times \frac{1 \text{ ml}}{1 \text{ mg}} = 100 \text{ ml}$$

The water compartments of the body can be estimated by using this same principle. Certain corrections, however, may have to be made in some in-

Figure 8-3 Dye dilution principle. An irregularly shaped vessel is filled with water. To determine the volume of the container, thoroughly mix in a known weight of a soluble dye or salt. Remove a sample for final concentration analysis.

stances because the injected material is metabolized or excreted. A corrected equation to account for such instances would be:

$$\text{Volume of compartment (ml)} = \frac{\text{amount injected minus amount removed}}{\text{final concentration}}$$

Total Body Water. Three substances are most frequently used to measure total body water. These are antipyrine and two isotopes of water, known as deuterium oxide (D_2O), and tritiated (radioactivated) water (H_3O).

Since the principle is identical for each material regardless of which substance is used, we shall consider only the use of antipyrine as an example. Suppose we inject 4 ml of a 100 mg/ml solution of antipyrine into a 70-kg (154-lb) man. After an equilibration time of 2 hours, we draw a blood sample and analyze the plasma for antipyrine concentration. Let us assume that the plasma concentration is 0.008 mg/ml. During the equilibration process, 18 percent of the administered dose was lost by excretion and metabolism. Substituting these values in the equation, we find:

Volume of compartment =
$$\frac{(4 \times 100 \text{ mg}) - (4 \times 100 \text{ mg} \times 0.18)}{0.008 \text{ mg/ml}}$$

$$= \frac{400 \text{ mg} - 72 \text{ mg}}{0.008 \text{ mg/ml}}$$

$$= \frac{328 \text{ mg}}{0.008 \text{ mg/ml}}$$

$$= 41,000 \text{ ml} = 41.0 \text{ liters}$$

41 liters weigh 41 kg

$$\text{Percent of body weight} = \frac{41 \text{ kg}}{70 \text{ kg}} = 58.6$$

Measurement of Extracellular Fluid Volume. The extracellular fluid compartment volume cannot be estimated very precisely. In order to measure this water compartment, it is necessary to find a chemical that is readily soluble in water, can diffuse through capillary walls easily, can enter every minute crevice and space between all cells, and yet does *not* appear *inside* any cell. In addition, the chemical should not be metabolized or excreted rapidly. Of course, the chemical should not be toxic to the body nor should it alter the water distribution between compartments. Unfortunately, no such ideal chemical has been discovered. Also, there is considerable controversy as to what should be included in the extracellular water compartment. For example, should it include the water in bone and the transcellular fluids previously mentioned? Some of the chemicals used to measure

extracellular water enter digestive juices; and some, such as sodium and chloride, enter cells to a slight extent.

Insulin, raffinose, sucrose, mannitol, thiosulfate, radiosulfate, thiocyanate, radiochloride, and radio-sodium are substances used to measure the extra-cellular fluid compartment. They produce results ranging from 12 to 21 liters for a 70-kg man. Radiosulfate seems the best compromise of good and bad features of all the substances and seems to be the substance used in most investigative laboratories. It gives values for the size of the extracellular compartment of about 16 liters for the standard 70-kg man.

Measurement of Intracellular Fluid Volume. The volume of the intracellular water compartment cannot be measured directly. It is calculated by subtracting the volume of the extracellular fluid compartment from the total water volume of the body. The value for the volume of the intracellular fluid volume is, then, only as reliable as the value for the extracellular fluid compartment.

Measurement of Plasma Fluid Volume. The volume of water in plasma is determined by the dye dilution technique. Two substances most frequently used are a dye, Evans Blue, and radioiodinated serum albumin (RISA). The Evans Blue dye bonds to the albumin molecules of the plasma and is easily distributed throughout the plasma compartment. The RISA is albumin labeled with iodine-131, and when injected in trace amounts, is diluted uniformly in the subject's plasma. After a half hour of equilibration, samples of plasma are removed and the final concentration of the dye or RISA is determined. Plasma volumes calculated for a 70-kg man average 3.15 liters.

Other techniques for measuring plasma volumes involve labeling the red cells with radioactive phosphorus (^{32}P) or chromium (^{51}Cr). When the volume percentage of cells in blood is known, the plasma volume can easily be calculated.

Now that we have taken a look at the distribution and measurement of body fluids, we can examine some of the chemicals present in them. First we shall see how the concentration of ions in body fluids is expressed.

IONS IN BODY FLUIDS

Concentration

The concentrations of ions in body fluids are expressed in units termed *milliequivalents per liter* (*meq/liter*). Chemists have determined that 1 equivalent of hydrogen ions contains 6.023×10^{23} (the Avogadro number) separate units of hydrogen ions, and their total weight is equal to 1.0008 g. This weight of hydrogen ions, termed one *equivalent weight,* will react with, and electrically neutralize, one equivalent weight of chloride ions. However, the total weight of the chloride ions is 35.453 g. Neutralization of one equivalent of hydrogen cations requires exactly one equivalent of anions, in this case chloride anions. The result of this neutralization is the formation of $1.008 + 35.453 = 36.461$ g of hydrochloric acid (1 mole of HCl). Since ions such as hydrogen (H^+) and chloride (Cl^-) each carry one unit of charge, they are termed *univalent ions.* Ions such as calcium (Ca^{2+}) and magnesium (Mg^{2+}) carry two unit charges and are termed *divalent ions.* When ions neutralize each other, the total number of oppositely charged ions must balance each other; that is, the total number of positive charges must equal the total number of negative charges. Accordingly, the equivalent weight of divalent ions contains half as many particles (3.013×10^{23}) as an equivalent weight of a univalent ion.

The concentrations of ions in body fluids are so small that it is more convenient to express them in *milliequivalents.* This enables us to express the concentrations in whole numbers, rather than fractional ones. One milliequivalent of a univalent ion is equal to 0.001 of an equivalent and consists of 6.023×10^{20} individual ionized particles. One milliequivalent of hydrogen ions weighs 1.008 mg, and one milliequivalent of chloride ions weighs 35.453 mg. The number of milliequivalents of an ion in each liter of solution is expressed by the following equation:

Milliequivalents per liter =
$$\frac{\text{milligrams of ion per liter of solution} \times \text{number of charges on one ion}}{\text{atomic weight of ion}}$$

The *atomic weight* of an element indicates how heavy it is when compared with the element carbon. Dividing the total weight of the solute by its atomic weight tells us how many ions we have. For instance, the atomic weight of calcium is 40, whereas that of sodium is 23. Calcium is therefore a heavier element, and 100 g of calcium contains fewer atoms than 100 g of sodium. The atomic weights of the elements have been calculated by scientists and can be found in a periodic table.

Using the preceding formula, we can calculate the milliequivalents per liter for calcium (Ca^{2+}). In 1 liter of plasma there are normally 100 mg of calcium. Thus, by substituting this value in the formula, we arrive at:

$$\text{meq/liter} = \frac{100 \times \text{number of charges}}{\text{atomic weight}}$$

The atomic weight of calcium is 40, and its number of charges is 2. By substituting these values we arrive at:

$$\text{meq/liter} = \frac{100 \times 2}{40} = 5 \text{ meq/liter for calcium}$$

The milliequivalents per liter for sodium can also be calculated. The number of milligrams of Na^+ per liter of plasma equals 3.300 mg/liter; number of charges equals 1; and the atomic weight equals 23. Therefore:

$$\text{meq/liter} = \frac{3.300 \times 1}{23} = 143.0 \text{ meq/liter for sodium}$$

Comparing the milliequivalents per liter of sodium with that of calcium, we can see that, even though calcium has a greater number of charges (two) than sodium (one), the body retains many more sodium ions than it does calcium ions. Therefore, the milliequivalent for sodium in plasma is higher.

Movements

Figure 8-4 illustrates the ionic composition of the three major fluid compartments. Of the ions shown, only the proteins, because of their large size, remain within their compartment. Proteins are considered *nondiffusible anions.* The other ions that are capable of crossing plasma membranes under varying conditions are referred to as *diffusible ions.*

Gibbs-Donnan Rule: Ionic Distributions

The distribution of the easily diffusible ions is affected by the presence of the nondiffusible protein anions. The ultimate distribution of *all* the ions (diffusible and nondiffusible) occurs according to what is called the *Gibbs-Donnan rule.* This rule states that when nondiffusible ions are present only on one side of a semipermeable membrane, the distribution of the diffusible ions across the membrane will become equal. The final distribution of *all* ions will satisfy three requirements:

1. The total number of cations and anions on the same side of the semipermeable membrane will be equal.
2. On the side containing the protein anions, the number of diffusible anions present will be less than and the number of diffusible cations will be greater than on the other side of the semipermeable membrane.
3. The osmotic pressure on the side containing the protein anions, which is called the *oncotic pressure,* will be slightly greater than on the side without protein anions. The oncotic pressure due to the protein anions needs to be balanced by some other mechanism to prevent water from shifting. Water does not move into capillaries because there is a

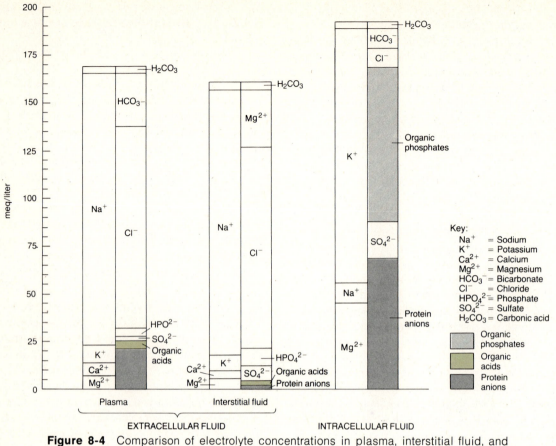

Figure 8-4 Comparison of electrolyte concentrations in plasma, interstitial fluid, and intracellular fluid. The height of each column represents the total electrolyte concentration.

counteracting *hydrostatic* or *blood pressure* equal to the oncotic pressure.

In order to illustrate the requirements for the operation of the Gibbs-Donnan rule, let us consider a simple closed system, such as the one shown in Figure 8-5. Two compartments of equal size are separated by a semipermeable membrane. Sodium and chloride ions can move freely through the membrane but the protein anions cannot. Initially, 10 Na$^+$ and 10 Cl$^-$ ions are placed in the left-hand compartment (L.H.C.) and 5 Na$^+$ and 5 protein$^-$ ions are placed in the right-hand compartment (R.H.C.) as shown in Figure 8-5a. In the final condition, after diffusion of the ions has taken place, an equilibrium is achieved (Figure 8-5b). At this point the three requirements of the Gibbs-Donnan rule are satisfied:

1. The number of cations and anions on the same side of the semipermeable membrane is equal. In the final state, the total number of anions in the L.H.C. = 6 and the total number of cations in the L.H.C. = 6. The total number of anions in the

R.H.C. = 9 (5 + 4) and the total number of cations in the R.H.C. = 9 (5 + 4).

2. In the compartment containing protein anions (R.H.C.), the number of *diffusible* anions = 4, which is less than the number of diffusible anions on the other side (6). The number of diffusible cations in the protein-anion-containing R.H.C. = 9 (5 + 4), which is greater than the number of diffusible cations on the other side (6).

3. The oncotic pressure on the side containing the protein anions (R.H.C.) is higher than that on the L.H.C. because the *total* number of osmotically active particles is 18 on the R.H.C., versus 12 on the L.H.C.

The Gibbs-Donnan rule predicts that at equilibrium the product of the concentrations of *any* pair of *diffusible* cations and anions on one side of a semipermeable membrane will equal the product of the concentrations of the same pair of ions on the other side. For the above example, the following simple calculation verifies the prediction. The diffusible pair of ions considered in the example are the sodium and

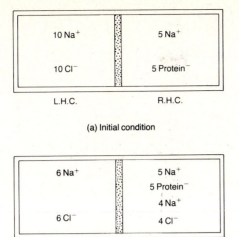

Figure 8-5 Illustration of the Gibbs-Donnan rule.

chloride ions. The product of their concentrations in the L.H.C. is $6 \times 6 = 36$. The product of their concentrations in the R.H.C. is $9 \times 4 = 36$. A further observation is that the L.H.C. has $6 + 6 = 12$ diffusible particles, whereas the R.H.C. has $4 + 9 = 13$ diffusible particles. These particles are osmotically active, and even if we ignore the presence of the protein anions in the R.H.C., the R.H.C. has a higher oncotic pressure than the L.H.C.

This difference in osmotic pressures would normally cause a movement of water from the L.H.C. to the R.H.C., but because our example employs *closed* compartments, there is no net (greater) movement of water. Later on in our discussion of the *capillary* or *microcirculation*, which is an open system, you will see that the movement of water does occur. *It is important to realize that the greater osmotic pressure present in the compartment with the proteins is due*

not only to the proteins themselves but also to the greater number of diffusible ions present on the side with the protein.

Extracellular Fluid Ions

The extracellular fluid compartment is conveniently studied by separating it into two subcompartments. These, as noted earlier, are plasma and interstitial fluid. Plasma is the fluid portion of blood and is confined within blood vessels. Interstitial fluid is the immediate environment surrounding all the cells of the body.

Plasma. Sodium (Na^+) is the major cation in blood plasma, whereas the major anion is chloride (Cl^-). It is important to reemphasize that the sum of the milliequivalent concentrations of the anions exactly equals the sum of the concentrations of the cations. The ionic concentrations in plasma and interstitial water are shown in Exhibit 8-2. While the concentrations of anions and cations must exactly balance each other, the concentrations of the individual ions vary ± 3 percent. The most significant difference between plasma and interstitial fluid is the concentration of protein anions in plasma. Whereas plasma has 17 meq/liter, interstitial fluid has 0 meq/liter.

Interstitial Fluid. The interstitial fluid is best considered an *ultrafiltrate* of plasma. A simple filtrate of plasma would exclude the red and white blood cells, whereas the ultrafiltrate of plasma would also exclude the large protein molecules. There may be traces of protein in the interstitial fluid in some regions of the body where the capillaries are more permeable or "leaky." The walls of the capillaries consists of closely packed simple squamous cells glued together and

Exhibit 8-2 IONIC COMPOSITION OF PLASMA AND INTERSTITIAL FLUID

Ionic Component	Plasma Water meq/liter	Interstitial Water meq/liter
Sodium (Na^+)	151.0	144.0
Potassium (K^+)	4.3	4.0
Calcium (Ca^{2+})	5.4	2.5
Magnesium (Mg^{2+})	3.2	1.5
Total cationic concentration	163.9	152.0
Chloride (Cl^-)	109.7	114.0
Bicarbonate HCO_3^-)	28.7	30.0
Phosphate (PO_4^{3-})	2.1	2.0
Sulfate (SO_4^{2-})	1.1	1.0
Organic acids (R—COOH)	5.3	5.0
Protein	17.0	0.0 (trace)
Total anionic concentration	163.9	152.0

rolled into a hollow tube called *endothelium*. It is through these endothelial tubes or capillaries that the continual exchange of materials between plasma and interstitial fluid occurs. In this way, the internal environment retains a composition suitable for the life processes of all cells. This is homeostasis.

Intracellular Fluid Ions

The ionic composition of this fluid is very difficult to describe because the interior of cells varies so greatly. Moreover, the analyses of the cellular contents involve numerous problems. It is obvious that cells contain different types and amounts of proteins for their own specialized functions. The types and amounts of various ions also differ. However, it is possible to set forth certain similar qualitative features. (See Figure 8-4.) Potassium (K^+) and magnesium (Mg^{2+}) ions are the most abundant cations, and proteins and organic phosphates (AMP, ADP, ATP, and creatine phosphate) are the major anions of the intracellular fluid. Added to the problem of measuring intracellular ionic concentrations is the disagreement about whether the ions inside the cells are free in solution or bound to large molecules or membranes.

MOVEMENT OF BODY FLUIDS

Osmotic Pressure

The movement and distribution of water into the various compartments of the body is determined in large part by osmotic forces. It is therefore important to have a clear understanding of the meaning of osmotic pressure.

The principle of osmotic pressure is based upon the fact that every chemical distributes itself *uniformly* through any compartment in which it is enclosed. Consider the illustration in Figure 8-6. The large container is filled with a pure solvent such as water. Suspended in the large container is a hollow cylinder fitted with a frictionless piston, and the end of the cylinder is closed with a semipermeable membrane. The cylinder is filled with a 5 percent solution, for example protein in water. The pure solvent and the solution are thus separated by a semipermeable membrane that permits free passage of water but confines the protein to the interior of the suspended cylinder.

Each chemical will tend to distribute itself uniformly throughout its compartment. The protein is thus distributed equally throughout the water inside the cylinder. It would not matter where one sampled the solution in the cylinder—top, bottom, middle, or side—the concentration of protein (the solute) would

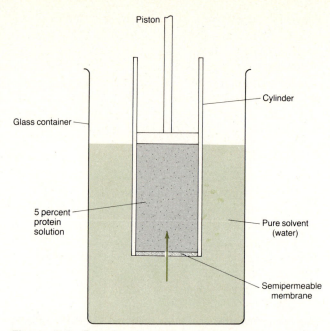

Figure 8-6 Experimental setup to demonstrate osmotic pressure.

be identical (5 percent) in any area sampled. However, the water demonstrates an interesting phenomenon. In the large container, water outside the cylinder is distributed uniformly. It is 100 percent pure solvent (water). But the water, being freely diffusible through the semipermeable membrane, exists, in reality, in a much larger compartment that includes the interior of the cylinder as well as the large vessel. Water therefore is *not* uniformly distributed throughout its compartment. It exists only as a 95 percent solution inside the cylinder but as 100 percent water outside. Therefore, water tends to distribute itself so as to equalize the concentration and make both the inside and outside concentrations the same. The result is that water flows into the cylinder. The frictionless piston is thus lifted by the inflowing water to reflect this movement of water. The inflowing water dilutes the 5 percent solution, striving to bring the water concentration to that of the large vessel. Theoretically, all the water should end up inside the cylinder because only then will the water be in a compartment where it is uniformly distributed.

In reality, however, as the water flows into the cylinder, the rising level of water in the cylinder generates an opposing force due to its weight acting downward with the force of gravity. Therefore, at some point, an equilibrium height is reached where the force of the inflowing water (**osmotic pressure**) is balanced by the weight of water acting through gravity (a hydrostatic pressure). It is possible to determine the osmotic pressure by measuring the equal but opposite hydrostatic pressure. An alternate way would be to measure the amount of additional pressure on the piston just necessary to keep it from rising (that is, to keep solvent water from entering into the solution cylinder).

Cell membranes are permeable to water but are often impermeable to many other chemicals. Thus they behave as tiny osmometers.

Fluid Distribution

The movement and distribution of water between the intracellular and extracellular compartments is governed by their concentration of osmotically active particles. To demonstrate the movements of water and ion concentrations that accompany the addition or removal of water or salt to one compartment, the following examples are presented. A standard 70-kg man is the subject of these four experiments. In the first experiment he will drink 2 liters of distilled water; in the second he will drink 200 ml of a 10 percent sodium chloride solution; in the third experiment he will be infused by an intravenous injection with 1 liter of isotonic saline; and in the fourth experiment, sodium chloride will be removed from his extracellular compartment. Figure 8-7 shows the extracellular and intracellular water compartments as rectangles. The horizontal axis represents the volume of the compartment in liters, and the vertical axis represents the ionic concentration in milliosmoles per liter (mOsm/liter).* The initial condition is outlined in solid lines and the changes that occur are outlined with broken lines.

Ingestion of Distilled Water

A standard 70-kg man is 60 percent water (42 liters). The intracellular volume is 25 liters and the extracellular volume is 17 liters. The ionic concentration of the fluids is 200 mOsm/liter.

Our imaginary subject drinks 2 liters of water, and for our purposes, absorbs it rapidly and does not excrete any during the time of the experiment. The water distributes uniformly throughout the total body water, increasing total volume and decreasing concentration. The final fluid concentration in the extracellular and intracellular compartments is $300 \times 42/(42 + 2) = 286$ mOsm/liter. The water distributes itself proportionately between the two compartments. The extracellular compartment receives $17/42 \times 2 = 0.81$ liters, and the intracellular compartment receives

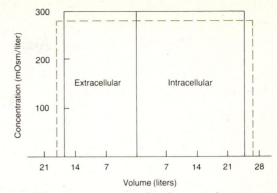

Figure 8-7 Demonstration of volume and concentration changes upon ingestion of 2 liters of distilled water. Initial concentration and volume are outlined in solid lines. Final concentration and volume are indicated with broken lines.

$25/42 \times 2 = 1.19$ liters (Figure 8-7). In reality, if a normal subject performed such an experiment, the increase in volume and the decrease in concentration would be less because excretion of water would have to occur before it was all absorbed from the digestive tract.

Infusion of 10 Percent Sodium Chloride

Suppose our standard 70-kg man is given 200 ml of 10 percent sodium chloride (that is, 20 g of sodium chloride). The sodium chloride is retained in the extracellular fluid compartment, thereby increasing its ionic concentration. As a consequence, water shifts from the intracellular compartment, decreasing its volume, and moves into the extracellular compartment, increasing its volume. The resultant increase in concentration is 20 g/58.5 × 2* osmotically charged particles = 0.684 Osm = 684 mOsm. These particles seem to be distributed in 42 liters, thus increasing the concentration by $684/42 = 16.3$ mOsm/liter. The ionic concentration of the extracellular and intracellular compartments is increased $300 + 16.3 = 316.3$ mOsm/liter. Since the concentration has increased in both compartments, it would seem that the sodium chloride was proportionately distributed. The added salt, however, actually remains in the extracellular compartment, and water leaves the cells. It is possible to calculate how much water would shift from the cells to the extracellular compartment. Initially, the cells contain 25 liters of water with an ionic concentration of 300 mOsm/liter. This translates to 300 × 25 or 7500 mOsm of osmotically active particles. In order to increase the concentration to 316.3 mOsm/liter, the size of the intracellular compartment must

---

* Osmolar concentrations are obtained by summing the molar concentrations of the solute particles; thus a solution of 0.15M NaCl would be reported as a 0.30 osmolar (Osm) concentration or a 300 milliosmolar (mOsm) concentration.

* One mole of sodium chloride exerts an osmotic effect of 2 Osm, since it dissociates into two osmotically active particles (NaCl → Na+ + Cl⁻).

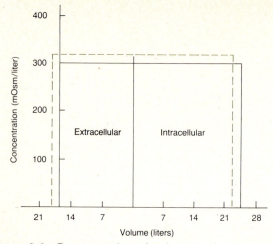

Figure 8-8 Demonstration of volume and concentration changes after ingestion of 200 ml of a 10 percent sodium chloride solution. Initial concentration and volume are outlined in solid lines. Final concentration and volume are indicated with broken lines.

be reduced to 7500/316.3 = 23.7 liters. This means that 25 − 23.7 or 1.3 liters of water shifted from inside the cells to the extracellular compartment (Figure 8-8).

Infusion of an Isotonic Saline Solution

One liter of an isotonic (300 mOsm/liter) solution is infused into our subject by intravenous injection. The concentration of the solution infused is identical to that of the intracellular and extracellular water, and therefore no concentration changes result. The sodium chloride is confined to the extracellular compartment, so that compartment increases in volume by the 1.0 liter of saline infused (Figure 8-9).

Removal of Sodium Chloride

It is possible, by indirect means, to remove sodium chloride from the extracellular fluid without taking out any water. The removal of sodium chloride decreases the ionic concentration of the extracellular

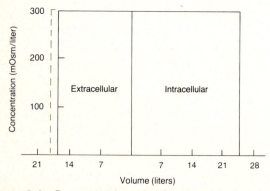

Figure 8-9 Demonstration of volume and concentration changes after an infusion of 1,000 ml of isotonic saline. Initial concentration and volume are outlined in solid lines. Final concentration is indicated with a broken line.

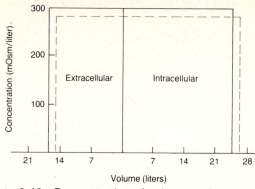

Figure 8-10 Demonstration of volume and concentration changes occurring after removal of only sodium chloride from extracellular fluid. Initial concentration and volume are outlined in solid lines. Final concentration and volume are indicated with broken lines.

fluid, and water shifts into the cells thereby decreasing the ionic concentration of the intracellular water. The loss of water from the extracellular compartment decreases its volume and increases that of the intracellular volume by the same amount (Figure 8-10).

MOVING MATERIALS ACROSS PLASMA MEMBRANES

The mechanisms whereby substances move across the plasma membrane are important to the life of the cell. Certain substances, such as oxygen and nutrients, must move into the cell to support life, whereas others, such as carbon dioxide and waste materials that may be harmful, must be moved out of the cell. The processes involved in such movements may be divided into two broad categories, depending upon whether the cell participates in the process by expending energy. Accordingly, the processes may be classified as either passive or active. In *passive processes,* substances move across plasma membranes without any energy from the cell. The substances move, because of their own kinetic energy, from an area where their concentration is greater to an area where their concentration is less. The substances can also be pushed through the cell membrane by pressure from an area where the pressure is higher to an area where it is lower. In *active processes,* by contrast, the cell contributes energy and assumes a role in moving the substance across the membrane. We shall first consider the passive processes.

Passive Processes

Bulk Flow and Diffusion

Two examples of passive processes are bulk flow and diffusion. In order to distinguish them, consider the

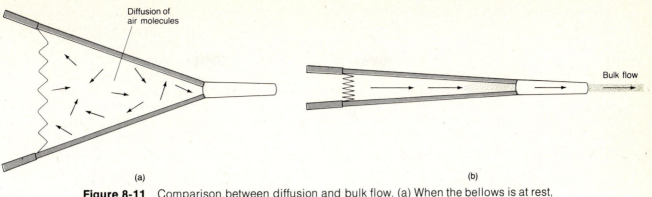

Diffusion of
air molecules

Bulk flow

(a)

(b)

Figure 8-11 Comparison between diffusion and bulk flow. (a) When the bellows is at rest, the air within moves randomly by diffusion. (b) When the bellows is compressed, the air is forced out in bulk flow.

air inside the bellows shown in Figure 8-11. When the bellows is at rest, the air molecules inside the bellows are moving rapidly and randomly, colliding and bouncing off each other like minuscule rubber balls. This type of movement is called *diffusion*. It is temperature-dependent and increases with a rise in temperature. If, however, the bellows is squeezed and the air is forced out, the molecules that were inside flow out from the nozzle as a unit. This type of movement is called *bulk flow*. The driving force for bulk flow is a pressure gradient (difference). When the bellows was squeezed, the pressure on the air inside became greater than the outside air pressure, so the air flowed out through the nozzle. The rate of bulk flow is directly proportional to the pressure gradient.

The flow of blood is an example of bulk flow in the body. As the heart contracts, the pressure created forces a volume of blood to move in bulk fashion into and through the blood vessels. Air flowing out of the lungs is another example. The contraction of the breathing muscles increases the pressure on the air in the lungs, forcing it to be exhaled.

The passive process called *diffusion* occurs when there is a *net* (greater) movement of molecules or ions from a region of high concentration to a region of low concentration. The movement from high to low concentration continues until the molecules are evenly distributed. This even distribution is called *equilibrium*. The difference between the higher and lower concentrations is called the *concentration gradient*. Molecules moving from the high-concentration area to the low-concentration area are said to move *down* or *with* the concentration gradient.

Consider the following example. If a dye pellet is placed in a beaker filled with water, the color of the dye is seen immediately around the pellet. At increasing distances from the pellet, the color becomes lighter (Figure 8-12). If the beaker is observed some time later, however, the water solution will be a uni-

form color. This happens because the dye molecules possess *kinetic energy,* which causes them to move about at random, dispersing them throughout the entire area. The dye molecules move down the concentration gradient from an area of high concentration to an area of low concentration. The water molecules also move from a high-concentration area to a low-concentration area. When dye molecules and water molecules are evenly distributed, equilibrium is reached and diffusion ceases, even though molecular movements still continue.

As another example of diffusion, consider what would happen if you opened a bottle of perfume in a room. The perfume molecules would diffuse until an equilibrium was reached between the perfume molecules and the air molecules in the room.

Five factors influence the rate of diffusion. These are: (1) the concentration or diffusion gradient, (2) the cross-sectional area through which diffusion occurs, (3) the temperature, (4) the molecular weight of the diffusing substances, and (5) the distance through which diffusion occurs.

1. The concentration gradient refers to the differences

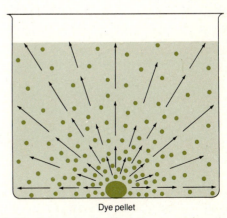

Dye pellet

Figure 8-12 Principle of diffusion.

in concentration of the diffusing substances measured at two points. The greater the difference, the greater the rate of diffusion.

2. The cross-sectional factor is visualized by considering the two cylinders in Figure 8-13. The rate of diffusion in cylinder (b) is greater since it permits more freedom of molecular movement from the higher concentration to the lower.

3. The greater the temperature, the more rapid the molecular movement and hence the more rapid the rate of diffusion. This is because the higher the temperature, the greater the kinetic energy.

4. The rate of diffusion of the molecular substance varies inversely with the square root of its molecular weight. That is, the smaller the value (the lighter the molecule) the greater the rate of diffusion.

5. The shorter the diffusion distance, the greater the rate of diffusion.

Osmosis

Another passive process by which materials move across membranes is called *osmosis*. As described earlier, osmosis is a special type of diffusion. Osmosis specifically concerns solvent movement, and in living systems, the solvent is water. Water molecules move through a semipermeable membrane from a region where they are in higher concentration to a region where they are in lower concentration. (See Figure 8-6.)

Filtration

Another passive process involved in moving materials across membranes is *filtration*. It is actually a specialized type of bulk flow. It is the movement of solvents such as water, and dissolved substances such as sugar, across a semipermeable membrane because of a pressure gradient. Such a movement is always from an area of higher pressure to an area of lower pressure and continues as long as a pressure difference exists. Most small-to-medium-sized molecules can be forced through a cell membrane by pressure. An example of filtration in the body occurs in the kidneys, where the blood pressure supplied by the heart forces water and urea (a waste) through thin cell membranes of blood capillaries into the kidney tubules. In this basic process, protein molecules are retained by the body since they are too large to be forced through the cell membranes. Harmful substances such as urea, however, are small enough to be forced through and eliminated.

Dialysis

The final passive process to be considered is dialysis, the process by which the artificial kidney works. *Dialysis* is the separation of small molecules from large molecules by diffusion of the smaller molecules through a semipermeable membrane. For example, assume that a solution containing molecules of various sizes is placed in a tube that is permeable only to the smaller molecules. The tube is then placed in a beaker of distilled water. Eventually, the smaller molecules will move from the tube into the water in the beaker, and the larger molecules will be left behind. The principle of dialysis is employed in artificial kidneys. In the operation of an artificial kidney, the blood of the patient is passed into a dialysis tube outside the patient's body. The dialysis tube takes the place of the kidneys. As the blood moves through the tube, waste products pass from the blood into a solution surrounding the dialysis tube. At the same time, certain nutrients are passed from the solution into the blood. The blood is then returned to the body.

We can now turn our attention to active processes, by which materials move across plasma membranes.

Active Processes

Active Transport

Many important molecules required by cells, such as glucose and amino acids, are excluded from the intracellular environment by the continuous lipid core of the cell membrane. However, glucose and amino acids must be available for the energy requirements and synthetic activity of the cell. Sometimes the concentration of these nutrients in the extracellular environment is very low, yet the cells are still obtaining these materials even against a considerable gradiet.

Active transport is an energy-requiring movement of a metabolite or an inorganic ion across a semi-

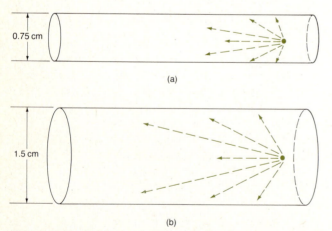

Figure 8-13 Diffusion and cross-sectional area. The larger cross-sectional diameter of cylinder (b) permits more freedom for molecular movement.

permeable membrane against a concentration gradient. Active transport is a great consumer of the cell's energy pool. In specialized tissues such as the kidney, up to 70 percent of the available metabolic energy is consumed for active transport processes.

In addition to the transport of fuels and essential nutrients into the cell, active tranport systems maintain the internal concentration of organic molecules relatively constant even during wide variations in the concentrations of the external environment. Active transport processes maintain constant and optimal internal concentrations of inorganic electrolytes (ions), especially K^+ and Ca^{2+}. They also help to maintain osmotic relationships so that cells neither swell nor shrink. Active transport systems are integral components in the transmission of information by the nervous system and in the excitation-contraction-relaxation cycle of muscle tissues. They also function in the absorption of nutrients through the intestinal epithelium and in the secretory functions of the kidneys and many glandular organs.

Mediated Membrane Transport

During passive processes, molecules are never altered either chemically or physically, and they enter cells without associating with other molecules as they pass through the membrane. These processes are the result of simple diffusion and are also termed *nonmediated membrane transport processes*.

Mediated (facilitated) transport processes, however, are quite different and have their own set of criteria. The first is that mediated transport systems exhibit *saturation kinetics*. As the concentration of substances outside the cell increases, the rate of entry into the cell increases up to a point and gradually levels off despite further increases in concentration of the substances outside the cell (Figure 8-14). These findings suggest that the rate of entry is regulated by a carrier molecule (system) that reaches a maximum level of operation which it cannot surpass (that is, saturated).

A second criterion of mediated transport is termed *specificity*. The transport molecule has a specific attraction for and complements exactly the shape of the molecule being transported. The combination of the transported molecule and the carrier depends upon their fitting together, much like two pieces of a jigsaw puzzle. Only one or very few molecules can fit and therefore be transported by the carrier.

Another criterion for mediated transport is that the transport process can be specifically inhibited. In *competitive inhibition*, another molecule with a similar structure can be transported by fitting into the receptor site of the transport molecule. In *noncompetitive inhibition*, a molecule attaches to the transport molecule at some site other than the normal receptor position. This causes the transport molecule to assume a new shape, altering the receptor site so that it can no longer accept molecules for transport.

The above criteria strongly indicate that the transport process involves proteins that can bind and release specific molecules after transferring them through the membrane.

Active and Passive Mediated Transport

Mediated transport systems can be divided into two classes, *active* and *passive,* depending upon the direction of the transport, the need for energy, and whether it operates against or with the gradient.

The transport of a molecule against a gradient, that is, moving a molecule from the dilute side of the membrane to the concentrated side, is possible only in an *active* transport process. To use this criterion for establishing the presence of an active transport system requires precise knowledge about the concentrations of the molecule on both sides of the membrane. It is often very difficult or impossible to obtain this information.

Another criterion utilized to establish whether the transport process is active or passive is the requirement for energy. If metabolic energy is required for the transport process, the system under investigation is considered an active one. The need for energy is often determined by using specific inhibitors that poison the energy-producing enzymes. With metabolic energy unavailable to the cell, it can be noted whether transport continues or not. If it ceases, then the process is dependent on energy and is therefore an active transport process. If it continues, then the process is passively mediated.

Active transport systems are also *unidirectional;* that is, they transport the substrate molecule across the membrane in one direction only. Passively medi-

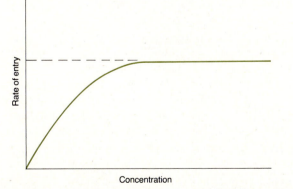

Figure 8-14 Saturation kinetics of a mediated transport system.

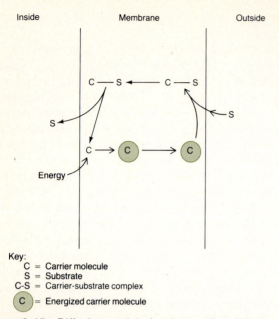

Inside   Membrane   Outside

Key:
C = Carrier molecule
S = Substrate
C-S = Carrier-substrate complex
ⓒ = Energized carrier molecule

Figure 8-15 Diffusion model of active mediated transport. The carrier molecule is energized on the inner surface of the membrane and then diffuses to the outer surface of the membrane. There, it joins with the substrate to form a carrier-substrate complex. This diffuses to the inner surface, where the substrate is released and the carrier is re-energized.

ated systems, by contrast, carry molecules in either direction across the membrane, with the gradient.

The Sodium-Potassium Transporting ATPase System

The intracellular concentration of K^+ is maintained at a high and relatively constant level, whereas the intracellular Na^+ level is much lower. (See Figure 8-4.) When we find that the reverse situation exists in the extracellular fluid, where the Na^+ level is higher and the K^+ level is lower, it becomes apparent that there is a large gradient for these ions across the membrane. These gradients exist because of an energy-requiring process that "pumps" Na^+ out of the cell and replaces it with K^+. The transport system involved is a remarkable process and is slowly being unraveled. This so-called *Na^+-K^+-ATPase transport system* seems to be essential not only for regulating concentrations of sodium and potassium ions, but also for the operation of other so-called "pumps," such as for glucose and amino acids.

The high concentrations of intracellular K^+ are needed for at least three vital processes. One is the synthesis of ribosomes. A second involves its participation in regulating the proper activity of certain enzymes. The third process is related to the electrical activity of excitable tissues. The so-called *action potential* or electrical discharge that occurs when certain cells are stimulated relates to the movement of Na^+ and K^+ across the cell membrane (Chapter 9).

Models of Active Mediated Transport

Actual transport molecules have been isolated, purified, and reconstructed. However, their method of intramembranous action remains unknown. Four types of models are generally discussed: diffusion, rotation, conformational change, and relay system.

1. **Diffusion.** According to this model, a small carrier molecule exists in the membrane, and when energized on the membrane's inner surface, it diffuses to the outer surface. Here it bonds to a substrate, and the carrier-substrate complex diffuses back to the inner surface, where the substrate is released. Then, the carrier is reenergized and repeats the cycle (Figure 8-15).

2. **Rotation.** According to the rotation model, a large carrier molecule that extends through the membrane attaches to the substrate and rotates 180° so that substrate is now exposed to inner surface of the membrane (Figure 8-16). Energy is supplied and the substrate is discharged to the interior of the cell and the molecule again rotates 180° to face the exterior and binds with another substrate molecule. The system is analogous to a revolving door.

3. **Conformational change.** In this proposed model, a large, coiled carrier molecule in the membrane projects a receptor site to the outside of the membrane and complexes with the substrate molecule. A conformational twist occurs, causing the carrier molecule to reorient itself so that the receptor site

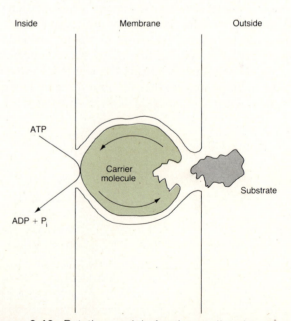

Inside   Membrane   Outside

ATP

Carrier molecule

ADP + P_i

Substrate

Figure 8-16 Rotation model of active mediated transport. The substrate binds to a large carrier molecule. Energy from ATP causes the carrier to rotate 180° and discharge the substrate to the interior.

of the carrier and the substrate now projects to the interior of the cell (Figure 8-17). The substrate is released, and in the presence of ATP, the carrier molecule returns to its original conformation.

4. **Relay system.** A fourth model of active mediated transport is based upon the operation of a bucket brigade (Figure 8-18). Certain fixed molecular groups of carriers pass the substrate from one to another through the membrane. Energy is used to keep the molecules oriented so that the transfers take place.

Endocytosis

In this type of active transport, material is introduced into the cell in bulk form. If the material entering the cell is solid, the process is called *phagocytosis* (*phago* = I eat). If the material is fluid, the process is termed *pinocytosis* (*pino* = I drink). With the aid of the electron microscope, the similarities of these two processes were uncovered, and the term *endocytosis* (*endo* = within) has been coined to cover both phagocytosis and pinocytosis. Since endocytosis can be observed only at the electron microscopic level, the term *micropinocytosis* is also used to describe this process.

Pinocytosis. In *pinocytosis*, the cell membrane forms channels by an infolding (invagination) process. These channels are then cut into short, small segments, forming vesicles (Figure 8-19a). In this process, the fluid to be taken into the cell is enclosed by a portion

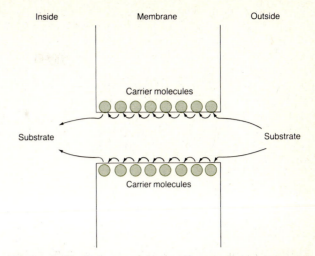

Figure 8-18 Relay model of active mediated transport. Groups of carrier molecules transfer the substrate through the membrane. Each carrier molecule moves the substrate a short distance to the next carrier molecule, and so on.

of the plasma membrane. An interesting example of pinocytosis occurs in bone marrow. The erythroblast cell—the cell that will mature into a red blood cell—obtains its iron-containing molecules by pinocytosis. Other cells that are active in pinocytosis are found in the liver, kidneys, and reticuloendothelial system. It is likely that pinocytosis is a common event that occurs to some degree in most cells. Extensive studies of the process in the amoeba indicate that the following sequence of events occurs in pinocytosis.

The first step requires the presence of protein inducer molecules, which adhere to the mucopolysaccharide coat on the exterior of the cell membrane. The next step is the formation of the invading channel by membrane infolding. The third step involves the fragmentation of the channel into small vesicles which then move into the interior of the cell. The last step involves the digestion of ingested material. That pinocytosis is an energy-requiring process is easily demonstrated by the use of inhibitors.

Phagocytosis. Phagocytosis is most likely a very similar process to pinocytosis. In *phagocytosis*, solid particles rather than fluids are engulfed. The process involves attachment of the particles to the cell surface coat, followed by complete engulfment (Figure 8-19b).

Now that we have surveyed the nature of the cellular environment and the various ways that substances are transported within the environment, we shall take a look at a very special type of cell, the nerve cell. Because of their ability to conduct impulses, nerve cells assume a key role in the maintenance of homeostasis.

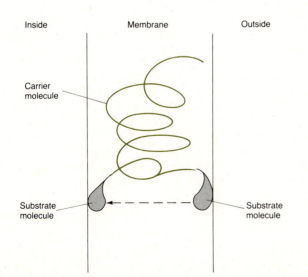

Figure 8-17 Conformational change model of active mediated transport. The large, coiled carrier molecule in the membrane has its tail facing to the outside to accept substrate molecules. The carrier molecule undergoes a conformational change so that the tail is flipped to the inside surface, where the substrate is released.

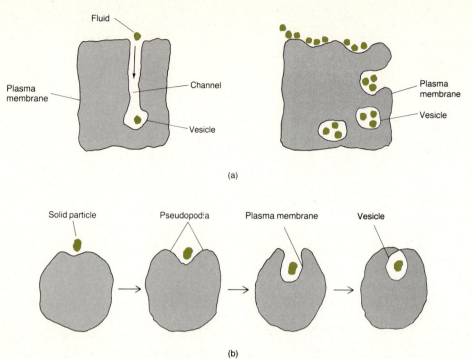

Figure 8-19   Endocytosis. (a) Pinocytosis. In the variation on the left, the ingested substance enters a channel formed by the plasma membrane and becomes enclosed in a vacuole at the base of the channel. In the variation on the right, the ingested substance becomes enclosed in a vacuole that forms and detaches at the surface of the cell. (b) Phagocytosis.

CHAPTER SUMMARY OUTLINE

WATER

1. Life as we know it on earth is not possible without water. Water comprises 60 to 90 percent of the weight of most living organisms, and possesses some very valuable properties.

Physical Properties of Water

1. A calorie is the amount of heat required to raise the temperature of 1 g of water from 15° to 16°C.
2. Specific heat is the number of calories required to raise the temperature of 1 g of a substance 1°C.
3. Heat of vaporization is the heat required to convert water at 100°C to steam at 100°C.
4. Heat of fusion is the amount of heat necessary to convert 1 g of ice at 0°C to water at 0°C.

Structure of a Water Molecule

1. A water molecule is shaped like an isosceles triangle.
2. The oxygen atom of the molecule exhibits very strong electronegative properties which create a permanent dipole moment within the molecule; that is, the oxygen atom has a negative charge and the two hydrogen atoms have slight positive charges.

Body Water

1. Water is the most abundant component of the human body, ranging from 45 to 75 percent of the body weight.
2. Fifty-two percent of body weight of young adult females is water, whereas 63 percent of the body weight of young adult males is water.
3. The intracellular water compartment consists of water contained within body cells.
4. The extracellular water compartment consists of the water distributed outside the body cells.

IONS IN BODY FLUIDS

Concentration

1. The concentrations of ions in body fluids are expressed in units termed milliequivalents per liter (meq/liter).

Movements

1. Proteins, because of their large size, remain within their compartment, and are considered nondiffusible ions, whereas other ions, which are capable of crossing plasma membranes, are referred to as diffusible ions.

MOVEMENT OF BODY FLUIDS

Osmotic Pressure

1. Osmotic pressure results from the fact that every chemical attempts to distribute itself uniformly throughout any compartment in which it is enclosed.

Fluid Distribution

1. The movement and distribution of water between the intracellular and extracellular compartments is governed by the different concentrations of osmotically active particles in the compartments.
2. When 2 liters of distilled water are ingested, the water will distribute uniformly throughout the total body water compartment, increasing the total water volume but decreasing the solute concentration.
3. If 200 ml of 10 percent sodium chloride is ingested, the sodium chloride is retained in the extracellular fluid compartment, increasing its ionic concentration. Water then shifts from the intracellular compartment, decreasing its volume, and moves into the extracellular compartment, increasing its volume.
4. When 1 liter of an isotonic solution is infused by intravenous injection, no concentration changes result since the concentration of the solution infused is identical to that of the intracellular and extracellular water.

MOVING MATERIALS ACROSS PLASMA MEMBRANES

1. The processes involved in moving various materials across plasma membranes may be divided into two broad categories, depending upon whether the cell participates in the process of expending energy.

Passive Processes

1. In passive processes, substances move across plasma membranes without any energy contribution from the cell. The substances move because of their own kinetic energy from an area where their concentration is greater to an area where their concentration is less.
2. Diffusion is temperature-dependent, increasing with a rise in temperature, and is due to the rapid and random movement of the molecules.
3. Bulk flow occurs when molecules move as a unit. The driving force for bulk flow is a pressure gradient.
4. Osmosis specifically concerns solvent movement. In living systems, the solvent is water. Water molecules move through a semipermeable membrane from a region where they are in higher concentration to a region where they are in lower concentration.
5. Filtration is the movement of solvents such as water, and dissolved substances such as sugar, across a semipermeable membrane because of a pressure gradient.
6. Dialysis is the separation of small molecules from large molecules by diffusion of the smaller molecules through a semipermeable membrane.

Active Processes

1. In active processes, the cell contributes energy and assumes a role in moving the substance across the membrane.
2. Active transport is an energy-requiring movement of a metabolite or an inorganic ion across a semipermeable membrane against a concentration gradient.
3. Mediated transport systems can be divided into two classes, active and passive, depending upon the direction of the transport, the need for energy, and whether the system operates against or with the gradient.

REVIEW QUESTIONS

1. Explain why water is so important to living organisms.
2. Define each of the following: calorie, specific heat, heat of vaporization, and heat of fusion.
3. Describe in detail how the structure of a water molecule creates the cohesive property of water.
4. Correlate the amount of body water with obesity.
5. What are the two major water compartments of the human body? Describe each one and give specific examples.

6. How are body fluid compartments measured?
7. How are ions in body fluids measured? Give one example.
8. What is Avogadro's number?
9. Explain the Gibbs-Donnan rule.
10. Which force plays the greatest role in the movement and distribution of water into the various compartments of the body? Explain your answer.
11. Correlate the movement and distribution of water between the intracellular and extracellular compartments with the concentration of osmotically active particles.
12. How is fluid distribution affected if 2 liters of distilled water are ingested?
13. What happens to water movement and distribution if 200 ml of 10 percent sodium chloride is ingested?
14. Suppose 1 liter of an isotonic solution is infused by intravenous injection. What effects would this have on water movement and water distribution?
15. What are the main differences between passive and active processes in moving substances across plasma membranes?
16. Define and give an example of each of the following: diffusion, bulk flow, osmosis, filtration, and dialysis.
17. Of the active processes, define and give an example of each of the following: active transport, mediated transport, endocytosis, pinocytosis, and phagocytosis.

Chapter 9

THE FUNCTIONING NERVE CELL

STUDENT OBJECTIVES

After reading this chapter, you should be able to

- Describe the function of the nervous system in maintaining homeostasis
- Classify the organs of the nervous system into central and peripheral portions
- Contrast the histological characteristics and functions of neuroglia and neurons
- Categorize neurons by shape and function
- Describe the necessary conditions for the regeneration of nervous tissue
- Explain some basic electrical principles
- Correlate these basic principles with the sequence of events involved in the initiation and transmission of a nerve impulse
- Describe the sodium and potassium pump
- Discuss in detail action potentials and the all-or-none law
- List the factors that determine the speed of impulse transmission
- List the factors involved in the conduction of an action potential across the synaptic gap
- Compare the mechanisms and functions of the excitatory postsynaptic potential with those of the inhibitory postsynaptic potential

- Describe summation
- Define the role of acetylcholine in the transmission of an action potential across a synaptic gap
- Describe the roles and locations of other chemical transmitters such as norepinephrine, serotonin, dopamine, histamine, gammaaminobutyric acid, and glutamate
- List several factors that affect impulse transmission

It is ironic that as we are learning about the functions of the human nervous system, we are employing the nervous system itself for that very purpose. In addition, when we are aware or conscious of our learning, we can experience one or several of a myriad of emotions ranging from exhilaration to fear or anger. An incredible evolutionary product, the human nervous system is probably the last and greatest frontier of our understanding. It functions to control, integrate, and maintain all the conscious and subconscious activities necessary for life. In simplest terms, the job of the nervous system is to communicate.

For the trillions of cells of the body to survive, the internal environment must be properly maintained. The nervous system operates to detect changes in this environment and to bring about appropriate cellular responses to restore the environment to the optimal state. In addition, the nervous system keeps the human organism aware of the external environment. At a basic level, for example, humans must be aware of and be able to respond to courtship. We also must be alert to danger and be able to secure and consume food. The life of a human being is clearly filled with varied and rewarding experiences, but the rewards of being human are obtained only by the efficient operation of an exquisitely designed nervous system.

NERVOUS SYSTEM

The nervous system is the control center and communications network of the body. In humans it performs three broad functions. First, it stimulates movements that are vital to life as well as movements that simply make life easier and more enjoyable. Second, it shares responsibility with the endocrine system for the maintenance of homeostasis. Third, it allows us to express uniquely human traits. Human life simply cannot exist without a functioning nervous system. For instance, skeletal muscle cells cannot contract until they are stimulated by a nerve impulse. If the intercostal muscles and diaphragm do not contract, we cannot breathe. If the digestive glands are not stimulated to release their secretions, food cannot be digested. Even if skeletal muscles and glands could contract and secrete by themselves, we could not live very long without our nervous system. This is because of the second great function of the nervous system — keeping the body in homeostasis, that is, the maintenance of a constant internal environment. The nervous system continually *senses* changes that occur both inside the body and in the external environment. It then analyzes all its information and decides on a course of action. This property is called *integration*. After deciding which action to take, it causes a *response* to occur by sending impulses to the appropriate muscles and glands. The third broad function of the human nervous system is to provide the uniquely human pleasures of thinking, feeling, and acting upon our thoughts and feelings. It is primarily this third function that provides human beings with our unique qualities that separate us from higher animals. In all probability, only we possess the capacity to have an awareness of fellowship and of self, we are the only species with a sense of past, present, and future of self and of our kind. This attribute of the human species appears to be a function of the enlarged portion of the nervous system known as the brain. There is a two-way flow of information about our internal and external world that is collected so that corrective actions can be made.

ORGANIZATION

The nervous system may be divided into two principal portions: the central nervous system and the peripheral nervous system (Figure 9-1).

The **central nervous system (CNS)** consists of the brain and spinal cord. All body sensations must be relayed from receptors to the central nervous system if they are to be interpreted and acted on. All the nerve impulses that stimulate muscles to contract and glands to secrete must also pass from the central system. The central nervous system constitutes the major portion of the nervous system and functions as the control center for the entire system. It processes incoming information; it initiates responses and sends them to the correct effector organ, that is, the organ that responds. In the central nervous system, the so-called higher functions such as learning and emotions also take place.

The various nerve processes that connect the brain and spinal cord with receptors, muscles, and glands comprise the **peripheral nervous system (PNS).** The peripheral nervous system may be classified into an afferent system and an efferent system. The *afferent system* consists of nerve cells that convey information from receptors in the periphery of the body to the central nervous system. These nerve cells, called afferent (sensory) neurons, are the first cells to pick up incoming information. The *efferent system* consists of nerve cells that convey information from the central nervous system to muscles and glands. These nerve cells are called efferent (motor) neurons. The efferent system is subdivided into a somatic nervous system and an autonomic nervous system.

The *somatic nervous system* or *SNS* (*soma* = body) consists of efferent neurons that conduct impulses from the central nervous system to skeletal muscle tissue. Since the somatic nervous system produces movement only in skeletal muscle tissue, it is under conscious control and therefore voluntary.

The *autonomic nervous system* or *ANS*, by contrast, contains efferent neurons that convey impulses from the central nervous system to smooth muscle tissue, cardiac muscle tissue, and glands. The autonomic system produces responses only in involuntary muscles and glands. Thus the autonomic system is usually considered to be involuntary. With few excep-

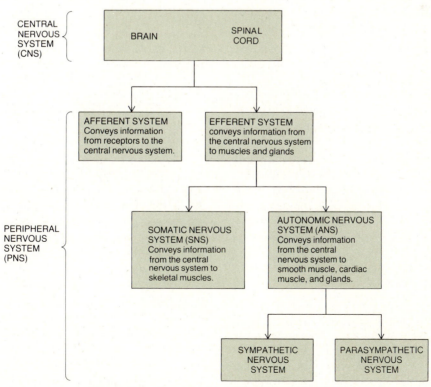

Figure 9-1 Organization of the nervous system.

tions, the viscera receive nerve fibers from the two divisions of the autonomic nervous system: the sympathetic division and the parasympathetic division.

HISTOLOGY

Despite the organizational complexity of the nervous system, it consists of only two principal kinds of cells. The first of these, the neurons, make up the nervous tissue that forms the structural and functional portion of the system. Neurons are highly specialized for impulse conduction and for all special functions attributed to the nervous system: thinking, controlling muscle activity, regulating glands. The second type of cell, the neuroglia, serves as a special supporting and protective component of the nervous system.

Neuroglia

The cells of the nervous system that perform the functions of support and protection are called **neuroglia** or **glial cells** (*neuro* = nerve; *glia* = glue). Many of the glial cells form a supporting network by twining around the nerve cells in the brain and spinal cord. Other glial cells bind nervous tissue to supporting structures and attach the neurons to their blood vessels. A few glial cells also serve specialized functions. For example, many nerve fibers are coated with a thick, fatty sheath produced by a particular type of neuroglia. Certain small glial cells are phagocytotic. They protect the central nervous system from disease by engulfing invading microbes and clearing away debris. These various functions performed by the neuroglia are divided among several different kinds of glial cells. Neuroglia are of clinical interest because they are a common source of tumors of the nervous system.

It has been estimated that in the central nervous system, about 90 percent of the cells are neuroglial cells. However, the neuroglial cells are so much smaller than neurons that overall, they comprise only about half of the total volume of the nerve tissue. Neuroglial cells are divided into: (1) astrocytes, (2) oligodendrocytes, and (3) microglia. The ependymal cell, a special kind of epithelial cell, is also considered one of the glial types.

Astrocytes

The astrocytes (*astro* = star; *cyte* = cell), with diameters of 8 to 10 μm, are the largest of the neuroglial cells. They are star-shaped cells with numerous long processes. Some of the processes have a terminal expansion or *pedicle* that attaches to the walls of blood vessels. There are two types of astrocytes found in the central nervous system, *protoplasmic,* found in the gray matter, and the *fibrous,* found in the white matter of the brain and spinal cord (Figure 9-2a). The astrocytes twine around nerve cells and form a supporting network. Some investigators think that these cells participate in the exchange of fluids, nutrients, wastes, oxygen, and carbon dioxide between the neurons and the blood or cerebrospinal fluid.

Oligodendrocytes

Oligodendrocytes (*oligo* = few; *dendro* = tree) are 6 to 8 μm in diameter and have fewer processes than the astrocytes (Figure 9-2b). They are found in gray and white matter. In gray matter, they cluster around the cell body of a neuron and are sometimes called satellite cells. It has been suggested that a symbiotic relationship exists involving transfers of RNA between glial and neuronal cells. In white matter, the oligodendrocytes lie adjacent to nerve fibers, and the processes extend toward and wrap around the nerve fibers. This layered wrapping is called *myelin* and serves to insulate the neuron and speed up nerve impulse conduction.

Microglia

The cell bodies of microglia (*micro* = small) are about 3 μm in diameter, dense, and elongated. Microglia have short processes that possess numerous small side branches giving them a prickly appearance (Figure 9-2c). These cells are not very numerous but are found associated with the blood vessels of nervous tissue. Normally stationary, they can become amoeboid and migrate to injured areas to serve as phagocytes within the central nervous system.

Ependymal Cells

Ependymal cells (*ependy* = to put on over, or to cover) line the cavities of the brain and the spinal cord. Therefore, these cells are bathed in the cerebrospinal fluid that fills these cavities. The ependymal cells form a true epithelium consisting of a single layer of closely packed columnar cells. The cells have tapered bases from which long, threadlike processes emerge and branch out into the surrounding nervous tissue. Cilia are present on the free surface during embryonic life, and a few may be present in the adult brain. The ependymal cells form a very selective barrier between the nervous tissue and the cerebrospinal fluid. The composition of the cerebrospinal fluid may be controlled in large measure by the secretory or absorption activity of the ependymal cells.

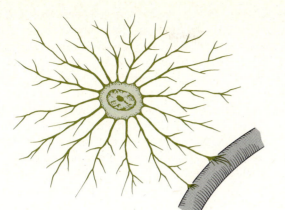
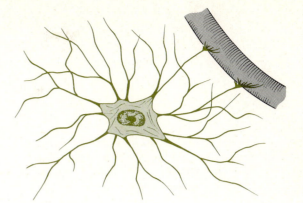

(a)

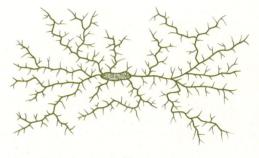

(b)   (c)

Figure 9-2 Neuroglia cells. (a) Protoplasmic astrocyte (left) and fibrous astrocyte (right). (b) Oligodendrocytes. (c) Microglial cell.

Neurons

The *neuron* or *nerve cell* differs greatly from any other cell type in the human body. The most obvious differences relate to shape and size. Neuronal cells have many extensions or processes, some of which can be over 1 m (3 ft) long, making neurons the longest cells in the body. The nerve cell processes make contact with other nerve cells and thus establish very extensive and complicated networks over which electrical impulses can travel. Some nerve cell processes make contact with other cells such as epithelial, glandular, or muscular cells. Not only do these anatomical peculiarities set the nerve cells apart, but their physiological and biochemical properties also make them very different from other cells of the body. For example, the cell membrane of nerve cells has the ability to initiate an electrical discharge called a *nerve impulse* and transmit this electrical event rapidly along its surface to the ends of the longest cellular processes. This property of *conduction* is highly developed. At the terminations of the long extensions, a potent chemical is released that can serve to transmit the signal to the adjacent cell, whether it be another neuron, a secretory cell, or a muscle cell. By means of such contacts, neurons are constantly receiving and conducting messages from and to all parts of the body.

Structure

A neuron consists of three structurally and functionally distinct portions: (1) the cell body, (2) dendrites, and (3) an axon (Figure 9-3). The *cell body* or *perikaryon* contains a well-defined nucleus and nucleolus surrounded by a granular cytoplasm. Within the cytoplasm are typical organelles such as mitochondria, Golgi apparatus, and endoplasmic reticulum. Also located in the cytoplasm are structures characteristic of neurons: Nissl bodies and neurofibrils. *Nissl bodies* are orderly arrangements of granular (rough) ER and free ribosomes whose function is protein synthesis. Newly synthesized proteins pass from the perikaryon into the neuronal processes, mainly the axon, at the rate of about 1 mm (0.0394 inch)/day. These proteins replace those lost during metabolism and are used for growth of neurons and regeneration of peripheral nerve fibers. *Neurofibrils* are long, thin fibrils composed of microtubules. They may assume a function in support.

The cell body receives a stimulus from the *dendrites,* which are highly branched extensions of the

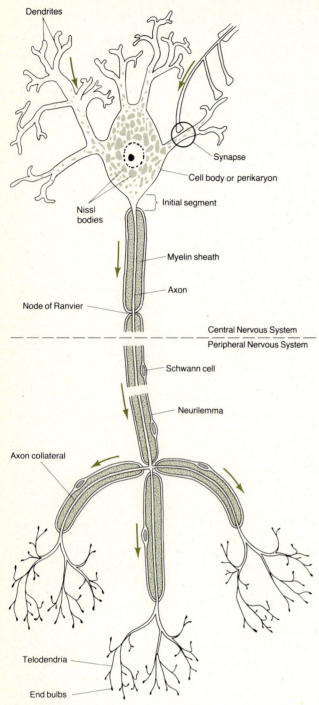

Figure 9-3 labels:
Dendrites

Synapse

Cell body or perikaryon

Nissl bodies

Initial segment

Myelin sheath

Axon

Node of Ranvier

Central Nervous System

Peripheral Nervous System

Schwann cell

Neurilemma

Axon collateral

Telodendria

End bulbs

Figure 9-3 Motor neuron. Arrows show the direction of the nerve impulse.

cytoplasm of the perikaryon. Their essential function is to pick up a stimulus and transmit it to the perikaryon. Nerve cells have large numbers of dendrites that provide the nerve cell with a large receiving area. A single neuron can make contact with great numbers of terminal processes called *end bulbs* from other nerve cells. The Purkinje cell, a special type of nerve cell found in the cerebellum, has been estimated to make contact with up to 200,000 end bulbs (Figure 9-4). Cells with only one dendrite (bipolar neurons) are rare and are found in special sites such as in the middle of the retina. Dendrites become thinner as they progressively branch.

The *axon*, or *nerve fiber*, by contrast, is a single, highly specialized, and relatively long process that conducts impulses away from the perikaryon to another neuron or to an organ of the body. Axons vary tremendously in length. Some of the axons inside the brain may be a half-inch in length. Those that run between the spinal cord and toes may be over 1 m (3 ft) long. Axons also vary in diameter, and as you will see later, this variation is related to the rate of impulse conduction. Axons contain one or more side branches called *axon collaterals*. The axon and its collaterals each terminate by branching into many fine filaments called *telodendria*. The termination of the telodendria are usually expanded into bulblike structures, the *end bulbs* or *boutons terminau*, which are important in the transmission of information to the next cell in the chain. The first portion of an axon plus the part of the cell body where the axon is joined is referred to as the *initial segment*. The cell membrane of the axon is often termed the *axolemma* (*lemma =*

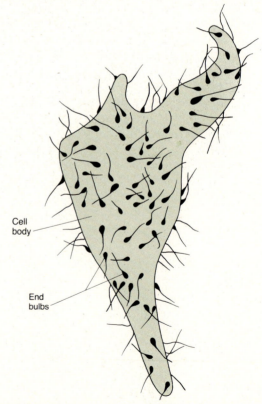

Cell body

End bulbs

Figure 9-4 End bulbs terminating on the body of a nerve cell.

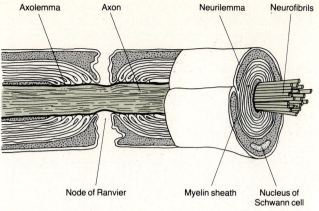

Axolemma Axon Neurilemma Neurofibrils

Node of Ranvier Myelin sheath Nucleus of Schwann cell

Figure 9-5 Cross section and longitudinal section through a myelinated axon.

the myelin sheath around axons outside the central nervous system. Oligodendrocytes make up a similar sheath on some of the axons inside the central nervous system. Each Schwann cell produces a segment of the sheath by first encircling an area of the axon until the ends of the encircling cell meet and overlap (Figure 9-6). The cell then winds itself around the axon a second or third time. The cytoplasm of the Schwann cell is gradually squeezed into a very thin layer that just separates the opposite sides of the cell membrane. The result is that the axon becomes surrounded by several double layers of cell membrane. The number of cytoplasmic layers in this concentric wrapping of the myelin sheath varies from six to eight, to nearly 50. Between the segments of sheath created by the Schwann cells are unmyelinated gaps called the *nodes of Ranvier* (see Figure 9-5). An electron micrograph of the myelin sheath is shown in Figure 9-6e. The function of the myelin layer is to insulate the axon. Never fibers that are wrapped in the sheath are called *myelinated axons*. Those that are not protected by the sheath are called *unmyelinated axons*. The *neurilemma* is the outermost layer of Schwann cell that now contains most of the cellular contents that have been squeezed to this region by the wrapping process. It contains most of the Schwann

husk or sheath) and its contents is called the *axoplasm*. Axons have a constant diameter along their entire length to the level of the telodendria.

Figure 9-5 shows a cross section of an axon. The axons of many neurons are surrounded by a *myelin sheath*, a covering that is made mostly of a lipid and shows interruption along its length. The myelin sheath is composed of flattened cells, called *Schwann cells*, that encircle the nerve fiber. Schwann cells become

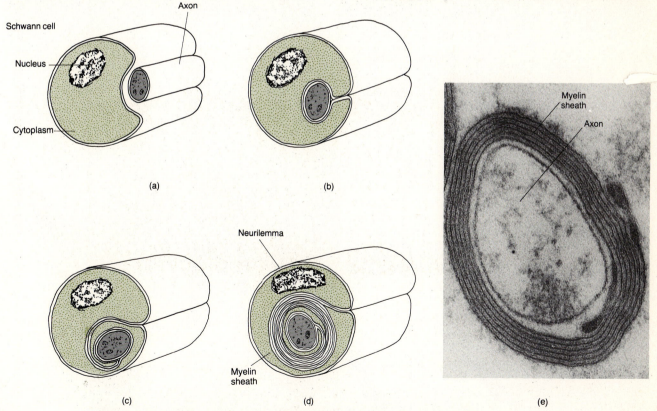

Schwann cell

Nucleus

Cytoplasm

Axon

(a)

(b)

Neurilemma

Myelin sheath

(c)

(d)

Myelin sheath

Axon

(e)

Figure 9-6 Myelin sheath. (a)–(d) Myelin formation in a peripheral nerve fiber. (e) Electron micrograph of a myelin sheath at a magnification of 111,000×. (Courtesy of F. Van Hoof, Université Catholique de Louvain.)

cell cytoplasm, with mitochondria, Golgi apparatus, ribosomes, and the nucleus. The essential function of the neurilemma is to assist in the regeneration of injured axons.

Classification

The many different kinds of neurons found in the body are classified according to structure and function. The structural classification is based upon the number of processes extending from the cell body (Figure 9-7). For example, *multipolar neurons* have several dendrites and one axon. Most neurons in the brain and spinal cord are of this type. *Bipolar neurons* have one dendrite and one axon and are found in the retina of the eye and in the inner ear. The third structural type of neuron is a *pseudounipolar neuron,* so called because it appears to have only one process extending from the cell body. The single process divides into a central branch, which functions as an axon, and a peripheral branch, which is an axon in structure but functions as a dendrite. Pseudounipolar neurons originate in the embryo as bipolar neurons, but during development, the origins of the axon and dendrite fuse into a common single process.

The functional classification of neurons is based upon the direction in which they carry impulses. *Sensory neurons,* also referred to as *afferent neurons,* are involved in the *reception* of stimuli from the external and internal environment. Impulses generated by receptor organs of the body are carried into the central nervous system over sensory neurons. *Motor neurons,* also referred to as *efferent neurons,* control a variety of effector organs such as exocrine and endocrine glands and muscles. Impulses departing the central nervous system travel over motor neurons on their way to the effector organ. Other neurons called *interneurons, internuncial,* or *association neurons* are located within the central nervous system and make the connections between the sensory and motor cells. These connections most often establish

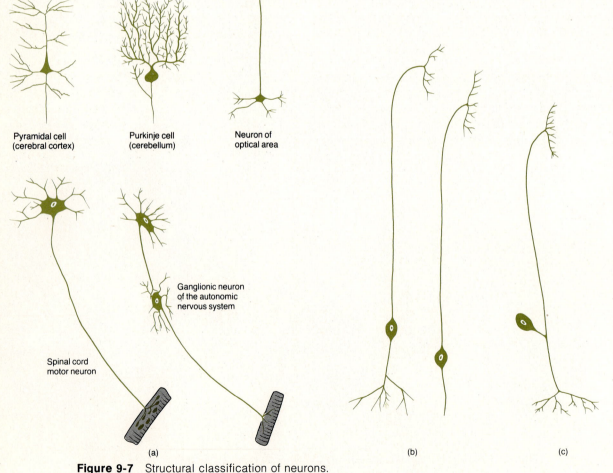

Pyramidal cell
(cerebral cortex)

Purkinje cell
(cerebellum)

Neuron of
optical area

Ganglionic neuron
of the autonomic
nervous system

Spinal cord
motor neuron

(a)

(b)

(c)

Figure 9-7 Structural classification of neurons.
(a) Multipolar. (b) Bipolar. (c) Pseudounipolar.

very complex circuits that distribute the sensory input to several motor neurons at different levels of the central nervous system so that at times the response to a simple stimulus results in a very complex response.

Nerve fibers outside of the central nervous system that travel to the same location are collected together and packaged to form a *nerve*. The individual fibers in a nerve retain their identity because they are held apart from each other by an elaborate packing of connective tissue called the *endoneurium* (Figure 9-8). Bundles of individual fibers and their packaging are gathered together into *fasciculi* which, in turn, are separated from each other by additional connective tissue called the *perineurium*. The number of fasciculi present within a nerve varies according to the size of the nerve. The fasciculi and all the component nerve fibers and connective tissues are completely enclosed within a tough outer sheath of connective tissue called the *epineurium. Spinal nerves* arise from the central nervous system in the region of the spinal cord, and the *cranial nerves* arise from the central nervous system from the bottom of the brain. The spinal and cranial nerves are large nerves that contain both myelinated and unmyelinated fibers. Many of these large nerves are said to be *mixed nerves* because they possess both sensory and motor fibers.

In the central nervous system, the nerve cell bodies or perikaryons are isolated or collected into regions termed the *gray matter*. The so-called *white matter,* on the other hand, excludes the perikaryons of nerve cells and consists of the neuronal processes, the dendrites and axons. The perikaryons of neurons of the peripheral nervous system are clustered or gathered together in usually ovoid structures covered with dense connective tissue called *ganglia.*

PHYSIOLOGY

Two striking features of nervous tissue are its limited ability to regenerate and its highly developed ability to send electrical messages called nerve impulses.

Regeneration

Unlike the cells of epithelial tissue, neurons have a limited ability to regenerate. Before or around the time of birth, the perikaryons of most nerve cells lose their ability to divide and multiply. This means that if a nerve cell dies, it cannot be replaced by the division of some other cell.

In the central nervous system, if a process is severed, the possibility for regeneration of a new process is very limited. If it occurs, it does so by growth through synthetic reactions occurring in the healthy perikaryon. In the peripheral nervous system, the process or fiber can be regenerated quite readily if the perikaryons are intact. When a single nerve cell dies, those cells that connect to it remain intact and healthy.

In contrast to the nerve cells, the glial cells of the central nervous system and the Schwann cells of the peripheral nervous system are capable of mitotic division. Often, when a nerve cell dies, these cells divide and occupy the vacant space left by the dead neuron.

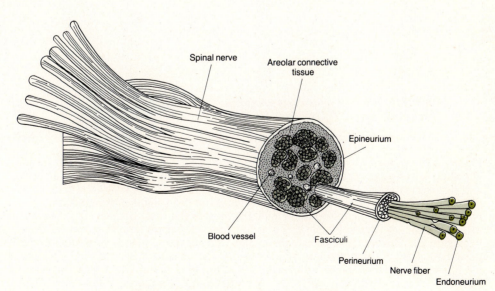

Figure 9-8 Coverings of a peripheral nerve.

The wide and vast distribution of nerves makes them very susceptible to injury. When nerves are cut, the proximal segment, which is that portion still connected with the perikaryon, frequently can regenerate a new process. The distal portion, separated from the perikaryon, dies, degenerates, and is eventually removed by the tissue phagocytes.

In the process of regeneration, the proximal stub of the cut nerve regresses temporarily, waits until the degenerating distal section is cleared away, and then begins to grow again (Figure 9-9). Meanwhile, the Schwann cells divide and increase in number to form a solid cylinder of cells connecting to the effector. As the nerve fiber stub grows, branches are directed toward the cylinder of Schwann cells. When one branch pierces and enters the column of Schwann cells, it continues to grow, and guided by the Schwann cells, reaches the effector. If the gap between the proximal and distal segments is great or if the distal segment is removed completely as in an amputation, the newly

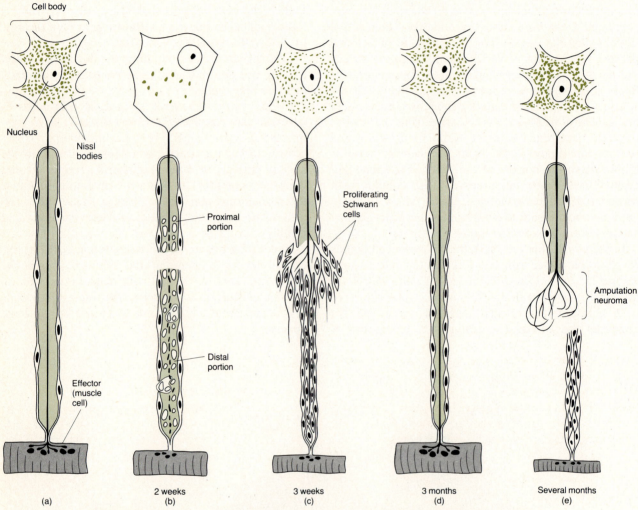

Figure 9-9 Process of nerve regeneration. (a) Normal nerve fiber connected to an effector (skeletal muscle cell). Note the location of the nucleus of the cell and the amount and distribution of the Nissl bodies. (b) When the nerve fiber is severed, the distal end degenerates and the debris is removed by phagocytic cells. In the proximal portion, the nucleus moves to the periphery of the cell and most of the Nissl bodies melt away into the cytoplasm. (c) Schwann cells begin to proliferate and form a solid column of cells into which the growing proximal stub penetrates. The effector originally innervated by the nerve cell shows considerable disuse atrophy. (d) The column of Schwann cells guides the regenerating nerve cell to the muscle cell at a rate of about 1 to 4 mm/day. When the connection is made, the effector is again activated and grows to its original size. (e) If the growing proximal end of the nerve cell misses the column of Schwann cells, the cell branches into a multitude of small disorganized extensions that may form an amputation neuroma.

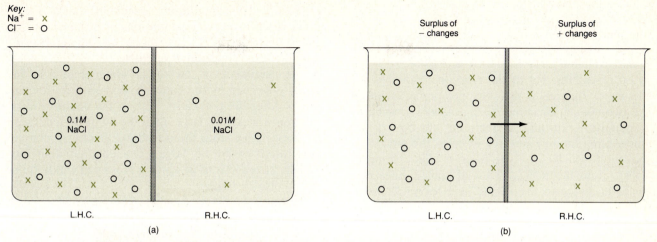

Figure 9-11 Voltage generated by diffusion of unequal concentrations of Na+ and Cl− through a permeable membrane. (a) Initial setup. Two solutions of sodium chloride separated by a membrane freely permeable to both Na+ and Cl− The L.H.C. contains 0.1*M* or ten times the concentration of Na+ and Cl− in the R.H.C. (b) Intermediate stage. After more Na+ than Cl− moves from the L.H.C. into the R.H.C., the R.H.C. has a surplus of positive charges and the L.H.C. has a surplus of negative charges. A voltage is generated.

sion differentials is shown in Figure 9-12. A solution of KCl in the L.H.C. has the same concentration as the NaCl in the R.H.C. In this instance, the concentrations of Cl− on both sides of the membrane are equal, and there is no net diffusion of Cl− from one side of the membrane to the other. Sodium ions, however, diffuse from the R.H.C. into the L.H.C., and potassium ions diffuse from the L.H.C. into the R.H.C. But the rates of diffusion are not identical.

Potassium ions diffuse more rapidly through the membrane than the sodium ions. As a result, the R.H.C. gains more K+ than it loses Na+ and therefore accumulates more positively charged ions than the L.H.C. The L.H.C. has lost more K+ than it has gained Na+ and therefore has an excess of negative charge. The resulting unequal distribution of electrical charges, more positive in the R.H.C. and more negative in the L.H.C., generates a voltage.

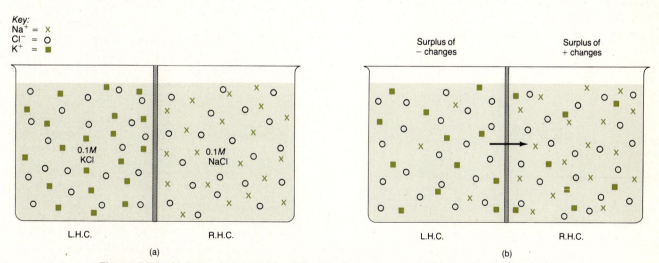

Figure 9-12 Voltage generated by diffusion of equal concentrations of Na+, Cl−, and K+ through a permeable membrane. (a) Initial setup. Two solutions of identical concentration are separated by a membrane permeable to Na+, K+, and Cl− The L.H.C. initially contains 0.1*M* KCl and the R.H.C. contains 0.1*M* NaCl. (b) Intermediate stage. After a time, the greater diffusibility of K+ from the L.H.C. into the R.H.C. causes the L.H.C. to have a surplus of negative charges and the R.H.C. to have a surplus of positive charges. A voltage is generated.

Membrane Permeable Only to Potassium

A third example that approximates the situation of living excitable cells is shown in Figure 9-13. The setup resembles that of the preceding example with the major difference that the membrane separating the two compartments is *semipermeable* or *selective*. It permits K^+ to diffuse freely but not Na^+ or Cl^-. Potassium, the only ion free to cross the membrane, moves down its concentration gradient from the L.H.C. into the R.H.C. This, of course, adds more positive charge to the R.H.C. and leaves an excess of negative charges in the L.H.C. Again, a voltage is generated across the membrane; the R.H.C. is the positive pole and the L.H.C. is the negative pole. To pursue this process to its conclusion, consider that as the K^+ moves from left to right, the R.H.C. becomes increasingly positive and the L.H.C. increasingly negative. As a result, it becomes increasingly more difficult for each successive K^+ to cross into the R.H.C. It would be leaving a negative region to which it would be strongly attracted and would be heading into a positive region where it would be strongly repelled. Potassium ions in the L.H.C. are therefore subjected to two forces; the first drives the K^+ down its concentration gradient into the R.H.C., and the second attempts to retain the K^+ in the L.H.C. due to electrical forces that begin retarding its movement into the L.H.C. At some point the two forces acting on the K^+ become equal. At this time, the movement of K^+ into the R.H.C. is balanced by a return of K^+ into the L.H.C., and no further changes in concentrations will occur. A measurement of the voltage at this time is referred to as the *equilibrium potential*. The equilibrium potential exists when there is no further *net* gain of ions on one side or the other because the forces acting on the freely diffusible ions are equal and opposite.

The magnitude of the equilibrium potential is directly governed by the concentration gradient for the freely diffusible potassium ions. If the concentration difference was very large, then the diffusion tendency is very great and the resulting electrical charge separation would be large. The counteracting but equally strong electrical voltage present at equilibrium would also be large.

Resting membrane voltages generated under these conditions are due almost entirely to the free diffusion of potassium ions and have values around -70 to -80 mV, the inside of the membrane being negative. The values for voltages across living cell membranes use the interior of the cell as the reference side. That is, the *minus* 70 mV is read as: the voltage across the membrane is 70 mV and the *inside* is negative. Similarly, a *plus* 35 mV would be read as: the voltage across the membrane is 35 mV with the *inside* positive.

Membrane Permeable Only to Sodium

In this example (Figure 9-14) suppose that the membrane is freely permeable only to Na^+ and not K^+ or Cl^-. A discussion similar to the preceding one for potassium would follow for sodium. The Na^+ diffuses down its gradient, that is, moves from the R.H.C. into the L.H.C. until the electrical gradient built up has a force equal to that of the concentration gradient. An equilibrium potential can then be measured. Neurons have sodium equilibrium potentials around $+60$ mV.

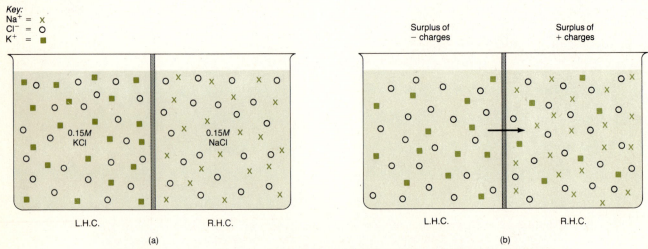

Key:
Na^+ = x
Cl^- = o
K^+ = ■

(a) L.H.C. 0.15M KCl   R.H.C. 0.15M NaCl

(b) Surplus of − charges   Surplus of + charges   L.H.C.   R.H.C.

Figure 9-13 Voltage generated by diffusion of K^+ through a semipermeable membrane. (a) Initial setup. Two solutions of identical concentrations are separated by a membrane permeable to K^+, but not Na^+ or Cl^-. (b) Intermediate stage. After a time, K^+ moves from the L.H.C. into the R.H.C. and this causes the R.H.C. to have a surplus of positive charges and the L.H.C. to have a surplus of negative charges. A voltage is generated.

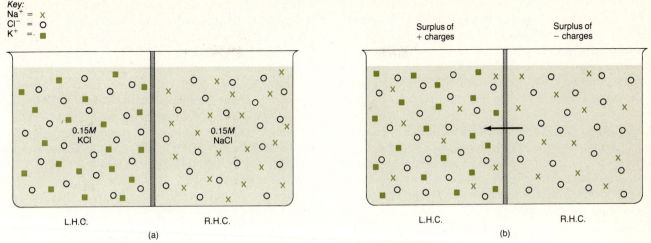

Figure 9-14 Voltage generated by diffusion of Na⁺ through a semipermeable membrane. (a) Initial setup. Two solutions of identical concentrations are separated by a membrane permeable to Na⁺ but not K⁺ or Cl⁻ (b) Intermediate stage. After a time, Na⁺ moves from the R.H.C. into the L.H.C. and this causes the L.H.C. to have a surplus of positive charges and the R.H.C. to have a surplus of negative charges. A voltage is generated.

From the above examples, we can conclude that the equilibrium potential *for each ion* is determined by the selective permeability characteristic of the membrane and by the magnitude of the concentration gradient.

Resting Membrane Potentials of Nerve Cells

The predicted value for the resting membrane potential of nerve cells, when calculated *using the potassium concentrations* across the membrane, is around -87 mV. Actual experimental measurements show the value to be nearer -70 mV. The discrepancy is due to the fact that the cell membrane is not totally impermeable to sodium. Sodium leaks into the cell continually but at a very slow rate. Addition of positive charges to the interior of the cell reduces the magnitude of the potential difference and produces the lower value of about -70 mV for the resting potential. As a consequence of this inward sodium leakage, the membrane voltage is below that of the potassium equilibrium of -87 mV. Therefore, potassium ions continue to diffuse out, trying to achieve the K⁺ equilibrium potential of -87 mV.

Na⁺-K⁺-ATPase ACTIVE TRANSPORT SYSTEM

The continual, although small, leakage of sodium ions inward and the resultant diffusion of potassium ions outward would eventually destroy the ionic gradient and the membrane potential. But this does not occur because of the presence of the *Na⁺-K⁺-ATPase active transport system*, also called the *sodium and potassium pump*. This system operates continually to extrude Na⁺ from the interior of the cell and to replace the sodium ions with K⁺ from the outside of the cell. Originally determined from crab nerves, this pump system has since been observed in a large variety of excitable cells such as brain, nerve, and muscle cells.

The Na⁺-K⁺-ATPase pump is a protein component of the cell membrane that requires energy expenditure by the cell in the form of ATP in order to operate. The enzyme complex has two component parts. One is a large protein molecule that appears to extend through the entire thickness of the membrane. It has receptor sites for K⁺ on its outer surface and receptor

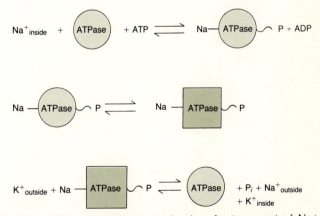

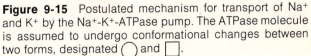

Figure 9-15 Postulated mechanism for transport of Na⁺ and K⁺ by the Na⁺-K⁺-ATPase pump. The ATPase molecule is assumed to undergo conformational changes between two forms, designated ◯ and ▢.

sites for Na⁺ and ATP on its inner surface. A second, smaller, component is a carbohydrate-protein complex whose exact contribution to the transport process is unknown at this time.

The operation of the Na^+-K^+-ATPase pump is believed to be a two-step mechanism that operates as shown in Figure 9-15. On the inner surface, the Na⁺ activates the ATPase, which splits the ATP into $ADP + P_i$. The P_i is retained by the ATPase enzyme, energizing it so that when K⁺ is added to the exterior site, a conformational change in the enzyme molecule occurs. This causes the phosphate to be released, and K⁺ appears inside the membrane while Na⁺ is released on the outside.

ELECTROTONIC POTENTIALS

When a nerve cell preparation is punctured with a microelectrode and a second microelectrode is situated just outside the membrane, a resting membrane potential or voltage is recorded of about −70 mV. This voltage is predictable when the ionic compositions inside and outside the cell are known. Experiments have shown that when the voltage across the membrane is imposed by a pair of clamping electrodes (Figure 9-16), the ionic composition of the fluids inside and outside the membrane change.

In voltage clamping experiments, the tip of one of the *clamping* electrodes is inserted just inside the

nerve membrane and the other electrode is placed in the extracellular fluid surrounding the nerve. A second set of microelectrodes similarly deployed is used for *recording* the voltage across the membrane very near the clamping electrodes. The instrument used to impose the clamping voltage is called a *current generator*. If the clamping electrode inserted into the cell is connected to the *negative* side of the generator, additional negative charges can be added to the inside of the membrane. The recording microelectrode will measure the voltage change that results. Figure 9-17 illustrates these changes. *Hyperpolarization* of the membrane results when the negative charges are added to the interior of the cell. The amount of additional negative charge that can be sustained by membranes is between two and four times the normal resting membrane potential.

A similar experiment can be repeated with the stimulating microelectrode connected to the *positive* side of the generator. In this instance, positive charges are added to the interior and the recording microelectrode will measure the resultant *less negative* value of resting membrane voltage. In this instance the membrane is said to be *depolarized*.

The resultant potential changes recorded in the vicinity of the stimulating electrode are called *electrotonic potentials*. The electrotonic potentials are localized at the region on the membrane where they are imposed. They are not conducted by the nerve. The recorded value decreases quickly as the recording electrode is moved further from the point of stimulus. Every millimeter distant from the point of stimulation decreases the recorded value by about half. For example, if the electrotonic potential is set to −60 mV and the recording electrodes are 1 mm away from the point of application, the recorded voltage will be about −30 mV. If the recording electrodes are 2 mm away from the point of application, the recorded voltage will be about −15 mV, and so on.

ACTION POTENTIALS

If the resting membrane at a given point on the nerve cell membrane is depolarized by addition of positive charges to the interior of the cell so that the membrane potential is reduced to approximately −50 mV, a point of electrical instability is reached, the so-called *threshold*. At this point an abrupt, spontaneous, and large depolarization occurs that causes an actual reversal of the voltage across the membrane. This electrical phenomenon, characteristic of all excitable cells, is called an *action potential* or *nerve impulse*. At the peak of the action potential (Figure 9-18), the interior record-

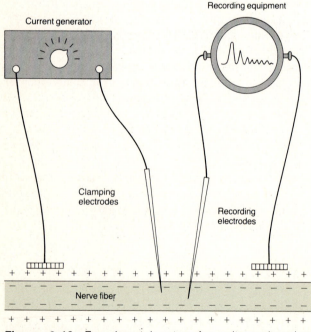

Figure 9-16 Experimental setup for voltage-clamping type experiments.

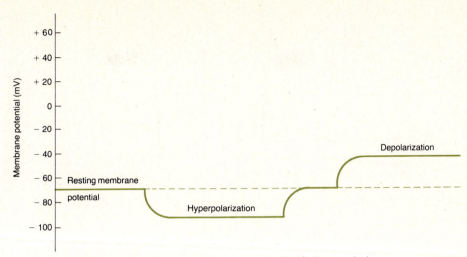

Figure 9-17 Development of hyperpolarization and depolarization.

ing electrode measures voltages around +40 to +50 mV. The peak voltage is reached in about 0.5 milliseconds and is immediately followed by a more gradual decline back to the resting potential. This action potential occurs because of a sequence of rapid changes in the membrane permeability, but only if depolarization reaches the threshold value. The size of

the action potential is never related to the strength of the stimulus. Nerve (and muscle) cells obey what is called the *all-or-none law,* which states that excitable cells, under a given set of conditions, respond to a stimulus with a maximal response or no response at all. An analogy may help in understanding this law. If a long trail of gun powder were spilled along the ground, it could be ignited at one end to send a blazing signal down the entire length of the trail. It would not matter how tiny or great the triggering spark or flame or explosion was; the blazing signal would be just the same, a maximal one (or none at all if it never started).

Latent and Refractory Periods

The time interval between the start of stimulation and the peak of the action potential is known as the *latent period* (see Figure 9-18). Depending on the nerve cell and the ionic conditions, there is another period of time of about 0.4 to 4 milliseconds following the onset of the action potential during which a stimulus, no matter how strong, cannot evoke a second action potential. This interval is called the *absolute refractory period* (Figure 9-19). Immediately following the absolute refractory period there is a *relative refractory period,* the time during which the cell will not respond to a normal-strength stimulus but will respond to a greater-than-normal stimulus (Figure 9-19).

Ionic Flow as a Basis for Action Potentials

During the action potential, membrane permeability changes occur very rapidly and reversibly. During the resting state, the membrane is 50 to 75 times more permeable to K^+ than to Na^+. It is because of this fact that the resting membrane potential is said to be a po-

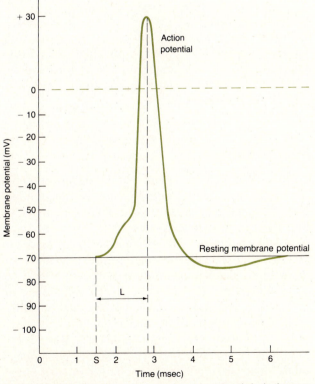

Figure 9-18 Changes in membrane potential during an action potential. At time S, the stimulus is applied. The latent period is shown by time L.

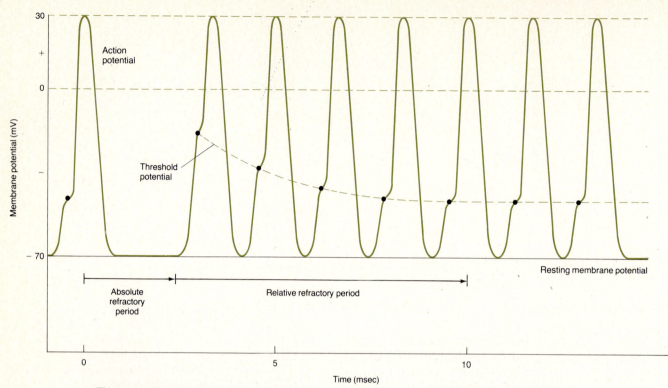

Figure 9-19 Absolute and relative refractory periods. The magnitude of a single, second stimulus necessary to generate a second action potential during the refractory period is greater than the initial stimulus because the threshold potential is farther away from the normal resting potential.

tassium potential, since its diffusion generates the potential difference. When the membrane is stimulated and depolarized to evoke an action potential, the membrane permeability to sodium suddenly increases almost 600 times. Thus, sodium ions move down their gradient into the cell in much greater numbers than do the potassium ions in diffusing outward. With the great influx of positive charges carried by the sodium ions, the inside of the membrane becomes less and less negative until it passes zero and begins to become positive.

In reality, the process produces a peak voltage of about +30 to +40 mV when it suddenly begins to reverse and return to its original resting value. The abrupt reversal occurs for two reasons. First, the 600-fold increase in sodium permeability suddenly ends and returns to its normal value. Second, the permeability to potassium increases from around 50 to 75 times that of sodium to almost 300 times. These relationships are shown in Figure 9-20. The rapid sodium influx slows gradually due to the increasing positivity of the internal environment of the cell, but ends with the change in permeability. The increase in K$^+$ permeability causes a rapid outflow of potassium ions down its gradient. The inactivation of the sodium permeability, combined with the great increase in potassium

permeability, quickly restores the membrane potential to its normal resting value, that is, to approximately the potassium diffusion potential of -85 mV. Often, the action potential shows a prolonged afterperiod of hyperpolarization due to the greater-than-normal K$^+$ permeability.

Recovery Activity

At this point, you may think that during an action potential the shifts of K$^+$ and Na$^+$ are as though floodgates have opened, and the extracellular sodium bursts into the cell in vast quantities while the intracellular potassium pours out into the extracellular space. In fact, an almost negligible amount of extracellular sodium enters the cell, and only about one out of every 1,000,000 potassium ions leaves the cell during an action potential. Such a small shift does not affect the ionic concentrations within and around the cell. However, if thousands of action potentials were transmitted along the nerve, the shifts of ions could eventually reduce the gradients to a point where a membrane potential would be almost nonexistent. This, of course, does not occur because of the continuous pumping activity of the Na$^+$-K$^+$-ATPase transport system.

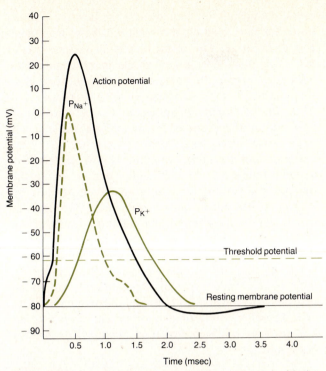

Figure 9-20 Graph showing the action potential with the permeability changes that occur for sodium and potassium superimposed. A local response shows the depolarization when it reaches the threshold. The action potential reaches a positive value of about 25 mV. When the action potential returns toward the normal resting value, it runs past to a more negative value before it slowly returns to the original value. Note the time sequencing of the permeability changes. The great influx of sodium reaches its peak value before the action potential does, and the permeability to sodium is declining when the action potential is at its peak. Potassium begins its rapid efflux just at the moment that the sodium begins to decrease and reaches its peak when the action potential is about half returned to its normal value.

SODIUM PERMEABILITY AND POSITIVE FEEDBACK

Voltage clamping experiments demonstrate that when the membrane is hyperpolarized, the permeability to sodium decreases. When the membrane is depolarized, the permeability to sodium increases. How or what the membrane does to cause these changes is still unknown, but the fact that the permeability does change is of greatest importance. This phenomenon is intriguing if we follow the consequences of the depolarization.

During depolarization, permeability increases and sodium enters more easily. The consequence of the entry of all this sodium is the same as adding *more* *positive* charges to the interior of the cell. This addition, of course, further depolarizes the membrane.

When the membrane is depolarized further, the membrane permeability to sodium increases more, raising further the rate of entry of positively charged sodium ions. The effect of the positive charges is to depolarize further the membrane, and the process snowballs faster and faster as the cycle repeats over and over. Thus the process can become "explosive." The rising phase of the action potential occurs because of the ever-increasing permeability to sodium. While it lasts, the process is an excellent example of a *positive feedback system*. Figure 9-21 summarizes the sequence of events in the positive feedback system. The ever-increasing permeability to sodium (often termed *sodium activation*) is halted abruptly (*sodium inactivation*) by another yet unknown mechanism. The mechanism for the switch to an increased potassium permeability is likewise still a mystery.

IMPULSE CONDUCTION

The function of the nervous system is to communicate with other parts of the body in order to maintain homeostasis. Action potentials are the means by which these communications are completed. These electrical phenomena are transmitted along the neuronal membranes in bursts of varying numbers and frequencies. They become the code for relaying information from one part of the nervous system to another and from the nervous system to the rest of the body. The question to be answered at this point is how is an action potential transmitted along the surface of the mem-

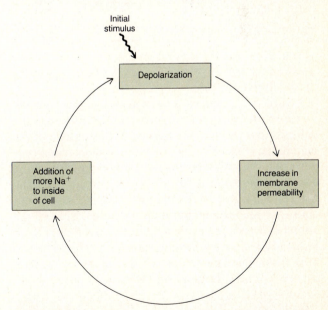

Figure 9-21 Summary of the sequence of events in the positive feedback system that occurs during the sodium activation phase of the action potential.

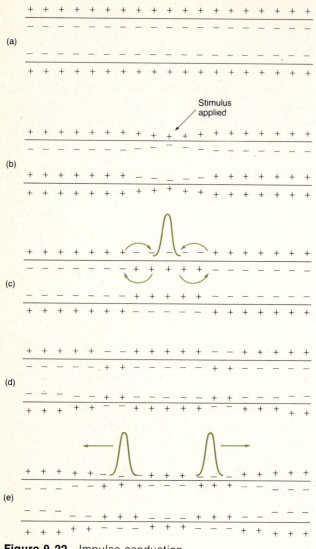

(a)

(b)

(c)

(d)

(e)

Stimulus applied

Figure 9-22 Impulse conduction.

brane the negative region attracts the positive charges on either side, creating a current that flows from positive to negative. The result of this *local current* is to draw the positive charges to the central stimulated area and to leave the adjacent areas depleted of positive charge, that is, in a negative state (Figure 9-22d). Thus, on the outside, the adjacent areas are now negative, and on the inside of the membrane, the adjacent areas are positive. The potential difference *across* the membrane in the adjacent regions is now decreased to a point where an action potential can occur. In this manner, the action potential at any given point on the surface of the membrane creates local currents that cause depolarizations of the adjacent areas of the membrane which trigger other action potentials (Figure 9-22e). The action potentials travel in both directions from the point of stimulation to the end of the nerve membrane. Under normal physiological conditions, neurons are almost always stimulated at one end of the cell. Therefore, only one action potential results, and it is conducted in one direction, toward the other end of the cell.

Conduction Velocity

The speed of impulse conduction along excitable cell membranes is dependent on three main factors: (1) the temperature of the cell, (2) the diameter of the conducting fiber, and (3) the presence or absence of myelin.

When warmed, nerve fibers conduct impulses at higher speeds; when cooled, they conduct impulses at lower speeds. Localized cooling of a nerve can block nerve conduction. Pain resulting from injured tissue can be reduced by the application of cold because the nerve fibers carrying the pain sensations are partially blocked.

The relationship of conduction velocity to the diameter of the nerve fiber is a linear one (Figure 9-23). The nerve fibers have been classified into three main types referred to as the A, B, and C groups. The A

brane? Figure 9-22 shows the series of events leading to impulse conduction.

Figure 9-22a shows a resting membrane with the inside of the nerve cell negative and the outside of the cell positive. A stimulus applied to the surface of the membrane (Figure 9-22b) increases sodium membrane permeability, which results in an action potential at the site of the stimulus. At the peak of the action potential, the cell's interior is positive with respect to the outside. At this instant, the areas on either side of the stimulated part are at their normal potentials. Figure 9-22c shows current flowing from the positive charges now inside the cell to the negative areas adjacent to it.* In a similar fashion, on the outside of the membrane

* The usual convention is to say that electric current flows from positive to negative potentials. This is contrary to the direction in which electrons flow in metallic conductors or in which negative ions flow in an electrolyte solution. Current appears to move with positive ions down their gradient.

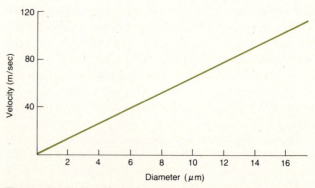

Figure 9-23 Relationship of velocity of nerve impulse conduction to nerve fiber diameter. The velocity varies directly with the diameter of the fiber.

group includes nerve fibers with a diameter of 1 to 20 μm and with velocities of 5 to 100 m/second. These are myelinated fibers and they function either as sensory or motor neurons. The sciatic nerve contains large numbers of such fibers. The B fibers are also myelinated, have diameters of between 1 and 3 μm and conduct impulses at rates of 3 to 14 m/second. These fibers are found only in the preganglionic autonomic nerves (see Chapter 16). The smallest fibers are the C group. They are unmyelinated, have diameters of less than 1 μm, and conduct impulses at a rate less than 2 m/second. They are found in the sensory nerves of the skin that relay touch, pressure, and heat sensations.

Unmyelinated nerves must have large diameters to conduct impulses rapidly. The giant nerve axon of the squid, which is unmyelinated, has a diameter of 600 to 1000 μm and conducts nerve impulses at a rate of 20 m/second. In contrast, a vertebrate myelinated nerve fiber with only a 15 μm diameter can conduct impulses three to four times faster than the giant squid axon. The reason for the greater speed is the difference in the manner of nerve impulse conduction. Myelinated fibers exhibit what has been termed *saltatory conduction* (*saltare* = to jump and dance). The action potential in myelinated nerves appears to jump or leap from one node of Ranvier to the next. The myelin is wrapped around the axon at regular intervals (Figures 9-3 and 9-5), leaving gaps where the axonal membrane is exposed to the extracellular environment. The local currents generated by the action potential extend under the membrane and over the myelin to a point where the next node is depolarized. Therefore, a new action potential results at the next node which, in turn, affects the next node. Thus, the impulse seems to reappear at each successive node down the length of the axon (Figure 9-24). This method of conduction has the advantage of being not only very rapid but also much more economical energetically. The pumping of sodium ions needs to occur only at the nodes.

SYNAPSES

A neuron transmits an action potential from the cell body down the axon to its terminal. This by itself is of no value to the organism. But, when this impulse is used to influence another cell, either to excite or inhibit it, then the possibilities of communication between cells are of great value. The point of contact between the axon terminals of one neuron and another neuron is called a *synapse* or *junction*. The nerve cell that carries the action potential *toward* the synapse is called the *presynaptic* neuron. The neuron that receives or is influenced by the electric impulse is the *postsynaptic* neuron. Figure 9-25 illustrates how a neuron in a long chain of communicating elements can be both presynaptic and postsynaptic to a different neuron. Neuron B is postsynaptic to neuron A but is presynaptic to neuron C.

When the action potential reaches the ends of the axon terminals, it is confronted with a gap of 200 to 1000 Å over to the next neuron. This gap is called the *synaptic gap* or *synaptic cleft*. Though not a great distance, it still offers too much electrical resistance for the small current to jump across. However, before the current disappears, it triggers the release of a powerful chemical from the presynaptic terminal. This chemical, called a transmitter, diffuses rapidly across the synaptic gap and reacts with receptor sites on the membrane of the postsynaptic neuron (Figure 9-26). Depending on its nature, the chemical released can either hyperpolarize the region of the postsynaptic membrane under the synaptic knob (*inhibitory*) or it can slightly depolarize the region under the synaptic knob (*excitatory*).

Unidirectional Transmission

Synapses that release chemical transmitters are *unidirectional* because only the terminals of the pre-

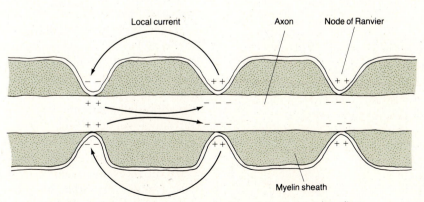

Figure 9-24 Saltatory conduction. Local currents flow around the myelin sheath, leaping from node to node.

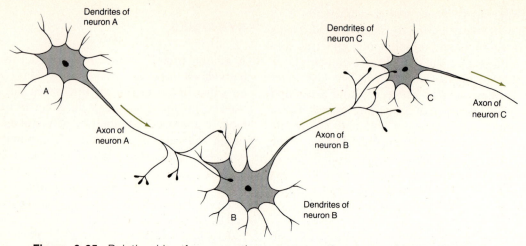

Figure 9-25   Relationship of presynaptic to postsynaptic neurons. Neuron B is postsynaptic to neuron A but is presynaptic to neuron C.

synaptic cell can release chemical transmitters. The direction of transmission can take place in only one direction and continues in the same direction along the postsynaptic fiber. This normal direction is said to be *orthodromic*. If the postsynaptic cell is electrically active, the synapse forms a barrier to any backward or *antidromic* transmission. Chemical transmitters are not released from the dendrites or cell bodies to affect the presynaptic cell. Thus, there is no possible transmission of the electrical signal in the opposite or reverse direction.

Synaptic Delay

Chemically mediated synapses are characterized by a time delay that interrupts the continuous propagation

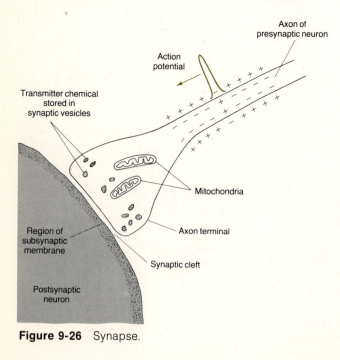

Figure 9-26   Synapse.

of the nerve impulse from one neuron to another. This interruption occurs during synaptic transmission and is called the *synaptic delay*. Synaptic delay includes the time necessary for the release of the chemical transmitter from the axon terminals, the diffusion time across the synaptic gap, plus the time for the transmitter to begin its effect on the postsynaptic membrane. This time usually lasts about 0.5 millisecond.

The Excitatory Postsynaptic Potential (EPSP)

The excitatory postsynaptic potential (EPSP) is the state of near excitation that is present when depolarization of the postsynaptic cell membrane occurs, so the membrane potential is brought *closer* to the threshold value. When the resting potential and threshold values are close, the addition of any other slight stimulus will further depolarize the membrane potential to the threshold level and trigger an action potential. Researchers believe that the chemical transmitters released by nerve endings *increase* the permeability of the postsynaptic membrane to positive ions. These cations flow mostly according to their gradient distribution. Therefore, there is a simultaneous flood of sodium ions into the cell and a small outward movement of potassium. (The membrane mechanism for this change is unknown.) As a consequence of the cation movements, more positive ions enter the cell than leave. This movement depolarizes the postsynaptic cell membrane and brings it closer to the threshold voltage (Figure 9-27). This new lowered voltage is called the *excitatory postsynaptic potential (EPSP)*. This lowering of the membrane voltage toward the threshold occurs only in the small area of the postsynaptic membrane and diminishes with distance from the area of the postsynaptic membrane (Figure 9-28).

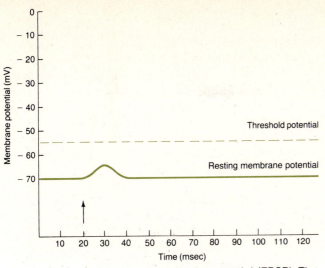

Figure 9-27 Excitatory postsynaptic potential (EPSP). The chemical transmitter increases the permeability of the membrane to positive ions. Sodium ions enter the cell in large numbers, causing the membrane potential to become less positive. The membrane potential thus is moved closer to the threshold value (depolarization).

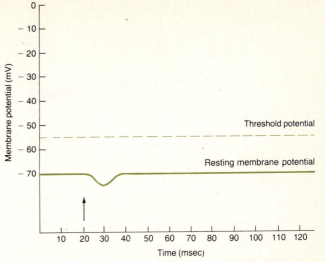

Figure 9-29 Inhibitory postsynaptic potential (IPSP). The result of many inhibitory postsynaptic potentials is to produce a significant depolarization of the membrane. The chemical transmitter increases membrane permeability only to K+ and Cl− ions. The loss of K+ from the cell makes the cell interior more negative. The membrane potential is thus moved farther from the threshold value (hyperpolarization).

The Inhibitory Postsynaptic Potential (IPSP)

The effect of the chemical transmitter released by other presynaptic cells can cause the postsynaptic cell membrane to hyperpolarize. The chemical transmitter released from such presynaptic nerve endings increases the permeability only to potassium and chloride, but not to sodium. The result is that positive charges in the form of potassium leave the cell and make the interior more negative (Figure 9-29). This increased interior negativity hyperpolarizes the area under the postsynaptic membrane and generates an *inhibitory postsynaptic potential (IPSP)*.

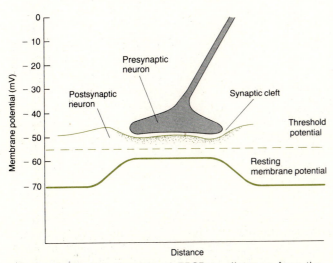

Figure 9-28 Relationship of EPSP to distance from the postsynaptic membrane.

The increase in chloride permeability tends to buffer the membrane, so it remains at the normal membrane resting potential. This buffering action of the chloride becomes important when a given nerve cell is receiving many stimuli from presynaptic nerve endings. With many EPSPs and IPSPs arriving simultaneously, the increase in chloride permeability due to the IPSPs operates to keep the membrane at the normal −70 mV resting level, and thus the membrane resists both depolarization and hyperpolarization.

Summation

At any given instant, it is likely that hundreds of synapses are acting on any single neuron, each one generating a local EPSP or IPSP. The magnitude of voltage change generated by each postsynaptic potential in a motor neuron is about 0.5 mV. Since it requires a total membrane depolarization of about 15 to 20 mV to initiate an action potential, it follows that many EPSPs are required to produce a sufficient depolarization. Also, because each postsynaptic potential only lasts about 100 milliseconds, the timing of EPSPs to arrive simultaneously is essential.

Consider the following experimental setup (Figure 9-30). The figure depicts a single postsynaptic neuron with three presynaptic connections labeled a, b, and c. Assume that a and c are excitatory inputs and b is an inhibitory input. A microelectrode is inserted into the postsynaptic neuron to record the changes in membrane potential that result as the axons to a, b, and c are stimulated.

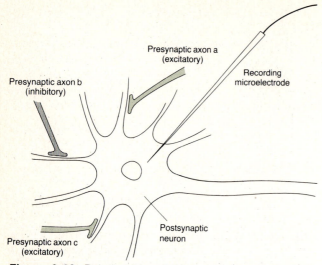

Figure 9-30 Relationship of three presynaptic terminations (a, b, c) to a postsynaptic neuron.

In the first experiment, we stimulate only excitatory axon a by applying stimuli closer and closer together in time. Figure 9-31 illustrates the results of the experiment. When the two stimuli to the same axon are applied sufficiently close together, it is possible to generate the second EPSP while the first is still in effect. The EPSPs added together depolarize the membrane much more than either could have alone, perhaps suf-

ficiently to reach the threshold potential and initiate an action potential. Adding together the effects of successive EPSPs generated by the same presynaptic terminal is called *temporal summation.*

In the second experiment (Figure 9-32), axons a and c are stimulated simultaneously. Each axon produces an EPSP at different places on the postsynaptic cell membrane. These EPSPs also summate, but because they are spatially separated, this is termed *spatial summation.*

In a similar experiment, it is possible to show that stimulating axon b hyperpolarizes the membrane (Figure 9-33). If axons a and b are stimulated simultaneously, the resulting EPSP and IPSP tend to cancel each other. Experiments show that inhibitory neurons generate IPSPs that also can be summated temporally and spatially.

The threshold voltage of a neuron varies over the surface of the cell. The beginning portion of the axon plus the region of the cell body that gives rise to the axon is called the *initial segment.* The initial segment possesses a threshold much closer to the resting potential of the cell than do other parts of the cell (Figure 9-34). When summation of the postsynaptic potentials results in an overall depolarization, the region of the initial segment reaches the threshold first and is activated first. An action potential is sent simultaneously down the axon and backward over the cell body,

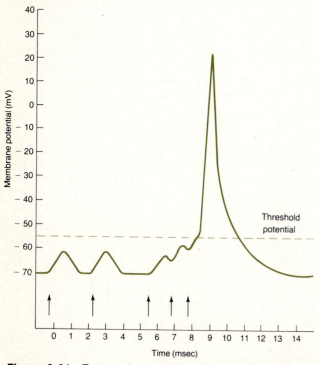

Figure 9-31 Temporal summation. Stimuli applied to the same axon sufficiently close together in time add together to depolarize the membrane. Depolarization may reach the threshold potential, triggering an action potential.

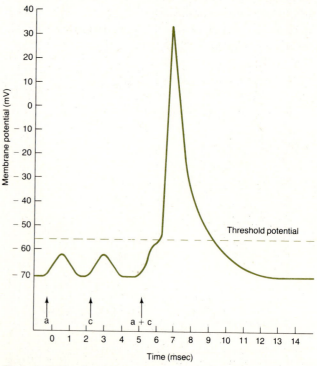

Figure 9-32 Spatial summation. Two axons stimulated simultaneously add together to depolarize the membrane. Depolarization can reach threshold and trigger an action potential.

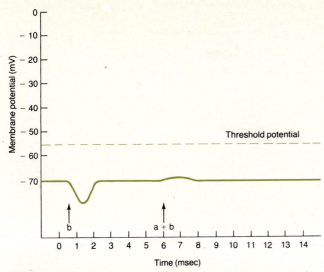

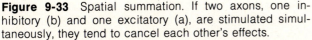

Figure 9-33 Spatial summation. If two axons, one inhibitory (b) and one excitatory (a), are stimulated simultaneously, they tend to cancel each other's effects.

where it overwhelms the many small postsynaptic potentials. Thus the membrane is wiped clean and starts fresh. The action potential is quite short-lived, about 1 to 2 milliseconds, whereas the postsynaptic potentials last about 100 milliseconds. This means that, after the sweep of the action potential, the membrane is still influenced by the lingering postsynaptic potentials. It is often possible that their sum is sufficient to maintain depolarization above the threshold level and thus evoke a second action potential. Indeed, if the summation of the depolarizing EPSPs is great

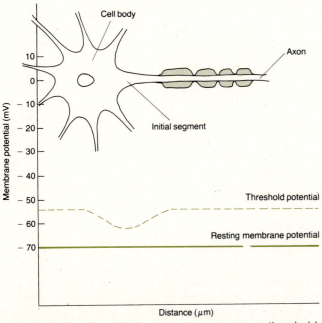

Figure 9-34 The initial segment possesses a threshold closer to the membrane resting potential.

enough, a series of impulses can be sent down the axon.

Also, the synapses closest to the initial segment exert greater influence on the response of the cell than those that end in the dendrites. The anatomical placement of the synapse on the cell body therefore is an important design feature in the control of nerve cell responses.

CHEMICAL TRANSMITTERS

Chemicals that operate as transmitter substances have been extracted from nervous tissue. Their simple chemical structure has enabled many laboratories to synthesize and market the chemicals for use by scientists and people in the health professions. Some of the chemicals thought to function as transmitter substances are described next.

Acetylcholine

In the nervous system, *acetylcholine (ACh)* is synthesized in the terminals of the presynaptic neurons from choline and acetylcoenzyme A. Acetylcholine is stored in synaptic vesicles and released by the arrival of an action potential. Calcium ions are required for the release of the transmitter substance. The ACh diffuses across the synaptic cleft to the receptor sites on the postsynaptic membrane. There, a transmitter-receptor site interaction causes a change in postsynaptic membrane permeability that leads to the production of an EPSP. This process, in turn, can develop an action potential that is transmitted down the axon of the postsynaptic cell (Figure 9-35). Nerve fibers that release acetylcholine from their terminals are called *cholinergic fibers.*

Neuromuscular transmission is mediated by acetylcholine as well as *all* preganglionic fibers of the autonomic nervous system. Postganglionic fibers of the parasympathetic division of the autonomic nervous system also employ acetylcholine as the transmitter substance. In the central nervous system, acetylcholine is distributed in varying amounts in different regions. There is little ACh in the cerebellum and optic nerves; whereas the brain stem, thalamus, and cerebral cortex have moderate levels. The highest concentrations are present in the retina and caudate nucleus.

Norepinephrine

In the peripheral nervous system, *norepinephrine (NE)* is present in the postganglionic fibers of the sympathetic division of the autonomic nervous system. In

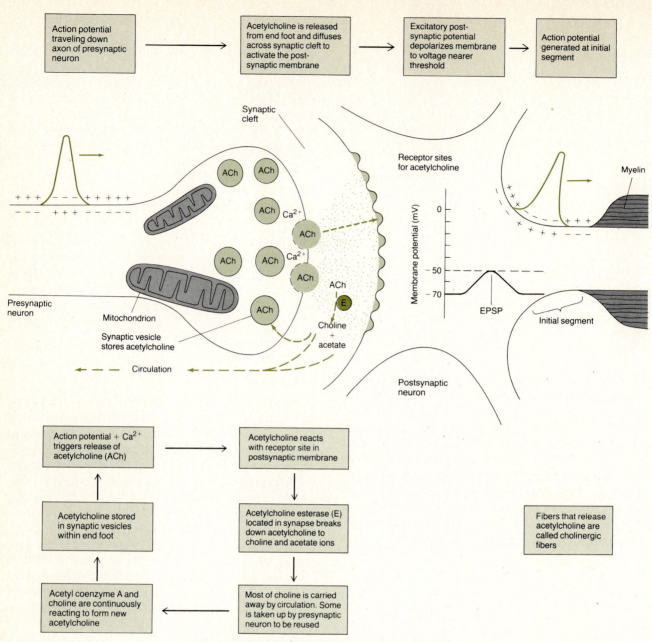

Figure 9-35 Synthesis and action of acetylcholine.

the central nervous system, it is present in moderate concentrations in the medulla oblongata, pons, and midbrain. The highest concentrations are localized in the hypothalamus and superior cervical ganglion. Nerve fibers containing NE are numerous in the olfactory bulb, retina, median eminence, limbic system, and the nuclei of the cranial nerves.

Nerve fibers that release norepinephrine from their terminals are called *adrenergic fibers*. Figure 9-36 illustrates the activities that occur in the adrenergic synapse. It shows that NE is synthesized from the amino acid tyrosine, metabolized through intermediary forms of dihydroxyphenylalanine (DOPA) and dopamine. In contrast to acetylcholine, NE is inactivated very slowly, mainly by diffusing back into the pre-

synaptic terminal for reuse. Small amounts are inactivated by the enzyme *catechol-o-methyl transferase (COMT)* and some is oxidized by the enzyme *monoamine oxidase (MAO)*.

Serotonin

Serotonin, also known as *5-hydroxytryptamine (5-HT)*, is a proposed transmitter substance whose activity is confined to the central nervous system. Highest concentrations are located in the hypothalamus, mesencephalon, and medulla. In the brain stem, high levels of serotonin inhibit the reticular activating system to produce sleep.

Dopamine

Dopamine is one of the intermediates formed in the synthetic pathway of norepinephrine and may function as a transmitter substance in some parts of the brain. It is found in high concentration in the region of the brain called the caudate nucleus.

Histamine

Histamine also occurs in brain tissue with highest concentration in the hypothalamus.

GABA

GABA (gammaaminobutyric acid) is of great interest to neurophysiologists because of its apparent action as an inhibitory transmitter. It is synthesized only in the central nervous system with highest concentrations in the colliculi, diencephalon, and occipital lobes. It seems to be inhibitory by its action in stabilizing membrane potentials near their resting values perhaps by regulating Cl^- permeability in some unknown way.

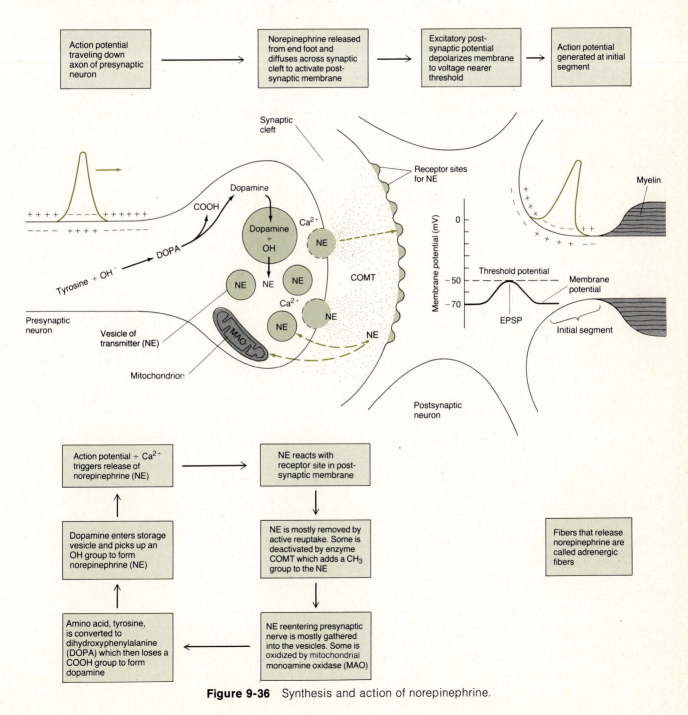

Figure 9-36 Synthesis and action of norepinephrine.

Glutamate and Glycine

Glutamate, an amino acid, seems to have a rapid and powerful excitatory effect on some central nervous system neurons. On the other hand, the amino acid *glycine* acts as an inhibitory transmitter in the spinal cord.

Other Factors that Affect Impulse Transmission

Impulse transmission and conduction may also be inhibited or blocked by a decrease in temperature, by pressure, and by certain drugs. For example, chemical transmitters become inactive at temperatures of 32°F or lower. If excessive or prolonged pressure is applied to a nerve, impulse transmission is interrupted and a part of the body may go to sleep. Removal of the pressure results in the "prickly" feeling when impulse transmission is reestablished. Among the drugs that may alter impulse transmission are anesthetics.

Anesthetics are chemical substances that produce partial or complete loss of sensation. They probably act by altering the permeability of the nerve cell membrane so that it is unable to generate an impulse.

Considerable space and much of your time has been consumed in order to explain in detail the structure, design, and operation of the basic functional unit of the nervous system—the neuron. It is hoped that by truly understanding the mechanism for nerve conduction, and visualizing the fantastic complexity of the possible interconnections between neurons, and the responses possible at all such connections, you have gained a special insight into the grand design and operation of the nervous system. After all, the nervous system is just made up of neurons relaying information to other neurons.

In the next chapter we deal with a second type of excitable cell—the muscle cell. It shares similar membrane phenomena with the nerve cell and indeed only functions when in contact with a viable nerve fiber.

CHAPTER SUMMARY OUTLINE

NERVOUS SYSTEM

1. The nervous system controls and integrates all body activities by sensing changes, interpreting them and reacting to them.

ORGANIZATION

1. The central nervous system consists of the brain and spinal cord.
2. The peripheral nervous system is classified into an afferent system and an efferent system.
3. The efferent system is subdivided into a somatic nervous system and an autonomic nervous system.
4. The somatic nervous system consists of efferent neurons that conduct impulses from the central nervous system to skeletal muscle tissue.
5. The autonomic nervous system contains efferent neurons that convey impulses from the central nervous system to smooth muscle tissue, cardiac muscle tissue, and glands.

HISTOLOGY

Neuroglia

1. Neuroglia are the cells of the nervous system that function to support and protect.

2. Neuroglial cells are divided into: astrocytes, oligodendrocytes, microglia, and ependymal cells.

Neurons

1. Neurons, or nerve cells, are the longest cells in the body. They consist of a cell body or perikaryon, dendrites that pick up stimuli and transmit them to the cell body, and usually a single axon or nerve fiber that conducts impulses away from the perikaryon to another neuron or to an organ of the body.
2. On the basis of structure, neurons are multipolar, bipolar, and pseudounipolar.
3. On the basis of function, sensory (afferent) neurons transmit impulses toward the central nervous system; motor (efferent) neurons transmit impulses to effector organs; and interneurons make the connections between the sensory and motor cells.

PHYSIOLOGY

Regeneration

1. Before or around the time of birth, the perikaryons of most nerve cells lose their ability to divide and multiply.

2. In the central nervous system, if a nerve process is severed, the possibility for regeneration of a new process is very limited.
3. In the peripheral nervous system, the process or fiber can be regenerated readily if the perikaryons are intact.

ELECTRICAL CONCEPTS FOR NEUROPHYSIOLOGY

1. Electrical charges exist in two types, positive and negative.
2. Like electrical charges repel each other, while unlike electrical charges attract each other.
3. Electrons behave like particles that carry negative charges and protons behave like particles that carry positive charges.
4. When oppositely charged particles are separated, they have a potential for doing work. Voltage is a measure of the potential work possible when charges are separated from each other.
5. When a voltage exists and charges move from one point in the system to another, a current is said to be flowing. The ability of the current to flow depends on how much resistance is presented by the intervening substance.
6. Some materials present great resistance to current flow and are used as insulators to prevent current flow. Other materials offer negligible resistance to current flow and are known as conductors.
7. Body fluids are good conductors because ions can transport charges. Lipids, by contrast, cannot carry charges and present high electrical resistance.

TRANSMEMBRANE POTENTIAL

1. The inside of the cell is negatively charged with respect to the outside. The difference in voltage across the membrane is known as the resting membrane potential.

DIFFUSION POTENTIALS

1. Certain ions in solution diffuse more rapidly than others. Each ion carries its own charge, and the unequal distribution of electrical charges generates a voltage.
2. The equilibrium potential for each ion is determined by the selective permeability characteristic of the membrane and by the magnitude of the concentration gradient.

THE SODIUM AND POTASSIUM PUMP

1. The continual leakage of sodium ions inward and the resultant diffusion of potassium ions outward would eventually destroy the ionic gradient and the membrane potential. This does not occur because of the presence of the Na^+-K^+-ATPase active transport system.
2. This system operates continually to extrude Na^+ from the interior of the cell and to replace the sodium ions with K^+ from the outside of the cell.

ACTION POTENTIALS

1. An action potential or nerve impulse occurs when there is depolarization at a given point on the nerve cell membrane. The depolarization can cause an actual reversal of the voltage across the membrane.
2. The all-or-none law states that excitable cells, under a given set of conditions, respond to a stimulus with a maximal response or no response at all.

Latent and Refractory Periods

1. The time interval between the start of stimulation and the peak of the action potential is known as the latent period.
2. The absolute refractory period is the period of time following the onset of the action potential during which a stimulus, no matter how strong, cannot evoke a second action potential.

SODIUM PERMEABILITY AND POSITIVE FEEDBACK

1. The rising phase of the action potential occurs because of an ever-increasing permeability to sodium. While it lasts, the process is an excellent example of a positive feedback system.

IMPULSE CONDUCTION

1. The job of the nervous system is to communicate to maintain homeostasis. Action potentials are the means by which these communications are completed. These electrical phenomena are transmitted along the neuronal membranes in bursts of varying numbers and frequencies.

Conduction Velocity

1. The speed in impulse conduction along excitable cell membranes is dependent on three main factors:

(1) the temperature of the cell, (2) the diameter of the conducting fiber, and (3) the presence or absence of myelin.

SYNAPSES

1. The point of contact between the axon terminals of one neuron and another neuron is called a synapse.
2. When the action potential reaches the ends of the axon terminals, it is confronted with a gap over to the next cell. This gap is called the synaptic gap or synaptic cleft.
3. The small current cannot jump across this gap. However, before the current disappears, it triggers the release of a powerful chemical called a transmitter. The transmitter diffuses rapidly across the synaptic gap and reacts with receptor sites on the membrane of the next cell.

The Excitatory Postsynaptic Potential

1. The excitatory postsynaptic potential is the slight depolarization of the postsynaptic cell membrane that occurs with each quantum of transmitter released. With each depolarization, the membrane potential is brought closer to the threshold value.

The Inhibitory Postsynaptic Potential

1. The inhibitory postsynaptic potential occurs when the area under the postsynaptic membrane hyperpolarizes as a result of increased interior negativity.

Summation

1. When two stimuli to the same axon are applied sufficiently close together, it is possible to generate a second excitatory postsynaptic potential while the first is still in effect. The two potentials add together to depolarize the membrane much more than either could have alone. The addition of several successive EPSPs can be sufficient to evoke an action potential.
2. Adding together the effects of successive excitation postsynaptic potentials generated by the same presynaptic terminal is called temporal summation.

CHEMICAL TRANSMITTERS

1. Various chemicals that operate as transmitter substances have been extracted from nervous tissue.
2. Included among them are the following: acetylcholine, norepinephrine, serotonin, dopamine, histamine, gammaaminobutyric acid, glutamate, and glycine.

Other Factors that Affect Impulse Transmission

1. Impulse transmission and conduction may also be inhibited or blocked by a decrease in temperature, by pressure, and by certain drugs.

REVIEW QUESTIONS

1. How does the nervous system maintain homeostasis?
2. Distinguish between the central and peripheral nervous systems. Relate the terms *voluntary* and *involuntary* to the nervous system.
3. What are neuroglia? List their principal functions.
4. Define a neuron. Diagram and label a neuron. Next to each part list its function.
5. Discuss how neurons are classified by structure and function.
6. Define each of the following: nerve, ganglia, endoneurium, perineurium, epineurium, mixed nerves.
7. What determines neuron regeneration?
8. Explain the phrases "like charges repel each other," and "unlike charges attract each other."
9. Define a coulomb, voltage, and a volt.
10. Correlate the words *current* and *resistance* with the words *insulator* and *conductor*.
11. Outline the main steps in the origin and transmission of a nerve impulse.
12. Define resting membrane potential, threshold, depolarization, latent period, and refractory period.
13. Explain the sodium and potassium pumps in relation to an action potential.
14. What is the all-or-none law?
15. What is the relationship between sodium permeability and a positive feedback system?
16. What factors determine the speed of impulse conduction? What is saltatory conduction?
17. Correlate a presynaptic neuron, a synapse, and a postsynaptic neuron.
18. What events are involved in the transmission of a nerve impulse across the synaptic gap?

19. Compare the excitatory postsynaptic potential with the inhibitory postsynaptic potential.
20. Explain summation.
21. Define the role of acetylcholine in the transmission of an action potential across a synaptic gap.
22. Describe the roles and locations of the chemical transmitters norepinephrine, serotonin, dopamine, histamine, gammaaminobutyric acid, and glutamate.
23. List other factors that affect impulse transmission.
24. Support this statement: "The nerve impulse is the body's best means for rapid correction of a deviation that tends to disrupt homeostasis."

Chapter 10

THE FUNCTIONING MUSCLE CELL

STUDENT OBJECTIVES

After reading this chapter, you should be able to

- List the characteristics and functions of muscle tissue
- Compare the location, microscopic appearance, nervous control, and functions of the three kinds of muscle tissue
- Define epimysium, perimysium, endomysium, tendons, and aponeuroses, and list their modes of attachment to muscles
- Describe the relationship of nerves and blood vessels to skeletal muscles
- Identify the histological characteristics of skeletal muscle tissue
- Describe the physiology of contraction by listing the events associated with the sliding-filament theory
- Describe the biochemistry and molecular events involved in muscle contraction
- Explain the molecular events involved in muscle relaxation
- Describe the source of energy for muscular contraction
- Describe the physiological importance of the motor unit and the neuromuscular junction
- Contrast the kinds of skeletal muscle contractions called isotonic and isometric
- Compare isotonic versus isometric exercise
- Explain skeletal muscle tone and the types of fibers present in skeletal muscle
- Describe the various periods of a myogram
- Define summation of muscle contraction, tetanus, and treppe
- Contrast cardiac muscle tissue with visceral muscle tissue
- Compare oxygen debt and heat production as examples of muscle homeostasis
- Define fibrosis, fibrositis, muscular dystrophy, and myasthenia gravis as homeostatic imbalances of muscle tissue
- Compare spasms, cramps, convulsions, fibrillation, and tics as abnormal muscular contractions

When biologists list the minimum requirements for a living system, the ability to move *(motility)* is always included. Unicellular organisms such as the amoeba can move, and slime molds such as *Physarum polycephalum* are motile. Recent research strongly suggests that the basic machinery responsible for the motility and the shortening of parts or contractility appeared very early in the evolution of living systems. Basic contractile elements persist today in all eucaryotic animal cells in only slightly modified forms.

In higher forms of animals, the contractile machinery is elaborated and organized to its highest degree in the muscle cell. Every muscle cell has the ability to shorten its length by converting the chemical energy stored in special molecules into mechanical work. When large numbers of muscle cells are organized into functional organs termed *muscles,* a synchronized contraction of the muscle cells produces a strong, purposeful movement. The contractions of muscle cells are coordinated and controlled in such a way that they occur at a time that benefits the animal. The control and coordination of these movements are a major function of the nervous system.

The nervous system regulates the activity of different types of muscle tissue, each of which has properties suited to a particular function. For example, if pieces of muscle tissue from the arm and intestine were exchanged and connected to the existing nervous system, neither would be effective in its new location. Muscle tissue of the arm would not contribute to moving food through the intestine, and the intestinal muscle tissue would not aid in moving the arm.

CHARACTERISTICS AND FUNCTIONS

Although bones and joints provide leverage and form the framework of the body, they are not capable of moving the body by themselves. Motion is an essential body function that is made possible by the contraction and relaxation of muscles. Muscle tissue constitutes about 40 to 50 percent of the total body weight and is composed of highly specialized cells having four striking characteristics. One of these features, *excitability (irritability),* is the ability of muscle tissue to receive and respond to stimuli. Stimuli are changes in the external or internal environment strong enough to initiate nerve impulses. A second characteristic of muscle tissue is *contractility,* the ability to shorten and thicken (contract) when a sufficient stimulus is received. Muscle tissue also exhibits *extensibility,* which means that it stretches (extends) when pulled. Many skeletal muscles are arranged in opposing pairs. While one is contracting, the other is extending. The final characteristic of muscle tissue is *elasticity,* the ability of muscle tissue to return to its original shape after contraction or extension. Through contraction, muscle tissue performs three important functions: (1) motion, (2) maintenance of posture, and (3) heat production.

The most obvious kinds of motions of the body are walking, running, and moving from one place to another. Other movements, such as grasping a pencil or nodding the head, are limited to one or more parts of the body. These kinds of movement rely on the integrated functioning of the bones, joints, and muscles that are attached to the bones. Less noticeable kinds of movement produced by muscles are the beating of the heart, the passage of food through the intestines, the contraction of the gallbladder to release bile, and the contraction of the urinary bladder to expel urine.

In addition to performing the function of movement, muscle tissue also enables the body to maintain posture. The contraction of skeletal muscles holds the body in stationary positions, such as standing and sitting.

The third function of muscle tissue is heat production. Skeletal muscle tissue contractions produce heat and are important in maintaining normal body temperature.

Not all contractile processes are confined to muscle cells. Inside *every* cell there is movement of parts. Nuclei revolve, chromosomes separate and move to the poles of a dividing cell, cilia that project from the surface of certain cells beat rhythmically to paddle fluids across their surfaces, and certain cells crawl slowly through tissue spaces and engulf foreign materials and worn-out cells.

KINDS

Three types of muscle tissue are present in the human body: skeletal, visceral, and cardiac. Each is categorized by location, microscopic striations, and nervous control.

Skeletal muscle tissue is attached to bones and moves parts of the skeleton. Skeletal muscle tissue is composed of long, cylindrical cells that are between 10 and 100 μm in diameter and reach lengths of almost 30 cm. It is also called *straited* muscle tissue because striations, or bandlike structures, are visible when the tissue is examined under a microscope. It is a *voluntary* muscle tissue because it can be made to contract by conscious, or voluntary, control.

Visceral muscle tissue is involved with processes related to maintaining the internal environment. This type of muscle tissue is found in the wall of the intestines, blood vessels, the uterus, and the urinary bladder. The contractions of smooth muscle cells are not usually under conscious control but are regulated by the autonomic nervous system and by the local chemical environment. Thus, they are usually involuntary. Contractions of smooth muscle cells are much slower than those of the skeletal or cardiac cells. The structure and operation of smooth muscle tissue is not as well understood as the other two types, although much of the biochemistry seems to be identical. Smooth muscle cells are nonstriated, long, spindle-shaped cells about 5 to 10 μm in diameter and range from 30 to 200 μm in length.

Cardiac muscle tissue is present only in the wall of the heart and is named because of its location. Cardiac muscle tissue is also striated and usually involuntary.

In summary, muscle tissues are classified in the following ways: skeletal, striated, voluntary muscle; cardiac, striated, and usually involuntary muscle; and smooth, visceral, and usually involuntary muscle tissue.

SKELETAL MUSCLE TISSUE

To understand the fundamental mechanisms of muscle movement, some knowledge of its connective tissue components, its nerve and blood supply, and its histology, or microscopic structure, is essential.

Connective Tissue Components

Skeletal muscles are protected, strengthened, and attached to other structures by several connective tissue components. The entire muscle is usually wrapped with a substantial quantity of fibrous connective tissue called the *epimysium* (Figure 10-1). The epimysium is an extension of deep fascia, a dense connective tissue. When the muscle is cut in cross section, invaginations of the epimysium are seen to divide the muscle into bundles called *fasciculi* or *fascicles*. These invaginations of the epimysium are called the *perimysium*. Perimysium, like epimysium, is an extension of deep fascia. In turn, invaginations of the perimysium, called *endomysium*, penetrate into the interior of each fascicle and separate each muscle cell. Endomysium is also an extension of deep fascia. The epimysium, perimysium, and endomysium are all continuous with the connective tissue that attaches the muscle to another structure, such as bone or other muscle. All three elements may be extended beyond the muscle cells as a *tendon*—a cord of connective tissue that attaches a muscle to the periosteum of a bone. In other cases, the connective tissue elements may extend as a broad, flat band of tendons called an *aponeurosis*. This structure also attaches to the coverings of a bone or another muscle. When a muscle contracts, the tendon and its corresponding bone or muscle are pulled toward the contracting muscle. In this way skeletal muscles produce movement. Certain tendons, especially those of the wrist and ankle, are enclosed by tubes of fibrous connective tissue called *tendon sheaths*. They are lined by a synovial membrane that permits the tendon to slide easily within the sheath. The sheaths also prevent the tendons from slipping out of place.

Nerve and Blood Supply

Skeletal muscles are well supplied with nerves and blood vessels. This heavy innervation and vasculari-

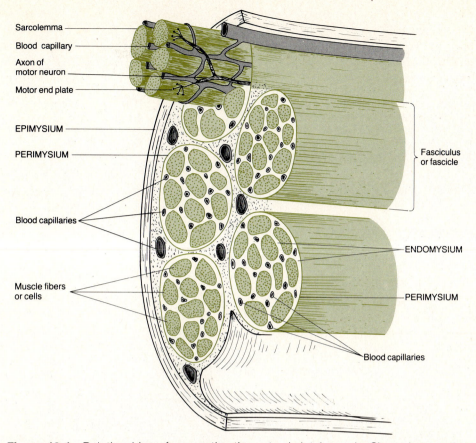

Sarcolemma
Blood capillary
Axon of motor neuron
Motor end plate
EPIMYSIUM
PERIMYSIUM
Fasciculus or fascicle
Blood capillaries
ENDOMYSIUM
Muscle fibers or cells
PERIMYSIUM
Blood capillaries

Figure 10-1 Relationships of connective tissue to skeletal muscle. Shown is a section of a skeletal muscle indicating the relative positions of the epimysium, perimysium, and endomysium. Compare the figure with the photomicrograph in Figure 10-2b.

zation is directly related to contraction, the chief characteristic of muscle. For a skeletal muscle cell to contract, it must first be stimulated by an impulse from a nerve cell. Muscle contraction also requires a good deal of energy—meaning large amounts of nutrients and oxygen. Moreover, the waste products of these energy-producing reactions must be eliminated. Thus muscle action depends on the blood supply.

Generally, an artery and one or two veins accompany each nerve that penetrates a skeletal muscle. The larger branches of the blood vessel accompany the nerve branches through the connective tissue of the muscle. Microscopic blood vessels called capillaries are arranged in the endomysium. Each muscle cell is thus in close contact with one or more capillaries. Each skeletal muscle cell usually makes contact with a portion of a nerve cell.

Microscopic Structure

When a typical skeletal muscle is teased apart and viewed microscopically, it can be seen that the muscle consists of thousands of elongated, cylindrical cells called *muscle fibers* (Figure 10-2a, b). These fibers lie parallel to each other and range from 10 to 100 μm in diameter. They can be as long as 30 cm. The cell membrane of the muscle fiber is known as the *sarcolemma* (*sarco* = flesh; *lemma* = sheath or husk). The sarcolemma encloses the intracellular fluid of the muscle fiber termed *sarcoplasm*. Within the sarcoplasm and lying close to the sarcolemma of each fiber are many nuclei. Filling up most of the interior space are contractile elements termed *myofibrils* (Figure 10-2c). The remaining sarcoplasm contains enzyme molecules and a variety of special high-energy molecules such as adenosine triphosphate (ATP), adenosine diphosphate (ADP), and creatine phosphate (CP). Located at regular intervals along the length of the myofibrils are numerous mitochondria. Muscle cells have a very special endoplasmic reticulum that is referred to as the *sarcoplasmic reticulum*. Its arrangement and role in the contractile mechanism will be described later in this chapter.

Myofibrils

The prefix *myo* means muscle. The *myofibrils* are structures arranged in parallel bundles that extend

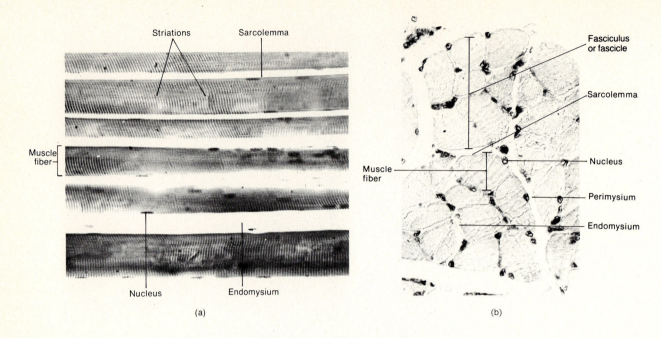

(a)

(b)

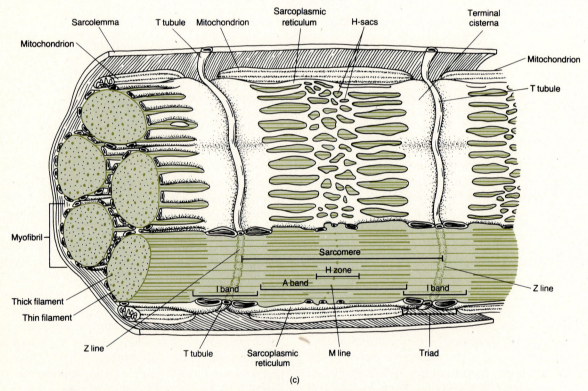

(c)

Figure 10-2 Histology of skeletal muscle tissue. (a) Photomicrograph of several muscle fibers in longitudinal section at a magnification of 640×. (b) Photomicrograph of several muscle fibers in cross section at a magnification of 640×. (c) Enlarged aspect of several muscle fibers based on an electron micrograph. (d) Details of a sarcomere showing thin and thick filaments and various internal zones. (e) Electron micrograph of several sarcomeres at a magnification of 35,000×. (Photomicrographs courtesy of Edward J. Reith, from *Atlas of Descriptive Histology,* by Edward J. Reith and Michael H. Ross, New York: Harper & Row, 1970. Electron micrograph courtesy of D. E. Kelly, from *Introduction to the Musculoskeletal System,* by Cornelius Rosse and D. Kay Clawson, New York: Harper & Row, 1970.)

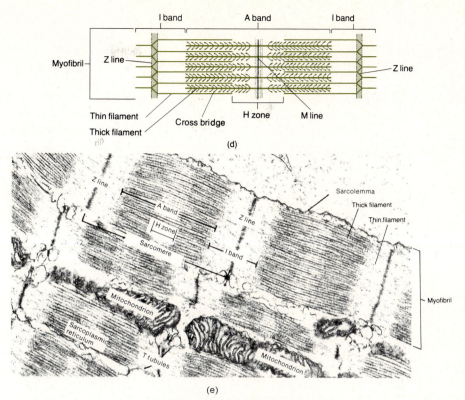

Figure 10-2 (Continued)

lengthwise within the muscle fiber. The myofibrils, like the fiber itself, are cylindrical in shape with a diameter of 1 to 2 μm. The myofibrils run lengthwise in the fiber and show a regular pattern of cross bands or striations that repeat about every 2.5 μm. The repeating unit is termed a *sarcomere* (Figure 10-2c, d). The electron microscope shows that, in skeletal muscle tissue, the banding patterns of adjacent myofibrils are in register. That is, the dark and light regions of the neighboring myofibrils are aligned so that the entire muscle cell takes on the appearance of having cross striations. The light regions of the repeating pattern are referred to as *I bands*. The dark regions are called *A bands*. When a section prepared from resting skeletal muscle is examined with the electron microscope, the A bands measure about 1.6 μm long and the I bands are about 1.0 μm long. A dark, thin line termed the *Z line* divides the I band into two equal sections. In the center of the A band is the *H zone*, an area of intermediate density about 0.5 μm wide. It is also divided into two equal parts by the *M line*. The repeating unit, the sarcomere, extends from Z line to Z line and is about 2.6 μm in length.

Thin and Thick Filaments

When myofibrils are studied with an electron microscope under high magnifications, their finer structure becomes visible (Figure 10-2e). Myofibrils consist of two principal components referred to as thin and thick *filaments*. Anchored in the Z line and projecting in both directions are the *thin filaments*. These thin filaments, extending in parallel array into the I band regions of the sarcomere, are about 6 nm in diameter. Overlapping the free ends of the thin filaments and occupying the A band region of the sarcomere are *thick filaments* with diameters of about 16 nm. In the region of filament overlap, *cross bridges* can be seen projecting from the sides of the thick filament, angled and pointing to the thin filaments. The cross bridges connect the thick filaments to the thin filaments during contraction.

A thick filament that has been isolated and enlarged is shown in Figure 10-3. The cross bridges are arranged in pairs and spiral around the main axis in 120° rotations from the adjacent pair. Each pair is 14.3 nm from the next pair.

If the myofibrils are cut in cross section, the orderly arrangement of thin and thick filaments can be seen (Figure 10-3b). In the region of filament overlap, each thick filament is surrounded by a hexagonal array of thin filaments, and in turn, each thin filament is enclosed by a triangle formed by three thick filaments. Figure 10-3b shows the arrangements of filaments in the overlapping portion of the A band.

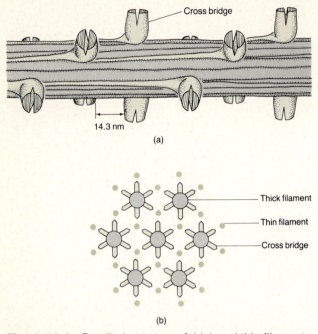

Figure 10-3 Detailed structure of thick and thin filaments. (a) Diagram of a thick filament showing the arrangement and orientation of the protruding cross bridges. (b) Cross section of a fibril illustrating the pattern of filament distribution. Note that each thick filament is surrounded by a hexagonal array of thin filaments.

Variations in Sarcomere Length Sliding-Filament Theory

The length of sarcomeres varies with the degree of muscle contraction (Figure 10-4). During a maximal contraction, the distance between Z lines can be shortened to 50 percent of the resting length. When muscle tissue is stretched, the sarcomere length can increase about another 20 percent of the resting length. The interesting feature of these observations is that the lengths of the thick and thin filaments remain constant. It has therefore been concluded that the interdigitating filaments slide along each other as shown in Figure 10-4. When contraction is taking place, the thin filaments approach each other from opposite ends and meet in the center in the region of the M line. In maximal contractions, the thin filaments may even slip past each other. The sliding process occurs when the cross bridges on the thick filaments angle outward and bond to the neighboring thin filaments. A series of very rapid pulls occurs as the cross bridges make and break bonds with the thin filaments.

Biochemistry of Contraction

The interior of the muscle fiber contains many proteins that are dissolved in its sarcoplasm. These are mostly enzymes that are involved with the early stages of glucose catabolism or breakdown.

The principal molecular elements inside the cell are the proteins that can be extracted from the thick and thin filaments. It has been shown that the thick filaments are composed of a protein named *myosin* and the thin filaments are composed mostly of a protein named *actin*. The protein myosin possesses enzymatic properties that enable it, in the presence of Ca^{2+}, to split a high-energy molecule of ATP into ADP and phosphate. When ATP was shown to be the immediate source for muscle contraction, the exact manner of energy utilization became greatly clarified.

Myosin

The *myosin* molecule is about 160 nm in length with a *globular head* at one end (Figure 10-5). The molecule consists of two so-called *heavy polypeptide chains* that are wound tightly around each other for most of their length. The individual polypeptide chains are the longest polypeptide chains yet found in living systems, each possessing about 1,800 amino acids. At one end of each chain, a free unwound portion is twisted and folded to form a globular head piece. Included in each of the globular head pieces are two additional smaller *light polypeptide chains*. A total of four light chains are thus present in the two globular heads.

Experiments that use enzymes to cut the long myosin molecules into pieces have shown that the ATP-splitting ability is localized in the globular head pieces and is a property of the light chains. In addition, the globular heads have been shown to possess the actin binding ability.

Electron microscopic examinations of the thick filaments have provided us with the information regarding the molecular arrangement of myosin molecules. Each thick filament contains about 400 myosin molecules arranged so that the cross section at any point contains 18 myosin molecules. The globular head pieces project from the filament in a spiral sequence corresponding to the cross-bridge pattern described earlier and shown in Figure 10-3a. Flexible regions, called *hinges,* are located at two places on the myosin molecule (Figure 10-6). These hinges allow parts of each myosin molecule to project from the thick filament and extend or reach toward the thin filament.

Two other binding-type proteins are part of the thick filaments. One is termed the *C-protein* and seems to be laced around the myosin molecules holding the thick filament together at regular intervals along its length. The other protein is located at the

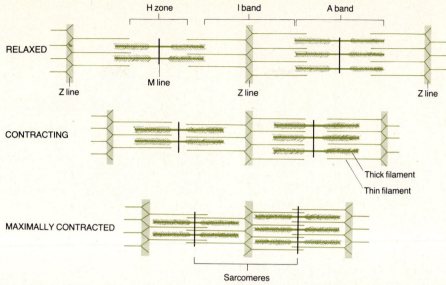

Figure 10-4 Sliding-filament theory of muscle contraction. Shown are the positions of the various parts of two sarcomeres in relaxed, contracting, and maximally contracted states. Note the movement of the thin filaments and the relative size of the H zone.

M line, which also seems to serve a binding function in this region.

Actin

The thin filaments are composed mostly of two strands of *fibrous* or *F-actin* wound around each other to form a double-stranded coil. Each F-actin strand is made up of globular or *G-actin* subunits strung together to form a long chain (Figure 10-7). Another protein, termed *α-actinin*, ties the actin molecules into the two lines. Also comprising the thin filaments are two other important molecules. The first is *tropomyosin*, another double-stranded, twisted molecule containing two different polypeptide chains. The tropomyosin molecules are arranged in a head-to-tail fashion and lie in the groove formed by the twisted F-actin strands. Each tropomyosin, however, is actually in contact with only one of the F-actin strands. The length of the tropomyosin molecule is about 40 nm and is just long enough to cover seven of the G-actin subunits.

Located every 40 nm along the thin filament at the end of each tropomyosin molecule is the second important molecule, *troponin*. Troponin is a large, globular-type protein made of several subunit polypeptides each with a very specific function. One subunit firmly binds two Ca^{2+} ions. Another serves an intermittent inhibitory role that prevents the globular head of myosin from attaching to the thin filament. The third subunit of troponin binds the entire troponin molecule to tropomyosin. Troponin anchors the actomyosintroponin complex to the actin.

Molecular Events in Muscle Contraction

The preceding sections have outlined the relationships of the component parts involved in the molecular contractile process. It is the utilization of the stored energy of ATP and its conversion into a power stroke that enables the cross bridges to pull the thin filaments, stroke by stroke, deeper into the channels between the thick filaments.

The description of the contractile process begins with the muscle fiber at rest; that is, the fiber is not contracting. In this state, the concentration of Ca^{2+} in the sarcoplasm is low, but the level of ATP is high. An examination of the muscle cell at this time with the electron microscope would show that the cross bridges are withdrawn or angled downward, close to the thick filaments. The globular heads are "loaded" or charged with two firmly bound ATP molecules.

In this low-Ca^{2+} environment, the troponin binding sites for calcium ions on the thin filament are vacant. As a consequence, the entire tropomyosin-troponin complex is oriented in such a way that the inhibitory component of troponin covers the potential binding site on the actin for the myosin globular head.

When a nerve impulse arrives at the muscle fiber, a sequence of events releases Ca^{2+} into the sarcoplasm. Free Ca^{2+} immediately binds to the two receptor sites on the troponin, causing a molecular rearrangement within the troponin. The consequent rearrangement causes the inhibitory component to withdraw from the binding site, and the charged globular head of myosin immediately binds to the exposed

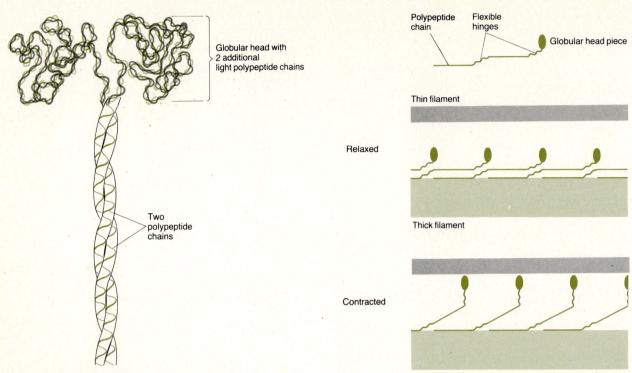

Figure 10-5 Structure of a myosin molecule.

Figure 10-6 Formation and attachment of cross bridges. Note the flexible hinges of the thick filament.

site on the actin. The binding of the myosin head to the actin in turn generates a conformational change in the hinge region behind the head piece. The conformational change in the hinge region constitutes the *power stroke*. The flexing of the head piece changes its angular relationship to the rest of the myosin molecule (Figure 10-8). This causes the thin filaments to slide past the thick filaments, drawing the Z lines closer together. Hence the sarcomere shortens (see Figures 10-4 and 10-6). At the end of the power stroke, the hydrolyzed ATP is released as ADP and P_i. The myosin quickly returns to its unenergized state, angled closely to the thick filament, and is recharged with two new ATP molecules. Repetitive power strokes occur, and the thin filaments continue sliding toward the central M line region. The contraction process ceases and the muscle relaxes when the active recapture and storage of Ca^{2+} into the sarcoplasmic reticulum takes place.

The Sarcoplasmic Reticulum

The endoplasmic reticulum (ER) of striated and cardiac muscle cells is of the smooth type and in these cells is known as the *sarcoplasmic reticulum*. These internal membranes are organized into three different parts. Wrapping around the Z line and spreading over part of the adjacent I band region, the sarcoplasmic reticulum is organized into *terminal cisternae*, which

appear as flattened sacs encircling each myofibril of the muscle cell. In the middle of each sarcomere in the H zone region, the sarcoplasmic reticulum is distributed as an irregular array of tubular channels called the *H sacs*. The third portion of the sarcoplasmic reticulum consists of *tubular channels* running lengthwise over the myofibril connecting the terminal cisternae and the H sacs (see Figure 10-2c).

Transverse Tubules: The T System

A close study of the surface of the muscle cell membrane reveals the openings to many repeating tubular invaginations of the membrane called *T tubules* (see Figure 10-2c). These T tubules protrude straight and deep into the interior of the muscle cell. Inside the cell, they branch into a pattern that encircles every myofibril and form a close association with the terminal cisternae of the sarcoplasmic reticulum.

The sarcolemma of the muscle cell is a polarized membrane similar to that of a nerve cell. When the sarcolemma is depolarized by the action of the nerve impulse, the wave of depolarization sweeps over the surface of the muscle cell and is carried down the T tubules to all the sarcomeres within the muscle cell.

The close association of the T tubules and the terminal cisternae has been examined with the electron microscope, and gap-type junctions seem to be present in the common surface connecting the two

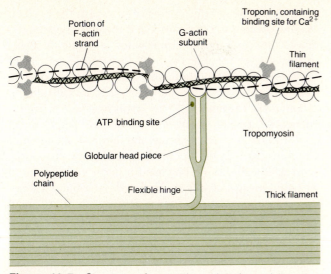

Figure 10-7 Structure of an actin molecule and its relationship to the globular head piece of a thick filament.

systems. The wave of depolarization traveling through the T tubular system is easily conducted through these junctions into the terminal cisternae of the sarcoplasmic reticulum. The arrival of the impulse onto the membranes of the sarcoplasmic reticulum causes a sudden increase in their permeability and a rapid outflow of stored calcium ions. These calcium ions then trigger the sequence of molecular interactions that lead to muscle shortening as described earlier.

Molecular Events in Muscle Relaxation

Shortening of the myofibrils ceases when nerve impulses to the muscle cell stop. At this moment, the sarcolemma restores its normal resting potential with the inside of the cell to about -60 mV. The sarcoplasmic reticulum, relieved of depolarizing currents, returns to its normal impermeable state. The calcium ions free in the sarcoplasm are quickly sequestered or gathered back into the interior of the sarcoplasmic reticulum. The contractile elements are thus deprived of the calcium ions needed for the operation of the trigger mechanism, and relaxation of the muscle results.

Figure 10-8 Power stroke of the myosin cross bridges. (a) The globular head piece of myosin has just been energized by the hydrolysis of ATP into ADP + P_i. Note the approach of the next ATP to be used. (b) The energized head piece makes contact with a G-actin subunit following activation of troponin by Ca^{2+}. (c) Following contact with the G-actin subunit, the myosin head piece undergoes a conformational change so that it displaces the actin molecule to the left (i.e., sliding the thin filament along the thick). (d) the power stroke ends with the release of ADP + P_i. The myosin head piece returns to its upright position and binds the next ATP molecule. The head piece will be again energized by the hydrolysis of the ATP to ADP + P_i.

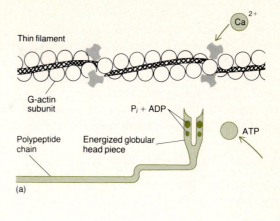
(a)

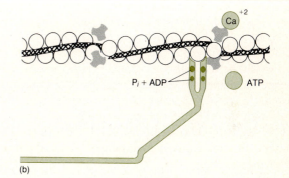
(b)

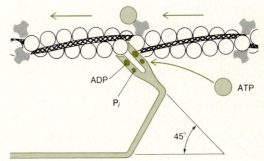
(c)

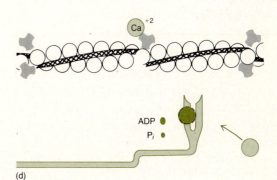
(d)

The transport protein called *calsequestrin* that gathers Ca^{2+} into the sarcoplasmic reticulum expends one ATP molecule for every two calcium ions that it collects. Clearly, then, relaxation also occurs at considerable energy cost. Experimental data indicate that, for one single contraction-relaxation cycle, two of every three ATP molecules used are spent in the contraction phase and one is spent in the relaxation phase.

The Energy for Muscular Activity

The immediate, direct source of energy for muscle contraction is ATP. The utilization of ATP by the myosin head piece has already been outlined. Skeletal muscle fibers function in a discontinuous manner. That is, they respond to work in bursts of activity and then rest for considerable periods. The supply of energy, therefore, must be able to accommodate these sudden bursts of activity. But the stores of ATP are limited and would be used up very quickly. However, a high-energy molecule, *phosphocreatine*, is not only present in muscle cells, but is in concentrations about five times that of ATP. The phosphocreatine serves as an energy source that converts ADP to ATP. The enzyme *creatine kinase* acts rapidly to bring about the following reaction:

$$Phosphocreatine + ADP \rightleftharpoons Creatine + ATP$$

When muscle activity is prolonged and even the supply of phosphocreatine is exhausted, then the source of energy for ATP regeneration comes directly from the catabolism of carbohydrates and fats. During this period of greatly increased catabolism, the uptake of oxygen may increase to 20 times that of resting muscle cell. Another result of the great increase in catabolism is the appearance of lactic acid in the blood. This carbohydrate results from the rapid but incomplete breakdown of glucose as the demand for energy becomes very great. At the end of a prolonged period of muscular activity, rapid and deep breathing continues for some time. The need for extra oxygen, the so-called *oxygen debt,* is required to complete the enzymatic breakdown of lactic acid to carbon dioxide and water. This restores the energy in the forms of ATP and phosphocreatine. This is discussed in detail at the end of the chapter.

Innervation of Skeletal Muscle

Nerve cells that carry impulses into muscle fibers are called *motor neurons.* The number of muscle cells innervated by the terminal feet of a single nerve cell varies greatly. In some instances, a nerve cell can connect to over 100 muscle cells, such as in the large postural muscles in the back. In other instances, a nerve cell may innervate only one or two muscle cells

as, for example, in the extrinsic muscles of the eyeball. Where precise control is needed for fine movements, fewer muscle cells are innervated by a nerve fiber. A nerve fiber together with all the muscle cells it innervates is called a *motor unit.* The degree of a muscle contraction depends on the number of motor units in the muscle responding. The smoothness of a skillful maneuver results from the sequential response of motor units as they reinforce and diminish the amount of contraction in the operating muscle.

Muscle cells, as is the case for nerve cells, respond in an *all-or-none* fashion. Therefore, to obtain a weak muscular contraction only a few motor units need respond. However, the motor unit is so designed that the muscle cells innervated by the nerve fiber are distributed throughout the muscle. Thus, the muscular response that occurs is a general one applied throughout the muscle as opposed to a local, sectional response.

The Neuromuscular Junction

The junction between a motor neuron and a skeletal muscle is known as a *neuromuscular* or *myoneural junction.* When a nerve fiber terminates within a muscle, it branches into many terminal feet, each with its ballooned terminus anchored firmly on the sarcolemma of the muscle cell (Figure 10-9). The portion of the muscle fiber membrane directly under the axon terminal is called a *motor end plate,* and covers a distance about two to three sarcomeres in length. These greatly expanded terminals are filled with many small synaptic vesicles containing the chemical transmitter *acetylcholine.* Viewed with the electron microscope, a very distinct gap of about 200 Å exists between the membranes of the nerve and muscle cells. This is the *synaptic cleft.* Normally filled with some amorphous material, this cleft is responsible for a time delay in the signal transmission from a neuron to a muscle fiber.

A nerve impulse arriving at a terminal foot causes the release of acetylcholine from many of the synaptic vesicles. A single nerve impulse releases one million or more acetylcholine molecules into the synaptic cleft. The molecules act together by binding to receptor sites on the surface of the muscle cell sarcolemma. This produces an *end plate potential* which, in turn, depolarizes the muscle cell membrane sufficiently to reach its threshold and initiate a propagated muscle action potential.

Kinds of Contractions

When a muscle cell receives stimuli and the protein myofilaments of the muscle cell begin to respond, a force is generated which is directed in a fashion

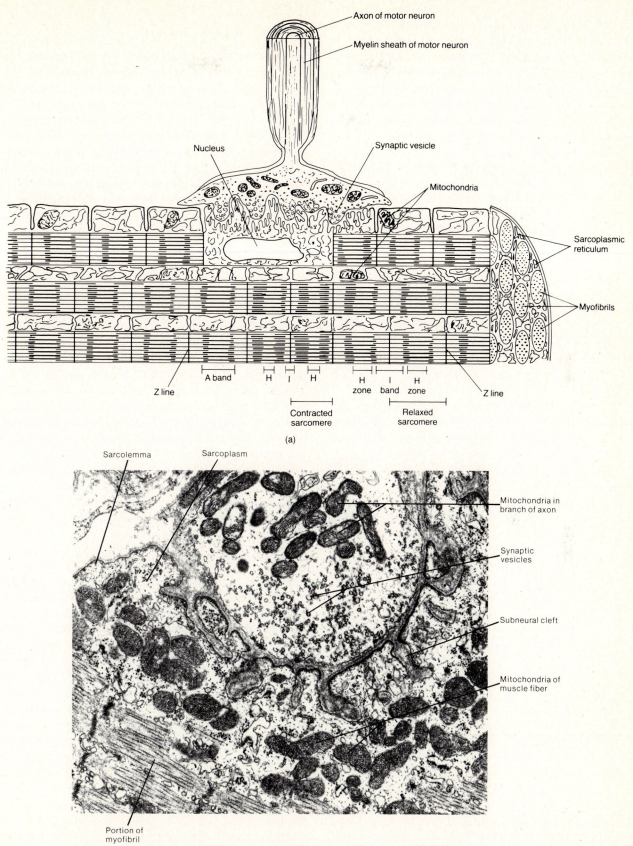

Axon of motor neuron

Myelin sheath of motor neuron

Nucleus

Synaptic vesicle

Mitochondria

Sarcoplasmic reticulum

Myofibrils

A band

H I H

H zone I band H zone

Z line

Contracted sarcomere

Relaxed sarcomere

Z line

(a)

Sarcolemma

Sarcoplasm

Mitochondria in branch of axon

Synaptic vesicles

Subneural cleft

Mitochondria of muscle fiber

Portion of myofibril

(b)

Figure 10-9 Motor end plate. (a) Diagrammatic representation. (b) Electron micrograph at a magnification of 30,000×. (Courtesy of Cornelius Rosse and D. Kay Clawson, from *Introduction to the Musculoskeletal System,* New York: Harper & Row, 1970.)

parallel to the length of the muscle cell. The force generated by many cells acting on some object is called *muscle tension.* The weight of the object is termed the *load.* Tension and load operate in opposite directions. If the tension generated by the muscle is greater than the load, the muscle will shorten and the object will move. Such contractions are called *isotonic contractions* (*iso* = equal; *tono* = tension).

When the load is greater than the tension, the contractile machinery attempts to shorten the muscle length, but is unable to and the object remains unmoved. Such contractions, which generate much tension but no shortening of the muscle, are called *isometric contractions* (*metric* = length).

Isotonic Versus Isometric Exercise

A comparison of the merits of isotonic and isometric training methods as a means of increasing muscular strength is complicated. Experiments designed specifically to compare the imposed work loads maintained during isometric and isotonic work sessions have been unsatisfactory. The direct comparisons are not reliable. However, each method has certain advantages.

Although both types of exercise are able to increase muscular strength in relatively short periods, those studies where direct comparisons were attempted tended to favor isotonic methods. The greatest advantage of isotonic exercise is that the increase in muscular strength is specific to the angle at which the resistance is encountered. Isotonic exercise works all the involved muscles over the entire range of a particular movement. Isometric exercise would require several maneuvers to encompass the same spectrum of muscular involvement.

There is also the psychological advantage of performing isotonic exercises. The performer has some satisfaction in seeing the exercise being accomplished, whereas since isometrics are static, they are less satisfying and often considered boring.

Some experiments indicate that greater muscle enlargement (hypertrophy) and endurance result from isotonic exercise.

The main advantages of isometric exercises are their ease of implementation and time-saving features. Very little equipment is needed, and therefore larger groups of participants work out in a shorter period of time.

One caution needs to be added. It is well documented that the systolic and diastolic blood pressures increase considerably during isometric maneuvers. Therefore, they are a potentially dangerous form of exercise for rehabilitating cardiac patients and for older adults.

Skeletal Muscle Tone

Each of us often stands or sits in an erect posture without any obvious skeletal muscle movements. The maintenance of posture, however, requires the steady, continual contractions of certain muscle fibers. The contractions are identical to those described earlier. The activity of skeletal muscles that are involved with maintaining some posture is called *tonus* or *tone.* Tone is a result of a continual but low-frequency volley of impulses delivered through only a few of the nerve cells to a muscle. The nerve fibers deliver the impulses in an asynchronous fashion so that the activations of the muscle cells are out of phase with each other. This assures a continual steady pull on the tendons rather than a series of surges.

When muscles in the back of the neck are in tonic contraction, they keep the head in the anatomical position and prevent it from slumping forward onto the chest. The term *flaccid* is applied to muscles with less than normal tone. Such a loss of tone may be the result of damage or disease of the nerve that conducts a constant flow of impulses to the muscle. If the muscle does not receive impulses for an extended period of time, its condition may progress from flaccidity to *atrophy,* which is a state of wasting away. Muscles may also become flaccid and atrophied if they are not used. For example, bedridden individuals and people with casts for broken bones may experience atrophy because the flow of impulses to the inactive muscle is greatly reduced. Upon resumption of normal activities, the atrophied muscles recover.

Types of Fibers Present in Skeletal Muscle

Dissection of the calf muscles in an animal or a human would reveal the *red* soleus muscle lying under the *white* gastrocnemius muscle. Skeletal muscles vary in color depending upon their content of a reddish protein called *myoglobin.* White muscle fibers have diameters about two to three times those of red muscle and have fewer mitochondria than the red-type cells. Red muscle is well adapted to maintaining activity over prolonged periods of time, a requirement of postural muscles. White muscles are phasic in nature, which means that they operate with bursts of activity for short periods of time. Red muscles are also supported in their prolonged activity by a rich blood supply.

Experiments have demonstrated that red muscles require fewer stimuli to maintain a sustained contraction than do the white muscles. A useful generalization is that all slowly contracting muscles are of the red type, but not all red muscles contract slowly. For example, the red soleus muscle contracts slowly, but the red masseter muscle that closes the mouth contracts quite rapidly.

Myograms

Figure 10-10 shows the *twitch* response of typical striated muscles to a single adequate stimulus. The time response for the muscles varies with the type of muscle and the temperature. The response to temperature, for instance, is to double the speed for every 10°C rise in temperature. The record of a contraction is called a *myogram*. Note that a brief period of time exists between the application of the stimulus and the beginning of contraction. This period of time, about 10 milliseconds in frog muscle, is called the *latent period*. The second phase, the *contraction period*, lasts about 40 milliseconds and is indicated by the upward part of the recording. The third phase, the *relaxation period*, lasts about 50 milliseconds and is indicated by the downward tracing. The duration of these periods varies with the muscle involved. For example, the latent period, contraction period, and relaxation period for the lateral rectus muscle that moves the eyeball are very short, whereas those of the leg muscles are longer.

A second stimulus applied to a muscle that is already responding to a first stimulus may or may not elicit a second contraction. Whether the second stimulus is effective or not depends on the time interval between the two stimuli. There exists a period of time during which even very strong stimuli are ineffective. This period of unresponsiveness is called the *absolute refractory period*. This time varies from 5 to 50 milliseconds depending on the muscle. The responsiveness of the muscles gradually returns. During this time, stimuli of greater than normal strength can elicit a response. This period is the *relative refractory*

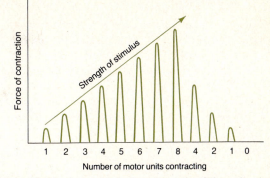

Figure 10-11 Increasing the strength of the stimulus increases the number of motor units responding.

period. Although these times vary from muscle to muscle, the absolute refractory periods and relative refractory periods are much shorter in fast-acting muscles.

Summation of Muscle Contraction

Stronger muscle contractions can be obtained in two ways. One is to increase the number of responding motor units in the muscle. The other is to increase the frequency of contractions by the individual motor units. Figure 10-11 shows the results of stimulating the motor nerve leading to a muscle. As the strength of the stimulus is increased stepwise, the number of motor units responding increases, and thus the strength of muscle contraction also increases. The nerve fibers in the motor nerve have different thresholds. The smaller-diameter nerve fibers are more easily stimulated than those of larger diameter. Therefore, as the strength of the stimulus to the nerve increases, additional larger nerve fibers respond, and of course, the muscle cells they innervate respond and shorten.

Figure 10-12 illustrates the phenomenon called *wave summation*. First a strong stimulus is used to obtain a muscle contraction. Then frequency of stimulation is increased so that the second stimulus arrives before the muscle has relaxed completely. The result is that each subsequent contraction adds to the preceding one. If the stimuli are applied with great rapidity, the additions are smoothed out and a *tetanic contraction* or a state of *tetanus* is obtained. This state of contraction represents the maximum contraction for that muscle.

Treppe

Treppe is a contraction in which a skeletal muscle contracts more forcefully after it has contracted

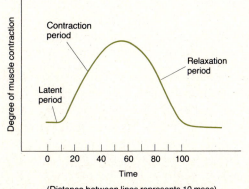

(Distance between lines represents 10 msec)

Figure 10-10 Myogram of a twitch contraction.

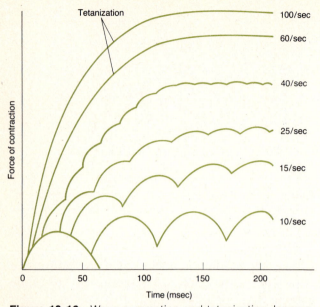

Figure 10-12 Wave summation and tetanization. Increasing the frequency of stimulation from 10/per second through 100/per second shows a greater force of contraction.

several times. Suppose that a series of stimuli are introduced into a muscle. If stimuli are repeated at intervals of 0.5 second, the first few tracings on the myogram will show an increasing height with each contraction. This is treppe, or the staircase phenomenon. It is the principle employed by athletes when warming up. After the first few stimuli, the muscle reaches its peak of performance and undergoes its strongest contraction.

The explanation for treppe is not yet complete. It partially involves the increase in calcium ions inside the muscle cell and perhaps a decrease in the concentration of potassium ions in the sarcolemma. Increased temperature may also be a factor.

CARDIAC MUSCLE TISSUE

The principal constituent of the heart wall is *cardiac muscle tissue*. It has a striated appearance similar to skeletal muscle tissue, but it is involuntary. The cells of cardiac muscle tissue are roughly quadrangular and have only a single nucleus (Figure 10-13). The individual fibers are covered by a thin, poorly defined sarcolemma, and the internal myofibrils produce the characteristic striations. Cardiac muscle cells have the same basic arrangement of actin, myosin, and sarcoplasmic reticulum that is found in skeletal muscle cells. In addition, they contain a system of transverse tubules similar to the T tubules of skeletal muscle. The nuclei in cardiac cells, however, are centrally located

compared to the peripheral location of nuclei in skeletal muscle.

While the groups of skeletal muscle fibers are arranged in a parallel fashion, those of cardiac muscle branch freely with other fibers to form two separate networks. The muscular walls and septum of the upper chambers of the heart (atria) compose one network. The muscular walls and septum of the lower chambers of the heart (ventricles) compose the other network. When a single fiber of either network is stimulated, all the fibers in the network become stimulated as well. Thus each network contracts as a functional unit. The fibers of each network were once thought to be fused together into a multinucleated mass called a syncytium. But it is now known that each fiber in a network is separated from the next fiber by an irregular transverse thickening of the sarcolemma called an *intercalated disc*. These discs strengthen the cardiac muscle tissue and aid in impulse conduction.

Under normal conditions, cardiac muscle tissue contracts rapidly, continuously, and rhythmically about 72 times a minute without stopping. This is a major physiological difference between cardiac and skeletal muscle tissue. Another difference is the source of stimulation. Skeletal muscle tissue ordinarily contracts only when stimulated by a nerve impulse. In contrast, cardiac muscle tissue can contract without extrinsic (outside) nerve stimulation. Its source of stimulation is a conducting tissue of specialized intrinsic (internal) muscle within the heart. About 72 times a minute, this tissue transmits electrical impulses that stimulate cardiac contraction. Extrinsic nerve stimulation merely causes the conducting tissue to increase or decrease its *rate* of discharge. A third difference is that cardiac muscle tissue has a long refractory period that extends into part of the relaxation period. As a result, even though the rate of the heartbeat can be drastically increased, the heart cannot undergo complete or incomplete tetanus. The anatomical and functional characteristics of the heart will be discussed in Chapter 22.

VISCERAL MUSCLE TISSUE

Visceral muscle tissue is a very heterogenous type. Its pattern of distribution, the manner in which it is laid down, and its physiological properties show great variation. For example, visceral muscle tissue is laid down in the form of sheets in structures such as the walls of the intestine and blood vessels. In some situations, single visceral muscle cells are present as myoepithelial cells in some glands. Small groups of visceral muscle cells are organized into a unit that resembles skeletal muscle in the tiny arrector pili

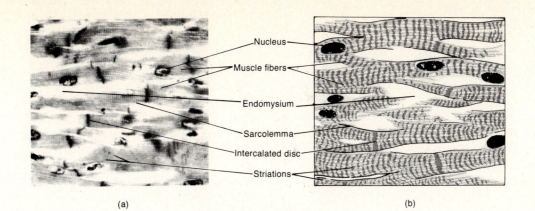

Nucleus

Muscle fibers

Endomysium

Sarcolemma

Intercalated disc

Striations

(a)

(b)

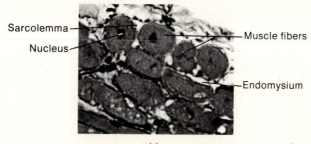

Sarcolemma

Nucleus

Muscle fibers

Endomysium

(c)

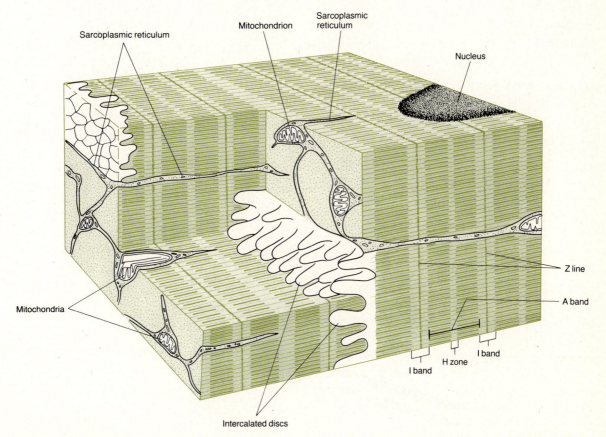

Sarcoplasmic reticulum

Mitochondrion

Sarcoplasmic reticulum

Nucleus

Z line

A band

I band

H zone

I band

Mitochondria

Intercalated discs

(d)

Figure 10-13 Histology of cardiac muscle tissue. (a) Photomicrograph of cardiac muscle tissue in longitudinal section at a magnification of 640×. (Courtesy of Edward J. Reith, from *Atlas of Descriptive Histology*, by Edward J. Reith and Michael H. Ross, New York: Harper & Row, 1970.) (b) Diagram of the photomicrograph. (c) Photomicrograph of cardiac muscle and tissue in cross section at a magnification of 100×. (Courtesy of Victor B. Eichler, Wichita State University.) (d) Diagram based on an electron micrograph.

muscle that produces a "goose bump" in our skin when it shortens. Visceral muscle tissue present in the walls of blood vessels and the vas deferens of the male reproductive system contract only when stimulated by their specific motor nerves. Visceral muscle tissue present in the intestinal walls shows spontaneous activity. The hormone norepinephrine relaxes the visceral muscle tissue in the intestinal wall, but causes the visceral muscle tissue in the blood vessel walls to contract.

Despite these diversities there are three physiological properties that appear to be common to all visceral muscle tissue. First, it contracts slowly and is able to maintain a prolonged contraction with a minimal expenditure of energy. Second, its motor innervation is entirely autonomic. Third, it exhibits muscular tone.

Visceral muscle tissue may be classified into two groups primarily on the basis of its response characteristics. The *unitary* or *visceral smooth muscle* tissue (Figure 10-14a) is characterized by spontaneous activity that arises from period fluctuations in the membrane potential in some key muscle cells. When action potentials are generated, they are quickly distributed uniformly over the entire muscle mass, causing all the cells to contract. The resting potential of the muscle cell is decreased when it is stretched, so the cell becomes more excitable. The excited regions or key muscle cells, therefore, can be anywhere in the muscle mass and can hop from one region to another from moment to moment.

The second group of visceral muscle tissue is the *multiunit muscle* tissue (Figure 10-14a,b). It is usually activated only by means of its autonomic innervation.

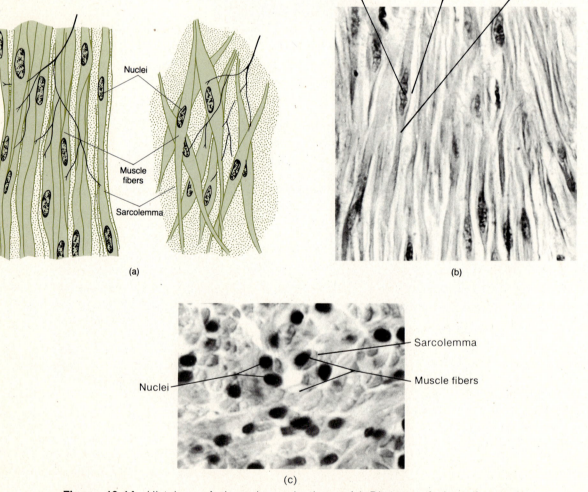

(a)

(b)

(c)

Figure 10-14 Histology of visceral muscle tissue. (a) Diagram of visceral smooth muscle tissue (left) and multiunit smooth muscle tissue (right). (b) Photomicrograph of several visceral muscle fibers in longitudinal section at a magnification of 640×. (Courtesy of Edward J. Reith, from *Atlas of Descriptive Histology,* by Edward J. Reith and Michael H. Ross, New York: Harper & Row, 1970.) (c) Photomicrograph of several visceral muscle fibers in cross section at a magnification of 200×. (Courtesy of Victor B. Eichler, Wichita State University.)

The tissue is composed of cells that function in "motor units." These groups of cells or "motor units" are diffuse and overlap each other considerably. Multiunit muscle tissue is generally less sensitive and unresponsive to being stretched. Multiunit muscle tissue is found in the walls of large blood vessels, in the arrector pili muscles, and in the intrinsic muscles of the eye—the iris, for example.

Discrete nerve endings have not been seen in visceral muscle tissue, and the chemical transmitter seems to be released at various places along the nerve axon. Visceral muscle cells close to the point of chemical release are stimulated directly, whereas those cells further from the nerve axon are activated by the diffusion of the transmitter. There is no neuromuscular junction such as is present in skeletal muscle tissue.

HOMEOSTASIS OF MUSCLE TISSUE

Muscle tissue has a vital role in maintaining the body's homeostasis. Two examples are the relationship of muscle tissue to oxygen and to heat production.

Oxygen Debt

The energy required to convert ADP into ATP is produced by the breakdown of digested foods. The primary nutrient that usually supplies this energy is the sugar glucose. The reaction proceeds as follows:

$$\text{Glucose} \longrightarrow \text{pyruvic acid} + \text{energy}$$

When the skeletal muscle is at rest, the breakdown of glucose is slow enough for the blood to supply sufficient oxygen to participate in the complete catabolism of pyruvic acid to the waste products carbon dioxide and water:

$$\text{Pyruvic acid} + O_2 \longrightarrow CO_2 + H_2O + \text{energy}$$

Reactions such as this that involve oxygen are called *aerobic reactions*.

When a skeletal muscle is contracting and ATP production increases, the breakdown of glucose occurs too rapidly for the blood to supply oxygen for the pyruvic acid to be completely catabolized into carbon dioxide and water. The pyruvic acid is only partly catabolized to lactic acid. The conversion of pyruvic acid to lactic acid proceeds *anaerobically* (without oxygen) as follows:

$$\text{Pyruvic acid} \longrightarrow \text{lactic acid}$$

Most of the lactic acid diffuses from the skeletal muscles and is transported to the liver. There it is eventually resynthesized to glycogen or glucose, which can be reused later. Some of the lactic acid accumulates in the muscle tissue, however. Physiologists suspect that this acid is responsible for the feeling of muscle fatigue. Ultimately, the lactic acid in the skeletal muscle, as well as that which has diffused into the blood, must be catabolized to CO_2 and H_2O. And for this conversion additional oxygen is needed—the **oxygen debt.** When vigorous activity is over, labored breathing continues in order to pay back the debt. Thus the accumulation of lactic acid causes hard breathing and sufficient discomfort to stop muscle activity until homeostasis is restored.

Heat Production

The production of heat by skeletal muscle is an important homeostatic mechanism for maintaining normal body temperature. Of the total energy released during muscular contraction, only 20 to 30 percent is used for mechanical work (contraction). The rest is released as heat, which is utilized to help maintain a normal body temperature.

Heat production by muscles may be divided into two phases: (1) *initial heat,* which is produced by the contraction and relaxation of a muscle, and (2) *recovery heat,* which is produced after relaxation. Initial heat is independent of O_2 and is associated with ATP breakdown. Recovery heat is associated with the anaerobic breakdown of glucose to pyruvic acid and the aerobic conversion of lactic acid to CO_2 and H_2O.

New discoveries always require us to modify and enlarge upon our current knowledge of a subject. Nowhere has this been more true than in our understanding of muscle contraction. The discovery and description of the components of the contractile machine, their functional relationships to each other, the precise point of energy utilization, and the complex trigger mechanism are all recent advances in muscle physiology. Even though this field of study has changed rapidly, we are confident that the information we have presented to you will remain a basis from which future discoveries will be understood.

HOMEOSTATIC IMBALANCES OF MUSCLE TISSUE

Disorders of the muscular system are related to disruptions of homeostasis. The disorders may involve a lack of nutrients, the accumulation of toxic products, disease, injury, disuse, or faulty nervous connections (innervations).

Fibrosis

The formation of fibrous tissue in locations where it normally does not exist is called **fibrosis.** Skeletal and cardiac muscle fibers cannot undergo mitosis, and dead muscle fibers are normally replaced with fibrous connective tissue. Fibrosis, then, is often a consequence of muscle injury or degeneration.

Fibrositis

Fibrositis is an inflammation of fibrous tissue. If it occurs in the lumbar region, it is termed *lumbago*. Fibrositis is a common condition characterized by pain, stiffness, or soreness of fibrous tissue, especially in the muscle coverings. It is not destructive or progressive. It may persist for years or spontaneously disappear. Attacks of fibrositis may follow an injury, repeated muscular strain, or prolonged muscular tension. *Myositis* is an inflammation of muscle cells.

Muscular Dystrophy

The term **muscular dystrophy** applies to a number of inherited myopathies, or muscle-destroying diseases. The word *dystrophy* means degeneration. The disease is characterized by degeneration of the individual muscle cells, which leads to a progressive atrophy (reduction in size) of the skeletal muscle. Usually the voluntary skeletal muscles are weakened equally on both sides of the body, whereas the internal muscles, such as the diaphragm, are not affected. Histologically, the changes that occur include the variation in muscle fiber size, degeneration of fibers, and deposition of fat.

Myasthenia Gravis

Myasthenia gravis is an atrophy of the skeletal muscles. It is caused by an abnormality at the neuromuscular junction that prevents the muscle fibers from contracting. Recall that motor neurons stimulate the skeletal muscle fibers to contract by releasing acetylcholine. Myasthenia gravis is caused by failure of the neurons to release acetylcholine or by the release from the muscle fibers of an excess amount of cholinesterase, a chemical that destroys acetylcholine. As the disease progresses, more neuromuscular junctions become affected. The muscle becomes increasingly weaker and may eventually cease to function altogether.

The cause of myasthenia gravis is unknown. It is more common in females, occurring most frequently between the ages of 20 and 50. The muscles of the face and neck are most apt to be involved. Initial symptoms include a weakness of the eye muscles and difficulty in swallowing. Later, the individual has difficulty chewing and talking. Eventually, the muscles of the limbs may become involved. Death may result from paralysis of the respiratory muscles, but usually the disorder does not progress to this stage.

Abnormal Contractions

One kind of abnormal contraction of a muscle is **spasm:** a sudden, involuntary contraction of short duration. **A cramp** is a painful spasmodic contraction of a muscle. It is an involuntary complete tetanic contraction. **Convulsions** are violent involuntary tetanic contractions of an entire group of muscles. Convulsions occur when motor neurons are stimulated by fever, poisons, hysteria, or changes in body chemistry due to withdrawal of certain drugs. The stimulated neurons send many bursts of seemingly disordered impulses to the muscle fibers. **Fibrillation** is the uncoordinated contraction of individual muscle fibers preventing the smooth contraction of the muscle. Cardiac muscle is particularly susceptible to this abnormality.

A **tic** is a spasmodic twitching made involuntarily by muscles that are ordinarily under voluntary control. Twitching of the eyelid and face muscles are examples. In general, tics are of psychological origin. They tend to develop in young individuals of nervous temperament.

CHAPTER SUMMARY OUTLINE

THE FUNCTIONING MUSCLE CELL

1. In higher forms of animals, the contractile machinery is elaborated and organized to its highest degree in the muscle cell.
2. The nervous system regulates the activity of different types of muscle tissue, each of which has properties suited to a particular function.

CHARACTERISTICS

1. Irritability is the ability of muscle tissue to receive and respond to stimuli.
2. Contractility is the ability to shorten and thicken, or contract.
3. Extensibility is the ability to be stretched or extended.

4. Elasticity is the ability to return to an original shape after contraction or extension.

FUNCTIONS

1. Through contraction, muscle performs the three important functions of motion, maintenance of posture, and heat production.

KINDS

1. Skeletal muscle is attached to bones. It is striated and voluntary.
2. Visceral muscle is located in viscera. It is non-striated and involuntary.
3. Cardiac muscle forms the wall of the heart. It is striated and involuntary.

SKELETAL MUSCLE TISSUE

Connective Tissue Components

1. The entire muscle is covered by the epimysium. Fasciculi are covered by perimysium. Fibers are covered by endomysium.
2. Tendons and aponeuroses attach muscle to bone.

Nerve and Blood Supply

1. Nerves convey impulses to muscle cells, and blood provides nutrients and oxygen for contraction.

Microscopic Structure

1. A skeletal muscle consists of thousands of cylindrical cells called muscle fibers. The fibers contain sarcoplasm, nuclei, and the sarcoplasmic reticulum.
2. The interior of the muscle fiber contains mostly contractile elements called myofibrils. The repeating units of myofibrils are called sarcomeres.
3. When myofibrils are studied with an electron microscope, they are seen to be constructed of thick and thin filaments.

Sliding-Filament Theory of Contraction

1. When contraction takes place, the thin filaments approach each other from opposite ends, and the cross bridges on the thick filaments angle outward and bond to the neighboring thin filaments.

Biochemistry of Contraction

1. The principal molecular elements inside the muscle cell are the proteins myosin and actin.
2. In addition, two other important molecules involved in contraction are tropomyosin and troponin.
3. When a nerve impulse arrives at the muscle fiber, a sequence of events releases calcium ions from the sarcoplasmic reticulum, triggering the contractile process.
4. The wave of depolarization traveling through the T tubule system is easily conducted through gap-type junctions into the terminal cisternae of the sarcoplasmic reticulum.

Molecular Events in Muscle Relaxation

1. Relaxation of a muscle results when the contractile elements are deprived of the calcium ions needed for the operation of the trigger mechanism.

The Energy for Muscular Activity

1. The immediate, direct source of energy is ATP.
2. Phosphocreatine serves as an energy source that converts ADP back to ATP.
3. When muscle activity is prolonged and the supply of phosphocreatine is exhausted, the source of ATP energy comes directly from the catabolism of carbohydrates and fats.

Innervation of Skeletal Muscle

1. A motor neuron transmits the impulse into muscle fibers for contraction.
2. The region of the sarcolemma specialized to receive the stimulus is the motor end plate.
3. The area of contact between a motor neuron and a muscle fiber is a neuromuscular junction.
4. A motor neuron and the muscle fibers it stimulates form a motor unit.
5. When a nerve impulse reaches the motor end plate, the chemical transmitter acetylcholine is released from many synaptic vesicles.

Kinds of Contractions

1. The various kinds of contractions are: isotonic, isometric, tonus, tetanus, and treppe.
2. A comparison of isotonic versus isometric exercise indicates that each method has certain advantages and disadvantages.
3. Skeletal muscles vary in color depending on their content of a reddish protein called myoglobin.

4. Red muscle is well adapted to maintaining activity over prolonged periods of time. White muscle can operate with bursts of activity for short periods of time.
5. A record of a muscle contraction is called a myogram. Three phases of a typical myogram include: the latent period, the contraction period, and the relaxation period.
6. Stronger muscle contractions can be obtained by: (1) increasing the number of motor units in the muscle responding, or (2) increasing the frequency of contractions by the individual motor units.
7. Treppe contraction is the principle employed by athletes when warming up.

CARDIAC MUSCLE TISSUE

1. This muscle is found only in the heart, and is striated and involuntary.
2. The cells are quadrangular and contain centrally placed nuclei.
3. The fibers form a continuous, branching network that contracts as a functional unit.
4. Intercalated discs provide strength and aid impulse conduction.

VISCERAL MUSCLE TISSUE

1. This muscle is found in viscera and blood vessels, and is nonstriated and involuntary.
2. Visceral muscle tissue is laid down in the form of sheets in structures such as the walls of the intestine and blood vessels.
3. Three physiological properties common to all visceral muscle are: (1) it contracts slowly and is able to maintain a prolonged contraction with a minimal expenditure of energy; (2) its motor innervation is entirely automatic; and (3) it exhibits muscular tone.
4. Visceral muscle tissue may be classified into two groups: the unitary or visceral smooth muscle tissue, and the multiunit muscle tissue.

HOMEOSTASIS

1. Muscle tissue has a vital role in maintaining the body's homeostasis. Two examples are the relationship of muscle tissue to oxygen and to heat production.

Oxygen Debt

1. Oxygen debt is the amount of oxygen needed to convert accumulated lactic acid into carbon dioxide and water. It occurs during strenuous exercise and is paid back by continuing to breathe rapidly after exercising.

Heat Production

1. The heat given off during muscular contraction maintains the homeostasis of body temperature.

HOMEOSTATIC IMBALANCES OF MUSCLE TISSUE

1. Disorders of the muscular system are related to disruptions of homeostasis.
2. Fibrosis is the formation of fibrous tissue in locations where it normally does not exist.
3. Fibrositis is an inflammation of fibrous tissue. If it occurs in the lumbar region it is called lumbago. Myositis is an inflammation of muscle cells.
4. Muscular dystrophy is a hereditary disease of muscles characterized by degeneration of individual muscle cells.
5. Myasthenia gravis is a disease exhibiting great muscular weakness and fatigability resulting from improper neuromuscular transmission.
6. Abnormal muscle contractions include: spasms, cramps, convulsions, fibrillation, and tics.

REVIEW QUESTIONS

1. How is the skeletal system related to the muscular system? What are the three basic functions of the muscular system?
2. What are the four characteristics of muscle tissue?
3. How can the three kinds of muscle tissue be distinguished?
4. Define epimysium, perimysium, endomysium, tendon, and aponeurosis. Describe the nerve and blood supply to a muscle.
5. Discuss the microcopic structure of skeletal muscle.
6. Contrast myofibrils with filaments.
7. Correlate the variations in sarcomere length with the sliding-filament theory.

8. In considering the contraction of skeletal muscle, describe the following: biochemistry of contraction; importance of tropomyosin and troponin; molecular events that occur.

9. Define: power stroke, sarcoplasmic reticulum, and the T tubules.

10. Describe the molecular events in muscle relaxation.

11. Where does the energy required for muscle contraction come from? Describe in detail all of the molecules involved in this process.

12. Describe in detail the motor unit and the neuromuscular junction.

13. Define each of the following contractions and state the importance of each: isotonic, isometric, tonus, tetanus, and treppe.

14. Compare isotonic and isometric exercise.

15. Distinguish between red muscles and white muscles.

16. What is a myogram? Describe the latent period, contraction period, and relaxation period of muscle contraction. Construct a diagram to illustrate your answer.

17. Define the refractory period. What is summation?

18. Compare cardiac and visceral muscle with regard to microscopic structure, functions, and locations.

19. Discuss fully each of the following as examples of muscle homeostasis: oxygen debt and heat production.

20. Define fibrosis. What is one of its causes? Define myositis.

21. What is muscular dystrophy?

22. What is myasthenia gravis? In this disease, why do the muscles not contract normally?

23. Define each of the following abnormal muscular contractions: spasm, cramp, convulsion, fibrillation, and tic.

Chapter 11

THE SPINAL CORD AND SPINAL NERVES

STUDENT OBJECTIVES

After reading this chapter, you should be able to

- Correlate the various ways that neural tissue is organized in the central and peripheral nervous systems
- Define white matter, gray matter, nerves, ganglia, tracts, nuclei, and horns
- Describe the principal structural features of the spinal cord
- Identify the structures responsible for protecting the spinal cord
- Describe the various areas of the spinal cord in cross section
- List the major functions of the spinal cord
- Contrast the functions of ascending and descending tracts of the spinal cord
- Identify the reflex center and reflex activities of the spinal cord
- Discuss the components of a reflex arc and its relationship to homeostasis
- List the basic components of a reflex arc
- Explain in detail how sensory information is processed from peripheral receptors
- Contrast the operation of a two-neuron and multineuron reflex arc
- Define a reflex
- Identify the relationship between reflexes and the maintenance of homeostasis
- Classify reflexes on the basis of organs stimulated and location of receptors
- List several clinically important reflexes
- Describe the composition and coverings of spinal nerves
- List the distribution of the 31 pairs of spinal nerves
- Define a plexus
- Define spinal cord injury, sciatica, and neuritis as homeostatic imbalances of the spinal cord and spinal nerves

Before we study the structure and functions of the spinal cord and spinal nerves, it might be helpful to examine the various ways that neural tissue is organized in the central and peripheral nervous systems.

GROUPING OF NEURAL TISSUE

The term **white matter** refers to aggregations of myelinated axons from many neurons supported by neuroglia. The lipid substance, myelin, has a whitish color that gives white matter its name. The gray areas of the nervous system are called **gray matter.** They contain either nerve cell bodies and dendrites or bundles of unmyelinated axons and neuroglia.

A **nerve** is a bundle of fibers located outside the central nervous system. Since the dendrites of somatic afferent neurons and axons of somatic efferent neurons of the peripheral nervous system are myelinated, most nerves are white matter. Nerve cell bodies that lie outside the central nervous system are generally grouped with other nerve cell bodies to form **ganglia** (*ganglion* = knot). Ganglia, since they are made up principally of unmyelinated nerve cell bodies, are masses of gray matter.

A **tract** is a bundle of fibers in the central nervous system. Tracts may run long distances up and down the spinal cord. Tracts also exist in the brain and connect parts of the brain with each other and with the spinal cord. The chief spinal tracts that conduct impulses up the cord are concerned with sensory impulses and are called *ascending tracts*. By contrast, spinal tracts that carry impulses down the cord are motor tracts and are called *descending tracts*. The major tracts consist of myelinated fibers and are there-

fore white matter. A **nucleus** is a mass of nerve cell bodies and dendrites in the central nervous system. It consists of gray matter. **Horns** or **columns** are the chief areas of gray matter in the spinal cord. The term *horn* describes the two-dimensional appearance of the organization of gray matter in the spinal cord as seen in cross section. The term *column* describes the three-dimensional appearance of the gray matter in longitudinal columns. Since the white matter of the spinal cord is also arranged in columns, we shall refer to the gray matter as being arranged in horns.

SPINAL CORD

General Features

The **spinal cord** is a cylindrical structure that is slightly flattened anteriorly and posteriorly. It begins as a continuation of the medulla oblongata, the inferior part of the brain stem, and extends from the foramen magnum of the occipital bone to the level of the second lumbar vertebra (Figure 11-1). The length of the adult spinal cord ranges from 42 to 45 cm (16 to 18 inches).

When the cord is viewed externally, two conspicuous enlargements can be seen. The superior enlargement, the *cervical enlargement,* extends from the fourth cervical to the first thoracic vertebra. Nerves that supply the upper extremities arise from the cervical enlargement. The inferior enlargement, called the *lumbar enlargement,* extends from the ninth to twelfth thoracic vertebra. Nerves that supply the lower extremities arise from the lumbar enlargement.

Below the lumbar enlargement, the spinal cord tapers to a conical portion known as the *conus medul-*

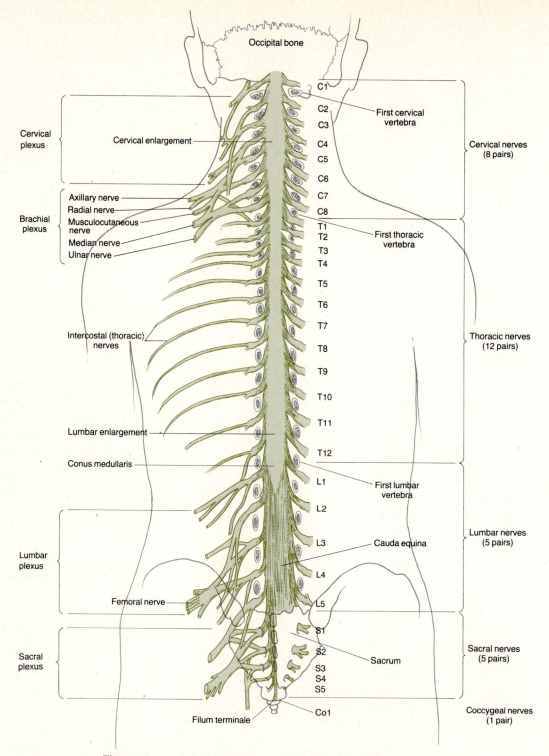

Occipital bone

Cervical plexus

Cervical enlargement

Brachial plexus

Axillary nerve
Radial nerve
Musculocutaneous nerve
Median nerve
Ulnar nerve

Intercostal (thoracic) nerves

Lumbar enlargement

Conus medullaris

Lumbar plexus

Femoral nerve

Sacral plexus

Filum terminale

C1
C2
C3
C4
C5
C6
C7
C8
T1
T2
T3
T4
T5
T6
T7
T8
T9
T10
T11
T12
L1
L2
L3
L4
L5
S1
S2
S3
S4
S5
Co1

First cervical vertebra

Cervical nerves (8 pairs)

First thoracic vertebra

Thoracic nerves (12 pairs)

First lumbar vertebra

Cauda equina

Lumbar nerves (5 pairs)

Sacrum

Sacral nerves (5 pairs)

Coccygeal nerves (1 pair)

Figure 11-1 Spinal cord and spinal nerves in posterior view.

laris. The conus medullaris lies at about the level of the first or second lumbar vertebra. Arising from the conus medullaris is the *filum terminale,* a nonnervous fibrous tissue of the spinal cord that extends inferiorly to the coccyx. The filum terminale consists mostly of pia mater, the innermost of three membranes that cover and protect the spinal cord and brain. Some nerves that arise from the lower portion of the cord

do not leave the vertebral column immediately. They angle inferiorly in the vertebral canal like wisps of coarse hair flowing from the end of the cord. They are appropriately named the *cauda equina* (horse's tail).

The spinal cord is a series of 31 segments, each giving rise to a pair of spinal nerves. *Spinal segment* refers to a region of the spinal cord from which a pair of spinal nerves arises. Reference to Figure 11-2 shows that the cord is divided into right and left sides by two grooves. One of these, the *anterior median fissure,* is a deep, wide groove on the anterior (ventral) surface. The other is the *posterior median sulcus,* a shallower, narrow groove on the posterior (dorsal) surface.

Protection and Coverings

The spinal cord is located in the vertebral canal of the vertebral column. The vertebral canal is formed by the foramina (holes) of all the vertebrae arranged on top of each other. Since the wall of the vertebral canal is essentially a ring of bone surrounding the spinal cord, the cord is well protected. A certain degree of protection is also provided by the meninges, the cerebrospinal fluid, and the vertebral ligaments.

Structure in Cross Section

The spinal cord consists of both gray and white matter. Figure 11-2 shows that the gray matter lies in an area shaped like an H. The gray matter consists primarily of nerve cell bodies, unmyelinated axons, dendrites of association and motor neurons. The white matter

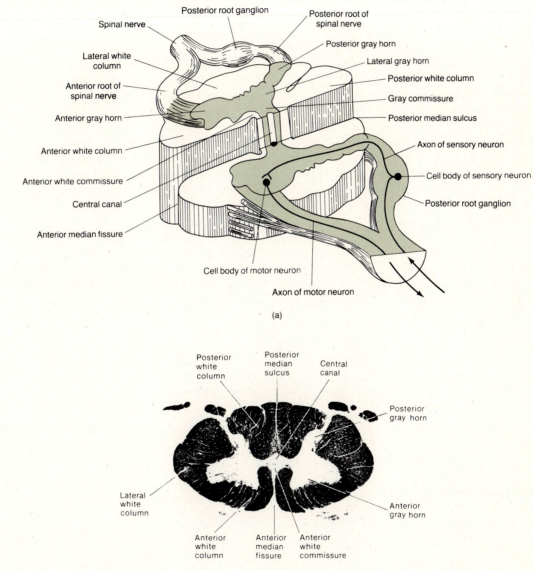

Figure 11-2   Spinal cord. (a) The organization of gray and white matter in the spinal cord as seen in a cross section of the spinal cord. The front side of the figure has been sectioned at a lower level than the back side so that you can see what is inside the posterior root ganglion, the posterior root of the spinal nerve, the anterior root of the spinal nerve, and the spinal nerve. (b) Photograph of the spinal cord at the seventh cervical segment, Weigert stain, at a magnification of 7×. (Courtesy of Murray L. Barr, *The Human Nervous System,* New York: Harper & Row, 1974.)

surrounds the gray matter and consists of bundles of myelinated axons of motor and sensory neurons.

In the center of the gray matter is a crossbar of the H called the *gray commissure,* connecting the right and left portions of the H. In the center of the gray commissure is a small space called the *central canal.* This canal runs the length of the spinal cord and is continuous with the fourth ventricle of the medulla. It contains cerebrospinal fluid. Anterior to the gray commissure is the *anterior (ventral) white commissure,* which connects the white matter of the right and left sides of the spinal cord. The upright portions of the H are further subdivided into regions. Those closer to the front of the cord are called *anterior (ventral) gray horns.* They represent the motor part of the gray matter. The regions closer to the back of the cord are referred to as *posterior (dorsal) gray horns.* They represent the sensory part of the gray matter. The regions between the anterior and posterior gray horns are *lateral gray horns.*

The gray matter of the cord also contains several nuclei that serve as relay stations for impulses and origins for certain nerves. Nuclei are clusters of nerve cell bodies and dendrites in the spinal cord and brain.

The white matter on each side of the cord, like the gray matter, is organized into regions. The anterior and posterior gray horns divide the white matter into three broad areas: *anterior (ventral) white column, posterior (dorsal) white column,* and *lateral white column.* Each column (or *funiculus*) in turn consists of distinct bundles of myelinated fibers that run the length of the cord. These bundles are called *fasciculi* or *tracts.* The longer *ascending tracts* consist of sensory axons that conduct impulses which enter the

spinal cord and pass upward to the brain. The longer *descending tracts* consist of motor axons that conduct impulses from the brain downward through the spinal cord and out to muscles and glands. Thus the ascending tracts are sensory tracts, and the descending tracts are motor tracts. Still other short tracts contain ascending or descending axons that convey impulses from one level of the cord to another.

Functions

A major function of the spinal cord is to convey sensory impulses from the periphery to the brain and to conduct motor impulses from the brain to the periphery. A second principal function is to provide reflexes.

Conduction Pathway

The vital function of conveying sensory and motor information is carried out by the ascending and descending tracts of the cord. The names of the tracts indicate the white column (funiculus) in which the tract travels, where the cell bodies of the tract originate, where the axons of the tract terminate, and the direction of impulse conduction within the tract. For example, the anterior spinothalamic tract is located in the anterior white column, it originates in the spinal cord, it terminates in the thalamus (a region of the brain), and it is an ascending (sensory) tract since it conveys impulses from the cord upward to the brain.

Important ascending and descending tracts are shown in Figure 11-3. Exhibit 11-1 summarizes the principal tracts.

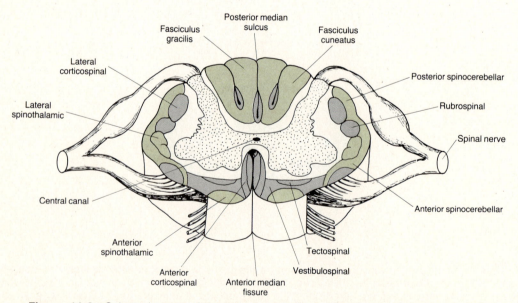

Figure 11-3 Selected tracts of the spinal cord. Ascending (sensory) tracts are indicated in color; descending (motor) tracts are shown in gray.

Exhibit 11-1   SELECTED ASCENDING AND DESCENDING TRACTS OF SPINAL CORD

Tract	Location (White Column)	Origin	Termination	Function
Ascending tracts				
Anterior (ventral) spinothalamic	Anterior (ventral) column	Posterior (dorsal) gray horn on one side of cord but crosses to opposite side	Thalamus; impulse eventually conveyed to cerebral cortex	Conveys sensations for crude touch and pressure from one side of body to opposite side of thalamus. Eventually sensations reach cerebral cortex.
Lateral spinothalamic	Lateral column	Posterior (dorsal) gray horn on one side of cord but crosses to opposite side	Thalamus; impulse eventually conveyed to cerebral cortex	Conveys sensations for pain and temperature from one side of body to opposite side of thalamus. Eventually sensations reach cerebral cortex.
Fasciculus gracilis and fasciculus cuneatus	Posterior (dorsal) column	Axons of afferent neurons from periphery that enter posterior (dorsal) column and rise to same side of medulla	Nucleus gracilis and nucleus cuneatus of medulla; impulse eventually conveyed to cerebral cortex	Convey sensations from one side of body to same side of medulla for fine touch; two-point discrimination (ability to distinguish that two points on skin are touched even though close together); proprioception (awareness of precise position of body parts and their direction of movement); stereognosis (ability to recognize size, shape, and texture of object); weight discrimination (ability to assess weight of an object); and vibrations. Eventually sensations may reach cerebral cortex.
Posterior (dorsal) spinocerebellar	Posterior (dorsal) portion of lateral column	Posterior (dorsal) gray horn on same side of cord	Cerebellum	Conveys sensations from one side of body to same side of cerebellum for subconscious proprioception.
Anterior (ventral) spinocerebellar	Anterior (ventral) portion of column	Posterior (dorsal) gray horn on one side of cord; tract contains both crossed and uncrossed fibers	Cerebellum	Conveys sensations from both sides of body to cerebellum for subconscious proprioception
Descending tracts				
Lateral corticospinal	Lateral column	Cerebral cortex on one side of brain but crosses in base of medulla to opposite side of cord	Anterior (ventral) gray horn	Conveys motor impulses from one side of cortex to anterior gray horn of opposite side. Eventually impulses reach skeletal muscles on opposite side of body that coordinate precise, discrete movements.
Anterior (ventral) corticospinal	Anterior (ventral) column	Cerebral cortex on one side of brain, uncrossed in the medulla, but crosses to the opposite side of cord	Anterior (ventral) gray horn	Conveys motor impulses from one side of cortex to anterior gray horn of opposite side. Eventually impulses reach skeletal muscles on opposite side of body that coordinate precise, discrete movements.
Rubrospinal	Lateral column	Midbrain (red nucleus) but crosses to opposite side of cord	Anterior (ventral) gray horn	Conveys motor impulses from one side of midbrain to skeletal muscles on opposite side of body that are concerned with muscle tone and posture.

Exhibit 11-1 *(Continued)*

Tract	Location (White Column)	Origin	Termination	Function
Tectospinal	Anterior (ventral) column	Midbrain but crosses to opposite side of cord	Anterior (ventral) gray horn	Conveys motor impulses from one side of midbrain to skeletal muscles on opposite side of body that control movements of head in response to auditory, visual, and cutaneous stimuli.
Vestibulospinal	Anterior (ventral) column	Medulla on one side of brain to same side of cord	Anterior (ventral) gray horn	Conveys motor impulses from one side of medulla to skeletal muscles on same side of body that regulate body tone in response to movements of head (equilibrium).

Reflex Center

The second principal function of the spinal cord is to serve as the center for certain reflexes. **Reflexes** are rapid responses to changes in the internal or external environment necessary to maintain homeostasis. Those that are carried out by the spinal cord alone are called *spinal reflexes*.

To understand what reflexes are, we must examine the general structure of spinal nerves. Examination of the cross section of the spinal cord in Figure 11-2a reveals that each pair of spinal nerves is connected to a segment of the cord by two points of attachment called roots. The **posterior** or **dorsal (sensory) root** contains sensory nerve fibers only and conducts impulses from the periphery to the spinal cord. These fibers extend into the posterior (dorsal) gray horn. Each dorsal root also has a swelling, the **posterior** or **dorsal (sensory) root ganglion,** which contains the cell bodies of the sensory neurons from the periphery. The other point of attachment of a spinal nerve to the cord is the **anterior** or **ventral (motor) root.** It contains motor nerve fibers only and conducts impulses from the spinal cord to the periphery. The cell bodies of the motor neurons are located in the gray matter of the cord. If the motor impulse supplies a skeletal muscle, the cell bodies are located in the anterior (ventral) gray horn. If, however, the impulse supplies smooth muscle, cardiac muscle, or a gland through the autonomic nervous system, the cell bodies are located in the lateral gray horn.

Reflex Arc and Homeostasis. The path an impulse follows from its origin in the dendrites or cell body of a neuron in one part of the body to its termination elsewhere in the body is called a *conduction pathway*. All conduction pathways consist of circuits of neurons. One pathway is known as a **reflex arc,** the func-

tional unit of the nervous system. A reflex arc contains two or more neurons over which impulses are conducted from a receptor to the brain or spinal cord and then to an effector. The basic components of a reflex arc are as follows:

1. **Receptor.** The distal end of a dendrite or a sensory structure associated with the distal end of a dendrite. Its role in the reflex arc is to respond to a change in the internal or external environment by initiating a nerve impulse in a sensory neuron.
2. **Sensory neuron.** The neuron that, once stimulated, passes the impulse from the receptor to its axonal termination in the central nervous system.
3. **Center.** A region, usually in the central nervous system, where an incoming sensory impulse generates an outgoing motor impulse. In the center, the impulse may be inhibited, transmitted, or rerouted. In the center of some reflex arcs, the sensory neuron directly generates the impulse in the motor neuron. The center may also contain an association neuron between the sensory neuron and the motor neuron leading to a muscle or a gland.
4. **Motor neuron.** Transmits the impulse generated by the sensory or association neuron in the center to the organ of the body that will respond.
5. **Effector.** The organ of the body, either muscle or gland, that responds to the motor impulse. This response is called a reflex action or reflex.

Processing Sensory Information. Sensory input comes from peripheral receptors, is transferred to the cord along sensory nerves, and is transmitted across a synapse to a motor nerve fiber at the same segmental level. Simultaneously, the information is relayed to higher centers for further processing. Signals from the higher centers descend the spinal cord to modify the motor response. The motor nerve that exits from the spinal cord carries the information *directly* to the muscle that will bring about the desired movement. This direct nonbranching, nonsynapsing connection from the spinal cord to the responding muscle is called the *final common path*.

It has been estimated that, in one segment of the human spinal cord, there are about 500,000 nerve cell bodies in the gray matter. These are the cells that make the vast and complex interconnections among themselves and are also responsible for the sensory-motor integration that occurs at each level. Over 25,000 dorsal root fibers function to deliver information into the spinal cord in one lumbar segment. But only about 13,000 ventral root fibers emerge from the same segment to deliver motor responses to the skeletal muscles. This discrepancy can be explained by the phenomena of *divergence* and *convergence*. Figure 11-4a illustrates the manner in which many sensory axons can branch after they enter the spinal cord. These collateral branches synapse with many association neurons of the gray matter. The arrangement is such that a single association neuron receives input from a wide variety of sensory neurons. Ultimately, the association neurons unload their information onto a single motor neuron; a typical motor neuron has about 5,000 synaptic connections. In this arrangement, the association neurons operate col-

lectively as an amplifier that can intensify or lessen the signals onto a final motor neuron.

Other association neuron arrangements are such that *recurrent loops* or *collaterals* can be added into the neuronal pathway (Figure 11-4b). In this situation, one of the collateral branches of a neuron may synapse to a second neuron that unloads its information back onto the first neuron. If the second neuron is an excitatory type, its input onto the original neuron will prolong its firing pattern. If the second neuron is an inhibitory type, it may inhibit the firing of the original neuron which will then show an interrupted pattern of firing. A third example of interneuronal influence is that of *signal inversion* or *reciprocal innervation*. Association neurons have the ability to change signals from excitatory to inhibitory. In Figure 11-4c, a simple example is presented. Impulses travel along two axonal branches (collaterals) of a sensory neuron. Each collateral makes a synapse with a different association neuron. One association neuron relays excitatory input onto a motor fiber, and the other relays inhibitory input onto a second motor fiber. The excitatory motor fiber will cause the prime mover around a joint to flex, while the other motor fiber will permit relaxation and extension of the antagonist muscle.

Since all neurons receive both excitatory and inhibitory synaptic input, for many other neurons they show the phenomena of temporal and spatial summation (Figure 11-4d). (See Chapter 9.) The membrane of each neuronal cell involved in a reflex is thus constantly algebraically summating the total input it receives.

Thus, one can understand the great difficulty in following the fate of one bit of sensory information that is sent into the central nervous system. The branching networks and great numbers of synapses, plus delay and facilitory circuits, hide every trace of an individual impulse.

Somatic Spinal Reflexes

As indicated earlier, reflexes carried out by the spinal cord alone are called spinal reflexes. On the other hand, reflexes that result in the contraction of skeletal muscles are known as *somatic reflexes*. We now examine a few somatic spinal reflexes: the stretch reflex, the flexor reflex, and the crossed extensor reflex.

Stretch Reflex. The **stretch reflex** is based on a *two-neuron*, or *monosynaptic*, *reflex arc*. Only two neurons are involved and there is only one synapse in the pathway (Figure 11-5). This reflex results in the contraction of a muscle when it is stretched. Slight

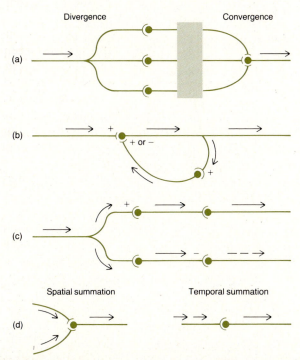

Figure 11-4 Processing information by different neuronal pathways. (a) Divergence and convergence. (b) Recurrent loops or collaterals, excitatory (+) and inhibitory (−). (c) Signal inversion. (d) Summation (spatial and temporal).

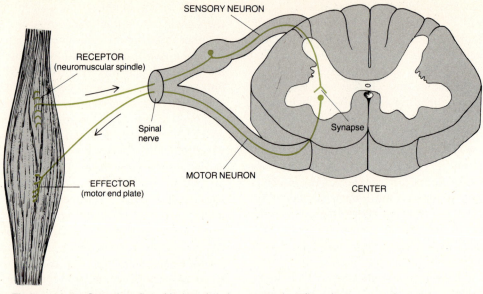

SENSORY NEURON

RECEPTOR
(neuromuscular spindle)

Spinal
nerve

EFFECTOR
(motor end plate)

MOTOR NEURON

Synapse

CENTER

Figure 11-5 Stretch reflex. Notice that in a stretch reflex there are only two neurons involved and there is only one synapse in the pathway. Thus, it is a monosynaptic reflex arc. Why is the reflex arc shown referred to as an ipsilateral reflex arc?

stretching of a muscle stimulates receptors in the muscle called *neuromuscular spindles*. The spindles monitor changes in the length of the muscle. Once the spindle is stimulated, an impulse is sent along a sensory neuron to the spinal cord. The sensory neuron lies in the posterior root of a spinal nerve and synapses with a motor neuron in the anterior gray horn. The sensory neuron generates an impulse at the synapse that is transmitted along the motor neuron. The motor neuron lies in the anterior root of the spinal nerve and terminates in a skeletal muscle. Once the impulse reaches the stretched muscle, it contracts. Thus, the stretch is counteracted by contraction. Since the sensory impulse enters the spinal cord on the same side that the motor impulse leaves the spinal cord, the reflex arc is called an *ipsilateral reflex arc*. All monosynaptic reflex arcs are ipsilateral. The stretch reflex is essential in maintaining muscle tone. Moreover, it is the basis for several tests used in neurological examinations. One such reflex is the *knee jerk,* or *patellar reflex*. This reflex is illustrated in Figure 11-5. This reflex is tested by tapping the patellar ligament (stimulus). Neuromuscular spindles in the quadriceps femoris muscle attached to the ligament send the sensory impulse to the spinal cord, and the returning motor impulse causes contraction of the muscle. The response is extension of the leg at the knee, or a knee jerk.

Flexor Reflex and Crossed Extensor Reflex. Reflexes other than stretch reflexes involve association neurons in addition to the sensory and motor neuron—they

are *polysynaptic reflex arcs*. One example of a reflex based on a polysynaptic reflex arc is the **flexor reflex,** or **withdrawal reflex** (Figure 11-6). Suppose you step on a tack. As a result of the painful stimulus, you immediately withdraw your foot. What has happened? A sensory neuron transmits an impulse from the receptor to the spinal cord. A second impulse is generated in an association neuron, which generates a third impulse in a motor neuron. The motor neuron stimulates the muscles of your foot and you withdraw it. Thus a flexor reflex is protective. It moves an extremity to avoid pain.

This stretch reflex is also ipsilateral. The incoming and outgoing impulses are on the same side of the spinal cord. The stretch reflex also illustrates another feature of reflex arcs. In the monosynaptic stretch reflex, the returning motor impulse affects only the quadriceps muscle of the thigh. When you withdraw your entire lower or upper extremity from a noxious stimulus, more than one muscle is involved. Therefore, several motor neurons are simultaneously returning impulses to several upper and lower extremity muscles at the same time. Thus a single sensory impulse causes several motor responses. This kind of reflex arc, in which a single sensory neuron splits into ascending and descending branches, each forming a synapse with association neurons at different segments of the cord, is called an *intersegmental reflex arc*. Because of intersegmental reflex arcs, a single sensory neuron can activate several motor neurons and thereby cause stimulation of more than one effector.

Something else may happen when you step on a

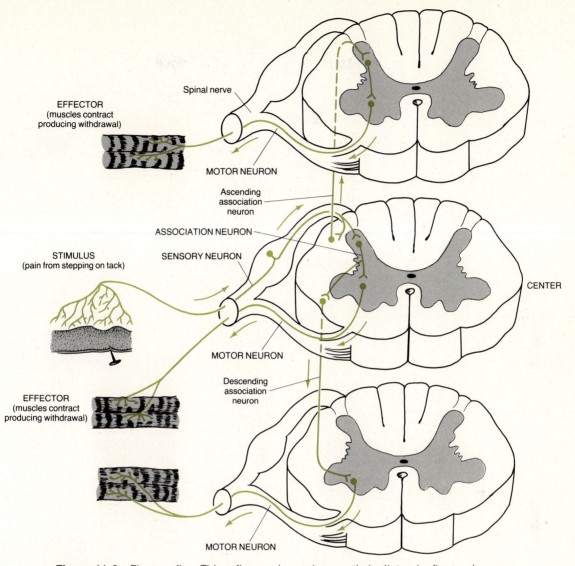

Figure 11-6   Flexor reflex. This reflex arc is a polysynaptic ipsilateral reflex arc because it involves more than one synapse. This is because it contains association neurons as well as sensory and motor neurons. Why is the reflex arch shown also an intersegmental reflex arc?

tack. You may lose your balance as your body weight shifts to the other foot. Then you do whatever you can to regain your balance so you do not fall. This means motor impulses are also sent to your unstimulated foot and both upper extremities. The motor impulses that travel to your unaffected foot cause extension at the knee so you can place your entire body weight on the foot. These impulses cross the spinal cord as shown in Figure 11-7. The incoming sensory impulse not only initiates the flexor reflex that causes you to withdraw, it also initiates an extensor reflex. The incoming sensory impulse crosses to the opposite side of the spinal cord through association neurons at that level and several levels above and below the point of sensory stimulation. From these levels, the motor neurons cause extension of the knee, thus maintaining balance. Unlike the flexor reflex, which passes over an ipsilateral reflex arc, the extensor reflex passes over a *contralateral reflex arc*—the impulse enters one side of the spinal cord and exits on the opposite side. The reflex just described in which extension of the muscles in one limb occurs as a result of flexion of the muscles of the opposite limb is simply called a **crossed extensor reflex.**

The flexor reflex and crossed extensor reflex also illustrate *reciprocal inhibition,* another feature of many reflexes. Reciprocal inhibition is the result of reciprocal innervation previously described. Reciprocal inhibition occurs when a reflex excites a muscle to cause its contraction and also inhibits another muscle

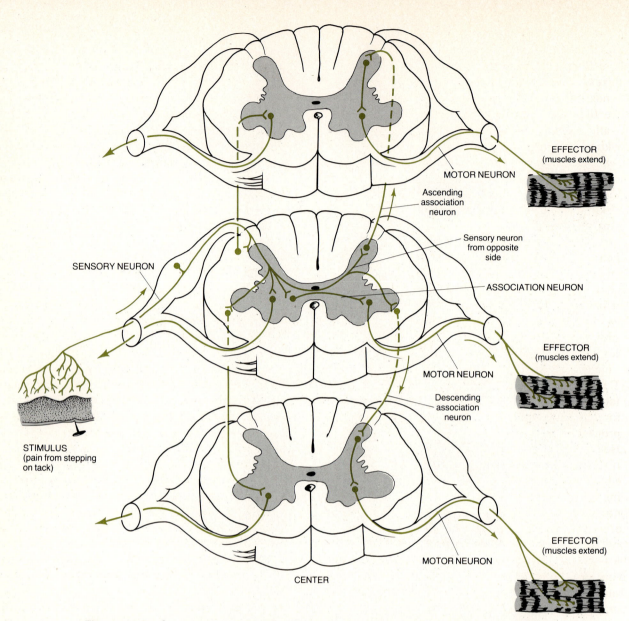

Figure 11-7 Crossed extensor reflex. Although the flexor reflex is shown on the left of the diagram so that you can correlate it with the crossed extensor reflex on the right, concentrate your attention on the crossed extensor reflex. Why is the crossed extensor reflex classified as a contralateral reflex arc?

to allow its extension. Thus, in this reflex, excitation and inhibition occur simultaneously.

In the flexor reflex, when the flexor muscles of your lower extremity are contracting, the extensor muscles of the same extremity are being extended. If both sets of muscles contracted at the same time, you would not be able to flex your limb because both sets of muscles would pull on the limb bones. But, because of reciprocal inhibition, one set of muscles contracts while the other is being extended.

In the crossed extensor reflex, reciprocal inhibition also occurs. While you are flexing the muscles of the limb that has been stimulated by the tack, the muscles of your other limb are producing extension to help

maintain balance. Reciprocal inhibition is vital in coordinating body movements. In flexing the forearm at the elbow, there is a prime mover, an antagonist, and a synergist. The prime mover (biceps) contracts to cause flexion, the antagonist (triceps) extends to yield to the action of the prime mover, and the synergist (deltoid) helps the prime mover perform its role efficiently.

Reflexes and Diagnosis

Reflexes are often used for diagnosing disorders of the nervous system and locating injured tissue. If a reflex ceases to function or functions abnormally, the physician may suspect that the damage lies somewhere along a particular conduction pathway. Visceral reflexes, however, are usually not practical tools for diagnosis. It is difficult to stimulate visceral receptors, since they are deep in the body. In contrast, many somatic reflexes can be tested simply by tapping or stroking the body.

Superficial reflexes are elicited by stroking the skin with a hard object, such as an applicator stick. The object is quickly passed over the skin once, depressing the skin but not producing a scratch.

Any skeletal muscle can normally be stimulated to contract by a slight sudden stretch of its tendon, which can be created by administering a light tap. Many muscle tendons are deeply buried, however, and cannot be readily tapped through the skin. Reflexes that involve a stretch stimulus to a tendon are called *deep tendon reflexes.* Deep tendon reflexes provide information about the integrity and function of the reflex arcs and spinal cord segments without involving the higher centers.

To obtain a substantial response when testing deep tendon reflexes, the muscle must be slightly stretched before the tap is administered. If it is stretched an appropriate amount, tapping the tendon elicits a muscle contraction.

If reflexes are weak or absent, *reinforcement* can be used. In this method, muscle groups other than those being tested are tensed voluntarily with isometric contractions to increase reflex activity in other parts of the body. For example, the person can be asked to hook the fingers together and then try to pull them apart. This action may increase the strength of reflexes involving other muscles. If the reflex can be demonstrated, it is certain that the sensory and motor nerve connections are intact between muscle and spinal cord.

Muscle reflexes can help determine the spinal cord's excitability. When a large number of facilitating impulses are transmitted from the brain to the spinal cord, the muscle reflexes become so sensitive that simply tapping the knee tendon with the tip of one's finger may cause the leg to jump a considerable distance. On the other hand, the cord may be so intensely inhibited by other impulses from the brain that almost no degree of pounding on the muscles or tendons can elicit a response.

Neurological impairment can be evaluated by using a stopwatch to time the reflex response. Sensitivity of sensory end organs in a muscle is demonstrated by stretching it by as little as 0.05 mm and for as short a duration as 0.05 second.

Among the reflexes of clinical significance are:

1. **Patellar reflex** (knee jerk). This reflex involves extension of the lower leg by contraction of the quadriceps femoris muscle in response to tapping the patellar ligament. The reflex is blocked by damaged afferent or efferent nerves to the muscle or reflex centers in the second, third, or fourth lumbar segments of the spinal cord. This reflex is also absent in people with chronic diabetes and neurosyphilis. The reflex is exaggerated in disease or injury involving the corticospinal tracts descending from cortex to spinal cord. This reflex may also be exaggerated by applying a second stimulus (a sudden loud noise) while tapping the patellar tendon.

2. **Achilles reflex** (ankle jerk). This reflex involves extension (plantar flexion) of the foot by contraction of the gastrocnemius and soleus muscles in response to tapping the Achilles tendon. Blockage of the ankle jerk indicates damage to the nerves supplying the posterior leg muscles or to the nerve cells in the lumbosacral region of the spinal cord. This reflex is also absent in people with chronic diabetes, neurosyphilis, alcoholism, and subarachnoid hemorrhages. An exaggerated Achilles reflex indicates cervical cord compression or a lesion of the motor tracts of the first or second sacral segments of the cord.

3. **Babinski reflex.** This reflex results from light stimulation to the outer margin of the sole of the foot. The great toe is extended, with or without fanning of the other toes. This phenomenon occurs in normal children under 1½ years of age and is due to incomplete development of the nervous system. The myelination of fibers in the corticospinal tract has not reached completion. A positive Babinski reflex after age 1½ is considered abnormal and indicates an interruption of the corticospinal tract as the result of a lesion of the tract, usually in the upper portion. The normal response after 1½ years of age is the *plantar reflex,* or negative Babinski—a curling under of all the toes accompanied by a slight turning in and flexion of the anterior part of the foot.

4. **Abdominal reflex.** This reflex compresses the abdominal wall in response to stroking the side of the abdomen. Two separate reflexes, the upper abdominal reflex and the lower abdominal reflex, are involved. The patient should be lying down and relaxed with arms at the sides and knees slightly flexed. The response is an abdominal muscle contraction that results in a lateral deviation of the umbilicus to the side opposite the stimulus. Absence of this reflex is associated with lesions of the corticospinal system. It may also be absent because of lesions of the peripheral nerves, lesions of reflex centers in the thoracic part of the cord, and multiple sclerosis.

SPINAL NERVES

Composition and Coverings

A **spinal nerve** has two points of attachment to the cord: a posterior root and an anterior root. A short distance from the spinal cord, the roots unite to form a spinal nerve. Since the posterior root contains sensory fibers and the anterior root contains motor fibers, a spinal nerve is a *mixed nerve*. The posterior (dorsal) root ganglion contains cell bodies of sensory neurons. The posterior and anterior roots unite to form the spinal nerve at the intervertebral foramen.

A spinal nerve contains many fibers surrounded by different coverings. The individual fibers, whether myelinated or unmyelinated, are wrapped in a connective tissue called the *endoneurium*. Groups of fibers with their endoneurium are arranged in bundles called fascicles, and each bundle is wrapped in connective tissue called the *perineurium*. The outermost covering around the entire nerve is the *epineurium*. The spinal meninges fuse with the epineurium as the nerve exits from the vertebral canal (see Figure 9-8).

Names

The 31 pairs of spinal nerves are named and numbered according to the region and level of the spinal cord from which they emerge (see Figure 11-1). The first

cervical pair emerges between the atlas and the occipital bone. All other spinal nerves leave the backbone from the intervertebral foramina between adjoining vertebrae. There are 8 pairs of cervical nerves, 12 pairs of thoracic nerves, 5 pairs of lumbar nerves, 5 pairs of sacral nerves, and 1 pair of coccygeal nerves. During fetal life, the spinal cord and vertebral column grow at different rates, the cord growing more slowly. Thus not all the spinal cord segments are in line with their corresponding vertebrae. Remember that the spinal cord terminates near the level of the first or second lumbar vertebra. Thus the lower lumbar, sacral, and coccygeal nerves must descend more and more to reach their foramina before emerging from the vertebral column. This arrangement constitutes the cauda equina.

Branches

Shortly after a spinal nerve leaves its intervertebral foramen, it divides into several branches (Figure 11-8). These branches are known as rami. The *dorsal ramus* innervates (supplies) the muscles and skin of the dorsal surface of the back. The *ventral ramus* of a spinal nerve innervates all the structures of the extremities and the lateral and ventral trunk. Except for thoracic nerves T2 to T12, the ventral rami of other spinal nerves enter into the formation of plexuses before supplying a part of the body. In addition to dorsal and ventral rami, spinal nerves also give off a

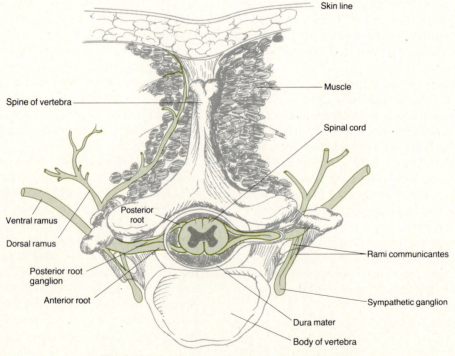

Figure 11-8 Branches of a typical spinal nerve.

meningeal branch. This branch supplies the vertebrae, vertebral ligaments, blood vessels of the spinal cord, and the meninges. Other branches of a spinal nerve are the *rami communicantes.*

Plexuses

The ventral rami of spinal nerves, except for T2 to T12, do not go directly to the structures of the body they supply. Instead, they form networks with adjacent nerves on either side of the body. Such networks are called **plexuses,** meaning braid. The principal plexuses are the cervical plexus, the brachial plexus, the lumbar plexus, and the sacral plexus. Emerging from the plexuses are nerves bearing names that are often descriptive of the general regions they supply or the course they take. Each of these nerves, in turn, may have several branches named for the specific structures they innervate.

HOMEOSTATIC IMBALANCES OF THE SPINAL CORD AND SPINAL NERVES

Spinal Cord Injury

The spinal cord may be damaged by fracture or dislocation of the vertebrae enclosing it or by wounds. All can result in **transection** — partial or complete severing of the spinal cord. Complete transection means that all ascending and descending pathways are cut. It results in loss of all sensation and voluntary muscular movement below the level of transection. In fact, individuals with complete cervical transections close to the base of the skull usually die of asphyxiation before treatment can be administered. This happens because impulses from the phrenic nerves to the breathing muscles are interrupted. If the upper cervical cord is partially transected, both the upper and lower extremities are paralyzed, and the patient is classified as *quadriplegic.* Partial transection between the cervical and lumbar enlargements results in paralysis of the lower extremities only, and the patient is classified as *paraplegic.*

In the case of partial transection, **spinal shock** lasts from a few days to several weeks. During this period, all reflex activity is abolished, a condition called *areflexia.* In time, however, there is a return of reflex activity. The first reflex to return is the knee jerk. Its reappearance may take several days. Next the flexion reflexes return. This may take up to several months. Then the crossed extensor reflexes return. Visceral reflexes such as erection and ejaculation are also affected by transection. Moreover, bladder and bowel function are no longer under voluntary control.

Sciatica

Sciatica is a type of neuritis characterized by severe pain along the path of the sciatic nerve or its branches. The term is commonly applied to a number of disorders affecting this nerve. Because of its length and size, the sciatic nerve is exposed to many kinds of injury. Inflammation of or injury to the nerve causes pain that passes from the back or thigh down its length into the leg, foot, and toes. Probably the most common cause of sciatica is a slipped or herniated disc. Other causes include irritation from osteoarthritis, back injuries, or pressure on the nerve from certain types of exertion. Other cases are idiopathic (unknown origin), and sciatica may be associated with diabetes mellitus, gout, or vitamin deficiencies.

Neuritis

Neuritis is inflammation of a single nerve, two or more nerves in separate areas, or many nerves simultaneously. It may result from irritation to the nerve produced by direct blows, bone fractures, contusions, or penetrating injuries. Additional causes include vitamin deficiency (usually thiamine) and poisons such as carbon monoxide, carbon tetrachloride, heavy metals, and some drugs. Neuritis exists in many forms and is usually considered a symptom rather than a disease. A thorough physical examination, along with laboratory studies, is necessary to discover its exact cause.

The portion of the central nervous system that exerts the major control of homeostasis is the brain. In the next chapter, we examine some of the anatomical features of the brain and cranial nerves and their functional relationships to homeostasis.

CHAPTER SUMMARY OUTLINE

GROUPING OF NEURAL TISSUE

1. White matter is an aggregation of myelinated axons and associated neuroglia.
2. Gray matter is a collection of nerve cell bodies and dendrites or unmyelinated axons along with associated neuroglia.
3. A nerve is a bundle of nerve fibers outside the central nervous system.
4. A ganglion is a collection of cell bodies outside the central nervous system.
5. A bundle of fibers of similar function in the central nervous system forms a tract.
6. A mass of nerve cell bodies and dendrites in the gray matter of the brain forms a nucleus.
7. A horn or column is an area of gray matter in the spinal cord.

SPINAL CORD

General Features

1. When the cord is viewed externally, two conspicuous enlargements can be seen: the superior, or cervical, enlargement; and the inferior, or lumbar, enlargement.
2. The spinal cord is a series of 31 segments, each giving rise to a pair of spinal nerves.
3. The spinal cord is protected by the vertebral canal, the meninges, the cerebrospinal fluid, and the vertebral ligaments.

Structure in Cross Section

1. The gray matter in the spinal cord is divided into horns and the white matter into funiculi or columns.
2. In the center of the spinal cord is the central canal, which runs the length of the spinal cord and contains cerebrospinal fluid.
3. There are ascending (sensory) tracts and descending (motor) tracts.

Functions and Reflexes

1. A major function of the spinal cord is to convey sensory impulses from the periphery to the brain and to conduct motor impulses from the brain to the periphery.
2. The ascending tracts consist of sensory axons that conduct impulses which enter the spinal cord and pass upward to the brain.
3. The descending tracts consist of motor axons that conduct impulses from the brain downward through the spinal cord and out to muscles and glands.
4. Another function of the spinal cord is to provide reflexes.
5. Sensory-motor integrations that occur within the spinal cord are called spinal reflexes.
6. A reflex arc is the functional unit of the nervous system, and is the shortest route that can be taken by an impulse from a receptor to an effector.
7. The basic components of a reflex arc are: receptor, sensory neuron, center, motor neuron, and effector.
8. Sensory input comes from peripheral receptors, is transferred to the cord along sensory nerves, and is transmitted across a synapse to a motor nerve fiber.
9. A two-neuron reflex arc contains one sensory and one motor neuron. Stretch reflexes such as the patellar reflex are all monosynaptic.
10. A multineuron reflex arc contains a sensory, association, and motor neuron. A withdrawal or flexor reflex such as pulling the hand away from a hot object is an example.
11. A reflex is a quick, involuntary response to a stimulus that passes along a reflex arc.
12. Reflexes represent the body's principal mechanisms for responding to changes in the internal and external environment to maintain homeostasis.
13. Two types of reflexes are the superficial reflexes and the deep tendon reflexes.
14. Among clinically important somatic reflexes are the patellar reflex, the Achilles reflex, the Babinski reflex, and the abdominal reflex.

SPINAL NERVES

Composition and Coverings

1. A spinal nerve has two points of attachment to the cord: a posterior root and an anterior root.
2. Since the posterior root contains sensory fibers, and the anterior root contains motor fibers, a spinal nerve is a mixed nerve.
3. A spinal nerve contains many fibers surrounded by different coverings. Connective tissue wrappings include the endoneurium, perineurium, and epineurium.
4. Thirty-one pairs of spinal nerves originate from the spinal cord. They are named and numbered ac-

cording to the region and level of the spinal cord from which they emerge.

5. The principal branches of the spinal nerves are the dorsal and ventral rami, and the visceral branches (rami communicantes)

6. The ventral rami of the spinal nerves, except for T2 to T12, do not go directly to the structures of the body they supply. Instead they form networks with adjacent nerves on either side of the body that are called plexuses.

HOMEOSTATIC IMBALANCES OF THE SPINAL CORD AND SPINAL NERVES

Spinal Cord Injury

1. Transection is the partial or complete severing of the spinal cord.

2. If the upper cervical cord is partially transected, both the upper and lower extremities are paralyzed and the patient is classified as quadriplegic.

3. Partial transection between the cervical and lumbar enlargements results in paralysis of the lower extremities only, and the patient is classified as paraplegic.

4. Spinal shock is the complete absence of all reflex activity.

Sciatica

1. Sciatica is a type of neuritis characterized by severe pain along the path of the sciatic nerve or its branches.

Neuritis

1. Neuritis is inflammation of a single nerve, two or more nerves in separate areas, or many nerves simultaneously.

REVIEW QUESTIONS

1. Define the following terms: white matter, gray matter, nerve, ganglion, tract, nucleus, and horn.
2. Where is the spinal cord located? Describe its general external appearance.
3. How is the spinal cord protected?
4. Diagram and label a cross section of the spinal cord and explain the function of each area.
5. What are the functions of the spinal cord?
6. Compare the ascending and descending tracts of the spinal cord with respect to location, origin, termination, and function.
7. Identify the reflex center, and describe its processing of information.
8. Define a reflex arc and list its components.
9. Explain briefly how sensory information is processed.
10. Distinguish between a monosynaptic and polysynaptic reflex arc.
11. Define a reflex. How are reflexes related to the maintenance of homeostasis?
12. Distinguish the following reflexes: somatic, visceral, ipsilateral, intersegmental, and contralateral.
13. Indicate the clinical importance of the following reflexes: patellar, Achilles, Babinski, and abdominal.
14. What is a spinal nerve? How is a spinal nerve related to the spinal cord? How are spinal nerves named?
15. Describe the coverings of a spinal nerve.
16. What are the four branches of a spinal nerve? Describe the distribution of each.
17. Define a plexus. What are the principal plexuses?
18. Define transection.
19. Compare quadriplegic with paraplegic.
20. What is spinal shock?
21. Contrast the disorders sciatica and neuritis.

Chapter 12

THE BRAIN AND THE CRANIAL NERVES

STUDENT OBJECTIVES

After reading this chapter, you should be able to

- Identify the structures responsible for protecting the brain and describe its coverings
- Describe the formation, circulation, and functions of cerebrospinal fluid in relation to homeostasis
- Define hydrocephalus
- Describe the structure and physiology of the blood-brain barrier
- Compare the components of the brain stem with regard to structure and function
- Identify the structural features of the cerebrum
- Compare the motor, sensory, and association functions of the cerebrum
- Describe the principle of an electroencephalograph and its significance in the diagnosis of certain disorders
- Identify the anatomical characteristics and functions of the cerebellum
- Identify by number and name the 12 pairs of cranial nerves
- Describe poliomyelitis, syphilis, cerebral palsy, Parkinsonism, multiple sclerosis, cerebrovascular accidents, dyslexia, and Tay-Sachs disease as homeostatic imbalances of the central nervous system

Now we shall consider how the brain is protected, what its principal parts are, how it is related to the spinal cord, and how it is related to the 12 pairs of cranial nerves.

BRAIN

Protection and Coverings

The **brain** of an average adult is one of the largest organs of the body, weighing about 1,300 g (3 lb). Figure 12-1 shows that the brain is mushroom-shaped. It is divided into four principal parts: brain stem, diencephalon, cerebellum, and cerebrum. The **brain stem,** the stalk of the mushroom, consists of the medulla oblongata, pons varolii, and midbrain. The lower end of the brain stem is a continuation of the spinal cord. Above the brain stem is the **diencephalon,** consisting primarily of the thalamus and hypothalamus. The **cerebrum** spreads over the diencephalon. The cerebrum constitutes about seven-eighths of the total weight of the brain and occupies most of the skull. Inferior to the cerebrum and posterior to the brain stem is the **cerebellum.**

The brain is protected by the cranial bones. Like the spinal cord, the brain is also protected by meninges. The *cranial meninges* surround the brain and are continuous with the spinal meninges. The cranial meninges have the same basic structure and bear the same names as the spinal meninges: the outermost *dura mater,* middle *arachnoid,* and innermost *pia mater* (see Figure 12-2).

Cerebrospinal Fluid

The brain, as well as the rest of the central nervous system, is further protected against injury by **cerebrospinal fluid.** This fluid circulates through the subarachnoid space around the brain and spinal cord and through the ventricles of the brain. The subarachnoid space is the area between the arachnoid and pia mater. The **ventricles** are cavities in the brain that communicate with each other, with the central canal of the spinal cord, and with the subarachnoid space. Each of the two *lateral ventricles* is located in a hemisphere (side) of the cerebrum under the corpus callosum (Figure 12-2). The *third ventricle* is a slit between and inferior to the right and left halves of the thalamus and between the lateral ventricles. Each lateral ventricle communicates with the third ventricle by a narrow, oval opening: the *interventricular foramen, or foramen of Monro.* The *fourth ventricle* lies between the inferior brain stem and the cerebellum. It communicates with the third ventricle via the *cerebral aqueduct (aqueduct of Sylvius),* which passes through the midbrain. The roof of the fourth ventricle has three openings: a *median aperture (foramen of Magendie)* and two *lateral apertures (foramina of Luschka).* Through these openings, the fourth ventricle also communicates with the subarachnoid space of the brain and cord.

The entire central nervous system contains about 125 ml (4 oz) of cerebrospinal fluid. It is a clear, colorless fluid of watery consistency. Chemically, it contains proteins, glucose, urea, and salts. It also contains some white blood cells. The fluid has two major functions related to homeostasis: (1) protection and (2) circulation. Its protective function is manifested in two ways. First, the cerebrospinal fluid acts as a shock absorbing medium to cushion the brain and spinal cord from jolts that would otherwise cause the soft nervous tissue to crash against the interior of the bony walls that enclose it. Second, the brain is buoyed by the cerebrospinal fluid so that it "floats"

inside the skull. Even though the human brain has a mass of about 1,300 g, it has been estimated that its in situ weight might be equivalent to about 50 g due to the buoyancy provided by the cerebrospinal fluid. To understand the importance of this buoyancy property, consider the process of *pneumoencephalography*. In this process, in order to visualize the ventricular cavities of the brain by x-ray examination, some of the

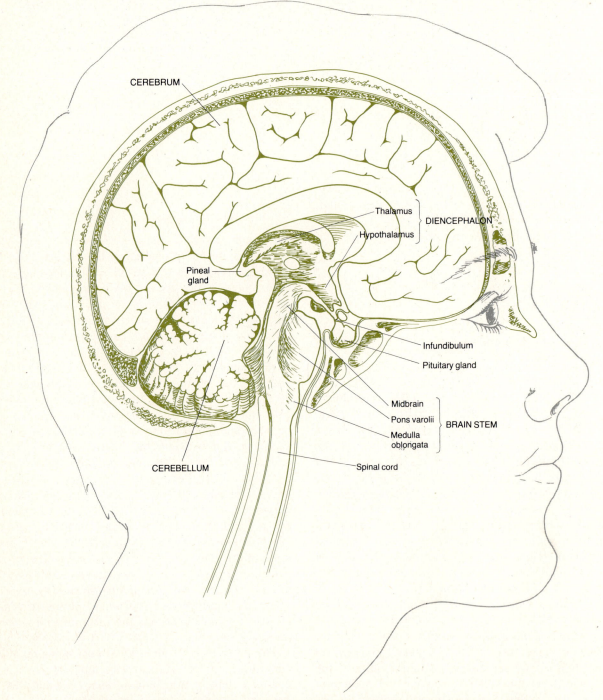

CEREBRUM

Thalamus

DIENCEPHALON

Hypothalamus

Pineal gland

Infundibulum

Pituitary gland

Midbrain

Pons varolii

BRAIN STEM

Medulla oblongata

CEREBELLUM

Spinal cord

Figure 12-1 Brain. Principal parts of the medial aspect of the brain seen in sagittal section. The infundibulum and pituitary gland are discussed in conjunction with the endocrine system in Chapter 17.

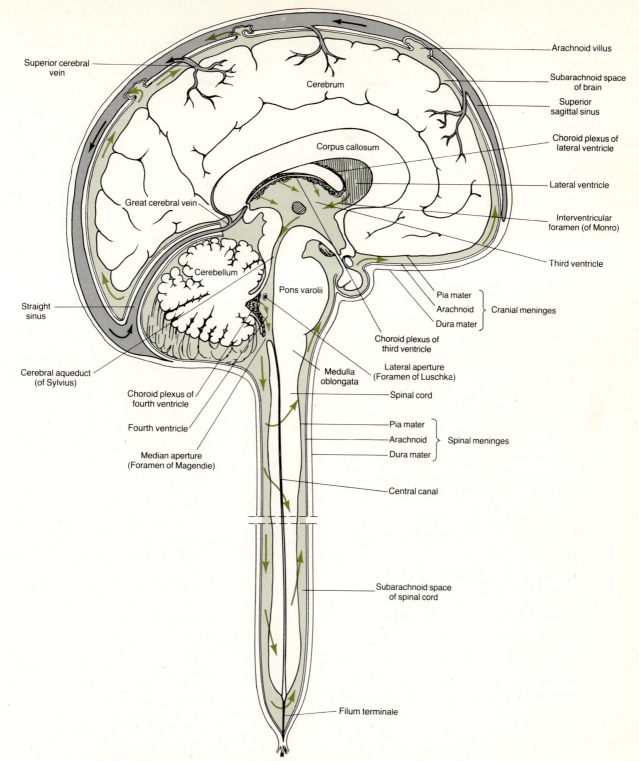

Figure 12-2 Brain and meninges seen in sagittal section. The direction of flow of cerebrospinal fluid is indicated by colored arrows.

Structures labeled in the figure:

- Superior cerebral vein
- Cerebrum
- Corpus callosum
- Great cerebral vein
- Cerebellum
- Pons varolii
- Straight sinus
- Cerebral aqueduct (of Sylvius)
- Choroid plexus of fourth ventricle
- Fourth ventricle
- Median aperture (Foramen of Magendie)
- Medulla oblongata
- Filum terminale
- Arachnoid villus
- Subarachnoid space of brain
- Superior sagittal sinus
- Choroid plexus of lateral ventricle
- Lateral ventricle
- Interventricular foramen (of Monro)
- Third ventricle
- Pia mater / Arachnoid / Dura mater — Cranial meninges
- Choroid plexus of third ventricle
- Lateral aperture (Foramen of Luschka)
- Spinal cord
- Pia mater / Arachnoid / Dura mater — Spinal meninges
- Central canal
- Subarachnoid space of spinal cord

cerebrospinal fluid is replaced with air. Considerable headache pain is experienced by the patient. The removal of the fluid permits the brain to sag slightly within its bony compartment and tug upon the blood vessels and meninges that connect to the surface of the brain. These structures are richly innervated with nerve endings that are sensitive to pain, and the pull exerted upon them generates the painful stimuli. The pain persists until the air is resorbed and the fluid is replaced by the normal secretory processes.

The cerebrospinal fluid circulates to provide a route for the elimination of metabolic by-products. It collects proteins, lactic acid, high-molecular-weight compounds, and toxic substances that may be produced by the brain. In other words, the continual flow of cerebrospinal fluid through the chambers of the central nervous system serves to wash away or "flush" the waste materials of the brain.

The cerebrospinal fluid is produced by the *choroid plexuses* in the ventricles of the brain. The active secretion of over 600 ml of fluid daily produces a pressure gradient that ensures the continual flow of fluid throughout all the spaces of the central nervous system. The choroid plexus is a tortuous network of capillaries in the ventricular walls. Surrounding these capillaries and separating them from the ventricular cavities is a layer of the pia mater and a layer of slightly modified ependymal (neuroglial) cells. There is considerable evidence that these cells are responsible for the active secretion of the fluid and its components into the chambers. The selective secretory activity of these cells ensures that the proper environment for nerve cell function remains constant. The rapid removal or turnover of the fluid also aids in the maintenance of the brain's environment. The flow of fluid starts in the lateral ventricles, where it is secreted by the ependymal cells around the choroid plexuses. The fluid flows out of the lateral ventricles via the interventricular foramen into the third ventricle (Figure 12-2). Additional fluid is produced by the ependymal cells of the third ventricle, and the total volume is directed through the cerebral aqueduct into the fourth ventricle. Further additions of fluid are secreted from the choroid plexus into this chamber. The fluid escapes from the fourth ventricle through the apertures in the roof of the chamber. The fluid then flows into the subarachnoid space around the back of the brain. It also passes downward to the subarachnoid space around the posterior surface of the spinal cord, up the anterior surface of the spinal cord, and around the anterior part of the brain. Some fluid is also produced by the ependymal cells in the central canal of the spinal cord. This fluid flows into the fourth ventricle and becomes part of the fluid exiting through the apertures. Fluid from the anterior part of the brain is gradually reabsorbed into the cerebral veins. Most of the fluid is collected by the superior sagittal sinus. The absorption actually occurs through *arachnoid villi,* which are fingerlike projections of the arachnoid that push into the superior sagittal sinus.

If an obstruction, such as a tumor, arises in the brain and interferes with the drainage of fluid from the ventricles into the subarachnoid space, large amounts of fluid accumulate in the ventricles. Fluid pressure inside the brain increases, and if the fontanels have not yet closed, the head bulges to relieve the pressure. This condition is called **internal hydrocephalus** (*hydro* = water; *cephalo* = head). If an obstruction interferes with drainage somewhere in the subarachnoid space

and cerebrospinal fluid accumulates inside the space, the condition is termed **external hydrocephalus.**

Blood Supply

The brain is well supplied with blood vessels, which provide oxygen and nutrients. Although the brain actually consumes less oxygen than most other organs of the body, it must receive a constant supply. If the blood flow to the brain is interrupted for even a few moments, unconsciousness may result. A 1- or 2-minute interruption may weaken the brain cells by starving them of oxygen. If the cells are totally deprived of oxygen for about 4 minutes, many are permanently injured. Occasionally during childbirth, the oxygen supply from the mother's blood is interrupted before the baby leaves the birth canal and can breathe. Often such babies are stillborn or suffer permanent brain damage that may result in mental retardation, epilepsy, and paralysis.

Blood supplying the brain also contains glucose, the principal source of energy for brain cells. Because carbohydrate storage in the brain is limited, the supply of glucose must be continuous. If blood entering the brain has a low glucose level, mental confusion, dizziness, convulsions, and even loss of consciousness may occur.

Glucose, oxygen, and certain ions pass rapidly from the circulating blood into brain cells. Other substances, such as creatinine, urea, chloride, insulin, and sucrose, enter quite slowly. Still other substances—proteins and most antibiotics—do not pass at all from the blood into brain cells. The differential rates of passage of certain materials from the blood into the brain suggest a concept called the **blood-brain barrier.** Electron micrograph studies of the capillaries of the brain reveal that they differ structurally from other capillaries. Brain capillaries are constructed of more densely packed cells and are surrounded by large numbers of glial cells and a continuous basement membrane. These features form a barrier to the passage of certain materials. Thus, substances that cross the barrier either are very small molecules or require the assistance of a carrier molecule to cross by active transport. The function of the blood-brain barrier is not known. It may assume a homeostatic role by protecting brain cells from harmful substances.

Brain Stem

Medulla Oblongata

The **medulla oblongata,** or simply **medulla,** is a continuation of the upper portion of the spinal cord and forms the inferior part of the brain stem. Its position in relation to the other parts of the brain may be noted in Figure 12-1. The medulla measures only 3 cm (about 1 inch) in length.

The medulla contains all ascending and descending tracts that communicate between the spinal cord and various parts of the brain. These tracts constitute the white matter of the medulla. Some tracts cross as they pass through the medulla. Let us see how this crossing occurs and what it means.

On the ventral side of the medulla are two roughly triangular structures called *pyramids* (Figure 12-3). The pyramids are composed of the largest motor tracts that pass from the outer region of the cerebrum (cerebral cortex) to the spinal cord. Just above the junction of the medulla with the spinal cord, most of the fibers in the left pyramid cross to the right side, and most of the fibers in the right pyramid cross to the left. This crossing is called the **decussation of pyramids.** The adaptive value, if any, of this phenomenon is unknown. The principal motor fibers that undergo decussation belong to the lateral corticospinal tracts. These tracts originate in the cerebral cortex and pass inferiorly to the medulla. The fibers cross in the pyramids and descend in the lateral columns of the spinal cord, terminating in the anterior gray horns. Here synapses occur with motor neurons that terminate in skeletal muscles. As a result of the crossing, fibers that originate in the left cerebral cortex activate muscles on the right side of the body, and fibers that originate in the right cerebral cortex activate muscles on the left side. Decussation explains why motor areas of one side of the cerebral cortex control muscular movements on the opposite side of the body.

The dorsal side of the medulla contains two pairs of prominent nuclei: the right and left *nucleus gracilis* and *nucleus cuneatus*. These nuclei receive sensory fibers from ascending tracts (right and left fasciculus gracilis and fasciculus cuneatus) of the spinal cord and relay the sensory information to the opposite side of the medulla. This information is conveyed to the thalamus and then to the sensory areas of the cerebral cortex. Nearly all sensory impulses received on one side of the body cross in the medulla or spinal cord and are perceived in the opposite side of the cerebral cortex.

In addition to its function as a conduction pathway for motor and sensory impulses between the brain and spinal cord, the medulla also contains an area of dispersed gray matter containing some white fibers. This region is called the **reticular formation.** Actually, portions of the reticular formation are located in the spinal cord, pons, midbrain, and diencephalon. The reticular formation functions in consciousness and arousal. Within the medulla are three vital reflex centers of the reticular system that assume key homeo-

static roles. The *cardiac center* regulates heartbeat; the *medullary rhythmicity area* adjusts the basic rhythm of breathing; and the *vasoconstrictor* or *vasomotor center* regulates the diameter of blood vessels. Other centers in the medulla coordinate swallowing, vomiting, coughing, sneezing, and hiccuping.

The medulla also contains the nuclei of origin for several pairs of cranial nerves. These are the cochlear and vestibular branches of the vestibulocochlear nerves (VIII), which are concerned with hearing and equilibrium (there is also a nucleus for the vestibular branches in the pons); the glossopharyngeal nerves (IX), which relay impulses related to swallowing, salivation, and taste; the vagus nerves (X), which relay impulses to and from many thoracic and abdominal viscera; the accessory nerves (XI), which convey impulses related to head and shoulder movements (a part of this nerve also arises from the first five segments of the spinal cord); and the hypoglossal nerves (XII), which convey impulses that involve tongue movements.

Pons Varolii

The relationship of the **pons varolii** or **pons** to other parts of the brain can be seen in Figures 12-1 and 12-3a. The pons, which means bridge, lies directly above the medulla and anterior to the cerebellum. It is about 2.5 cm (1 inch) long. Like the medulla, the pons consists of white fibers scattered throughout with nuclei. As the name implies, the pons is a bridge connecting the spinal cord with the brain and parts of the brain with each other. These connections are provided by fibers that run in two principal directions. The transverse fibers connect with the cerebellum through the middle cerebellar peduncles. The longitudinal fibers of the pons belong to the motor and sensory tracts that connect the spinal cord or medulla with the upper parts of the brain stem.

The nuclei for certain paired cranial nerves are also contained in the pons. These include the trigeminal nerves (V), which relay impulses for chewing and for sensations of the head and face; the abducens nerves (VI), which regulate certain eyeball movements; the facial nerves (VII), which conduct impulses related to taste, salivation, and facial expression; and the vestibular branches of the vestibulocochlear nerves (VIII), which are concerned with equilibrium.

Other important nuclei in the reticular formation of the pons are the *pneumotaxic area* and *apneustic area*. Together with the medullary rhythmicity area in the medulla, they help control respiration.

Midbrain

The **midbrain** or **mesencephalon** extends from the pons to the lower portion of the diencephalon (Figure 12-1).

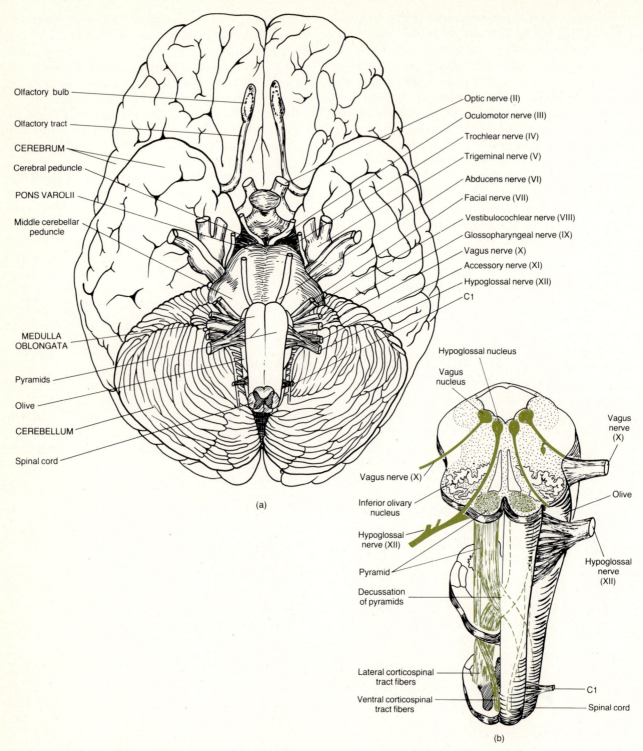

Olfactory bulb
Olfactory tract
CEREBRUM
Cerebral peduncle
PONS VAROLII
Middle cerebellar peduncle
MEDULLA OBLONGATA
Pyramids
Olive
CEREBELLUM
Spinal cord

Optic nerve (II)
Oculomotor nerve (III)
Trochlear nerve (IV)
Trigeminal nerve (V)
Abducens nerve (VI)
Facial nerve (VII)
Vestibulocochlear nerve (VIII)
Glossopharyngeal nerve (IX)
Vagus nerve (X)
Accessory nerve (XI)
Hypoglossal nerve (XII)
C1

(a)

Hypoglossal nucleus
Vagus nucleus
Vagus nerve (X)
Inferior olivary nucleus
Hypoglossal nerve (XII)
Pyramid
Decussation of pyramids
Lateral corticospinal tract fibers
Ventral corticospinal tract fibers

Vagus nerve (X)
Olive
Hypoglossal nerve (XII)
C1
Spinal cord

(b)

Figure 12-3 Structure of the brain. (a) Ventral view. (b) Details of the medulla.

It is about 2.5 cm (1 inch) long. The cerebral aqueduct passes through the midbrain and connects the third ventricle above with the fourth ventricle below.

The ventral portion of the midbrain contains a pair of fiber bundles referred to as *cerebral peduncles*. The cerebral peduncles contain many motor fibers that convey impulses from the cerebral cortex to the pons and spinal cord. They also contain sensory fibers that pass from the spinal cord to the thalamus. The cerebral peduncles constitute the main connection for tracts between upper parts of the brain and lower parts of the brain and the spinal cord.

The dorsal portion of the midbrain is called the *tectum* and contains four rounded eminences: the *corpora quadrigemina.* Two of the eminences are known as the *superior colliculi.* These serve as a reflex center for movements of the eyeballs and head in response to visual and other stimuli. The other two eminences are the *inferior colliculi.* They serve as reflex centers for movements of the head and trunk in response to auditory stimuli. The midbrain also contains the *substantia nigra,* a large, heavily pigmented nucleus near the cerebral peduncles.

A major nucleus in the reticular formation of the midbrain is the *red nucleus.* Fibers from the cerebellum and cerebral cortex terminate in the red nucleus. The red nucleus is also the origin of cell bodies of the descending rubrospinal tract. Other nuclei in the midbrain are associated with cranial nerves. These include the oculomotor nerves (III), which mediate some movements of the eyeballs and changes in pupil size and lens shape, and the trochlear nerves (IV), which conduct impulses that move the eyeballs.

A structure called the *medial lemniscus* is common to the medulla, pons, and midbrain. The medial lemniscus is a band of white fibers containing axons that convey impulses for fine touch, proprioception, and vibrations from the medulla to the thalamus.

Diencephalon

The **diencephalon** consists principally of the thalamus and hypothalamus. The relationship of these structures to the rest of the brain is shown in Figure 12-1.

Thalamus

The **thalamus** is a large oval structure above the midbrain (Figure 12-4a). It consists of two masses of gray matter covered by a thin layer of white matter. It measures about 3 cm (1 inch) in length and constitutes four-fifths of the diencephalon. The thalamus contains numerous nuclei organized into masses (Figure 12-4b). Some nuclei in the thalamus serve as relay stations for all sensory impulses, except smell, to the cerebral cortex. These include the *medial geniculate nuclei* (hearing), the *lateral geniculate nuclei* (vision), and the *ventral posterior nuclei* (general sensations and taste). Other nuclei are centers for synapses in the somatic motor system. These include the *ventral lateral nuclei* (voluntary motor actions) and *ventral anterior nuclei* (voluntary motor actions and arousal). The thalamus is the principal relay station for sensory impulses that reach the cerebral cortex from the spinal cord, brain stem, cerebellum, and parts of the cerebrum.

The thalamus also functions as an interpretation center. That is, some sensory impulses that enter the thalamus are interpreted there. At the thalamic level, one can have conscious recognition of pain and temperature and some awareness of crude touch and pressure. The thalamus also contains a *reticular nucleus* in its reticular formation and an *anterior nucleus* in the floor of the lateral ventricle.

Hypothalamus

The **hypothalamus** is a small portion of the diencephalon, and its relationship to other parts of the brain is shown in Figures 12-1 and 12-4a. The hypothalamus forms the floor and part of the lateral walls of the third ventricle. The hypothalamus is protected by the sphenoid bone and indirectly by the sella turcica of the sphenoid bone. Despite its small size, nuclei in the hypothalamus control many body activities, most of them related to homeostasis. The chief functions of the hypothalamus are these:

1. It controls and integrates the autonomic nervous system, which stimulates smooth muscle, regulates the rate of contraction of cardiac muscle, and controls the secretions of many glands. Through the autonomic nervous system, the hypothalamus is the main regulator of visceral activities. It regulates heart rate, movement of food through the digestive tract, and contraction of the urinary bladder.
2. It is involved in the reception of sensory impulses from the viscera.
3. It is the principal intermediary between the nervous system and endocrine system—the two major control systems of the body. The hypothalamus lies just above the pituitary, the main endocrine gland. When the hypothalamus detects certain changes in the body, it releases chemicals called releasing factors that stimulate or inhibit the anterior pituitary gland. The anterior pituitary then releases or holds back hormones that regulate carbohydrates, fats, proteins, certain ions, and sexual functions.
4. It is the center for the mind-over-body phenomenon. When the cerebral cortex interprets strong emotions, it often sends impulses along tracts that connect the cortex with the hypothalamus. The hypothalamus then directs impulses via the autonomic nervous system and also releases chemicals that stimulate the anterior pituitary gland. The result can be a wide range of changes in body activities. For instance, when you panic, impulses leave the hypothalamus to stimulate your heart to beat faster. Likewise, continued psychological stress can produce long-term abnormalities in body function that result is serious illness. These so-called psychosomatic disorders definitely are real.
5. It is associated with feelings of rage and aggression.
6. It controls normal body temperature. Certain cells of the hypothalamus serve as a thermostat—a

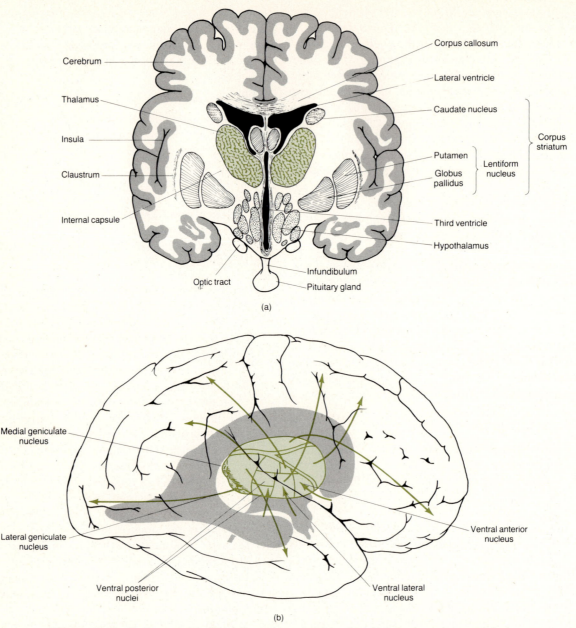

Figure 12-4 Thalamus. (a) Frontal section showing the thalamus and associated structures. (b) Right lateral view of the thalamic nuclei.

mechanism sensitive to changes in temperature. If blood flowing through the hypothalamus is above normal temperature, the hypothalamus directs impulses along the autonomic nervous system to stimulate activities that promote heat loss. Heat can be lost through relaxation of the smooth muscle in the blood vessels, causing vasodilation of cutaneous blood vessels and increased heat loss from the skin. Heat loss also occurs by sweating. Conversely, if the temperature of the blood is below normal, the hypothalamus generates impulses that promote heat retention. Heat can be retained through the contraction of cutaneous blood vessels, cessation of sweating, and shivering.

7. It regulates food intake through two centers. The *feeding center* is stimulated by hunger sensations from an empty stomach. When sufficient food has been ingested, the *satiety center* is stimulated and sends out impulses that inhibit the feeding center.

8. It contains a *thirst center*. Certain cells in the hypothalamus are stimulated when the extracellular fluid volume is reduced. The stimulated cells produce the sensation of thirst in the hypothalamus.

9. It is one of the centers that maintain the waking state and sleep patterns.

Cerebrum

Supported on the brain stem and forming the bulk of the brain is the **cerebrum** (Figure 12-1). The surface of the cerebrum is composed of gray matter 2 to 4 mm (0.08 to 0.16 inch) thick and is referred to as the *cerebral cortex* (*cortex* = rind or bark). The cortex, containing millions of cells, consists of six layers of nerve cell bodies. Beneath the cortex lies the *cerebral white matter.*

During embryonic development when there is a rapid increase in brain size, the gray matter of the cortex enlarges out of proportion to the underlying white matter. As a result, the cortical region rolls and folds upon itself. The upfolds are called *gyri* or *convolutions* (Figure 12-5b). The deep downfolds are referred to as *fissures;* the shallow downfolds are *sulci.* The most prominent fissure, the *longitudinal fissure,* nearly separates the cerebrum into right and left halves, or *hemispheres* (Figure 12-5b). The hemispheres, however, are connected internally by a large bundle of transverse fibers composed of white matter called the *corpus callosum.* Between the hemispheres is an extension of the cranial dura mater called the *falx cerebri.*

Lobes

Each cerebral hemisphere is further subdivided into four lobes by deep sulci or fissures (Figure 12-5). The *central sulcus,* or *fissure of Rolando,* separates the *frontal lobe* from the *parietal lobe.* A major gyrus, the *precentral gyrus,* is located immediately anterior to the central sulcus. The *lateral cerebral sulcus,* or *fissure of Sylvius,* separates the *frontal lobe* from the *temporal lobe.* The *parietooccipital sulcus* separates the *parietal lobe* from the *occipital lobe.* Another prominent fissure, the *transverse fissure,* separates the cerebrum from the cerebellum. The frontal lobe, parietal lobe, temporal lobe, and occipital lobe are named after the bones that cover them. A fifth part of the cerebrum, the *insula (island of Reil),* lies deep within the lateral cerebral fissure, under the parietal, frontal, and temporal lobes. It cannot be seen in an external view of the brain.

White Matter

The white matter underlying the cortex consists of myelinated axons running in three principal directions:

1. **Association fibers** connect and transmit impulses between gyri in the same hemisphere.

2. **Commissural fibers** transmit impulses from the gyri in one cerebral hemisphere to the corresponding gyri in the opposite cerebral hemisphere. Three important groups of commissural fibers are the *corpus callosum, anterior commissure,* and *posterior commissure.*

3. **Projection fibers** form ascending and descending tracts that transmit impulses from the cerebrum to other parts of the brain and spinal cord.

Basal Ganglia

The **basal ganglia** or **cerebral nuclei** are paired masses of gray matter in each cerebral hemisphere (Figure 12-6). The largest of the basal ganglia of each hemisphere is the *corpus striatum.* It consists of the *caudate nucleus* and the *lentiform nucleus.* The lentiform nucleus, in turn, is subdivided into a lateral portion called the *putamen* and a medial portion called the *globus pallidus.* Figure 12-4a shows the two divisions of the lentiform nucleus and a structure called the *internal capsule.* It is made up of a group of sensory and motor white matter tracts that connect the cerebral cortex with the brain stem and spinal cord. The portion of the internal capsule passing between the lentiform nucleus and the caudate nucleus and between the lentiform nucleus and thalamus is sometimes considered part of the corpus striatum.

The basal ganglia are interconnected by many fibers. They are also connected to the cerebral cortex, thalamus, and hypothalamus. The caudate nucleus and the putamen control large subconscious movements of the skeletal muscles — such as swinging the arms while walking. Such gross movements are also consciously controlled by the cerebral cortex. The globus pallidus is concerned with the regulation of muscle tone required for specific body movements.

Limbic System

Certain components of the cerebral hemispheres and diencephalon constitute the **limbic system.** It includes the following regions of gray matter:

1. **Limbic lobe.** Formed by two gyri of the cerebral hemisphere: the cingulate gyrus and the hippocampal gyrus.
2. **Hippocampus.** An extension of the hippocampal gyrus that extends into the floor of the lateral ventricle.
3. **Amygdaloid nucleus.** Located at the tail end of the caudate nucleus.
4. **Hypothalamus.** The regions of the hypothalamus that form part of the limbic system are the perifornical nuclei.
5. **Anterior nucleus of the thalamus.** Located in the floor of the lateral ventricle.

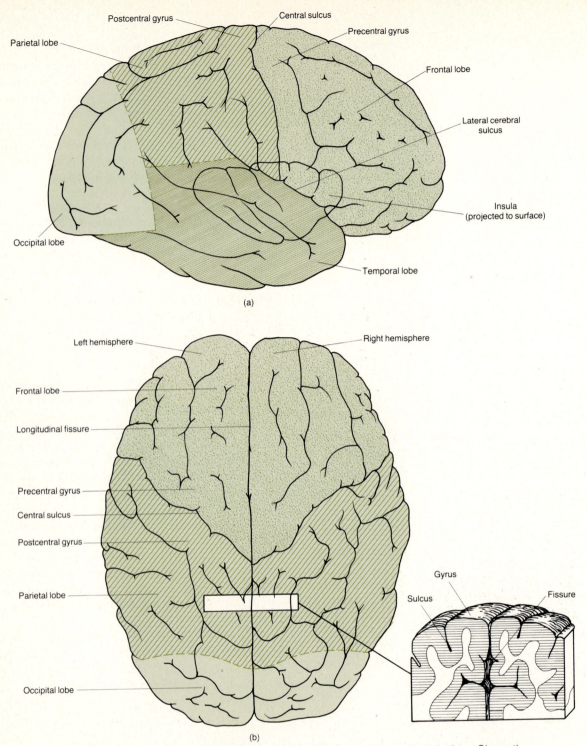

Figure 12-5 Lobes and fissures of the cerebrum. (a) Right lateral view. Since the insula cannot be seen externally, it has been projected to the surface. It can be seen in Figure 12-4a. (b) Superior view. The insert in (b) indicates the relative differences between a gyrus, sulcus, and fissure.

The limbic system functions in the emotional aspects of behavior related to survival. It also functions in memory. Although behavior is a function of the entire nervous system, the limbic system controls most of its involuntary aspects. Experiments on the limbic system of monkeys and other animals indicate that the amygdaloid nucleus assumes a major role in controlling the overall pattern of behavior.

Other experiments have shown that the limbic system is associated with pleasure and pain. When certain areas of the limbic system of the hypothalamus, thalamus, and midbrain are stimulated, experimental

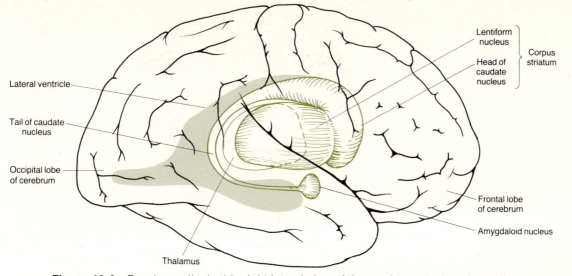

Figure 12-6 Basal ganglia. In this right lateral view of the cerebrum, the basal ganglia have been projected to the surface. Refer to Figure 12-4a to note the positions of the basal ganglia in the frontal section of the cerebrum.

animals indicate they are experiencing intense punishment. When other areas are stimulated, the animals' reactions indicate they are experiencing extreme pleasure. In still other studies, stimulation of the perifornical nuclei of the hypothalamus result in a behavioral pattern called *rage*. The animal assumes a defensive posture—extending its claws, raising its tail, hissing, spitting, growling, and opening its eyes wide. Stimulating other areas of the limbic system results in an opposite behavioral pattern: docility, tameness, and affection.

Functional Areas of Cerebral Cortex

The functions of the cerebrum are numerous and complex. In a general way, the cerebral cortex is divided into motor, sensory, and association areas. The **motor areas** control muscular movement. The **sensory areas** interpret sensory impulses. And the **association areas** are concerned with emotional and intellectual processes.

Sensory Areas. The *general sensory area* or *somesthetic area* is located directly posterior to the central sulcus of the cerebrum on the postcentral gyrus. It extends from the longitudinal fissure on the top of the cerebrum to the lateral cerebral sulcus. In Figure 12-7a the general sensory area is designated by the areas numbered 1, 2, and 3.* The general sensory area receives sensations from cutaneous, muscular, and visceral receptors in various parts of

* These numbers, as well as most of the others shown, are based on K. Brodmann's cytoarchitectural map of the cerebral cortex. His map, first published in 1909, is an attempt to correlate structure and function.

the body. Each point of the general sensory area receives sensations from specific parts of the body. Essentially the entire body is spatially represented in the general sensory area. The portion of the sensory area receiving stimuli from body parts is not dependent on the size of the part but on the number of receptors. For example, a greater portion of the sensory area receives impulses from the lips than from the thorax. The major function of the general sensory area is to localize exactly the points of the body where the sensations originate. The thalamus is capable of localizing sensations in a general way. That is, the thalamus receives sensations from large areas of the body but cannot distinguish between specific areas of stimulation. This ability is reserved to the general sensory area of the cortex.

Posterior to the general sensory area is the *somesthetic association area*. It corresponds to the areas numbered 5 and 7 in Figure 12-7a. The somesthetic association area receives input from the thalamus, other lower portions of the brain, and the general sensory area. The somesthetic association area integrates and interprets sensations. This area permits you to determine the exact shape and texture of an object without looking at it, to determine the orientation of one object to another as they are felt, and to sense the relationship of one body part to another. Another role of the somesthetic association area is the storage of memories of past sensory experiences. Thus you can compare sensations with previous experiences.

Other sensory areas of the cortex include:

1. **Primary visual area** (area 17). Located on the medial surface of the occipital lobe and occa-

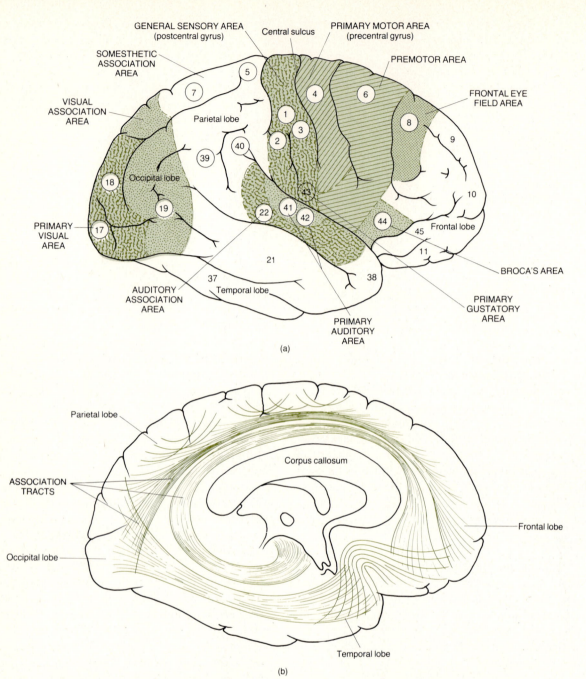

Figure 12-7 Functional areas of the cerebrum. (a) The lateral view indicates the sensory and motor areas. Although the right hemisphere is illustrated, Broca's area is in the left hemisphere of most people. (b) The sagittal section shows the association tracts.

sionally extends around to the lateral surface. It receives sensory impulses from the eyes and interprets shape and color.

2. **Visual association area** (areas 18 and 19). Located in the occipital lobe and receives sensory signals from the primary visual area and the thalamus. It relates present to past visual experiences with recognition and evaluation of what is seen.

3. **Primary auditory area** (areas 41 and 42). Located

in the superior part of the temporal lobe near the lateral cerebral sulcus. It interprets the basic characteristics of sound such as pitch and rhythm.

4. **Auditory association area** (area 22). Inferior to the primary area in the temporal cortex. It determines if a sound is speech, music, or noise. It also interprets the meaning of speech by translating words into thoughts.

5. **Primary gustatory area** (area 43). Located at the

base of the postcentral gyrus above the lateral cerebral sulcus in the parietal cortex. It interprets sensations related to taste.

6. **Primary olfactory area.** Located in the temporal lobe on the medial aspect and interprets sensations related to smell.

7. **Gnostic area** (areas 5, 7, 39, and 40). This *common integrative area* is located between the somesthetic, visual, and auditory association areas. The gnostic area receives impulses from these areas, as well as from the taste and smell areas, the thalamus, and lower portions of the brain stem. The gnostic area integrates all thoughts from the various sensory areas so that a common thought can be formed from the various sensory inputs. It then transmits signals to other parts of the brain to cause the appropriate response to the sensory signal.

Motor Areas. The *primary motor area* is located mainly in the precental gyrus of the frontal lobe (Figure 12-7a). This region is also designated as area 4. Like the general sensory area, the primary motor area consists of points that control specific muscles or groups of muscles. Stimulation of a specific point of the primary motor area results in a muscular contraction, usually on the opposite side of the body.

The *premotor area* (area 6) is anterior to the primary motor area. It is concerned with learned motor activities of a complex and sequential nature. It generates impulses that cause a specific group of muscles to contract in a specific sequence. An example of this is writing. Thus the premotor area controls skilled movements.

The *frontal eye field area* (area 8) in the frontal cortex is sometimes included in the premotor area. This area controls voluntary scanning movements of the eyes—searching for a word in a dictionary, for instance.

The *language areas* are also significant parts of the motor cortex. When you listen to someone speaking, sounds are relayed to the primary auditory area of the cortex. The sounds are then interpreted as words in the auditory association area. The words are interpreted as thoughts in the gnostic area. Written words are interpreted by the visual association area and converted into thoughts by the gnostic area. Thus you can translate speech or written words into thoughts.

The translation of thoughts into speech involves *Broca's area* or the *motor speech area*, designated as area 44 and located in the frontal lobe just superior to the lateral cerebral sulcus. From this area, a sequence of signals is sent to the premotor regions that control the muscles of the larynx, throat, and mouth. The

impulses from the premotor area to the muscles result in specific, coordinated contractions that enable you to speak. Simultaneously, impulses are sent from Broca's area to the primary motor area. From here, impulses reach your breathing muscles to regulate the proper flow of air past the vocal cords. The coordinated contractions of your speech and breathing muscles enable you to translate your thoughts into speech.

Broca's area is usually located in the left cerebral hemisphere of most individuals regardless of whether they are left-handed or right-handed. Injury to the sensory or motor speech areas results in *aphasia*, which is an inability to speak; *agraphia*, an inability to write; *word deafness*, an inability to understand spoken words; or *word blindness*, an inability to understand written words.

Association Areas. The *association areas* of the cerebrum are made up of association tracts that connect motor and sensory areas (see Figure 12-7b). The association region of the cortex occupies the greater portion of the lateral surfaces of the occipital, parietal, and temporal lobes and the frontal lobes anterior to the motor areas. The association areas are concerned with memory, emotions, reasoning, will, judgment, personality traits, and intelligence.

Brain Waves

Brain cells can generate electrical activity as a result of literally millions of action potentials of individual neurons. The electrical potentials are called **brain waves** and indicate activity of the cerebral cortex. Brain waves pass through the skull easily and can be detected by sensors called electrodes. A record of such waves is called an **electroencephalogram (EEG).** An EEG is obtained by placing electrodes on the head and amplifying the waves with an electroencephalograph. As indicated in Figure 12-8, four kinds of waves are produced by normal individuals:

1. **Alpha waves.** These rhythmic waves occur at a frequency of about 10 to 12 cycles/second. They are found in the EEGs of nearly all normal individuals when awake and in the resting stage. These waves disappear entirely during sleep.

2. **Beta waves.** The frequency of these waves is between 15 and 60 cycles/second. Beta waves generally appear when the nervous system is active—that is, during periods of sensory input and mental activity.

3. **Theta waves.** These waves have frequencies of 5 to 8 cycles/second. Theta waves normally occur in children and in adults experiencing emotional stress.

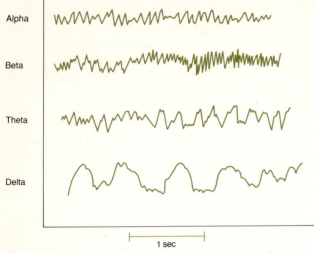

Alpha

Beta

Theta

Delta

|← 1 sec →|

Figure 12-8 Kinds of waves recorded in an electroencephalogram.

4. **Delta waves.** The frequency of these waves is 1 to 5 cycles/second. Delta waves occur during sleep. They are normal in an awake infant. When produced by an awake adult, they indicate brain damage.

Distinct EEG patterns appear in certain abnormalities. In fact, the EEG is used clinically in the diagnosis of epilepsy, infectious diseases, tumors, trauma, and hematomas. Electroencephalograms also furnish information regarding sleep and wakefulness.

Cerebellum

The **cerebellum** is the second-largest portion of the brain and occupies the inferior and posterior aspects of the cranial cavity. Specifically, it is below the posterior portion of the cerebrum and is separated from it by the *transverse fissure* (see Figure 12-1). The cerebellum is also separated from the cerebrum by an extension of the cranial dura mater called the *tentorium cerebelli*. The cerebellum is shaped somewhat like a butterfly. The central constricted area is the *vermis*, which means worm-shaped, and the lateral "wings" are referred to as *hemispheres* (Figure 12-9). Between the hemispheres is another extension of the cranial dura mater: the *falx cerebelli*. It passes only a short distance between the cerebellar hemispheres.

The surface of the cerebellum, called the *cortex*, consists of gray matter in a series of slender, parallel ridges called *gyri*. These gyri are less prominent than those located on the cerebral cortex. Beneath the gray matter are white matter tracts (*arbor vitae*) that resemble branches of a tree. Deep within the white matter are masses of gray matter: the *cerebellar nuclei*.

The cerebellum is a motor area of the brain that produces certain subconscious movements in the skeletal muscles. These movements are required for coordination, for maintenance of posture, and for balance. The cerebellar peduncles are the fiber tracts that allow the cerebellum to perform its functions.

Let us now see how the cerebellum produces coordinated movement. Motor areas of the cerebral cortex voluntarily initiate muscle contraction. Once the movement has begun, the sensory areas of the cortex receive impulses from nerves in the joints. The impulses provide information about the extent of muscle contraction and the amount of joint movement. The term *proprioception* is applied to this sense of position of one body part relative to another. The cerebral cortex uses the proprioceptive sensations to determine which muscles are required to contract next and what strength they are to contract in order to continue moving in the desired direction. Then a pattern of impulses is generated by the cerebral cortex

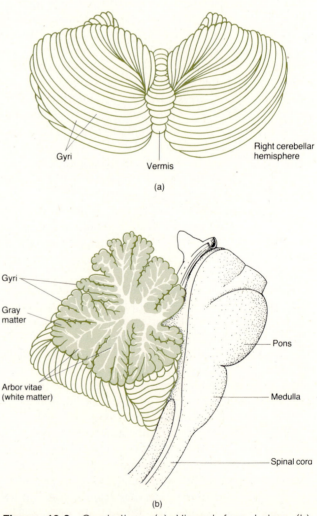

Gyri

Vermis

Right cerebellar hemisphere

(a)

Gyri

Gray matter

Arbor vitae (white matter)

Pons

Medulla

Spinal cord

(b)

Figure 12-9 Cerebellum. (a) Viewed from below. (b) Viewed in sagittal section.

along tracts to the pons and midbrain, which relay the impulses over the middle and superior cerebellar peduncles to the cerebellum. The cerebellum then generates subconscious motor impulses along the inferior cerebellar peduncles to the medulla and spinal cord. The impulses pass downward along the spinal cord and out the nerves that stimulate the prime movers and synergists to contract and that inhibit the contraction of the antagonists. The result is smooth, coordinated movement. A well-functioning cerebellum is essential for delicate movements such as playing the piano.

The cerebellum also transmits impulses that control postural muscles. That is, the cerebellum is required for maintaining normal muscle tone. The cerebellum also maintains body equilibrium. The inner ear contains structures that sense balance. Information such as whether the body is leaning to the left or right is transmitted from the inner ear to the cerebellum. The cerebellum then discharges impulses that cause the contraction of the muscles necessary for maintaining equilibrium.

Damage to the cerebellum through trauma or disease is characterized by certain symptoms involving skeletal muscles. There may be lack of muscle coordination, called *ataxia*. Blindfolded people with ataxia cannot touch the tip of their nose with a finger because they cannot coordinate movement with their sense of where a body part is located. Another sign of ataxia is a change in the speech pattern due to a lack of coordination of speech muscles. Cerebellar damage may also result in disturbances of gait in which the subject staggers or cannot coordinate normal walking movements.

CRANIAL NERVES

Of the 12 pairs of **cranial nerves,** 10 pairs originate from the brain stem, but all 12 pairs leave the skull through foramina in the base of the skull (see Figure 12-3a). The cranial nerves are designated in two ways —with Roman numerals and with names. The Roman numerals indicate the order in which the nerves arise from the brain (front to the back). The names indicate the distribution or function of the nerves. Some cranial nerves are termed *mixed nerves*. They contain both sensory and motor fibers. Other cranial nerves contain sensory fibers only. The cell bodies of sensory fibers are located in ganglia outside the brain. The cell bodies of motor fibers lie in nuclei within the brain.

Some motor fibers control subconscious movements, yet the somatic nervous system has been defined as a *conscious* system. The reason for this apparent contradiction is that some fibers of the auto-nomic nervous system leave the brain bundled together with somatic fibers of the cranial nerves. Therefore subconscious functions transmitted by the autonomic fibers are described along with the conscious functions of the somatic fibers of the cranial nerves. A summary of cranial nerves is presented in Exhibit 12-1.

HOMEOSTATIC IMBALANCES OF THE CENTRAL NERVOUS SYSTEM

Many disorders can affect the central nervous system. Some are caused by viruses or bacteria. Others are caused by damage to the nervous system during birth. The origins of many conditions, however, are unknown. Here we discuss the origins and symptoms of some common central nervous system disorders.

Poliomyelitis

Poliomyelitis, also known as **infantile paralysis,** is a viral infection that is most common during childhood. Onset of the disease is marked by fever, severe headache, a stiff neck and back, deep muscle pain and weakness, and loss of certain somatic reflexes. The virus may spread via the respiratory passages and blood to the central nervous system where it destroys the motor nerve cell bodies, specifically those in the anterior horns of the spinal cord and in the nuclei of the cranial nerves. Injury to the spinal gray matter is the basis for the name of this disease (*polio* = gray matter; *myel* = spinal cord). Destruction of the anterior horns produces paralysis. The first sign of bulbar polio is difficulty in swallowing, breathing, and speaking. Poliomyelitis can cause death from respiratory or heart failure if the virus invades the brain cells of the vital medullary centers. In recent years, an immunization against the disease has been used.

Syphilis

Syphilis is a venereal disease caused by the *Treponema pallidum* bacterium. Venereal diseases are infectious disorders that can be spread through sexual contact. The disease progresses through several stages: primary, secondary, latent, and sometimes tertiary. During the *primary stage*, the chief symptom is an open sore, called a chancre, at the point of contact. The chancre eventually heals. About 6 weeks later, symptoms such as a skin rash, fever, and aches in the joints and muscles usher in the *secondary stage*. At this stage, syphilis can usually be treated with antibiotics. Even if individuals do not undergo treatment, their symptoms will eventually disappear. Within a

Exhibit 12-1 SUMMARY OF CRANIAL NERVES

Number, Name, and Type	Location	Function and Comments
I. Olfactory (sensory)	Arises in olfactory mucosa, passes through olfactory bulb and olfactory tract, and terminates in the olfactory area of the cerebral cortex.	Smell. Fractures of the cribriform plate of the ethmoid may result in the loss of the sense of smell (*anosmia*).
II. Optic (sensory)	Arises in the retina of the eye, forms the optic chiasma, passes through the optic tracts, a nucleus in the thalamus, and terminates in the visual areas of the cerebral cortex.	Vision. Fractures and growths in the orbit and diseases of the nervous system may affect the optic nerve resulting in visual field defects and loss of visual acuity.
III. Oculomotor (mixed)	The motor portion originates in the midbrain and is distributed to the levator palpebrae superioris of the upper eyelid, four extrinsic eyeball muscles (superior rectus, medial rectus, inferior rectus, and inferior oblique), the ciliary muscle of the eyeball, and the sphincter muscle of the iris.	Movement of the eyelid and eyeball, accommodation of the lens for near vision, and constriction of the pupil.
	The sensory portion consists of afferent fibers from proprioceptors in the eyeball muscles terminating in the midbrain.	Muscle sense (proprioception).
		Complete paralysis of the oculomotor, trochlear, or abducent nerves is indicated by squints (*strabismus*) and seeing two objects instead of one (*diplopia*). In oculomotor paralysis, the affected eye is directed downward and outward, the upper eyelid droops (*ptosis*), the pupil is dilated and fixed, and there is a loss of accommodation for near vision.
IV. Trochlear (mixed)	The motor portion originates in the midbrain and is distributed to the superior oblique muscle.	Movement of the eyeball.
	The sensory portion consists of afferent fibers from proprioceptors in the superior oblique muscles to terminate in the midbrain.	Muscle sense (proprioception).
		In trochlear nerve paralysis, the individual may carry the head tilted to affected side.
V. Trigeminal (mixed)	The motor portion originates in the pons and terminates in the muscles of mastication.	Chewing. Injury results in paralysis of the muscles of mastication.
	The sensory portion consists of three branches: *ophthalmic*—contains sensory fibers from upper eyelid, eyeball, nasal cavity, side of nose, forehead, and anterior half of scalp; *maxillary*—contains sensory fibers from palate, pharynx, upper teeth, upper lip, cheek, and lower eyelid; *mandibular*—contains sensory fibers from anterior two-thirds of tongue, lower teeth, skin over mandible, and side of head in front of ear. The three branches terminate in the pons. There are also sensory fibers from proprioceptors in the muscles of mastication.	Conveys sensations for touch, pain, and temperature from the head, face, and pharynx. Injury results in a loss of sensation of touch and temperature.
		Injury to the motor portion results in paralysis of the muscles of mastication. Injury to a sensory division results in loss of sensation of touch and temperature.

Exhibit 12-1   *(Continued)*

Number, Name, and Type	Location	Function and Comments
		Neuralgia (pain) of one or more branches of the trigeminal nerve is called *tic douloureux* or *trigeminal neuralgia.*
VI. Abducens (mixed)	The motor portion originates in the pons and is distributed to the lateral rectus muscle.	Movement of the eyeball.
	The sensory portion consists of afferent fibers from proprioceptors in the lateral rectus and terminates in the pons.	Muscle sense (proprioception).
		In abducens nerve paralysis, the affected eye is directed inward.
VII. Facial (mixed)	The motor portion originates in the pons and is distributed to facial and scalp muscles and the sublingual and submandibular glands.	Facial expression and the secretion of saliva. Injury produces paralysis of the muscles of facial expression and reduced salivary secretion.
	The sensory portion arises from taste buds on the anterior two-thirds of the tongue, passes through the geniculate ganglion, a nucleus in the pons that sends fibers to the thalamus for relay to the cerebral cortex, and terminates in the thalamus. There are also sensory fibers from proprioceptors in the muscles of the face and scalp.	Taste. Injury produces loss of taste.
VIII. Vestibulocochlear (sensory)	Consists of a cochlear and a vestibular branch. The cochlear branch arises in the spiral organ of Corti, forms the spiral ganglion, passes through a nucleus in the medulla; and terminates in the thalamus. The vestibular branch arises in the semicircular canals, the saccule, and the utricle and forms the vestibular ganglion, and terminates in the medulla and pons.	The cochlear branch conveys impulses associated with hearing; the vestibular branch conveys impulses associated with equilibrium. Injury to the cochlear branch may cause ringing, buzzing, or deafness. The commonest symptoms of damage to the vestibular branch are dizziness and loss of balance.
IX. Glossopharyngeal (mixed)	The motor portion originates in the medulla and is distributed to the swallowing muscles of the pharynx and to the parotid gland.	Swallowing movements and the secretion of saliva. Injury results in pain during swallowing and reduced secretion of saliva.
	The sensory portion arises from taste buds on the posterior one-third of the tongue and from the carotid sinus, and terminates in the thalamus. There are also sensory fibers from proprioceptors in the muscles innervated by this nerve.	Taste and regulation of blood pressure. Injury results in loss of sensations in the throat and loss of taste from posterior part of tongue.
X. Vagus (mixed)	The motor portion originates in the medulla and terminates in the muscles of the pharynx, larynx, respiratory passageways, esophagus, heart, stomach, small intestine, most of large intestine, and gallbladder.	Visceral muscle movement and swallowing movements.
	The sensory portion arises from essentially the same structures supplied by the motor fibers and terminates in the medulla and pons. There are also sensory fibers from proprioceptors supplied by this nerve.	Sensations from organs supplied.
		Severing of both nerves in the upper body interferes with swallowing, paralyzes vocal cords, and interrupts sensations from many organs. Injury to both nerves in the abdominal area has little effect since the abdominal organs

Exhibit 12-1 *(Continued)*

Number, Name, and Type	Location	Function and Comments
XI. Accessory (mixed)	The motor portion consists of a bulbar portion and a spinal portion. The bulbar portion originates from the medulla and supplies voluntary muscles of the pharynx, larynx, and soft palate. The spinal portion originates from the anterior gray horn of the first five cervical segments of the spinal cord and supplies the sternocleido-mastoid and trapezius muscles.	are also supplied by autonomic fibers that arise from the spinal cord. The bulbar portion mediates swallowing movements; the spinal portion mediates movements of the head. Injury to the spinal portion results in an inability to turn the head or raise the shoulders.
	The sensory portion consists of afferent fibers from proprioceptors in the muscles supplied by the motor components.	Muscle sense (proprioception).
XII. Hypoglossal (mixed)	The motor portion originates in the medulla and supplies the muscles of the tongue.	Movement of the tongue during speech and swallowing.
	The sensory portion consists of fibers from proprioceptors in the tongue muscles that terminate in the medulla.	Muscle sense (proprioception).
		Injury results in difficulty in chewing, speaking, and swallowing. The tongue, when protruded, curls toward the affected side and the affected side becomes atrophied, shrunken, and deeply furrowed.

few years, the disease will cease to be infectious. The symptoms of the disease disappear, but a blood test is generally positive. During this later "symptomless" period, called the *latent stage*, the bacteria may invade and slowly destroy body organs. Untreated syphilis is considered dangerous for this reason. When organ degeneration appears, the disease is said to be in the *tertiary stage*. If the syphilis bacteria attack the organs of the nervous system, the tertiary stage is called *neurosyphilis*. Neurosyphilis may take different forms, depending on the tissue involved. For instance, about 2 years after the onset of the disease, the bacteria may attack the meninges, producing meningitis. The blood vessels that supply the brain may also become infected. In this case, symptoms depend on the parts of the brain destroyed by oxygen and glucose starvation. Cerebellar damage is manifested by uncoordinated movements as in writing. As the motor areas become extensively damaged, victims may be unable to control urine and bowel movements. Eventually, they may become bedridden, unable even to feed themselves. Damage to the cerebral cortex produces memory loss and personality changes that range from irritability to hallucinations.

Cerebral Palsy

The term **cerebral palsy** refers to a group of motor disorders caused by damage to the motor areas of the brain during fetal life, birth, or infancy. One cause is infection of the mother with German measles during the first 3 months of pregnancy. During early pregnancy, certain cells in the fetus are dividing and differentiating in order to lay down the basic structures of the brain. These cells can be abnormally changed by toxin from the measles virus. Radiation during fetal life, temporary oxygen starvation during birth, and hydrocephalus during infancy may also damage brain cells.

Cases of cerebral palsy are categorized into three groups depending on whether the cortex, the basal ganglia of the cerebrum, or the cerebellum is affected most severely. Most cerebral palsy victims have at least some damage in all three areas. The location and extent of motor damage determine the symptoms. The victim may be deaf or partially blind. About 70 percent of cerebral palsy victims appear to be mentally retarded. The apparent mental slowness, however, is often due to the person's inability to speak or hear well. Such individuals are often more mentally acute than they appear.

Cerebral palsy is not a progressive disease. Thus it does not worsen as time elapses. Once the damage is done, however, it is irreversible.

Parkinsonism

This disorder, also called **Parkinson's disease,** is a progressive malfunction of the basal ganglia of the cerebrum. The basal ganglia regulate subconscious contractions of skeletal muscles that aid activities desired by the motor areas of the cerebral cortex — swinging the arms when walking, for example. In Parkinsonism, the basal ganglia produce unnecessary skeletal movements that often interfere with voluntary movement. For instance, the muscles of the upper extremities may alternately contract and relax, causing the hands to shake. This shaking is called *tremor.* Other muscles may contract continuously, causing rigidity of the involved body part. *Rigidity* of the facial muscles gives the face a masklike appearance. The expression is characterized by a wide-eyed, unblinking stare and a slightly open mouth with uncontrolled drooling. Vision, hearing, and intelligence are unaffected by the disorder, indicating that Parkinsonism does not attack the cerebral cortex.

Parkinsonism seems to be caused by a malfunction at the neuron synapses. The motor neurons of the basal ganglia release the chemical transmitter acetylcholine. In normal people, the basal ganglia also produce a synaptic transmitter called dopamine, which quickly inactivates the acetylcholine and prevents continuous conduction across the synapse. People with Parkinsonism do not manufacture enough dopamine in their brains. As a result, stimulated basal ganglia neurons do not easily stop conducting impulses. Injections of dopamine are useless; the blood-brain barrier stops it. However, symptoms are somewhat relieved by a drug developed a few years ago, levodopa, and its successors carbidopa and bromocriptine — none without distressing side effects.

Multiple Sclerosis

Multiple sclerosis causes progressive destruction of the myelin sheaths of neurons in the central nervous system. The sheaths deteriorate to *scleroses*, which are hardened scars or plaques, in multiple regions — hence the name. The destruction of myelin sheaths interferes with the transmission of impulses from one neuron to another, literally short-circuiting conduction pathways. Multiple sclerosis is one of the most common disorders of the central nervous system. Usually the first symptoms occur between the ages of 20 and 40. Early symptoms are generally produced by the formation of a few plaques and are, consequently, mild. Plaque formation in the cerebellum may produce lack of coordination in one hand. The patient's handwriting becomes strained and irregular. A short-circuiting of pathways in the corticospinal tract may partially paralyze the leg muscles so that the patient drags a foot when walking. Other early symptoms include double vision and urinary tract infections. Following a period of remission during which the symptoms temporarily disappear, a new series of plaques develop and the victim suffers a second attack. One attack follows another over the years. Each time the plaques form, certain neurons are damaged by the hardening of their sheaths. Other neurons are uninjured by their plaques. The result is a progressive loss of function interspersed with remission periods during which the undamaged neurons regain their ability to transmit impulses.

The symptoms of multiple sclerosis depend on the areas of the central nervous system most heavily laden with plaques. Sclerosis of the white matter of the spinal cord is common. As the sheaths of the neurons in the corticospinal tract deteriorate, the patient loses the ability to contract skeletal muscles. Damage to the ascending tracts produces numbness and short-circuits impulses related to position of body parts and flexion of joints. Damage to either set of tracts also destroys spinal cord reflexes.

As the disease progresses, most voluntary motor control is eventually lost and the patient becomes bedridden. Death occurs anywhere from 7 to 30 years after the first symptoms appear. The usual cause of death is a severe infection resulting from the loss of motor activity. Without the constricting action of the urinary bladder wall, for example, the bladder never totally empties and stagnant urine provides an environment for bacterial growth. Bladder infection may then spread to the kidney, damaging kidney cells.

Multiple sclerosis may be caused by a virus. The occasional appearance of more than one case in a family suggests such an infectious agent, but the same circumstances also suggest a genetic predisposition. Like other demyelinating diseases, multiple sclerosis is incurable. Electrical stimulation of the spinal cord can improve function in certain patients, however.

Cerebrovascular Accidents

The most common brain disorder is a **cerebrovascular accident (CVA),** also called a **stroke** or **cerebral apoplexy.** A CVA is the destruction of brain tissue or infarction resulting from disorders in the vessels that supply the brain. Common causes of CVAs are intracerebral hemorrhage from aneurysms, embolism, and atherosclerosis of the cerebral arteries. An *intracerebral hemorrhage* is a rupture of a vessel in the pia mater or brain. Blood seeps into the brain and damages neurons by increasing intracranial fluid pressure. An *embolus* is a blood clot, air bubble, or bit of foreign material, most often debris from an inflammation, that becomes lodged in an artery and blocks circulation. *Atherosclerosis* is the formation of plaques in the artery walls. The plaques may slow down circulation by constricting the vessel. Both emboli and atherosclerosis cause brain damage by reducing the supply of oxygen and glucose needed by brain cells.

Many elderly people suffer mild CVAs as a result of short periods of reduced blood supply. Another cause is *arteriosclerosis,* or hardening of the arteries, which occurs with aging. Damage is generally undetectable or very mild. During these mild CVAs the individual may have a short blackout, blurred vision, or dizziness and does not realize anything serious has occurred. A CVA can also cause sudden, massive damage, however. Severe CVAs cause about 21 percent of all deaths from cardiovascular disease. The person who recovers may suffer partial paralysis and mental disorders such as speech difficulty. The malfunction depends on the parts of the brain that were injured. Vascular disorders are more common after age 40.

Dyslexia

Dyslexia (*dys* = difficulty; *lexis* = words) is unrelated to basic intellectual capacity, but it causes a mysterious difficulty in handling words and symbols. Apparently some peculiarity in the brain's organizational pattern distorts the ability to read, write, and count. Letters in words seem transposed, reversed, or even topsy-turvy — *dog* becomes *god; b* changes identity with *d;* sign saying "OIL" inverts into "710." Many dyslexics cannot orient themselves in the three dimensions of space and may show bodily awkwardness.

The cause of dyslexia is unknown, since it is unaccompanied by outward scars of detectable neurological damage and its symptoms vary from victim to victim. It occurs three times as often among boys as among girls. It has been variously attributed to defective vision, brain damage, lead in the air, physical trauma, or oxygen deprivation during birth. It remains an unsolved problem.

Tay-Sachs Disease

Tay-Sachs disease is a central nervous system affliction that brings death before age 5. The Tay-Sachs gene is carried by normal-appearing individuals descended from the Ashkenazi Jews of Eastern Europe. Approximately one in 3,600 of their offspring will be afflicted with Tay-Sachs disease. The disease is caused by the neuronal degeneration of the central nervous system because of excessive amounts of the sphingolipid known as ganglioside G_{m2} in the nerve cells of the brain. The afflicted child will develop normally until the age of 4 to 8 months. Then symptoms follow a course of progressive degeneration: paralysis, blindness, inability to eat, decubitus ulcers, and death from infection. There is no known cure for Tay-Sachs disease.

Now that you have a basic understanding of the structure and function of the brain and cranial nerves, and their relationship to homeostasis, we shall examine the various motor and integrative functions of the nervous system. You will see how sensory information is translated into motor responses that assume a key role in controlling homeostasis.

CHAPTER SUMMARY OUTLINE

BRAIN

Protection and Coverings

1. The brain is protected by the cranial bones, the meninges, and the cerebrospinal fluid.

Cerebrospinal Fluid

1. Cerebrospinal fluid circulates through the subarachnoid space around the brain and spinal cord, and through the ventricles of the brain.

2. The fluid has two major functions related to homeostasis: first, it acts as a shock absorbing medium to cushion the brain and spinal cord from jolting, and second, it provides a route for the elimination of metabolic by-products.

3. The cerebrospinal fluid is produced by the choroid plexuses in the ventricles of the brain.

4. The accumulation of cerebrospinal fluid in the head is called hydrocephalus. If the fluid accumulates in the ventricles, it is called internal hydrocephalus. If it accumulates in the subarachnoid space, it is called external hydrocephalus.

Blood Supply

1. Interruption of the oxygen supply to the brain can result in paralysis, mental retardation, epilepsy, or death.
2. Glucose deficiency may produce dizziness, convulsions, and unconsciousness.
3. The blood-brain barrier is a concept that explains the differential rates of passage of certain materials from the blood into the brain.

Brain Stem

1. The medulla oblongata is continuous with the upper part of the spinal cord. It contains nuclei that are reflex centers for regulation of heart rate, respiration rate, vasoconstriction, swallowing, coughing, vomiting, sneezing, and hiccuping.
2. The pons is superior to the medulla. It connects the spinal cord with the brain and links parts of the brain with each other. It relays impulses from the cerebral cortex to the cerebellum related to voluntary skeletal movements. The reticular formation of the pons contains the pneumotaxic and apneustic centers.
3. The midbrain connects the pons and cerebellum with the diencephalon. It conveys motor impulses from the cerebrum to the cerebellum and cord and conveys sensory impulses from cord to thalamus. It regulates auditory and visual reflexes.

Diencephalon

1. The thalamus is superior to the midbrain. It contains nuclei that serve as relay stations for all sensory impulses except smell. These include hearing, vision, general sensation, and taste. It also contains nuclei that are centers for synapses in the somatic motor system, such as voluntary motor actions and arousal. Conscious recognition of pain and temperature and general awareness of touch and pressure are also located in the thalamus.
2. The hypothalamus is anterior and inferior to the thalamus. It controls the autonomic nervous system, body temperature, food and fluid intake, waking state, and sleep.

Cerebrum

1. The cerebrum is the largest brain portion. Its surface (cortex) contains gyri, fissures, and sulci.
2. The functions of the cerebral cortex are motor (voluntary muscular movement), sensory (interpreting sensory impulses), and associational (emotional and intellectual processes). It contains the basal ganglia or cerebral nuclei in each cerebral hemisphere. Certain parts of these hemispheres constitute the limbic system—sometimes called the "visceral" or "emotional" brain.
3. The sensory areas of the cerebral cortex include the following: primary visual area, visual association area, primary auditory area, auditory association area, primary gustatory area, primary olfactory area, and gnostic area.
4. The motor areas of the cerebral cortex include the following: premotor area, frontal eye field area, language areas, and Broca's area or motor speech area.
5. The association areas of the cerebral cortex are concerned with memory, emotions, reasoning, will, judgment, personality traits, and intelligence.
6. Brain waves generated by the cerebral cortex may be recorded on an EEG. They are used to diagnose epilepsy, infections, and tumors.

Cerebellum

1. The cerebellum is the second-largest portion of the brain. It is posterior to the pons and medulla and inferior to the cerebrum. The cerebellum is the motor area of the brain that produces certain subconscious movements in the skeletal muscles. These movements are required for coordination, posture, and balance.

CRANIAL NERVES

1. Twelve pairs of cranial nerves originate from the brain.
2. The pairs are named primarily on the basis of distribution and numbered by order of attachment to the brain.

HOMEOSTATIC IMBALANCES OF THE CENTRAL NERVOUS SYSTEM

1. Poliomyelitis or infantile paralysis is a viral infection that is more common in children. The virus spreads via the respiratory passages and blood to the central nervous system, where it destroys the motor nerve cell bodies.
2. Syphilis is a venereal disease caused by bacteria. The disease progresses through several stages. When it reaches the fourth stage, called the tertiary stage, the disease is called neurosyphilis because the nervous system is affected.
3. Cerebral palsy refers to a group of motor disorders caused by damage to the motor areas of the brain during fetal life, birth, or infancy.

4. Parkinsonism, or Parkinson's disease, is a progressive malfunction of the basal ganglia of the cerebrum that regulate subconscious contractions of skeletal muscles.
5. Multiple sclerosis causes progressive destruction of the myelin sheaths of neurons in the central nervous system. This interferes with the transmission of impulses from one neuron to another, short-circuiting conduction pathways.
6. Cerebrovascular accident (CVA), also called stroke or apoplexy, is the most common brain disorder. Cerebrovascular accidents can be intracerebral hemorrhage, emboli, atherosclerosis, or arteriosclerosis.
7. Dyslexia causes a mysterious difficulty in handling words and symbols. The disorder distorts the individual's ability to read, write, and count.
8. Tay-Sachs disease is a central nervous system affliction that brings death before age 5. It is caused by the neuronal degeneration of the central nervous system.

REVIEW QUESTIONS

1. How is the brain protected?
2. Describe the composition, formation, and circulation of cerebrospinal fluid.
3. What are the functions of cerebrospinal fluid?
4. Describe the process of pneumoencephalography.
5. What is the difference between internal and external hydrocephalus?
6. Discuss the importance of a constant supply of oxygen and glucose to brain functioning. What is the blood-brain barrier?
7. What is the brain stem? Compare the medulla oblongata, pons varolii, and midbrain with regard to structure and function.
8. Define decussation. Give an example.
9. What are the functions of the thalamus?
10. Summarize the major functions of the hypothalamus.
11. What are the principal sulci and fissures of the cerebrum? How are they related to the lobes of the cerebrum? What is the cerebral cortex?
12. How are the names of the cranial bones related to the lobes of the cerebrum?
13. Compare the fibers in the white matter of the cerebrum with regard to direction and function.
14. What are basal ganglia? List two and describe their functions.
15. Discuss the motor, sensory, and association functions of the cerebrum. What is aphasia? Agraphia? Word deafness? Word blindness?
16. What is an electroencephalograph? What is its diagnostic value?
17. Describe the structure and functions of the cerebellum. What are cerebellar peduncles? With which body system is the cerebellum most closely associated in terms of coordination? Explain your answer.
18. How would damage to the cerebellum affect skeletal muscles?
19. What is ataxia? List several signs of ataxia.
20. What is a cranial nerve? Why are some cranial nerves classified as mixed nerves? How are cranial nerves named and numbered?
21. List the location and function of each cranial nerve.
22. What is poliomyelitis? What areas of the nervous system are mostly affected by it?
23. Describe the various stages of syphilis. Why is the last stage called neurosyphilis?
24. Define cerebral palsy. What are some causes of cerebral palsy?
25. What is Parkinson's disease? Correlate Parkinson's disease with neuronal synapses.
26. Describe the effects of multiple sclerosis on the central nervous system. Why do the symptoms of multiple sclerosis differ so much from one individual to another?
27. What is a CVA? Describe various CVAs and their possible harmful effects on the body.
28. Describe the disorder dyslexia. Give some examples of word distortion caused by dyslexia.
29. What is Tay-Sachs disease? Describe its progressive degeneration of the central nervous system.

Chapter 13

SENSORY PHYSIOLOGY

STUDENT OBJECTIVES

After reading this chapter, you should be able to

- Explain the relationship between the various types of sensory information and the human brain
- Define sensory information and follow its pathways to the central nervous system
- Define adequate stimulus, sensory unit, and convergence
- Explain the relationship between the intensity of the stimulus and the location of the stimulus
- Differentiate between the three types of sensory receptors on the basis of location and function
- Define mechanoreceptors, generator potential, and adaptation
- Give the name and describe the structure, location, and function of the receptors for touch, pressure, temperature, and the different types of pain
- Explain acupuncture and the "gate control" theory
- Identify the various sensory pathways

The complexity of the human nervous system is the result of an adaptive process that has had to cope with ever-increasing bits of information about the external and internal environments. The central nervous system, which consists of the brain and spinal cord, requires a continual flow of information in order to regulate the body's homeostasis and initiate appropriate responses to changes in the external and internal environments. The Canadian psychologist Donald Hebb found that when subjects were confined in a special laboratory and deprived of practically all sensory stimulation, many suffered profound psychotic illness. They suffered visual and auditory hallucinations, feelings of panic, and extreme agitation. He concluded that the central nervous system depends on a continual stream of information from the peripheral nervous system. Without such a flow, the brain conjures erroneous information on its own that can often lead to destructive behavioral responses.

Our brains receive and respond to many varieties of sensory information. The primary forms, or modalities, of sensation include touch, smell, temperature, light, pressure, and sound. Additional sensory information arising continually from within our bodies also bombards the brain. However, at any moment we are aware only of those sensations that we consciously focus upon. For example, we search anxiously with our eyes the crowd of people at an airport terminal for the face of a loved one. We may be oblivious to the weariness in our legs, the cold wind, or the jostling of the waiting crowd. We are tuning and straining our visual sense to the utmost. In a similar fashion, we "strain" our ears to hear a funny story told quietly above the loud din of a raucous party. We ignore any irrelevant sensation. We can appreciate the exquisite sensitivity of our fingertips for temperature and touch when we feel the fevered brow of a sick child or the swollen lymph glands in the neck. The point is that, when our attention is focused acutely to one type of sensory stimulus, the others are still functioning just as they normally would should we focus on their sensations. But the central nervous system has selected only those bits of information that are important for that moment, and it is only those bits that are brought to our conscious level. Indeed, we would collapse into nervous wrecks should our consciousness be forced to deal with all the information arriving from our sense receptors. Experiments on volunteers subjected to loud noises, intense flashing lights, and pungent odors have shown that, under such conditions, the subjects retreat into a kind of sleep. The conscious mind is turned off to protect itself from overstimulation. The "turn off" mechanism is a subconscious act. This is a good illustration of the operation of homeostasis for self-protection. The sense receptors operate faithfully to detect and relay the stimuli to the brain, where the information is processed. Because the inflow is overwhelming, the subconscious mind switches it off, and a kind of stupor results.

It is critical for the brain to receive continual information about the external and internal environments for its proper functioning in order to maintain homeostasis. But overwhelming the brain with sensory input triggers a defensive mechanism that protects against a nervous collapse. We shall now examine the mechanisms involved in generating sensory nerve impulses.

CONTENT OF SENSORY INFORMATION

Information from the environment reaching the central nervous system and rising to the level of conscious awareness is termed **sensory information.** The pathways for sensory information are the afferent nerve fibers, which are activated by sensory receptors. The structure of nerve fibers and how they carry impulses have been described in detail in Chapter 9. Their outstanding feature is the ability to conduct action potentials over considerable distances with a constant amplitude. These action potentials are able to carry information from the receptor to the central nervous system describing the kind, intensity, location, and duration of the incoming stimulus. The central nervous system interprets the pattern and the frequency of action potentials delivered by the nerve fiber. The brain never sees the image, hears the sound, or feels the sensation. It only receives volleys of action potentials from sensory receptors. By some unknown process, we then perceive the image, sound, or sensation.

Kinds of Stimuli

It is well known that receptors are designed to respond more readily to only one certain form of energy. For example, light energy striking the retina of the eye generates a visual sensation. That same light energy shone into the ear has no effect in generating sound. But other forms of energy can stimulate the eye and ear. A sharp blow to the eye or ear, that is, a form of mechanical energy, can produce a flash of color or a loud, thumping sound. However, for pressure to excite the retina or ear it needs to be sharp and intense with a total energy input far beyond that of the normal stimulus. An *adequate stimulus* is that form of stimulation to which a given receptor has the lowest threshold. Thus, for the retina, light energy is the adequate stimulus, and for the ear, sound energy is the adequate stimulus.

It becomes apparent, therefore, that different external environmental changes or stimuli are detected and coded by different sense organs. In order for this coded information to reach the appropriate center in the central nervous system, the afferent nerve fibers or pathway connecting the receptor to the central nervous system remain separated from other pathways relaying a different type of sensory information.

Sensory Unit

A single afferent nerve fiber plus the receptor units to which it is connected is termed a **sensory unit.** This situation is analogous to the motor unit you studied in Chapter 10. In some instances, the nerve fiber innervates only one receptor unit, but generally the afferent nerve fiber branches into several peripheral terminals and connects with several receptor units (Figure 13-1). The surface area covered by a single afferent fiber varies according to the extent of its branching. For example, if the branching is widely distributed, the afferent fiber can be activated by applying the adequate stimuli anywhere within a large area. If, on the other hand, the branching is limited, then the stimuli will have to be applied within a smaller area around the receptors. The *receptive field* of a neuron is that area which, if stimulated, leads to activity in the afferent neuron (Figure 13-1).

A nerve fiber that has been activated carries the action potentials into the central nervous system. The end of the afferent fiber that enters the central nervous system usually branches into a number of *central terminals,* which can make synapses with many association neurons in the central nervous system. This is referred to as *divergence* (Figure 13-2a). Since many other afferent fibers also enter the central nervous system at the same location, their central terminals may also synapse with the same association neurons. This overlap of the central terminals produces a phenomenon called *convergence* (Figure 13-2b). In the spinal cord, the association neurons are grouped into bundles arranged in a parallel array according to the sensory information they convey toward the brain. In the region of the brain called the thalamus, these ascending pathways make their final synapse before

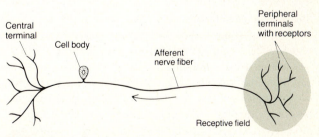

Figure 13-1 Components of a sensory unit.

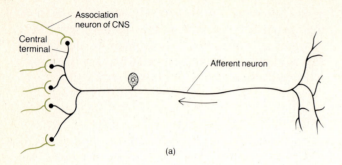

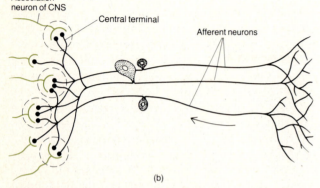

Figure 13-2 (a) Divergence of afferent neuron terminals. (b) Convergence of afferent neurons.

they are distributed to their ultimate destinations in the cerebral cortex.

Intensity of the Stimulus

Information about the intensity of the stimulus is relayed to the central nervous system in two fashions. First, the greater the intensity of the stimulus, the greater the number of sensory units that will be activated. Second, each responding unit indicates increasing stimulation by sending out a greater number of action potentials per unit time.

Location of the Stimulus

Since the pathway for information from the sensory unit to the central nervous system is relatively direct and fixed, the information arrives describing the location of stimulation. The accuracy of the location is determined on a differential basis that can be explained in the following way.

The sensitivity of the receptors in one sensory unit is not uniform. The threshold for stimulation of the receptors nearest the center of the receptive field is lower (more easily activated) than those toward the edges of the receptive field (Figure 13-3).

By visualizing an arrangement of overlapping termi-

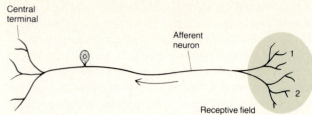

Figure 13-3 Threshold for stimulation of receptors nearest the center of the receptive field is lower than for those near the edge. 1 and 2 are the stimulus points.

nals (Figure 13-4), we can see that the point of stimulation occurs near the center of the receptive field for neuron 2 but in the periphery of the receptive fields for 1 and 3. The result is that neuron 2 fires a volley of impulses of greater frequency than those of neurons 1 and 3. Analysis of the incoming information enables the central nervous system to compare the frequency of the impulses and locate the center of the stimulation to be near the center of receptive field 2.

The analysis of this information for accurate location is a relatively simple task when the brain deals with sensations of touch, pressure, temperature, pain, and joint position—the so-called *somatic sensory systems*. Each *modality*, or type of sensory stimulus, has its own pathway up the spinal cord, through the brain stem, and to the thalamus. In the spinal cord or brain stem, the sensory pathways cross over to the opposite side of the central nervous system. Thus, the sensory information from the right side of the body crosses over to the left side and is projected to the left cerebral hemisphere. Similarly, the sensory information from the left side of the body crosses over to the right side and is projected to the right cerebral hemisphere. Areas on the surface of the cerebral hemispheres have been mapped out which represent the termination of sensory information from all parts

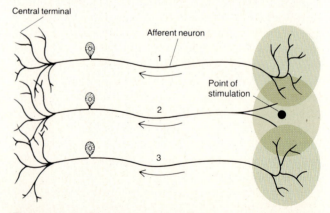

Figure 13-4 The point of stimulation occurs nearest the center of the receptive field for afferent neuron 2, but in the periphery of the receptive field for afferent neurons 1 and 3.

of the body. Figure 13-5 shows the location and areas of representation on the cerebral cortex. Note that the areas with greatest representation are those parts of the body with the largest number of sensory receptors. Some parts with greatest sensitivity are the fingers, lips, and tongue. The somatosensory area on one side of the cortex is matched by a corresponding strip on the opposite side.

CLASSIFICATION OF SENSORY RECEPTORS

The central nervous system detects changes in its environments through impulses sent from strategically located structures called **sensory receptors.** These structures may be as simple as a single bare nerve ending or as complex as the human eye. Sensory receptors are sometimes divided into three types on the basis of location:

1. **Exteroceptors** are receptors that are located at or near the surface of the body and deliver information to the central nervous system so that it finally reaches the level of consciousness. Thus, we are aware of these sensations. Exteroceptors respond to stimuli that arise from outside of the body. Included in this group are the eye, which responds to light; receptors in the skin, which respond to pressure and temperature changes; the ear, which responds to sound; and the taste (olfactory) receptors, which are stimulated by soluble and volatile molecules.

2. **Interoceptors** perform the vital role of monitoring the internal environment of our bodies. They are located in blood vessels and viscera. Some operate to bring certain sensations to the level of consciousness, and others operate reflexively to regulate internal functions. Hunger, thirst, respiration, circulation, and temperature are sensed by interoceptors and are controlled by the nervous system.

3. **Proprioceptors** are sensory receptors that gather information about the positions of the various parts of the body. They are located in skeletal muscles, tendons, and joints. Other proprioceptors respond to pressure changes that occur in the muscular wall of the heart and carotid sinus and in the walls of the gastrointestinal tract. The proprioceptors are specially designed mechanoreceptors that signal both the degree of stretch or tension present in a muscle and the speed of contraction when it shortens.

FUNCTION OF SENSORY RECEPTORS

Receptors are the specialized cells that transform the different kinds of energy that continually affect o[ur] bodies into action potentials. Receptors the[n] as transducers (transformers), [turn one] energy into another. In this reg[ard] assume a key role in helping [maintain] homeostasis. Each receptor is [sensitive] mainly to only one form of ene[rgy. A receptor is] simply the bare nerve endings of sens[ory nerves] or an elaborate complex of cells that connects to the sensory nerve.

The simplest examples of receptor organs are the pressure and touch receptors (see Figure 13-7). These *mechanoreceptors,* as they are called, transduce mechanical energy into electrical energy. Thorough studies of mechanoreceptors have led to certain general conclusions that apply to most sensory receptors.

Generator Potentials (Sensory Potentials)

When a receptor is stimulated, the nerve ending connected to the receptor demonstrates changes in its resting potential. This change may, in turn, initiate the all-or-none response of an action potential. The initial electrical activity at the nerve ending is called the **generator** or **sensory potential.**

In mechanoreceptors, the terminal, unmyelinated

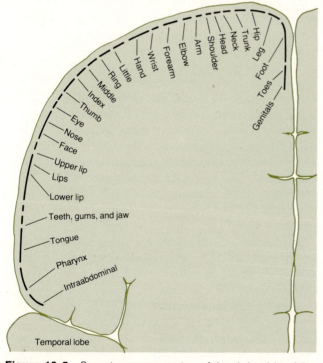

Figure 13-5 Somatosensory cortex of the right side of the brain showing areas of pathway terminations to various regions of the left side of the body.

portion of the afferent nerve seems to be very sensitive to stretch deformation. Application of pressure presumably causes some changes in the physical properties of the membrane. These changes increase membrane permeability so that the small ions inside and outside the cell flow down their gradients. As you will recall from Chapter 9, potassium ions are higher in concentration inside the cell, sodium ions are higher outside the cell, and the resting membrane potential is near −70 mV. Since sodium is further from its equilibrium potential than is potassium, there is a much greater inflow of sodium than an outflow of potassium. The net result is that the membrane is partially depolarized. The depolarizations that are taking place in the peripheral terminals of the afferent fiber are the generator potentials. In this unmyelinated region of the nerve, the membrane has a very high threshold, and the generator potential is unable to depolarize the nerve sufficiently to trigger an action potential. However, the terminal depolarization does set up local currents that draw charges from the adjacent regions. The first node region of the myelinated portion of the afferent fiber has a lower threshold. As charges are drawn from the first node, depolarization occurs at that point, and because the threshold is much lower there, it can easily be reached. Thus, an action potential is initiated and conducted toward the central nervous system.

Properties of Generator Potentials

...therefore act ...changing one type of ...sensory receptors ...the body to maintain ...designed to respond ...rgy. Receptors can be ...sensory nerve fibers ...action potentials. Instead, their amplitude varies directly in proportion to the strength of the stimulus. Their rate of recovery is much slower, and they persist for 5 to 10 milliseconds. The generator potential will maintain the local current and will continually depolarize the membrane at the first node. As long as this condition exists, action potentials will be generated and propagated along the sensory fiber toward the central nervous system. The pattern of impulses sent along the afferent fiber is determined by the amplitude and duration of the generator potential.

Factors Affecting Generator Potentials

The amplitude and duration of the generator potential vary with the intensity of the provoking stimulus and the manner in which the stimulus is applied: whether gradually or very abruptly. The generator potential

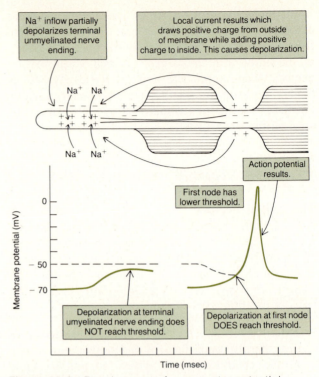

Figure 13-6 Development of a generator potential.

increases with greater intensity of stimulus because the membrane permeability increases proportionately and there is a greater influx of sodium ions. Thus, the depolarization is greater and the local current that results triggers a longer burst of action potentials at the first node. If the application of a stimulus is gradual, the burst of action potentials is not nearly as large. Apparently, the membrane can recover or prevent some of the permeability changes if the stimulus is applied slowly.

Some receptors signal the removal of a stimulus. For example, you can be totally unaware of the pressure of a hat that you have been wearing for a long time. But its removal would immediately trigger a response because the membranes are deformed as they change to a new shape. Thus, their permeability characteristics would be altered, eliciting the generator potential.

Summation. If stimuli are applied at sufficiently frequent intervals, their individual effects will accumulate. This phenomenon is called *summation* of generator potentials and can be explained as follows. Generator potentials persist for 5 to 10 milliseconds. If a second stimulus is applied to the receptor before the first generator potential has died away, the second generator potential adds to, or summates with, the first to produce a single but larger generator potential.

Adaptation. The phenomenon of adaptation is demonstrated by the drop in the frequency of action potentials along the afferent nerve fiber despite a constant, prolonged level of stimulation. The decrease in frequency, or *adaptation*, occurs for three reasons. First, though the level of stimulation may remain constant, the proportion reaching the receptor member decreases with time. In mechanoreceptors, the surrounding tissues adjust to the applied pressure and are redistributed, so the actual level of intensity on the receptor is less. Similar tissue adjustments can be described for other sensory modalities. A second factor in the adaptation phenomenon involves the receptor membrane itself. Its responsiveness to the pressure decreases with time. This may reflect some molecular adjustments in the membrane structure to decrease the permeability and thus lower the amplitude of the generator potential, which would decrease the number of action potentials sent forth from the first nodal region. A third factor, which is suggested by experimental evidence, is that the membrane in the region of the first node can also change and become less responsive. The threshold may rise slightly, thus reducing the number of action potentials sent forth even if the generator potential remains the same.

Exhibit 13-1 compares the properties of the generator and action potentials. The important fact linking the two types is that the *amplitude* of the generator potential determines the *number* of action potentials that will be sent down the afferent fiber.

Recruitment. To complete our description of how information about the stimulus strength is relayed to the central nervous system, we shall briefly discuss the process of *recruitment*. As the strength of stimulation increases, additional receptors adjacent to those first responding begin to respond. For example, if you touch your finger lightly with a pencil point, only those receptors under the point will respond. But if you press harder, the area of skin depressed will increase and additional receptors will be "recruited" to respond. This recruitment contributes information about the total increase in the energy of stimulation.

CUTANEOUS SENSORY SYSTEMS

The human skin is a sense organ that responds to a myriad of stimuli. Certain gentle stimuli, such as moving a hair on the back of your hand, trigger a sensation referred to as *touch*. If the hair is bent and then held in position, the sensation of touch disappears in a couple of seconds. If, however, a dull needle is pressed gently into the skin and held constant, the sensation is detected, persists for a minute or so, and then disappears. This is termed a *pressure* sensation. If more pressure is applied to the needle, receptors in the subcutaneous tissues respond, and this sensation will persist for several minutes. This sensation is termed *deep pressure*. If one of the tines of a driven tuning fork is placed against the skin, a *vibration* will be felt. And if a feather is gently drawn across the skin, a sensation called *tickle* is experienced. These are all harmless mechanical stimuli felt by the skin.

Very small increases in the temperature of the skin are sensed as *warmth*, whereas decreases in skin temperature are sensed as *cold*. The receptors in the skin can sense changes as small as 0.2°C. If however, skin temperature is elevated to 45°C or more, or dropped to a temperature of 10°C or less, the sensations become that of *pain*. Similarly, excessively strong mechanical and chemical stimuli that are destructive to skin tissue also create painful sensations. Another sensation associated with certain forms of skin damage is *itch*. Distinct from pain, itch accompanies inflammatory responses.

Exhibit 13-1 COMPARISON BETWEEN GENERATOR POTENTIAL AND ACTION POTENTIAL

Point of Comparison	Generator Potential	Action Potential
1. Response	Graded.	All-or-none.
2. Summation	Yes. If a second stimulus arrives before the generator potential of the first stimulus is over, the generator potential from the second stimulus is added to the depolarization from the first.	No.
3. Refractory period	No.	Yes, about 1 msec.
4. Conduction	Passive; decreases in magnitude with increasing distance along the nerve fiber.	Propagated without loss of amplitude along the length of the nerve fiber
5. Duration	Greater than 1 to 2 msec; varies.	1 to 2 msec.

This brief description serves to illustrate the array of sensations detected by skin receptors. Figure 13-7 shows various receptors found in the skin.

Touch

The sensation of touch is triggered by free nerve endings surrounding the hair follicles and by specialized receptors called *Meissner's corpuscles*. They are oval structures consisting of coiled nerve endings surrounded by masses of irregular-shaped cells (Figure 13-7a).

Pressure

The sensation of pressure is detected by *Pacinian corpuscles* (Figure 13-7b). Located in the deeper parts of the subcutaneous tissue, they are also able to respond to vibratory stimuli. When cut in section, the corpuscles resemble a sliced onion, consisting of a bare nerve ending surrounded by thin, concentric layers of connective tissue.

Temperature

Heat and cold receptors were once thought to be *end organs of Ruffini* and *end bulbs of Krause,* respectively (Figure 13-7c, d). The distribution of the receptors is uneven. The warmth detectors lie deep in the skin and are distributed generally over the body, but are most numerous in the skin on the sole of the foot. The cold receptors are more numerous in the skin generally, but are particularly concentrated in the mucous membranes of the mouth and genital organs.

Pain

Pain is a very complex sensation that always includes a reaction or response component. It is an indicator of a homeostatic imbalance. When a painful stimulus is applied, it causes tissue damage or sets up a potential situation that could result in damage. It triggers an immediate reflex withdrawal reaction and initiates a series of physiological responses that are usually accompanied by emotions such as rage, fear, or the "fight or flight" response. The physiological changes usually include an increase in heart rate and blood pressure, and an elevation in blood glucose concentration, probably due to an outpouring of epinephrine from the adrenal medulla. The adrenal response also reduces blood flow to the viscera and skin, reduces gastric secretion and motility, causes increased blood flow to the skeletal muscles, causes profuse sweating, and dilates the pupils. Pain sensations can be modified by past experiences, other sensory input, hypnosis, alcohol, and various drugs. The reactions to pain vary from person to person, and the same person may react differently under changing circumstances. Serious battle wounds may be ignored, whereas under quieter conditions, the same injuries could be incapacitating. Patients given morphine or patients with frontal lobe lesions seem able to report pain sensations, but the reactive, emotional component is no longer present. In such reports the patient senses pain but says, "It doesn't hurt."

The peripheral receptors for pain are high-threshold bare nerve endings that innervate areas up to one square centimeter (Figure 13-7e). These fine, unmyelinated terminals extend between cells and intermesh profusely with terminals from many other pain neurons. Pain endings do not respond selectively to only one type of painful stimulus. Rather, they respond nonselectively to any mechanical, thermal, or chemical stimulus if the intensity of stimulation is useful in protecting us from greater destruction. We learn to avoid future contact with pain-inducing situations. A patient's description of painful symptoms plays an important part in medical diagnosis.

Referred Pain

In most cases, the cortex accurately projects the pain back to the stimulated area. For example, if you burn your finger, you localize the pain experience in your finger. Similarly, an individual with inflammation of the lining of the pleural cavity experiences pain in the affected area. In the viscera, when tissue damage occurs, pain is often experienced in a location removed from the site of injury. The pain may be felt in or just under the skin that overlies the stimulated organ; or the pain may be felt in a surface area of the body distant from the stimulated organ. This phenomenon is called **referred pain**. In general, the area to which the pain is referred and the visceral organ involved receive their innervation from the same segment of the spinal cord. Consider the following example. Afferent fibers from the heart enter segments T1 to T4 of the spinal cord as do afferents from the skin over the heart and the skin over the medial surface of the left arm. Thus, pain resulting from heart damage is typically felt in the skin over the heart and along the medial surface of the left arm. Figure 13-8 illustrates cutaneous regions to which visceral pain may be referred.

Phantom Pain

A kind of pain frequently experienced by amputees is called **phantom pain.** The person experiences pain in

81 82 83 84 9 8 7 6 5 4 3 2 1

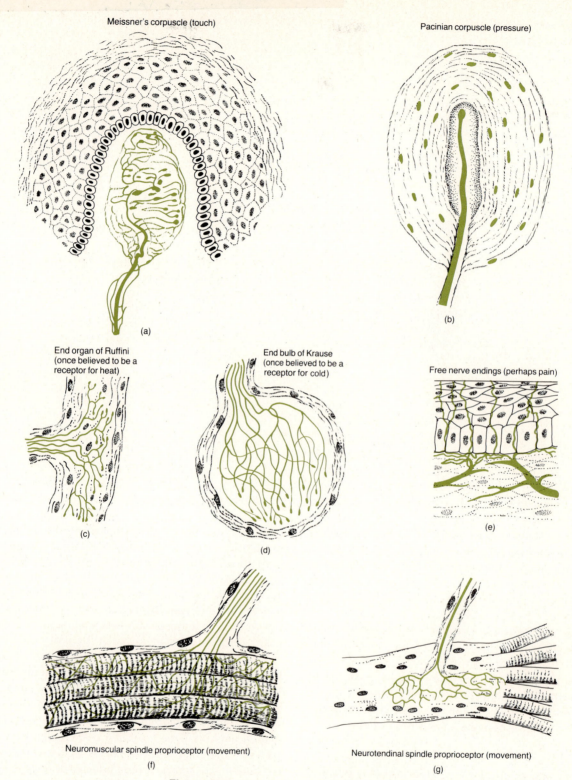

Meissner's corpuscle (touch)

Pacinian corpuscle (pressure)

(a)

(b)

End organ of Ruffini
(once believed to be a
receptor for heat)

End bulb of Krause
(once believed to be a
receptor for cold)

Free nerve endings (perhaps pain)

(c)

(d)

(e)

Neuromuscular spindle proprioceptor (movement)
(f)

Neurotendinal spindle proprioceptor (movement)
(g)

Figure 13-7 Types of sensory nerve endings.

a limb or part of a limb after it has been amputated. Here is how phantom pain occurs. Let us say that a foot has been amputated. A sensory nerve that origi- nally terminated in the foot is severed during the opera- tion, but repairs itself and returns to function within the remaining leg. From past experience, the brain

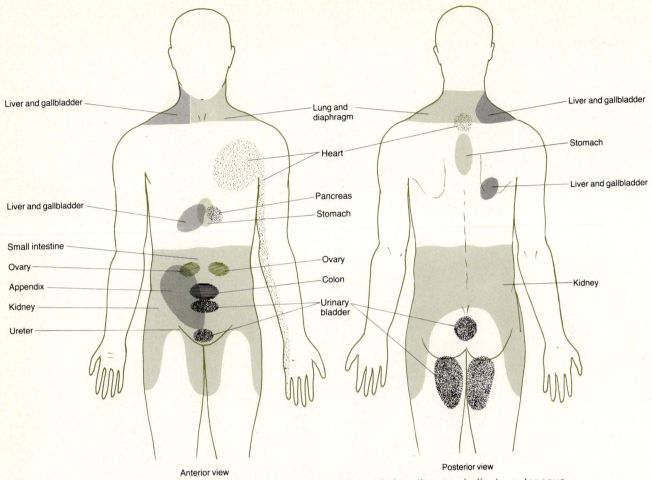

Figure 13-8 Referred pain. The shaded parts of the diagram indicate cutaneous areas to which visceral pain is referred.

has always projected stimulation of the neuron back to the foot. So, when the distal end of this neuron is now stimulated, the brain continues to associate the sensation with the missing part. Thus, even though the foot has been amputated, the patient still "feels" pain in the toes.

Most pain sensations respond to painkilling drugs. In a few individuals, however, pain may be controlled only by surgery. The aim of surgical treatment is to interrupt the pain impulse somewhere between the receptors and the interpretation centers of the brain. This is done by severing the sensory nerve, its spinal root, or certain tracts in the spinal cord or brain.

Acupuncture and Pain

Another method of inhibiting pain impulses is called **acupuncture** (*acus* = needle; *pungere* = to sting). Here is how the procedure is performed. Needles are inserted through selected areas of the skin. The needles are then twirled by the acupuncturist or by a battery-operated device. About 20 to 30 minutes after the

twirling starts, pain is deadened for 6 to 8 hours. The location of needle insertion varies with the part of the body the acupuncturist desires to anesthetize. For a tooth extraction, one needle is inserted in the web between the thumb and the index finger. For a tonsillectomy, one needle is inserted about 2 inches above the wrist. For removal of a lung, one needle is placed in the forearm, midway between the wrist and elbow.

There is no complete or totally satisfactory explanation of why acupuncture works. According to the "gate control" theory, the twirling of the acupuncture needle stimulates two sets of nerves that eventually enter the spinal cord and synapse with the same association neurons. One very fine nerve is the nerve for pain and the other, a much thicker nerve, is the nerve for touch. The speed of the impulse passing along the touch nerve is faster than that passing along the pain nerve. Recall that fibers with larger diameters conduct impulses faster than those with smaller diameters. Because the touch impulse reaches the dorsal horn of the spinal cord first, it has right of way over the pain

impulse. It thus "closes the gate" to the brain before the pain impulse reaches the cord. Since the pain impulse does not pass to the brain, no pain is felt. We should mention, however, that before acupuncture can be used as a routine procedure by American physicians, for the well-being of the public additional research and understanding of the process must take place.

SENSORY PATHWAYS

Sensory information transmitted from the spinal cord to the brain is conducted along two general pathways: the posterior column pathway and the spinothalamic pathway.

In the **posterior column pathway** to the cerebral cortex there are three separate sensory neurons. The *first-order neuron* connects the receptor with the spinal cord and medulla on the same side of the body. The cell body of the first-order neuron is in the posterior root ganglion of a spinal or cranial nerve. The first-order neuron synapses with a *second-order neuron*. The second-order neuron passes from the medulla upward to the thalamus. The cell body of the second-order neuron is located in the nuclei cuneatus and gracilis of the medulla. Before passing into the thalamus, the second-order neuron crosses to the opposite side of the medulla and enters the medial lemniscus, a projection tract that terminates at the thalamus. In the thalamus, the second-order neuron synapses with a *third-order neuron*. The third-order neuron terminates in the somesthetic sensory area of the cerebral cortex. The posterior column pathway conducts impulses related to proprioception, fine touch, two-point discrimination, and vibrations.

The **spinothalamic pathway** is composed of three orders of sensory neurons also. The first-order neuron connects a receptor of the neck, trunk, and extremities with the spinal cord. The cell body of the first-order neuron is in the posterior root ganglion also. The first-order neuron synapses with the second-order neuron, which has its cell body in the posterior gray horn of the spinal cord. The fiber of the second-order neuron crosses to the opposite side of the spinal cord and passes upward to the brain stem in the lateral spinothalamic tract or ventral spinothalamic tract. The fibers from the second-order neuron terminate in the thalamus. There, the second-order neuron synapses with a third-order neuron. The third-order neuron terminates in the somesthetic sensory area of the cerebral cortex. The spinothalamic pathway con-

veys sensory impulses for pain and temperature as well as crude touch and pressure.

The second-order neurons of the spinothalamic pathway enter the medulla, pons, and midbrain. Thus, the spinothalamic pathway conducts sensory signals that result in subconscious motor reactions. By contrast, second-order neurons of the posterior column pathway have a direct connection with the thalamus and cerebral cortex. Thus, the posterior column pathway conducts sensory information primarily into the conscious areas of the brain.

Now we can examine the physiology of specific sensory pathways—for pain and temperature, for crude touch and pressure, and for fine touch, proprioception, and vibration.

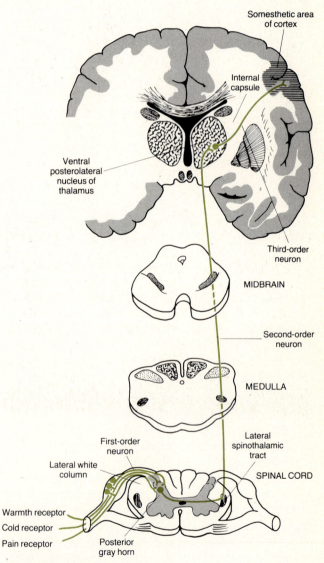

Figure 13-9 Sensory pathway for pain and temperature—the lateral spinothalamic pathway.

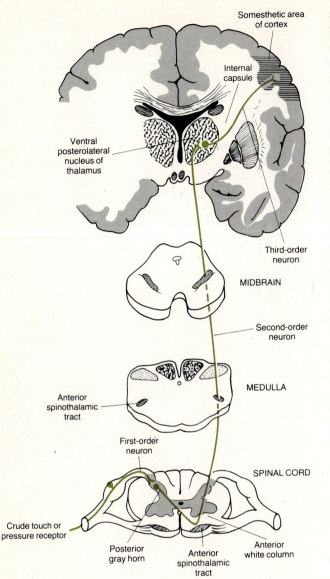

Figure 13-10 Sensory pathway for crude touch and pressure—the anterior spinothalamic pathway.

Pain and Temperature

The sensory pathway for pain and temperature is called the **lateral spinothalamic pathway** (Figure 13-9). The first-order neuron conveys the impulse for pain or temperature from the appropriate receptor to the posterior gray horn on the same side of the spinal cord. In the horn, the first-order neuron synapses with a second-order neuron. The axon of the second-order neuron crosses to the opposite side of the cord. Here it becomes a component of the *lateral spinothalamic tract* in the lateral white column. The second-order neuron passes upward in the tract through the brain stem to a nucleus in the thalamus called the ventral posterolateral nucleus. In the thalamus, conscious recognition of pain and temperature occurs. The

sensory impulse is then conveyed from the thalamus through the internal capsule to the somesthetic area of the cortex by a third-order neuron. The cortex analyzes the sensory information for the precise source, severity, and quality of the pain and heat stimuli.

Crude Touch and Pressure

The neural pathway that conducts impulses for crude touch and pressure is the **anterior (ventral) spinothalamic pathway** (Figure 13-10). By crude touch and pressure is meant the ability to perceive that something has touched the skin, although its exact location, shape, size, or texture cannot be determined. The first-order neuron conveys the impulse from a crude touch or pressure receptor to the posterior gray horn

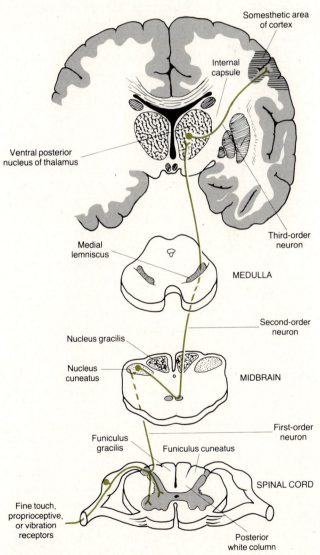

Figure 13-11 Sensory pathway for fine touch, proprioception, and vibration—the posterior column pathway.

on the same side of the spinal cord. In the horn, the first-order neuron synapses with a second-order neuron. The axon of the second-order neuron crosses to the opposite side of the cord and becomes a component of the *anterior spinothalamic tract* in the anterior white column. The second-order neuron passes upward in the tract through the brain stem to the ventral posterolateral nucleus of the thalamus. The sensory impulse is then relayed from the thalamus through the internal capsule to the somesthetic area of the cerebral cortex by a third-order neuron. Although there is some awareness of crude touch and pressure at the thalamic level, it is not fully perceived until the impulses reach the cortex.

Fine Touch, Proprioception, Vibration

The neural pathway for fine touch, proprioception, and vibration is called the **posterior column pathway** (Figure 13-11). This pathway conducts impulses that give rise to several discriminating senses:

1. Fine touch: the ability to recognize the exact location of stimulation and to distinguish that two points are touched, even though they are close together (two-point discrimination).
2. Stereognosis: recognizing the size, shape, and texture of an object.
3. Weight discrimination: the ability to assess the weight of an object.

4. Proprioception: the awareness of the precise position of body parts and directions of movement.
5. The ability to sense vibrations.

First-order neurons for the discriminating senses just noted follow a pathway different from those for pain and temperature and crude touch and pressure. Instead of terminating in the posterior gray horn, the first-order neurons from appropriate receptors pass upward in the fasciculus gracilis or fasciculus cuneatus in the posterior white column of the cord. From here the first-order neurons enter either the nucleus gracilis or nucelus cuneatus in the medulla, where they synapse with second-order neurons. The axons of the second-order neurons cross to the opposite side of the medulla and ascend to the thalamus through the medial lemniscus, a projection tract of white fibers passing through the medulla, pons, and midbrain. The second-order neuron axons synapse with third-order neurons in the ventral posterior nucleus in the thalamus. In the thalamus, there is no conscious awareness of the discriminating senses, except for a possible crude awareness of vibrations. The third-order neurons convey the sensory impulses to the somesthetic area of the cerebral cortex. It is here that you perceive your sense of position and movement and fine touch.

Now that you have an understanding of some of the important principles of sensory physiology, we shall take a look at the special senses such as taste, smell, sight, hearing, and equilibirum.

CHAPTER SUMMARY OUTLINE

CONTENT OF SENSORY INFORMATION

1. Information from the environment reaching the central nervous system and rising to the level of conscious awareness is termed sensory information.

Kinds of Stimuli

1. Different external environmental changes or stimuli are detected and coded by different sense organs.
2. The greater the intensity of the stimulus, the greater the number of sensory units that will be activated.
3. Each responding unit indicates increasing stimulation by sending out a greater number of action potentials.
4. In the spinal cord or brain stem the sensory pathways cross over to the opposite side of the central nervous system.

CLASSIFICATION OF SENSORY RECEPTORS

1. Exteroceptors are receptors that are located at or near the surface of the body and deliver information to the central nervous system so that it finally reaches the level of consciousness. Examples are the eye, ear, and taste receptors.
2. Interoceptors perform the vital role of monitoring the internal environment of our bodies. Examples include hunger and thirst.
3. Proprioceptors are sensory receptors that gather information about the position of the various parts of the body. They are located in skeletal muscles, tendons, and joints.

FUNCTION OF SENSORY RECEPTORS

1. Receptors are the specialized cells that transform different forms of energy that continually affect our bodies into action potentials. They help the body maintain homeostasis.

Generator Potentials

1. The initial electrical activity at the nerve ending is called the generator or sensory potential.
2. The amplitude of generator potentials varies directly in proportion to the strength of stimulus.
3. The amplitude and duration of the generator potential vary with the intensity of the provoking stimulus and the manner in which the stimulus is applied: whether gradually or very abruptly.

CUTANEOUS SENSORY SYSTEMS

Touch

1. The sensation of touch is triggered by free nerve endings surrounding the hair follicles and by specialized receptors called Meissner's corpuscles.

1. The sensation of pressure is detected by the Pacinian corpuscles.

Temperature

1. Heat and cold receptors are known as the end organs of Ruffini and the end bulbs of Krause, respectively.

Pain

1. Pain is a very complex sensation that always includes a reaction or response component. It is an indicator of a homeostatic imbalance.
2. Referred pain is pain that is felt in a surface area of the body distant from the stimulated organ.
3. Phantom pain occurs when an individual experiences pain in a limb or part of a limb after it has been amputated.
4. Acupuncture is another method of inhibiting pain impulses.

SENSORY PATHWAYS

1. In the posterior column pathway and the spinothalamic pathway there are first-order, second-order, and third-order neurons.
2. The sensory pathway for pain and temperature is the lateral spinothalamic pathway.
3. The neural pathway that conducts impulses for crude touch and pressure is the anterior spinothalamic pathway.
4. The neural pathway for fine touch, proprioception, and vibration is called the posterior column pathway.

REVIEW QUESTIONS

1. Explain in detail how the body is constantly receiving sensory information both externally and internally.
2. Correlate some different kinds of sensory information with the central nervous system response.
3. Define sensory information, adequate stimuli, sensory unit, receptive field, and convergence.
4. How is information about the intensity of the stimulus relayed to the central nervous system?
5. Why is sensory information from the right side of the body felt on the left side of the body and vice versa?
6. Define the three types of sensory receptors, and include examples or locations of each.
7. Define receptors, mechanoreceptors, and generator potentials.
8. How do generator and action potentials differ?
9. What are the factors that affect generator potentials?
10. What is adaptation? How does it occur?
11. Give the name and describe the structure and location of the receptors involved in touch, pressure, and temperature.
12. What is your understanding of pain? Why are pain receptors important?
13. Compare referred pain with phantom pain.
14. Explain the "gate control" theory of acupuncture.
15. What is the sensory pathway for pain and temperature?
16. Which pathway controls crude touch and pressure?
17. Which pathway is responsible for fine touch, proprioception, and vibrations?

Chapter 14

SPECIAL SENSES

STUDENT OBJECTIVES

After reading this chapter, you should be able to

- Identify the gustatory receptors and the four basic tastes, and describe the neural pathway for taste
- Locate the receptors for olfaction and describe the neural pathway for smell
- List the structures related to visual sensations
- Identify the structural divisions of the eyeball
- Discuss retinal image formation by describing refraction, accommodation, constriction of the pupil, convergence, and inverted image formation
- Diagram and discuss the rhodopsin cycle responsible for light sensitivity of rods
- Describe the afferent pathway of light impulses to the brain
- Define the anatomical subdivisions of the ear
- List the principal events in the physiology of hearing
- Identify the receptor organs for equilibrium
- Contrast cataracts, glaucoma, conjunctivitis, trachoma, Ménière's disease, and impacted cerumen as homeostatic imbalances of the special sense organs

The sensations of taste, smell, sight, hearing, and equilibrium are often referred to as the **special senses** because they have receptor organs that are highly complex. Of these, the sense of smell is the least specialized, whereas the sense of sight is considered the most complex. Like all sensory organs, the special senses allow us to detect changes in our environment in order to make homeostatic adjustments.

GUSTATORY SENSATIONS

Most people would support the idea that gustation or taste is a sensation that contributes enjoyment to life. The importance of this sensation in humans may be questioned, but in other animals it seems to have a very important function in selecting proper nutrients. It has been shown that animals with certain food deficiencies will preferentially select those foods that have a taste associated with the substance in which they are deficient.

Taste, as well as smell, is a chemical sense. Compared to other sensory structures, the taste receptors are highly adaptive, so the taste of any given substance can be modified by what the taste receptors experienced just a moment before. The taste sensation is rather complex, and the physiological mechanisms involved are not completely worked out.

Taste Buds

The primary chemical receptors for taste are the approximately 10,000 **taste buds,** most of which are located on the dorsal surface of the tongue. Others are distributed over the roof of the mouth and the mucosal linings of the soft palate, pharynx, lips, and cheeks. The taste buds of the tongue are located on the small connective tissue elevations called *papillae* that project out over the surface of the tongue. The papillae are of three types and differ in number, size, and distribution (Figure 14-1a). The smallest papillae, called the *filiform*, are most numerous and are distributed in parallel rows across the tongue. They rarely possess taste buds. The *fungiform* papillae project from the dorsal surface of the tongue like little mushrooms. They are not nearly as numerous as the filiform papillae. Their distribution is somewhat scattered, but with a higher concentration at the tip of the tongue, where they can be seen as red spots. Each of these papillae possesses 8 to 10 taste buds in the covering epithelium. There are only about 7 to 12 of the third type of papillae, called the *vallate* papillae. These are located along an inverted V-shaped line that separates the mucous membrane of the body of the tongue from the root portion. These papillae possess more than 200 taste buds each.

A taste bud looks like an oval wooden barrel and is 50 μm by 70 μm in size (Figure 14-1b–d). Each taste bud communicates with the fluids of the mouth through a *gustatory (taste) pore*. Two types of cells make up the taste bud: peripheral *supporting cells* and the elongated *gustatory (receptor) cells*. *Gustatory hairs (microvilli)* on the tips of the receptor cells project through the taste pore on the surface of the papilla. Each taste bud contains from 5 to 18 gustatory cells. The gustatory cells are innervated by terminal

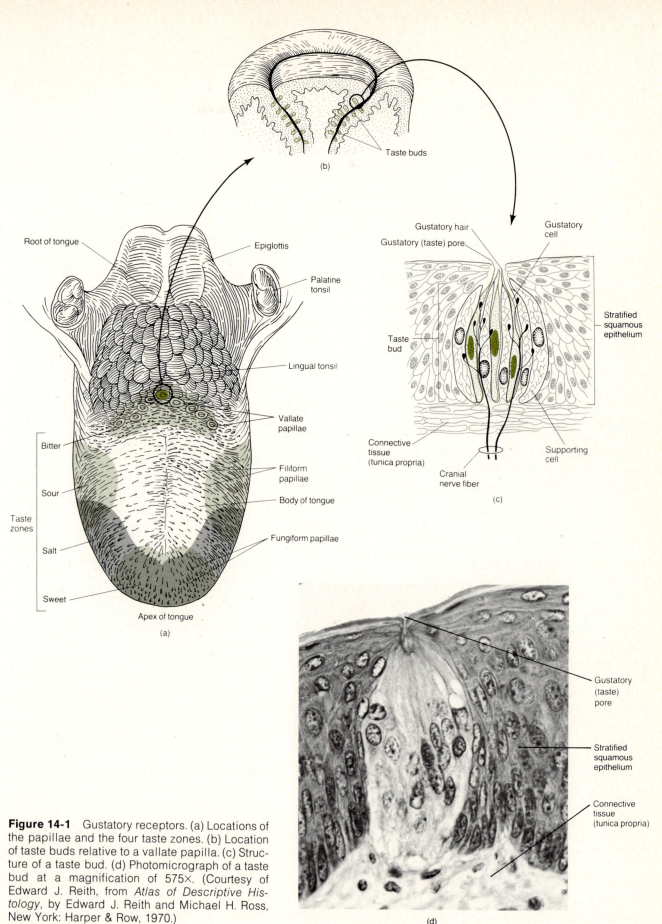

Figure 14-1 Gustatory receptors. (a) Locations of the papillae and the four taste zones. (b) Location of taste buds relative to a vallate papilla. (c) Structure of a taste bud. (d) Photomicrograph of a taste bud at a magnification of 575×. (Courtesy of Edward J. Reith, from *Atlas of Descriptive Histology,* by Edward J. Reith and Michael H. Ross, New York: Harper & Row, 1970.)

fibers that branch out and make contact with the basal surface of the gustatory cells. The gustatory cells of the taste bud are modified epithelial cells that are being constantly replaced every 10 to 15 days.

Basic Tastes

Only dissolved molecules are capable of triggering a taste sensation. The first step involved in taste probably involves the loose binding of the dissolved molecule with a specific receptor site on the gustatory cell membrane. Though we experience a considerable range and variety of taste sensations, we recognize four basic ones that combine in ways to express the much larger range. Sweet, sour, salty, and bitter are the four basic tastes. The receptors for these sensations are located in different parts of the tongue (see Figure 14-1a). Sweet is detected on the tip of the tongue, salt on the sides near the tip, sour on the sides nearer the root of the tongue, and bitter at the root of the tongue.

The existence of four basic tastes has led many investigators to search for four different types of taste buds, each able to respond better to certain tastes than to others. There is good evidence to support this notion. Other experiments suggest that each taste bud is capable of responding to all four basic tastes.

Innervation

The nerves that supply afferent fibers to taste buds are the facial (VII) nerve, which supplies the anterior two-thirds of the tongue; the glossopharyngeal (IX) nerve, which supplies the posterior one-third of the tongue; and the vagus (X) nerve, which supplies the epiglottis area of the throat. Taste impulses are conveyed from the gustatory cells in taste buds to the three nerves just cited. From these, the impulses enter the medulla, pass through the thalamus, and terminate in the parietal lobe of the cortex. It is here that the final interpretation and experience of taste occurs. Impulses from responding taste cells are summed, compared, and integrated with information received from other sensory receptors. It is known that the odor of a substance greatly influences its taste. The correct identification of a substance is also modified by its temperature and the tactile sensations it projects.

OLFACTORY SENSATIONS

For many other organisms, the sense of olfaction or smell is vital for self-preservation and for preservation of the species. For insects and lower animals, olfactory stimulation may be the most important signal for sexual activity. Though the sense of smell adds another dimension to life's experiences, it is certainly not as important in human activity.

Receptors

The **olfactory** or **smell receptors** are located in the nasal epithelium in the superior portion of the nasal cavity on either side of the nasal septum (Figure 14-2a). It has been estimated that the human has 10 million receptors located within a 5 cm² area. Since only about 2 percent of the inhaled air reaches the upper portions of the nasal cavity, only the more volatile molecules reach the olfactory receptor cells. Sniffing enhances the flow of air into these upper areas and enhances olfaction.

The nasal epithelium consists of two principal kinds of cells: supporting and olfactory (Figure 14-2b–c). The *supporting cells* are columnar epithelial cells of the mucous membrane lining the nose. Olfactory glands in the mucosa keep the mucous membrane moist. The *olfactory cells* are bipolar neurons whose cell bodies lie between the supporting cells. The distal (free) end of each olfactory cell contains six to eight dendrites, called *olfactory hairs*.

Innervation

The unmyelinated axons of the olfactory cells unite to form the *olfactory nerves*, which pass through foramina in the cribriform plate of the ethmoid bone. The olfactory nerves terminate in paired masses of gray matter called the *olfactory bulbs*. The olfactory bulbs lie beneath the frontal lobes of the cerebrum on either side of the crista galli of the ethmoid bone. The first synapse of the olfactory neural pathway occurs in the olfactory bulbs between the axons of the olfactory nerves and the dendrites of neurons inside the olfactory bulbs. Axons of these neurons run posteriorly to form the *olfactory tract*. From here, impulses are conveyed to the olfactory portion of the cortex. In the cortex, the impulses are interpreted as odor and give rise to the sensation of smell.

Odor Classification

There have been several attempts to classify odors in the hope that such grouping would provide clues to the mechanism of odor discrimination. To date, no progress in this direction has proved helpful in describing the underlying process of olfaction. Present research focuses on the molecular shapes of the odiferous molecules and receptor sites with matching

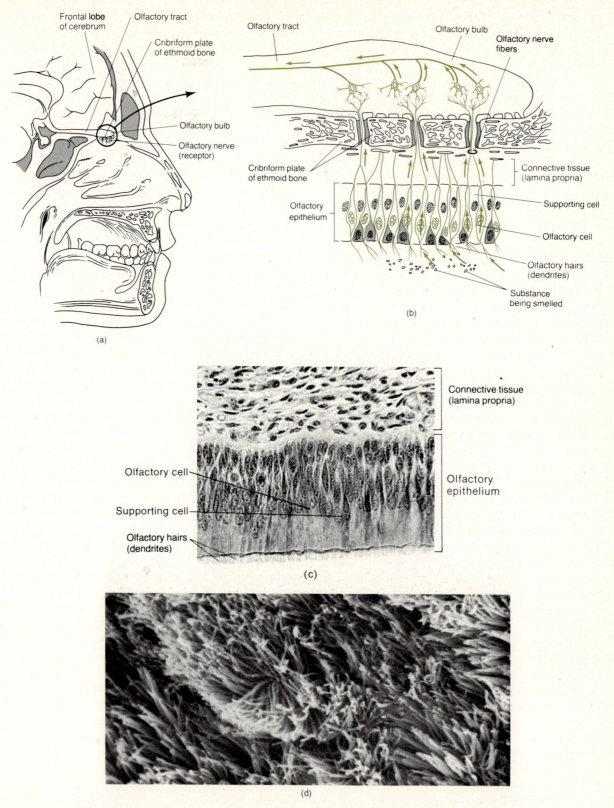

Figure 14-2 Olfactory receptors. (a) Location of receptors in nasal cavity. (b) Enlarged aspect of olfactory receptors. (c) Photomicrograph of the olfactory mucosa at a magnification of 400×. (Courtesy of Donald I. Patt, from *Comparative Vertebrate Histology*, by Donald I. Patt and Gail R. Patt, New York: Harper & Row, 1969.) (d) Scanning electron micrograph of the ciliated nasal mucosa at a magnification of 5,000×. (Courtesy of Fisher Scientific Company and S.T.E.M. Laboratories, Inc., Copyright 1975.)

shapes. The matching of molecule and receptor site might trigger action potentials in a manner that would indicate something of the quality of the odiferous molecule.

Despite 60 years of research, the operation of the olfactory mechanism is far from understood. It is known that different nerve fibers respond to different stimuli and the pattern of action potentials carries the information to the central nervous system, but the process that discriminates accurately the odors at the primary receptor site is still unknown.

VISUAL SENSATIONS

The structures related to visual sensations include the eyeball, which is the receptor organ for sight; the optic nerve; the visual area of the brain; and a number of accessory structures such as the eyebrows, eyelids, eyelashes, and the lacrimal apparatus. Details of the accessory structures may be found in any standard anatomy textbook.

Structure of Eyeball

The adult **eyeball** measures about 2.5 cm (1 inch) in diameter. Of its total surface area, only the anterior one-sixth is exposed. The remainder is recessed and protected by the orbit into which it fits. Anatomically, the eyeball can be divided into three layers: fibrous tunic, vascular tunic, and retina (Figure 14-3).

The **fibrous tunic** is the outer coat of the eyeball. It can be divided into two regions: the posterior sclera and the anterior cornea. The *sclera,* the "white of the eye," is a white coat of fibrous tissue that covers all the eyeball except the anterior colored portion. The sclera gives shape to the eyeball and protects its inner parts. Its posterior surface is pierced by the optic nerve. The anterior portion of the fibrous tunic is called the *cornea.* It is a nonvascular, nervous, transparent fibrous coat that covers the iris, the colored part of the eye. The outer surface of the cornea is covered by an epithelial layer that is continuous with the epithelium of the bulbar conjunctiva. At the junction of the sclera and cornea is a venous sinus known as the *canal of Schlemm.*

The **vascular tunic** is the middle layer of the eyeball and is composed of three portions: the posterior choroid, the anterior ciliary body, and the iris. The *choroid* is a thin, dark-brown membrane that lines most of the internal surface of the sclera. It contains numerous blood vessels and a large amount of pigment. The choroid absorbs light rays so they are not reflected back out of the eyeball. Through its blood

supply, it nourishes the retina. The optic nerve also pierces the choroid at the back of the eyeball. The anterior portion of the choroid becomes the *ciliary body.* It is the thickest portion of the vascular tunic. It extends from the *ora serrata* of the retina (inner tunic) to a point just behind the sclerocorneal junction. The ora serrata is simply the jagged margin of the retina. The second division of the vascular tunic contains the *ciliary muscle*—a smooth muscle that alters the shape of the lens for near or far vision. The *iris* is the third portion of the vascular tunic. It consists of circular and radial smooth muscle fibers arranged to form a doughnut-shaped structure. The black hole in the center of the iris is the *pupil,* the area through which light enters the eyeball. The iris is suspended between the cornea and the lens and is attached at its outer margin to the ciliary body. A principal function of the iris is to regulate the amount of light entering the eyeball. When the eye is stimulated by bright light, the circular muscles of the iris contract and decrease the size of the pupil. When the eye must adjust to dim light, the radial muscles of the iris contract and increase the pupil's size.

The third and inner coat of the eye, the **retina,** lies only in the posterior portion of the eye. Its primary function is image formation. It consists of a nervous tissue layer and a pigmented layer. The retina covers the choroid. At the edge of the ciliary body, it appears to end in a scalloped border, the ora serrata. This is where the outer nervous layer or visual portion of the retina ends. The inner pigmented layer extends anteriorly over the back of the ciliary body and the iris as the nonvisual portion of the retina. The inner nervous layer contains three zones of neurons. These three zones, named in the order in which they conduct impulses, are the photoreceptor neurons, bipolar neurons, and ganglion neurons.

The dendrites of the photoreceptor neurons are called rods and cones because of their shapes. These visual receptors are highly specialized for stimulation by light rays. **Rods** are specialized for vision in dim light. They also allow us to discriminate between different shades of dark and light and permit us to see shapes and movement. **Cones,** by contrast, are specialized for color vision and sharpness of vision *(visual acuity).* Cones are stimulated only by bright light. This is why we cannot see color by moonlight. It is estimated that there are 7 million cones and somewhere between 10 and 20 times as many rods. Cones are most densely concentrated in the *central fovea,* a small depression in the center of the macula lutea. The *macula lutea,* or yellow spot, is in the exact center of the retina. The fovea is the area of sharpest vision because of the high concentration of

cones. Rods are absent from the fovea and macula, but they increase in density toward the periphery of the retina.

When impulses for sight have passed through the photoreceptor neurons, they are conducted across synapses to the bipolar neurons in the intermediate zone of the nervous layer of the retina. From here the impulses are passed to the ganglion neurons.

The axons of the ganglion neurons extend posteriorly to a small area of the retina called the *optic disc,* or *blind spot.* This region contains openings through which the fibers of the ganglion neurons exit as the optic nerve. Since this area contains neither rods nor cones, and only nerve fibers, no image is formed on it. Thus it is called the blind spot.

In addition to the fibrous tunic, vascular tunic, and retina, the eyeball itself contains the lens, just behind the pupil and iris. The *lens* is constructed of numerous layers of protein fibers arranged like the layers of an onion. Normally, the lens is perfectly transparent. It is enclosed by a clear connective tissue capsule and held in position by the *suspensory ligament.*

The interior of the eyeball contains a large cavity divided into two smaller ones: the anterior cavity and the posterior cavity. They are separated from each other by the lens. The *anterior cavity,* in turn, has two subdivisions referred to as the anterior chamber and the posterior chamber. The *anterior chamber* lies behind the cornea and in front of the iris. The *posterior chamber* lies behind the iris and in front of the suspensory ligament and lens. The anterior cavity is filled with a watery fluid called the *aqueous humor.* The fluid is believed to be secreted into the posterior chamber by choroid plexuses of the ciliary processes of the ciliary bodies behind the iris. It is very similar to cerebrospinal fluid. From the posterior chamber, the fluid permeates the posterior cavity and then passes forward between the iris and the lens, through the pupil, into the anterior chamber. From the anterior chamber, the aqueous humor, which is continually produced, is drained off into the *canal of Schlemm* and then into the blood. The anterior chamber thus serves a function similar to the subarachnoid space around the brain and spinal cord. The canal of Schlemm is analogous to a venous sinus of the dura mater. The pressure in the eye, called *intraocular pressure,* is produced mainly by the aqueous humor. The intraocular pressure keeps the retina smoothly applied to the choroid so the retina may form clear images. Besides maintaining normal intraocular pressure, the aqueous humor is also the principal link between the circulatory system and the lens and cornea. Neither the lens nor the cornea has blood vessels.

The second, and larger, cavity of the eyeball is the *posterior cavity.* It lies between the lens and the retina and contains a jellylike substance called the *vitreous humor.* This substance contributes to intraocular pressure, helps to prevent the eyeball from collapsing, and holds the retina flush against the internal portions of the eyeball. The vitreous humor, unlike the aqueous humor, does not undergo constant replacement. It is formed during embryonic life and is not replaced thereafter.

Physiology of Vision

Before light can reach the rods and cones of the retina, it must pass through the cornea, aqueous humor, pupil, lens, and vitreous humor. Moreover, for vision to occur, light reaching the rods and cones must form an image on the retina. The resulting nerve impulses must then be conducted to the visual areas of the cerebral cortex. In discussing the physiology of vision, let us first consider retinal image formation.

Retinal Image Formation

The formation of an image on the retina requires four basic processes, all concerned with focusing light rays:

1. Refraction of light rays.
2. Accommodation of the lens.
3. Constriction of the pupil.
4. Convergence of the eyes.

Accommodation and pupil size are functions of the smooth muscle cells of the ciliary muscle and the dilator and sphincter muscles of the iris. They are termed *intrinsic eye muscles* since they are inside the eyeball. Convergence is a function of the voluntary muscles attached to the outside of the eyeball called the *extrinsic eye muscles.*

Refraction and Accommodation. When light rays traveling through a transparent medium (such as air) pass into a second transparent medium with a different density (such as water), they bend at the surface of the two media. This is *refraction* (Figure 14-4a). The eye has four such media of refraction: cornea, aqueous humor, lens, and vitreous humor. Light rays entering the eye from the air are refracted at the following points: (1) the anterior surface of the cornea as they pass from the lighter air into the denser cornea; (2) the posterior surface of the cornea as they pass into the less dense aqueous humor; (3) the anterior surface of the lens as they pass from the aqueous humor into the denser lens; and (4) the posterior surface of the lens as they pass from the lens into the less dense vitreous humor.

When an object is 6 m (20 ft) or more away from

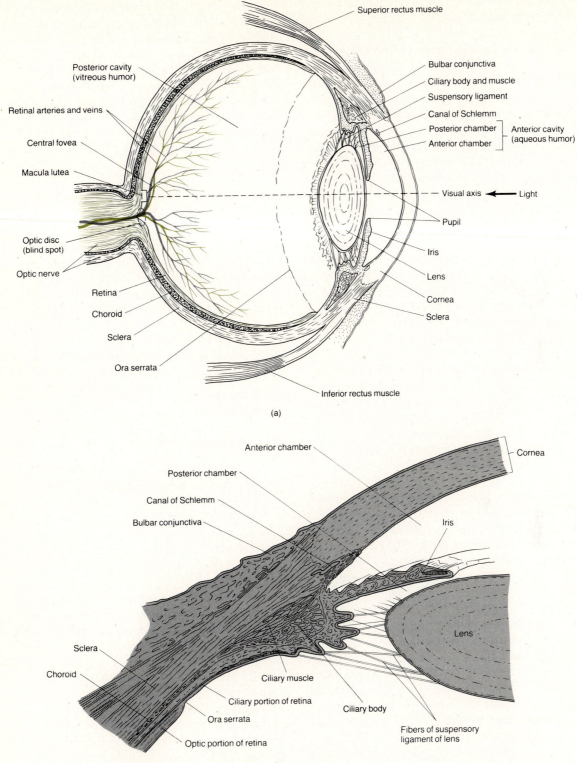

Figure 14-3 Structure of the eyeball. (a) Gross anatomy in sagittal section. (b) Section through the anterior part of the eyeball at the sclerocorneal junction. (c) Diagram of the microscopic structure of the retina. (d) Photomicrograph of the posterior wall of the eyeball, showing the layers of the retina at a magnification of 100×. (Courtesy of Donald I. Patt, from *Comparative Vertebrate Histology,* by Donald I. Patt and Gail R. Patt, New York: Harper & Row, 1969.)

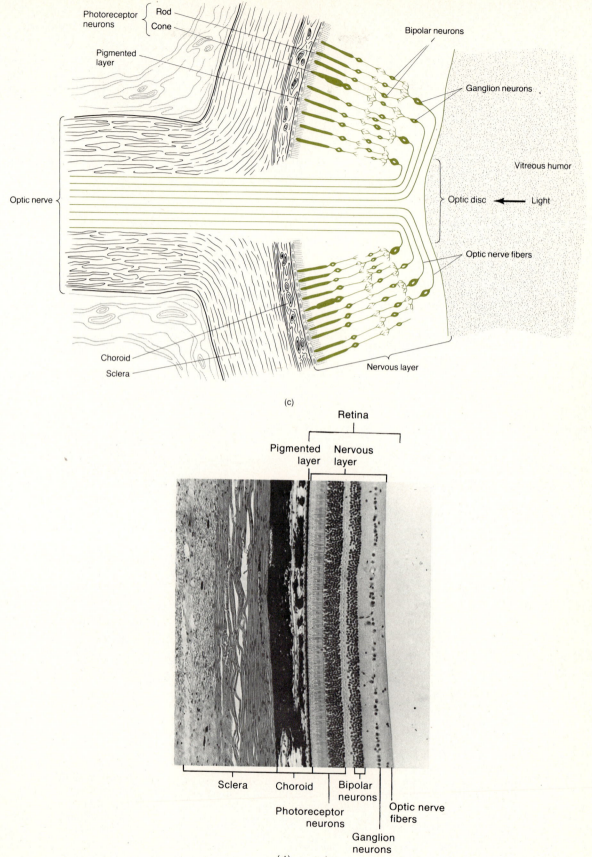

Photoreceptor neurons { Rod / Cone

Pigmented layer

Bipolar neurons

Ganglion neurons

Optic nerve

Vitreous humor

Optic disc ← Light

Optic nerve fibers

Choroid

Sclera

Nervous layer

(c)

Retina

Pigmented layer / Nervous layer

Sclera

Choroid

Photoreceptor neurons

Bipolar neurons

Ganglion neurons

Optic nerve fibers

(d)

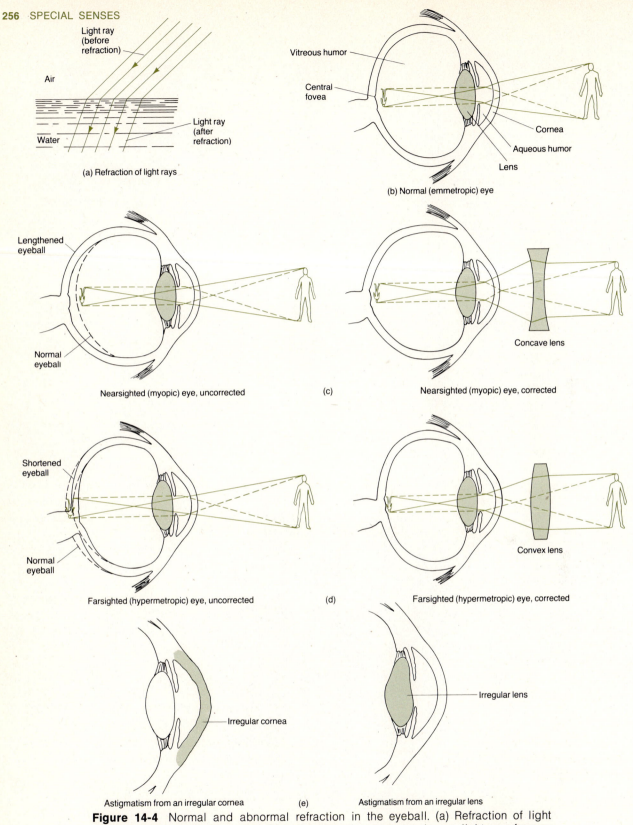

Figure 14-4 Normal and abnormal refraction in the eyeball. (a) Refraction of light rays passing from air into water. (b) In the normal or emmetropic eye, light rays from an object are bent sufficiently by the four refracting media and converged on the central fovea. A clear image is formed. (c) In the nearsighted or myopic eye, the image is focused in front of the retina. The condition may result from an elongated eyeball or a thickened lens. Correction is by use of a concave lens, which diverges entering light rays. (d) In the farsighted or hypermetropic eye, the image is focused behind the retina. The condition

the viewer, the light rays reflected from the object are nearly parallel to one another. The degree of refraction that takes place at each surface in the eye is very precise. Therefore the parallel rays are sufficiently bent to fall exactly on the central fovea, where vision is sharpest. However, light rays that are reflected from nearby objects are divergent rather than parallel. As a result, they must be refracted toward each other to a greater extent. This change in refraction is brought about by the lens of the eye.

If the surfaces of a lens curve outward, as in a convex lens, the lens will refract the rays toward each other so they eventually intersect. The more the lens curves outward, the more acutely it bends the rays toward each other. Conversely, when the surfaces of a lens curve inward, as in a concave lens, the rays bend away from each other. The lens of the eye is biconvex. Furthermore, it has the unique ability to change the focusing power of the eye by becoming moderately curved at one moment and greatly curved the next. When the eye is focusing on a close object, the lens curves greatly in order to bend the rays toward the central fovea. This increase in the curvature of the lens is called *accommodation* (Figure 14-5). The ciliary muscle contracts, pulling the ciliary body and choroid forward toward the lens. This action releases the tension on the lens and suspensory ligament. Due to its elasticity, the lens shortens, thickens, and bulges. In near vision, the ciliary muscle is contracted and the lens is bulging. In far vision, the ciliary muscle is relaxed and the lens is flatter. With aging, the lens loses elasticity and, therefore, its ability to accommodate.

The normal eye, known as an *emmetropic eye*, can sufficiently refract light rays from an object 6 m (20 ft) away to focus a clear object on the retina. Many individuals, however, do not have this ability because of abnormalities related to improper refraction. Among these abnormalities are *myopia* (nearsightedness), *hypermetropia* (farsightedness), and *astigmatism* (irregularities in the surface of the lens or cornea). The conditions are illustrated and explained in Figure 14-4c–e.

Constriction of Pupil. The muscles of the iris also assume a function in the formation of clear retinal images. Part of the accommodation mechanism con-

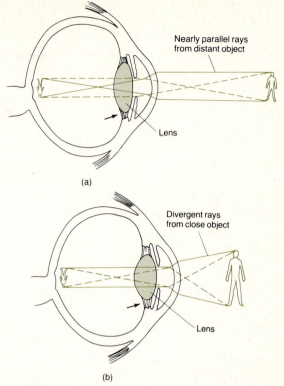

(a)

(b)

Figure 14-5 Accommodation. (a) For objects 6 m (20 ft) or more away. (b) For objects nearer than 6 m.

sists of the contraction of the circular muscle fibers of the iris to constrict the pupil. Constricting the pupil means narrowing the diameter of the hole through which light enters the eye. This action occurs simultaneously with accommodation of the lens and prevents light rays from entering the eye through the periphery of the lens. Light rays entering at the periphery would not be brought to focus on the retina and would result in blurred vision. The pupil, as noted earlier, also constricts in bright light to protect the retina from sudden or intense stimulation.

Convergence. Birds see a set of objects off to the left through one eye and an entirely different set off to the right through the other. This characteristic doubles their field of vision and allows them to detect predators behind them. In human beings, both eyes focus on only one set of objects—a characteristic called *single binocular vision*.

may be the result of the eyeball being too short or the lens being too thin. Correction is by a convex lens, which refracts entering light rays. (e) Astigmatism. In this condition, the curvature of the cornea or lens is uneven. As a result, horizontal and vertical rays are focused at two different points on the retina. Suitable glasses correct the refraction of an astigmatic eye. On the left, astigmatism resulting from an irregular cornea. On the right, astigmatism resulting from an irregular lens. The image is not focused on the area of sharpest vision of the retina. This results in blurred or distorted vision.

Single binocular vision occurs when light rays from an object are directed toward corresponding points on the two retinas. When we stare straight ahead at a distant object, the incoming light rays are aimed directly at both pupils and are refracted to identical spots on the retinas of both eyes. But as we move close to the object, our eyes must rotate medially—that is, become "crossed"—for the light rays from the object to hit the same points on both retinas. The term *convergence* refers to this medial movement of the two eyeballs so they are both directed toward the object being viewed. The nearer the object, the greater the degree of convergence necessary to maintain single binocular vision. Convergence is brought about by the coordinated action of the extrinsic eye muscles.

Inverted Image. Images are focused upside down on the retina. They also undergo mirror reversal. That is, light reflected from the right side of an object hits the left side of the retina and vice versa. Note in Figure 14-4b that reflected light from the top of the object crosses light from the bottom of the object and strikes the retina below the central fovea. Reflected light from the bottom of the object crosses light from the top of the object and strikes the retina above the central fovea. The reason we do not see a topsy-turvy world is that the brain learns early in life to coordinate visual images with the exact locations of objects. The brain stores memories of reaching and touching objects and automatically turns visual images right-side-up and right-side-around.

Stimulation of Photoreceptors

After an image is formed on the retina by refraction, accommodation, constriction of the pupil, and convergence, light impulses must be converted into nerve impulses by the rods and cones. Both rods and cones contain special photosensitive substances. These *visual pigment molecules,* as they are called, are able to absorb quanta of light and undergo structural changes that lead to a decrease in membrane permeability and hyperpolarization. This is contrary to what happens in all other sensory cell membranes. This may uninhibit or release nerve fibers in the visual pathway to fire action potentials into the central nervous system.

There are four visual pigment molecules. It has been estimated that each rod cell contains 70 million molecules of a visual colored pigment called *rhodopsin,* which is sensitive to low levels of illumination. The cone cells contain one of three other visual pigments known as *erythrolabe, chlorolabe,* or *cyanolabe.* These pigments are responsive to wavelengths in the red, green, and blue regions, respectively.

All four pigment molecules are constructed in a similar way. They consist of a protein called *opsin* and a *chromophore.* Opsin, which is different for each of the four visual pigments, determines the frequency of light to which each pigment will respond. It determines whether the visual pigment will respond to all light or just to red, green, or blue light. The chromophore, chemically known as *retinene,* is the same in all four visual pigments.

In the following discussion of how photoreceptors are stimulated, we shall describe only what happens to rhodopsin in rods since the same basic changes occur in the other visual pigments in cones. *Rhodopsin,* also known as *visual purple,* consists of the specific protein *scotopsin* and the chromophore *retinene.* The retinene is a derivative of vitamin A and exists in a form called *cis-retinene.* The importance of this form will be noted shortly. In the first step in the *rhodopsin cycle* (Figure 14-6), light strikes the rhodopsin and is absorbed by it. Consequently, the *cis*-retinene portion of rhodopsin is changed to an *all-trans* form of retinene (Figure 14-7). Whereas the *cis*-retinene is a curved molecule and the all-*trans*-retinene has a straight configuration, the all-*trans*-retinene no longer fits with the scotopsin. The result of the change in configuration is a decomposition or splitting of rhodopsin into scotopsin and all-*trans*-retinene. As the separation begins, the rhodopsin is successfully converted into *prelumirhodopsin, lumirhodopsin,* and *metarhodopsins I and II.* It is at this point that the pigment is bleached (loses color) and the visual excitatory event is believed to occur. When separation is complete, the products are scotopsin and all-*trans*-retinene. The last step in the rhodopsin cycle leading

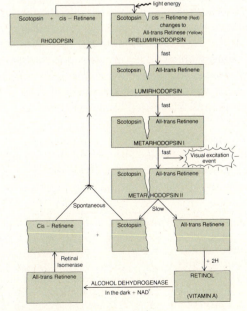

Figure 14-6 The rhodopsin cycle.

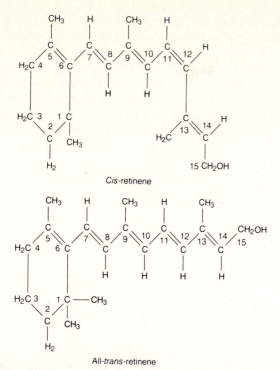

Figure 14-7 Comparison between *cis*-retinene and all-*trans*-retinene. Note the rotation that occurs around the carbon numbered 12. The structural change from *cis*-retinene to all-*trans*-retinene triggers the decomposition of rhodopsin.

to the breakdown of the pigment molecule to its component scotopsin and all-*trans*-retinene is very slow and, therefore, is considered not to contribute directly to visual excitation. The point of excitation must be linked to one of the preceding steps. In the visual sensation, the wavelengths of light serve only to trigger the changes in the shape of the retinene. The next changes are related to the cell membrane alterations, and these are poorly understood.

Following decomposition of rhodopsin into scotopsin and all-*trans*-retinene in the presence of light, rhodopsin is once again re-formed. In this process, the all-*trans*-retinene is converted back to *cis*-retinene, in the presence of an enzyme, *retinene isomerase*. Then *cis*-retinene combines with scotopsin to re-form rhodopsin. This compound is stable until its decomposition is again triggered by light energy.

Rhodopsin is highly sensitive to even small amounts of light. For that reason the rods, which contain rhodopsin, are the photoreceptors that respond during conditions of poor illumination, such as during the nighttime. Their responses to the light stimulation generate colorless images, and objects are seen only in shades of gray. Their *acuity*, or ability to see clearly and to distinguish two points close together in space, is very poor. On the other hand, the pigments of the cones are much less sensitive to light and therefore require a much higher intensity to trigger the

molecular changes in the chromophore. Therefore, the cones are the photoreceptors that are used for daytime vision. Visual acuity of cones is very high, and their responses produce colored images.

Impulses from both rods and cones are transmitted through bipolar neurons to ganglion cells (see Figure 14-3c). The cell bodies of the ganglion cells lie in the retina, and their axons leave the eye via the optic nerve. The synaptic connections for the rods and cones differ. Cones connect to the central nervous system by much more direct pathways. Bipolar cells receive impulses from few cones, and the ganglion cells similarly receive input from few bipolar cells. In this way, precise information about the image projected on the retina is conducted accurately to the brain. On the other hand, many rods connect with one bipolar cell, and many bipolar cells transmit impulses to one ganglion cell. This greatly lowers visual acuity, but it permits summation effects to occur so that low levels of light can stimulate a ganglion cell that would not respond had it been connected directly to a cone. The synaptic connections thus contribute much to the difference in visual acuity and light sensitivity.

In the dim light of a theater, objects are unclear and seen in shades of gray. The low light intensity does not affect the pigment molecule in the cones, so they do not respond. But the rhodopsin pigment in the rods does respond, so the rods produce their action potentials. Therefore, vision at low levels of light is due to the properties and pathways that connect the rods. In daylight, the rods are of only limited help. This is because rhodopsin breaks down very quickly in the bright light and is re-formed very slowly. In bright light, rhodopsin is destroyed faster than it can be manufactured. In dim light, production can keep pace because of the slower rate of breakdown. These characteristics of rhodopsin are responsible for the dark-adaptation process that occurs when one moves from the bright sunshine into a darkened room. The period for complete adjustment requires about 20 to 30 minutes, and during this time, the destroyed rhodopsin

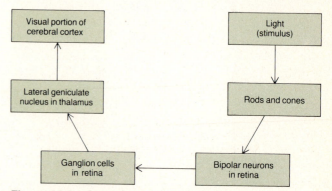

Figure 14-8 Cells in the visual pathway.

is re-formed. *Night blindness* is the lack of normal night vision that persists after the dark-adaptation period. It is most often caused by a deficiency of vitamin A.

Afferent Pathway to the Brain

The synaptic connections delivering visual sensations to the brain are shown in Figure 14-8. In the retina, the light rays activate the receptor cells, which transmit impulses to the ganglion cells via the bipolar cells. The axons of the ganglion cells pass through the *optic chiasma*, a crossing point of the optic nerves (Figure 14-9). Some fibers cross to the opposite side; others remain uncrossed. Upon passing through the optic chiasma, the fibers, now part of the *optic tract*, enter the brain and terminate in the *lateral geniculate nucleus*, which is part of the thalamus. Here the fibers

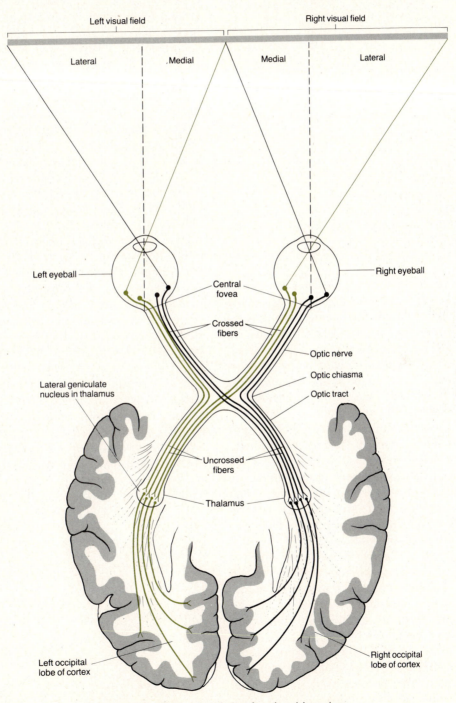

Figure 14-9 Afferent pathway for visual impulses.

synapse with neurons whose axons pass to the visual cortex located in the occipital lobes of the cerebral cortex.

Analysis of the afferent pathway to the brain reveals that the visual field of each eye is divided into two regions. These are referred to as the *medial* (or *nasal*) *half* and the *lateral* (or *temporal*) *half*. For each eye, light rays from an object in the nasal half of the visual field fall on the temporal half of the retina. Light rays from an object in the temporal half of the visual field fall on the nasal half of the retina. Note that in the optic chiasma, nerve fibers from the nasal halves of the retinas cross and continue on to the thalamus. Also note that nerve fibers from the temporal halves of the retinas do not cross but continue directly to the thalamus. As a result, the visual center in the cortex of the right occipital lobe "sees" the left side of an object via impulses from the temporal half of the retina of the right eye and the nasal half of the retina of the left eye. The cortex of the left occipital lobe interprets visual sensations from the right side of an object via impulses from the nasal half of the right eye and the temporal half of the left eye. Blind spots in the field of vision may indicate a brain tumor along one of the afferent pathways. For instance, a symptom of tumor in the right optic tract might be an inability to see the left side of a normal field of vision without moving the eyeball.

AUDITORY SENSATIONS AND EQUILIBRIUM

In addition to containing receptors for sound waves, the ear also contains receptors for equilibrium. Anatomically, the ear is divided into three principal regions: the external or outer ear, the middle ear, and the internal or inner ear.

External or Outer Ear

The **external** or **outer ear** is structurally designed to collect sound waves and direct them inward (Figure 14-10a). It consists of the pinna, the external auditory canal, and the tympanic membrane, also called the eardrum. The *pinna*, or *auricle*, is a trumpet-shaped flap of elastic cartilage covered by thick skin. The rim of the pinna is called the helix; the inferior portion is the lobe. The pinna is attached to the head by ligaments and muscles. The *external auditory canal* or *meatus* is a tube about 2.5 cm (1 inch) in length that lies in the external auditory meatus of the temporal bone. It leads from the pinna to the eardrum. The *tympanic membrane* or *eardrum* is a thin, semi-transparent partition of fibrous connective tissue between the external auditory meatus and the middle ear.

Middle Ear

Also called the **tympanic cavity,** the **middle ear** is a small, epithelial-lined, air-filled cavity hollowed out of the temporal bone (Figure 14-10a and b). The cavity is separated from the external ear by the eardrum and from the internal ear by a thin bony partition that contains two small openings: the oval window and the round window. The posterior wall of the cavity communicates with the mastoid cells of the temporal bone through a chamber called the *tympanic antrum*. The anterior wall of the cavity contains an opening that leads directly into the *Eustachian tube*, also called the *auditory tube* or *meatus*. The Eustachian tube connects the middle ear with the nose and nasopharynx of the throat. The function of the tube is to equalize air pressure on both sides of the tympanic membrane. Abrupt changes in external or internal air pressure might otherwise cause the eardrum to rupture. Since the tube opens during swallowing and yawning, these activities allow atmospheric air to enter or leave the middle ear until the internal pressure equals the external pressure. Any sudden pressure changes against the eardrum may be equalized by deliberately swallowing.

Extending across the middle ear are three exceedingly small bones called **auditory ossicles.** These are called the malleus, incus, and stapes. According to their shapes, they are commonly named the hammer, anvil, and stirrup, respectively. The "handle" of the *malleus* is attached to the internal surface of the tympanic membrane. Its head articulates with the base of the incus. The *incus* is the intermediate bone in the series and articulates with the stapes. The base or footplate of the *stapes* fits into a small opening between the middle and inner ear called the *fenestra vestibuli*, or *oval window*. Directly below the oval window is another opening, the *fenestra cochlea*, or *round window*. This opening, which also separates the middle and inner ears, is enclosed by a membrane called the secondary tympanic membrane. The auditory ossicles are attached to the tympanic membrane, to each other, and to the oval window by means of ligaments and muscles (Figure 14-10a–c).

Internal or Inner Ear

The **internal** or **inner ear** is also called the **labyrinth** (Figure 14-11) because of its complicated series of canals. Structurally, it consists of two main divisions:

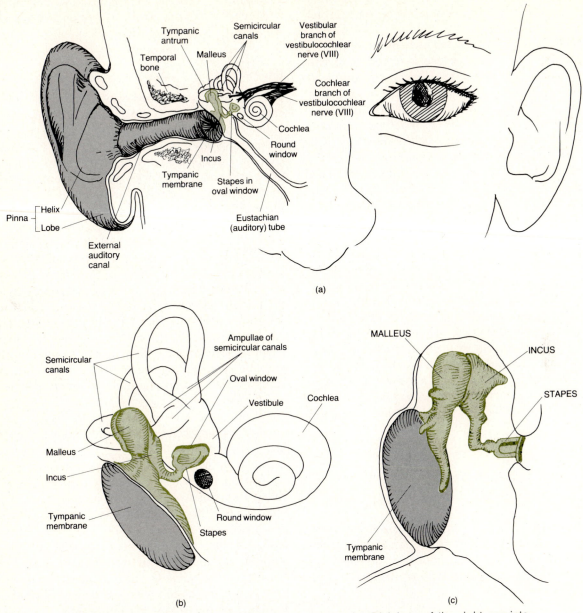

Figure 14-10 Structure of the auditory apparatus. (a) Divisions of the right ear into external, middle, and internal portions seen in a coronal section through the right side of the skull. The middle ear is shown in color. (b) Details of the middle ear and the bony labyrinth of the internal ear. (c) Ossicles of the middle ear.

(1) a bony labyrinth and (2) a membranous labyrinth that fits in the bony labyrinth. The *bony labyrinth* is a series of cavities in the petrous portion of the temporal bone. It can be divided into three areas named on the basis of shape: the vestibule, cochlea, and semicircular canals. The bony labyrinth is lined with periosteum and contains a fluid called *perilymph*. This fluid surrounds the *membranous labyrinth*, a series of sacs and tubes lying inside and having the same general form as the bony labyrinth. Epithelium lines

the membranous labyrinth and contains a fluid called *endolymph*.

The *vestibule* constitutes the oval central portion of the bony labyrinth. The membranous labyrinth in the vestibule consists of two sacs called the *utricle* and *saccule*. These sacs are connected to each other by a small duct.

Projecting upward and posteriorly from the vestibule are the three bony *semicircular canals*. Each is arranged at approximately right angles to the other

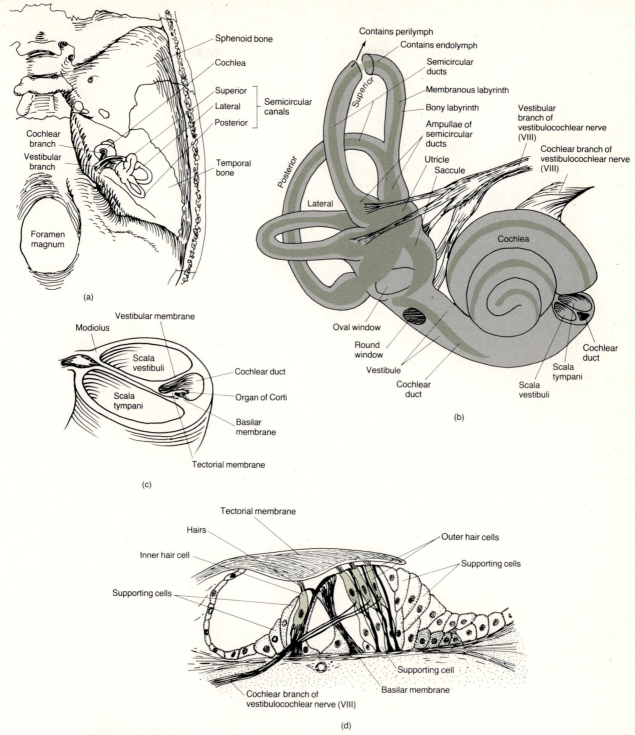

Figure 14-11 Details of the internal ear. (a) Relative position of the bony labyrinth projected to the inner surface of the floor of the skull. (b) The outer, gray area belongs to the bony labyrinth. The inner, colored area belongs to the membranous labyrinth. (c) Cross section through the cochlea. (d) Enlargement of the spiral organ or organ of Corti.

two. On the basis of their positions, they are called the superior, posterior, and lateral canals. One end of each canal enlarges into a swelling called the *ampulla*. In-

side the bony semicircular canals lie portions of the membranous labyrinth: the *semicircular ducts* or *membranous semicircular canals*. These structures

are almost identical in shape to the bony semicircular canals and communicate with the utricle of the vestibule.

Lying in front of the vestibule is the *cochlea,* so designated because of its resemblance to a snail's shell. The cochlea consists of a bony spiral canal that makes about 2¾ turns around a central bony core called the *modiolus.* A cross section through the cochlea shows the canal is divided by partitions into three separate channels resembling the letter Y on its side. The stem of the Y is a bony shelf that protrudes into the canal. The wings of the Y are composed of the bony labyrinth. The channel above the bony partition is the *scala vestibuli;* the channel below is the *scala tympani.* The cochlea adjoins the wall of the vestibule, into which the scala vestibuli opens. The scala tympani terminates at the round window. The perilymph of the vestibule is continuous with that of the scala vestibuli. The third channel (between the wings of the Y) is the membranous labyrinth: the *cochlear duct.* The cochlear duct is separated from the scala vestibuli by the *vestibular membrane,* also called *Reissner's membrane.* It is separated from the scala tympani by the *basilar membrane.* Resting on the basilar membrane is the *spiral organ,* or *organ of Corti,* the organ of hearing. The organ of Corti is a series of epithelial cells on the inner surface of the basilar membrane. It consists of a number of supporting cells and hair cells, which are the receptors for auditory sensations. The inner hair cells are medially placed in a single row and extend the entire length of the cochlea. The outer hair cells are arranged in several rows throughout the cochlea. The hair cells have long hairlike processes at their free ends that extend into the endolymph of the cochlear duct. The basal ends of the hair cells are in contact with fibers of the cochlear branch of the vestibulocochlear nerve (VIII). Projecting over and in contact with the hair cells of the organ of Corti is the *tectorial membrane,* a delicate and flexible gelatinous membrane.

Physiology of Hearing

Sound Waves

Sound waves result from the alternate compression and decompression of air. They originate from a vibrating object and travel through air in much the same way that waves travel over the surface of water. The sound waves that reach either the external ear or pinna are channeled into the external auditory canal to the tympanic membrane (eardrum) (see Figure 14-10a). Here, the energy of the sound waves is converted to a mechanical vibratory energy as the eardrum starts oscillating inward and outward. The total area of the eardrum is about 70 mm². Attached to the inside of the eardrum by its handle is the malleus (Figure 14-12). Its head end is connected to the second of the auditory ossicles, the incus, which in turn is connected to the stapes, which is anchored firmly in the oval window of the cochlea. Calculations done on the length of the auditory ossicles and their arrangement show that a mechanical advantage of about 1.3 is realized. Small muscles of the middle ear, the *stapedius* and *tensor tympani* muscles, can significantly change the efficiency of transmission of low-frequency sound. When tension is increased on the bones of the middle ear, the energy transmitted to the oval window is reduced, thus protecting the inner ear from sudden surges of energy. Since the area of the footplate of the stapes is 3 mm² and the effective area of the eardrum is 40 mm², the mechanical advantage gain is 13 to 1. This advantage is necessary to overcome the greater impedance of propagation through the internal medium of the perilymph as compared to the external medium of air. The sound energy is then carried down the perilymph of the scala vestibuli and around the 2¾ turns of the cochlea to its tip. The energy wave is then returned down the scala tympani to the round window. The movements of the windows are synchronized so that as the oval window bulges inward, the round window bulges outward in compensation. Movements of the stapes cause simultaneous displacements of the basal parts of the basilar and vestibular membranes (see Figure 14-11b and c). The basilar membrane is quite rigid at the base, where it begins, but becomes increasingly more flexible along its length so that, at its tip, it is about 100 times less rigid. Simultaneous with the increase in flexibility is an increase in breadth of the basilar membrane.

Sound entering the cochlea stimulates the basilar membrane at its base so that it begins to propagate a wave that moves from the base toward the tip. As the vibrations are sent out along the membrane, the position where the maximum deflection occurs moves further down the basilar membrane to a position determined by the frequency of the stimulus. Sound frequencies between 20 and 30 Hertz (Hz) set the entire membrane into motion so that it all moves in phase upward into the scala vestibuli and then down toward the scala tympani (Figure 14-13a). For sound frequencies of up to about 200 Hz, there are at least two regions moving in opposite directions. The point where the wave crosses the membrane varies with the frequency of stimulation (Figure 14-13b). As sound frequencies increase above 200 Hz, the higher frequencies cause only the stiffer base end of the basilar membrane to vibrate up and down. Also, as the fre-

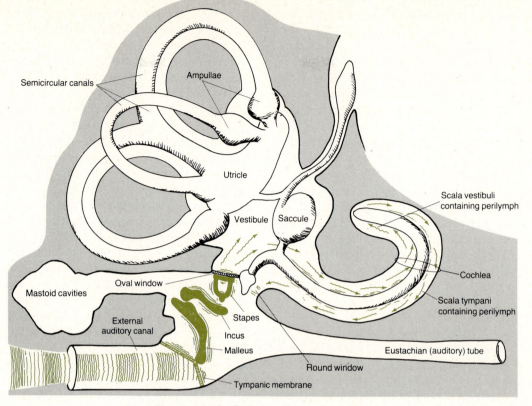

Figure 14-12 Pathway of sound waves.

quency gets higher, less of the total length responds. For complex sounds, the motion of the basilar membrane is also complex. The places where the waves cross the basilar membrane and the amplitude or the amount of excursion of the membrane changes as rapidly as the sound characteristics. You can gain some insight into the process by envisioning a violin string caused to vibrate more or less by the pressure

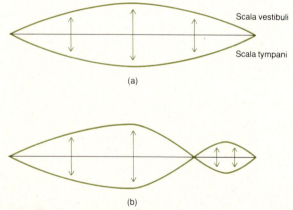

Figure 14-13 Relationship of sound waves to the basilar membrane. (a) Sound frequencies between 20 and 30 Hz set the entire membrane into motion. (b) Sound frequencies up to 200 Hz generate two waves moving in opposite directions. The point where the wave crosses the membrane varies with the frequency of stimulation.

of the drawn bow while the wave vibration pattern changes as fingers are rapidly changing the length of the bowed string.

Sound Energy Transduction

The next step in the sensory process deals with the way in which the vibrating basilar membrane triggers the generator potentials in sensory receptors. A closer examination of a cross section of the cochlear duct is necessary at this point (see Figure 14-11d). The *spiral organ of Corti* is the specialized end organ of hearing. It is a long, multicellular structure that lies on the floor of the cochlear duct and runs the entire length of the basilar membrane. The thin *tectorial membrane* is of jellylike consistency and is connected to the bony spiral lamina by the cells of the periosteum. In this position it hangs over, and barely touches, the hairs of the hair cells that lie directly underneath. The hair cells are arranged into two groups: three or more rows of *outer* hair cells and a single row of *inner* hair cells. It has been estimated that the human ear has about 15,000 outer hair cells and about 3,500 inner hair cells. All the hair cells are bathed in endolymph, which is similar in composition to the perilymph.

When the basilar membrane responds to sound stimuli and begins moving, the hair cells begin to rock. The hairs of the hair cells are in contact with the more

stationary tectorial membrane and are deflected in response to the rolling motion of the basilar membrane. In other words, as the basilar membrane moves up and down, hairs on top of the hair cells rub on the tectorial membrane and are bent. The bending of the hairs distorts the structure of the cell membrane, which results in excitation of the cell. The fibers of the cochlear branch of the vestibulocochlear nerve (VIII) connect with many hair cells. The possibility for convergence and divergence of excitation from the hair cells is suggested by such an anatomical arrangement.

The afferent fibers from the organ of Corti enter the medulla in the cochlear branch of the vestibulocochlear nerve and then divide into two groups running to the dorsal and ventral cochlear nuclei. Here some impulses cross to the opposite side and finally travel to the auditory area of the temporal lobe of the cerebral cortex.

Physiology of Equilibrium

Under normal living conditions, most of us are unaware of the sensations served by the vestibular mechanisms. A brief ship ride on a rough sea will certainly alert most of us to certain disturbances detected by this very exquisite system. The vestibular mechanism keeps us informed of the nature of our movements and our orientation in space (equilibrium). The receptors within this system respond to acceleration forces: The semicircular canals detect *angular* accelerations, while the utricles and saccules detect *linear* accelerations. The functions of the vestibular apparatus are very important for maintenance of equilibrium and homeostasis as we move about in space. Proper orientation of our head, limbs, and entire body is achieved by reflex signals originating from the vestibular apparatus. Cooperation of the vestibular senses with the other senses, particularly vision, enables us to perceive the world around us from a knowing perspective.

The receptor organs for equilibrium are the semicircular canals, the saccule, and utricle (see Figure 14-12). These organs are essentially a series of interconnected tubes and bulbous sacs filled with a fluid-like material called *endolymph*. The tubelike portions are the *semicircular canals,* each of which forms an incomplete circular loop. The three canals are oriented in the three dimensions of space to detect basically side-to-side, front-to-back, and up-and-down movements. All three semicircular canal loops open at both ends into an enlarged saclike chamber called the *utricle*. A second, and slightly smaller, chamber called the *saccule* connects to the utricle by means of a short duct. The hollow bony cavity in which the utricle and saccule reside is termed the *vestibule*. The vestibule is the inner space separated from the middle ear cavity by the round and oval windows. The vestibule extends in two directions—somewhat anteriorly to form the coiled bony cavity of the cochlea, and more posteriorly into three bony tubes—each housing one incomplete circle of the membranous canals.

Utricle and Saccule

Suspended in the perilymphatic fluid of the vestibule are the enlarged sacs of the membranous network, the utricle and saccule. Their walls are composed of a thin sheet of connective tissue lined on the inside with a

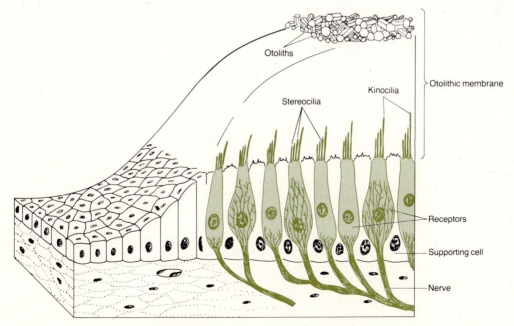

Figure 14-14 Structure of the macula.

simple squamous-type epithelium. Each contains a small, flat, plaquelike region called a *macula* (Figure 14-14).

Microscopically, the structure of the macula resembles the organ of Corti. It consists of differentiated neuroepithelial cells that are innervated by the vestibular branch of the vestibulocochlear nerve (VIII). The two maculae are anatomically located in planes perpendicular to one another. They possess two kinds of cell types: *receptors* and *supporting cells*. Two shapes of receptor cells have been identified: one is more flask-shaped and the other is more cylindrical. Both show long extensions of the cell membrane called *stereocilia* (they are actually microvilli) and one *kinocilium* anchored firmly to its basal body and extending beyond the longest microvilli. Some of the microvilli reach lengths of over 100 μm, and some receptor cells have over 80 such projections. It is easy to understand why these are also named *hair cells*.

The supporting cells of the macula are scattered between the receptor cells, are columnar, and have short microvilli on their exposed surface. Floating directly over the hair cells is a thick, gelatinous, glycoprotein layer, probably secreted by the supporting cells, called the *otolithic membrane*. A layer of calcium carbonate crystals, called *otoliths*, extends over the entire surface of the otolithic membrane. The specific gravity of these otoliths is about three, which makes them much denser than the endolymph fluid that fills the rest of the utricle.

The entire otolithic structure sits on top of the macula like a discus on a greased cookie sheet. If you were to tilt your head forward, the otolithic membrane (the discus in our analogy) would slide downhill over the stereocilia in the direction determined by the tilt of your head. In the same manner, if you were sitting upright in a cart that was suddenly jerked forward, the otolithic membrane, due to its inertia, would slide backwards and stimulate the receptor cells that, in turn, would signal the forward movement. Research has made it clear that the utricle functions in this manner to inform us of linear acceleration and gravity directions. The saccule's role in such balance functions is not as clear. While it may perform similar functions, some studies suggest it may be more important in sensing low-frequency vibration.

Semicircular Canals

The walls of the membranous *semicircular canals* are constructed just like those of the utricle and saccule: connective tissue plus a lining of squamous epithelium. These fluid-filled tubes are situated within the bony canals of the inner ear, but occupy only a small portion of the hollow space (Figure 14-15). The canals lie in contact with the outer rim of the bony canal. Connective tissue strands anchor the membranous tubes to the other parts of the bony canal.

The receptor cell area for the semicircular canals is located in their ampullae (Figure 14-15). Extending from the outer, or centrifugal wall of each ampulla in the form of a transverse ridge is the *crista*. This extended shelflike projection is made up of receptor

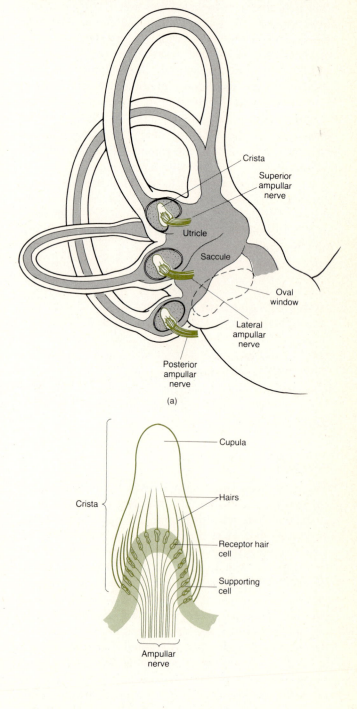

Figure 14-15 Semicircular ducts and dynamic equilibrium. (a) Position of the cristae relative to the membranous ampullae. (b) Enlarged aspect of a crista.

(hair) cells, supporting cells, nerve fibers, and capillaries. Its receptive structure is very similar to that of the macula, but the gelatinous covering differs in that it is heaped up over the crista to form an extended amorphous topping called the *cupula*. The cupula, which lacks the calcium carbonate crystals, extends from the crista across the entire ampulla to form a flap that acts like a swinging door. When you turn your head quickly, the fluid of the horizontal semicircular canal tends not to move because of its inertia. But the cristae, with the hairs of the receptor cells embedded in the cupula, do move with the direction of your head. As a result, the gelatinous cupula is bent in the direction opposite to the rotation of your head. Thus the hairs embedded in the cupula are pulled and trigger the excitation of the receptor cell. The afferent nerve fibers are activated and information about the rotation is transmitted, via the vestibular branch of the vestibulocochlear nerve (VIII), to the temporal lobe of the cerebrum. If the head rotation is abrupt, it is likely that a reflex response will send impulses to skeletal muscles, causing them to contract or relax as necessary for the body to balance itself by assuming a new position.

HOMEOSTATIC IMBALANCES OF THE SPECIAL SENSE ORGANS

The special sense organs can be altered or damaged by numerous disorders. The causes of disorder can range from congenital origins to the effects of old age. Here we discuss a few common disorders of the eyes and ears.

Cataract

The most prevalent disorder resulting in blindness is **cataract** formation. This disorder causes the lens or its capsule to lose its transparency. Cataracts can occur at any age, but we shall discuss the type that develops with old age. As a person gets older, the cells in the lenses may degenerate and be replaced with nontransparent fibrous protein. Or the lenses may start to manufacture nontransparent protein. The main symptom of cataract is a progressive, painless loss of vision. The degree of loss depends on the location and extent of the opacity. If vision loss is gradual, frequent changes in glasses may help maintain useful vision for a while. Eventually, though, the changes may be so extensive that light rays are blocked out altogether. At this point, surgery is indicated. Essentially, the surgical procedure consists of removing the opaque lens and substituting an artificial lens by means of eyeglasses or by implanting a plastic lens inside the eyeball to replace the natural one.

Glaucoma

The second most common cause of blindness, especially in the elderly, is **glaucoma.** This disorder is characterized by an abnormally high pressure of fluid inside the eyeball. The aqueous humor does not return into the bloodstream through the canal of Schlemm as quickly as it is formed. The fluid accumulates and, by compressing the lens into the vitreous humor, puts pressure on the neurons of the retina. If the pressure continues over a long period of time, it destroys the neurons and brings about blindness. It can affect a person of any age, but 95 percent of the victims are over 40. Glaucoma affects the eyesight of more than one million people in this country.

Conjunctivitis

Many different eye inflammations exist, but the most common type is **conjunctivitis (pinkeye)** — an inflammation of the membrane that lines the insides of the eyelids and covers the cornea. Conjunctivitis can be caused by microorganisms — most often the pneumococci or staphylococci bacteria. In such cases, the inflammation is very contagious. It can also be caused by a number of irritants, in which case the inflammation is not contagious. Irritants include dust, smoke, wind, air pollution, and excessive glare. The condition may be acute or chronic. The epidemic type in children is extremely contagious, but normally it is not serious.

Trachoma

This chronic, contagious conjunctivitis is caused by an organism called the TRIC agent. The organism has characteristics of both viruses and bacteria. **Trachoma** is characterized by many granulations or fleshy projections on the eyelids. If untreated, these projections can irritate and inflame the cornea and reduce vision. The disease produces an excessive growth of subconjunctival tissue and the invasion of blood vessels into the upper half of the front of the cornea. The disease progresses until it covers the entire cornea, bringing about a loss of vision because of corneal opacity.

Ménière's Disease

An important cause of deafness and loss of equilibrium in adults is **Ménière's disease** of the inner ear. It is a

disturbance or malfunction of any part of the inner ear. It can be the result of many causes:

1. Infection of the middle ear.
2. Trauma from brain concussion producing hemorrhage or splitting of the labyrinth.
3. Cardiovascular diseases, such as arteriosclerosis and blood vessel disturbances.
4. Congenital malformation of the labyrinth.
5. Excessive formation of endolymph.
6. Allergy.

The last two causes can produce an increase in pressure in the cochlear duct and vestibular system. This pressure, in turn, causes a progressive atrophy of the hair cells of the cochlear or semicircular ducts.

If the cochlear duct is injured, typical symptoms are hissing, roaring, or ringing in the ears and deafness. If the semicircular ducts are involved, the person feels dizzy and nauseous. The dizzy spells can last from a few minutes to several days. Ménière's disease is a chronic disorder. It affects both sexes equally and usually begins in late middle life.

Impacted Cerumen

Some people produce an abnormal amount of cerumen, or earwax, in the external auditory canal. Here it becomes impacted and prevents sound waves from reaching the tympanic membrane. The treatment for **impacted cerumen** is usually periodic ear irrigation or removal of wax with a blunt instrument.

Now that you have a basic understanding of the nature of sensations and the role of the special senses in maintaining homeostasis, we shall take a look at the motor and integrative functions of the nervous system in relation to the internal and external environments.

CHAPTER SUMMARY OUTLINE

GUSTATORY SENSATIONS

1. Receptors in the taste buds send impulses to the cranial nerves, thalamus, and cortex.
2. The four basic tastes are sweet, sour, salty, and bitter.

OLFACTORY SENSATIONS

1. Receptor cells in the nasal epithelium send impulses to the olfactory bulbs, olfactory tracts, and cortex.

VISUAL SENSATIONS

Structure of Eyeball

1. The eye is constructed of three coats: (1) fibrous tunic (sclera and cornea); (2) vascular tunic (choroid, ciliary body, and iris); and (3) retina, which contains the rods and cones.
2. The anterior cavity contains aqueous humor. The posterior cavity contains vitreous humor.

Physiology of Vision

1. The formation of an image on the retina requires four basic processes: refraction of light rays; accommodation of the lens; constriction of the pupil; and convergence of the eyes.
2. Improper refraction may result from myopia (nearsightedness), hypermetropia (farsightedness), and astigmatism (irregularities in the surface of the lens or cornea).
3. After an image is formed on the retina by refraction, accommodation, constriction of the pupil, and convergence, light impulses must be converted into nerve impulses by the rods and cones.
4. Both rods and cones contain special photosensitive substances. There are four visual pigment molecules.
5. It has been estimated that each rod cell contains 70 million molecules of a visual colored pigment called rhodopsin, which is sensitive to low levels of illumination.
6. Cone cells contain one of three other visual pigments known as erythrolabe, chlorolabe, or cyanolabe.
7. All four pigment molecules consist of a protein called opsin and a chromophore chemically known as retinene. Opsin is different for each of the four visual pigments, but retinene is the same in all four visual pigments.
8. In the first step of the rhodopsin cycle, light strikes the rhodopsin and is absorbed by it. The rhodopsin splits and loses color, and the visual excitatory event occurs.
9. Following decomposition of rhodopsin in the pres-

ence of light, rhodopsin is re-formed and is stable until its decomposition is again triggered by light energy.

Afferent Pathway to the Brain

Impulses from rods and cones are conveyed through the retina to the optic nerve, the optic chiasma, the optic tract, the thalamus, and the cortex.

AUDITORY SENSATIONS AND EQUILIBRIUM

1. The ear consists of three anatomical subdivisions: (1) the outer ear (pinna, external auditory canal, and tympanic membrane); (2) the middle ear (ossicles, oval window, round window, and opening into the Eustachian tube); and (3) the inner ear (bony labyrinth and membranous labyrinth).

Physiology of Hearing

1. Sound waves result from the alternate compression and decompression of air.
2. Waves enter the external auditory canal, strike the tympanic membrane, pass through the ossicles, strike the oval window, set up waves in the perilymph, strike the vestibular membrane and scala tympani, increase pressure in the endolymph, strike the basilar membrane, and stimulate hairs on the spiral organ. A sound impulse is then initiated.

Physiology of Equilibrium

1. The receptor organs for equilibrium are the semicircular canals, the saccule, and utricle.

2. When your head is turned quickly, the afferent nerve fibers in receptor cells are activated and information about the rotation is transmitted via the vestibular branch of the vestibulocochlear nerve (VIII), to the temporal lobe of the cerebrum.
3. If the head rotation is abrupt enough, a reflex sends impulses to skeletal muscles, causing them to contract or relax as necessary in order for the body to balance itself by assuming a new position.

HOMEOSTATIC IMBALANCES OF THE SPECIAL SENSE ORGANS

1. The most prevalent disorder resulting in blindness is cataract formation. This disorder causes the lens or its capsule to lose its transparency.
2. The second most common cause of blindness, especially in the elderly, is glaucoma. This disorder is characterized by an abnormally high pressure of fluid inside the eyeball.
3. Conjunctivitis, or pinkeye, is the most common type of eye inflammation. It can be caused by microorganisms or any number of irritants, and can be very contagious.
4. Trachoma is characterized by many granulations or fleshy projections on the eyelids. It is caused by the TRIC organism, which has characteristics of both viruses and bacteria.
5. Ménière's disease is an important cause of deafness and loss of equilibrium in adults. It has many causes, affects the inner ear, is a chronic disorder affecting both sexes equally, and begins in late middle life.
6. Impacted cerumen or earwax prevents sound waves from reaching the tympanic membrane, producing hearing difficulties.

REVIEW QUESTIONS

1. How are papillae related to taste buds? Describe the structure and location of the papillae. Discuss how an impulse for taste travels from a taste bud to the brain.
2. Discuss the origin and path of an impulse that results in smelling.
3. By means of a labeled diagram, indicate the three principal anatomical structures of the eyeball.
4. Explain how the retina is adapted to its function.
5. Describe the location and contents of the chambers of the eye. What is intraocular pressure?

How is the canal of Schlemm related to this pressure?
6. Explain how each of the following events is related to the physiology of vision: (a) refraction of light, (b) accommodation, (c) constriction of the pupil, (d) convergence, and (e) inverted image formation.
7. Distinguish emmetropia, myopia, hypermetropia, and astigmatism by means of a diagram.
8. What are the four visual pigment molecules?
9. How is a light stimulus converted into an impulse?

Relate your discussion to the rhodopsin cycle by means of a diagram.

10. What is night blindness? What causes it?
11. Describe the path of a visual impulse from the optic nerve to the brain.
12. Define visual field. Relate the visual field to image formation on the retina.
13. Diagram the principal parts of the outer, middle, and inner ear. Describe the function of each part labeled.
14. Explain the events involved in the transmission of sound from the pinna to the spiral organ.
15. What is the afferent pathway for sound impulses from the vestibulocochlear nerve to the brain?
16. What are the receptor organs for equilibrium? Describe their locations and functions.
17. Describe the series of events that occur if you turn your head quickly. Trace them all the way to the skeletal muscles.
18. What is cataract? Explain what happens to the lens of the eye as an individual gets older.
19. Describe what happens to the eye in glaucoma. How many people are affected with glaucoma in this country?
20. What is pinkeye? Explain the different ways an individual can get pinkeye. When is it considered very contagious?
21. Describe trachoma. What is the TRIC agent?
22. What is Ménière's disease? List some of its causes and possible effects on the individual.
23. What is cerumen? What happens if there is an unusual buildup of this material in an individual?

Chapter 15

MOTOR AND INTEGRATIVE FUNCTIONS

STUDENT OBJECTIVES

After reading this chapter, you should be able to

- Explain how our motor response system integrates incoming sensory information along the various pathways of the central nervous system
- Correlate sensory input with motor unit activation of simple or complex muscle responses or secretion of glands
- Describe the various locations within the central nervous system where the major sensory-motor linkages take place
- Differentiate between conscious sensation and input that never gives rise to conscious sensation
- Explain in detail the myotatic reflex as proposed by C. S. Sherrington
- Define and explain proprioception and its relationship to homeostasis
- Correlate proprioception with the muscle spindle and Golgi tendon organs
- Describe the afferent pathway for proprioception
- Identify the various motor pathways
- Distinguish between the three main tracts of the pyramidal system
- Distinguish between the three main extrapyramidal pathways
- Explain the integrative functions of the cerebrum such as memory, sleep and wakefulness, and emotions

LINKAGE OF SENSORY INPUT AND MOTOR RESPONSES

In preceding chapters we have described the sensory systems in detail. These systems function to keep us aware of our external and internal environment. Our active responses to this information result from the activity of **motor systems.** The motor systems enable us to move about and change our relationship to the world around us. As sensory information is conveyed into the central nervous system, it becomes part of a large pool of sensory input. That is, we do not necessarily respond actively to every bit of input the central nervous system receives. Rather, the incoming information is added to, subtracted from, or integrated with other information arriving from all other operating sensory receptors. The integration process occurs not just once but at many stations along the pathways of the central nervous system. It occurs within the spinal cord, in the brain stem, the cerebellum, and the cerebral cortex. As a result, a motor response to make a muscle contract or a gland secrete can be initiated at any of these levels.

The design of this motor response system is very precise. Each skeletal muscle is connected to its own group of nerve fibers, and these nerve fibers innervate *only* fibers of that muscle. These nerve fibers are called *skeletomotor fibers.* This arrangement ensures that, when the central nervous system initiates a response, only the desired muscle contracts. Recall the description of the motor unit in Chapter 10.

Muscle actions can vary greatly in complexity, and the pattern of activation of motor units therefore also varies in complexity. Motor units can respond simultaneously to produce a sudden, forceful movement, or they can be orchestrated to respond asynchronously in a time sequence so that the force of contraction builds up more gradually. When more than one joint is involved, the sequential patterns of motor unit activation involving all the joints can be very complicated indeed. Yet, when complicated movements are performed by an Olympic gymnast, for example, the beauty of the act belies the underlying complexity of the myriad of neuromuscular interactions.

Some muscular responses are simple and occur subconsciously. The quick withdrawal of a hand when it touches a hot object or the knee jerk that results when a percussion hammer taps the patellar tendon are simple responses. Playing a piano is a much more complex response. Simple or very complex, all muscle responses involve both contraction and relaxation of muscle groups. The innervation to antagonistic (opposing) muscles always acts in a *reciprocal fashion.* One nerve fiber set causes the contraction of the prime movers and the other set of fibers causes relaxation of the antagonists. The complexity of the muscular movements involved in piano playing can be astonishing. Sensory input is rapidly and continually changing as the eyes scan the music score. Each bit of information initiates and modifies the reciprocal pattern of muscle innervation to the fingers, within milliseconds. Additional information from sensory receptors relating to touch, position, pressure, speed, and sound also affect the reciprocal innervation program. This almost instantaneous analysis of the total sensory input generates the incredible array of motor responses to produce beautiful music. The linking of the sensory input

and motor response occurs at four major levels in the central nervous system, and each level can add richness and variety to the repertoire of movements.

Figure 15-1 illustrates the locations within the central nervous system where the major sensory-motor linkages take place. As the location of the sensory-motor linkage climbs to higher levels in the central nervous system—from spinal cord to brain stem to cerebellum to basal ganglia to cerebral cortex—additional contributions are introduced that enrich the growing inventory of motor responses. The figure indicates that all information about each sensory modality arises from the peripheral nervous system and is directed into the spinal cord. At the level of entry, a synapse is made that directs the input of information to higher centers in the central nervous system. When information reaches the highest level, the cerebral cortex, conscious sensation results. Most conscious sensations originate from fibers connected to skin receptors.

Other inputs to the central nervous system never give rise to conscious sensation. These operate to coordinate and modulate the contractions of muscle groups. Nerve fibers that connect to special structures within the muscles and tendons deliver input to the central nervous system about the present state of the muscle length, and its speed and strength of contraction. These fibers enter the spinal cord, synapse, and project their input up the cord to the cerebellum.

The number of interconnections between the regions of the central nervous system is so great that it is impossible to define the final locus of sensory input. Some sensory input arrives at the cerebral cortex and also is delivered to the cerebellum. In other cases, the sensory input is delivered to the cerebellum, which then relays some of the input to the cerebral cortex.

When the input reaches the highest center, the process of sensory-motor integration occurs. This involves not only the utilization of information contained within that center but also the information impinging upon that center delivered from other centers within the central nervous system. After integration occurs, the output of the center is sent down the spinal cord in the *descending motor pathways*. These pathways originate from two locations: the brain stem and the cerebral cortex.

MYOTATIC REFLEX

Over 50 years ago, C. S. Sherrington, one of the most famous pioneers in the field of neurophysiology, showed that if a muscle was abruptly stretched or jerked, its response was to contract immediately. It reacted in a manner that tended to return it immediately to its original length. From many additional experiments, he finally concluded that the behavior was the result of afferent nerve signals that originated in the stretched muscle and reached the spinal cord; efferent motor signals returned from the cord to the muscle, causing the quick contraction. Sherrington termed this rapid response the **myotatic** (muscle stretching) **reflex**. Scientists have been able to describe in detail the nature of the stretch receptors that initiate the afferent impulses. It is now also known that, in the spinal cord, the afferent nerve fibers make a direct synapse onto the motor neuron that innervates the stretched muscle. Since the myotatic reflex makes only one synapse in its loop, it is classed as a monosynaptic reflex. In fact, the stretch reflexes are the only monosynaptic reflex loops known.

PROPRIOCEPTION

Proprioception is the ability to recognize the location and rate of movement of one body part in relation to other parts. It allows us to estimate weight and to de-

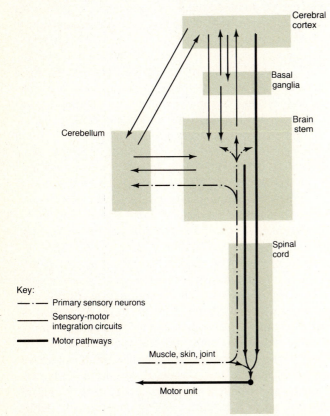

Figure 15-1 Sensory-motor linkages of the central nervous system. Primary sensory neurons project to the spinal cord, brain stem, and cerebellum. Sensory-motor integration circuits interconnect all major subdivisions of the central nervous system. Descending motor pathways to the spinal cord originate in the brain stem and the cerebral cortex.

Cerebral cortex

Basal ganglia

Brain stem

Cerebellum

Spinal cord

Key:
— · — Primary sensory neurons
——— Sensory-motor integration circuits
━━━ Motor pathways

Muscle, skin, joint

Motor unit

termine the muscular work necessary to perform a task. With the proprioceptive sense, we can judge the position and movements of our limbs when we walk, type, play a musical instrument, or undress even in the dark. The proprioceptive system relies on input from the vestibular apparatus of the inner ear (Chapter 14) plus information that arrives from receptors in the muscles, tendons, and joints.

All muscles in the human body possess not only a few pain and deep pressure sensory receptors, but also special types of receptors that contribute to the proprioceptive sense. *Muscle spindles* are scattered throughout the belly of skeletal muscles and operate continually to monitor the overall length of muscle fibers and the speed at which they contract. *Golgi tendon organs*, which are enmeshed in the tendons of skeletal muscle, measure the tension generated by the contracting muscle.

The Muscle Spindle

The **muscle spindle** is a fusiform structure (shaped like an elongated football) that is about 4 to 7 mm long and 100 to 200 μm wide. Many such spindles are distributed within a skeletal muscle, and each spindle has its own special nerves that relay information to the spinal cord and back again to the spindle. The spindle is constructed of six or seven modified muscle cells, called *intrafusal fibers*, enveloped in a thin, fibrous sheath.

The regular contractile cells of the skeletal muscle are known as *extrafusal fibers*. The intrafusal fibers are long, thin cells with cross striations evident over most of their length; they are absent only near the middle of the cell. Intrafusal fibers are further divided into two groups called *nuclear bag fibers* and *nuclear chain fibers* (Figure 15-2). The nuclear bag fibers, usually two in number, are longer and thicker and possess many nuclei that are generally collected in the center of the cell. The nuclear chain fibers, usually four or five in number, are shorter and thicker cells with a few nuclei distributed lengthwise in the middle region of the cell.

Muscle spindles are inserted between the extrafusal fibers of the muscle in a more or less parallel fashion. Each spindle is fixed in position by anchoring itself with very long, threadlike, fibrous extensions that insert into the tendon fibers at each end of the muscle. Some spindles can attach one end to the endomysial component of the muscle tissue. Normally, the intrafusal muscle fibers within the spindle are under tension, but if the extrafusal fibers were to shorten, the tension on the intrafusal fibers would be removed. If, on the other hand, the extrafusal fibers were stretched, greater tension would be placed on the intrafusal fibers.

Sensory Innervation

The innervation of the intrafusal fibers includes both sensory (afferent) and motor (efferent) fibers (Figure

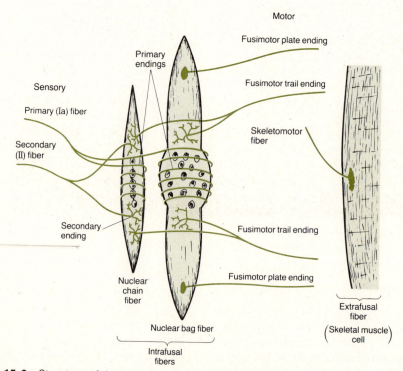

Figure 15-2 Structure of the muscle spindle, showing sensory and motor innervations.

15-2). The sensory nerves leave the intrafusal fibers in two groups. *Primary (Ia) fibers* seem to originate from a coil around the central region of both the nuclear bag and the nuclear chain fibers. This coiled origin is referred to as the *primary ending*. The primary fibers are 12 to 15 μm in diameter and have the fastest conduction velocity of any nerve group, of more than 90 m/second.

The other group of afferent fibers, called *secondary (II) fibers*, also arise from the intrafusal fibers. Their only point of origin is called the *secondary ending*, and they arise from the regions on either side of the central clear area of the intrafusal fibers. The nuclear chain fibers appear to be the main source of these secondary afferent neurons.

Motor Innervation

Intrafusal muscle fibers can contract when stimulated by *fusimotor neurons* that connect to the polar ends of the fibers. When fusimotor neurons fire, the polar end portions of the intrafusal fibers shorten, thus putting tension on the central region. The sensory afferents that arise from this central region are thereby stimulated to increase their rate of discharge. It is important to distinguish the two types of motor nerves that are present in muscle tissue. The *skeletomotor neurons* are those which innervate the normal skeletal muscle fibers (extrafusal) and cause them to contract, thus causing a joint to move. Skeletomotor efferent fibers and the fusimotor fibers both have their cell bodies in the ventral gray horn of the spinal cord. Both motor sets exit from the spinal cord via the ventral root and travel together in the peripheral muscle nerve. The skeletomotor fibers are larger in diameter (around 12 to 15 μm) and conduct impulses at a speed of 70 to 90 m/second, whereas the fusimotor fibers are about 4 to 7 μm in diameter and have conduction velocities between 18 and 40 m/second.

Fusimotor fibers terminate on the intrafusal fibers in two different patterns. The *fusimotor plate endings* form connections like the end plates of the myoneural junctions described in Chapter 10. They terminate only on the polar ends of the nuclear bag fibers (Figure 15-2). *Fusimotor trail endings* make connections that are more trailing and branched and confined to the regions midway between the center and ends of both the nuclear chain and bag fibers.

Operation of the Spindle

The sensory fibers leaving the spindle can be activated in either of two ways, or both. First, if the skeletal fibers of the muscle are stretched, the intrafusal fibers, which are in parallel array, are also stretched and the sensory endings are deformed to produce a rapid series of action potentials. Second, the sensory endings of the intrafusal fibers can also be made to discharge if the polar ends of the intrafusal fibers shorten. This stretches the central region where the sensory endings are located and they are triggered to fire. Properly orchestrated, these two mechanisms produce a smooth-running spindle that sends sensory information into the spinal cord about the overall contractile state of the muscle under every possible condition.

The response of the spindle sensory fibers can be viewed in two ways: (1) when the muscle is "resting" and (2) when the muscle is contracting. When the muscle is at any constant "resting" length, the spindles discharge at some fixed rate. The greater the length of the muscle, the greater the stretch on the spindle and the greater the rate of sensory discharge. Under these constant length conditions, both types of sensory endings, primary and secondary, send impulses into the cord. However, during a fast stretch of the skeletal muscle, the primary endings respond immediately to the extension and fire a rapid volley of action potentials toward the spinal cord. The secondary endings do not respond to the rapid change but *slowly* adjust the firing frequency to the new length of the muscle. Similarly, the primary endings abruptly stop firing when a stretched muscle is released, whereas the secondary endings *gradually* decrease their rate of firing to a slower frequency, indicating the shorter length. The primary endings that are able to indicate these sudden changes in length are described as possessing a *dynamic-response component*.

In their continual search for a deeper understanding of how our bodies function, scientists have come upon an example of homeostasis that can even *anticipate* changes that will need to be made. Consider the following sequence of events involving the muscle spindle and its controlled muscle. The muscle is stretched by an outside force (such as a percussion hammer), which also stretches the intrafusal fibers and the sensory endings. The sensory nerve fiber is activated to discharge with greater frequency. In the spinal cord, a monosynaptic relay to the skeletomotor neurons of the muscle of origin is made. This causes a contraction of the skeletal muscle and removes the stretch on the spindle so that it is returned to firing at its original rate. Thus, the system operates to restore the muscle to its original length. In a similar way, shortening the skeletal muscles reduces tension on the spindle sensors, and the skeletomotor response is a reflex extension of the shortened muscle. In this way, the reflex system serves to maintain the muscles at a constant length when *small* deviations from the desired length are imposed. Impulses descending from higher regions of the central nervous system can influence the fusi-

motor neurons that innervate the intrafusal spindle fibers. Such influences can increase or decrease the tension on the sensory fibers of the spindle, which will change the length of the skeletal muscle to a new, constant length. In many homeostatic illustrations, it is pointed out that fluctuations around the "desired" or "set" point are a common feature of control systems. For muscle length control, fluctuations around a desired length would be totally useless because this would be a very unstable situation. It appears that the primary spindle fibers are able to predict the extent of the imposed muscle extension by the velocity with which it is imposed. Thus, these primary receptors are able to *predict* the necessary amount of opposing contraction required to restore the muscle to the constant length. This *dynamic* or *velocity sensitive* ability of the primary sensory fibers prevents the fluctuations that are characteristic of many other homeostatic systems.

Function of the Fusimotor Reflexes

Fusimotor reflexes are polysynaptic, which means they have one or more association neurons between the sensory fiber and the fusimotor fiber. Recall that the reflex action initiated by stretching a muscle results in a powerful jerk response of the same muscle. The sensory fibers of the spindle in the muscle make a direct monosynaptic connection to the skeletomotor fiber of the same muscle to elicit the strong response. The same sensory fibers exert only very weak influences on the fusimotor fibers of the spindle.

It has long been known that the fusimotor nerve fibers operate to keep the intrafusal fibers under a constant degree of stretch. Because the muscle spindles are parallel to the extrafusal fibers, the spindle fibers would become slack if the extrafusal fibers shortened. The consequence of such a shortening would be to remove any stimulus on the sensory fibers of the spindle. Thus, the spinal cord would lose all input about the speed and degree of the contraction. To prevent and correct any slack, the fusimotor fibers trigger an immediate contraction of the intrafusal fibers which takes up the slack and keeps the sensory fibers operating. In this manner, the spindle is constantly kept in operation, sending impulses into the spinal cord.

As described earlier, the primary fibers of the spindle connect directly with the skeletomotor neurons to complete a monosynaptic reflex. If the fusimotor response were absent, any muscle contraction would remove the tension from the spindle, and the primary fibers would become silent and the monosynaptic reflex inoperative. The extrafusal fiber would be deprived of further reflex stimulus for additional

contraction, and any purposeful movement would be greatly hindered. The operation of the fusimotor fibers keeps tension "on," and this greatly aids the motor systems in their quick and powerful response.

In summary, the fusimotor fibers offer a twofold contribution to motor activity. First, they keep tension on the sensory endings of the intrafusal fibers, which enables the proprioceptors to relay information into the central nervous system. The central nervous system can then better monitor whether the muscle contraction is occurring in the appropriate fashion. Second, by keeping the primary fibers "turned on," the spindles are constantly informed about the activity of the skeletomotor fibers. This rapid reflex activity enables the spindle to participate throughout the entire contractile process by smoothing out and dampening any fluctuations in the muscular contraction.

Golgi Tendon Organs

Another group of receptors that contribute to proprioception are Golgi tendon organs. The *Golgi tendon organs* are encapsulated structures usually located in the tendon near its connection to the muscle. The structure of the Golgi tendon organ is simple in comparison to that of the muscle spindle previously described. This receptor organ consists of the terminal branches of thick, myelinated nerve fibers (Ib type fibers) wrapped around a cluster of tendon fascicles or bundles (Figure 15-3). The structure is one-tenth the length of the muscle spindle (about 700 μm) and about 200 μm wide, which, on the average, is wider than the muscle spindle. The capsule, which consists of several layers of connective tissue fibers, extends inward and divides the interior into several long, nar-

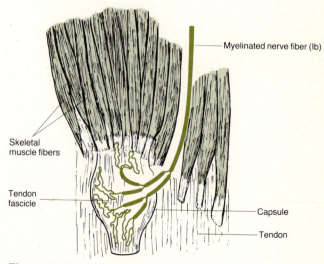

Figure 15-3 Structure of a Golgi tendon organ.

row compartments. The capsule of the Golgi tendon organ anchors the entire structure in position since it is continuous with the connective tissue of the tendon.

Sensory Innervation

The Ib type sensory fibers of the Golgi tendon organ conduct impulses with a velocity of almost 80 m/second, which is somewhat slower than the primary (Ia) sensory fibers of the muscle spindle. The large myelinated Ib fiber penetrates the capsule of the receptor organ and branches rapidly into secondary and tertiary myelinated branches. These terminal branches attach to the tendon fascicles, which are stretched during the muscle contraction. It is thought that deformation of the terminals of the Ib sensory fibers triggers the action potentials.

Operation of the Golgi Tendon Organ

In contrast to the muscle spindle, which was said to be parallel with the extrafusal fibers, the Golgi tendon organ is connected in series (Figure 15-4). When connected "in series," the Golgi tendon organ is able to signal muscle *tension*. When the muscle contracts, pull is exerted directly on the tendon organ, causing it to increase its firing frequency. Recall that the muscle spindle, which is connected parallel, is unloaded when the muscle contracts. The sensory input from the tendon organ serves to keep the central nervous system constantly informed about the tension in the muscle and also serves to *inhibit* the skeletomotor neuron to the same muscle.

One might think that it would make no difference from which direction the Golgi tendon organ was stretched. It would be stretched when the muscle

contracted, because pull on the tendon stretches the tendon organ. Similarly, if the muscle were stretched, pull would be exerted on the tendon organ. It is possible to show that the tendon organ fires more readily, that is, has a lower threshold, when the muscle contracts than when it is passively stretched. When the muscle is passively stretched, much of the stretching force is absorbed by the muscle fibers as they elongate. This means less stretch is detected by the tendon organ. Thus, fewer impulses are sent to the central nervous system.

An interesting anatomic relationship has recently been pointed out. Muscle spindles are present in the same muscle fascicles that possess Golgi tendon organs in the tendon anchoring the muscle. Although this arrangement suggests some functional cooperation between the proprioceptors, the exact meaning and relative contributions of each is still unknown.

Afferent Pathway for Proprioception

The afferent pathway for muscle sense consists of impulses sent from proprioceptors into cranial and spinal nerves. Impulses for conscious proprioception travel in ascending tracts in the cord, where they are relayed to the thalamus and cerebral cortex (see Figure 13-11). The sensation is registered in the general sensory area in the parietal lobe of the cerebral cortex posterior to the central fissure. Proprioceptive impulses that result in reflex action travel to the cerebellum through spinocerebellar tracts.

The *posterior spinocerebellar tract* is an uncrossed tract that conveys impulses concerned with subconscious muscle sense. That is, the tract assumes a role in reflex adjustments for posture and muscle tone. The nerve impulses originate in neurons that run between proprioceptors in muscles, tendons, and joints and the posterior gray horn of the spinal cord. Here the neurons synapse with afferent neurons that pass to the ipsilateral lateral white column of the cord to enter the posterior cerebellar tract. The tract enters the inferior cerebellar peduncles from the medulla and ends at the cerebellar cortex. In the cerebellum, synapses are made that ultimately result in the transmission of impulses back to the spinal cord to the anterior gray horn to synapse with the lower motor neurons leading to skeletal muscles.

The *anterior spinocerebellar tract* also conveys impulses for subconscious muscle sense. It, however, is made up of both crossed and uncrossed nerve fibers. Sensory neurons deliver impulses from proprioceptors to the posterior gray horn of the spinal cord. Here a synapse occurs with neurons that make up the anterior spinocerebellar tracts. Some fibers cross to the oppo-

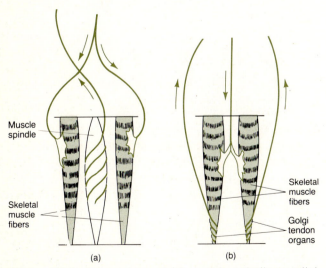

Figure 15-4 Comparison of (a) muscle spindle parallel with skeletal muscle fibers and (b) Golgi tendon organs in series with skeletal muscle fibers.

Muscle spindle

Skeletal muscle fibers

(a)

Skeletal muscle fibers

Golgi tendon organs

(b)

site side of the spinal cord in the anterior white commissure. Others pass laterally to the ipsilateral anterior spinocerebellar tract and move upward, through the brain stem, to the pons to enter the cerebellum through the superior cerebellar peduncles. Here again, the impulses for subconscious muscle sense are registered.

We have already noted in Chapter 11 that sensory fibers terminating in the spinal cord can bring about spinal reflexes. These reflexes do not require immediate action by the brain in order to occur. Sensory fibers that terminate in the lower brain stem can bring about much more complex motor reactions than simple reflexes. When sensory information reaches the lower brain stem, it causes subconscious motor reactions. Sensory signals that reach the thalamus can be localized crudely in the body. In fact, at the thalamic level, sensations are distinguished on the basis of their modality. That is, the sensations are sorted by type. When sensory information reaches the cerebral cortex, we experience very precise localization. It is at this level that memories of previous sensory information are stored and perception of the sensation occurs on the basis of past experiences.

MOTOR PATHWAYS

The principal parts of the brain concerned with skeletal muscle control are the cerebral motor cortex, basal ganglia, reticular formation, and cerebellum. The motor cortex assumes the major role for controlling precise, discrete muscular movements. The basal ganglia largely integrate semivoluntary movements such as walking, swimming, and laughing. The cerebellum, although not a control center, assists the motor cortex and basal ganglia by making body movements smooth and coordinated. Voluntary motor impulses are conveyed from the brain through the spinal cord by way of two major pathways: the pyramidal pathways and the extrapyramidal pathways.

Pyramidal Pathways

Voluntary motor impulses are conveyed from the motor areas of the brain to somatic efferent neurons leading to skeletal muscles via the **pyramidal pathways**. Most pyramidal fibers originate from cell bodies in the precentral gyrus. They descend through the internal capsule of the cerebrum and cross to the opposite side of the brain. They terminate in nuclei of cranial nerves that innervate voluntary muscles or in the anterior gray horn of the spinal cord. A short connecting neuron probably completes the connection of

the pyramidal fibers with the motor neurons that activate voluntary muscles.

The pathways over which the impulses travel from the motor cortex to skeletal muscles have two components: *upper motor neurons (pyramidal fibers)* and *lower motor neurons (peripheral fibers)*. Here we consider three tracts of the pyramidal system:

1. **Lateral corticospinal tract** *(pyramidal tract proper)*. This tract begins in the motor cortex, descends through the internal capsule of the cerebrum, the cerebral peduncle of the midbrain, and then the pons on the same side as the point of origin (Figure 15-5). In the medulla, the fibers decussate to the opposite side to descend through the spinal cord in the lateral white column in the lateral corticospinal tract. Thus, the motor cortex of the right side of the brain controls muscles on the left side of the body and vice versa. The upper motor neurons of the lateral corticospinal tract probably synapse with short association neurons in the anterior gray horn of the cord. These then synapse in the anterior gray horn with lower motor neurons that exit all levels of the cord via the ventral roots of spinal nerves. The lower motor neurons terminate in skeletal muscles.

2. **Anterior corticospinal tract.** About 15 percent of the upper motor neurons from the motor cortex do not cross in the medulla. These pass through the medulla and continue to descend on the same side to the anterior white column to become part of the *anterior (straight or uncrossed) corticospinal tract*. The fibers of these upper motor neurons decussate and probably synapse with association neurons in the anterior gray horn of the spinal cord of the side opposite the origin of the anterior corticospinal tract. The association neurons in the horn synapse with lower motor neurons that exit the cervical and upper thoracic segments of the cord via the ventral roots of spinal nerves. The lower motor neurons terminate in skeletal muscles that control muscles of the neck and part of the trunk.

3. **Corticobulbar tract.** The fibers of this tract begin in upper motor neurons in the motor cortex. They accompany the corticospinal tracts through the internal capsule to the brain stem, where they decussate and terminate in the nuclei of cranial nerves in the pons and medulla. These cranial nerves include the trigeminal (V), abducens (VI), facial (VII), glossopharyngeal (IX), vagus (X), accessory (XI), and hypoglossal (XII). The corticobulbar tract conveys impulses that largely control voluntary movements of the head and neck.

The various tracts of the pyramidal system convey

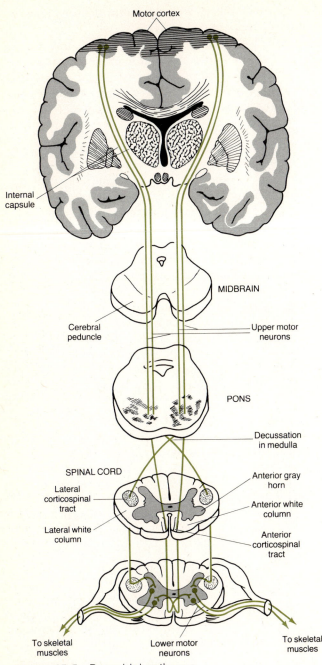

Figure 15-5 Pyramidal pathways.

impulses from the cortex that result in precise muscular movements.

Extrapyramidal Pathways

The **extrapyramidal pathways** include all descending tracts other than the pyramidal tracts. Generally, these include tracts that begin in the basal ganglia and reticular formation. The main extrapyramidal tracts are as follows:

1. **Rubrospinal tract.** This tract originates in the red nucleus of the midbrain (after receiving fibers from the cerebellum), crosses over to descend in the lateral white column of the opposite side, and terminates in the anterior gray horns of the cervical and upper thoracic segments of the cord. The tract transmits impulses to skeletal muscles concerned with tone and posture.

2. **Tectospinal tract.** This tract originates in the superior colliculus of the midbrain, crosses to the opposite side, descends in the anterior white column, and enters the anterior gray horns in the cervical segments of the cord. Its function is to transmit impulses that control movements of the head in response to auditory, visual, and cutaneous stimuli.

3. **Vestibulospinal tract.** This tract originates in the vestibular nucleus of the medulla, descends on the same side of the cord in the anterior white column, and terminates in the anterior gray horns, mostly in the cervical and lumbosacral segments of the cord. It conveys impulses that regulate muscle tone in response to movements of the head. This tract, therefore, plays a major role in equilibrium.

Only one motor neuron carries the impulse from the cerebral cortex to the cranial nerve nuclei or spinal cord: an *upper motor neuron.* Only one motor neuron in the pathway actually terminates in a skeletal muscle: the *lower motor neuron.* This neuron, a somatic efferent neuron, always extends from the central nervous system to the skeletal muscle. Since it is the final transmitting neuron in the pathway, it is also called the *final common pathway.* This neuron is important clinically. If it is damaged or diseased, there is neither voluntary nor reflex action of the muscle it innervates, and the muscle remains in a relaxed state —a condition called *flaccid paralysis.* Injury or disease of upper motor neurons in a motor pathway is characterized by varying degrees of continued contraction of the muscle *(spasticity)* and exaggerated reflexes. Another characteristic is the sign of Babinski. Stroking the plantar surface along the outer border of the foot produces a slow dorsiflexion of the great toe accompanied by fanning of the lateral toes. The normal response is plantar flexion of the toes.

Lower motor neurons are subjected to stimulation by many other presynaptic neurons. Some signals are excitatory; others are inhibitory. The algebraic sum of the opposing signals determines the final response of the lower motor neuron. It is not just a simple matter of the brain sending an impulse and the muscle always contracting.

Association neurons are of considerable importance in the motor pathways. Impulses from the brain are conveyed to association neurons before being received by lower motor neurons. These association

neurons are integrators. They integrate the pattern of muscle contraction.

The basal ganglia have many connections with other parts of the brain. Through these connections, they help to control subconscious movements. The caudate nucleus controls gross intentional movements. The caudate nucleus and putamen, together with the cortex, control patterns of movement. The globus pallidus controls positioning of the body for performing a complex movement. The subthalamic nucleus is thought to control walking and possibly rhythmic movements. Many potential functions of the basal ganglia are held in check by the cerebrum. Thus, if the cerebral cortex is damaged early in life, a person can still perform many gross muscular movements.

The role of the cerebellum is significant also. The cerebellum is connected to other parts of the brain that are concerned with movement. The vestibulocerebellar tract transmits impulses from the equilibrium apparatus in the ear to the cerebellum. The olivocerebellar tract transmits impulses from the basal ganglia to the cerebellum. The corticopontocerebellar tract conveys impulses from the cerebrum to the cerebellum. The spinocerebellar tracts relay proprioceptive information to the cerebellum. Thus, the cerebellum receives considerable information regarding the overall physical status of the body. Using this information, the cerebellum generates impulses that integrate body responses.

Take tennis, for example. To make a good serve, you must bring your racket forward just far enough to make solid contact. How do you stop at the exact point without swinging too far? This is where the cerebellum comes in. It receives information about your body status while you are serving. Before you even hit the ball, the cerebellum has already sent information to the cerebral cortex and basal ganglia informing them that your swing must stop at an exact point. In response to cerebellar stimulation, the cortex and basal ganglia transmit motor impulses to your opposing body muscles to stop the swing. The cerebellar function of stopping overshoot when you want to zero in on a target is called its *damping function*. The cerebellum also helps you to coordinate different body parts while walking, running, and swimming. Finally, the cerebellum helps you maintain equilibrium.

INTEGRATIVE FUNCTIONS

We turn now to a fascinating, though poorly understood, function of the cerebrum: integration. The **integrative functions** include cerebral activities such as memory, sleep and wakefulness, and emotions. The role of the limbic system in emotional behavior was discussed in Chapter 12.

Memory

Memory may be defined as the ability to recall thoughts originally established by incoming sensory impulses. Memory may be classified into two kinds: short-term and long-term.

Short-term memory is the ability to recall bits of information. One example is finding a number in the phone book and then dialing it. If the number has no special significance, it is usually forgotten in a few seconds. Short-term memories leave no permanent imprint on the brain. One theory of short-term memory claims that memories may be caused by reverberating neuronal circuits—an incoming impulse stimulates the first neuron, which stimulates the second, which stimulates the third, and so on. Branches from the second and third neurons synapse with the first, sending the impulse back through the circuit again and again. Thus the output neuron generates continuous impulses. Once fired, the output signal may last from a few seconds to many hours, depending on the arrangement of neurons in the circuit. If this pattern is applied to short-term memory, an incoming thought—the phone number—continues in the brain even after the initial stimulus is gone. Thus you can recall the memory only for as long as the reverberation continues.

The concept of a reverberating circuit does not, however, explain long-term memory. *Long-term memory* is the persistence of an incoming impulse for years. One theory explains long-term memory on the basis of another principle: facilitation at synapses. When an incoming thought enters a neuronal circuit, the synapses in the circuit become facilitated for the passage of a similar signal later on. Thus, an incoming signal facilitates the synapses in the circuit used for that signal over and over, and you recall the memory. Such a neuronal circuit is called an *engram*.

The first incoming thought leading to long-term memory lasts only a brief time. How then does a short-term memory result in a long-term memory? One explanation is that the reverberating circuit of a short-term memory may persist for up to an hour after the initial thought. This reverberation establishes the engram. Later, another incoming thought can cause facilitation of the neurons in the engram, and the result is long-term memory. Another theory suggests that long-term memory is reated to protein synthesis by RNA. According to this notion, the storage of memories results from the production of RNA that synthesizes specific proteins for each memory stored. The theory also suggests that when a memory is re-

called, the protein is destroyed but the RNA template remains so the memory is not lost.

The portions of the brain thought to be associated with memory include: (1) the association cortex of the frontal, parietal, occipital, and temporal lobes; and (2) parts of the limbic system, especially the hippocampus.

Sleep and Wakefulness

Humans sleep and awaken in a fairly constant 24-hour rhythm called a *circadian rhythm*. This cycle occurs even when there is total light or total darkness. Since neuronal fatigue precedes sleep and the signs of fatigue disappear after sleep, fatigue is apparently one cause of sleep. Moreover, EEG recordings indicate that during wakefulness the cerebral cortex is very active, sending impulses continuously through the body. During sleep, however, fewer impulses are transmitted by the cerebral cortex. The activity of the cerebral cortex is thought to be related to the reticular formation.

The reticular formation has numerous connections with the cerebral cortex (Figure 15-6). Stimulation of portions of the reticular formation results in increased cortical activity. Thus the reticular formation is also known as the **reticular activating system (RAS).** One part of the system, the mesencephalic part, is composed of areas of gray matter of the pons and midbrain. When this area is stimulated, many impulses pass upward into the thalamus and disperse to widespread areas of the cerebral cortex. The effect is a generalized increase in cortical activity. The other part of the RAS, the thalamic part, consists of gray matter in the thalamus. When the thalamic part is stimulated, signals from specific parts of the thalamus cause activity in specific parts of the cerebral cortex. Apparently the mesencephalic part of the RAS causes general wakefulness (consciousness) and the thalamic part causes *arousal* — that is, awakening from deep sleep.

For the *arousal reaction* to occur, the RAS must be stimulated by input signals. Almost any sensory input can activate the RAS: pain stimuli, proprioceptive signals, bright light, an alarm clock. Once the RAS is activated, the cerebral cortex is also activated and you experience arousal. Signals from the cerebral cortex can also stimulate the RAS. Such signals may originate in the somesthetic cortex, the motor cortex, or the limbic system. When the signals activate the RAS, the RAS activates the cerebral cortex and arousal occurs.

Following arousal, the RAS and cerebral cortex continue to activate each other through a feedback

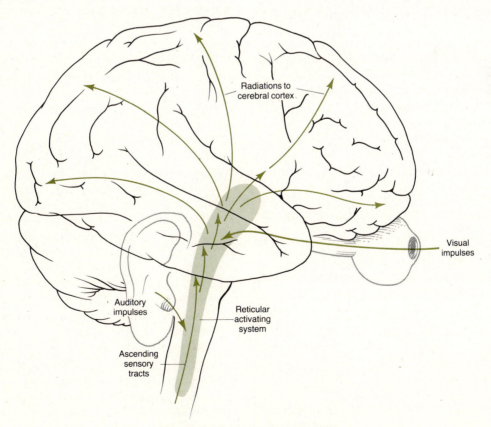

Figure 15-6 Reticular formation.

system consisting of many circuits. The RAS also has a feedback system with the spinal cord that is composed of many circuits. Impulses from the activated RAS are transmitted down the spinal cord and then to skeletal muscles. Muscle activation causes proprioceptors to return impulses that activate the RAS. The two feedback systems maintain activation of the RAS, which in turn maintains activation of the cerebral cortex. The result is a state of wakefulness called *consciousness*. Since humans experience different levels of consciousness (alertness, attentiveness, relaxation, nonattentiveness), it is assumed that the level of consciousness depends on the number of feedback circuits operating at the time. In fact, consciousness may be altered by various factors. Amphetamines probably activate the RAS to produce a state of wakefulness and alertness. Meditation produces a lack of consciousness. Anesthetics produce a state of consciousness called anesthesia. Damage to the nervous system, as well as disease, can produce a lack of consciousness called coma. And drugs such as LSD can alter consciousness also.

If this theory of wakefulness is accepted, how then does sleep occur if the activating feedback systems are in continual operation? One explanation is that the synapses in the feedback circuits eventually undergo fatigue. The feedback system slows tremendously or is inhibited. Inactivation of the RAS produces a state known as *sleep*.

Just as there are different levels of consciousness, there are different levels of sleep. Normal sleep consists of two types: nonrapid eye movement sleep (non-REM) and rapid eye movement sleep (REM).

Non-REM sleep consists of four stages. Each has been identified by EEG recordings:

Stage 1. The person is relaxing with eyes closed. During this time, respirations are regular, pulse is even, and the person has fleeting thoughts. If awakened during stage 1, the person will frequently deny having been asleep. During stage 1, alpha waves appear on the EEG.

Stage 2. It is a little harder to awaken the person. Fragments of dreams may be experienced, and the eyes may slowly roll from side to side. During stage 2, the EEG shows *sleep spindles:* sudden, short bursts of sharply pointed alpha waves that occur 14 to 16 cycles/second.

Stage 3. The person is very relaxed. Body temperature begins to fall and blood pressure decreases. During this stage, it is difficult to awaken the person and the EEG shows a mixture of sleep spindles and delta waves. This stage occurs about 20 minutes after falling asleep.

Stage 4. Deep sleep occurs. The person is very relaxed and responds slowly if awakened. Bed-wetting and sleepwalking may occur during this stage. The EEG of stage 4 is dominated by delta waves.

In a typical 7 or 8 hour sleep period, a person goes from stages 1 to 4 of non-REM sleep. Then the person ascends to stages 3 and 2 and then to REM sleep within 50 to 90 minutes.

In **REM sleep,** the EEG readings are similar to those of stage 1 of non-REM sleep. There are significant physiological differences, however. During REM sleep, respirations and pulse rate increase and are irregular. Blood pressure also fluctuates considerably. It is during REM sleep that most dreaming occurs. Following REM sleep, which lasts about 10 to 20 minutes, the person descends again to stages 3 and 4 of non-REM sleep. This cycle repeats itself from three to five times during the entire sleep period. Each time the person returns to REM sleep, however, more time is spent at this stage. As much as 50 percent of an infant's sleep is REM sleep as contrasted with 20 percent for adults. Most sedatives significantly decrease REM sleep.

Emotions

It was noted in Chapter 12 that the limbic system assumes a function in emotions such as rage, pleasure, anger, fear, sorrow, sexual feelings, docility, and affection. In this respect, the limbic system is referred to as the emotional brain. The details of the role of the limbic system in emotional behavior may be reviewed by referring to Chapter 12.

We shall now conclude our discussion of the nervous system by considering the autonomic nervous system.

CHAPTER SUMMARY OUTLINE

LINKAGE OF SENSORY INPUT AND MOTOR RESPONSES

1. Incoming sensory information is added to, subtracted from, or integrated with other information arriving from all other operating sensory receptors.
2. The integration process occurs at many stations along the pathways of the central nervous system such as within the spinal cord, in the brain stem, the cerebellum, and the cerebral cortex.
3. A motor response to make a muscle contract or a gland secrete can be initiated at any of these stations or levels.
4. The motor response system is very precise so that when the central nervous system initiates a response only the desired muscle contracts.
5. The pattern of activation of motor units varies in complexity from the quick knee jerk to playing a piano.
6. As the location of the sensory-motor linkage climbs to higher levels in the central nervous system, additional contributions are introduced that enrich the growing inventory of motor responses.
7. Input to the central nervous system may or may not give rise to conscious sensation.
8. When the input reaches the highest center, the process of sensory-motor integration occurs. This involves not only the utilization of information contained within that center but also the information impinging upon that center delivered from other centers within the central nervous system.

MYOTATIC REFLEX

1. The myotatic reflex shows that if a muscle is abruptly stretched or jerked, its response is to contract immediately. Since the myotatic reflex makes only one synapse in its loop, it is classed as a monosynaptic reflex.

PROPRIOCEPTION

1. Proprioception is the ability to recognize the location and rate of movement of one body part in relation to other parts.

The Muscle Spindle

1. Muscle spindles are scattered throughout the belly of skeletal muscles and operate continually to monitor the overall length of muscle fibers and the speed at which they contract.
2. The extrafusal fibers of muscle spindles have both sensory and motor innervations.

Golgi Tendon Organs

1. Another group of receptors that contribute to proprioception are Golgi tendon organs.
2. In contrast to the muscle spindle, which is said to be in parallel with the extrafusal fibers, the Golgi tendon organ is connected in series and is able to signal muscle tension.

Afferent Pathway for Proprioception

1. The afferent pathway for muscle sense consists of impulses sent from proprioceptors into cranial and spinal nerves.
2. Impulses for conscious proprioception travel in ascending tracts in the cord, where they are relayed to the thalamus and cerebral cortex.
3. Proprioceptive impulses that result in reflex action travel to the cerebellum through spinocerebellar tracts.
4. The posterior spinocerebellar tract is an uncrossed tract that conveys impulses concerned with subconscious muscle sense.
5. The anterior spinocerebellar tract also conveys impulses for subconscious muscle sense; however, it is made up of both crossed and uncrossed nerve fibers.

MOTOR PATHWAYS

Pyramidal Pathways

1. Voluntary motor impulses are conveyed from the motor areas of the brain to somatic efferent neurons leading to skeletal muscles via the pyramidal pathway.
2. The pathways over which the impulses travel from the motor cortex to skeletal muscles have two components: upper motor neurons (pyramidal fibers), and lower motor neurons (peripheral fibers).
3. Three important tracts of the pyramidal system are: (1) the lateral corticospinal tract, (2) the anterior corticospinal tract, and (3) the corticobulbar tract.

Extrapyramidal Pathways

1. The extrapyramidal pathways include all descending tracts other than the pyramidal tracts.

2. The main extrapyramidal tracts are: (1) the rubrospinal tract, (2) the tectospinal tract, and (3) the vestibulospinal tract.

INTEGRATIVE FUNCTIONS

Memory

1. Memory is defined as the ability to recall thought originally established by incoming sensory impulses.

2. Sleep and wakefulness are integrative functions.
3. Non-REM sleep consists of four stages identified by EEG recordings.
4. The limbic system assumes a function in emotions such as rage, pleasure, anger, fear, sorrow, sexual feelings, docility, and affection. In this respect, the limbic system is referred to as the emotional brain.

REVIEW QUESTIONS

1. What happens to incoming sensory information in the central nervous system?
2. What are the different stations along the pathways of the central nervous system where the integration process occurs?
3. Explain the statement, "The design of the motor response system to skeletal muscle is very precise."
4. Explain how the pattern of activation of motor units varies in complexity. Give examples.
5. Why doesn't all sensory input to the central nervous system give rise to conscious sensation?
6. What happens when the input reaches the highest center in the central nervous system?
7. Explain in detail the myotatic reflex.
8. What is proprioception? Explain your answer. Where are the receptors for this sense located?
9. Relate proprioception to the maintenance of homeostasis.
10. Define muscle spindles, Golgi tendon organs, and joint kinesthetic receptors.
11. Describe the operation of the muscle spindle.
12. What are the contributions of the fusimotor fibers to motor activity?
13. Describe the operation of the Golgi tendon organ.
14. Describe the structure and function of the posterior spinocerebellar tract. Do the same for the anterior spinocerebellar tract.
15. What are the three main tracts of the pyramidal pathway?
16. Describe these three pathways in detail.
17. What are the extrapyramidal pathways? Describe the three main extrapyramidal tracts.
18. Define memory. What are the two kinds of memory?
19. Explain circadian rhythm.
20. What are the four stages of non-REM sleep?

Chapter 16

THE AUTONOMIC NERVOUS SYSTEM

STUDENT OBJECTIVES

After reading this chapter, you should be able to

- Explain the relationship between the autonomic nervous system and homeostasis
- Compare the structural and functional differences between the somatic efferent and autonomic portions of the nervous system
- Identify the structural features of the autonomic nervous system
- Compare the sympathetic and parasympathetic divisions of the autonomic nervous system in terms of structure and physiology
- Compare in detail the activities of the sympathetic versus the parasympathetic divisions of the autonomic nervous system
- Explain the functions of transmitter substances in the autonomic nervous system
- Contrast cholinergic versus adrenergic transmission as to location and different chemical transmitter substances
- Outline the contradictory effects of catecholamines
- Explain the role of the hypothalamus and its relationship to the sympathetic and parasympathetic divisions
- Explain the relationship between biofeedback and the autonomic nervous system
- Describe the relationship between meditation and the autonomic nervous system

The portion of the nervous system that regulates the activities of smooth muscle, cardiac muscle, and glands is the **autonomic nervous system.** Structurally, the system consists of visceral efferent neurons organized into nerves, ganglia, and plexuses. Functionally, it usually operates without conscious control. Physiologists originally thought the system functioned autonomously with no control from the central nervous system—hence its name, the autonomic nervous system. In truth, the autonomic system is neither structurally nor functionally independent of the central nervous system. It is regulated by centers in the brain, in particular by the cerebral cortex, hypothalamus, and medulla oblongata. However, the autonomic nervous system does differ from the somatic efferent nervous system in some ways. For convenience of study the two are separated.

SOMATIC EFFERENT AND AUTONOMIC NERVOUS SYSTEMS

Whereas the somatic efferent nervous system produces conscious movement in skeletal muscles, the autonomic nervous system (visceral efferent nervous system) regulates visceral activities. And it generally does so involuntarily and automatically. Examples of visceral activities regulated by the autonomic nervous system are changes in the size of the pupil, accommodation for near vision, dilation and constriction of blood vessels, adjustment of the rate and force of the heartbeat, emptying of the urinary bladder by contraction of its smooth muscle, movements of the gastrointestinal tract, formation of gooseflesh, and secretion by most glands. These activities usually lie beyond conscious control. They are automatic.

The autonomic nervous system is entirely motor. All its axons are efferent fibers, which transmit impulses from the central nervous system to visceral effectors. Autonomic fibers are called **visceral efferent fibers. Visceral effectors** include cardiac muscle, smooth muscle, and glandular epithelium. This does not mean there are no afferent (sensory) impulses from visceral effectors, however. Visceral sensations pass over visceral afferent neurons that have cell bodies located outside but close to the central nervous system. Some functions of these afferent neurons were described with the cranial and spinal nerves. However, the hypothalamus, which largely controls the autonomic nervous system, also receives impulses from the visceral sensory fibers.

The autonomic nervous system consists of two principal divisions: the **sympathetic** and the **parasympathetic.** Many organs innervated by the autonomic nervous system receive visceral efferent neurons from both components of the autonomic system—one set from the sympathetic division, another from the parasympathetic division. In general, impulses transmitted by the fibers of one division stimulate the organ to start or increase activity, whereas impulses from the other division decrease or halt the organ's activity. Organs that receive impulses from both sympathetic and parasympathetic fibers have *dual innervation*. In the somatic efferent nervous system, only one kind of motor neuron innervates an organ, which is always a skeletal muscle. When the somatic neurons stimulate the cells of the skeletal muscle, the muscle becomes active. When the neuron ceases to stimulate the

muscle, contraction stops altogether. Skeletal muscle cells of each motor unit contract only when stimulated by their motor neuron. When the impulse stops, contraction stops.

STRUCTURE

The sympathetic and parasympathetic divisions of the autonomic nervous system are also referred to as the **thoracolumbar** and **craniosacral** divisions, respectively. Let us see what this means by discussing the general features applicable to both divisions.

Visceral Efferent Neurons

Autonomic visceral efferent pathways always consist of two neurons. One runs from the central nervous system to a ganglion. The other runs directly from the ganglion to the effector.

The first of these visceral efferent neurons in an autonomic pathway is called a **preganglionic neuron** (Figure 16-1). Preganglionic neurons have their cell bodies in the brain or spinal cord. Their myelinated axons, called **preganglionic fibers,** pass out of the central nervous system as part of a cranial or spinal nerve. At some point, they leave these nerves and run to autonomic ganglia, where they synapse with the dendrites or cell bodies of postganglionic neurons.

Postganglionic neurons, the second visceral efferent neurons in an autonomic pathway, lie entirely outside the central nervous system. Their cell bodies and dendrites (if they have dendrites) are located in the autonomic ganglia, where the synapse with the preganglionic fibers occurs. The axons of postganglionic neurons are called **postganglionic fibers.** Postganglionic fibers are nonmyelinated, and they terminate in visceral effectors.

Thus preganglionic neurons convey efferent impulses from the central nervous system to autonomic ganglia. Postganglionic neurons relay the impulses from the autonomic ganglia to visceral effectors.

Preganglionic Neurons

In the sympathetic division, the preganglionic neurons have their cell bodies in the lateral gray horns of the thoracic segments and first two lumbar segments of the spinal cord (Figure 16-2). It is for this reason that the sympathetic division is also called the **thoracolumbar division** and the fibers of the sympathetic preganglionic neurons are known as the **thoracolumbar outflow.**

The cell bodies of the preganglionic neurons of the parasympathetic division are located in nuclei in the brain stem and in the lateral gray horns of the second through fourth sacral segments of the spinal cord (Figure 16-2)—hence the synonymous term **craniosacral division.** The fibers of the parasympathetic preganglionic neurons are referred to as the **craniosacral outflow.**

Autonomic Ganglia

Autonomic pathways also include **autonomic ganglia,** where synapses between visceral efferent neurons occur. Autonomic ganglia differ from posterior root ganglia. The latter contain cell bodies of sensory neurons and no synapses occur in them. The autonomic ganglia may be divided into three general groups (Figure 16-3). The *sympathetic trunk* or *vertebral chain ganglia* are a series of ganglia that lie in a horizontal row on either side of the vertebral column extending from the base of the skull to the coccyx. They are also known as *paravertebral* or *lateral ganglia.* They receive preganglionic fibers only from the thoracolumbar (sympathetic) division (Figure 16-2).

A second kind of ganglion of the sympathetic division of the autonomic nervous system is called a

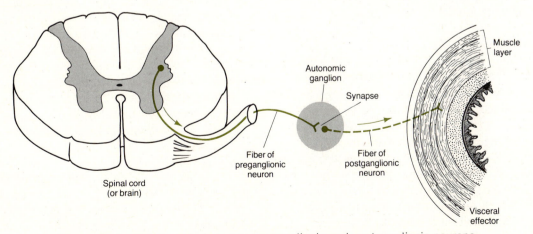

Figure 16-1 Relationship between preganglionic and postganglionic neurons.

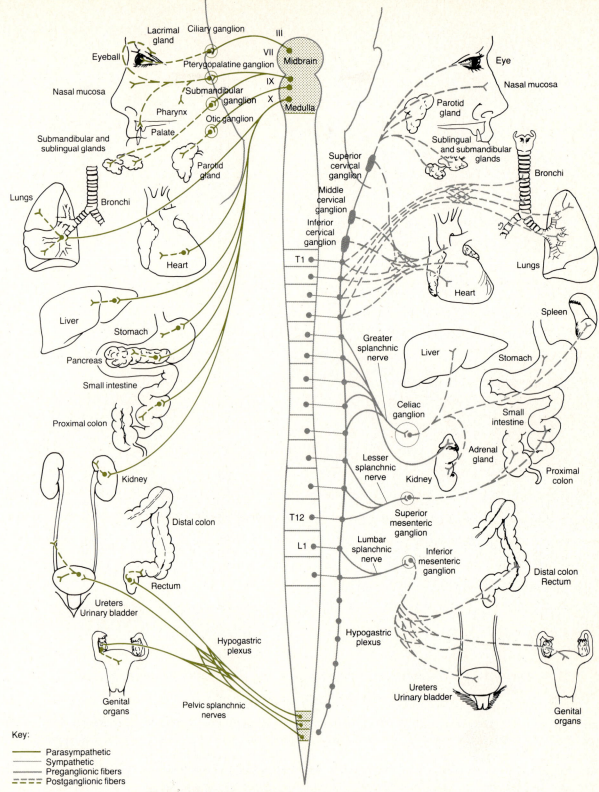

Figure 16-2 Structure of the autonomic nervous system.

prevertebral or *collateral ganglion* (Figure 16-3). The ganglia of this group lie anterior to the spinal column and close to the large abdominal arteries from which their names are derived. Examples of prevertebral ganglia so named are the celiac ganglion, on either side

of the celiac artery just below the diaphragm; the superior mesenteric ganglion, near the beginning of the superior mesenteric artery in the upper abdomen; and the inferior mesenteric ganglion, located near the beginning of the inferior mesenteric artery in the middle

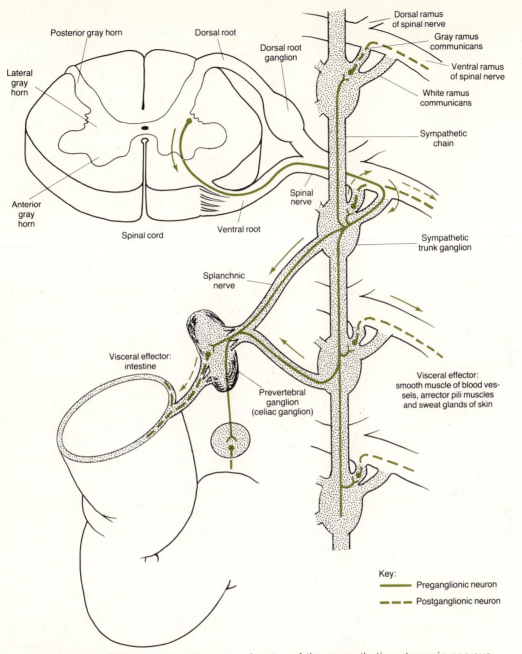

Figure 16-3 Ganglia and rami communicantes of the sympathetic autonomic nervous system.

of the abdomen (Figure 16-2). Prevertebral ganglia receive preganglionic fibers from the thoracolumbar (sympathetic) division.

The third kind of autonomic ganglion belongs to the parasympathetic division and is called a *terminal* or *intramural ganglion*. The ganglia of this group are located at the end of a visceral efferent pathway very close to visceral effectors or within the walls of visceral effectors. Terminal ganglia receive preganglionic fibers from the craniosacral (parasympathetic) division. The preganglionic fibers do not pass through sympathetic trunk ganglia (Figure 16-2).

In addition to autonomic ganglia, the autonomic nervous system also contains **autonomic plexuses.** Slender nerve fibers from ganglia containing postganglionic nerve cell bodies arranged in a branching network constitute an autonomic plexus.

Postganglionic Neurons

Axons from preganglionic neurons of the sympathetic division pass to ganglia of the sympathetic trunk. Either they can synapse in the sympathetic chain ganglia with postganglionic sympathetics, or

they can continue, without synapsing, through the chain ganglia to end at a prevertebral ganglion where synapses with the postganglionic sympathetics can take place. Each sympathetic preganglionic fiber synapses with several postganglionic fibers in the ganglion, and the postganglionic fibers pass to several visceral effectors. Upon exiting their ganglia, the postsynaptic fibers innervate their visceral effectors.

Axons from preganglionic neurons of the parasympathetic division pass to terminal ganglia near or within a visceral effector. In the ganglion, the presynaptic neuron usually synapses with only four or five postsynaptic neurons to a single visceral effector. Upon exiting their ganglia, the postsynaptic fibers supply their visceral effectors.

With this background in mind, we can now examine some specific structural features of the sympathetic and parasympathethic divisions of the autonomic nervous system.

Sympathetic Division

The preganglionic fibers of the sympathetic division have their cell bodies located in the lateral gray horn of the spinal cord in the thoracic and first two lumbar segments (Figure 16-2). The preganglionic fibers are myelinated and leave the spinal cord through the ventral root of a spinal nerve along with the somatic efferent fibers of the same segmental levels. After exiting through the intervertebral foramina, the preganglionic sympathetic fibers enter a white ramus to pass to the nearest sympathetic trunk ganglion on the same side. Collectively, the white rami are called the **white rami communicantes.** Their name indicates that they contain myelinated fibers. Only thoracic and upper lumbar nerves have white rami communicantes. The white rami communicantes connect the ventral ramus of the spinal nerve with the ganglia of the sympathetic trunk.

When a preganglionic fiber of a white ramus communicans enters the sympathetic trunk, it may terminate (synapse) in several ways. Some fibers synapse in the first ganglion at the level of entry. Others pass up or down the sympathetic trunk for a variable distance to form the fibers on which the ganglia are strung. These fibers, known as *sympathetic chains* (Figure 16-3), may not synapse until they reach a ganglion in the cervical or sacral area. Some postganglionic fibers leaving the sympathetic trunk ganglia pass directly to visceral effectors of the head, neck, chest, and abdomen. Most, however, rejoin the spinal nerves before supplying peripheral visceral effectors such as sweat glands, and the smooth muscle in blood vessels and around hair follicles. The **gray ramus communicans** is the structure containing the postganglionic

fibers that run from the ganglion of the sympathetic trunk to the spinal nerve (Figure 16-3). The term *gray* refers to the fact that the fiber is unmyelinated. All spinal nerves have gray rami communicantes. Gray rami communicantes outnumber the white rami since there is a gray ramus leading to each of the 31 pairs of spinal nerves.

In most cases, a sympathetic preganglionic fiber terminates by synapsing with a large number of postganglionic cell bodies in a ganglion, usually 20 or more. Often the postganglionic fibers then terminate in widely separated organs of the body. Thus an impulse that starts in a single preganglionic neuron may affect several visceral effectors. For this reason, most sympathetic responses have widespread effects on the body.

Parasympathetic Division

The preganglionic cell bodies of the parasympathetic division are found in nuclei in the brain stem and the lateral gray horn of the second through fourth sacral segments of the spinal cord (Figure 16-2). Their fibers emerge as part of a cranial nerve or as part of the ventral root of a spinal nerve. The **cranial parasympathetic outflow** consists of preganglionic fibers that leave the brain stem by way of the oculomotor nerves (III), facial nerves (VII), glossopharyngeal nerves (IX), and vagus nerves (X). The **sacral parasympathetic outflow** consists of preganglionic fibers that leave the ventral roots of the second through fourth sacral nerves. The preganglionic fibers of both the cranial and sacral outflows end in terminal ganglia where they synapse

Exhibit 16-1 COMPARISON OF SYMPATHETIC AND PARASYMPATHETIC DIVISIONS

Sympathetic	Parasympathetic
Forms thoracolumbar outflow.	Forms craniosacral outflow.
Contains sympathetic trunk and prevertebral ganglia.	Contains terminal ganglia.
Ganglia are close to the CNS and distant from visceral effectors.	Ganglia are near or within visceral effectors.
Each preganglionic fiber synapses with many postganglionic neurons that pass to many visceral effectors.	Each preganglionic fiber usually synapses with four or five postganglionic neurons that pass to a single visceral effector.
Distributed throughout the body, including the skin.	Distribution limited primarily to head and viscera of thorax, abdomen, and pelvis.

with postganglionic neurons. Since the terminal ganglia are close to their visceral effectors, postganglionic parasympathetic fibers are short. Postganglionic sympathetic fibers are relatively long.

The salient structural features of the sympathetic and parasympathetic divisions are compared in Exhibit 16-1.

PHYSIOLOGY

Chemical Transmitters

In the autonomic nervous system, impulses are relayed across synapses by means of *transmitter substances*. These chemicals are released by the preganglionic fibers in the ganglia and by the postganglionic fiber onto the effector. The two transmitter substances important in the operation of the autonomic reflexes are *acetylcholine (ACh)* and *norepinephrine (NE)*. The term *cholinergic* is used to describe fibers that use acetylcholine as the transmitter substance, and *adrenergic* describes fibers that utilize epinephrine as the transmitter substance.

Cholinergic Transmission

Cholinergic fibers include *all* preganglionic fibers, the postganglionic fibers of all parasympathetic neurons, and the sympathetic postganglionic fibers that innervate the sweat glands. Recall from Chapter 10 that the innervation of skeletal muscle is also mediated by acetylcholine.

Another important situation involves the *adrenal gland*. This gland consists of two major divisions, a *medulla* and a *cortex*. The medulla is under autonomic nervous control, whereas the cortex is under endocrine control (Chapter 17). The medulla is innervated by the *splanchnic* nerve, which belongs to the sympathetic division. The innervation reaching the adrenal medulla corresponds to preganglionic fibers, and the cells of the medulla correspond to the postganglionic cells (which, of course, have no axons). The preganglionic fibers release ACh from their terminals when they are stimulated. The transmitter substance then acts upon the *chromaffin cells* of the adrenal medulla, which release a mixture of norepinephrine ($\frac{1}{5}$) and epinephrine ($\frac{4}{5}$) into the bloodstream.

Acetylcholine is synthesized and stored in an inactive form in the terminals of the cholinergic fibers. The small membrane-enclosed synaptic vesicles that store the ACh can be seen with the electron microscope. The ACh is synthesized rapidly within these vesicles from choline and acetyl coenzyme A. The biosynthesis is carried out by an enzyme named *choline-acetylase* (Figure 16-4). The release of ACh from the terminals requires Ca^{2+} ions to be present. The release of the chemical comes in bursts of thousands of molecules each. The ACh then diffuses the short distance across the synaptic cleft to bind with a receptor substance on the postsynaptic membrane. The postsynaptic membrane is quickly depolarized and the ACh is quickly inactivated by an enzyme *acetylcholinesterase*. This enzyme can be inhibited by a variety of drugs. Physostigmine (also known as eserine) and neostigmine are reversible inhibitors. As the body removes the inhibitor, the enzyme regains its functional ability and removes the ACh. Other drugs such as di-isopropyl fluorophosphate (DFP) combine permanently with the enzyme, and thus the effect of ACh is prolonged until the level of new enzyme surpasses the level of the inhibitor.

Adrenergic Transmission

Norepinephrine is released from most of the postganglionic sympathetic fibers; exceptions include the postganglionic sympathetic fibers that innervate the sweat glands which release ACh and the postganglionic fibers that innervate the uterus and the pilomotor muscles.

Norepinephrine and epinephrine, which are secreted by the adrenal medulla, belong to a class of compounds called *catecholamines*. Included in this class of compounds is a variety of structurally related mediators, all of which can be synthesized from the amino acid phenylalanine. Norepinephrine is synthesized and stored in the synaptic vesicles located in the

Presynaptic neuron synaptic vesicles   Synaptic cleft   Postsynaptic neuron

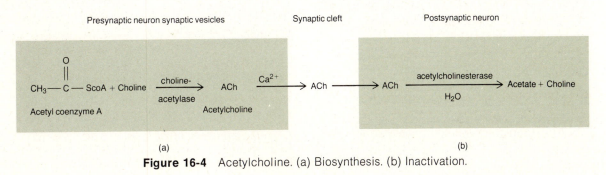

(a)   (b)

Figure 16-4   Acetylcholine. (a) Biosynthesis. (b) Inactivation.

axon terminals of the adrenergic fibers. When an action potential reaches the axon terminal, a rapid extrusion of norepinephrine occurs into the synaptic cleft. The molecules diffuse across the cleft and combine with specific receptor sites on the postsynaptic membrane to elicit the effector response.

The effect of the transmitter is short-lived, as it is quickly removed by a number of possible mechanisms. The primary method of inactivation is by the active reuptake of the norepinephrine by the presynaptic fiber, which then stores the transmitter in synaptic vesicles. The enzyme *monoamine oxidase (MAO)* is a component of the outer membrane of mitochondria and is therefore distributed throughout the body. In the axon terminal, it acts slowly to inactivate norepinephrine and epinephrine. Another enzyme present, called *catechol-o-methyl transferase (COMT)* also inactivates these compounds.

Contradictory Effects of Catecholamines

The contradictory effects of catecholamines are due to the presence of two different types of receptor systems on the surface of cell membranes. Receptors designated as *α receptors* and *β receptors* combine with the transmitter molecules and cause changes in membrane permeability. Most cells have only one type of adrenergic receptor, but some possess both types. In general, sites that are excited by catecholamines are the *α* receptors and those inhibited by catecholamines are the *β* receptors. Exhibit 16-2 shows the type of receptor present on the cells and their response to autonomic stimulation. There are, however, exceptions to the general rule describing *α* and *β* receptors. For example, heart muscle has *β* receptors that cause the heart muscle to contract more forcibly. Other body cells possess both types of receptors. The *β* receptors greatly outnumber the *α* receptors in the smooth muscle cells of the blood vessels that flow through skeletal muscles. Therefore, an injection of *norepinephrine* onto the surface of such smooth muscle cells would cause a *vasoconstriction* because norepinephrine reacts mainly on the *α* type receptor. An injection of *epinephrine,* however, would result in a *vasodilation* because epinephrine acts equally on both types of receptors. But since there is a preponderance of *β* receptors, the effect is that of the *β* type receptor, that is, vasodilation due to *relaxation* of the smooth muscle cells.

The simple rule presently in operation is that norepinephrine stimulates only *α* receptors, that isopropyl norepinephrine (a synthetic derivative of norepinephrine also called isoproterenol) stimulates only *β* receptors, and that epinephrine stimulates both *α* and *β* receptors.

Also, depending upon the organ, it has been shown that the *α* receptors are 2 to 10 times more responsive to epinephrine than norepinephrine and that the *β* receptors are 2 to 10 times more responsive to isopropyl norepinephrine than epinephrine.

In general, stimulation of the *α* type receptor produces membrane *depolarization* whereas the *β* type receptor produces membrane *hyperpolarization*.

Certain drugs are used to produce the identical effects of sympathetic stimulation. Such drugs are said to mimic the sympathetic nerves and are, in fact, referred to as being *sympathomimetic*. Included in this class of drugs are epinephrine, phenylephrine, methoxamine, isoproterenol, terbutaline, and many others. These drugs differ in their potencies, in their duration of action, and in their major targets. Other drugs produce sympathomimetic effects because they cause the release of stored norepinephrine. Included in this category are the amphetamines, tyramine, and ephedrine.

When it is necessary to block the action of the sympathetic transmitters, various drugs are available that act in different ways to prevent the sympathetic stimulation. Reserpine belongs to a class of inhibitors that prevents the synthesis and storage of norepinephrine in the axon terminals, thereby depleting the nerves of their chemical transmitter. Another group of drugs blocks the release of norepinephrine that is stored in the axon terminals. Drugs such as guanethidine and bethanidine belong to this group. Specific blockage of the *α* receptors can be achieved by the use of drugs such as dibenamine, phenoxybenamine, and phentolamine, while drugs such as propranolol and sotalol are specific for the *β* type receptors.

Activities of the Autonomic System

Because of its body-wide distribution, the autonomic nervous system is able to exert prime control in maintaining the constancy of the internal environment. It effectively combats stresses that arise from either inside or outside the body. The composition, temperature, volume, and distribution of the body fluid compartments is regulated by the action of autonomic nerve fibers acting upon circulatory, respiratory, excretory, endocrine, and exocrine organs. Most of the organs controlled by the autonomic nervous system receive innervation from both divisions, and since the two divisions produce opposite effects, they can maintain a fine degree of control at all times. It is possible to show that both divisions exert a constant or *tonic* effect on a given organ. For example, sectioning the branch of the vagus nerve to the heart causes the heart to speed up since the effect of the sympathetic nerves to the heart is now unchecked. The opposite

Exhibit 16-2 RESPONSES OF EFFECTOR ORGANS TO AUTONOMIC NERVE IMPULSES

| Effector Organs | ADRENERGIC IMPULSES | | CHOLINERGIC IMPULSES |
	RECEPTOR TYPE	RESPONSES	RESPONSES
Eye			
Radial muscle iris	α	Contraction (mydriasis)	_____
Sphincter muscle iris		_____	Contraction (miosis)
Ciliary muscle	β	Relaxation for far vision (slight effect)	Contraction for near vision
Heart			
S-A node	β	Increase in heart rate	Decrease in heart rate; vagal arrest
Atria	β	Increase in contractility and conduction velocity	Decrease in contractility and (usually) increase in conduction velocity
Blood vessels			
Coronary		Dilation	Dilation
Skin and mucosa	α	Constriction	Dilation
Skeletal muscle	α, β	Constriction; dilation	Dilation
Cerebral	α	Constriction (slight)	Dilation
Pulmonary	α	Constriction	Dilation
Abdominal viscera	α, β	Constriction; dilation	_____
Salivary glands	α	Constriction	Dilation
Lung			
Bronchial muscle	β	Relaxation	Contraction
Bronchial glands		Inhibition(?)	Stimulation
Stomach			
Motility and tone	β	Decrease (usually)	Increase
Sphincters	α	Contraction (usually)	Relaxation (usually)
Secretion		Inhibition(?)	Stimulation
Intestine			
Motility and tone	α, β	Decrease	Increase
Sphincters	α	Contraction (usually)	Relaxation (usually)
Secretion		Inhibition(?)	Stimulation
Gallbladder and ducts		Relaxation	Contraction
Urinary bladder			
Detrusor	β	Relaxation (usually)	Contraction
Trigone and sphincter	α	Contraction	Relaxation
Ureter			
Motility and tone		Increase (usually)	Increase(?)
Uterus	α, β	Variable	Variable
Sex organs		Ejaculation	Erection
Skin			
Pilomotor muscles	α	Contraction	_____
Sweat glands	α	Slight, localized secretion	Generalized secretion
Spleen capsule	α	Contraction	_____
Adrenal medulla		_____	Secretion of epinephrine and norepinephrine
Liver		Glycogenolysis	_____
Pancreatic acini		_____	Secretion
Salivary glands	α	Thick, viscous secretion	Profuse, watery secretion
Lacrimal glands		_____	Secretion
Nasopharyngeal glands		_____	Secretion

Source: From L. S. Goodman and A. Gilman, eds., *The Pharmacological Basis of Therapeutics*, 3rd ed., New York: Macmillan, 1965. Copyright © 1965, Macmillan Publishing Co., Inc.

experiment, sectioning the sympathetic nerves to the heart, results in a marked slowing of the heart due to the now unchecked influence of the vagus nerve.

Sympathetic Division

The sympathetic division is primarily concerned with the *total* well-being of the individual. It influences many homeostatic responses: blood sugar levels, heart rate, body temperature, muscular contraction, and blood flow distribution. The sympathetic division does respond in selective instances to adjust single variables such as blood pressure. But, in emergencies, the system responds totally. It not only increases the frequency of impulses in fibers that are tonically active, but also activates those fibers that were quiescent. For example, it activates sympathetic fibers to the

adrenal medulla and vasodilator fibers to the skeletal muscles.

Similar to this emergency response are the directives sent out by the sympathetic division during vigorous muscular exercise. The increased need for energy and removal of waste products requires that blood flow to the exercising muscles be greatly increased. The sympathetic division causes a vasodilation of blood vessels in the skeletal muscles but a vasoconstriction of blood vessels in the splanchnic circulation (blood flow to the liver, spleen, and intestines). Sympathetic effects on the heart cause an increase in heart rate and force of contraction that produce a greater volume of blood to be pumped into the arteries every minute (that is, increases cardiac output). In the lungs, the sympathetic nerves cause the air passages (bronchioles) to dilate, which improves the oxygen delivery to the blood. The adrenal medulla is activated to release epinephrine and norepinephrine into the blood, and their effects are mainly metabolic ones. That is, the epinephrine reaching the liver cells causes the breakdown of stored glycogen so that glucose is released into the blood, and in muscle cells the rate of glucose utilization for energy production is greatly accelerated. Fat stored in body depots is also broken down more rapidly to produce free fatty acids which are able to provide additional energy for the contracting muscles. Another activity promoted by the sympathetic nerves during vigorous exercise relates to the elimination of heat produced by the muscular activity. An increase in the flow of blood to the skin and sweat glands causes water evaporation from the body surface, thus cooling the body.

The importance of the role of the sympathetic nerves in regulating the caliber of the arterioles must be stressed. Increasing the tone in sympathetic nerves leading to the arterioles causes constriction of most of these blood vessels. The effect of such a generalized arteriolar constriction is a rise in the blood pressure in the arteries. The effect of releasing epinephrine from the adrenal medulla, added to the effect of the sympathetic nerves acting on the heart, causes an increase in cardiac output that also produces a rise in arterial blood pressure. In this manner, the body ensures that blood is delivered to the vital organs (brain and heart) at the expense of less vital regions.

Parasympathetic Division

In contrast to the sympathetic division, which demonstrates general widespread effects, the parasympathetic division produces *specific* effects that are directed to one organ. Cranial nerves III, VII, and IX (see Figure 16-2) supply the parasympathetic control of structures in the head region. Part of the supply of III innervates the ciliary muscle that permits the pupil to contract and thus accommodate for near vision. Other parasympathetic fibers innervate the lacrimal glands, which cause tear secretion when stimulated, while other fibers innervate the salivary glands to secrete saliva in great amounts when they are activated.

The vagus nerves (X) are the source of parasympathetic innervation to the thoracic and abdominal viscera. Increased tone in fibers to the heart results in cardiac slowing; in fibers to the lungs, it causes the bronchioles to constrict and secrete when they are stimulated.

The vagus also has fibers innervating the gastrointestinal tract. Here, an increase in parasympathetic tone results in increased secretion and motility, contributing factors to peptic ulcer formation.

The sacral outflow of parasympathetic nerves innervates the urinary bladder and the genital organs. Increased tone in the fibers to the bladder causes contraction of the bladder musculature and relaxation of the internal sphincter muscles. In the genitals, activation of the parasympathetic nerves results in erectile tissue becoming engorged with blood.

It is possible to demonstrate that, in many instances, the two divisions of the autonomic nervous system have antagonistic effects. Important exceptions include the effects on the peripheral blood vessels. The sympathetic nerves produce a large increase in peripheral resistance, which causes an increase in arterial blood pressure. Parasympathetic stimulation has little effect in reducing the peripheral resistance since its distribution is restricted to only small, discrete areas of the body. Thus, this vasodilation effect does not contribute significantly to reducing the total peripheral resistance. In order to reduce arterial blood pressure, it is much more effective to diminish the sympathetic tone rather than attempt to increase that of the parasympathetic.

The two systems differ in the character of their activity. The parasympathetic division exhibits specific effects, whereas the sympathetic division exhibits coordinated but widespread effects.

The activities of the sympathetic and parasympathetic divisions of the autonomic nervous system may be reviewed by examining Exhibit 16-2.

CONTROL OF THE AUTONOMIC NERVOUS SYSTEM BY THE CENTRAL NERVOUS SYSTEM

The regulation of the autonomic reflexes, as with the motor reflexes, is divided among the different levels of the central nervous system. Regulation of the caliber of blood vessels, initiation of sweating, control of the

secretory activity and motility of the gastrointestinal tract, and adjustments in respiratory and cardiac activity are all reflex mechanisms. Sensory input arises from the viscera and certain somatic areas and is carried into the central nervous system where synapses are located. After many synapses, the motor output emerges from the same segment of the spinal cord to innervate the effector organ. Within the cord, neurons from various levels synapse with the association neurons to modify the final output. Suprasegmental controls also operate to change the peripheral reflexes. For example, autonomic centers in the cerebral cortex are connected to autonomic centers in the thalamus. These, in turn, are connected to the hypothalamus. In this hierarchy of control, the thalamus functions by sorting out incoming impulses before they reach the cerebral cortex. The cerebral cortex then turns over the control and integration of visceral activities to the hypothalamus. It is at the level of the hypothalamus that the major control and integration of the autonomic nervous system is exerted.

The hypothalamus is connected to both the sympathetic and parasympathetic divisions of the autonomic nervous system. The posterior and lateral portions of the hypothalamus appear to control the sympathetic division. When these areas are stimulated, there is an increase in visceral activities such as an increase in heart rate, a rise in blood pressure due to a vasoconstriction of blood vessels, an increase in the rate and depth of respiration, dilation of the pupils, and inhibition of the digestive tract. On the other hand, the anterior and medial portions of the hypothalamus seem to control the parasympathetic division. Stimulation of these areas results in a decrease in heart rate, a lowering of blood pressure due to a vasodilation of blood vessels, constriction of the pupils, and increased activity of the digestive tract.

The control of the autonomic nervous system by the cerebral cortex occurs primarily during emotional stress. During extreme anxiety, the cerebral cortex influences the hypothalamic centers which, in turn, causes the heart rate to increase and blood pressure to rise. Students often experience this reaction just before taking an examination. Another example of cortical control of the autonomic nervous system via the hypothalamus is a person's fainting upon viewing a tragic or unpleasant event. The visual cortex relays input to the hypothalamus, which causes vasodilation of blood vessels and slowing of the heart. The consequent sudden drop in blood pressure causes fainting to occur.

It may seem contradictory to refer to the autonomic nervous system as *autonomic* when it is, in fact, subject to control by the central nervous system. Unfortunately, the naming of the system preceded our current knowledge of it, and attempts to establish a more appropriate name have not been successful. Results from biofeedback and meditation experiments provide additional support for the idea that the autonomic nervous system can be influenced by other parts of the nervous system.

Biofeedback

In its simplest terms, **biofeedback** is a process in which the subject receives signals, or feedback, regarding the level of operation of certain body functions such as heart rate, blood pressure, and muscle tension. Special monitoring devices display this information to the subject who then attempts *consciously* to alter the level of operation of the visceral function.

In recent years, some simpler monitoring devices have been used in experimental situations. One of the more simple monitoring devices operates as follows. The device is connected to the subject so that the heart rate is displayed either by a red light to indicate a normal rate, or a green light to indicate a faster rate. When the green light is on, the subject is made aware of the too-rapid heart rate and so attempts to relax by thinking of something pleasant. If the subject is successful, the light changes to red, indicating the slower heart rate. In this technique, subjects are thus taught to control their heart rhythm.

Alleviation of migraine headache pain has also been reported by investigators. Using monitoring devices that indicate skin temperatures, sufferers have been taught to raise their skin temperatures by increasing blood flow to their extremities. The basis of treatment is the finding that it is the *distension* of blood vessels in the head that causes the excruciating pain. The patient consciously tries to dilate the blood vessels in the skin in the hope that this will cause the blood flowing into the head vessels to be shunted to the skin. The successful redistribution of blood is indicated by the elevation of the patient's skin temperature. The procedure used by workers at the Menninger Foundation has shown considerable success. Patients are taught how to monitor the temperature of the skin of their right index finger. In addition, they are provided with suggested phrases they can repeat to help them relax.

I feel quite quiet. . . . I am beginning to feel quite relaxed. . . . My feet feel heavy and relaxed. . . . My ankles, my knees, and my hips, feel heavy, relaxed, and comfortable. . . . My solar plexus, and the whole central portion of my body, feel relaxed and quiet. . . . My hands, my arms, and my shoulders, feel heavy, relaxed, and comfortable. . . . My neck, my jaws, and my forehead feel relaxed. . . . They

feel comfortable and smooth. . . . My whole body feels quiet, heavy, comfortable and relaxed.

Then another set of phrases helps them concentrate on increasing the blood flow to the hands.

I am quite relaxed. . . . My arms and hands are heavy and warm. . . . I feel quite quiet. . . . My whole body is relaxed and my hands are warm, relaxed and warm. . . . My hands are warm. . . . Warmth is flowing into my hands, they are warm. . . . warm.

The subjects are instructed to practice raising their skin temperature at home for 5 to 15 minutes a day. When the skin temperature rises, the monitor produces a high-pitched sound. In this way, the subjects can practice controlling blood flow into their hands, and on alternate days they attempt the exercise without the monitor. In time, the monitor becomes unnecessary.

In still other experiments, it has been shown that biofeedback can be of value during childbirth. Apprehensive and worried mothers can be taught to relax by biofeedback techniques. The electrical conductivity of the skin and the degree of muscle tension of the patient can be easily monitored by simple connection to the fingers and arms. Both parameters increase with nervousness and make labor more difficult. In these experiments, an increase in the level of muscle tension caused a high-pitched, sirenlike sound to become louder. An increase in skin conductivity caused a louder popping, crackling static noise. With the aid of these biofeedback mechanisms, women were able to focus on relaxing exercises that ultimately resulted in a shorter, easier birth process that required less medication.

The ultimate direction and extent to which biofeedback will be utilized by medicine is difficult to predict. It has shown us that the autonomic nervous system is not so autonomic and that dedicated *conscious* efforts can alter certain physiological processes. Its ultimate usefulness may lie in the treatment of a great number of pyschosomatic disorders and in helping us to understand better the mind-body interactions.

Meditation

Yoga, which literally means *union*, is defined by some practitioners as a higher consciousness that is achieved when the body is completely rested and relaxed and the mind is fully awake, but also relaxed. One technique that is widely employed for reaching this level of higher consciousness is called *transcendental meditation*. In this procedure, the person sits upright in a comfortable position with eyes closed, and concentration is focused on a suitable sound or thought.

Physiological measurements taken during the time a person is meditating have shown that some dramatic changes can occur. Metabolic rates can drop drastically as evidenced by a drop in oxygen consumption and carbon dioxide production. Lung ventilation and heart rate both decrease. Brain wave patterns show an intensification of the alpha waves, and there is an increase in the electrical resistance of the skin. All these responses are indicative of a relaxed state of mind. Recall that the alpha waves are characteristic patterns obtained from normal individuals when they are awake and in a resting state, but disappear during sleep.

The physiological changes that occur when a person is practicing transcendental meditation have been referred to as an *integrated response*. Essentially, the response represents a *hypo*metabolic state due to a great reduction in the tone of the sympathetic outflow of the autonomic nervous system. This is exactly the opposite response that occurs in the "fight or flight" response described earlier.

The integrated response further supports the concept that the autonomic nervous system can be controlled by other parts of the nervous system.

The social impact of some discoveries will require the general public to be much more enlightened about research trends and the implications of the discoveries as they unfold. Will we allow mind control to be used on a large scale? As we learn more about brain mechanisms, will we develop super computers? For what purpose? Will we be able to develop ways to control our own internal environment by thought processes? Will we be able to heighten pleasurable sensations without drugs or electrode implants? In the same regard, will we be able to remove or lessen painful experiences? The answers to these and many other questions will evolve from discussions that cross many scientific, artistic, philosophical, legal, and political disciplines. Each citizen of the world will be affected by these answers.

CHAPTER SUMMARY OUTLINE

SOMATIC EFFERENT AND AUTONOMIC NERVOUS SYSTEMS

1. The somatic efferent nervous system produces conscious movement in skeletal muscles. The autonomic nervous system (visceral efferent nervous system) regulates visceral activities.
2. The autonomic nervous system automatically regulates the activities of smooth muscle, cardiac muscle, and glands.
3. It usually operates without conscious control.
4. It is regulated by centers in the brain, in particular by the cerebral cortex, the hypothalamus, and the medulla oblongata.

STRUCTURE

1. The autonomic nervous system consists of visceral efferent neurons organized into nerves, ganglia, and plexuses.
2. It is entirely motor. All autonomic axons are efferent fibers.
3. Efferent neurons are preganglionic (with myelinated axons) and postganglionic (with unmyelinated axons).
4. The autonomic system consists of two principal divisions: the sympathetic and the parasympathetic (also called the thoracolumbar and craniosacral divisions).
5. Autonomic ganglia are classified as sympathetic trunk ganglia (on sides of spinal column), prevertebral ganglia (anterior to spinal column), and terminal ganglia (near or inside visceral effectors).
6. Sympathetic responses are widespread and, in general, concerned with energy expenditure. Parasympathetic responses are restricted and are typically concerned with energy restoration and conservation.

PHYSIOLOGY

Chemical Transmitters

1. In the autonomic nervous system, impulses are relayed across synapses by means of transmitter substances.
2. The two transmitter substances important in the operation of the autonomic reflexes are acetylcholine and norepinephrine.
3. The term cholinergic is used to describe fibers that use acetylcholine as the transmitter substance, and adrenergic is used for fibers that utilize epinephrine as the transmitter substance.
4. Cholinergic fibers include all preganglionic fibers, the postganglionic fibers of all parasympathetic neurons, and the sympathetic postganglionic fibers that innervate the sweat glands.
5. Acetylcholine is synthesized and stored in an inactive form in the terminals of the cholinergic fibers.
6. Norepinephrine is released from most of the postganglionic sympathetic fibers except the postganglionic sympathetic fibers that innervate the sweat glands, and the postganglionic fibers that innervate the uterus and pilomotor muscles.
7. Norepinephrine and epinephrine are secreted by the adrenal medulla and belong to a class of compounds called catecholamines.
8. The contradictory effects of catecholamines are due to the presence of two different types of receptor systems on the surface of cell membranes.

Activities of the Autonomic Systems

1. The autonomic nervous system exerts prime control in maintaining the constancy of the internal environment of the body.
2. The composition, temperature, volume, and distribution of the body fluid compartments is regulated by the action of autonomic nerve fibers acting upon circulatory, respiratory, excretory, endocrine, and exocrine organs.
3. Most of the organs controlled by the autonomic nervous system receive tonic innervation from both divisions, producing a fine degree of control at all times.
4. The sympathetic division is primarily concerned with the total well-being of the individual, while the parasympathetic division operates in specific effects that are directed to one organ.
5. In many instances, the two divisions of the autonomic nervous system have antagonistic effects. One important exception is the effect on the peripheral blood vessels.

CONTROL OF THE AUTONOMIC NERVOUS SYSTEM BY THE CENTRAL NERVOUS SYSTEM

1. The hypothalamus controls and integrates the autonomic nervous system.

2. The hypothalamus is connected to both the sympathetic and the parasympathetic divisions.
3. The control of the autonomic nervous system by the cerebral cortex occurs primarily during emotional stress.
4. Biofeedback is a process in which the subject receives signals, or feedback, regarding the level of operation of certain body functions such as heart rate, blood pressure, and muscle tension.
5. Yoga is a higher consciousness achieved through a fully rested and relaxed body and a fully awake and relaxed mind.
6. Transcendental meditation produces the following physiological responses: decrease in oxygen consumption and carbon dioxide elimination, reduction in metabolic rate, decrease in heart rate, increase in the intensity of alpha brain waves, sharp decrease in the amount of lactic acid in the blood, and increase in the skin's electrical resistance.

REVIEW QUESTIONS

1. What are the principal components of the autonomic nervous system? What is its general function? Why is it called involuntary?
2. What is the principal anatomical difference between the voluntary nervous system and the autonomic nervous system?
3. Relate the role of visceral efferent fibers and visceral effectors to the autonomic nervous system.
4. Distinguish the following with respect to locations and function: preganglionic neurons and postganglionic neurons.
5. What is an autonomic ganglion? Describe the location and function of the three types of autonomic ganglia. Define white and gray rami communicantes.
6. On what basis are the sympathetic and parasympathetic divisions of the autonomic nervous system differentiated anatomically and functionally?
7. Discuss the distinction between cholinergic and adrenergic fibers of the autonomic nervous system.
8. Give examples of the antagonistic effects of the sympathetic and parasympathetic divisions of the autonomic nervous system.
9. Summarize the principal functional differences between the voluntary nervous system and the autonomic nervous system.
10. What are the functions of transmitter substances of the autonomic nervous system? Name two important chemical transmitters.
11. Where are acetylcholine and norepinephrine synthesized?
12. Explain the contradictory effects of catecholamines.
13. What is the main function of the autonomic nervous system?
14. Explain the statement, "Organs controlled by the autonomic nervous system exhibit a fine degree of control at all times."
15. Describe how the hypothalamus controls and integrates the autonomic nervous system.
16. Define biofeedback. Explain how it could be useful.
17. What is transcendental meditation? Can it be useful? Explain your answer with a specific example.

Chapter 17

ENDOCRINE PHYSIOLOGY

STUDENT OBJECTIVES

After reading this chapter, you should be able to

- Compare the nervous and endocrine systems as functional communicating and coordinating systems
- Define an endocrine gland and differentiate between the three different classes of hormones
- Identify the relationship between tropic hormones and target organs
- Define the anatomical and physiological relationship between the pituitary gland and the hypothalamus
- Explain the control of glandular secretion by hormone levels
- Differentiate the four distinct mechanisms by which hormones change the metabolic patterns of cells
- List the hormones of the adenohypophysis, their target organs, and their functions
- Define hypophysectomy and list its effects on the bodies of young animals
- Describe the many physiological effects of the human growth hormone on the body, as well as its control
- Define the source of hormones stored by the neurohypophysis, their target organs, and their functions
- Discuss how thyroxine is synthesized and stored by thyroid follicles and transported
- Identify the physiological effects and regulation of secretion of thyroxine and thyrocalcitonin
- Describe the physiological effects of the parathyroid hormone
- Distinguish the effects of adrenal cortical mineralocorticoids, glucocorticoids, and gonadocorticoids on physiological activities
- Identify the function of the adrenal medullary secretions as supplements of sympathetic responses
- Compare the roles of glucagon and insulin in the control of blood sugar level
- Identify the physiological effects of the hormones secreted by the pineal gland
- Describe the importance of the thymus gland
- Define the general adaptation stress syndrome
- Identify the body reactions during the alarm, resistance, and exhaustion stages of stress
- Explain the disorders of the human growth hormone, comparing hyposecretion with hypersecretion
- Explain the disorders of the neurohypophysis
- Define the various disorders of the thyroid gland
- Identify the principal effects of abnormal secretion of the parathyroid hormone on calcium metabolism
- Explain the disorders of the adrenal cortex
- Explain the disorders of the adrenal medulla
- Describe the disorders of the pancreas

The need for communication and coordination in order to achieve homeostasis within the human body is apparent. Each organ has its requirements not only for glucose and oxygen, but also for those substances peculiar to its function. Therefore, the human body with its complex organization of parts must monitor closely the needs and operation of each part. Each organ must cooperate with other organs if the total organism is to survive. The kidneys must decrease the excretion of water when the other parts of the body begin to require water. The heart must beat faster and more vigorously when the skeletal muscles are performing vigorous exercise. The secretion of digestive juices must be turned on when food is present in the stomach and intestine. These, and many other related processes, are successful because the human body possesses efficient communication systems. Two systems function primarily as communicating and coordinating systems—the nervous system, described in preceding chapters, and the endocrine system, which is the topic of this chapter.

The two communicating systems also interact with each other. The two systems have, at times, been likened to two somewhat more familiar communication systems—the telephone and postal systems. The telephone system is analogous to the nervous system, which responds with great rapidity and pinpoint accuracy and connects points separated by considerable territory. The postal system corresponds to the endocrine system, which is considerably slower, has a somewhat bulkier effect on its targets, but does cover the same territory.

Interaction between the nervous and endocrine system in animals occurs in the hypothalamus, one of the major controlling centers of the autonomic nervous system. The hypothalamus regulates a number of important functions including appetite, temperature, water balance, sleep, sexual activity, and blood vessel diameters. Many of these controls are mediated in great part by the endocrine system, and the hypothalamus is the link between the two communicating networks.

ENDOCRINE GLANDS

Definition

Distributed within the body are several glandular structures that operate as a team and function to regulate the rates of growth and development, the activity of certain tissues, and the rates of the metabolic processes within the body. The glandular structures are termed **endocrine glands** (*endo* = within; *krin* = to separate) because they are ductless glands and their products are delivered into the blood or lymph. The secretory products of the endocrine glands are called **hormones** (*hormon* = to set in motion), and they have the special property of exerting physiological control over other cells of the body. The cells, tissues, or other organs influenced by the hormone are referred to as the *target* cells, tissues, or organs.

Chemical Structure

Over 35 hormones have been identified in the human body, and other substances are always being studied

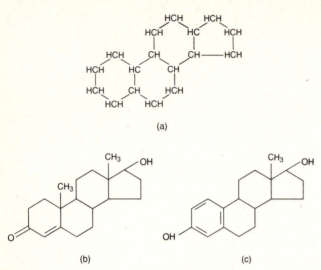

(a)

(b)   (c)

Figure 17-1 Steroid hormones. (a) Steroid ring structure. (b) Male hormone, testosterone. (c) Female hormone, estradiol.

to determine whether they also qualify for classification as hormones. The chemical structure of many hormones is known, and, of these, many have been synthesized in laboratories. Hormones are molecules that fall into three classes of compounds—steroids, proteins, and catecholamines.

The ring structure depicted in Figure 17-1a is the basic structure common to the *steroid hormones*. Figures 17-1b and c show the structure of estradiol, a female sex hormone, and testosterone, a male sex hormone, both of which are steroid hormones. Though their structural differences are slight, their physiological effects are very different—one is feminizing and the other masculinizing. One common characteristic shared by steroid hormones is that they are secreted by glands whose embryonic origin is related to the same mesodermal tissues. Included in the group of glands that secrete steroid hormones are the ovaries, testes, and adrenal cortex.

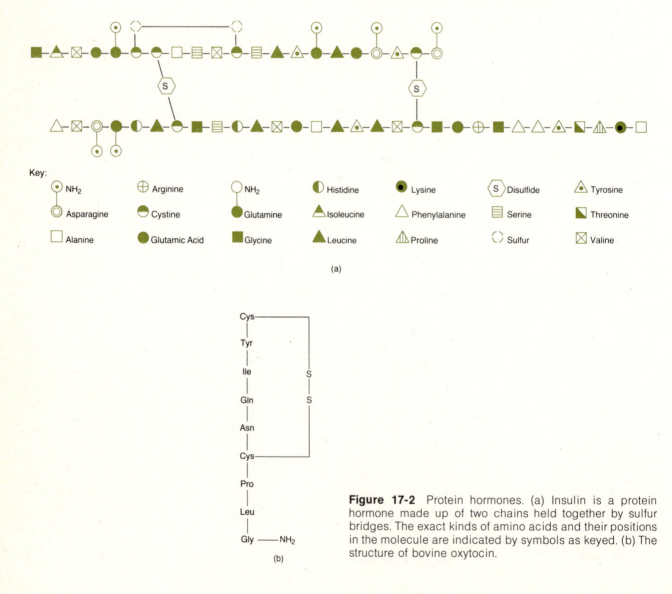

Key:

⊙ NH$_2$	⊕ Arginine	○ NH$_2$	◐ Histidine	● Lysine	⬡ Disulfide	△ Tyrosine
◎ Asparagine	◓ Cystine	● Glutamine	△ Isoleucine	△ Phenylalanine	▤ Serine	◣ Threonine
☐ Alanine	● Glutamic Acid	■ Glycine	▲ Leucine	△ Proline	◯ Sulfur	⊠ Valine

(a)

Cys ——
Tyr
Ile   S
Gln   S
Asn
Cys ——
Pro
Leu
Gly —— NH$_2$

(b)

Figure 17-2 Protein hormones. (a) Insulin is a protein hormone made up of two chains held together by sulfur bridges. The exact kinds of amino acids and their positions in the molecule are indicated by symbols as keyed. (b) The structure of bovine oxytocin.

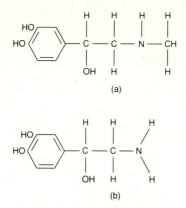

Figure 17-3 Catecholamines. (a) Epinephrine. (b) Norepinephrine.

Protein hormones can be large molecules of considerable complexity as illustrated by the insulin molecule, or they can be simple polypeptides such as oxytocin (Figure 17-2). Protein hormones are secreted by endocrine glands that develop embryonically from the endoderm of the digestive tract. The only exception to this endodermal origin is the anterior portion of the pituitary, which arises from an outpouching of ectodermal tissue that originates from the roof of the oral cavity. Other endocrine glands that secrete protein hormones include the thyroid, parathyroid, and pancreas.

Catecholamines are structurally the simplest of the hormone molecules—essentially modified amino acids. Epinephrine and norepinephrine (Figure 17-3) are examples of catecholamines. This type of hormone is secreted by endocrine glands whose cells originate from the same kind of ectoderm that gives rise to the nervous system. The adrenal medulla is an example.

ENDOCRINE REGULATION

In order to explain the way in which the endocrine system regulates certain functions, it is necessary to describe two related but separate processes. Hormones are the products of glandular tissues, and their synthesis and release are one important aspect of the control mechanism. The second process relates to the manner in which the hormones trigger the specific biochemical events that take place in the target tissues. We shall now examine these two important processes.

Control of Glandular Secretion by Tropic Hormones

Endocrine secretions are controlled either by hormonal feedback systems or by some nervous activity generally channeled through the hypothalamus. The anterior pituitary, which will be described in detail in a later section, functions as a *master gland*—it regulates the activity of several of the other endocrine glands. It does so by liberating so-called *tropic hormones* which are hormones that stimulate one specific endocrine gland to secrete its hormone. *Thyroid stimulating hormone (TSH), follicle stimulating hormone (FSH), luteinizing hormone (LH),* and *adrenorcorticotropic hormone (ACTH)* are examples of tropic hormones released by the anterior pituitary.

The target endocrine glands, in turn, respond to the specific tropic hormone by releasing their own hormone. For example, in response to the effect of thyroid stimulating hormone (TSH) or thyrotropin, the thyroid gland releases its hormone called *thyroxine.* Thyroxine then exerts its effect on all body tissues by increasing their metabolic rate. As the level of thyroxine rises in the body fluids, the hypothalamus detects the rising level. After assessing the body's need for thyroxine, the hypothalamus is able to reduce the output of the thyroid gland by its control over the anterior pituitary. The hypothalamus regulates the activity of the anterior pituitary by means of chemical messengers called *hypothalamic regulating factors (hormones).* The hypothalamic regulating factors may function as *releasing factors,* which stimulate secretion, or *inhibiting factors,* which inhibit secretion. The regulating factors are synthesized within the nerve cells of the hypothalamus and are selectively discharged into the blood vessels that connect the hypothalamus directly to the anterior pituitary. The cells of the anterior pituitary can thus be selectively stimulated or inhibited. In the example being described, the rise in thyroxine level causes a decrease in the liberation of *thyrotropin releasing factor* (TRF) from the hypothalamus into the blood vessels to the anterior pituitary. This, in turn, removes the stimulation from cells of the anterior pituitary that secrete the thyroid stimulating hormone (thyrotropin). The generalized feedback loop or circuit is illustrated in Figure 17-4.

In a normal person, when metabolic functions are balanced, there is homeostasis and the level of thyroxine in body fluids is kept fairly constant. It is maintained at a fixed level by the constant output of TSH from the anterior pituitary and thyrotropin releasing factor (TRF) from the hypothalamic cells. The key element in the feedback control system is the action of thyroxine on the hypothalamus. If there is an abrupt uptake of thyroxine by the body cells that lowers the circulating level of the hormone in the body fluids, the hypothalamus increases its output of TRF. This triggers a greater output of thyroid stimulating hormone from the anterior pituitary, and the thyroid

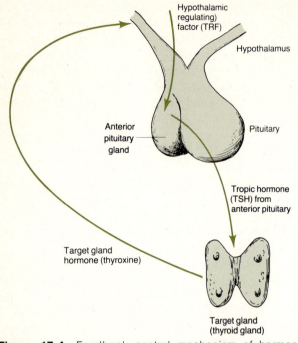

Figure 17-4 Feedback control mechanism of hormone secretion. The control of target hormone production depends on stimulation by an anterior pituitary tropic hormone, which is in turn controlled by hypothalamic regulating factors. The hypothalamus "senses" the quantity of target hormone in the blood and produces its regulating factor in relation to normal target hormone levels. Too much target hormone turns off the hypothalamus and too little turns it on more.

gland, in turn, releases more thyroxine. Thus, the circulating level of thyroxine is restored to the required level. The hypothalamus functions to set the appropriate level of circulating hormone. Under changing circumstances, the body's need will vary and the set level for the hormone can be raised or lowered.

Control of Glandular Secretion by Hormone Levels

All endocrine glands, however, are not regulated by tropic hormones from the pituitary. Some glands are controlled by the level of the substance they seek to regulate. For example, the pancreatic hormones, *insulin* and *glucagon*, are released by the fluctuations in blood glucose level. Following a meal, the blood glucose level begins to rise. The beta cells of the pancreas detect the rising level and respond by releasing insulin. The circulating insulin promotes the uptake of glucose from the blood stream by the cells of the body. On the other hand, a drop in blood glucose levels not only shuts off the release of insulin but also causes the alpha cells of the pancreas to release glucagon. This hormone functions to elevate the

blood glucose to a normal value. Thus the endocrine portion of the pancreas detects the level of the metabolite it regulates (glucose) and releases the appropriate hormone in order to maintain the blood glucose level within the prescribed limits that it determines.

The blood calcium ion level is controlled by a similar system. The parathyroid glands regulate the circulating level of Ca^{2+} by two hormones, *parathormone* and *calcitonin*. When the blood Ca^{2+} level drops, the parathyroid glands release parathormone, which acts to elevate the blood Ca^{2+} level. Conversely, if the blood Ca^{2+} level is too high, calcitonin is released by the parathyroid glands and it acts to lower the circulating level of Ca^{2+}.

Hormone levels can also be under the direct control of the autonomic nervous system. The cells of the adrenal medulla function as postganglionic nerve cells. For example, when a generalized autonomic discharge results because of a crisis situation (recall the "fight or flight" response), the adrenal medulla responds to the nervous stimulation by releasing epinephrine and norepinephrine. The origin of the generalized nervous outburst is the hypothalamus, a response further emphasizing the nervous control of hormone release.

Nerve impulses arising from the hypothalamus also control the release of hormones stored in the cells of the posterior pituitary. For example, dehydration is detected by the hypothalamus, which then stimulates the release of *antidiuretic hormone (ADH)* from the posterior pituitary. This hormone acts on the tubules of the kidney to conserve water and discharge a more concentrated urine.

The hypothalamus is intimately involved in hormonal regulation. Since it receives information from all parts of the nervous system, it responds to a variety of stress situations. Emotional stress, which is a homeostatic imbalance, can often cause the hypothalamus to react in a way that severely disturbs the balance of hormonal interplay.

MECHANISMS OF HORMONAL ACTION

Hormones bring about changes in the metabolic patterns of cells in a variety of ways. They can activate certain suppressed genes, increase the rate of translation by ribosomes, enhance enzyme activity, and promote active transport across the cell membrane.

Gene Activation by Hormones

Steroid hormones secreted by the adrenal cortex, the ovaries, and the testes mediate their effect by activating genes that were previously silent. To illustrate the operation of a steroid hormone and how it "turns on" particular genes, let us examine one group of female hormones produced by the ovary called estrogens. These hormones have the primary function of stimulating development of the female reproductive system and the secondary sex characteristics. The most potent estrogen is called *estradiol.*

At the onset of puberty, the ovary begins to produce increasing amounts of estrogens, such as estradiol. All the cells that are going to respond to the effect of the estradiol possess special chemicals called *receptor molecules.* For example, the cells of the female breast, the hip and thigh muscles, and blood vessels in the uterus possess such receptor molecules. However, it is believed that the receptor molecules in these different tissues differ somewhat in their appearance. That is, they only share the characteristic of being able to bind to the hormone by some "lock and key" arrangement (Figure 17-5a). Therefore it is thought that, though the molecules are unlike in their overall three-dimensional appearance, they all possess in some region of their anatomy the identical combining site for the estradiol. When the hormone combines with the receptor molecule of the responding cells, the total hormone-receptor complex moves into the nucleus of the cell. There, with a new shape, the complex "fits" into a complementary location on the DNA molecule to unlock a specific gene. Thus, even though every cell in the body contains the same genetic information, estradiol will unlock different genes in different tissues. In the cells of the breast it can cause development of glandular and duct tissue, whereas in the uterus, it promotes the proliferation and growth of blood vessels.

Figure 17-5b shows how the same hormone can unlock two different genes by combining with two different receptor molecules. It is even possible that one cell could have more than one type of receptor molecule and thus be able to respond in more than one fashion.

Stimulation of Protein Synthesis

At present, there is evidence supporting the idea that the primary focus of action of *human growth hormone (HGH)* is the ribosome. In the presence of HGH, ribosomes speed up their rate of translation of mRNA and thus turn out protein molecules in rapid succession. The burst of additional enzymes (which are protein molecules) enables the cell to function faster. In general, most of the body cells respond to the action of human growth hormone, but the cells do not respond to produce *all* proteins. The cells speed up the

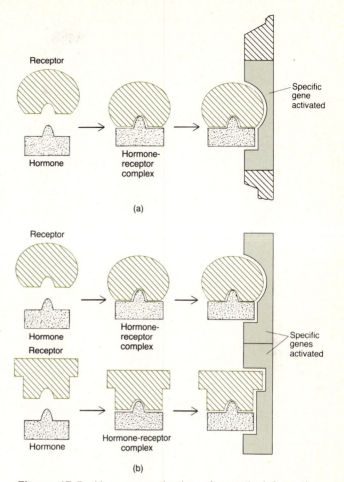

Figure 17-5 Hormone activation of genetic information. (a) The receptor "recognizes" the specific shape of the hormone, and the result is the formation of a hormone-receptor complex. The complex then interacts specifically with a gene activator. (b) One hormone activates two different genes by interacting with two different receptors.

rate of production of only certain proteins, and different cells produce different sets of proteins in response to the human growth hormone. Investigators think that receptor molecules may also operate in this process by somehow directing the hormone to speed up translation of only certain protein molecules.

Activation of the Cyclic AMP System

This mode of hormonal operation was described by Earl W. Sutherland, and for his outstanding accomplishment he was awarded the Nobel prize for medicine in 1971. Though in his research he sought to elaborate the mechanism of action of the hormone epinephrine on liver cells, the system he described seems to have widespread application in biological systems.

Epinephrine acts to produce changes in many locations, but its effect on blood glucose level serves to illustrate its molecular mode of action. For years, epinephrine has been known to cause liver cells to break down stored glycogen to glucose, which is released into the blood to raise the circulating glucose level. In skeletal muscles, the stored glycogen is broken down to glucose, which remains in the cell for the production of energy.

Glycogen is a large molecule that consists of many glucose molecules linked together. Glucose is obtained from glycogen by the action of the enzyme *phosphorylase a*. This enzyme removes the glucose

units one by one from the ends of the highly branched glycogen molecule. The operation of this cyclic AMP system is extremely rapid because of its so-called *amplification cascade action.*

Figure 17-6a illustrates the cascading sequence of events in a liver cell following the arrival of epinephrine. When the concentration of epinephrine reaches a value of about $10^{-9}M$, it begins to bind to specific epinephrine receptor sites on the surface of the liver cell membrane. The binding process most likely causes a conformational change in the membrane structure that results in the activation of the

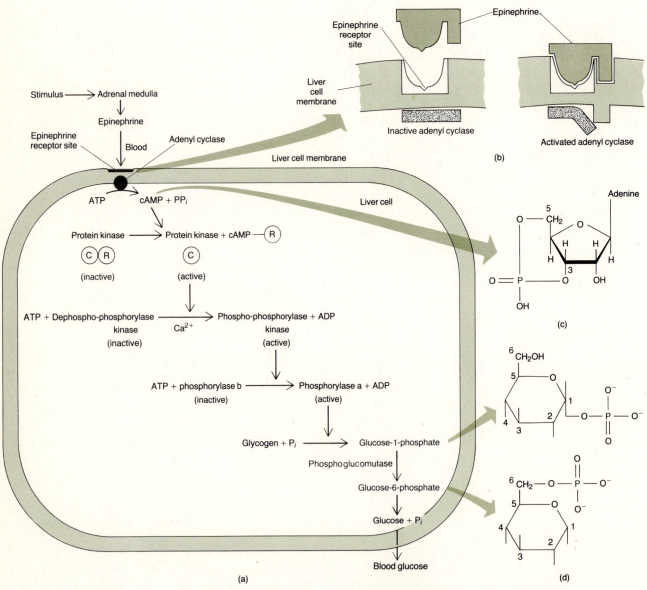

Figure 17-6 Amplification cascade action of the cyclic AMP system. (a) Cascading sequence in a liver cell following the arrival of epinephrine. (b) Activation of adenyl cyclase. (c) Structure of cyclic AMP. (d) Conversion of glucose-1-phosphate to glucose-6-phosphate by phosphoglucomutase.

enzyme called *adenyl cyclase* (Figure 17-6b). This enzyme is located on the inner surface of the cell membrane and, when activated, converts ATP into what is known as *cyclic adenosine-3', 5'-monophosphate (cyclic AMP or cAMP)* (Figure 17-6c). Located in the cytoplasm of the cell is another key protein molecule—an enzyme called *protein kinase*. This enzyme is *allosteric*, which means it has two different combining sites. When this enzyme is not functioning, it exists as a combination of two subunits designated C and R. The C stands for the catalytic subunit and R stands for the regulatory subunit. When they are together, the R subunit prevents any action of the catalytic subunit.

When the cyclic AMP is formed in large amounts (it can reach $10^{-6}M$ concentrations), it splits the protein kinase into the two subunits, thus releasing the active C subunit. The catalytic subunit now acts upon another enzyme called *phosphorylase kinase*. This enzyme is activated by the addition of a phosphate group donated by an ATP molecule. This activated (now phospho-) phosphorylase kinase acts upon yet another inactive enzyme known as *phosphorylase b*. By the addition of four phosphate groups donated by four ATP molecules, phosphorylase b is converted to an active phosphorylase a enzyme. This is the active enzyme that attacks glycogen and produces, with the aid of inorganic phosphate (P_i), the glucose-1-phosphate units. The conversion of glucose-1-phosphate to glucose-6-phosphate by the enzyme phosphoglucomutase transfers the phosphate from the carbon number one position to the carbon number six position (Figure 17-6d). This step readies the molecule for *glucose-6-phosphatase*, an enzyme present only in liver cells, which removes the phosphate group and delivers the freed glucose into the blood.

The sequence of steps cascade one after another, each step amplifying the result of the preceding step. A low concentration of epinephrine molecules acting on the surface of the cell is rapidly amplified to produce a large release of free glucose in the blood. The system continues to operate as long as epinephrine is secreted by the adrenal medulla. Once it stops, the cascading events within the cell also cease and the active molecules revert to inactive form.

It seems, at this time, that many hormones exert their effect by utilizing a similar cyclic AMP system. Adrenocorticotropic hormone (ACTH), luteinizing hormone (LH), thyroid stimulating hormone (TSH), melanocyte stimulating hormone (MSH), antidiuretic hormone (ADH), parathormone, insulin, glucagon, epinephrine, norepinephrine, and in some instances, cortisol, are thought to utilize cyclic AMP, at least in part, in producing their effects.

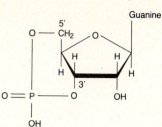

Figure 17-7 Structure of cyclic guanosine-3', 5'-monophosphate (cGMP).

Activation of Membrane Transport Systems

Insulin is a hormone that is released from the beta cells of the pancreas into the blood. Its most observable effect is an immediate drop in blood glucose level. It is believed that insulin acts to speed up the active transport of glucose across the cell membranes of muscle and fat cells for storage as glycogen in muscle cells and for storage as fat in fat cells. It is known that insulin binds to receptor molecules in the membrane of muscle and fat cells. Interestingly, insulin causes a decrease in the intracellular level of cyclic AMP in these cells, and another cyclic nucleotide called *cyclic guanosine-3', 5'-monophosphate (cyclic GMP or cGMP)* is found to increase (Figure 17-7). How the speedup of glucose transport occurs is not known. Indeed, the molecular events of any transport process remain only conjecture and require much more research.

PITUITARY (HYPOPHYSIS)

The **pituitary gland,** or **hypophysis,** as it is also called, regulates so many body activities that it has been referred to as the "master gland." It is surprising therefore, that despite its great regulating power, it is a very small structure. Measuring about 1.3 cm in diameter and weighing about 0.5 g, it lies in a protective bony cavity of the sphenoid bone called the sella turcica. It is attached to the hypothalamus of the brain via a stalklike structure called the infundibulum (Figure 17-8).

The pituitary is divided structurally and functionally into an anterior lobe and a posterior lobe, both of which are physically connected to the hypothalamus. The *anterior lobe* contains many glandular epithelial cells that synthesize and release seven different hormones. A system of blood vessels, called the *hypothalamic-hypophyseal portal system,* connects the anterior lobe with the hypothalamus. The *posterior lobe* contains axons of neurons that form the neural part of

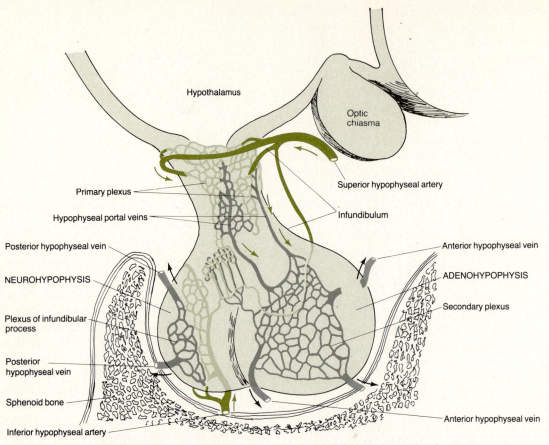

Figure 17-8 Divisions and blood supply of the pituitary gland.

the pituitary. The bodies of the neurons are located in two nuclei in the hypothalamus. The axons that leave the hypothalamus to terminate in the posterior lobe form the *hypothalamic-hypophyseal tract* in the infundibulum (see Figure 17-12).

The Adenohypophysis

The anterior lobe of the pituitary is also called the *adenohypophysis*. It releases its hormones into the circulation upon receipt of signals from the hypothalamus. These signals arrive via the hypothalamic-hypophyseal portal system and are special excitatory or inhibitory molecules called regulating factors. The design of the communication system between the hypothalamus and anterior pituitary makes use of two capillary beds arranged in series. Branches of the internal carotid and posterior communicating arteries give rise to several *superior hypophyseal arteries* (see Figure 17-8). These arteries branch abruptly to form a network or plexus of capillaries called the *primary plexus*. These capillaries not only maintain the nerve cells around the base of the hypothalamus but also serve to receive the regulating factors secreted by the

hypothalamic cells. The plexus drains into a series of veins known as the *hypophyseal portal veins* that course down the infundibulum. Upon reaching the adenohypophysis, the veins abruptly branch to form another capillary network, the *secondary plexus*, which not only maintains the cells of the anterior pituitary with nutrients, but also delivers the regulating factors from the hypothalamus to every cell. The secondary plexus drains into the *anterior hypophyseal veins* which carry, in addition to the cellular wastes, the hormones released by the cells of the anterior pituitary.

The anterior pituitary consists of three cell types, each determined by certain staining properties (Figure 17-9). These are the *acidophils* which stain with acid dyes, the *basophils* which stain with basic dyes, and the *chromophobes* which tend to resist staining. The acidophils synthesize and secrete two of the anterior pituitary hormones: *human growth hormone (HGH)*, or *somatotropin (STH)* as it is also called, which controls body growth; and *prolactin (PR)*, which initiates milk secretion in breast tissue. The basophils are known to secrete at least five other hormones. These are the *thyroid stimulating hormone*

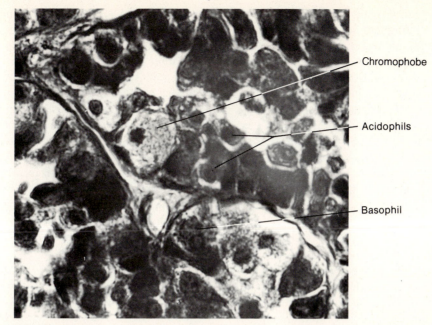

Chromophobe

Acidophils

Basophil

Figure 17-9 Histology of the adenohypophysis. (Courtesy of Victor B. Eichler, Wichita State University.)

(TSH), which controls the thyroid gland; *adrenocorticotropic hormone (ACTH)*, which regulates the activity of the adrenal cortex; *follicle stimulating hormone (FSH)*, which stimulates production of eggs by the ovaries and sperm by the testes; *luteinizing hormone (LH)* in the female or *interstitial cell stimulating hormone (ICSH)* in the male, which stimulate other sexual and reproductive activities; and finally the *melanocyte stimulating hormone (MSH)*, which regulates skin pigmentation. The chromophobes may have some slight ACTH producing activity but generally are thought to be precursors of the other two types.

Except for growth hormone and melanocyte stimulating hormone, all the anterior pituitary secretions are referred to as *tropic hormones* (*trophe* means to nourish), which indicates that the tropic hormones stimulate the growth and function of some other endocrine gland. Prolactin, follicle stimulating hormone, and luteinizing hormone are also classed as *gonadotropic hormones* because they regulate the functions of the gonads. The gonads (ovaries and testes) are the endocrine glands that produce sex hormones.

Hypophysectomy

Much of the basic information about the function of the anterior pituitary hormones was obtained in the early 1900s. Experimental studies involving the removal of the pituitary *(hypophysectomy)* from dogs demonstrated the far-reaching effects of the pituitary gland. It was shown that the hypophysis was not essential to life, for hypophysectomized animals, given no substitute hormonal treatment, were able to survive for months and years. The experimental animals required special treatment such as a constant warm temperature, simple paste diets that were easy to digest, and a regimen of antibiotics to prevent infections.

The most obvious effects of the removal of the pituitary from *young* animals was the decrease in the rate of growth. The exact reason for the slowing of growth is not known, but scientists think it involves a number of mechanisms. Force-feeding animals by stomach tubes enables them to grow and gain considerable weight. The removal of the pituitary abolishes the source of the tropic hormones, and therefore the animal suffers the secondary loss of function from the adrenal cortex, thyroid, and the gonads.

The loss of ACTH from the adenohypophysis causes the adrenal cortex to atrophy and greatly reduce the output of its own hormones. The lack of the glucose regulating hormones from the adrenal cortex generates the greatest potential problem for survival. (The effects of adrenal cortical insufficiency will be described later.)

The lack of thyroid stimulation reduces the output of thyroxine from the thyroid gland and leads to a condition of secondary hypothyroidism. The thyroid gland regresses and its synthetic activity operates at a greatly reduced rate.

The loss of the gonadotropins (FSH, LH, and prolactin) results in a complete loss of reproductive capability. The ovaries and testes do not develop, and as a consequence, the primary and secondary sex characteristics never appear.

The loss of melanocyte stimulating hormone (MSH) may result in cutaneous pigmentation deficiency.

It seems appropriate to describe the role and activity of the tropic hormones in the section that describes the function of their target glands. The following discussion of human growth hormone focuses on its metabolic interactions and control.

Human Growth Hormone (HGH) Action

To fully understand the relationship of human growth hormone to the growth of the body, it is necessary to focus our attention on the process of *cellular* growth and proliferation. It is at this level that we can describe the synthesis of additional new cellular components that result from the uptake of amino acids. A complete description of how human growth hormone functions must relate to the stimulus it imparts to the utilization of these molecules. Human growth hormone is the only known product of the anterior pituitary that operates without an intermediary to produce its widespread effect.

Human growth hormone from a number of mammalian species has been isolated, and its biochemical composition for each species has been compared (Exhibit 17-1). The average molecular weight of 22,000 indicates that it is a relatively small protein molecule.

The amino acid sequence for human growth hormone is also known and is shown in Figure 17-10. A comparison of the structure of human growth hormone with other mammalian species shows many identical amino acid sequences and structural similarities. By focusing on the similarities, scientists will soon be able to identify functional requirements of the growth hormone molecule and thus explain its molecular mode of action with the target cells.

Human growth hormone causes body growth by stimulating growth of those tissues that are able to grow. Its mode of operation is to: (1) stimulate protein synthesis in body cells, (2) inhibit the cellular utilization of glucose as an energy source, but to (3) stimulate the utilization of stored body fat as an energy source.

The stimulation of protein synthesis by human growth hormone occurs at three points along the pathway of *protein anabolism (ana* = up, again; *metabol* = to change). First, it has been shown by the use of radioactively tagged amino acids that their uptake into the cell is greatly enhanced by even minute amounts of human growth hormone. It has been suggested by some scientists that this uptake is the rate-limiting step in protein synthesis. Thus, by increasing the intracellular concentration of these protein building blocks, the cellular machinery can operate at a faster rate. Second, there is evidence that human growth hormone and certain other hormones can stimulate ribosomes to initiate the translation of mRNA molecules at a greater rate. The result is the formation

Exhibit 17-1   AMINO ACID COMPOSITION OF EIGHT MAMMALIAN GROWTH HORMONES

Amino Acid	Human	Bovine		Ovine	Equine		Porcine	Canine		Rabbit	Rat	
Lysine	9	11	12	13	8	11	11	11	12	10	10	10
Histidine	3	3	3	3	3	3	3	3	3	3	3	3
Arginine	11	13	13	13	11	13	12	13	12	12	11	8
Aspartic acid	20	16	16	16	14	14	15	17	18	14	13	16
Threonine	10	12	12	12	7	7	7	8	10	7	7	7
Serine	18	13	12	12	13	13	14	15	17	12	12	11
Glutamic acid	26	24	25	25	21	21	24	26	24	22	22	22
Proline	8	6	6	8	7	7	7	8	11	6	6	8
Glycine	8	10	10	10	9	9	8	8	10	9	7	10
Alanine	7	15	13	14	14	15	16	18	19	13	15	13
Half-cystine	4	4	4	4	4	2	4	4	4	3	3	2
Valine	7	6	6	7	7	7	8	8	7	6	4	8
Methionine	3	4	4	4	3	2	3	3	3	3	5	3
Isoleucine	8	7	7	7	6	5	6	6	6	4	6	7
Leucine	26	27	24	22	21	21	24	26	24	21	19	17
Tyrosine	8	6	6	6	6	5	7	7	7	6	6	5
Phenylalanine	13	13	13	13	9	9	12	13	12	10	10	8
Tryptophan	1	1	1	1		2	1			1	1	

Source: Modified from A. E. Wilhelmi. Chemistry of growth hormone. *Handb. Physiol. Endocrinol.* IV, Part 2:59, 1974.
Note: Two sets of values are found in the literature for bovine, equine, canine and rat growth hormone.

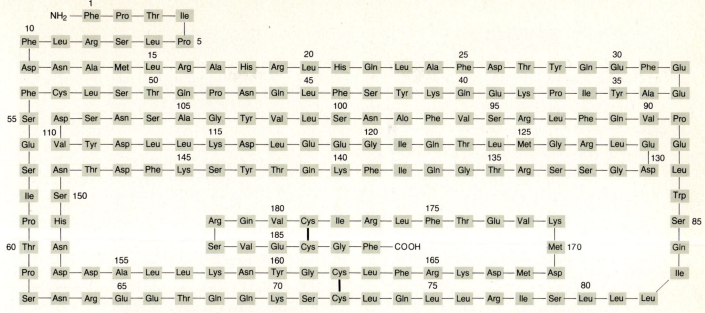

Figure 17-10 Amino acid sequence of human growth hormone, a single polypeptide chain with two disulfide bridges.

of greater numbers of polysomes, and thus multiple copies of proteins are produced. Third, a more long-term effect of additional human growth hormone is the enhanced rate of RNA transcription in the nucleus. This will produce more of the cytoplasmic machinery that is required for protein synthesis.

The inhibition of glucose utilization for cellular energy by human growth hormone occurs mainly because of its powerful action in stimulating fat catabolism (*katabol* = throw down) in adipose tissue. This causes production of large amounts of an intermediary metabolite called acetyl coenzyme A. With an excess of this compound, the utilization of glucose is reduced to zero. In fact, the glucose is stored in cells as glycogen. When the cellular stores are filled with glycogen, the addition of more glucose from the diet causes an increase in the circulating level of blood glucose. Levels of 200 mg percent can easily be obtained. This elevation of blood glucose is called the *diabetogenic effect* of growth hormone. With the elevation in blood glucose comes the increased demand for more insulin to restore the blood glucose level to a more normal value. Prolonged demand for large amounts of insulin can cause exhaustion of the beta cells of the islets of Langerhans in the pancreas that produce insulin. When this point is reached, the condition of *diabetes mellitus* results. Its symptoms will be described later.

Control of Human Growth Hormone

Though the physiological effects of human growth hormone are most profound during the early growth years, its release continues throughout adult life. Indeed, the levels detected in the blood of an adult, though normally lower than that of a child (3×10^{-9} g/ml versus 5×10^{-9} g/ml) can reach levels of 50×10^{-9} g/ml. The day-to-day, minute-to-minute levels of human growth hormone fluctuate according to the level of stress, exercise, and nutrition demonstrated by the person. Any condition that lowers blood glucose promotes human growth hormone release. Recently it has also been demonstrated that a good correlation exists between the release of human growth hormone and the level of cellular protein. During prolonged fasting or starvation, the depletion of cellular proteins coincides with the rising level of human growth hormone in the circulation. The control of human growth hormone, therefore, seems intimately linked with the nutritional status of the individual. What it is or how the feedback signal operates is still unknown. Some speculation centers on the hypothalamus, which can release either the *human growth hormone releasing factor (HGHRF)* or the *human growth hormone inhibiting factor (HGHIF)*, also known as *somatostatin* (Figure 17-11). The hypothalamus may have access to information regarding the total nutritional state of the body and regulate the level of human growth hormone by factors it releases.

The Neurohypophysis

In a strict sense, the posterior lobe, or *neurohypophysis*, is not an endocrine gland since it does not synthesize hormones. The posterior lobe consists of supporting cells called *pituicytes*, which are similar in appearance to the neuroglia of the nervous system. It

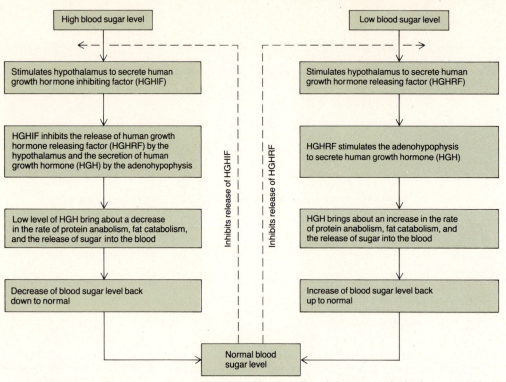

Figure 17-11 Regulation of the secretion of human growth hormone. As is the case with other hormones of the adenohypophysis, the secretion of HGH is controlled by regulating factors. As with most hormones of the body, HGH secretion and inhibition may involve negative feedback systems.

also contains neuron fibers that establish an important connection between the hypothalamus and neurohypophysis. The cell bodies of the neurons originate in two nuclei in the hypothalamus—the supraoptic and paraventricular nuclei (Figure 17-12). The fibers project from the hypothalamus, form the *hypothalamic-hypophyseal tract*, and terminate on blood capillaries in the neurohypophysis. The cell bodies located in the supraoptic nucleus synthesize the hormone *antidiuretic hormone (ADH)*, while the cell bodies located in the paraventricular nucleus synthesize *oxytocin*. Following their production, the hormones are transported in the neuron fibers into the neurohypophysis and are stored in the axon terminals resting on the capillaries. Later, when the hypothalamus is properly stimulated, it sends impulses over the neurons. The impulses cause the release of the hormones from the axon terminals into the blood.

The blood supply to the neurohypophysis is from the *inferior hypophyseal arteries*, derived from the internal carotid arteries. Within the neurohypophysis, the inferior hypophyseal arteries form a plexus of capillaries called the *plexus of the infundibular process*. From this plexus, hormones stored in the neurohypophysis pass into the posterior hypophyseal veins for distribution to all parts of the body (see Figure 17-8).

Oxytocin Action and Control

Oxytocin stimulates the contraction of the smooth muscle cells in the pregnant uterus and the contractile cells around the ducts of the mammary glands. It is released in large quantities just prior to giving birth. When labor begins, the uterus and vagina distend. This distension initiates afferent impulses to the hypothalamus. These impulses stimulate the synthesis of more oxytocin in the cell bodies of the paraventricular nucleus. The oxytocin migrates down the nerve fibers from the hypothalamus to the neurohypophysis. The impulses also cause the neurohypophysis to release stored oxytocin into the blood. It is then carried by the blood to all parts of the body. It acts on the cells of the uterus to reinforce uterine contractions. The effect of oxytocin on milk ejection is as follows. Milk formed by the glandular cells of the breasts is stored until the baby begins active suckling. From about 30 seconds to 1 minute after nursing begins, the baby receives no milk. During this latent period, nerve impulses from the nipple are transmitted to the hypothalamus. The hypothalamus sends impulses down the neurosecretory neurons, which release oxytocin from their axonic ends in the neurohypophysis. Oxytocin then flows from the neurohypophysis via the blood to the breasts, where it stimulates smooth muscle cells to contract and eject milk out of the mammary glands.

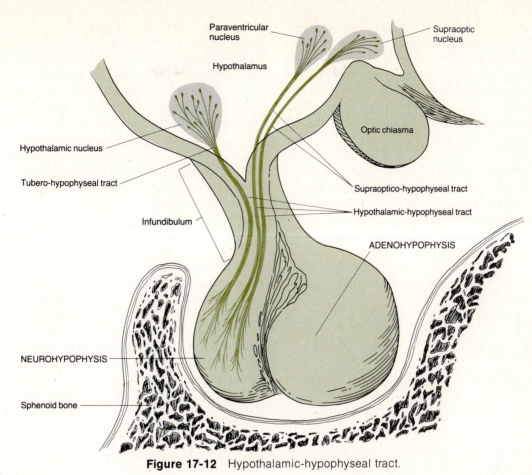

Paraventricular nucleus

Supraoptic nucleus

Hypothalamus

Optic chiasma

Hypothalamic nucleus

Tubero-hypophyseal tract

Supraoptico-hypophyseal tract

Hypothalamic-hypophyseal tract

Infundibulum

ADENOHYPOPHYSIS

NEUROHYPOPHYSIS

Sphenoid bone

Figure 17-12 Hypothalamic-hypophyseal tract.

Oxytocin is inhibited by progesterone but works together with estrogens.

Antidiuretic Hormone (ADH) Action and Control

The more important physiological activity of ADH is its effect on urine volume. ADH causes the kidneys to remove water from newly forming urine and return it to the blood. Since water is the chief constituent of urine, ADH decreases urine volume. However, in the absence of ADH, urine output may be increased 10 times. An *antidiuretic* is any chemical substance that prevents excessive urine production. Another minor physiological activity of ADH is to cause a rise in blood pressure by bringing about constriction of arterioles. This effect is noted when large quantities of the purified hormone are injected. In only rare instances does the body secrete enough hormone to significantly affect blood pressure. The amount of ADH normally secreted varies with the needs of the body. For example, when the body is dehydrated, the concentration of water in the blood falls below normal limits. As the salt-to-water ratio changes, receptors in the hypothalamus detect the low water concentration in the plasma and stimulate the hypothalamus to produce ADH. The hormone travels

down the fibers to the neurohypophysis. It is then released into the bloodstream and transported to the kidneys. The kidneys respond by reabsorbing water and decreasing urine output. ADH also decreases the rate at which perspiration is produced during dehydration. By contrast, if the blood contains a higher-than-normal water concentration, the hypothalamic receptors detect the increase, and the secretion of the hormone is stopped. The kidneys then release large quantities of dilute urine, and the volume of body fluid is brought down to normal.

Secretion of ADH can also be altered by a number of special conditions. Pain, stress, acetylcholine, and nicotine all stimulate secretion of the hormone. Alcohol inhibits secretion and thereby increases urine output. This is why thirst is one of the symptoms of a hangover. The exact mechanism by which ADH actually regulates water volume of the kidneys is discussed in Chapter 26.

THYROID

The endocrine organ located just below the larynx is called the **thyroid gland.** The right and left lateral lobes lie one on either side of the trachea and are connected by a mass of tissue called an *isthmus* that lies in front of the trachea just below the cricoid cartilage (Figure 17-13). The thyroid gland weighs about 25 g (almost 1 oz) and has a very rich blood supply, receiving about 80 to 120 ml of blood/minute. Histologically, the thyroid is composed of spherical sacs called thyroid follicles (Figure 17-14). The walls of each follicle consist of cells that reach the surface of the lumen of the follicle (principal cells) and cells that do not reach the lumen (parafollicular cells). The principal cells manufacture the hormones *thyroxine* (T_4 or tetraiodothyronine) and *triiodothyronine* (T_3). Together, these hormones are referred to as the thyroid hormones. Thyroxine is considered to be the major hormone produced by the principal cells. The parafollicular cells produce the hormone thyrocalcitonin. The interior of each thyroid follicle is filled with *thyroid colloid,* a stored form of the thyroid hormones.

Physiology of the Thyroid

One of the unique features of the thyroid gland is its ability to store its hormones and release them in a steady flow over a long period of time. For example, the principal hormone, thyroxine, is synthesized from iodine and an amino acid called tyrosine. Synthesis usually occurs on a fairly continuous basis (Figure 17-15). The production begins in the gland with an *iodide trapping* process. This term is used to describe the concentrating capability of the follicular cells. In some yet unknown manner, these cells are able to selectively remove I^- *(iodide)* from the plasma and actively transport it across the membrane into the cytoplasm of the cell. Under normal conditions, the concentration inside the cell is 25 times that of the plasma. Administration of iodide after periods of an iodine-free diet shows that these cells have the ability to concentrate iodide to over 300 times that in the plasma. Once inside the cell, the iodide is oxidized to form *iodine* (I_2). This process is catalyzed by the enzyme *peroxidase*. The iodine is then secreted into the follicular colloid along with enzymes and a very unusual glycoprotein molecule named *thyroglobulin*. This molecule has a molecular weight of about 680,000 and contains 25 tyrosine amino acid residues. In the follicular colloid, enzymes begin a series of attacks to progressively iodinate the tyrosine residues. First,

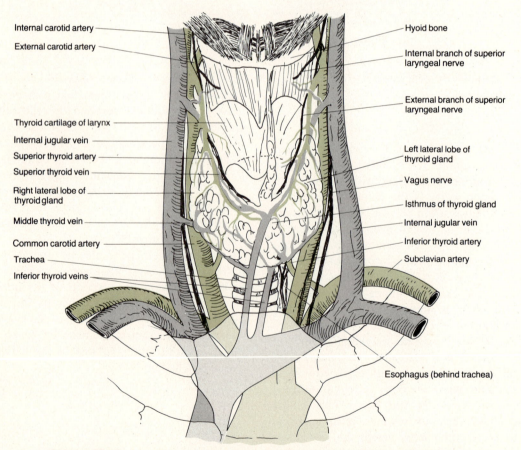

Internal carotid artery

External carotid artery

Thyroid cartilage of larynx

Internal jugular vein

Superior thyroid artery

Superior thyroid vein

Right lateral lobe of thyroid gland

Middle thyroid vein

Common carotid artery

Trachea

Inferior thyroid veins

Hyoid bone

Internal branch of superior laryngeal nerve

External branch of superior laryngeal nerve

Left lateral lobe of thyroid gland

Vagus nerve

Isthmus of thyroid gland

Internal jugular vein

Inferior thyroid artery

Subclavian artery

Esophagus (behind trachea)

Figure 17-13   Location and blood supply of the thyroid gland in anterior view.

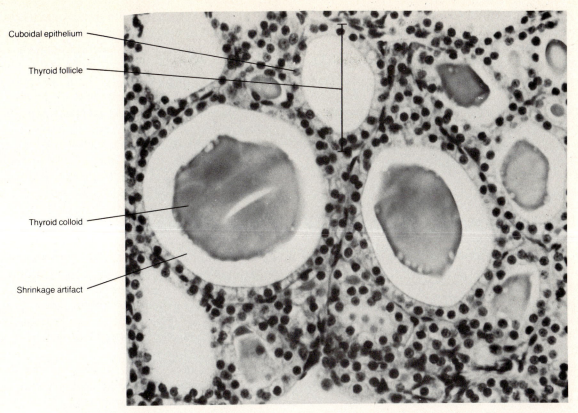

Cuboidal epithelium

Thyroid follicle

Thyroid colloid

Shrinkage artifact

Figure 17-14 Histology of the thyroid gland at a magnification of 90×. (Courtesy of Victor B. Eichler, Wichita State University.)

the tyrosine molecules are converted to monoiodotyrosines (MIT). Most of these molecules are iodinated a second time to form diiodotyrosine (DIT) molecules. By a condensing type of reaction, two of the diiodotyrosine units link together to form one molecule of 3,5,3',5'-*tetraiodothyronine* or *thyroxine* (T₄). A molecule of the amino acid alanine is formed and released into the follicular pool. In some instances a

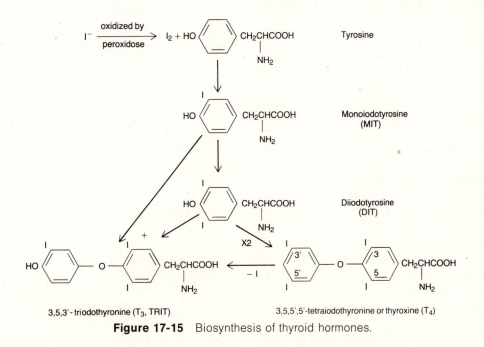

Figure 17-15 Biosynthesis of thyroid hormones.

diiodotyrosine links with a monoiodotyrosine to form one molecule of 3,5,3'-*triiodothyronine* (T_3). At the conclusion of the enzymatic attacks, the thyroglobulin molecule contains two or three active hormone molecules. Overall production results in 90 percent thyroxine and 10 percent triiodothyronine. Thyroglobulin molecules can persist in the colloidal matrix for months before they are called upon to give up the active portions. The hormones are released into the circulation by a process involving pinocytosis and digestion.

First, a small section of the follicle cell membrane reaches into the colloid to surround a small portion and form a small internal vesicle. A lysosome then fuses with the vesicle, thus adding its enzymatic complement to the combined vesicle. Enzymes then cut out the active thyroxine and triiodothyronine portions, which are then released to diffuse out of the cell into the blood.

In the blood, the hormones are immediately bound to plasma proteins and are transported to all parts of the body. In this combined form, the hormone is referred to as *protein-bound iodine (PBI)*. Under normal circumstances, the amount of PBI in the blood is fairly constant—4 to 8 μg PBI/100 ml blood. This amount can easily be measured. PBI is a good index of thyroid hormone secretion and is often used as a tool to diagnose suspected thyroid malfunction. Some of the thyroxine in the blood is deiodinated to triiodothyronine, which is 5 to 10 times more potent a stimulator of cellular metabolism than the thyroxine. However, it is calculated that because thyroxine has a longer duration of action and is present in greater concentration, its overall effectiveness extended over its life span is about equal to that of triiodothyronine over a similar period.

Cells take up the active hormone molecules by removing them from their plasma carrier molecules. Once inside the cell, however, they again combine with a protein molecule and are stored in this fashion.

Thyroxine Action and Control

The major function of thyroxine is to control the rate of metabolism by regulating the catabolic or energy-releasing processes. Thyroxine increases the rate at which carbohydrates are burned. It stimulates cells to break down proteins for energy instead of using them for building processes. At the same time, thyroxine decreases the breakdown of fats. The overall effect, though, is to increase catabolism. Thus it produces energy and raises body temperature as heat energy is given off. This is called the *calorigenic effect*. Thyroxine is also an important factor in the regulation of tissue growth and in the development of tissues. It works with human growth hormone to accelerate body growth, especially the growth of nervous tissue. Thyroxine deficiency during fetal development can result in fewer and smaller neurons, defective myelination of nerve fibers, and mental retardation. Hyposecretion of thyroxine during the early years of life causes small stature and failure of some organs to develop. Thyroxine also acts as a diuretic, increases the reactivity of the nervous system, and increases the heart rate.

The secretion of thyroxine seems to be brought on by any of several factors (Figure 17-16). For instance, if levels of thyroxine in the blood fall below normal, chemical sensors in the hypothalamus detect the change in the blood chemistry and stimulate the hypothalamus to release thyrotropin releasing factor (TRF). TRF stimulates the adenohypophysis to secrete thyroid stimulating hormone (TSH). TSH acts upon the thyroid, stimulating it to release thyroxine until the metabolic rate is returned to normal. Conditions that increase the body's need for energy, such as a cold environment, or high altitude, or pregnancy, also trigger this feedback system and increase the secretion of thyroxine.

When the amount of thyroxine in the blood is returned to the required level, the secretion of TSH stops, and thyroxine release is reduced. Thyroid activity can also be slowed down by a number of other factors. For instance, when very large amounts of the sex hormones—estrogens and androgens—are circulating within the blood, TSH secretion is decreased. Aging slows the activities of most glands. Thus thyroid production decreases as an individual gets older.

Thyrocalcitonin Action and Control

The hormone produced by the parafollicular cells of the thyroid gland is *thyrocalcitonin* or *calcitonin*. It is involved in the homeostasis of blood calcium level.

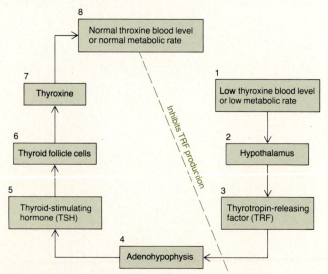

Figure 17-16 Regulation of thyroxine secretion.

Bones are continually remolded during adult life and part of this process consists of the breakdown of osseous tissue by osteoclasts with a consequent release of calcium into the blood. The other part of the process is the deposition of calcium in bones and the subsequent production of new ossified tissue by osteoblasts. Thyrocalcitonin lowers the amount of calcium in the blood by inhibiting bone breakdown and by accelerating the absorption of calcium by the bones. Thyrocalcitonin appears to exert its influence by antagonizing a number of bone resorption agents such as vitamin D, vitamin A, and the parathyroid hormone (PTH). If thyrocalcitonin is administered to a person with normal blood calcium levels, it causes *hypocalcemia,* or low blood calcium level. Hypocalcemia is also a complication of magnesium deficiency. If thyrocalcitonin is given to an individual with *hypercalcemia* (high blood calcium level), the level returns to normal. It is suspected that the blood calcium level directly controls the secretion of calcitonin.

PARATHYROIDS

Structure

Embedded on the posterior surfaces of the thyroid gland are from one to three pairs of small, yellowish-brown masses of tissue called the **parathyroid glands.** About 75 percent of the population possess two pairs — the superior and inferior parathyroids — one of each pair attached to each lateral thyroid lobe (Figure 17-17). They measure 3 to 8 mm in length and 3 to 4 mm in breadth and are about 1 to 2 mm thick. Despite the variability in their number, the total mass of parathyroid tissue in the human seems to be quite constant, around 120 mg. The blood supply to the glands is chiefly from the inferior thyroid supply with additional anastomotic connections to arteries supplying the larynx and pharynx.

Histologically, the parathyroids contain two kinds of epithelial cells (Figure 17-18). The first kind, called the *principal* or *chief cell,* is believed to be the major synthesizer of the parathyroid hormone called *parathyroid hormone (PTH)* or *parathormone.* Some researchers believe that the other kind of cell, called an *oxyphil cell,* synthesizes a reserve capacity of hormone.

Parathyroid Hormone (PTH) and 1,25-Dihydroxycholecalciferol

There are three important hormones that function to regulate the metabolism of Ca^{2+} and phosphate in the human body. They are thyrocalcitonin described in the preceding section, *parathyroid hormone,* and *1,25-dihydroxycholecalciferol.*

Control and Action of Parathyroid Hormone

Parathyroid hormone is secreted into the blood in increasing amounts whenever the Ca^{2+} concentration of the blood falls below 10 mg/100 ml of plasma. This

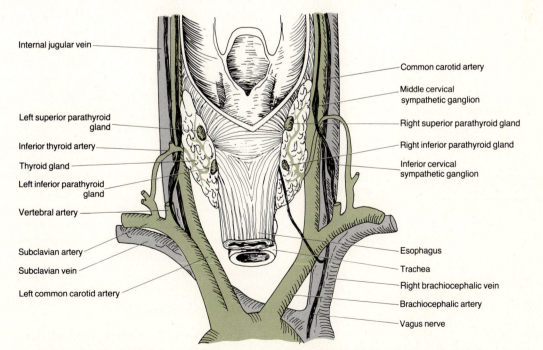

Figure 17-17 Location and blood supply of the parathyroid glands in posterior view.

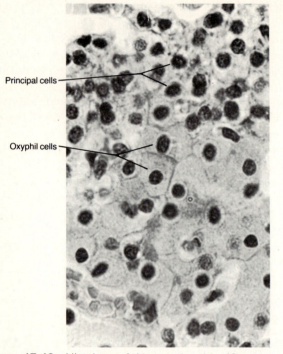

Principal cells

Oxyphil cells

Figure 17-18 Histology of the parathyroids. (Courtesy of Victor B. Eichler, Wichita State University.)

polypeptide hormone contains 84 amino acid residues and acts to increase the blood Ca^{2+} concentration in three ways. First, it stimulates the release of Ca^{2+} from bone by increasing the number of functional osteoclastic cells. Second, the hormone stimulates Ca^{2+} reabsorption by the kidney tubular cells, while it inhibits the reabsorption of phosphate. Third, the parathyroid hormone acts as a tropic hormone causing the formation of another hormone called 1,25-dihydroxycholecalciferol in the kidneys from its precursor molecule, 25-hydroxycholecalciferol. This hormone, in turn, promotes the absorption of Ca^{2+} by the intestinal epithelial cells.

The body obtains its calcium from the foodstuffs that enter the digestive tract. Milk and milk products are the major sources not only of Ca^{2+} but also phosphate. However, it should be noted that the divalent calcium cation, Ca^{2+}, is poorly absorbed by the intestinal cells whereas phosphate is very easily taken up. Almost 90 percent of the dietary intake of calcium is not absorbed but is excreted in the feces primarily as an insoluble calcium phosphate salt.

Production and Action of 1,25-Dihydroxycholecalciferol

The hormone 1,25-dihydroxycholecalciferol is the most active form of the vitamin D compounds. Its synthesis (Figure 17-19) begins with 7-dehydrocholesterol, which is present in the skin. There, in the

presence of the ultraviolet rays from sunlight, it is converted to vitamin D_3 or cholecalciferol. This is the normal manner in which most people acquire vitamin D. Dietary vitamin D comes almost entirely from fish-liver oils. An adult requires about 20 μg of vitamin D daily, and much can be stored in the liver to supply the body's needs for weeks. In the liver, the cholecalciferol is converted to 25-hydroxycholecalciferol and then released into the circulation. In the kidney, the 25-hydroxycholecalciferol is converted to 1,25-dihydroxycholecalciferol, a reaction promoted by parathyroid hormone. The active hormone, 1,25-dihydroxycholecalciferol, works directly on its targets: the epithelial cells of the small intestine and the osteoclasts in bone. It has been shown that the 1,25-dihydroxycholecalciferol stimulates the epithelial cells in the intestine to synthesize additional molecules of a specific receptor protein that binds and transports calcium ions across the intestinal mucosa into the circulation. Thus, when the plasma level of Ca^{2+} begins to fall, the parathyroid glands respond by releasing parathyroid hormone, which stimulates the kidney to excrete phosphate and to convert 25-hydroxycholecalciferol to 1,25-dihydroxycholecalciferol.

ADRENALS (SUPRARENALS)

The body has a pair of **adrenal** or **suprarenal glands**—flattened, yellowish masses of tissue—each located on the superior border of one kidney (Figure 17-20). The average dimensions of the adult adrenal gland are about 50 mm long, 30 mm wide, and 10 mm thick. The adrenals, like the thyroid, are among the most vascular organs of the body. The gland is covered with a thick layer of fatty connective tissue and an outer, thin, fibrous capsule.

Structurally and functionally, each adrenal is differentiated into two sections: the outer adrenal cortex and the inner adrenal medulla (Figure 17-21). The cortex arises embryonically from two masses of mesodermal tissue in the dorsal mesentery. These tissues secrete steroid type hormones that are similar in chemical structure to those synthesized by the testes and ovaries.

During embryonic development, the mesodermal masses combine with and envelop ectodermal cells that pinched off from the developing neural tissue. This inner neuroendocrine tissue becomes the adrenal medulla.

The Adrenal Cortex

Histologically, the cortex is subdivided into three zones (Figure 17-21). Each zone has a different cellu-

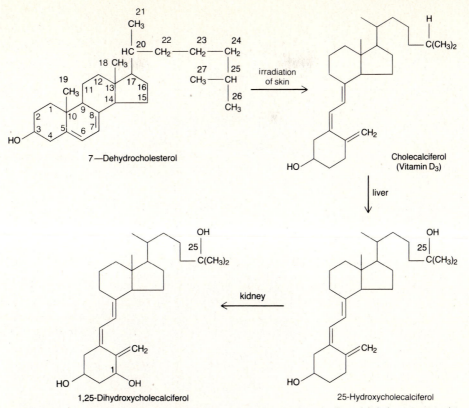

Figure 17-19 Conversion of 7-dehydrocholesterol to vitamin D_3, which is converted to 1,25-dihydroxycholecalciferol.

lar arrangement and secretes different hormones. The outer zone, directly underneath the connective tissue covering, is referred to as the *zona glomerulosa*. Its cells are arranged in arched loops or round balls, and they secrete a group of hormones called *mineralocorti-*

coids. The middle zone, or *zona fasciculata*, is the widest of the three zones and consists of cells arranged in long, straight cords. The zona fasciculata secretes *glucocorticoid* hormones. The inner zone, the *zona reticularis*, contains cords of cells that branch

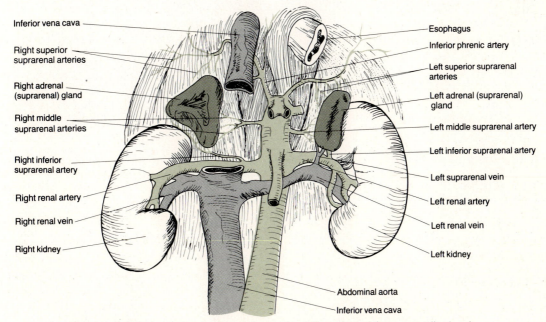

Figure 17-20 Location and blood supply of the adrenal (suprarenal) glands.

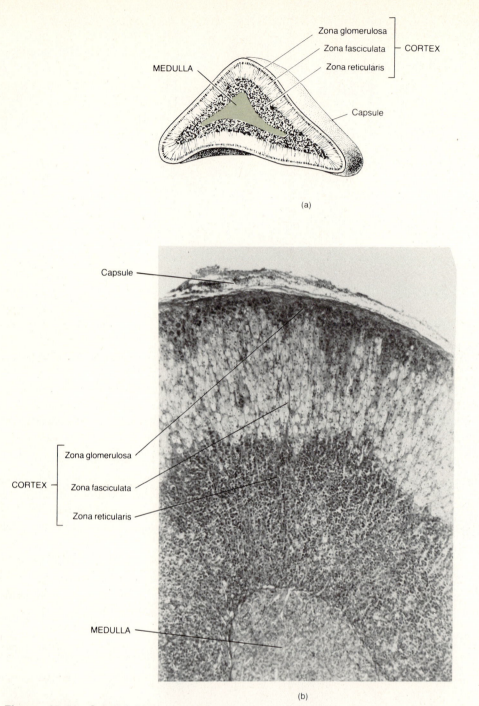

Figure 17-21 Subdivisions and histology of the adrenal (suprarenal) glands. (a) Diagram of the subdivisions. (b) Histology of the subdivisions at a magnification of 15×. (Courtesy of Bio-Art.)

freely. This zone synthesizes *sex hormones,* chiefly male hormones called *androgens.*

Action of Mineralocorticoids

Mineralocorticoids help control electrolyte homeostasis, particularly the concentrations of sodium and po-

tassium. Although the adrenal cortex secretes at least three different hormones classified as mineralocorticoids, the one called *aldosterone* is responsible for about 95 percent of the mineralocorticoid activity. Aldosterone acts on the tubule cells in the kidneys and causes them to increase their reabsorption of sodium, with the result that sodium ions are removed from the urine and returned to the blood. In this manner, aldosterone prevents rapid depletion of sodium from the body. On the other hand, aldosterone decreases reabsorption of potassium. Large amounts of potassium are moved from the blood into the urine and excreted from the body.

These two basic functions—conservation of sodium and elimination of potassium—cause a number of secondary effects. For example, a large proportion of the sodium reabsorption occurs through an exchange reaction whereby positive hydrogen ions pass into the urine to take the place of the positive sodium ions. Since this mechanism removes hydrogen ions, it makes the blood less acidic and thus acts to prevent acidosis. The movement of Na^+ ions begins to create an excess of positive charge in the blood vessels. As a result, negatively charged chloride and bicarbonate ions are attracted out of the urine and returned to the blood. The increase in the ionic concentration of the blood finally causes water to move by osmosis from the urine into the blood. In summary, aldosterone stimulates potassium excretion and sodium reabsorption. The sodium reabsorption leads to the elimination of H^+ ions, the retention of Na^+, Cl^-, and HCO_3^-, and a consequent retention of water.

Aldosterone operates at a molecular level in a manner characteristic of the steroid hormones. Being lipid soluble, the molecule easily penetrates into the cytoplasm of the cell, where it then reacts and hydrogen bonds with a specific receptor protein. The complex of aldosterone-receptor is able to enter the nucleus of the cell, where it then attaches to DNA and derepresses a specific gene. Transcription of *m*RNA results and this messenger diffuses back to the cytoplasm, where it is translated to a specific protein. Current speculation is that the protein is an enzyme that energizes a membrane transport system by hydrolyzing ATP. The transport system operates selectively to reabsorb sodium ions.

With the sequence of steps required to elicit a response, a time delay of about 1 hour usually occurs between the time of administration of the hormone and the onset of increased reabsorption.

Control of Aldosterone

The control of aldosterone is rather complex and not completely understood. While it regulates the ionic composition of body fluids more directly, it is also part of the water-regulating mechanism. For example, when the body is dehydrated, the blood contains less water. This means that a smaller volume of blood is circulating through the vessels and blood pressure is lowered. The blood pressure drop is detected by pressure receptors in the largest arteries, which then direct nerve signals to the hypothalamus, where a thirst sensation is produced. The drop in renal arterial blood pressure stimulates certain kidney cells called *juxtaglomerular cells* to secrete into the blood an enzyme called *renin*. In this sequence of reactions, called the *renin-aldosterone pathway* (Figure 17-22), renin converts a substance called *angiotensinogen*, which is continually secreted by the liver into the circulation, to a substance called *angiotensin I*. Angiotensin I circulates through the lungs, where it is converted to *angiotensin II* by the removal of two amino acids. The angiotensin II is carried by the blood to the adrenal cortex, which responds to produce more aldosterone which, in turn, promotes active Na^+ reabsorption and passive water reabsorption in the kidney tubules. This reduces greatly the water loss in urine which, coupled with the thirst sensation, works toward restoring fluid balance and blood volume.

It is thought by some investigators that the elevation in potassium ions is the most important stimulator of aldosterone. Cells in the zona glomerulosa of the adrenal cortex are very sensitive to slight increases in the concentration of K^+. The potassium ion stimulates an early key step in the biosynthetic pathway of aldosterone, speeding up its rate of synthesis. A drop in sodium ion concentration also speeds up this same key enzyme, but the drop in Na^+ level must be considerable before the effect is noticed.

It should be pointed out that ACTH has little or no direct effect on the adrenal cortex to produce aldosterone.

Action of Glucocorticoids

The *glucocorticoids* are a group of hormones that are largely concerned with normal metabolism and the ability of the body to resist stress. Three examples of glucocorticoids are *hydrocortisone (cortisol), corticosterone,* and *cortisone.* Of the three, hydrocortisone is the most abundant and is responsible for about 95 percent of the glucocorticoid activity. The glucocorticoids have the following effects on the body:

1. Glucocorticoids work with other hormones in promoting normal metabolism. Their role is to make sure that enough energy is provided to meet the total body needs. Cortisol increases the rate at which amino acids are transported to the liver, where they can be converted to carbohydrates for energy purposes. Alternatively, amino acids may be used for synthesis of new proteins, such as the enzymes that are needed for the metabolic reactions. Cortisol also promotes both the removal of protein from skeletal muscle and its transport to the liver, where the protein can be used for syn-

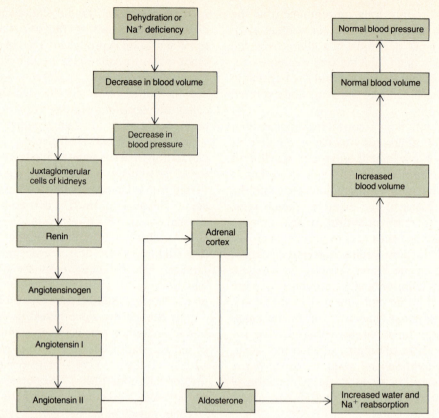

Figure 17-22 Proposed mechanism for the regulation of the secretion of aldosterone by the renin-angiotensin pathway.

thetic purposes. Glucocorticoids reduce the rate of glucose utilization by cells and, in the liver, stimulate glycogen synthesis. In addition, the hormones encourage the movement of fats from storage depots to all the cells, where they are catabolized for energy. The overall picture of the effect of glucocorticoids is that they promote gluconeogenesis (protein to carbohydrate) and glycogen synthesis (storage of carbohydrate).

2. Glucocorticoids work in many ways to provide resistance to stress. One of the more obvious ways is their rapid protein mobilizing capability. A sudden increase in available amino acids makes the body cells more able to repair tissue damage. Additional protein also gives the body energy for combating a range of stressors, such as fright, temperature extremes, high altitude, bleeding, and infection. Glucocorticoids also make the blood vessels more sensitive to vessel-constricting chemicals. In this manner, they raise blood pressure. This is advantageous if the stressor happens to be blood loss, which causes a drop in blood pressure.

3. Glucocorticoids decrease the blood vessel dilation and hence the edema associated with inflammations. Thus they are anti-inflammatory agents. Unfortu-

nately, they also decrease connective tissue regeneration and are thereby responsible for slow wound healing.

Control of Glucocorticoids

The control of glucocorticoid secretion is another example of a typical negative feedback mechanism (Figure 17-23). Extreme stress and low blood levels of glucocorticoids trigger the response mechanism. Stress could include emotional stress or physical damage, such as that produced by contusions, broken bones, disease, or tissue destruction. The stress may directly stimulate the hypothalamus. For example, the hypothalamus itself could detect and respond to a low blood pressure resulting from excessive bleeding. Alternatively, the stimulus may be detected by some other part of the nervous system and the information relayed to the hypothalamus. In any case, either extreme stress or abnormally low levels of glucocorticoids stimulate the hypothalamus to secrete a regulating factor called *adrenocorticotropin hormone releasing factor (ACTHRF)* that initiates the release of ACTH from the anterior lobe of the pituitary. ACTH is carried through the blood to the adrenal cortex, where it stimulates glucocorticoid secretion.

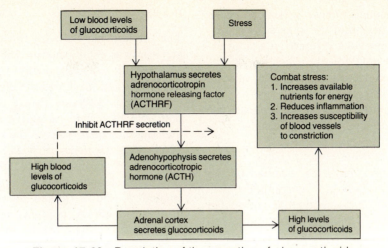

Figure 17-23 Regulation of the secretion of glucocorticoids.

Gonadocorticoids

The adrenal cortices secrete both male and female *gonadocorticoids,* or *sex hormones.* These are estrogens and androgens. But the amount of sex hormones normally secreted by the adrenals is usually so small that it is insignificant.

The Adrenal Medulla

The adrenal medulla consists of hormone-producing cells, called *chromaffin cells,* which surround large blood-containing sinuses (see Figure 17-21). Chromaffin cells develop from the same embryonic tissue as the postganglionic cells of the sympathetic division of the nervous system. They are directly innervated by preganglionic cells of the sympathetic division of the nervous system and indeed may be regarded as postganglionic nerve cells that do not conduct nerve impulses but are specialized only to secrete. In all other visceral effectors, preganglionic sympathetic fibers first synapse with postganglionic neurons before innervating the effector. In the adrenal medulla, however, the preganglionic fibers pass directly to the chromaffin cells of the gland. The secretion of hormones from the chromaffin cells is directly controlled by the autonomic nervous system, and innervation by the preganglionic fibers allows the gland to respond very rapidly to a stimulus.

Action of Medullary Hormones

The adrenal medulla synthesizes two important hormones, *norepinephrine* and *epinephrine,* by the biosynthetic pathway outlined in Figure 17-24. Tyrosine is an amino acid easily obtained from the diet. Each of the intermediates in the pathway has been found to be present in nervous tissue. DOPA has been synthe-

sized in laboratories and used therapeutically for the treatment of Parkinson's disease. It is effective because it can cross the blood-brain barrier into the brain. There, it is converted to dopamine, which is a very important neurotransmitter in the brain. Epinephrine constitutes about 80 percent of the total secretion of the gland and is more widespread in its action than is norepinephrine. Both hormones are *sympathomime-*

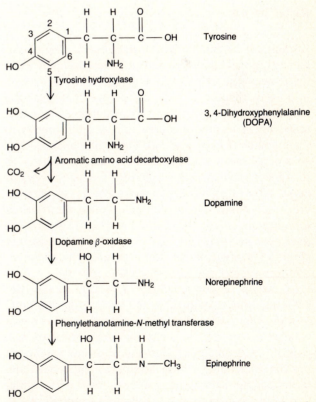

Figure 17-24 Biosynthesis of norepinephrine and epinephrine.

tic. That is, they produce effects similar to those brought about by the action of the sympathetic division of the autonomic nervous system. To a large extent, they are responsible for the "fight or flight" response. Like the glucocorticoids of the adrenal cortices, these hormones help the body resist stress situations. However, unlike the cortical hormones, the medullary hormones are not essential for life.

Control of Medullary Hormones

Under a stress condition, impulses received by the hypothalamus are conveyed to sympathetic preganglionic neurons, which act upon the chromaffin cells so they increase their output of epinephrine and norepinephrine. Epinephrine increases blood pressure by increasing the heart rate and by constricting the blood vessels. It also accelerates the rate of respiration, dilates respiratory passageways, decreases the rate of digestion, increases the efficiency of muscular contractions, increases blood sugar level, and stimulates cellular metabolism. Hypoglycemia may also stimulate medullary secretion of epinephrine and norepinephrine.

The mechanism of action of epinephrine was described earlier in the chapter in the section on mechanisms of hormone action.

PANCREAS

Because of its functions, the **pancreas** can be classified as both an endocrine and an exocrine gland. Since the exocrine functions of the gland will be discussed in a later chapter, in this section we shall describe only its endocrine functions. The pancreas is a flattened organ located posterior and slightly inferior to the stomach (Figure 17-25a). The adult pancreas consists of a head, body, and tail. Its average length is 12 to 15 cm (5 to 6 inches), and its average weight is about 85 g (3 oz). The endocrine portion of the pancreas consists of clusters of cells called the *islets of Langerhans* (Figure 17-25b). Two kinds of cells are found in the islets: (1) *alpha cells,* which constitute about 25 percent of the islet cells and which synthesize and secrete the hormone *glucagon;* and (2) *beta cells,* which constitute about 75 percent of the islet cells and which synthesize and secrete the hormone *insulin.* The islets are surrounded by numerous blood capillaries and by the cells that form the exocrine part of the gland (acini). Both glucagon and insulin are peptide molecules whose structures have been determined. Both hormones are concerned with regulation of carbohydrate, fat, and protein metabolism.

Insulin

The relationship between the pancreas and diabetes mellitus was established in 1889 by Von Mering and Minkowski, who surgically removed the pancreas from dogs and produced the diabetic condition. However, it took 32 years before Banting and Best were first able to isolate an active preparation of insulin that functioned to relieve the diabetic symptoms. In 1955, after 10 years of work, Sanger described the structure of the insulin molecule (see Figure 17-2a). The molecule consists of two polypeptide chains designated A and B, linked together by two disulfide bridges. Chain A contains an intrachain disulfide bridge that encloses a series of six amino acids.

The biosynthesis of insulin by the islet tissue has recently been worked out (Figure 17-26). The molecule of insulin is synthesized on the ribosomes as a molecule of proinsulin—a molecule containing some 81 to 86 amino acid residues. This molecule contains both the A and B chains of the insulin molecule and an interconnecting chain C. The proinsulin molecule has the three disulfide bridges characteristic of native insulin. The proinsulin molecule is transported to the Golgi complex via the channels of the rough endoplasmic reticulum. The proinsulin is then attached by enzymes and the C chain portion is removed from the molecule. The insulin is packaged into Golgi vesicles along with Zn^{2+} and the C chain to form crystalline insulin. The vesicles release their contents into the capillaries when the blood glucose level rises to values over 80 to 90 mg/100 ml plasma. The insulin molecule has a half-life in plasma of only about 3 to 4 minutes.

Action of Insulin

Insulin release causes the level of glucose in the blood to drop. It accomplishes this by promoting the transport of glucose from blood into muscle and fat cells, where the glucose is then stored as glycogen and fat. Insulin also increases the rate of synthesis of proteins from amino acids.

In general, insulin operates at the level of the cell membrane, promoting the increased entry not only of glucose but also of amino acids, lipids, and potassium ions. The regulation of insulin secretion is directly determined by the level of sugar in the blood and is based upon a negative feedback system (Figure 17-27). However, other hormones can indirectly affect insulin production. For instance, human growth hormone raises blood glucose level, and the rise in glucose level triggers insulin secretion. ACTH, by stimulating the secretion of glucocorticoids, produces hyperglycemia and also directly stimulates the release

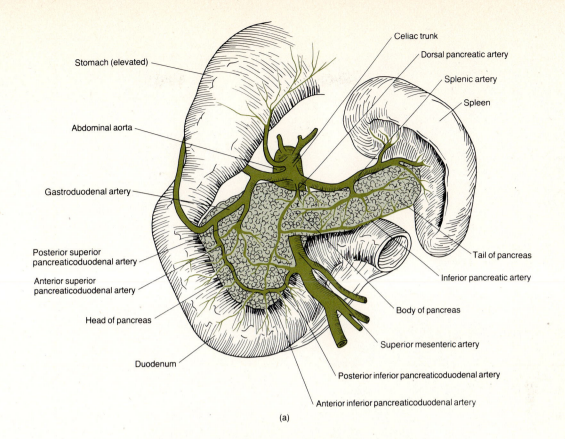

Stomach (elevated)

Celiac trunk

Dorsal pancreatic artery

Splenic artery

Spleen

Abdominal aorta

Gastroduodenal artery

Posterior superior pancreaticoduodenal artery

Anterior superior pancreaticoduodenal artery

Head of pancreas

Tail of pancreas

Inferior pancreatic artery

Body of pancreas

Superior mesenteric artery

Duodenum

Posterior inferior pancreaticoduodenal artery

Anterior inferior pancreaticoduodenal artery

(a)

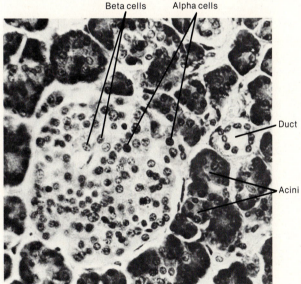

Beta cells   Alpha cells

Duct

Acini

(b)

Figure 17-25   The pancreas. (a) Location and blood supply of the pancreas. (b) Histology of the pancreas. Enlarged aspect of a single islet of Langerhans and surrounding acini at a magnification of 100×. (Courtesy of Victor B. Eichler, Wichita State University.)

of insulin. Epinephrine is also an insulin antagonist. Recent experiments have shown that insulin binds to specific receptor molecules located in the cell membrane. However, unlike other protein hormones described, it seems to *decrease* the level of intracellular cyclic AMP, while it causes the increase of another

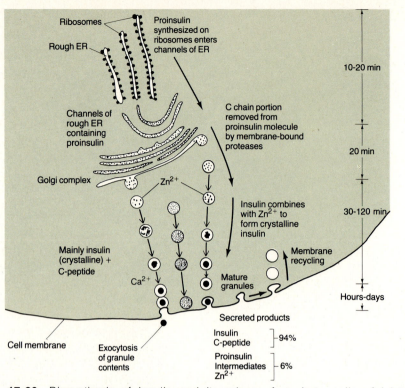

Figure 17-26 Biosynthesis of insulin and its release from beta cells of islets of Langerhans. Time relationships are also noted. Mature granules are not released until blood glucose levels are significantly elevated.

cyclic nucleotide, cGMP (cyclic guanosine-3′,5′-monophosphate) (see Figure 17-7). Some scientists believe that perhaps a reciprocal relationship operates between the two cyclic nucleotides — one promoting one response and the other promoting the opposite response — a hypothesis tentatively referred to as the *yin-yang hypothesis*.

Glucagon

The product of the alpha cells is *glucagon*, a hormone whose principal physiological activity is to increase the blood glucose level. It has the opposite effect to that of insulin. Glucagon is a polypeptide hormone containing 29 amino acid residues but is

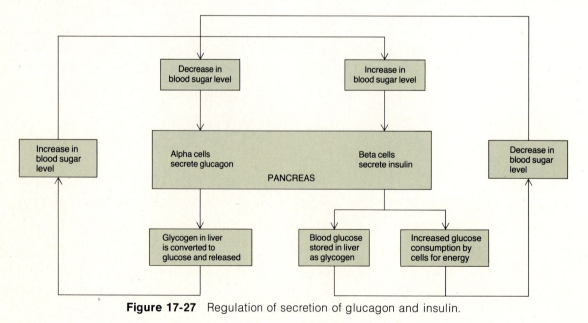

Figure 17-27 Regulation of secretion of glucagon and insulin.

synthesized as *proglucagon,* an inactive precursor (Figure 17-28) that has eight additional amino acid residues at the C terminal end.

Glucagon raises blood glucose level by acting primarily on the liver to accelerate the breakdown of glycogen into glucose. In a manner like epinephrine, glucagon manifests its effect by binding to a receptor protein on the liver cell membrane. The resultant complex on the outer cell membrane surface activates the adenyl cyclase enzyme on the inner surface of the membrane to produce cyclic AMP. Through the same *amplification cascade* sequence outlined for epinephrine, the concentration of *phosphorylase a* increases inside the liver cell, leading to a rapid conversion of glycogen to blood glucose.

Secretion of glucagon is directly controlled by the level of blood sugar according to a negative feedback system (see Figure 17-27). For example, when the blood sugar level falls below normal, chemical sensors in the alpha cells of the islets stimulate the cells to secrete glucagon. When blood sugar rises, the cells are no longer stimulated and production slackens. If for some reason the self-regulating device fails and the alpha cells secrete glucagon continuously, hyperglycemia may result. Glucagon secretion can be inhibited by somatostatin from the adenohypophysis.

PINEAL

The cone-shaped gland located in the roof of the third ventricle is known as the **pineal gland,** or epiphysis cerebri (see Figure 12-1). The gland is about 5 to 8 mm (0.2 to 0.3 inch) long and 9 mm wide. It weighs about 0.2 g. It is covered by a capsule formed by the pia mater. It consists of masses of parenchymal and glial cells. Around the cells are scattered preganglionic sympathetic fibers. The pineal gland starts to degenerate at about age 7, and in the adult it is largely fibrous tissue.

Although many anatomical facts concerning the pineal gland have been known for years, its physiology is still somewhat obscure. One hormone secreted by the pineal gland is *melatonin,* which appears to affect the secretion of hormones by the ovaries. It has been known for years that light stimulates the sexual endocrine glands. Researchers have also discovered that blood levels of melatonin are low during the day and high at night. Putting these observations together, some investigators now believe that melatonin inhibits the activities of the ovaries. During daylight hours, light entering the eye stimulates neurons to transmit impulses to the pineal that inhibit melatonin secretion. Without melatonin intereference, the ovaries are free to step up their hormone production. But at night, the pineal gland is able to release melatonin, and ovarian function is slowed. One function of the pineal gland might very well be regulation of the activities of the sexual endocrine glands, particularly the menstrual cycle.

Some evidence also exists that the pineal secretes a second hormone called *adrenoglomerulotropin.* This hormone may stimulate the zona glomerulosa of the adrenal cortex to secrete aldosterone. Still other functions attributed to the pineal gland are the secretion of a growth-inhibiting factor and the secretion of a hormone called *serotonin* that is involved in normal brain physiology.

THYMUS

Usually a bilobed organ, the **thymus gland** is located in the upper mediastinum posterior to the sternum and between the lungs (Figure 17-29). The gland is conspicuous in the infant, and during puberty when it reaches its absolute maximum size it weighs 28–35 g. (1 to 1.3 oz). After puberty, the thymic tissue, which consists primarily of lymphocytes, is replaced by fat. By the time the person reaches maturity, the gland has atrophied.

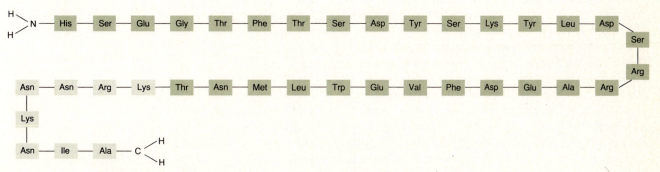

Figure 17-28 Structure of bovine glucagon (darker boxes) and bovine proglucagon (darker and lighter boxes).

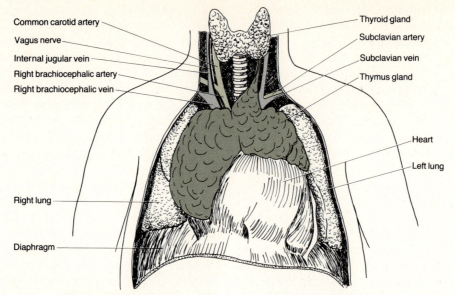

Common carotid artery

Vagus nerve

Internal jugular vein

Right brachiocephalic artery

Right brachiocephalic vein

Right lung

Diaphragm

Thyroid gland

Subclavian artery

Subclavian vein

Thymus gland

Heart

Left lung

Figure 17-29 Location of the thymus gland in a young child.

The importance of the thymus gland is related to the production of a special type of sensitized lymphocyte cell that is very significant in cellular immunity. This will be fully explained in Chapter 21.

STRESS AND HOMEOSTASIS

Throughout this text, we have emphasized the concept of homeostasis—the maintenance of the body's internal physiological environment in response to stresses that originate within the body or in the external environment. Essentially, homeostasis may be viewed as specific responses of the body to specific stimuli. For instance, when blood calcium goes up, the rise stimulates the thyroid gland to release thyrocalcitonin. When blood calcium falls, thyrocalcitonin secretion is inhibited, and parathyroid hormone secretion is stimulated. Homeostatic mechanisms "fine tune" the body. If the mechanisms are successful, the internal environment maintains a uniform chemistry, temperature, and pressure.

Homeostatic mechanisms are geared toward counteracting the everyday stresses of living. However, if a stress is extreme or unusual, the normal ways of keeping the body in balance may not be sufficient. In this case, the stress triggers a wide-ranging set of bodily changes called the **general adaptation stress syndrome.** Unlike the homeostatic mechanisms, the general adaptation stress syndrome does not maintain a constant internal environment. In fact, it does just the opposite. For instance, during the general adaptation stress syndrome, blood pressure and blood sugar level are raised above normal. The purpose of these changes in the internal environment is to gear up the body for meeting an emergency. Here is how it works.

The hypothalamus can be called the watchdog of the physical and psychological state of the body. It has sensors that detect changes in the chemistry, temperature, and pressure of the blood. It is made aware of increasing emotional feelings via nerve tracts that connect it with the emotional centers of the cerebral cortex. When the hypothalamus realizes that a severe or unusual stress is occurring, it initiates a chain of reactions that produce the general adaptation stress syndrome. The severe or unusual stresses that produce the syndrome are called *stressors.* A stressor may be almost any severe disturbance such as extreme heat or cold, environmental poisons, poisons given off by bacteria during a raging infection, heavy bleeding from a wound or surgical procedure, or a strong emotional reaction.

When a stressor is sensed, it stimulates the hypothalamus to initiate the syndrome through two pathways. The first pathway is stimulation of the sympathetic nervous system and adrenal medulla. This produces an immediate set of responses called the alarm reaction. The second pathway, called the resistance reaction, involves the anterior pituitary gland and adrenal cortex. The resistance reaction is slower to start, but its effects are longer lasting.

Alarm Reaction

The *alarm reaction,* or "fight or flight" response, is the initial reaction of the body to any stressor (Figure 17-30a). It is actually a complex of reactions initiated when the hypothalamus stimulates the sympathetic nervous system and the adrenal medulla. The responses of the visceral effectors are immediate, but short-lived. They are designed to counteract danger by mobilizing the body's resources for immediate

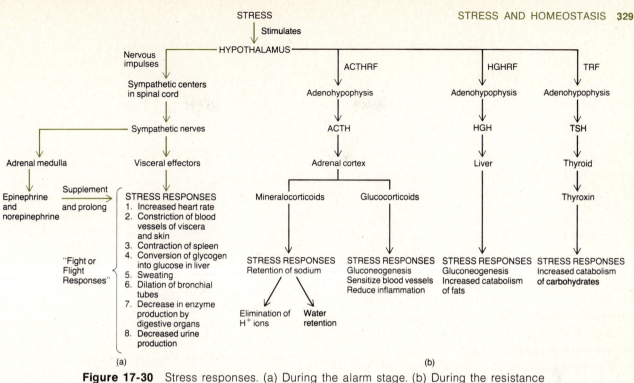

Figure 17-30 Stress responses. (a) During the alarm stage. (b) During the resistance stage. Colored arrows indicate immediate reactions. Black arrows indicate long-term reactions.

physical activity. In essence, the alarm reaction brings tremendous amounts of glucose and oxygen to the organs that are most active in warding off danger. These are the brain, which must become highly alert; the skeletal muscles, which may have to fight off a strong attacker; and the heart, which must work furiously to pump enough materials to the brain and muscles. Among the stress responses that characterize the alarm stage are the following:

1. The heart rate and the strength of cardiac muscle contraction increase. This circulates substances in the blood very quickly to areas where they are needed to combat the stress.
2. Blood vessels supplying the skin and viscera, except the heart and lungs, undergo constriction. At the same time, blood vessels supplying the skeletal muscles and brain undergo dilation. These responses route more blood to organs that are active in the stress responses and decrease the blood supply to organs that do not assume an immediate, active role.
3. The spleen contracts and discharges stored blood into the general circulation to provide additional blood. Moreover, red blood cell production is accelerated, and the ability of the blood to clot is increased. These preparations are made by the body for combating possible bleeding.
4. The liver transforms large amounts of stored glycogen into glucose and releases it into the bloodstream. The glucose is broken down by the active

cells to provide the energy needed to meet the stressor.
5. An increase in sweat production also occurs. This response helps to maintain a normal body temperature, which starts to rise as circulation increases and body catabolism increases. Profuse sweating also helps to eliminate wastes produced as a result of accelerated catabolism.
6. The rate of breathing increases and the respiratory passageways widen in order to accommodate larger volumes of air. This response enables the body to acquire more oxygen, which is needed in the oxidation reactions of catabolism. It also allows the body to eliminate more carbon dioxide, an end product of catabolism.
7. A decrease in the production of saliva, stomach enzymes, and intestinal enzymes occurs. This takes place since digestive activity is not essential for counteracting the stress.
8. Sympathetic impulses to the adrenal medulla increase its secretion of epinephrine. This hormone supplements and prolongs many sympathetic responses, such as increasing heart rate and strength, constricting blood vessels, accelerating the rate of breathing, widening respiratory passageways, increasing blood sugar level, increasing the rate of catabolism, and decreasing the rate of digestion.

If the stress responses of the alarm stage are grouped according to general functions, you will note that they are designed to rapidly increase circulation, promote catabolism for energy production, and decrease non-essential activities.

Resistance Reaction

The second stage in the stress responses is the *resistance reaction* (Figure 17-30b). Unlike the short-lived alarm reaction that is initiated by nervous impulses from the hypothalamus, the resistance reaction is initiated by regulating factors secreted by the hypothalamus and is a long-term reaction. The regulating factors are ACTHRF, HGHRF, and TRF.

ACTHRF stimulates the adenohypophysis to increase its secretion of ACTH. ACTH stimulates the adrenal cortex to secrete some of its hormones. The adrenal cortex is also indirectly stimulated by the alarm reaction. During the alarm reaction, activity of the kidney is reduced because of decreased blood circulation to the kidneys. Kidney function is not essential for meeting sudden danger. The resultant decrease in urine production stimulates the secretion of mineralocorticoids.

The mineralocorticoids secreted by the adrenal cortex promote the reabsorption of sodium ions by the kidney. A secondary effect of sodium reabsorption is the elimination of H^+ ions. The H^+ ions accumulate in the blood as a result of increased catabolism and tend to make the blood more acidic. Thus, during stress, a lowering of body pH is prevented. Sodium retention also leads to water retention—thus maintaining the high blood pressure that is typical of the alarm reaction. It also helps make up for fluid lost through severe bleeding.

The glucocorticoids, which are produced in high concentrations during stress, bring about the following reactions:

1. The glucocorticoids accelerate protein catabolism and the conversion of amino acids into glucose so that the body has a large supply of energy long after the immediate stores of glucose have been used up. The glucocorticoids also stimulate the removal of proteins from cell structures and stimulate the liver to break them down into amino acids. The amino acids can then be rebuilt into enzymes that are needed to catalyze the increased chemical activities of the cells or converted to glucose.
2. The glucocorticoids make blood vessels more sensitive to stimuli that bring about their constriction. This response counteracts a drop in blood pressure caused by bleeding.
3. The glucocorticoids also inhibit the production of fibroblasts, which develop into connective tissue cells. Injured fibroblasts release chemicals that play a role in stimulating the inflammatory response.

Thus the glucocorticoids reduce inflammation and prevent it from becoming disruptive rather than protective. Unfortunately, through their effect on fibroblasts, the glucocorticoids also discourage connective tissue formation. Wound healing is therefore slow during a prolonged resistance stage.

Two other regulating factors are secreted by the hypothalamus in response to a stressor: TRF and HGHRF. TRF causes the adenohypophysis to secrete TSH. HGHRF causes it to secrete HGH. TSH stimulates the thyroid to secrete thyroxine, which increases the catabolism of carbohydrates. HGH stimulates the catabolism of fats and the conversion of glycogen to glucose. The combined actions of TSH and HGH increase catabolism and thereby supply additional energy for the body.

The resistance stage of the general adaptation syndrome allows the body to continue fighting a stressor long after the effects of the alarm reaction have dissipated. It increases the rate at which life processes occur. It also provides the energy, functional proteins, and circulatory changes required for meeting emotional crises, performing strenuous tasks, fighting infection, or resisting the threat of bleeding to death. During the resistance stage, blood chemistry returns to nearly normal. The cells use glucose at the same rate it enters the bloodstream. Thus blood sugar level returns to normal. Blood pH is brought under control by the kidneys as they excrete more hydrogen ions. However, blood pressure remains abnormally high because the retention of water increases the volume of blood.

All of us are confronted by stressors from time to time, and we have all experienced the resistance stage. Generally, this stage is successful in seeing us through a stressful situation, and our bodies then return to normal. Occasionally, the resistance stage fails to combat the stressor, however, and the body "gives up." In this case, the general adaptation syndrome moves into the stage of exhaustion.

Exhaustion

One of the primary causes of exhaustion is loss of potassium ions. When the mineralocorticoids stimulate the kidey to retain sodium ions, potassium and hydrogen ions are exchanged for sodium ions and are secreted in the urine. As the chief positive ions within cells, potassium is partially responsible for controlling the water concentration of the cytoplasm. As the loss of potassium increases, cell function becomes less and less effective, until the cells finally start to die. This is called the *stage of exhaustion*. Unless the condition is rapidly reversed, vital organs cease functioning, and death occurs. Another cause of exhaustion is depletion of the adrenal cortical hormones. This, in turn, causes the blood glucose level to suddenly fall, and therefore the cells do not receive enough nutrients. A final cause is a weak organ. A long-term or strong resistance reaction puts heavy demands on the body, particularly on the heart, blood vessels, and adrenal cortex. They may not be up to handling the demands, or they may suddenly fail under the strain. In this respect, ability to handle stressors is determined to a large degree by previous health.

In summary, the general adaptation stress syndrome is a mechanism for adapting the body to severe stress. Unlike the homeostatic mechanisms, it changes the internal environment. In addition, the general adaptation stress syndrome is not a specific response to a specific stimulus. Instead, it produces wide-ranging changes in the body. The same set of changes occur, no matter what the particular stressor happens to be. Generally, this is advantageous, but occasionally it is not. For instance, just as fear of an immediate danger can produce the general adaptation stress syndrome, so can intense worry, tension, and resentment. Some investigators say that people who live under constant psychological stress seem to be in a continual state of mild resistance response. It has been suggested that this often causes chronic high blood pressure and contributes to heart disease by increasing wear and tear on the circulatory system.

HOMEOSTATIC IMBALANCES OF ENDOCRINE GLANDS

Disorders of the endocrine system, in general, are based upon underproduction or overproduction of hormones. The term **hyposecretion** describes an underproduction, whereas the term **hypersecretion** means an overproduction.

Disorders of the Adenohypophysis

The anterior pituitary gland produces many hormones. All these hormones, with the exception of the human growth hormone, directly control the activities of other endocrine glands. Therefore, it is hardly surprising that hyposecretion or hypersecretion of an anterior pituitary hormone produces widespread and complicated abnormalities.

Among the clinically interesting disorders related to the adenohypophysis are those involving human growth hormone. Human growth hormone builds up cells, particularly those of bone tissue. If the hormone is hyposecreted during the growth years, bone growth is slow, and the epiphyseal plates close before normal height is reached. This is the condition called **pituitary dwarfism.** Other organs of the body also fail to grow, and the pituitary dwarf is childlike in many physical respects. Treating the condition requires administration of human growth hormone during childhood, before the epiphyseal plates close.

If secretion of human growth hormone is normal during childhood, but lower than normal during adult life, a rare condition called **pituitary cachexia (Simmond's disease)** occurs. The tissues of a person with Simmond's disease waste away, or atrophy. The victim becomes quite thin and shows signs of premature aging. For instance, as the connective tissue degenerates, it loses its elasticity, and the skin hangs and becomes wrinkled. The atrophy occurs because the person is not receiving enough human growth hormone to stimulate the protein-building activities that are required for replacing cells and cell parts.

Hypersecretion of the human growth hormone produces completely different disorders. For example, hyperactivity during childhood years results in **giantism,** which is due primarily to an abnormal increase in the length of long bones. Hypersecretion during adulthood results in a condition called **acromegaly.** Since there is no possible further *lengthening* of the long bones, because the epiphyseal plates are already closed, in acromegaly, the bones of the hands, feet, cheeks, and jaws *thicken.* Other tissues also grow. For instance, the eyelids, lips, tongue, and nose enlarge, the skin thickens and furrows, especially on the forehead and soles of the feet.

Disorders of the Neurohypophysis

The principal abnormality associated with dysfunction of the neurohypophysis is **diabetes** *(passing through)* **insipidus** *(tasteless).* This disorder should not be confused with diabetes mellitus (sugar), which is a disorder of the pancreas and is characterized by sugar in the urine. Diabetes insipidus is the result of a hyposecretion of ADH, usually caused by damage to the neurohypophysis or to the hypothalamus. Symptoms of the disorder include excretion of large amounts of urine and subsequent thirst. Diabetes insipidus is treated by administering ADH.

Disorders of the Thyroid

Hyposecretion of thyroxine during the growth years results in a condition called **cretinism.** Two outstanding clinical symptoms of the cretin are dwarfism and mental retardation. The first is caused by failure of the skeleton to grow and mature. The second is caused by failure of the brain to develop fully. Recall that one of the functions of thyroid hormone is to control tissue growth and development. Cretins also exhibit retarded sexual development and a yellowish skin color. Flat pads of fat develop, giving the cretin the characteristic round face and thick nose; a large, thick, protruding tongue; and a protruding abdomen. Because the energy-producing metabolic reactions are so slow, the cretin has a low body temperature and demonstrates a continual lethargic behavior. Carbohydrates are stored rather than utilized. Heart rate is also slow. If the condition is diagnosed early, and therapy is promptly instituted with thyroid hormone, the condition can be reversed.

Hypothyroidism during the adult years produces a disorder called **myxedema.** The name refers to the fact that thyroxine is a diuretic. Lack of thyroxine causes the body to retain water. And one of the hallmarks of myxedema is an edema that causes the facial tissues to swell and look puffy. Another symptom caused by the retention of water is an increase in blood volume that frequently causes high blood pressure. Like the cretin, the person with myxedema also suffers from slow heart rate, low body temperature, muscular weakness, general lethargy, and a tendency to gain weight easily. The long-term combination of a slow heart rate and high blood pressure may overwork the heart muscles, causing the heart to enlarge. Because the brain has already reached maturity, the person with myxedema does not experience mental retardation. However, in moderately severe cases nerve reactivity may be dulled so that the person lacks mental alertness. Myxedema occurs eight times more frequently in females than in males. Its symptoms are abolished by the administration of thyroxine.

Hypersecretion of thyroxine gives rise to a condition called **exophthalmic goiter.** This disease, like myxedema, is also more frequent in females, affecting eight females to every one male. One of its primary symptoms is an enlarged thyroid, called a **goiter.** The two other characteristic symptoms are an edema behind the eye, which causes the eye to "pop out" (*exophthalmos*), and an abnormally high metabolic rate. The high metabolic rate produces a range of effects that are generally opposite to those of myxedema. The person has an increased pulse. The body temperature is high, and the skin is warm, moist, and flushed. Weight loss occurs and the person is usually full of "nervous" energy. The thyroxine also increases the responsiveness of the nervous system. Thus a person with exophthalmic goiter may become irritable and may exhibit tremors of the fingers when they are extended. The usual methods for treating hyperthyroidism are administering drugs that suppress thyroxine synthesis or surgically removing a part of the gland.

The term *goiter* simply means an enlargement of the thyroid gland, and it is a symptom of many thyroid disorders. Goiter may occur if the gland does not receive enough iodine to produce sufficient thyroxine for the body's needs. The follicular cells, which enlarge in a futile attempt to produce more thyroxine, secrete large quantities of colloid. This condition is called **simple goiter,** and it is most often caused by a lower than average amount of iodine in the diet. However, it may also develop if iodine intake is not increased during certain conditions that put a high demand on the body for thyroxine. Such conditions include: frequent exposure to cold, high-fat and high-protein diets.

Disorders of the Parathyroids

A constant level of Ca^{2+} in the body fluids is important to maintenance of the resting potential of excitable cells. **Hypoparathyroidism** is usually a result of damage incurred during thyroid surgery. Its signs are a drop in plasma Ca^{2+} from 10 mg percent to values of 6 or 7 mg percent and a rise in plasma phosphate. The most dramatic sign is **tetany,** which involves muscle spasms that are **tonic** (sustained contractions) and **clonic** (alternating contractions and relaxations). The muscles of the larynx can also constrict forcibly, causing death by asphyxiation unless immediate intervention occurs. Lowered levels of Ca^{2+} can be detected by clinical tests such as the Trousseau and Chvostek signs. Trousseau's sign is observed when the tightening of a blood pressure cuff around the upper arm produces contraction of the fingers and inability to open the hand. The Chvostek sign is a contracture of the facial muscles elicited by tapping the facial nerves located at the angle of the jaw.

Treatment of hypoparathyroidism would obviously be to introduce additional hormone. However, the hormone, which is difficult to obtain and is very expensive, has a short life span, and can be antigenic. Therefore, the usual treatment is to give large doses of vitamin D and put the patient on a high-calcium, low-phosphate diet.

Hyperparathyroidism is usually due to a tumor of the parathyroids. In this condition a continual secretion of hormone occurs even if the level of Ca^{2+} in the circulation is already high. The clinical signs include muscular weakness, slowing of the heart, kidney stones, and stomach ulcers due to stimulation of acid secretion. These symptoms result from the rise in the threshold level of excitable cells such as nerve and muscle cells. Therefore, depolarization is more difficult to achieve and fewer action potentials result. Demineralization of bone, a condition known as **osteitis fibrosa cystica,** also results because areas of destroyed bone tissue are replaced with fibrous tissue.

The bones thus become deformed and are highly susceptible to fracture. Treatment of hyperparathyroidism is to surgically remove the parathyroid tumor—a very difficult procedure due to the small size of the glands and their unpredictable location.

Disorders of the Adrenal Cortex

Hypersecretion of the **mineralocorticoid** aldosterone results in a decrease in the body's potassium concentration. Recall how important potassium movement is in the transmission of nerve impulses. Consequently, if potassium depletion is great enough, neurons cannot depolarize and muscular paralysis results. Hypersecretion of the adrenal cortex also brings about excessive retention of sodium and water. Water retention increases the volume of the blood and causes high blood pressure. It also increases the volume of the interstitial fluid, producing edema.

Disorders associated with **glucocorticoid** secretion include Addison's disease and Cushing's syndrome. Hyposecretion of glucocorticoids results in the condition called **Addison's disease.** Clinical symptoms include hypoglycemia, which leads to muscular weakness, mental lethargy, and weight loss. In addition, increased potassium blood levels and decreased sodium blood levels lead to low blood pressure and dehydration. **Cushing's syndrome** is a hypersecretion of glucocorticoids, especially hydrocortisone and cortisone. The condition is characterized by the redistribution of fat. This results in spindly legs accompanied by a characteristic "moon face," "buffalo hump" on the back, and pendulous abdomen. The facial skin is flushed and the skin covering the abdomen develops stretch marks. The individual also bruises easily, and wound healing is poor.

The **adrenogenital syndrome** results from overproduction of **sex hormones,** particularly the male androgens, by the adrenal cortex. Hypersecretion in male infants and young male children results in an enlarged penis. In young boys, it also causes premature development of male sexual characteristics. Hypersecretion in male adults is characterized by overgrowth of body hair, enlargement of the penis, and increased sexual drive. Hypersecretion in young girls results in premature sexual development. Hypersecretion in both girls and women usually produces a receding hairline, baldness, an increase in body hair, deepening of the voice, muscular arms and legs, small breasts, and an enlarged clitoris.

Disorders of the Adrenal Medulla

Tumors of the chromaffin cells of the adrenal medulla, called **pheochromocytomas,** cause hypersecretion of the medullary hormones. The oversecretion causes high blood pressure, high levels of sugar in the urine and blood, and elevated basal metabolic rate, nervousness, and sweating. Since the medullary hormones create the same effects as does sympathetic stimulation, hypersecretion puts the individual into a prolonged version of the "fight or flight" response. Needless to say, this eventually exhausts the body, and the individual eventually suffers from general weakness.

Disorders of the Pancreas

Hyposecretion of insulin produces a number of clinical symptoms referred to as **diabetes mellitus** (*diabe* = to pass through; *meli* = honey). The name of the illness suggests a sweet urine. Typically an inherited disease, it is caused by destruction or malfunction of the beta cells and a consequent reduction of available insulin. Therefore, blood glucose levels rise, sometimes reaching levels as high as 3000 mg of glucose per 100 ml of blood, (3000 mg percent). When blood glucose levels approach and surpass 300 mg percent, glucose begins to appear in the urine because the cells of the kidney tubules are unable to reabsorb excessive amounts. The glucose molecules remaining in the tubular fluid act as osmotically active particles and reduce the reabsorption of water so that large volumes of water, as well as glucose, are lost in the urine. Thus, the diabetic is deprived of energy that could be extracted from glucose, and a weight loss soon becomes apparent. The diabetic patient also becomes weak because of the consequent breakdown of protein and fat stores for energy purposes. This causes extreme hunger and thirst, so the diabetic eats and drinks large amounts.

Increased fat metabolism results in the production of fatty acid derivatives called *ketone bodies*. These molecules are acids, and when they accumulate in body fluids, the pH of extracellular fluids begins to drop. A dangerous *acidosis* results, and the situation can be disastrous unless counteracting measures are taken. Diabetic coma and death can be the tragic consequences.

Diabetes cannot be cured but can be controlled by insulin injections, diet, and certain supportive drugs. The life of a diabetic, though greatly prolonged because of control measures, is not free from secondary problems. With proper management, a person with symptomatic diabetes may remain free of serious complications. Secondary complications that may occur and require treatment include: (1) cholesterol deposition in blood vessels, leading to atherosclerosis; (2) kidney failure, producing hypertension; (3) minute aneurysms of capillaries on the retina of the eye, which may lead to blindness.

Exhibit 17-2 DIFFERENTIAL DIAGNOSIS OF COMA DUE TO DIABETIC ACIDOSIS AND HYPOGLYCEMIA

	Acidosis (due to too little insulin)	Hypoglycemia (due to too much insulin)
Onset	Hour to days	Minutes
Background events	Intercurrent disease, omission of insulin	Exercise, omission of meal
Symptoms	Thirst, polyuria (excessive urination), nausea, vomiting, abdominal pain	Hunger, headache, perspiration, confusion, stupor
Physical findings	Kussmaul respirations (labored breathing), dehydration, flushed face, fast pulse; person appears ill	Normal pulse, respiration; person appears well
Typical laboratory findings		
Urine		
Glucose	5 percent	0 to 5 percent
Ketones	Strongly positive	0 to positive
Serum		
Glucose	400 mg/100 ml. or more	Less than 40 mg/100 ml.
Ketones	Positive, 1:8	Negative
HCO_3^-	Less than 10 meq/liter	26 meq./liter
Response to 50 percent glucose, I.V.	None	Dramatic

Hyperinsulinism is much rarer than hyposecretion and can be the result of a tumor of the islet cells. The excessive release of insulin produces **hypoglycemia** (lower than normal levels of glucose in the blood). The most frequent cause of hypoglycemia seen by doctors is due to insulin overdose by diabetics. Whatever the cause, the symptoms of hypoglycemia follow a characteristic pattern. Mild hypoglycemia produces hunger, tremor, perspiration, weakness, blurred vision, and impaired mental function. Mental confusion often leads to bizarre behavior, and the diabetic patient may even resist assistance. Neuromuscular impairment occurs, and a resultant staggering walk and sometimes hostile behavior are misinterpreted as alcoholic drunkenness. If the hypoglycemia is severe, then epileptic type seizures can result. If the condition is untreated, permanent brain damage can occur and death sometimes results.

It is, of course, very important to be able to diagnose the cause of coma in a diabetic patient, since coma can result from too much insulin or too little insulin. Exhibit 17-2 makes the distinctions.

We shall now turn our attention from endocrine physiology to the closely related topic, reproductive physiology. In the next chapter, you will have an opportunity to study the close relationship between reproductive structures and hormones.

CHAPTER SUMMARY OUTLINE

1. The human body possesses two systems that function primarily as communicating and coordinating systems: the nervous system and the endocrine system.
2. The hypothalamus is the link between the two communicating networks.

ENDOCRINE GLANDS

1. Endocrine glands are ductless and secrete hormones directly into the blood.

2. Hormones are molecules that fall into one of three classes of compounds: steroids, proteins, or catecholamines.
3. Tropic hormones are hormones that stimulate one specific endocrine gland to secrete its hormone.
4. Organs or cells that respond to hormones are called the targets.
5. The anterior pituitary gland is called the master gland because it regulates the activities of several endocrine glands.
6. The hypothalamus receives information from all

parts of the nervous system and responds to a variety of stress situations.

MECHANISMS OF HORMONAL ACTION

1. Hormones bring about changes in the metabolic patterns of cells in a variety of ways.
2. They can activate certain suppressed genes, increase the rate of translation by ribosomes, enhance enzyme activity, and promote active transport across the cell membrane.

PITUITARY (HYPOPHYSIS)

1. This gland is differentiated into the adenohypophysis (the anterior lobe and glandular portion) and the neurohypophysis (the posterior lobe and nervous portion).

The Adenohypophysis

1. The adenohypophysis releases its hormones into the circulation upon receipt of signals from the hypothalamus.
2. The adenohypophysis secretes tropic hormones and gonadotropic hormones.
3. Hormones of the adenohypophysis are (a) human growth hormone (HGH), or somatotropin (STH), which regulates body growth and is controlled by HGHRF; (b) thyroid stimulating hormone (TSH), which controls the thyroid gland; (c) adrenocorticotropic hormone (ACTH), which regulates the activity of the cells in the cortex of the adrenal glands; (d) follicle stimulating hormone (FSH), which stimulates production of ova and spermatozoa in the reproductive organs; (e) luteinizing hormone (LH) in the female, or interstitial cell stimulating hormone (ICSH) in the male, which stimulate other sexual and reproductive activities; and finally (f) the melanocyte stimulating hormone (MSH), which regulates skin pigmentation.
4. Hypophysectomy is the removal of the pituitary gland, and the most obvious effect of its removal from young animals is decrease in the rate of growth.
5. Human growth hormone causes body growth by stimulating growth of those tissues that are able to grow.
6. The stimulation of protein synthesis by human growth hormone occurs at three points along the pathway of protein anabolism.
7. Human growth hormone elevates blood glucose, and this is called the diabetogenic effect.
8. The control of human growth hormone is intimately linked with the nutritional status of the individual.

The Neurohypophysis

1. Hormones of the neurohypophysis are oxytocin, which stimulates the contraction of the pregnant uterus and the ejection of milk, and ADH, which stimulates water reabsorption by the kidneys.

THYROID

1. This gland synthesizes thyroxine, triiodothyronine, and thyrocalcitonin.
2. The major function of thyroxine is to control the rate of cellular metabolism by regulating the catabolic or energy-releasing reactions.
3. Thyrocalcitonin lowers the amount of calcium in the blood by inhibiting bone breakdown and by accelerating the absorption of calcium by the bones.

PARATHYROIDS

1. Parathyroid hormone regulates the homeostasis of calcium and phosphate by stimulating osteoclasts in response to hypocalcemia.

ADRENALS (SUPRARENALS)

1. These glands consist of an outer cortex and inner medulla.
2. Cortical secretions include the mineralocorticoids, such as aldosterone, that regulate sodium reabsorption and potassium excretion; the glucocorticoids, which are essential to normal metabolism and resistance to stress; and the gonadocorticoids (male and female sex hormones).
3. Medullary secretions are norepinephrine and epinephrine. Both hormones are sympathomimetic, which means that they produce effects similar to those brought about by the sympathetic division of the autonomic nervous system.

PANCREAS

1. The pancreas has both an endocrine and an exocrine function.
2. Alpha cells of the pancreas secrete glucagon, which increases blood glucose level.
3. Beta cells secrete insulin, which decreases blood glucose level.

PINEAL

1. This gland secretes melatonin, which appears to affect the secretion of hormones by the ovaries.

2. It also secretes adrenoglomerulotropin, which may stimulate the adrenal cortex, and a third hormone called serotonin that is involved in normal brain physiology.

THYMUS

1. The thymus gland is related to the production of a special type of sensitized lymphocyte cell that is very important in cellular immunity.

STRESS AND HOMEOSTASIS

1. Stress is a condition of the body produced in response to extreme stimuli.
2. If the stress is extreme or unusual it triggers a wide-ranging set of bodily changes called the general adaptation stress syndrome.
3. Unlike the homeostatic mechanisms, this syndrome does not maintain a constant internal environment.
4. The stresses that produce the general adaptation stress syndrome are called stressors.
5. Stressors include surgical operations, poisons, infections, and strong emotional reactions.

Alarm Reaction

1. The alarm reaction, or "fight or flight" response is the initial reaction of the body to any stressor.
2. It is actually a complex of reactions initiated by the hypothalamic stimulation of the sympathetic nervous system and the adrenal medulla.
3. The responses are immediate but short-lived, and they are designed to counteract danger by mobilizing the body's resources for immediate physical activity.
4. They are designed to rapidly increase circulation, promote catabolism for energy production, and decrease nonessential activities.

Resistance Reaction

1. The second stage in the stress responses is the resistance reaction, and it is initiated by regulating factors secreted by the hypothalamus.
2. The regulating factors are ACTHRF, HGHRF, and TRF.
3. ACTHRF stimulates the adenohypophysis to increase its secretions of ACTH, which in turn stimulates the adrenal cortex to secrete hormones.
4. Resistance reactions are long-term and accelerate catabolism to provide energy to counteract stress.
5. Glucocorticoids are produced in high concentra-

tions during stress, and they creat[e] physiological effects.
6. The stage of exhaustion results from dramatic changes during alarm and resistance reactions.
7. If stress is too great, exhaustion may lead to death.

HOMEOSTATIC IMBALANCES OF ENDOCRINE GLANDS

Disorders of the Adenohypophysis

1. The term hyposecretion describes an underproduction, whereas the term hypersecretion means an overproduction.
2. Hyposecretion of the human growth hormone during the growth years results in pituitary dwarfism, while a lower than normal secretion during adult years results in pituitary cachexia.
3. Hypersecretion of the human growth hormone during childhood results in giantism, but during adulthood produces acromegaly.

Disorders of the Neurohypophysis

1. The principal abnormality associated with dysfunction of the neurohypophysis is diabetes insipidus.

Disorders of the Thyroid

1. Hyposecretion of thyroxine during the growth years results in a condition called cretinism, while an undersecretion during the adult years produces myxedema.
2. Hypersecretion results in a condition called exophthalmic goiter.

Disorders of the Parathyroids

1. Hypoparathyroidism produces tetany, which involves muscle spasms that are tonic and clonic.
2. Hyperparathyroidism results in the condition known as osteitis fibrosa cystica.

Disorders of the Adrenal Cortex

1. Hypersecretion of aldosterone produces muscular paralysis, retention of sodium and water, high blood pressure, and edema.
2. Hyposecretion of glucocorticoids results in Addison's disease, while hypersecretion of glucocorticoids produces Cushing's syndrome.
3. An overproduction of sex hormones results in many effects in males and females collectively called the adrenogenital syndrome.

Disorders of the Adrenal Medulla

1. Tumors of the adrenal medulla, called pheochromocytomas, cause hypersecretion of the medullary hormones, which produce many effects, including a prolonged version of the "fight or flight" response.

Disorders of the Pancreas

1. Hyposecretion of insulin causes a number of clinical symptoms referred to as diabetes mellitus, while hyperinsulinism results in hypoglycemia.

REVIEW QUESTIONS

1. Compare the nervous and endocrine systems as functional communicating and coordinating systems.
2. Define an endocrine gland. What is the relationship between an endocrine gland and a target organ?
3. What is a hormone? Distinguish between tropic and nontropic hormones.
4. Differentiate between the three classes of hormones, and give an example of each.
5. Describe the anatomical and physiological relationship between the pituitary gland and the hypothalamus. Correlate this relationship with endocrine glands and hormones.
6. Give a specific example of an endocrine gland regulated by a tropic hormone from the pituitary.
7. Give a specific example of an endocrine gland controlled by the level of the substance it seeks to regulate.
8. Differentiate between the four distinct mechanisms by which hormones change the metabolic patterns of cells.
9. In what respect is the pituitary gland actually two glands? Describe the histology of the adenohypophysis. Why does the anterior lobe of the gland have such an abundant blood supply?
10. What hormones are produced by the adenohypophysis? What are their functions?
11. Define hypophysectomy and list its effects on the bodies of young animals.
12. Describe the effects of human growth hormone on body growth. Correlate human growth hormone with protein anabolism.
13. What is the diabetogenic effect of growth hormone?
14. Discuss the histology of the neurohypophysis and the function and regulation of its hormones.
15. Describe the location and histology of the thyroid gland. How are the thyroid hormones made, stored, and secreted? Explain PBI.
16. What is the calorigenic effect of thyroxine? What is the function of thyrocalcitonin?
17. Where are the parathyroids located? What is their histology? What are the functions of the parathyroid hormone (PTH)? How is it regulated?
18. Compare the adrenal cortex and adrenal medulla with regard to location and histology.
19. Describe the renin-aldosterone pathway in detail.
20. Describe the hormones produced by the adrenal cortex in terms of type, normal function, and control.
21. What relationship does the adrenal medulla have to the autonomic nervous system? What is the action of adrenal medullary hormones?
22. Describe the location of the pancreas and the histology of the islets of Langerhans. What are the actions of glucagon and insulin?
23. Where is the pineal gland located? What are its assumed functions?
24. Describe the location of the thymus gland, and give its function.
25. Define the general adaptation stress syndrome. What is a stressor? Give examples of different kinds of stressors.
26. Outline the reactions of the body during the alarm reaction, resistance reaction, and stage of exhaustion when placed under stress. What is the central role of the hypothalamus during stress?
27. With the knowledge you have acquired in this chapter, respond to the following statement: "The combined activities of the nervous and endocrine systems are essential for maintaining homeostasis and overcoming stress."
28. Explain the disorders of the human growth hormone, comparing hyposecretion and hypersecretion in early childhood and adult life.
29. Explain the main abnormality associated with dysfunction of the neurohypophysis.
30. Describe the disorders of the thyroid gland. Define goiter.
31. Identify the main effects of abnormal secretion of the parathyroid hormone on calcium metabolism.
32. Identify the main disorders of the adrenal cortex, including the adrenogenital syndrome.
33. What is a pheochromocytoma? Explain its effects on the body.
34. Explain in detail diabetes mellitus and its management. What is hyperinsulinism?

Chapter 18

REPRODUCTIVE PHYSIOLOGY

STUDENT OBJECTIVES

After reading this chapter, you should be able to

- List the organs that comprise the male and female systems of reproduction
- Describe the role of the scrotum in protecting the testes
- Describe the testes as glands that produce sperm and the male hormone testosterone
- Trace the course of sperm cells through the system of ducts that lead from the testes to the exterior
- Contrast the functions of the seminal vesicles, prostate gland, and bulbourethral glands in secreting constituents of seminal fluid
- Describe the chemical composition of seminal fluid
- Describe the penis as the organ of copulation
- Describe the physiological effects of testosterone
- Explain the relationships between the male reproductive hormones, the pituitary gland, and the hypothalamus
- Describe the regulation of FSH and ICSH
- Describe the ovaries as glands that produce ova and female sex hormones
- Describe the function of the uterine tubes in transporting a fertilized ovum to the uterus
- Identify the anatomical portions of the uterus and its three layers of tissue
- Describe the role of the vagina in the menstrual flow and copulation
- Identify the components of the vulva
- Describe the structure and development of the mammary glands
- Describe the principal events of the menstrual and ovarian cycles
- Describe the physiological effects of estrogens and progesterone
- Correlate the activities of the menstrual and ovarian cycles
- Discuss the hormonal interactions that control the menstrual and ovarian cycles and menopause
- Describe the symptoms and causes of disorders of the reproductive systems including venereal diseases, prostate disorders, infertility, and menstrual abnormalities
- Describe ovarian cysts, endometriosis, and leukorrhea
- Discuss breast cancer and cervical cancer
- Describe special detection procedures such as mammography and thermography

Sex seems to have preoccupied human endeavor from the earliest recorded times, as evidenced by cave drawings and unearthed sculptures dating from prehistoric times. Most likely these were primitive attempts to support and promote the fertility of the tribe or clan. From these prehistoric eras many expressions of sexual organs have been discovered. For example, they have been found decorating pottery and weapons and they have been carved into amulets for good luck purposes.

The male sexual organs, very obvious because of their external location, were easier to comprehend than those of the female. The effects of castration, in both men and animals, were realized very early in human history. In some cultures, castration was used as punishment. In others it was used to provide "safe" guards for the harems of powerful rulers, and up until the late 1800s castration was also used to maintain the soprano voices of boys in male choirs.

Among the earliest of the scientific writings relating to sexual physiology were those of Aristotle. Though he never had a microscope he often referred to sperm as the "male seed." He described the female menstrual cycle in considerable detail and was accurate in his detailing of the course of pregnancy. The details of hormonal interactions were not understood, however, for centuries. Indeed it was only 60 to 70 years ago that the extracts containing reproductive hormones were utilized. Since then, the momentum of our discoveries has been constantly accelerating. Today not only do we have a great deal of knowledge about the actions and interactions of the sex hormones, but the mechanisms of control are also well understood. In-

deed, this knowledge forms the basis for understanding much about human behavior.

FUNCTIONAL ANATOMY OF THE MALE REPRODUCTIVE SYSTEM

The organs of the male reproductive system (Figure 18-1) are the testes, or male gonads, which produce sperm; a number of ducts that either store or transport sperm to the exterior; accessory glands that add secretions comprising the semen; and several supporting structures, including the penis.

Scrotum

The **scrotum** is a pouching of the abdominal wall consisting of loose skin and superficial fascia (see Figure 18-1) and is the supporting structure for the testes. Internally, the scrotum is divided by a septum into two sacs, each of which contains a single testis. The testes are the organs that produce sperm. Sperm production and survival require a temperature that is lower than body temperature. Because the scrotum is isolated from the body cavities, it is not as warm as the rest of the abdominal cavity and in fact it provides the testes with an environment about 3°F below body temperature. If the testes are too cold, a contraction of smooth muscle fibers elevates the testes closer to the abdomen where they can absorb body heat. This causes the skin of the scrotum to appear more wrinkled. Exposure to warmth reverses the process. Another muscle in the spermatic cord, the cremaster

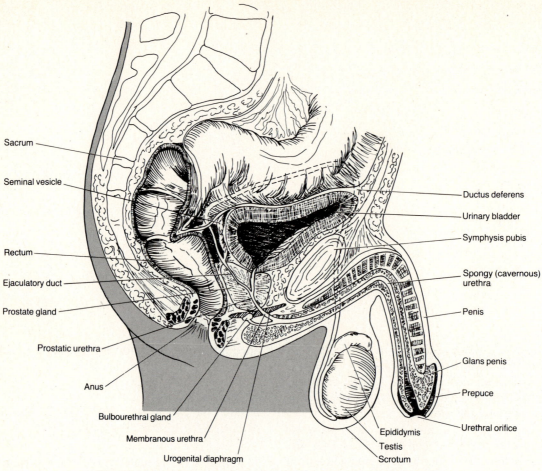

Sacrum

Seminal vesicle

Rectum

Ejaculatory duct

Prostate gland

Prostatic urethra

Anus

Bulbourethral gland

Membranous urethra

Urogenital diaphragm

Ductus deferens

Urinary bladder

Symphysis pubis

Spongy (cavernous) urethra

Penis

Glans penis

Prepuce

Urethral orifice

Epididymis

Testis

Scrotum

Figure 18-1   Male organs of reproduction seen in sagittal section.

muscle, also elevates the testes upon exposure to cold and upon sexual arousal.

The Testes

The **testes** are paired oval glands measuring about 5 cm in length and 2.5 cm in diameter. They weigh between 10 and 15 g (Figure 18-2a). During most of fetal life, they are located in the pelvic cavity, but about 2 months prior to birth they descend into the scrotum. If the testes do not descend, the condition is referred to as *cryptorchidism.* Cryptorchidism results in sterility because the sperm cells are destroyed by the higher body temperature of the pelvic cavity. Undescended testes can be placed in the scrotum prior to puberty with only slightly reduced fertility by administering hormones or by surgical means.

The testes are covered by a dense layer of white fibrous tissue, the *tunica albuginea,* that extends inward and divides each testis into a series of internal compartments called *lobules.* Each lobule contains one to three tightly coiled tubules, the *seminiferous tubules,* that produce the sperm by a process called *spermatogenesis.* A cross section through a semi-

niferous tubule reveals that it is packed with sperm cells in various stages of development (Figure 18-2b). The most immature cells, the *spermatogonia,* are located against the basement membrane. Moving toward the lumen in the center of the tube are layers of progressively more mature cells. In order of advancing maturity, these are called primary spermatocytes, secondary spermatocytes, and spermatids. By the time a *sperm cell,* or *spermatozoan,* has reached maturity, it is in the lumen of the tubule and begins to be moved through a series of ducts. Embedded between the developing sperm cells of the seminiferous tubules are *Sertoli cells* that probably produce secretions for supplying nutrients to the spermatozoa. Outside the seminiferous tubules are clusters of *interstitial cells of Leydig.* These cells secrete the male hormone *testosterone.* The functions of the testes are to produce sperm and testosterone.

Spermatozoa

Spermatozoa, once ejaculated, have a life expectancy of about 48 hours. A spermatozoan is highly adapted for reaching and penetrating a female ovum. It is composed of a head, a middle piece, and a tail (Figure

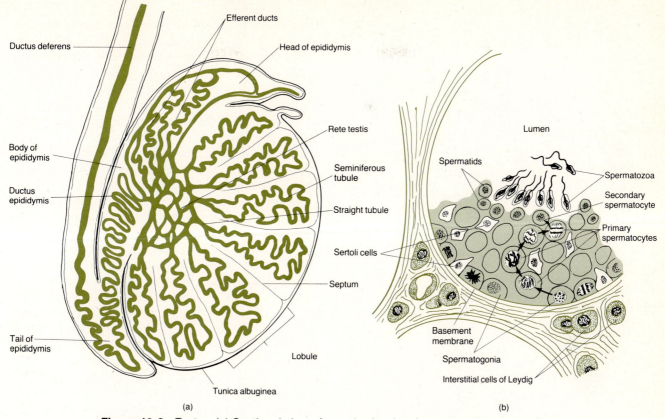

Figure 18-2 Testes. (a) Sectional view of a testis showing the system of ducts. (b) Cross section of a seminiferous tubule showing the stages of spermatogenesis.

18-3). The head contains the nuclear material and is capped by the acrosome, which contains chemicals that effect penetration of the sperm cell into the ovum. The middle piece contains numerous mitochondria, which carry on the catabolism that provides energy for locomotion. The tail, a typical flagellum, propels the sperm.

Ducts

When the sperm mature, they are moved through the seminiferous tubules to the **straight tubules** (see Figure 18-2a). The straight tubules form a network of ducts in the center of the testis called the **rete testis.** Some of the cells lining the rete testis possess cilia that probably help to push the sperm along. The sperm are next transported out of the testis through a series of coiled **efferent ducts** that empty into a single tube called the epididymis. At this point, the sperm are morphologically mature.

Epididymis

The **epididymis** is a highly coiled tube, measuring about 6 m (20 ft) in length and 1 mm in diameter, that lies tightly packed within the scrotum. An epididymis attaches to each testis, via the efferent ducts, and then descends along the posterior side of the testis, makes a loop, and then ascends (see Figures 18-1 and 18-2). Functionally, the epididymis is the site of sperm maturation. It stores spermatozoa in the tail portion in anticipation of ejaculation.

Vas Deferens

The terminal portion of the tail of the epididymis is less coiled and considerably thicker. At this point it is referred to as the **vas (ductus) deferens** (see Figure 18-2a). The vas deferens, about 45 cm long, ascends along the posterior border of the testis, penetrates the inguinal canal, and enters the abdomen, where it loops over the top and down over the posterior surface of the bladder. Peristaltic contractions of the muscular coat propel the spermatozoa toward the urethra during ejaculation.

Traveling alongside the vas deferens are the testicular artery, autonomic nerves, veins that drain the testes, lymphatics, and a small circular band of skeletal muscle called the *cremaster muscle.* These structures together constitute the *spermatic cord,* a supporting structure of the male reproductive system. The spermatic cord passes through the *inguinal canal,* a slitlike passageway in the anterior abdominal wall just superior to the medial half of the inguinal ligament.

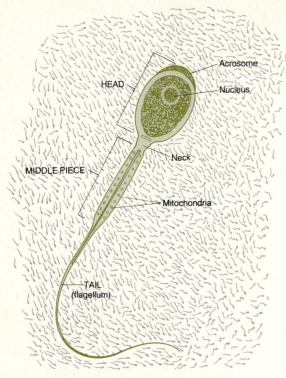

Figure 18-3 Parts of a spermatozoan.

The area of the inguinal canal and spermatic cord represents a weak spot in the abdominal wall, and it is frequently the site of a *hernia*. Very simply, a hernia is a rupture or separation of a portion of the abdominal wall resulting in the protrusion of a part of the viscera.

Ejaculatory Duct

Posterior to the urinary bladder, each vas deferens joins its **ejaculatory duct** (Figure 18-4). Each duct is about 2 cm long. Both ejaculatory ducts eject spermatozoa into the prostatic urethra. The urethra is the terminal duct of the system, serving as a common passageway for both spermatozoa and urine.

Urethra

In the male, the **urethra** passes through the prostate gland, the urogenital diaphragm, and the penis. This tube measures about 20 cm in length and is subdivided into three parts (see Figures 18-1 and 18-4). The *prostatic portion* is about 2 to 3 cm long and passes through the prostate gland. It continues inferiorly as the membranous portion as it passes through the urogenital diaphragm, a muscular partition between the two ischiopubic rami. The *membranous portion* is about 1 cm in length. After passing through the urogenital diaphragm, it is known as the *spongy (cavernous) portion* of the urethra. This portion is about 15 cm long. The spongy urethra enters the bulb of the penis and terminates at the external urethral orifice.

Accessory Glands

Whereas the ducts of the male reproductive system store and transport sperm cells, the **accessory glands** secrete the liquid portion of semen. The first of the accessory glands are the paired **seminal vesicles** (see Figure 18-4). These glands are convoluted pouchlike structures, about 5 cm in length, lying posterior to and at the base of the urinary bladder, in front of the rectum. They secrete an alkaline, viscous fluid, rich in the sugar fructose, into the ejaculatory duct. The seminal vesicles contribute about 60 percent of the volume of semen.

The **prostate gland** is a single, doughnut-shaped gland about the size of a chestnut (see Figure 18-4). It measures about 3.7 cm from side to side, 2.5 cm from front to back, and 3.5 cm in height. It is inferior to the urinary bladder and surrounds the upper portion of the urethra. The prostate secretes an alkaline fluid that constitutes about 13 to 33 percent of the semen into the prostatic urethra. Liquid semen coagulates rapidly due to a clotting enzyme synthesized in the prostate that acts on a substance produced in the seminal vesicle. This clot liquefies in a few minutes due to another enzyme produced by the prostate gland. It is not clear why semen coagulates and then liquefies. In older men, the prostate sometimes enlarges to the point where the "doughnut hole" portion narrows and compresses the urethra and obstructs urine flow. At this stage, surgical removal of part or the entire gland (prostatectomy) usually is indicated. The prostate gland is also a relatively common tumor site in older males.

The paired **bulbourethral** or **Cowper's glands** are about the size and shape of peas. They are located beneath the prostate on either side of the urethra (see Figure 18-4) and secrete an alkaline fluid into the spongy urethra.

Semen

Semen, or **seminal fluid,** is a mixture of sperm and the secretions of the seminal vesicles, the prostate gland, and the bulbourethral glands. The average volume of semen for each ejaculation is 2.5 to 6 ml, and the number of spermatozoa ejaculated is between 50 and 100 million spermatozoa/ml. If the number of spermatozoa per milliliter falls below 20 million, the male is likely to be physiologically sterile. Even though only a single spermatozoan fertilizes an ovum, fertilization requires the combined action of millions of spermatozoa. The ovum is enclosed by membranes and follicle cells that form a barrier against the sperm. The acrosomes of sperm secrete an enzyme called hyaluronidase, which is believed to dissolve inter-

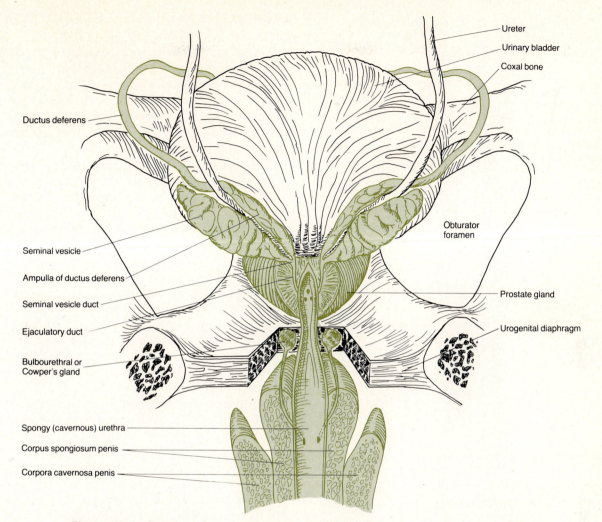

Figure 18-4 Relationships of some male reproductive organs as seen in a posterior view of the urinary bladder.

cellular materials of the cells covering the ovum, giving the sperm a passageway into the ovum. Apparently vast numbers of sperm are required to secrete an effective amount of the enzyme.

Semen has a pH range of 7.35 to 7.50; that is, it is slightly alkaline. The prostatic secretion gives semen a milky appearance, and fluids from the seminal vesicles and bulbourethral glands give it a mucoid consistency. Semen provides spermatozoa with a transportation medium and nutrients. It also acts as a buffer to neutralize the acid environment of the female reproductive system and the male urethra from urine. Semen also activates sperm with enzymes after ejaculation.

Penis

The **penis** is used to introduce spermatozoa into the female vagina (Figure 18-5). The distal end of the penis has a slightly enlarged region called the *glans*, which means shaped like an acorn. Covering the glans

is the loosely fitting skin, called the *prepuce* or *foreskin*. Internally, the penis is composed of three cylindrical masses of tissue bound together by fibrous tissue. The two dorsally located masses are called the *corpora cavernosa penis,* and the smaller ventral mass, the *corpus spongiosum penis,* contains the spongy urethra. All three masses of tissue are sponge-like and contain venous sinuses. Under the influence of sexual excitation, the arteries supplying the penis dilate and large quantities of blood enter the venous sinuses. Expansion of these spaces compresses the veins draining the penis so most entering blood is retained. These vascular changes result in an *erection*. The penis returns to its flaccid state when the arteries constrict and the pressure on the veins is relieved. During ejaculation, a smooth muscle sphincter at the base of the urinary bladder is closed due to the higher pressure in the urethra caused by expansion of the corpus spongiosum penis. Thus, urine is not expelled during ejaculation and semen does not enter the urinary bladder.

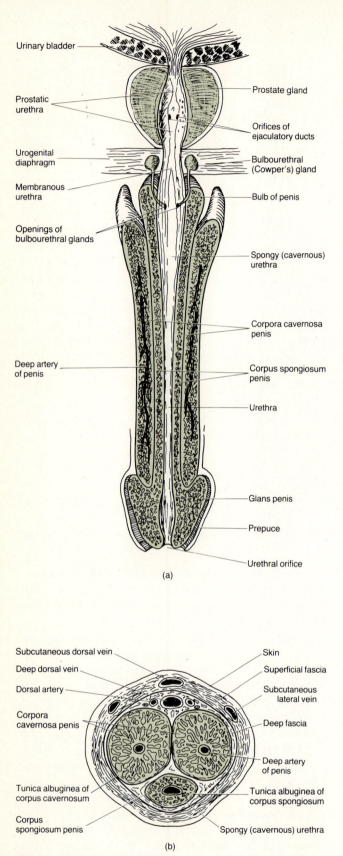

(a)

(b)

Figure 18-5 Internal structure of the penis. (a) View from above. (b) Cross section.

ANDROGENS—THE MALE SEX HORMONES

Androgens are the male sex hormones secreted primarily by the interstitial cells of Leydig. These cells, which make up about 20 percent of the weight of the testes, are located outside the seminiferous tubules and receive a very rich blood supply. Androgens are also produced in the adrenal cortex, but the amount secreted there is normally physiologically insignificant. Only if a tumor of the androgen synthesizing cells occurs does the quantity reach levels to exert physiological effects such as masculinizing changes in women.

The androgenic hormones are steroid in nature, and the structures of three are shown in Figure 18-6. The most important of these steroid hormones is testosterone, which accounts for almost 90 percent of the total circulating male hormones.

Testosterone is carried in the circulation bound to plasma proteins. From there, either it is bound to receptor molecules within responsive cells or it is metabolized by the liver to an inactive steroid such as androsterone, which is then attached to sugars or sulfates and excreted from the body.

Testosterone that is bound to an intracellular carrier is converted to a much more potent steroid molecule, *dihydrotestosterone* (Figure 18-6). This new complex of receptor-hormone enters the nucleus of the responsive cell where it derepresses (activates) certain genes for transcription into new messenger RNA molecules. These messengers are then translated into new cellular enzymes that enable the cell to

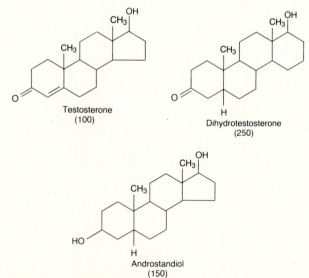

Figure 18-6 Androgens. The relative biologic potency of each is shown in parentheses.

manifest the correct sexual expression. The male androgenic hormones cause changes in most tissues of the body. In fact, their influence begins during fetal development, when they stimulate differentiation of the hypothalamus and external genitalia. Prior to birth, the testes descend from their location in the pelvis into the scrotum. The initiation of the process requires elevated levels of testosterone at that time. During the teen years, testosterone is important in promoting muscular development and rapid growth in height. It also stimulates the development, growth, and maintenance of the male sex organs, including the penis, scrotum, prostate, and seminal vesicles. Normal sexual behavior and the development of secondary male sex characteristics also result from androgen stimulation. The secondary sex characteristics, in addition to muscular development, include body hair patterns such as pubic hair, axillary and chest hair (within hereditary limits), facial hair, and temporal hairline recession; and enlargement of the thyroid cartilage, producing deepening of the voice.

REGULATION OF THE MALE REPRODUCTIVE HORMONES

Normal reproductive functions are the result of a very delicate interplay between the hypothalamus, the anterior pituitary, and the gonads. In the male, this relationship is often referred to as the **hypothalamo-pituitary-tubular axis.** The central position of the pituitary was shown by classical experiments involving removal of the gland. In young animals a failure of the gonads to mature results, and in adults the gonads regress and eventually atrophy. Extracts prepared from pituitaries have been shown to restore and ameliorate both conditions. Two active principles or gonadotropins have been isolated from the pituitary: (1) follicle stimulating hormone (FSH) and (2) interstitial cell stimulating hormone (ICSH). The latter hormone is identical to luteinizing hormone (LH) in the female. Both pituitary hormones are glycoproteins with molecular weights of about 25,000 for FSH and 29,000 for ICSH.

The activation of the testes to secrete the androgenic hormones normally occurs at about the age of 14 plus or minus 3 years. Precocious development prior to 10 years of age has been observed. This period of time when the hormonal changes begin is a gradual process in the male, and the period of **puberty** begins with the first ejaculate that contains mature sperm. The continual sequence of changes that follow, including the gradual molding of body shape and size, the development of the secondary sex characteristics, and personality and psychological changes, constitute a period of life called **adolescence.**

The initiating process lies within the functional properties of the hypothalamus. It presently appears that some stage of "maturation" is reached within the hypothalamus and it begins to secrete regulating factors into the hypophyseal portal system. Regulating factors for FSH and ICSH are called follicle stimulating hormone releasing factor (FSHRF) and interstitial cell stimulating hormone releasing factor (ICSHRF), respectively. They stimulate the anterior pituitary to release FSH and ICSH into the systemic circulation. In the male, the FSH acts on the seminiferous tubules to stimulate and maintain sperma-

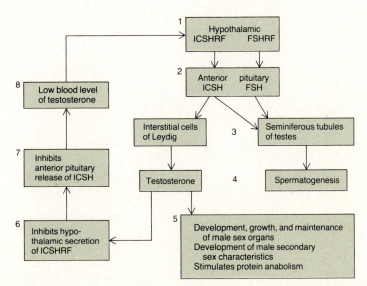

Figure 18-7   Secretion, physiological effects, and inhibition of testosterone.

togenesis (Figure 18-7). For normal sperm maturation, testosterone is also required. The action of ICSH is to stimulate the interstitial cells of Leydig to produce the androgenic hormone testosterone.

Regulation of FSH and ICSH

The interaction of the anterior pituitary hormone ICSH and testosterone illustrates again the operation of negative feedback mechanisms (see Figure 18-7). ICSH stimulates the production of testosterone, but once the testosterone concentration in the blood is increased to a certain level, it inhibits the release of ICSHRF by the hypothalamus. This inhibition, in turn, inhibits the release of ICSH by the anterior pituitary. Thus, testosterone production is decreased. A low blood level of ICSH puts another set of reactions into operation. Low blood levels of testosterone, in turn, stimulate the release of ICSHRF by the hypothalamus, which stimulates the anterior pituitary secretion of ICSH. As the blood level of ICSH increases, the testes are stimulated to produce testosterone. Thus, the stimulatory-inhibitory cycle is complete.

The level of circulating testosterone has no feed-back effect on the release of FSH. In some yet unclear manner, the rate of spermatogenesis in the testes controls the level of FSH. If spermatogenesis begins to slow or falter, there is an increase in the amount of FSH released by the anterior pituitary. Some investigators suggest that the active seminiferous tubules may release a substance that inhibits FSH.

FEMALE REPRODUCTIVE SYSTEM

The female organs of reproduction (Figure 18-8) include the ovaries, which produce ova; the uterine tubes (fallopian tubes), which transport the ova to the uterus (womb); the vagina; and external organs that constitute the vulva or pudendum. The mammary glands, or breasts, also are considered part of the female reproductive system, even though anatomically they are part of the integument.

Ovaries

The **ovaries,** or female gonads, are paired glands resembling unshelled almonds in size and shape. They are about 3 cm long, 2 cm wide, and 1 cm thick. They

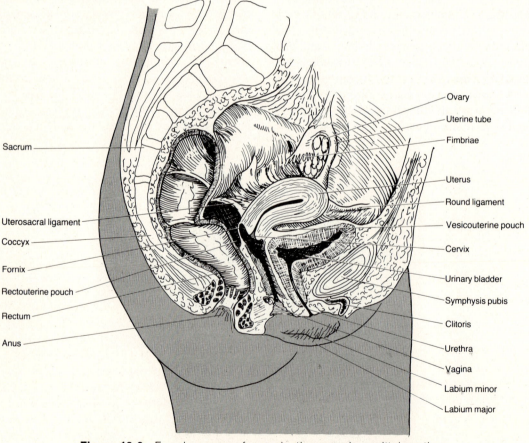

Figure 18-8 Female organs of reproduction seen in sagittal section.

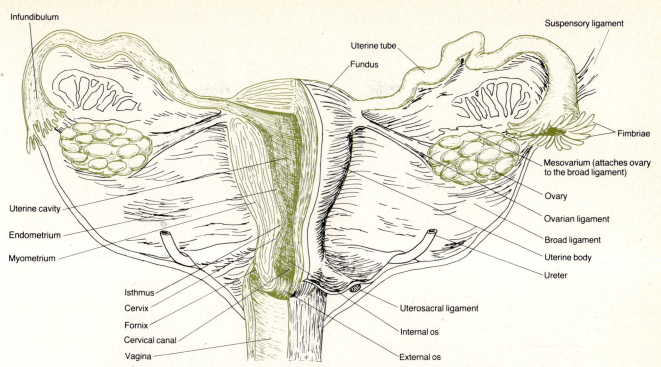

Figure 18-9 Uterus and associated female reproductive structures. The left side of the figure has been sectioned to show internal structures.

are positioned in the upper pelvic cavity, one on each side of the uterus (Figure 18-9).

In microscopic view, it can be seen that each ovary consists of the following parts (Figure 18-10):

1. **Germinal epithelium.** This is a layer of simple cuboidal epithelium that covers the free surface of the ovary.
2. **Tunica albuginea.** This is a capsule of collagenous connective tissue immediately deep to the germinal epithelium.
3. **Stroma.** This is a region of connective tissue deep to the tunica albuginea. The stroma is composed of an outer, more dense layer called the *cortex* and an inner, looser layer known as the *medulla*. The cortex contains ovarian follicles.
4. **Primary ovarian follicles.** Each of these consists of an oocyte surrounded by a single layer of flattened epithelium-like cells. These progress through various stages of development and mature into ova.
5. **Graafian follicle.** This is a mature ovum and its surrounding tissues. It secretes estrogens.
6. **Corpus luteum.** This is a glandular body developed from a Graafian follicle after extrusion of an ovum (ovulation). The corpus luteum produces the hormone progesterone.

The ovaries produce ova and secrete the female sexual hormones. The ovaries are analogous to the testes of the male.

Uterine Tubes

The female body contains two **uterine** or **fallopian tubes,** which transport the ova from the ovaries to the uterus (see Figures 18-8 and 18-9). Measuring about 10 cm long, each tube has a funnel-shaped open end called the *infundibulum.* This end, located very close to the ovary but not attached to it, is surrounded by a fringe of fingerlike projections called *fimbriae.* From the infundibulum the tube extends inward and downward and attaches to the upper side of the uterus.

About once a month, a mature ovum is released from the surface of one ovary near the infundibulum of the uterine tube—a process called **ovulation.** The ovum adheres to the surface of the ovary until it is swept off by the ciliary action of the epithelium of the infundibulum into the uterine tube. The ovum is then moved along by ciliary action supplemented by the wavelike contractions of the smooth muscle of the uterine tube. Fertilization, if it occurs, takes place in the upper one-third of the uterine tubes. This may occur at any time up to 24 hours following ovulation. The ovum, fertilized or unfertilized, descends into the uterus within 3 to 4 days. Sometimes an ovum is fertilized outside the uterine tube and is not drawn into the uterine tubes but instead becomes implanted in the pelvic cavity. Pelvic implantations fail because the developing fertilized ovum does not make satisfactory vascular connections with the maternal blood supply. Also, on occasion, a fertilized ovum fails to descend

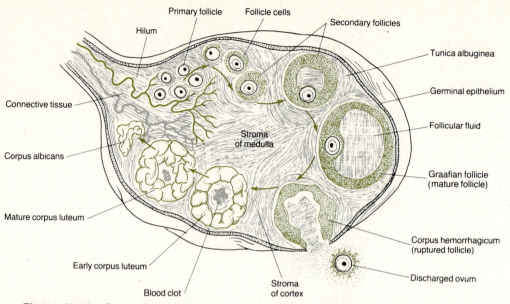

Figure 18-10 Ovary. Parts of the ovary seen in sectional view. The arrows indicate the sequence of developmental stages that occur as part of the ovarian cycle.

to the uterus. Instead it implants in a uterine tube and then signals this unusual condition by causing severe pain. In this instance, the pregnancy must be terminated surgically before the tube ruptures. Both pelvic and tubular implantations are referred to as *ectopic pregnancies*.

Uterus

The organ of the female reproductive system that assumes a role in menstruation, implantation of a fertilized ovum, development of the fetus during pregnancy, and labor is the **uterus.** Situated between the bladder and the rectum, the uterus is an inverted-pear-shaped organ (see Figure 18-9). Before the first pregnancy, the adult uterus measures approximately 7.5 cm long, 5 cm wide, and 1.75 cm thick. Anatomical subdivisions of the uterus include the dome-shaped portion above the uterine tubes called the *fundus,* the major, tapering, central portion called the *body,* and the inferior, narrow portion called the *cervix.* Between the body and the cervix is a constricted region about 1 cm long, the *isthmus.* The interior of the body of the uterus is called the *uterine cavity,* and the interior of the narrow cervix is called the *cervical canal.* The junction of the uterine cavity with the cervical canal is the *internal os;* the *external os* is the place where the cervix opens into the vagina.

Normally the uterus is flexed between the uterine body and the cervix (see Figure 18-8). In this position, the body of uterus projects forward and slightly upward over the urinary bladder. The cervix projects

downward and backward, joining the vagina at nearly a right angle.

Histologically, the uterus consists of three layers of tissue (Figure 18-11). The outer layer, derived from the peritoneum, is referred to as the *serous layer* or *serosa.* It is part of the parietal peritoneum.

The middle layer of the uterus, the *myometrium,* forms the bulk of the uterine wall. This layer consists of smooth muscle fibers and is thickest in the fundus and thinnest in the cervix. During childbirth, coordinated contractions of the muscles dilate the cervix and help to expel the fetus from the body of the uterus.

The inner layer of the uterus, the *endometrium,* is a mucous membrane composed of two principal layers. The *stratum functionalis* is the layer closer to the uterine cavity and is the layer shed during menstruation from the fundus and body. The other layer, the *stratum basalis,* is maintained during menstruation and produces a new functionalis following menstruation. The endometrium contains numerous glands.

Blood is supplied to the uterus by branches of the internal iliac artery called *uterine arteries.* Branches of the uterine arteries, called *arcuate arteries,* are arranged in a circular fashion underneath the serosa and give off branches, called *radial arteries,* that penetrate the myometrium (Figure 18-11). Just before these branches enter the endometrium, they divide into two kinds of arterioles. One, the *straight arteriole,* terminates in the basalis and supplies it with the materials necessary to regenerate the functionalis. The other branch, the *spiral arteriole,* penetrates the functionalis and changes markedly during the men-

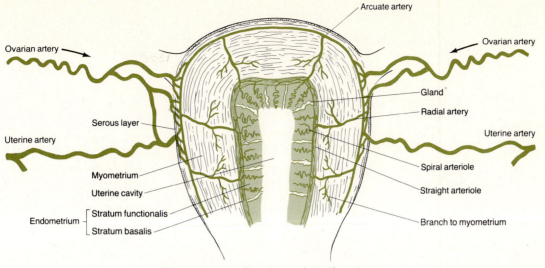

Figure 18-11 Blood supply to the uterus.

strual cycle. The uterus is drained by the uterine veins.

Vagina

The **vagina** serves as a passageway for the menstrual flow; as the receptacle for the penis during coitus, or sexual intercourse; and as the lower portion of the birth canal. It is a muscular, tubular organ lined with mucous membrane, and it measures about 10 cm in length (Figure 18-8 and 18-9). Situated between the bladder and the rectum, it is directed upward and backward, where it attaches to the uterus. A recess called the *fornix* surrounds the vaginal attachment to the cervix. The dorsal recess, called the posterior fornix, is larger than the ventral and two lateral fornices. The fornices make possible the use of contraceptive diaphragms. The mucosa of the vagina consists of stratified squamous epithelium overlying connective tissue that is arranged in a series of transverse folds, the *rugae*. The vagina is capable of distension. The muscle layer is composed of smooth muscle that can stretch considerably. This distension is important because the vagina receives the penis during sexual intercourse and also serves as the lower portion of the birth canal. At the lower end of the vaginal opening (*vaginal orifice*) is a thin fold of vascularized mucous membrane called the *hymen*, which forms a border around the orifice, partially closing it (Figure 18-12). Sometimes the hymen completely covers the orifice, a condition called *imperforate hymen*, and surgery is required to open the orifice to permit the discharge of the menstrual flow. The mucosa of the vagina contains large amounts of glycogen that, upon decomposition, produce organic acids. These acids create a low-pH environment in the vagina, which retards microbial growth. However, the acidity is also injurious to sperm cells. For this reason, the buffering action of semen is important. Semen neutralizes the acidity of the vagina to ensure survival of the sperm.

Vulva

The term **vulva** or **pudendum** is a collective designation for the external genitalia of the female (Figure 18-12).

The *mons pubis (veneris)* is an elevation of adipose tissue covered by coarse pubic hair situated over the symphysis pubis. It lies in front of the vaginal and urethral openings. From the mons pubis, two longitudinal folds of skin, the *labia majora*, extend downward and backward. The labia majora, the female homologue of the scrotum, contain an abundance of adipose tissue and sebaceous and sweat glands; they are covered by hair on their upper outer surfaces. Medial to the labia major are two folds of skin called the *labia minora*. Unlike the labia majora, the labia minora are devoid of hair and have relatively few sweat glands. They do, however, contain numerous sebaceous glands.

The *clitoris* is a small, cylindrical mass of erectile tissue and nerves. It is located just behind the junction of the labia minora. A layer of skin called the *prepuce, or foreskin*, is formed at the point where the labia minora unite, and it covers the body of the clitoris. The exposed portion of the clitoris is referred to as the *glans*. The clitoris is homologous to the penis of the male, in that it is capable of enlargement upon tactile stimulation and assumes a role in sexual excitement of the female.

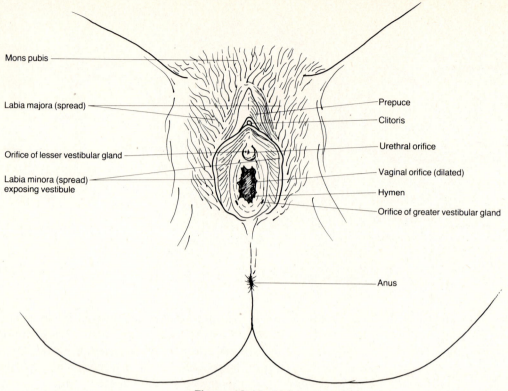

Mons pubis

Labia majora (spread)

Orifice of lesser vestibular gland

Labia minora (spread) exposing vestibule

Prepuce

Clitoris

Urethral orifice

Vaginal orifice (dilated)

Hymen

Orifice of greater vestibular gland

Anus

Figure 18-12   Vulva.

The cleft between the labia minora is called the *vestibule*. Within the vestibule are the hymen, vaginal orifice, urethral orifice, and the openings of several ducts. The vaginal orifice occupies the greater portion of the vestibule and is bordered by the hymen. In front of the vaginal orifice and behind the clitoris is the *urethral orifice*. Behind and to either side of the urethral orifice are the openings of the ducts of the *lesser vestibular* or *Skene's glands*. These glands secrete mucus and are the female homologue of the prostate gland. On either side of the vaginal orifice itself are two small glands, the *greater vestibular* or *Bartholin's glands*. These glands open by a duct into the space between the hymen and labia minora and produce a mucoid secretion that serves to supplement lubrication during sexual intercourse. Whereas the lesser vestibular glands are homologous to the male prostate, the greater vestibular glands are homologous to the male bulbourethral or Cowper's glands.

Mammary Glands

The **mammary glands** are branched tubuloalveolar glands that lie over the pectoralis major muscles and are attached to them by a layer of connective tissue (Figure 18-13). Internally, each mammary gland consists of 15 to 20 *lobes,* or compartments, separated by adipose tissue. The amount of adipose tissue present

is the principal determinant of the size of the breasts. However, the size of the breasts has nothing to do with the amount of milk they produce. Within each lobe are several smaller compartments, called *lobules,* that are composed of connective tissue in which milk-secreting cells, referred to as *alveoli,* are embedded. Between the lobules are strands of connective tissue called the *suspensory ligaments of Cooper.* These ligaments run between the skin and deep fascia and support the breast. Alveoli are arranged in grape-like clusters. They convey the milk into a series of *secondary tubules.* From here, the milk passes into the *mammary ducts.* As the mammary ducts approach the nipple, they expand into sinuses called *ampullae,* where milk may be stored. The ampullae continue as *lactiferous ducts* that terminate in the *nipple.* Each lactiferous duct conveys milk from one of the lobes to the exterior, although some ducts may join before reaching the surface. The circular pigmented area of skin surrounding the nipple is called the *areola.* It appears rough because it contains modified sebaceous glands.

At birth, both male and female mammary glands are undeveloped and appear as slight elevations on the chest. With the onset of puberty, the female breasts begin to develop—the mammary ducts elongate, extensive fat deposition occurs, and the areola and nipple grow and become pigmented. These changes

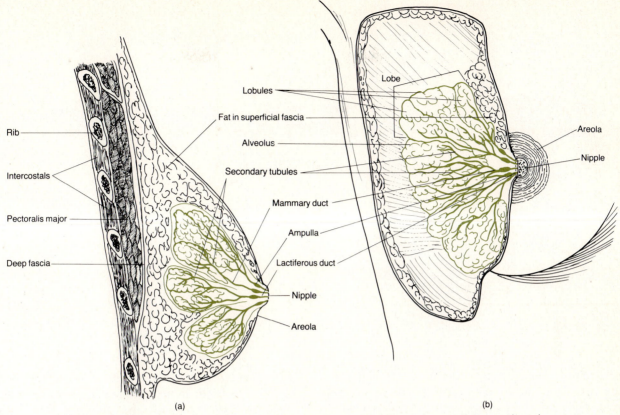

Figure 18-13 Mammary glands. (a) Sagittal section. (b) Front view partially sectioned.

are correlated with an increased output of estrogens by the ovary. Further mammary development occurs at sexual maturity with the onset of ovulation and the formation of the corpus luteum. During adolescence, lobules are formed and fat deposition continues, increasing the size of the glands. Although these changes are associated with estrogens and progestins secreted by the ovaries, remember ovarian secretion is ultimately controlled by FSH. The essential function of the mammary glands is milk secretion or lactation.

Endocrine Relations: Ovarian and Menstrual Cycles

The principal events of the menstrual cycle can be correlated with those of the ovarian cycle and changes in the endometrium. These are hormonally controlled events.

The term **menstrual cycle** refers to a series of changes that occur in the endometrium of a nonpregnant female. Each month the endometrium is prepared to receive a fertilized ovum. An implanted ovum eventually develops into a fetus and normally remains in the uterus until delivery. If no fertilization occurs, a portion of the endometrium is shed. The

ovarian cycle is a monthly series of events associated with the maturation of an ovum.

Gonadotropic hormones (FSH and LH) of the anterior pituitary gland initiate the menstrual cycle, ovarian cycle, and other changes associated with puberty in the female (Figure 18-14). The release of the gonadotropic hormones is controlled by regulating factors from the hypothalamus called the follicle stimulating hormone releasing factor (FSHRF) and the luteinizing hormone releasing factor (LHRF). FSH acts on the ovarian follicles and stimulates the secretion of estrogens by the follicles. Another anterior pituitary hormone, the luteinizing hormone (LH), acts to stimulate the further growth of ovarian follicles, brings about ovulation, and stimulates production of estrogens and progesterone by the follicle. The female sex hormones, estrogens and progesterone, affect the body in different ways. **Estrogens** have four main functions. First is the development and maintenance of female reproductive organs, especially the endometrium, secondary sex characteristics, and the breasts. The secondary sex characteristics include fat distribution to the breasts, abdomen, and hips; voice pitch; and hair pattern. Second, they control fluid and electrolyte balance. Third, they increase protein anabolism, and in this regard, estrogens are syner-

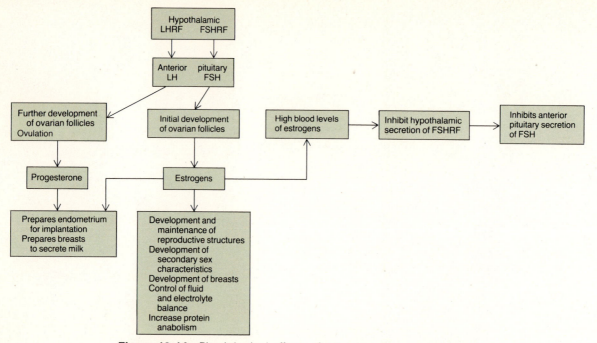

Figure 18-14 Physiological effects of estrogens and progesterone.

gistic with human growth hormone (HGH). Fourth, they may help cause an increase in the female sex drive. High levels of estrogens in the blood inhibit the secretion of FSH by the anterior pituitary gland. **Progesterone,** the other female sex hormone, works with estrogens to prepare the endometrium for implantation and to prepare the breasts for milk secretion.

Phases of the Menstrual and Ovarian Cycles

The duration of the menstrual cycle is variable among different females, normally ranging from 24 to 35 days. For purposes of our discussion, we shall assume that the average duration is 28 days. Events occurring during the menstrual cycle may be divided into three phases: (1) the menstrual phase, (2) the preovulatory or proliferative phase, and (3) the postovulatory or secretory phase (Figure 18-15).

The Menstrual Phase. The **menstrual phase,** also called **menstruation** or **the menses,** is the periodic discharge of about 25 to 65 ml of blood, tissue fluid, mucus, and epithelial cells. It lasts approximately for the first 5 days of the cycle. The discharge is associated with endometrial changes in which the functionalis layer degenerates and patchy areas of bleeding develop. The spiral arterioles of the functionalis layer constrict and shut off the flow of blood to the endometrium. As a consequence, the tissue begins to die and small bleeding areas result. These small areas of the functionalis detach one at a time (total detachment would result in hemorrhage), the uterine glands discharge their contents and collapse, and tissue fluid is discharged. The menstrual flow passes from the uterine cavity to the cervix, through the vagina, and ultimately to the exterior. Generally the flow terminates by the fifth day of the cycle. At this time the entire functionalis is shed, and the endometrium is very thin because only the basalis remains.

During the menstrual phase, the ovarian cycle is also in operation. Ovarian follicles, called *primary follicles,* begin their development (see Figure 18-10). At birth, each ovary contains about 400,000 such follicles, each consisting of an ovum surrounded by a layer of cells. During the early part of each menstrual phase, four or five primary follicles start to grow and differentiate. A clear membrane, the zona pellucida, develops around the ova in 4 to 5 days. The primary follicles develop into *secondary follicles* and start to produce low levels of estrogens. As the cells of the surrounding layer increase in number, they differentiate and secrete a fluid called the follicular fluid. When this fluid forces the ovum to the edge of the follicle, it becomes known as a tertiary or Graafian follicle. The production of estrogens by the secondary follicle elevates the estrogen level of the blood slightly. Ovarian follicle development is the result of FSH and LH production by the anterior pituitary, and during this part of the ovarian cycle, FSH secretion is maximal, while LH secretion is low. Although a num-

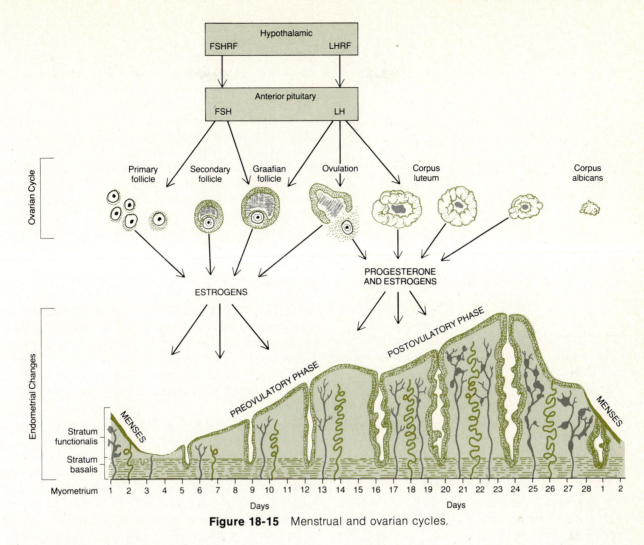

Figure 18-15   Menstrual and ovarian cycles.

ber of follicles begin development each cycle, only one will mature into a Graafian follicle.

The Preovulatory Phase. The **preovulatory phase,** the second phase of the menstrual cycle, is the period between menstruation and ovulation. This phase of the menstrual cycle is more variable in length than are the other phases. It lasts from day 6 to day 13 in a 28-day cycle. During the preovulatory phase, usually only one of the secondary follicles in the ovary matures into a *Graafian follicle,* a follicle ready for ovulation. The remaining follicles shrink and regress into a nonviable structure and become known as *atretic follicles.* During the maturation phase, the one Graafian follicle increases its production of estrogens. This increase of estrogens results in a *positive feedback* response that causes the pituitary to release a large surge of luteinizing hormone (LH). Luteinizing hormone acting on the growing follicular cells causes them to further increase secretion of estrogens. This

reinforcing cycle raises the levels of both hormones rapidly.

Though at least six estrogenic hormones have been isolated from the plasma of human females, only three, *β-estradiol, estrone,* and *estriol* are present in physiologically significant levels (Figure 18-16). Of these three, the β-estradiol has an estrogenic potency about 10 times that of estrone and about 100 times that of estriol, so in effect, β-estradiol is *the* estrogenic hormone.

During the phase of rising levels of estrogens, the hormones promote the repair of the endometrium. Cells of the stratum basalis undergo mitosis and produce a new stratum functionalis. As the endometrium thickens, the endometrial glands are short and straight, and the arterioles coil and lengthen as they penetrate the functionalis. Because the proliferation of endometrial cells occurs during the preovulatory phase, the phase is also referred to as the *proliferative phase.* Still another name for this phase is the *follicular phase*

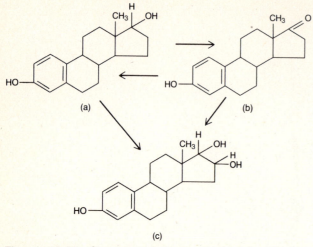

Figure 18-16 Structure of three principal female hormones. (a) β-estradiol. (b) Estrone. (c) Estriol.

because of the increasing secretion of estrogens by the developing follicle. Functionally, estrogens are the dominant ovarian hormones during this phase of the menstrual cycle.

Ovulation. Ovulation begins on day 14 of the 28-day cycle. About 18 hours before ovulation, there is a sixfold to tenfold increase in the plasma concentration of LH, accompanied by a twofold rise in the plasma FSH. The sudden elevation in the ratio of these hormones is the apparent trigger for ovulation. Also at this time, small amounts of progesterone are secreted from the cells of the Graafian follicle—probably initiated by the rising LH levels. The location of the Graafian follicle at this time is obvious, causing a protrusion on the surface of the ovary. The follicular surface ruptures, releasing the follicular fluid which floats the ovum into the abdominal cavity. The ovum is still surrounded by supporting layers of cells that form the so-called *corona radiata.* Following ovulation, the Graafian follicle collapses, and blood within it forms a clot called the *corpus hemorrhagicum.* The clot is eventually absorbed by the remaining follicular cells. In time, the follicular cells enlarge, change character, and form the *corpus luteum,* or *yellow body.*

The Postovulatory Phase. The **postovulatory phase** of the menstrual cycle is fairly constant in duration and lasts from day 15 to day 28 in a 28-day cycle. It is the period between ovulation and the onset of the next menses. Following ovulation, the level of estrogens in the blood drops slightly, and LH secretion stimulates the development of the corpus luteum. The corpus luteum then secretes increasing quantities of estrogens and progesterone, the latter being responsible

for the preparation of the endometrium to receive a fertilized ovum. Preparatory activities include the filling of the endometrial glands with secretions that cause the glands to appear tortuously coiled, vascularization of the superficial endometrium, thickening of the endometrium, and an increase in the amount of tissue fluid. These preparatory changes are maximal about 1 week after ovulation, and they correspond to the anticipated arrival of the fertilized ovum. During the postovulatory phase, FSH secretion again gradually increases and LH secretion gradually decreases. The functionally dominant ovarian hormone during this phase is progesterone.

If fertilization and implantation do not occur, the rising levels of progesterone and estrogens inhibit LH secretion, and as a result the corpus luteum degenerates and becomes the *corpus albicans.* The decreased secretion of progesterone and estrogens by the degenerating corpus luteum then initiates another menstrual period. In addition, the decreased levels of progesterone and estrogens in the blood bring about a new output of the anterior pituitary hormones, especially FSH, and a new ovarian cycle is initiated. A summary of these hormonal interactions is presented in Figure 18-17.

If, however, fertilization and implantation do occur, the corpus luteum is maintained for about 3 months during which time it continues to supply larger amounts of estrogens and progesterone. During this period, the maintenance of the corpus luteum gradually becomes the responsibility of a new hormone called *human chorionic gonadotropin (HCG),* which is produced by the trophoblastic cells of the forming placenta. HCG begins to appear in increased amounts around the fourth week of pregnancy and reaches and maintains a high level from about day 40 to day 80 (Figure 18-18). The level then drops to a maintenance level for the rest of the pregnant period. HCG is a glycoprotein molecule that structurally and physiologically resembles luteinizing hormone. It functions to maintain the corpus luteum, which not only secretes estrogens and progesterone, but also supports the continual growth of the uterus. When the placenta is completely formed, it assumes the role of producing the estrogens and progesterone, and the corpus luteum no longer plays a vital role in the continuation of the pregnancy. If the fetus is male, the HCG stimulates the interstitial cells of the fetal testes to begin secreting their androgenic hormones. These hormones in turn cause the proper development of the male sex organs. HCG is also used clinically to bring about the descent of testes in cases of cryptorchidism.

If the ovum is not fertilized, the corpus luteum

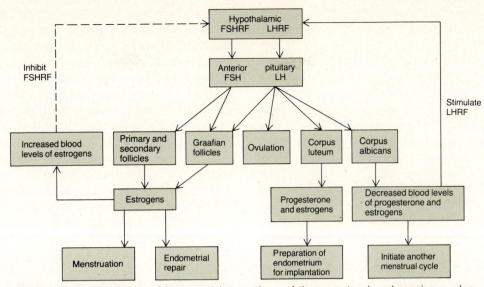

Figure 18-17 Summary of hormonal interactions of the menstrual and ovarian cycles.

begins to regress. Since a placenta is not begun then there is no HCG to maintain the corpus luteum. The level of LH from the anterior pituitary is also insufficient to support the corpus luteum beyond a few days. There is evidence that when the levels of estrogens and progesterone reach a certain point, they inhibit the further release of LH from the anterior pituitary. The regression of the corpus luteum brings about a marked drop in the circulating levels of estrogens and progesterone, which then deprives the endometrium of its supportive stimuli. As a consequence, the endometrial glands collapse and regress, the blood flow through the capillaries markedly decreases, and

ischemia of tissue results. Necrosis of the tissue follows, and as a result of cellular death, an anticoagulant substance is released into the tissue spaces. The spiral arteries constrict and this reinforces the ischemia. Then, individually, the arterioles relax, and the return of blood into the damaged vessels causes them to burst, producing menstrual bleeding. The blood vessels again constrict, and the functionalis layer of the endometrium is slowly but completely shed over a 5-day period.

Concurrently with the drop in levels of estrogens and progesterone, the inhibitory feedback on the anterior pituitary is removed and the anterior pituitary begins to secrete larger amounts of FSH, and the cycle begins again.

The menstrual cycle normally occurs once each month from *menarche* (the first menses) to *menopause* (the last menses). Just before menopause, there is a period of time lasting from several months to several years during which the sexual cycles become irregular and gradually stop. This is called the *female climacteric*. Its onset typically occurs between 40 and 50 years of age, and results from the failure of the ovaries to respond to the stimulation of gonadotropic hormones from the anterior pituitary.

In this period, the female must readjust her life from one that has been physiologically stimulated by estrogens and progesterone to one devoid of these feminizing hormones. The secretion of estrogens decreases rapidly, and essentially no progesterone is secreted after the last ovulatory cycle. The loss of the estrogens often causes marked physiologic changes in the function of the body. These changes include: (1) "hot flashes" characterized by extreme flushing of the

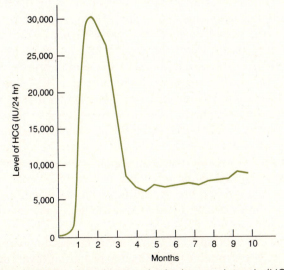

Figure 18-18 Mean human chorionic gonadotropin (HCG) levels during normal pregnancy. The time scale begins from the first day of the last period.

skin, (2) psychic sensations of dyspnea, (3) irritability, (4) fatigue, (5) anxiety, and (6) occasionally various psychotic states.

The cause of menopause (cessation of the sexual cycles) apparently is "burning out" of the ovaries. The pituitary is not involved, as it is with puberty. FSH is produced in great quantities but for some reason the ovaries fail to respond. Throughout a woman's sexual life, many of the primary ovarian follicles grow into Graafian follicles with each sexual cycle. Eventually most of the follicles have either degenerated or ovulated. Therefore, at the age of approximately 45, greatly reduced numbers of primary follicles still remain to be stimulated by FSH and LH. Therefore, the production of estrogens by the ovary decreases. In the postmenopausal woman there is some degree of atrophy of the ovaries, uterine tubes, uterus, vagina, external genitalia, and breasts.

Many specialists regard menopause as an estrogen deficiency disorder that can be treated. They believe treatment may relieve the discomforting symptoms of irritability and hot flashes; prevent osteoporosis (softening of the bones, which is a normal occurrence in postmenopausal women); and lessen the future possibility of arteriosclerosis, hypertension, or heart disease. Before menopause, it appears, women are less likely than men to suffer hypertension or heart disease. However, after menopause, women develop these diseases at roughly the same rate as men. It is believed that the normal balance of female hormones before menopause acts as a protection.

Many doctors prescribe estrogens to replace the deficiency. For practical purposes, estrogens are often given for several months—or several years—and the results are reviewed periodically. All agree, however, that estrogens should be given cautiously to normal individuals who have been properly screened and are carefully followed up.

HOMEOSTATIC IMBALANCES OF THE REPRODUCTIVE SYSTEM

In discussing the disorders that occur in the reproductive systems, we shall consider venereal diseases first. Then we shall look at some common diseases involving the male and female reproductive tracts, respectively.

Venereal Diseases

The term *venereal* comes from Venus, the goddess of love. **Venereal diseases** are a group of infectious diseases that are spread primarily through sexual inter-

course. With the exception of the common cold, venereal diseases are ranked as the number one communicable diseases in the United States. Gonorrhea and syphilis are the two most common diseases.

Gonorrhea, more commonly known as "clap," is an infectious disease that primarily affects the mucous membrane of the urogenital tract, the rectum, and occasionally the eyes. The disease is caused by the bacterium *Neisseria gonorrhoeae*. Symptoms in the male include a yellowish, puslike discharge from the urethra—usually within a week after contraction. It is accompanied by a burning sensation during urination. Fibrosis sometimes occurs in an advanced stage of gonorrhea, causing stricture of the urethra. There also may be involvement of the epididymis and prostate gland. In females, infection may occur in the urethra, vagina, and cervix, and there may be a discharge of pus. If untreated, the infection may spread to the uterus and uterine tubes. If the condition remains untreated and develops into a chronic infection, scar tissue may form and close off the tubes, leading to infertility. Females commonly can harbor the disease and be asymptomatic.

Discharges from the involved mucous membranes are the source of infection, and the bacteria are transmitted by sexual contact. The bacteria may be transmitted to the eyes of a newborn when the baby passes through the birth canal. If so, the infant initially develops conjunctivitis, but later, other structures of the eye may become involved and blindness may result. Administration of a 1 percent silver nitrate solution or penicillin in the eyes of the infant is very effective in preventing their infection. Penicillin is also the drug of choice for the treatment of gonorrhea in adults.

Syphilis is an infectious disease caused by the bacterium *Treponema pallidum*. It also is acquired through sexual contact. The early stages of the disease primarily affect the organs that are most likely to have made sexual contact—the genital organs, the mouth, and the rectum. The point where the bacteria enter the body is marked by a hard, round ulcer with raised edges called a *chancre* (pronounced shank-er). In males, it usually occurs on the penis, and in females, it usually occurs in the vagina or cervix. The chancre, which is usually painless, heals without scarring. Following the initial infection, the bacteria enter the bloodstream and are spread throughout the body. In some individuals, an active secondary stage of the disease occurs. It is characterized by a skin rash, which soon disappears, and sometimes is accompanied by vague symptoms such as a sore throat, fever, and muscle or joint pain. The signs of the secondary stage also go away without medical treatment. During the next several years, the disease progresses without symptoms and is said to be in a *latent phase*. When symptoms again appear, anywhere from 5 to 40 years after the initial infection, the person is said to be in the tertiary stage of the disease. Tertiary or late syphilis occurs in about one-third of the untreated patients, apparently because they are unable to generate a sufficient immune response. As a result, certain long-term destructive changes progress through the years involving the circulatory system, skin, bones, viscera, and central nervous system. The underlying destructive mechanism arises from inflammation of the small arteries, which then results in an obstructive coagulation and necrosis of the supplied tissue. The skin and mucous membranes eventually show gross nodular, ulcerating lesions. Pain in the bones is also a common symptom. Involvement of the central nervous system also is manifested as arterioles become inflamed and obstructed. Partial paralysis, seizures, loss of position sense, impotence, incontinence, vision impairment, delusions, hallucinations, and general paralysis that progresses rapidly are all possible symptoms of the tertiary syphilitic patient. Without medical treatment, the patient dies. Treatment with penicillin even during the tertiary stages is beneficial, but it does not cure.

Syphilis can also be acquired congenitally. A fetus can become infected *in utero* since the bacterium can cross the placenta. This is the basis in many states for the mandatory blood tests prior to issuance of a marriage license. Treatment of the mother with penicillin during the first months of pregnancy can prevent congenital syphilis. Treatment of congenital syphilis in early infancy with penicillin can alleviate many of the potential problems.

Another venereal disease is **genital herpes.** The sexual transmission of herpes virus is well established. Increased awareness by both patient and physician, better techniques of diagnosis of all venereal diseases, and a true rise in the incidence has brought herpes genitalis into the spotlight. Unlike syphilis and gonorrhea, it is uncurable. In one study, genital herpes was seen seven times more frequently than primary syphilis. Herpes simplex I is the virus that causes the typical lip cold sore, and herpes simplex II causes painful genital blisters, sometimes high up in the vaginal vault. The blisters cyclically disappear and reappear, but the disease itself remains present in the body.

Genital herpes virus infection causes a considerable amount of discomfort in both sexes, and there is an extraordinarily high rate of recurrent disease. Women who have genital herpes during the last trimester of pregnancy may infect the newborn, which could cause its death. The disease has been linked recently to cervical cancer. There is no cure for the disease, but discomfort can be alleviated with medication, rest, and other therapy. For pregnant women with genital herpes, a cesarean section will usually prevent complications in the child.

Herpes seems to be more common and severe in uncircumcised men. However, we do not know whether this has any influence on the etiology of cancer in these men or in their sex partners. Medical people are very concerned about the possibility of cancer in chronic recurring viral infections.

Treatment involves symptomatic therapy including pain medication, saline compresses to the intact bullae or open ulcers, and abstinence for the duration of the eruption.

The microorganism *Trichomonas vaginalis* causes **trichomoniasis.** Symptoms include a vaginal discharge and severe vaginal itch in women. Men can have it without symptoms, but can transmit it to women. Both partners must be treated simultaneously.

Chlamydial infections can cause very painful urination in men, while many women have no symptoms at all. It is caused by an organism midway between a virus and bacterium. In women who may not know they are carriers, it can cause serious pelvic disease, arthritis, and heart trouble, and during birth, the organism may enter the baby's eyes, causing conjunctivitis. Routine use of drugs in the eyes of newborns cures it, and tetracycline is the main treatment for men and women.

Nonspecific urethritis, or NSU, also known as **nongonococcal urethritis,** is inflammation of the urethra. Chlamydias are known to cause many cases, but others stem from other organisms. It occurs in both men and women. The urethra swells and narrows, impeding the flow of urine. Both urination and the urgency to urinate increase. Urination is accompanied by burning pain, and there may be a purulent discharge. In men, there is usually an associated infection of the prostate and adjacent structures. Untreated, the disease may seriously damage the male or female urethra along its entire length. Urethritis usually responds to treatment with antibiotics or sulfonamides.

The fungus *Candida albicans* causes **candidiasis,** sometimes called **vaginal moniliasis.** This infection is common in women, especially diabetic women. The fungus may respond to treatment and, like the *Trichomonas* organism, may lodge in the male prostate gland.

Male Disorders

Disorders of the male reproductive system involve the prostate gland and testes or else are related to sexual malfunctions. The latter will be examined in detail in the next chapter, which deals with the physiology of sex.

The prostate gland is susceptible to infection, enlargement, and benign and malignant tumors. Because the prostate surrounds the urethra, any of these disorders can cause obstruction to the flow of urine. Prolonged obstruction also may result in serious changes in the bladder, ureters, and kidneys.

Acute and chronic infections of the prostate gland are common in postpubescent males, many times in association with inflammation of the urethra. In **acute prostatitis** the prostate gland becomes swollen and very tender. Appropriate antibiotic therapy, bed rest, and above-normal fluid intake are effective in treatment.

Chronic prostatitis is one of the most common chronic infections in men of the middle and later years. On examination, the prostate gland feels enlarged, soft, and extremely tender. The surface outline is irregular and may be hard. This disease frequently produces no symptoms, but the prostate is believed to harbor infectious microorganisms responsible for some allergic conditions, arthritis, and inflammations of nerves (neuritis), muscles (myositis), and the iris (iritis).

An **enlarged prostate** gland occurs in approximately one-third of all males over 60 years of age. The enlarged gland is two to four times larger than normal. The cause is unknown, and the enlarged condition usually can be detected by rectal examination.

Tumors of the male reproductive system primarily involve the prostate gland. Carcinoma of the prostate is the second leading cause of death from cancer in men in the United States. It is responsible for approximately 19,000 deaths annually. Its incidence is related to age, race, occupation, geography, and ethnic origin. Both benign and malignant growths are common in elderly men. Both types of tumors put pressure on the urethra, making urination painful and difficult. At times, the excessive back pressure destroys kidney tissue and gives rise to increased susceptibility to infection. Therefore, even if the tumor is benign, surgery is indicated to remove the prostate or parts of it if the tumor is obstructive and perpetuates urinary tract infections.

Infertility, or **sterility,** is an inability to fertilize the ovum and does not imply impotence. Male fertility requires viable spermatozoa, adequate production of spermatozoa by the testes, unobstructed transportation of sperm through the seminal tract, and satisfactory deposition within the vagina. The tubules of the testes are sensitive to many factors—x-rays, infections, toxins, malnutrition, and others—that may cause degenerative changes and produce male sterility.

If inadequate spermatozoa production is suspected, a sperm analysis should be performed. Analysis includes measuring the volume of semen, counting the number of sperm per milliliter, evaluating sperm motility at 4 hours after ejaculation, and determining the percentage of abnormal sperm forms (not to exceed 20 percent).

Female Disorders

Common disorders of the female reproductive system include menstrual abnormalities, ovarian cysts, leukorrhea, infertility, breast tumors, and cervical cancer.

Abnormalities of Menstruation

Disorders of the female reproductive system frequently include menstrual disorders. This is hardly surprising because proper menstruation reflects not only the health of the uterus, but the health of the glands that control it, that is the ovaries and the pituitary gland.

Amenorrhea is the absence of menstruation in a woman. If the woman has never menstruated, the condition is called *primary amenorrhea.* Primary amenorrhea can be caused by endocrine disorders, most often in the pituitary gland and hypothalamus, or by genetically caused abnormal development of the ovaries or uterus. *Secondary amenorrhea* is cessation of uterine bleeding in women who have previously menstruated. The first cause considered is pregnancy. If that is ruled out, various endocrine disturbances are considered.

Dysmenorrhea is painful menstruation caused by contractions of the uterine muscles. A primary cause is believed to be low levels of progesterone. Recall that progesterone prevents uterine contractions. It can also be caused by pelvic inflammatory disease, uterine tumors, cystic ovaries, or congenital defects.

Abnormal uterine bleeding includes menstruation of excessive duration and/or excessive amount, too-frequent menstruation, intermenstrual bleeding, and postmenopausal bleeding. These abnormalities may be caused by disordered hormonal regulation, emotional factors, and systemic diseases.

Ovarian Cysts

Ovarian cysts are tumors of the ovary that contain fluid. Follicular cysts may occur in the ovaries of elderly women, in ovaries that have inflammatory diseases, and in menstruating females. They have thin walls and contain a serous albuminous material. Cysts may also arise from the corpus luteum or the endometrium.

Endometriosis

Endometriosis occurs when the endometrial tissue is found growing any place other than inside the uterus. Remember that the endometrium is the inner lining of the uterus that is sloughed off in menstruation. The tissue may be found in any of a dozen possible sites, including on the ovaries, cervix, abdominal wall, and bladder. Causes are unknown. Endometriosis is common in women 30 to 40 years of age. Symptoms include premenstrual or unusual menstrual pain. If a women who has endometriosis plans to have children, she should do so as soon as possible because the disorder can cause infertility. Treatment usually consists of hormone therapy or surgery. Endometriosis does disappear by itself at menopause or when the ovaries are removed.

Leukorrhea

Leukorrhea is a nonbloody vaginal discharge that may occur at any age and affects most women at some time. It is not a disease; it is a symptom of infection or congestion of some portion of the reproductive tract. It may be a normal discharge in some women. If it is evidence of an infection, it may be caused by a protozoan microorganism called *Trichomonas vaginalis,* or a yeast, a virus, or a bacterium.

Diseases of the Breasts

The breasts are highly susceptible to cysts and tumors. Men are also susceptible to breast tumor, but certain breast cancers are 100 times more common in women than in men. Usually these growths can be detected early by the woman who inspects and palpates her breasts regularly. To palpate means to feel or examine by touch. Unfortunately, so few women practice periodic self-examination that many growths are discovered by accident and often too late for proper treatment.

In the female, the benign **fibroadenoma** is a common tumor of the breast. It occurs most frequently in young women. Fibroadenomas have a firm, rubbery consistency and are easily moved about within the mammary tissue. The usual treatment is excision of the growth. The breast itself is not removed.

Breast cancer has the highest fatality rate of all cancers affecting women, but it is rare in men. In the female, breast cancer is rarely seen before age 30 and its occurrence rises rapidly after menopause.

Breast cancer is generally not painful until it becomes quite advanced. Often it is not discovered early, or if it is noted, it is ignored. Any lump, be it ever so small, should be reported to a doctor at once. If there is no evidence of *metastasis* (the spread of cancer cells from one part of the body to another or from one organ to another), the treatment of choice could be a *lumpectomy,* to remove a localized tumor, or a *modified* or *radical mastectomy.* A radical mastectomy involves removal of the affected breast, along with the underlying pectoral muscles and the axillary lymph nodes. Metastasis of cancerous cells is usually through the lymphatics or blood. Radiation treatments may follow the surgery to ensure the destruction of any remaining stray cancer cells.

The mortality from breast cancer has not improved significantly in the last 50 years. Early detection is still the most promising method to increase the survival rate, and two methods of detection are breast self-examination and mammography.

Examination of the breasts is an essential part of the physical examination of all women since 95 percent of breast cancer is first detected by the women themselves. It deserves more than a casual approach, and whether it is done by the woman or a physician, the breasts should be completely and thoroughly examined for lumps, puckering of the skin, or discharge. Self-examination should be done each month after the menstrual period.

Mammography is a sophisticated breast x-ray technique that is being used increasingly for breast cancer detection purposes. Breast cancers can be detected by mammography even before axillary lymph nodes have been affected. This means that with mammography, women may have a better than average prognosis. The purpose of mammography is to detect breast masses and to determine whether they are malignant. The examination consists of two x-ray, right-angle views of each breast. It is a painless, brief examination.

Mammographic diagnoses of breast masses have proven to be 80 percent reliable. In addition to their role in breast cancer detection, mammographic x-ray prints are used as a guide to surgeons performing mastectomies. As an aid in analyzing mammographic findings, a relatively new x-ray image processing technique, *xeroradiography,* is being used. This is a photoelectric (rather than photochemical) method in which the x-ray is reproduced on paper instead of film. Xeroradiography provides excellent soft-tissue detail and requires less radiation than film mammography.

Modern x-ray films, xeroradiography, and special x-ray machines have reduced the problem of mammographic radiation to a minimum. Ovaries are not exposed to radiation during mammography, and the technique can be used safely on pregnant women.

Most cancer experts agree that mammography should be used as follows:

1. To evaluate breast complaints, especially pain, and to check questionable masses.
2. When there is a bloody or serous discharge from the nipple.
3. Where there is a family history of breast cancer.
4. To evaluate the opposite breast in patients with known breast cancer, at the time of surgery and as a follow-up.
5. To reassure the cancerphobic patient.
6. To search for the primary malignancy in patients with peripheral metastases or those having axillary nodes but no palpable breast mass.

Mammography should be used only after a careful clinical examination. *Thermography,* a method of measuring and graphically recording heat radiation emitted by the breast, is also frequently used in conjunction with mammography. This is important because tumors, both benign and malignant, emit more heat than nonaffected areas.

Cervical Cancer

Another common disorder of the female reproductive system is cancer of the uterine cervix. It ranks third in frequency after breast and skin cancers. **Cervical cancer** starts with a change in the shape of the cervical cells called *cervical dysplasia.* Cervical dysplasia is not a cancer in itself, but the abnormal cells tend to become malignant if untreated.

In theory, invasive cancer of the cervix is a preventable disease since early neoplastic changes in the epithelium of the cervix can be recognized and eradicated in almost 100 percent of women. Nevertheless, approximately 8,000 women in the United States will die this year of this type of malignancy. Not too many years ago the deaths from cervical cancer in this country numbered 13,000 each year.

Cervical cancer, for the most part, is a venereal disease with a long incubation period. It occurs primarily in women who have had sexual intercourse. The early onset of sexual activity and promiscuity are factors that predispose to cervical cancer. Inciting factors are not as yet known, but herpes virus type II has recently become suspect. Smegma* and the DNA of spermatozoa have also been implicated. Cancer of the cervix (except for adenocarcinoma) does not occur in celibate women, and for unknown reasons, it is rare in Jewish women.

The death rate from this type of malignancy is steadily declining, due in considerable part to early diagnosis and treatment. Deaths from cancer of the cervix would be virtually eliminated if: (1) every woman were to report regularly for a pelvic examination, (2) every physician were to properly examine and sample the cervix for cytologic and histologic findings, (3) every cytopathologist were to properly interpret the findings, and (4) appropriate treatment were then administered.

Early diagnosis of cancer of the uterus is accomplished by the *Papanicolaou test,* or "Pap" smear. In this generally painless procedure, a few cells from the vaginal fornix (that part of the vagina surrounding the cervix) and the cervix are removed with a swab and examined microscopically. Malignant cells have a characteristic appearance and are indicative of an early stage of cancer, even before any symptoms occur. Estimates indicate that the "Pap" smear is more than 90 percent reliable in detecting cancer of the cervix. Treatment of cervical cancer may involve complete or partial removal of the uterus, called a hysterectomy, or radiation treatments.

---

* Smegma is a thick, cheesy, ill-smelling secretion, consisting chiefly of desquamated epithelial cells, found under the prepuce and around the labia minora.

CHAPTER SUMMARY OUTLINE

MALE REPRODUCTIVE SYSTEM

1. The scrotum is a pouching of the abdominal wall that provides an appropriate temperature for the testes.
2. The major functions of the testes are sperm production and the secretion of testosterone.
3. Sperm cells are conveyed from the testes to the exterior through the convoluted seminiferous tubules, straight tubules, rete testis, efferent ducts, ductus epididymis, ductus deferens, ejaculatory duct, and urethra.
4. The seminal vesicles, prostate, and bulbourethral glands secrete the liquid portion of semen.
5. Semen is a mixture of sperm and secreted liquids.
6. The penis serves as the organ of copulation.

ANDROGENS—THE MALE SEX HORMONES

1. Androgens, the male sex hormones, are secreted primarily by the interstitial cells of Leydig.
2. The most important androgen is testosterone, which accounts for almost 90 percent of the total circulating male hormones.
3. Testosterone promotes muscular development, rapid growth in height, and the development, growth, and maintenance of the male reproductive organs.
4. Normal reproductive functions are the result of a very delicate interplay between the hypothalamus, the anterior pituitary, and the gonads.
5. FSH initiates spermatogenesis, and ICSH stimulates the secretion of testosterone.
6. The hypothalamus secretes regulating factors that control the secretion of FSH and ICSH.

FEMALE REPRODUCTIVE SYSTEM

1. The ovaries produce ova and secrete estrogens and progesterone.
2. The uterine tubes convey ova from the ovaries to the uterus and are the sites of fertilization.
3. The uterus is associated with menstruation, implantation of a fertilized ovum, development of the fetus, and labor.
4. The vagina serves as a passageway for the menstrual flow, as the receptacle for the penis, and as the lower portion of the birth canal.
5. The vulva is a collective designation for the external genitalia of the female.
6. The mammary glands function in the secretion of milk.

Endocrine Relations: Ovarian and Menstrual Cycles

1. The function of the menstrual cycle is to prepare the endometrium each month for the reception of a fertilized egg.
2. The ovarian cycle produces a mature ovum each month.
3. FSH and FSHRF, and LH and LHRF control the ovarian cycle. Estrogens and progesterone control the menstrual cycle.
4. The female climacteric is the time immediately before menopause.
5. Menopause is the cessation of the sexual cycles.

HOMEOSTATIC IMBALANCES OF THE REPRODUCTIVE SYSTEM

Venereal Diseases

1. Venereal diseases are a group of infectious diseases that are spread primarily through sexual intercourse.
2. With the exception of the common cold, venereal diseases are ranked as the number one communicable diseases in the United States.
3. Gonorrhea or "clap" is an infectious disease that mostly affects the mucous membrane of the urogenital tract, the rectum, and occasionally the eyes. It is caused by the bacterium *Neisseria gonorrhoeae.*
4. Syphilis is an infectious disease caused by the bacterium *Treponema pallidum.* The early stage produces a chancre, the secondary stage produces systemic effects, and the tertiary stage produces severe destructive changes involving the circulatory system, skin, bones, viscera, and central nervous system.
5. Genital herpes is a venereal disease transmitted by the herpes virus and is considered incurable.
6. Trichomoniasis is caused by the microorganism *Trichomonas vaginalis,* and symptoms include a vaginal discharge and severe vaginal itch in women.
7. Chlamydial infections are caused by organisms midway between viruses and bacteria. They cause painful urination in men, and in women they can cause serious pelvic disease, arthritis, heart trouble, and during birth, conjunctivitis in the newborn.
8. Nonspecific urethritis, or NSU, is inflammation of the urethra caused by Chlamydias and other organisms. Untreated, the disease may seriously damage the male or female urethra.
9. *Candida albicans* is a common fungal infection that causes candidiasis or vaginal moniliasis in women.

Male Disorders

1. The prostate gland is susceptible to infection, enlargement, and benign and malignant tumors.
2. In acute prostatitis the prostate gland becomes swollen and very tender. Chronic prostatitis is one of the most common chronic infections in men of the middle and late years.
3. Carcinoma of the prostate is the second leading cause of death from cancer in men in the United States, causing 19,000 deaths annually.
4. Infertility or sterility is an inability to fertilize the ovum and does not imply impotence.

Female Disorders

1. Abnormalities of menstruation include amenorrhea and dysmenorrhea.
2. Ovarian cysts are tumors of the ovary that contain fluid, while endometriosis occurs when the endometrial tissue is found growing any place other than inside the uterus.
3. Leukorrhea is a nonbloody, vaginal discharge that may occur at any age and affects most women at some time.
4. Breast cancer has the highest fatality rate of all cancers affecting women, but it is rare in men. Early detection is still the most promising method to increase the survival rate, and two methods of detection are self-examination and mammography.
5. Another common disorder of the female is cancer of the uterine cervix. It ranks third in frequency after breast and skin cancers. Early diagnosis of cancer of the uterus is accomplished by the Papanicolaou test, or "Pap" smear.

REVIEW QUESTIONS

1. List the male and female organs of reproduction.
2. Describe the function of the scrotum in protecting the testes from temperature fluctuations.
3. Describe the internal structure of a testis. Where are the sperm cells produced? What are Sertoli cells?
4. Identify the principal parts of a spermatozoan.
5. Trace the course of a sperm cell through the male system of ducts from the seminiferous tubules through the urethra.
6. What is the spermatic cord?
7. Briefly explain the functions of the seminal vesicles, prostate gland, and bulbourethral glands.
8. What is seminal fluid? What is its function?
9. How is the penis structurally adapted as an organ of copulation?
10. Describe the physiological effects of testosterone. How is the testosterone level in the blood controlled?
11. Explain the effects of FSH and ICSH on the male reproductive system.
12. Explain the hypothalamus and its regulating factors.
13. Describe the functions of the ovaries. What is ovulation?
14. What is the function of the uterine tubes? Define an ectopic pregnancy.
15. Diagram the principal parts of the uterus. Describe the three layers of tissue of the uterus.
16. Discuss the blood supply to the uterus. Why is an abundant blood supply important?
17. What are the functions of the vagina?
18. List the parts of the vulva. Compare the anatomy of the penis with that of the clitoris.
19. Describe the passage of milk from the alveolar cells of the mammary gland to the nipple.
20. Define menstrual cycle and ovarian cycle. What is the function of each?
21. Briefly outline the major events of the menstrual cycle and correlate them with the events of the ovarian cycle.
22. Prepare a labeled diagram of the principal hormonal interactions involved in the menstrual and ovarian cycles.
23. Define female climacteric and menopause. What is the cause of menopause?
24. What are the two most common venereal diseases? How are they usually transmitted? Name the causative microorganisms.
25. Correlate lip cold sores and genital herpes.
26. Compare trichomoniasis, chlamydial infections, NSU, and candidiasis as to causative microorganisms and possible symptoms.
27. Describe some pathologies of the prostate gland. Why is prostate cancer a serious disorder?
28. Define infertility and describe some of its possible causes.
29. Compare amenorrhea with dysmenorrhea.
30. Describe ovarian cysts, leukorrhea, and endometriosis.
31. What type of cancer has the highest fatality rate in women? Outline the different steps that women can take toward early detection of this cancer.
32. Describe cervical cancer. List some possible causes of this disorder, and identify the main early detection procedure.

Chapter 19

PHYSIOLOGY OF SEX

STUDENT OBJECTIVES

After reading this chapter, you should be able to

- Define coitus and orgasm
- Explain the physiological basis of sexual arousal
- Define foreplay and the sexual response phenomena
- Compare the four successive phases of the sexual response cycle of females and males
- Describe, in detail, the phases of the female sexual cycle
- Describe, in detail, the phases of the male sexual cycle
- Explain homeostatic imbalances related to sex physiology in the male such as impotence and premature ejaculation
- Explain homeostatic imbalances related to sex physiology in the female such as orgasmic dysfunction and vaginismus
- Contrast the various kinds of birth control and their effectiveness
- Compare surgical sterilization methods of females and males
- Define abortion

Despite the fact that the physiology of sex has been of universal interest from the earliest recorded times, scientific information on this topic has been very slow to accumulate. Investigations of sex had always been viewed as inappropriate for scholarly pursuit or scientific investigation. Consequently, the scientific community, reflecting the mores of society, had shunned this area of human physiology until the mid 1950s. But now it is quite common to view on television or read in weekly periodicals the latest therapy for sexual problems.

The total sexual experience cannot be explained as a biological sequence of events. The total enjoyment (or nonenjoyment) of sex is a multicomponent experience that also includes psychological and sociological forces. All these ramifications are beyond the scope of this chapter. Instead, we shall describe only the more biological components as they are currently understood. A review of the anatomy of the reproductive system as presented in Chapter 18 would be of great value in understanding the physiology of sex.

COITUS AND ORGASM

The term **coitus** comes from the Latin word *coitio*, meaning a coming together or uniting. In sexual behavior, it refers to insertion of the erect penis into the vagina. Two frequently used synonyms are **sexual intercourse** (*intercurs(us)* = a running between) and **copulation** (*copulatus* = bound together).

The word **orgasm** comes from the Greek word *orgasmos*, which means to be lustful or to swell. For the human being, it is one of the most intense experiences possible. Orgasm is the culmination of a mounting excitement and tension brought on by sexual activity.

Orgasm is a phenomenon that is common to human males and females and is usually experienced as a very intense physical pleasure. The physiological expression of orgasm is different in males and females. In physically mature males, orgasm includes **ejaculation** as part of the experience. In females, there is no ejaculation component. A common component of orgasm in both sexes is a neuromuscular discharge that occurs after a buildup of sexual tensions.

Many sexual experiences never produce this culmination, and the frequency and manner in which successful attempts are achieved vary greatly. They vary not only from person to person but for the same person at different times, under different circumstances, and at different stages of life. The term **climax** is often used in place of the word orgasm.

SEXUAL AROUSAL

Sexual arousal is a basic instinct since it initiates a sequence of events that may lead to perpetuation of the species. Although an individual can survive without sex, the human species obviously cannot. Humans have inherited responses and behaviors similar to other mammals, and these traits have evolved over millions of years. Sexual arousal is a normal, healthy characteristic of all beings. For humans the variations in degree and intensity of sexual arousal experienced during a lifetime are almost limitless. Sometimes the arousal culminates in orgasm, other times it lingers comfortably and then fades away, while at other times it generates severe frustration and anxiety.

Sexual arousal can result from stimulation by almost any sense modality, the most important being touch, followed to a lesser degree by sight. There are areas of the body called the **erogenous zones** that are especially susceptible to touch and often play a significant role in generating sexual arousal. Though these zones vary in sensitivity for different people, they normally include: the ear (especially the lobes); the mouth (lips, tongue, and the whole interior); the breasts (particularly the nipples); the buttocks; the inner surface of the thighs; the area between the anus and the genitals; the anus; the glans penis, but not the shaft of the penis; the clitoris; the labia minora and the space they enclose (vestibule), but not the vaginal canal.

The other sense modalities—smell, taste, hearing, and vision—also can be sexually arousing. These modalities are contributory in a conditioned manner. Certain visions, scents, tastes, and sounds become associated with sexual connotations and can be powerful or weak adjuncts to sexual arousal. This diversity of associations means that some individuals can be greatly aroused by a certain stimulus while others are completely immune to the same signal. Overall, the sexual responsiveness of an individual depends on his or her past conditioning. The brain, which stores past experiences in the memory, can generate excitement by recalling pleasant situations and thus enhance or enrich a current episode.

Foreplay

Foreplay, or petting, is a term used to describe the touching that occurs purposely to heighten sexual arousal. As sexual stimulation progresses, the level of sexual arousal increases, and the focus of attention becomes increasingly self-centered as external factors and concerns become less obvious. It is often possible during these early stages to control the rate of sexual arousal to some extent.

The rate at which arousal occurs varies with age. In younger people the rate is faster than in older ones. Differences of sexual arousal between males and females are not as significant as might be expected. Experiments using visual stimuli of films featuring male and female masturbation, petting, and coitus proved to be equally stimulating to both sexes. Men tended to respond more readily than women, but the variation within the groups was much greater than the difference between the groups, indicating that some

women responded sexually to a greater degree than the "average" man.

In this same context, interesting data are accumulating that show that women reach peaks of sexual arousal and arousability between the ages of 36 and 40, whereas men reach peaks during the late teens. These are average figures and the variation within groups is very large. More recent data suggest the age of peak arousal for women is getting lower.

The approach to orgasm, if prolonged and maintained at a high level of arousal, is often in itself a satisfying experience, and a climax may never be achieved. In other circumstances, if the heightened arousal is not culminated by orgasm, a high level of frustration and sometimes pain results.

SEXUAL RESPONSE CYCLES

The physiological changes occurring in males and females during sexual intercourse were first described in detail by Masters and Johnson in 1966. Their publication contained information derived from studying 619 females and 654 males and 10,000 sexual response cycles. Their data were the first to report measurements of the sexual organs during different phases of sexual arousal. Intrauterine electrodes measured uterine contractions while simultaneous recordings were obtained on heart activity, blood pressure, and breathing changes. The information was obtained from: (a) couples during as-normal-as-possible intercourse, (b) the same individuals during masturbation, and (c) women during artificial coitus. For artificial coitus an electronic plastic penis was utilized that permitted direct visualization and filming of the intravaginal physiological responses. The response cycles for both men and women were basically independent of the manner in which they were aroused.

Masters and Johnson divided the sexual response cycle of men and women into four successive phases: (1) the excitement phase, (2) the plateau phase, (3) the orgasmic phase, and (4) the resolution phase. While it is convenient to divide the sexual response cycle into phases, it must be emphasized that these successive phases blend into each other from the initial stages through orgasm. The graphs in Figures 19-1 and 19-2 represent data collected from many responses of males and females.

Individual variations can be considerable. In males, the duration of the phases shows large variability, but the intensity response pattern seems to be typical. In females, the response cycle also includes variable intensity experiences. Three general alternative patterns of response are possible for females. Though

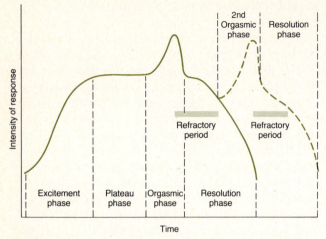

Figure 19-1 The sexual response cycle in males. (Adapted from Masters and Johnson, *Human Sexual Response,* Boston: Little, Brown and Company, 1966.)

these graphs indicate the most likely responses, they do not begin to illustrate the broad repertoire of responses demonstrated by females. A female can experience any one of many patterns during any given sexual episode.

The sexual responses of males and females share two basic physiological processes. First, there is a reaction called *vasocongestion.* This describes the engorgement of blood vessels with increased volumes of blood that causes the tissues to swell and enlarge. This accumulation of blood occurs particularly in the penis, clitoris, and labia minora, and also to a lesser degree in the breasts, skin, earlobes, and nose.

The second reaction is called *myotonia* and refers to the increase in skeletal muscle tension. During sexual arousal, both sexes show an increase in involuntary muscle contractions. Facial muscles tighten and produce grimacing and scowling expressions. Myotonia increases during the sexual response cycle and

reaches its peak at orgasm. At this point, involuntary rhythmic contractions of the penile muscles and outer third of the vaginal musculature occur.

Both sexes also show rhythmic contractions of the gluteal muscles during orgasm, and during the height of the response, clenching movements of both hands and feet occur.

Phases of the Female Sexual Cycle

Excitement Phase

As a result of sexual stimulation during this period, initial increases in heart rate and blood pressure occur. The nipples become erect and the breasts increase slightly in size. Vasocongestion begins in various organs. The clitoris and labia minora become engorged with blood and increase in size. Over 75 percent of females show a *sex flush*—a kind of blushing of the skin that covers the shoulders, neck, back, breasts, and lower abdomen. During this phase, the inner end of the vagina expands and lengthens and the labia majora withdraw by flattening out and rising slightly. The vagina produces a clear lubrication fluid—most likely a result of the vasocongestion occurring in the vaginal wall. This moistening of the vaginal wall is now recognized as the first positive indication of sexual arousal and can occur within 10 to 30 seconds after erotic stimulation. Due to the intense vasocongestion, the vagina changes from the normal purple-red color to a darker, more intense coloration. Initially patchy in distribution, the intensity spreads as the vasocongestion and excitement increase.

Plateau Phase

During this phase, myotonia, vasocongestion, and breast size increase. The clitoral glans (the very sensitive tip of the clitoris) and the shaft begin to

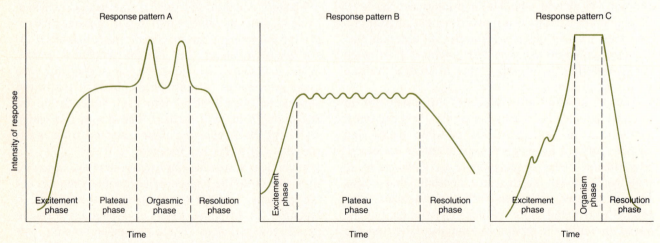

Figure 19-2 The sexual response cycle in females. (Adapted from Masters and Johnson, *Human Sexual Response,* Boston: Little Brown and Company, 1966.)

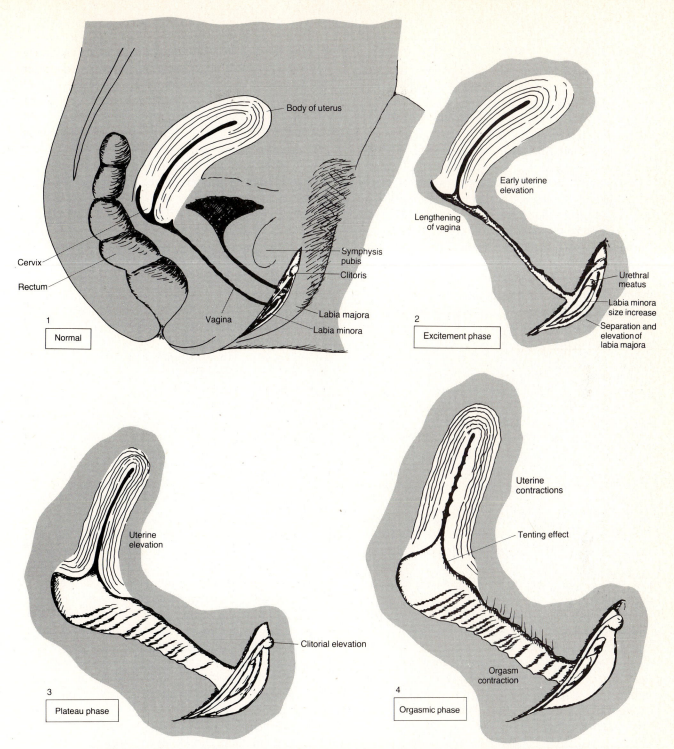

Figure 19-3 Anatomical and physiological changes within the female pelvic organs accompanying the sexual response cycle. (1) Normal female pelvis, lateral view. (2) Female pelvis in the excitement phase. (3) Plateau phase. (4) Orgasmic phase.

recede or withdraw into the upper folds (hood) of the labia minora. Also during this phase, the outer third of the vagina becomes very congested so that the vaginal lumen is reduced to about two-thirds its relaxed size. This vasocongestion occurring in the outer third of the vaginal canal is termed the *orgasmic platform*. It is here that the rhythmic muscle contractions of orgasms occur most intensely. In addition,

during this phase, the deepest part of the vagina enlarges to reach its greatest expansion. This ballooning in size is referred to as the *tenting effect*. The uterus also responds to the sexual stimulation. The uterus, which when relaxed normally leans forward (Figure 19-3), now is elevated to a more vertical position, and the cervix withdraws from the vaginal canal, contributing to the tenting effect.

Orgasmic Phase

As orgasm approaches, the physiological changes described in the preceding phases intensify. The breasts achieve their greatest size and the nipples are erect. Vasocongestion is at its peak and the clitoris is completely hidden. The deepest portions of the vaginal canal are expanded to their greatest volume, while the wall of the vagina is "sweating" profusely. The outer third of the vagina, the orgasmic platform, has reached peak congestion and is forming a tight closure upon the male penis. Breathing and heart rate reach maximum values while certain skeletal muscle contractions become most intense. The orgasmic phase is relatively short, lasting between 3 and 10 seconds. The onset is indicated by contractions of the orgasmic platform, the uterine musculature, and the anal sphincter. These contractions occur every 0.8 seconds about 5 to 12 times. Blood pressure rises abruptly during orgasm, sometimes reaching twice the normal values.

Females show considerable variation in the physiology and subjective feelings of orgasm. Masters and Johnson have described the three most common response patterns in the graph shown in Figure 19-2.

The B pattern seems most characteristic of young or sexually inexperienced females. The sequence shows a rising level of excitement that levels out in the plateau phase with several minor fluctuations toward orgasm. The level of sexual excitement then slowly declines to the normal, prearousal level.

The C pattern, as interpreted by Masters and Johnson, shows a rapidly climbing arousal state that reaches an extreme intensity of orgasm that resolves itself into a total feeling of release and complete gratification.

The A pattern shows that some females achieve multiple orgasms. The sexual intensity subsides between the peak orgasmic episodes only to the plateau level. There is, however, no refractory period between orgasms. This means that following one orgasm, a female can respond immediately to continual erotic stimulation and achieve successive orgasms. Some females can experience six or seven such orgasms.

The patterns of response for females change over time and can vary from one sexual episode to the next even with the same partner.

Resolution Phase

Normally, within 5 to 10 minutes following orgasm, body functions return to their prearousal level or condition. This phase is characterized by feelings of great relaxation. Females may laugh, cry, or fall asleep. A considerable percentage demonstrate a postorgasmic perspiration response that covers most of their skin surface. Others show more specific area responses in only one or two areas such as the palms of the hands and soles of the feet.

Phases of the Male Sexual Cycle

The sexual response cycle of males bears more similarities to the female response than differences. Vasocongestion and myotonia are the major physiological changes that are also basic to the male response.

Excitement Phase

One of the earliest signs is erection of the penis. Vasocongestion occurring in the three cylinders of erectile tissue (corpus spongiosum and corpora cavernosa) that run the length of the penis cause the penis to enlarge greatly and assume a protruding position. Penile erection is a reflex phenomenon that begins with any of a variety of erotic stimuli. These stimuli form the afferent limb of the reflex and travel to a reflex center in the sacral portion of the spinal cord. The efferent limb of the reflex is composed of both parasympathetic and sympathetic fibers that innervate the arteriolar blood vessels supplying the spongy erectile tissues. The parasympathetic fibers are activated and cause the arterioles to dilate, permitting a rapid flow of blood into the corpora cavernosa and corpus spongiosum of the penis. The simultaneous inhibition of the sympathetic nerve fibers removes any of their vasoconstrictive action. The sudden surge of blood into the erectile bodies causes compression of the veins of the penis and occludes the outflow of blood from the spongy tissue. The rapid vasocongestion is followed by stiffening and erection of the penis.

Return of the penis to its normal, flaccid condition results from increased tone of the sympathetic fibers to the arterioles and decreased tone of the parasympathetic fibers. Constriction of the arterioles results, and the inflow of blood into the erectile tissue is markedly reduced. Drainage through the veins becomes possible, and the penis slowly returns to its normal, prearousal position and state.

During the excitement phase, the increased tone of the parasympathetic fibers also causes elevation and flattening of the scrotum, and elevation and considerable increase in the size of the testes. Vasocongestion occurring in the skin of the scrotum is a factor in scrotal elevation, whereas an involuntary contraction of muscles near the spermatic cords causes them to pull and thus elevate the testes.

This phase can be quite short, but usually is of considerable duration. It is during this period that a male can easily lose his erection if sexual stimulation or interest declines.

Plateau Phase

The penis reaches its maximum state of erection during the preceding phase and now shows an increase in engorgement primarily in the corona of the glans penis. The penis achieves a state of more permanent erection and the male is able to be temporarily diverted from sexual activity without loss of tumescence (swelling). The testes are elevated to the maximum extent and have increased in size about 50 percent. If the plateau phase is prolonged for an extended period of time without achieving orgasm, the testes may increase in size 100 percent. Heart rate, blood pressure, breathing rate, and muscle tension all increase.

Orgasmic Phase

Orgasm in the male, as in the female, involves the total body. Myotonia reaches a peak, causing a sustained muscular tension so that most of the body is rigid. A rhythmic throbbing of the pelvic muscles is often experienced as the sex organs respond.

The response of the sex organs begins internally with the rhythmic contractions of the prostate, seminal vesicles, and vas deferens. The contractions extend quickly to include the penis (Figure 19-4). The contractions are spaced at intervals of 0.8 seconds and become weaker after the first several powerful contractions. A feeling that ejaculation is about to occur or is "coming" is sensed, and the male continues directly to a second stage wherein there is a forcible ejection of fluid from the urethra. The fluid, called *semen* or *seminal fluid,* normally contains millions of spermatozoa suspended in the fluid secreted by the prostate gland and seminal vesicles. The volume of fluid ejected varies, but 3 ml, or about a teaspoonful, is most common.

Ejaculation, like erection, is a reflex mechanism. The ejaculation center is in the spinal cord located between T12 and L3. The afferent limb of the reflex arises from nerves in the penis, and their tone increases as penile stimulation increases.

There are two stages to ejaculation. The first is termed the *emission* or *first stage*. During this stage, the sperm and secretions from the prostate gland and seminal vesicles are moved toward the urethra by rhythmic contractions of the smooth muscles in the wall of the vas deferens and the glands (Figure 19-4). The sphincter at the origin of the urethra where it connects to the bladder closes tightly. This prevents urine from entering the urethra and also prevents semen from entering the urinary bladder.

The *expulsion* or *second stage* occurs when the semen is ejected forcibly from the urethra mainly by contraction of the bulbocavernosus muscle located around the base of the penis. The initial ejaculation can cause semen to spurt out a great distance, but usually it is ejected only a short distance or simply oozes out over the tip of the penis. Subsequent contractions are less forceful and the semen flows out quite gently.

For males, ejaculation is *part* of orgasm and should not be considered equal to or the same as orgasm.

Resolution Phase

Return to the normal, prearousal state occurs during resolution. The penile erection decreases in two stages. Initially, there is rapid loss of vasocongestion and the tumescence rapidly disappears so that the penis reaches

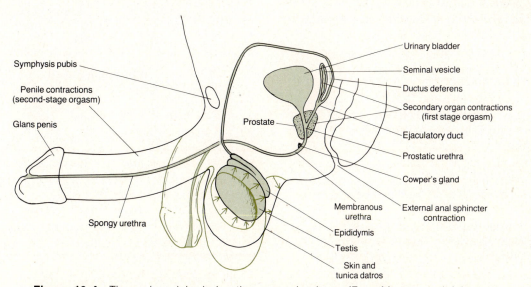

Figure 19-4 The male pelvis during the orgasmic phase. (From Masters and Johnson, *Human Sexual Response,* Boston: Little, Brown and Company, 1966.)

a semierect state. A more gradual stage follows and the penis slowly returns to its normal state. About one-third of all males show a perspiration response during this phase. Adult males however, differ from females in this phase because males experience a *physiological refractory period*. Immediately following orgasm there is a period of time during which sensitivity to erotic stimulation is greatly reduced. The duration of this refractory period varies, and because of its existence, males are seldom able to experience multiple orgasms.

Exhibits 19-1 and 19-2 summarize the reactions of sex organs and body during the sexual response cycles of males and females.

HOMEOSTATIC IMBALANCES RELATED TO SEX PHYSIOLOGY

Sex organs and their functioning are subject to a variety of disorders. Simple infections of the female reproductive organs are treated with drugs and are usually relatively easy to cure. Venereal diseases have been discussed in Chapter 18, and these also can be treated successfully with antibiotics if caught early. Cancers that involve the sex organs strike both males and females, and early detection and treatment can save lives. These disorders have a physical basis and so can deprive, lessen, temporarily interrupt, or prevent a satisfactory sexual experience. Many other sexual disorders have their basis in psychological or sociological or specific interpersonal interactions and conflicts. It has been estimated that over half of the married couples in the United States have some form of sexual problem.

Among males the most common difficulties are *impotence* and *premature ejaculation*. In females the

Exhibit 19-1 REACTIONS OF SEX ORGANS DURING THE SEXUAL RESPONSE CYCLE

Male	Female
Excitement phase	
Penile erection (within 3–8 seconds)	Vaginal lubrication (within 10–30 seconds)
As phase is prolonged: Thickening, flattening, and elevation of scrotal sac	*As phase is prolonged:* Thickening of vaginal walls and labia
As phase is prolonged: Partial testicular elevation and size increase	*As phase is prolonged:* Expansion of inner $\frac{2}{3}$ of vagina and elevation of cervix and corpus
	As phase is prolonged: Tumescence of clitoris

Exhibit 19-1 *(Continued)*

Male	Female
Plateau phase	
Increase in penile coronal circumference and testicular tumescence (50–100 percent enlarged)	Orgasmic platform in outer $\frac{1}{3}$ of vagina
Full testicular elevation and rotation (orgasm inevitable)	Full expansion of $\frac{2}{3}$ of vagina, uterine and cervical elevation
Purple hue on corona of penis (inconsistent, even if orgasm is to ensue)	"Sex skin": discoloration of labia minora (constant, if orgasm is to ensue)
Mucoid secretion from Cowper's gland	Mucoid secretion from Bartholin's gland
	Withdrawal of clitoris
Orgasmic phase	
Ejaculation	*Pelvic response (no ejaculation)*
Contractions of accessory organs of reproduction: vas deferens, seminal vesicles, ejaculatory duct, prostate	Contractions of uterus from fundus toward lower uterine segment
Relaxation of external bladder sphincter	Minimal relaxation of external cervical opening
Contractions of penile urethra at 0.8 second intervals for 3–4 contractions (slowing thereafter for 2–4 more contractions)	Contractions of orgasmic platform at 0.8 second intervals for 5–12 contractions (slowing thereafter for 3–6 more contractions)
Anal sphincter contractions (2–4 contractions at 0.8 second intervals)	External rectal sphincter contractions (2–4 contractions at 0.8 second intervals)
	External urethral sphincter contractions (2–3 contractions at irregular intervals, 10–15 percent of subjects)
Resolution phase	
Refractory period with rapid loss of pelvic vasocongestion	Ready return to orgasm with retarded loss of pelvic vasocongestion
Loss of penile erection in primary (rapid) and secondary (slow) stages	Loss of "sex skin" color and orgasmic platform in primary (rapid) stage
	Remainder of pelvic vasocongestion as secondary (slow) stage
	Loss of clitoral tumescence and return to position

Source: Courtesy of H. A. Katchadaurian and D. T. Lunde, *Fundamentals of Human Sexuality*, New York: Holt, Rinehart and Winston, 1972.

Exhibit 19-2   GENERAL BODY REACTIONS DURING THE SEXUAL RESPONSE CYCLE

Male	Female
Excitement phase	
Nipple erection (30 percent)	Nipple erection (consistent)
	Sex-tension flush (25 percent)
Plateau phase	
Sex-tension flush (25 percent)	Sex-tension flush (75 percent)
Carpopedal spasm	Carpopedal spasm
Generalized skeletal muscle tension	Generalized skeletal muscle tension
Hyperventilation	Hyperventilation
Heart rate increase 100–160 beats per minute	Heart rate increase 100–160 beats per minute
Orgasmic phase	
Specific skeletal muscle contractions	Specific skeletal muscle contractions
Hyperventilation	Hyperventilation
Heart rate increase 100–180 beats per minute	Heart rate increase 110–180 beats per minute
Resolution phase	
Sweating reaction (30–40 percent)	Sweating reaction (30–40 percent)
Hyperventilation	Hyperventilation
Heart rate decrease 150–80 beats per minute	Heart rate decrease 150–80 beats per minute

Source: Courtesy of H. A. Katchadaurian and D. T. Lunde, *Fundamentals of Human Sexuality*, New York: Holt, Rinehart and Winston, 1972.

most common malfunctions are the inability to reach orgasm (*orgasmic dysfunction*) and *vaginismus*.

Impotence

This is a term that comes from the Latin word *impotens* meaning without power over oneself or others. In the sexual context, impotence is defined as the inability to have an erection. Impotence is separated into two categories. **Primary impotence** is the condition in which a male has never been able to have an erection sufficient for coitus to take place. **Secondary impotence** is a condition in which the ability to achieve an erection is lost after successful attempts have been achieved. Some males have secondary impotence in only certain situations, and many males occasionally experience an episode of impotence.

Impotence occurs at a rate of about 1 percent in males under 35 years of age and, of these, very few are totally and chronically incapable. By age 70, about 25 percent of males are impotent. There are suggestions that impotence is increasing. The fear of failure in sexual intercourse has been implicated as the greatest factor in causing impotence. Since 90 percent of the cases of impotence are psychologically based, the treatment usually requires the analysis of sexual attitudes and past inhibitory experiences with the aid of an expert counselor.

Premature Ejaculation

Definition of *premature ejaculation* focuses on the word *premature*. Masters and Johnson recognized the presence of the disorder in males who could not control ejaculation during penile-vaginal confinement for a long enough time to satisfy their partners at least 50 percent of the time. More clinically oriented investigators define a premature ejaculator as a male who cannot control his ejaculation for 30 to 60 seconds after penile-vaginal penetration.

When expectations of performance and worry about premature ejaculation increase in the male, the degree of frustration by the female partner also increases. Thus, often a destructive cycle is set into motion. The studies of Masters and Johnson indicate that premature ejaculators result from preconditioning during their early sexual experiences. Certain situations require that the sexual episode be completed quickly. Repeated episodes of this nature during early experiences reinforce the conditioning process.

Treatment for premature ejaculators is best carried out in a sex therapy clinic where the therapists are properly trained and the histories of both partners can be studied.

Orgasmic Dysfunction

Females who have never reached an orgasm are classed as *primary* sufferers, whereas women who have had at least one orgasm, either from masturbation or coitus, but who are thereafter unable to achieve orgasm are classed as *situational* sufferers. Though some cases result from anatomical abnormalities of the genitalia, from abuse of drugs and alcohol, and occasionally from hormonal irregularities, most *orgasmic dysfunctions* result from psychological forces. A variety of forces may precondition a female to feel guilt or fear when she is about to experience a sexual episode, or she may have expectations of performance and peaks of arousal that are unrealistic. The nature of the psychological forces and factors that are most predisposing to orgasmic failure in females needs much more careful research.

Present therapy for orgasmic dysfunction requires good sexual counseling from qualified therapists. Attitudes of both partners toward sex need to be revealed, examined, and restructured.

Vaginismus

In *vaginismus*, involuntary contractions of the muscles surrounding the outer third of the vagina make insertion of the erect penis impossible or at least very difficult. This phenomenon occurs in some females as a result of certain kinds of preconditioning. The preconditioning can be so powerful a force that the mere thought of sexual intercourse triggers the involuntary contraction of the pelvic muscles.

Forces, such as those generated in youth due to strong sexual inhibitions related to a sexually repressive upbringing, violent sexual episodes, or humiliation from incestuous experiences, have been known to produce vaginismus. Females who have suffered painful intercourse for any of a number of reasons—infections, torn pelvic ligaments, hormonal imbalances—can also develop vaginismus.

Treatment is oriented in two directions. Psychological counseling is needed to overcome the sexual fears of the female, and a series of exercises to dilate the vagina are instituted.

BIRTH CONTROL

Couples who enjoy intercourse but who do not wish to have a baby can take certain precautions to reduce the likelihood of pregnancy. Many methods of birth control have been used from the times of remote antiquity. One of the earliest mentions of birth control is in a document found in Egypt and believed to date from between 1900 and 1100 B.C. The technique of birth control now called *coitus interruptus* is described in the Bible and Talmud. A brief chronology sequencing some of the developments in birth control is presented in Exhibit 19-3.

People give many reasons for practicing some form of birth control, and each individual has his or her own personal and private reason. Most reasons given revolve around a desire to limit family size because of economic, physical, or mental health considerations.

Methods of Birth Control

Methods of birth control fall into four classes or techniques. (1) *Abstinence* (from the Latin "to hold away from"), in the birth control context, means to avoid

Exhibit 19-3 LANDMARK EVENTS IN THE DEVELOPMENT OF SOPHISTICATED CONTRACEPTION

Date	Development
Late 1700s	Casanova (1725–1798) popularizes and publicizes use of the sheath or "English Riding Coat."
1798	Malthus urges "moral restraint."
1840s	Goodyear vulcanizes rubber. Production of rubber condoms soon follows.
1883	Mensigna invents the diaphragm.
1893	Harrison performs the first vasectomy.
1909	Richter uses the intrauterine silkworm gut.
1910–1920	Margaret Sanger pioneers in New York City; the term "birth control" is coined.
1930	Graffenberg publishes information documenting his 21 years of experience with the ring (silver and copper) and catgut.
1930–1931	Knaus and Ogino elucidate "safe and unsafe" periods of the woman's menstrual cycle: the rhythm method.
1934	Corner and Beard isolate progesterone.
1937	Makepeace demonstrates that progesterone inhibits ovulation.
1950s	Abortions are utilized extensively in Japan.
1950–1960	Hormonal contraceptive research results in F.D.A. approval for the use of "The Pill" as a contraceptive in 1960 (Pincus and Rock).
1960s	Many Western nations liberalize abortion laws. Modern IUDs become available.
	Contraceptive sterilization becomes more acceptable.
	Laparoscopic tubal ligation technique is developed.
1973	U.S. Supreme Court rules on abortion.
	First "mini-pill" or low-dose progestin wins FDA approval.
	"The Shot" is provided in over 50 nations.

Source: R. A. Hatcher, G. K. Stewart, F. Guest, R. Finkelstein, and C. Godwin, *Contraceptive Technology, 1976–1977*, 8th ed., page 9. New York: Wiley.

intercourse during the time when ovulation and fertilization are most likely to occur. (2) *Contraception* (from the Latin "against taking in") describes any method that prevents fertilization or implantation from taking place. Such methods utilize physical barriers or chemicals that prevent the sperm from reaching or penetrating the ovum, or they utilize the presence of a foreign body within the uterus which prevents implantation. (3) *Sterilization* techniques usually require surgery to remove or tie off a portion of a duct or organ so that the reproductive function is blocked. These, contrary to some claims, should be regarded as permanent conditions. (4) *Abortion* involves removal of the contents of the pregnant uterus—the fetus, placenta, and enlarged endometrial tissue on the

Exhibit 19-4 COMPARISON OF APPROXIMATE FAILURE RATES OF VARIOUS FORMS OF BIRTH CONTROL (PREGNANCIES PER 100 WOMAN YEARS)

	Theoretical Failure Rate	Actual Use Failure Rate
Abortion	0+	0+
Abstinence	0	?
Hysterectomy	0.0001	0.0001
Tubal ligation	0.04	0.04
Vasectomy	Less than 0.15	0.15
Oral contraceptives (combined)	Less than 1.0	2–5
I.M. long-acting progestin	Less than 1.0	5–10
Condom + spermicidal agent	1.0	5
Low-dose oral progestin	1–4	5–10
IUD	1–5	6
Condom	3	15–20
Diaphragm	3	20–25
Spermicidal foam	3	30
Coitus interruptus	15	20–25
Rhythm (calendar)	15	35
Lactation for 12 months	15	40
Chance (sexually active)	80	80

Source: From *Our Bodies, Ourselves.* Copyright © 1971, 1973, 1976 by The Boston Women's Health Book Collective, Inc. Reprinted by permission of Simon & Schuster, a division of Gulf & Western Corporation.

Note: Extensive references, often conflicting, are available on the complicated subject of contraceptive effectiveness, including 14 references in *Contraceptive Technology, 1973–1974.*

uterine wall. Methods used for abortion vary depending on the age and size of the fetus.

Exhibit 19-4 shows the approximate failure rates for various forms of birth control.

Abstinence

This is a perfectly acceptable method of birth control. Careful determination of a female's fertile periods by methods such as calendar calculation, observation of cervical mucus, and observation of basal body temperature is called the *rhythm method* (newer terms for this method include *natural family planning* and *fertility awareness*) of birth control. Abstinence is practiced during those days calculated to be the female's fertile period.

Somewhat related to the technique of abstinence is *coitus interruptus.* This involves withdrawal of the penis from the vagina prior to ejaculation so that the semen is ejected far from the vagina. This is not a very reliable method of birth control because often the drops of fluid that flow from the penis when it reaches full erection contain healthy sperm. Also, the probability of late withdrawal or premature ejaculation raises the possibility of pregnancy occurring. Coitus interruptus has a pregnancy rate of about 20 per 100 women per year in the United States.

Contraception

Contraceptive techniques include physical barriers that prevent the sperm from reaching the ovum, chem-icals that either inactivate sperm or prevent release of the ovum from the ovary, and contraimplantation devices.

Physical barrier types include the *condom* (*condus* = a receptacle) used by the male and the *diaphragm* used by the female.

The condom is a thin, cylindrical sheath of strong latex rubber with a domed or ballooned tip that is pulled over the penis. Its purpose is to capture the ejaculate and prevent the sperm from entering the vaginal canal of the female. When used every time during sexual intercourse, condoms are about 97 percent effective. In actual use, because of carelessness, their effectiveness drops to about 80 to 85 percent.

A diaphragm is also made of latex rubber but is shaped in the form of a dome or shallow cup (Figure 19-5). The latex is stretched over a flexible metal spring ring. Diaphragms are always used in combination with spermicidal cream or jelly. Diaphragms come in varying sizes with diameters from 45 to 105 mm and are fitted with the aid of a trained clinician. When fitted correctly, the diaphragm fits snugly over the cervix. It is inserted prior to intercourse and is left in place at least six hours after intercourse. When fitted correctly and used with spermicidal cream or jelly, the effectiveness of the diaphragm is, in theory, about 97 percent. In actual use however, its effectiveness drops to between 75 and 80 percent, depending upon how well the diaphragm is fitted and the knowledge and motivation of its user (Exhibit 19-4).

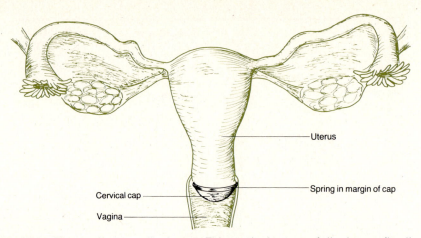

Figure 19-5 The cervical cap diaphragm. This particular type of diaphragm fits directly over the cervix of the uterus. The thin spring around the margin of the diaphragm opens out, presses against the wall of the vagina, and stretches across the cervix.

The *intrauterine device* (IUD) is a small piece of plastic that comes in a variety of different shapes and sizes (Figure 19-6). It is carefully inserted by a trained clinician. One or two strings attached to the IUD lead out into the upper end of the vaginal canal and are used to check that the device is still in place and to remove the IUD. The method by which the IUD prevents pregnancy is unknown. Some researchers suspect that a local inflammatory reaction occurs that prevents the fertilized ovum from implanting. Some newer IUDs contain an added active substance, usually copper or progesterone. These seem to be more effective and well tolerated but need to be replaced when the active ingredient has been consumed. Pregnancy rates for IUDs now used are about 6 per 100 women per year (Exhibit 19-4).

Chemical means of contraception include the use of various foams, creams, jellies, suppositories, and douches that make the vagina and cervix unfavorable for sperm survival. Of the newer chemical means, *oral contraceptives — the pill —* have found rapid and widespread use. The pills most widely used are combination types containing both estrogens and progesterones in varying amounts. The estrogens and progestins function to inhibit the release of FSH from the anterior pituitary gland. With lowered levels of FSH, no ovum matures and therefore ovulation does not occur until the pill regimen concludes. There is evidence that elevated estrogen levels also inhibit implantation of the fertilized ovum.

It is estimated that some form of contraceptive pill is now used by 100 million females throughout the world, including 15 million in the United States. While the theoretical effectiveness of the combination pill approaches 100 percent, pregnancy rates among pill users are actually about 4 per 100 females per year. The most serious side effect of the pill is an increased incidence of blood clotting within blood vessels. A clot that occludes the vessel in which it forms is called a thrombus. A clot that is carried to some other part of the circulation is called an embolism. The danger of the clot is that it can obstruct the blood flow to a vital region such as the brain, heart, or lung. Hypertension is also a possible side effect of estrogen treatment. Estrogen causes an increased secretion of renin which, in turn, causes increased levels of angiotensin II. The result is increased vasoconstriction and an increase in blood pressure.

Sterilization

Birth control methods that are permanent in nature are commonly called *sterilization*. Increasing in favor, these permanent methods are the most commonly used forms of fertility control in couples over 30 years of age. The permanent methods of contraception for both males and females include surgical and nonsurgical ones.

For females, nonsurgical methods may include: *immunological control*, which is presently only in the experimental stage; *hormonal control*, which involves a time release process from either a pill or an intramuscular injection and is also still only experimental; *irradiation*, which can be used to terminate ovulation, but which also eliminates production of most of the estrogen and progesterone and is a method that should never be used; *tubal occlusion* by means of transcervical insertion of a variety of chemicals (silver nitrate, zinc chloride, phenol), cauterization with heat or cryosurgery, placement of plastic occlusives, or tissue adhesives. These have the advantages of speed of implementation, and they are potentially very effective.

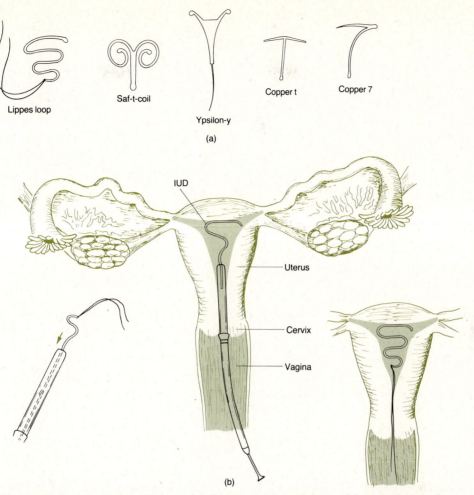

Lippes loop

Saf-t-coil

Ypsilon-y

Copper t

Copper 7

(a)

IUD

Uterus

Cervix

Vagina

(b)

Figure 19-6 Intrauterine devices. (a) Representative designs of intrauterine devices. (b) Procedure for insertion. IUDs are inserted into a slightly open cervix. The device is threaded into a long, narrow-bore tube that is passed through the cervix. Once in position in the uterus, it spreads out to its former shape. Most IUDs have a thread or chain projecting into the vagina that may be detected by a finger (indicating the device is still in place) and that may be used for removal.

Surgery to produce sterilization in females can be approached in a variety of ways, so a female can make a selection. Oophorectomy, surgical removal of the ovary, removes the source of ova, producing permanent infertility. However, it also removes the major source of estrogens, which can cause physical symptoms of menopause such as hot flashes and vaginal dryness if not replaced by oral administration. This procedure is unusual and not recommended in the absence of hysterectomy. *Hysterectomy* is surgical removal of the uterus. Sometimes a "partial" hysterectomy is performed in which the body of the uterus is removed but the cervical portion is retained. Since the ovaries and tubes are left, ovulation occurs regularly but there are no menstrual periods. A *radical total hysterectomy* is a more extreme procedure in which the complete uterus, cervix, and upper vagina are also removed. This procedure is generally done because of the presence of disease such as cancer and not for the primary purpose of sterilization. The most common sterilization procedure in females is *tubal ligation* (Figure 19-7a). In this procedure, the uterine tubes are located and rendered impassable by tying off and then by cutting out and/or cauterizing a section of the tubes. The procedures involved in closing the tubes are varied and many, but there are only two commonly used approaches: an incision through the abdominal wall and an incision through the vaginal wall. In both these approaches, the tubes are visually identified and ligated.

The introduction of endoscopic instruments (long, flexible, tubelike telescopic instruments) has permitted sterilization surgery to be done using local anesthetics, frequently on an outpatient basis.

For males, *vasectomy* is the only method of sterilization that is used and in fact is the single most

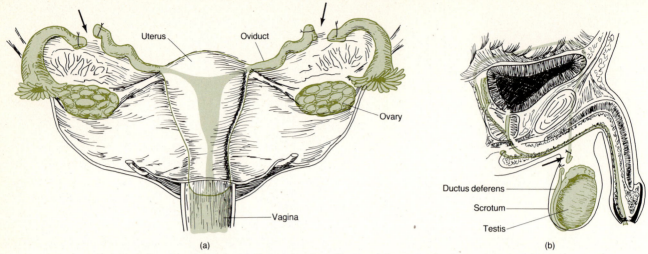

Figure 19-7 Sterilization. (a) Tubal ligation. Each uterine tube is cut and tied after an incision is made into the abdomen. (b) Vasectomy. The ductus deferens of each testis is cut and tied after an incision is made into the scrotum.

popular method of permanent contraception (Figure 19-7b). Through a small (2 cm long) incision in the scrotum, the surgeon cuts, ties, and cauterizes the vas deferens. Vasectomy done on a physically and psychologically healthy male will not affect his sexual responsiveness.

Abortion

The premature expulsion from the uterus of the products of conception is termed an *abortion*.

A *miscarriage* is a *spontaneous abortion* that occurs most frequently during the first trimester of pregnancy. In about 50 percent of the miscarriages, the fetus is defective in some way. Only about 15 percent of the miscarriages have an apparent reason — illness, malnutrition, trauma, or some other known factor affecting the mother. The other 85 percent occur from unknown causes.

An *induced abortion* is the result of a procedure done specifically for the purpose of terminating the pregnancy. In an induced abortion, the contents of the uterus (fetus, placenta, and the supportive tissue of the uterus) are removed. There are several methods used to perform an abortion, and the selection of the method is determined in large part by the size and age of the fetus.

Despite the legalization of abortion, the topic continues to be most controversial in our society. Moral aspects related to the destruction of an unborn fetus are hotly debated. Can abortion be strictly a private matter that concerns only the pregnant woman and her physician? Undoubtedly, these and other aspects of abortion will be argued for a long time.

We shall now turn our attention to a discussion of development and inheritance. Here we shall examine the events associated with gamete development, fertilization, implantation, embryonic growth, fetal growth, labor, and lactation. We shall also examine some elementary principles related to inheritance.

CHAPTER SUMMARY OUTLINE

COITUS AND ORGASM

1. In sexual behavior the word coitus refers to insertion of the erect penis into the vagina.
2. Sexual intercourse is the most frequently used synonym.
3. Orgasm is the culmination of a mounting excitement and tension brought on by sexual activity.
4. Orgasm is common to both males and females and is usually experienced as a very intense physical pleasure.

SEXUAL AROUSAL

1. Sexual arousal is a basic instinct that initiates a sequence of events that may lead to perpetuation of the species.

2. Sexual arousal is a normal, healthy characteristic of all beings.
3. There are areas of the body, called the erogenous zones, that are especially susceptible to touch and often play a large role in generating sexual arousal.

Foreplay

1. Foreplay, or petting, is a term used to describe the caressing and stroking that occurs purposely to heighten sexual arousal.
2. The base line for sexual excitement must include heart rate acceleration and heavy breathing, or else it can be assumed that there is no sexual arousal.
3. In a totally sexually aroused person every body system participates in some fashion in the sexual response.
4. Data are accumulating that show that females reach peaks of sexual arousal and arousability between the ages of 36 and 40, whereas males reach peaks during the late teens.
5. More recent data, however, suggest the age of peak arousal for females is lower.

SEXUAL RESPONSE CYCLES

1. The sexual response cycle of males and females is divided into four successive phases: the excitement phase, the plateau phase, the orgasmic phase, and the resolution phase.
2. The sexual responses of males and females share two basic physiological processes: vasocongestion and myotonia.

Phases of the Female Sexual Cycle

1. In the excitement phase there are increases in heart rate and blood pressure, the nipples become erect, and the breasts increase slightly in size. Over 75 percent of women show a sex flush, and the vagina produces a clear lubricating fluid.
2. During the plateau phase, myotonia, vasocongestion, and breast size increase. The vagina exhibits the orgasmic platform and a tenting effect.
3. The orgasmic phase intensifies the previously mentioned physiological changes, and there are three most common response patterns possible during this phase.
4. Following one orgasm, some women can respond immediately to continual stimulation and achieve successive orgasms.
5. Normally, within 5 to 10 minutes following orgasm, body functions return to their prearousal level or

condition. This resolution phase is characterized by feelings of great relaxation.

Phases of the Male Sexual Cycle

1. In the excitement phase of the male one of the earliest signs is erection of the penis. There is also scrotal and testes elevation, and considerable increase in the size of the testes.
2. During the plateau phase the penis reaches its maximum state of engorgement primarily in the corona of the glans. Heart rate, blood pressure, breathing rate, and muscle tension all increase.
3. The orgasmic phase in the male, as in the female, involves the total body. The response of the sex organs begins internally with the rhythmic contractions of the prostate, seminal vesicles, vas deferens, and penis. These contractions result in the two-stage phenomenon of ejaculation.
4. The resolution phase is a return to the normal, prearousal state, with the penis returning to its normal state in two stages. Immediately following orgasm there is a period of time during which sensitivity to erotic stimulation is greatly reduced. As a result men are seldom able to experience multiple orgasms.

HOMEOSTATIC IMBALANCES RELATED TO SEX PHYSIOLOGY

1. Many sexual disorders have their basis in psychological, sociological, or specific interpersonal interactions and conflicts. It has been estimated that over half of the married couples in the United States have some form of sexual difficulty.

Impotence

1. Impotence is defined as the inability to have an erection.
2. Primary impotence is the condition in which a man has never been able to have an erection sufficient for coitus to take place.
3. Secondary impotence is a condition in which the ability to achieve an erection is lost after successful attempts have been achieved.
4. The fear of failure in sexual intercourse has been implicated as the greatest factor in causing impotence.

Premature Ejaculation

1. A premature ejaculator is a male who cannot control his ejaculation for 30 to 60 seconds after penile-vaginal penetration.

2. Treatment for premature ejaculators is best carried out in a sex therapy clinic where the therapists are properly trained and the proper histories of both partners can be studied.

Orgasmic Dysfunction

1. Females who have never reached an orgasm are classed as primary sufferers, whereas women who have had at least one orgasm, but who are thereafter unable to achieve orgasm, are classed as situational sufferers.
2. Present therapy for orgasmic dysfunction requires good counseling from qualified therapists.

Vaginismus

1. In vaginismus, involuntary contractions of the muscles surrounding the outer third of the vagina make insertion of the erect penis impossible or at least very difficult.
2. Treatment involves psychological counseling and a series of exercises to dilate the vagina.

BIRTH CONTROL

Methods of Birth Control

1. Methods of birth control fall into four classes or techniques: abstinence, contraception, sterilization, and abortion.
2. Abstinence means to avoid intercourse during the time when ovulation and fertilization are most likely to occur.
3. Contraception describes any method that prevents fertilization or implantation from taking place. It includes physical barriers such as the condom and diaphragm, the IUD, and chemical means including "the pill."
4. Sterilization usually requires surgery to remove or tie off a portion of a duct or organ so that the reproductive function is blocked.
5. Abortion involves removal of the contents of the pregnant uterus—the fetus, placenta, and built-up tissue on the uterine wall. Abortions include miscarriages, or spontaneous abortions, induced abortions, very early abortions and late abortions.

REVIEW QUESTIONS

1. Define coitus and orgasm.
2. Explain sexual arousal. What are erogenous zones?
3. Define foreplay. What responses constitute the base line for sexual excitement?
4. Compare females with males as to peaks of arousal, using age as a criterion.
5. What are the four successive phases of the sexual response cycle of males and females?
6. Which two basic physiological processes are shared by both males and females as sexual responses? Explain them.
7. What are the main anatomical and physiological happenings during each of the four phases of the female sexual cycle?
8. What are the main anatomical and physiological happenings during each of the four phases of the male sexual cycle?
9. Compare females and males as to orgasmic frequency.
10. Define sex flush, orgasmic platform, and tenting effect.
11. Contrast the three most common response patterns of orgasm in females as described by Masters and Johnson.
12. Explain why penile erection and ejaculation are said to be reflex phenomena.
13. What is the physiological refractory period in the male?
14. Define impotence and differentiate between primary and secondary impotence.
15. What is premature ejaculation?
16. Describe orgasmic dysfunction in females.
17. Define vaginismus and its treatment.
18. Methods of birth control fall into four classes or techniques. What are they?
19. Explain abstinence. What is the rhythm method of birth control?
20. Distinguish between the condom, diaphragm, and IUD as methods of contraception.
21. List several examples of functions of chemical contraceptives. Include the advantages and disadvantages.
22. What are the most commonly performed surgical sterilization procedures of both females and males?

Chapter 20

EMBRYOLOGY: DEVELOPMENT AND INHERITANCE

STUDENT OBJECTIVES

After reading this chapter, you should be able to

- Define chromosome number, gametes, homologous chromosomes, meiosis, diploid number, and haploid number
- Contrast the events of spermatogenesis and oogenesis
- Describe the activities associated with fertilization and implantation
- Discuss the formation of the primary germ layers, embryonic membranes, and placenta
- List the body structures produced by the three primary germ layers
- Describe the function of the embryonic membranes
- Describe the roles of the placenta and umbilical cord during embryonic and fetal growth
- Discuss the principal body changes associated with fetal growth
- Compare the sources and functions of the hormones secreted during pregnancy
- Describe the three stages of labor
- Explain prenatal photography
- Describe the physiology of lactation
- Define inheritance and genetics, compare

- genotype and phenotype, and explain a Punnett square
- Describe the inheritance of PKU and define dominant, recessive, homozygous, and heterozygous
- Explain inheritance of sex, color blindness, and X-linked inheritance
- Describe some dominant and recessive traits in human beings
- Describe the procedure of amniocentesis and its value in preventing disease in the newborn
- Describe the cause and symptoms of Down's syndrome

In the preceding chapter it was pointed out that not all sexual activity is related to procreation. Indeed, in our modern society, sexual activity is pursued increasingly for personal enjoyment. But sexual activity is the natural way for initiating reproduction, and reproduction is necessary for the survival of the human species.

FORMATION OF GAMETES

Chromosome Number

Each human being develops from the union of an ovum and a sperm. Ova and sperm are collectively called **gametes,** and they differ radically from almost all the other cells in the body in that they have only half the normal number of chromosomes in their nuclei. **Chromosome number** is the number of chromosomes contained in each nucleated cell that is not a gamete. Chromosome numbers vary from species to species. The human chromosome number is 46, which means that every somatic cell contains 46 chromosomes in its nucleus. In other words, there are 23 pairs of chromosomes in each cell other than a gamete. Two chromosomes that belong to a pair are called **homologous chromosomes.** The ovum or spermatozoan has only one-half of each pair. Another word for chromosome number is **diploid number** (*di-* = two), symbolized as $2n$. The duplicate set of chromosomes has an important genetic significance, but now we shall look only at how it relates to the development of the gametes.

Suppose that a sperm containing 46 chromosomes fertilizes an egg that also contains 46 chromosomes. You would expect the first generation of offspring to have 92 chromosomes, the second generation 184 chromosomes, and so on. In reality, the chromosome number does not double with each generation because of a different kind of cellular division called **meiosis.** Meiosis occurs only in sex cells before they mature. It causes a developing sperm or ovum to relinquish its duplicate set of chromosomes so that the mature gamete has only 23. This is called the **haploid** (one-half) **number,** and is symbolized as n.

In the testes, the formation of haploid spermatozoa by meiosis is called **spermatogenesis.** In the ovary, the formation of a haploid ovum by meiosis is referred to as **oogenesis.**

Spermatogenesis

In humans, the process of spermatogenesis takes about $2\frac{1}{2}$ months. The seminiferous tubules are lined with immature cells called *spermatogonia* or sperm mother cells (Figure 20-1). Spermatogonia contain the diploid chromosome number and are the precursor cells for all the spermatozoa that the male will produce. At puberty, when the anterior pituitary secretes FSH in response to FSHRF from the hypothalamus, the spermatogonia embark on a lifetime of continual mitotic division. As a result of their active mitosis, some of the daughter cells are pushed toward the lumen of the seminiferous tubule. These cells lose contact with the basement membrane of the seminiferous tubule and undergo a series of developmental changes. The cells first become known as *primary spermatocytes* and, like spermatogonia, are diploid. The other daughter cells formed by mitosis of the spermatogonia remain near the basement membrane and function as a reservoir of spermatogonial cells.

Each primary spermatocyte then undergoes a re-

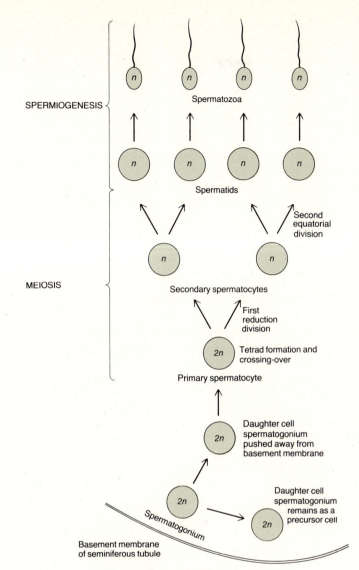

Figure 20-1 Spermatogenesis.

ing crossing-over, the spindle forms and the spindle threads attach to the centromeres of the paired chromosomes.

Of the 23 pairs of chromosomes present in these cells, 22 pairs are known as the *autosomes,* and the twenty-third pair consists of the *sex chromosomes,* so called because they have evolved to function as determiners of sex. Only in male cells does this pair of chromosomes *not* form a matching set. One chromosome is longer than its partner and is called the X chromosome; its shorter partner is designated the Y chromosome.

As the 23 pairs of chromosomes separate, one member of each pair migrates to the opposite pole of the dividing cell. In this way, one-half of the cells formed by the first nuclear division possesses an X chromosome and one-half a Y chromosome. The cells with only 23 single chromosomes (the haploid number) are called *secondary spermatocytes.* Each chromosome, however, is made up of two chromatids. Moreover, the chromosomes of secondary spermatocytes are rearranged as a result of crossing-over.

The second cellular division of meiosis is called an *equatorial division.* In this division, there is no replication of DNA. The chromosomes (each of two chromatids) line up in single file around the equatorial plane, and the chromatids of each chromosome separate from each other. The cells thus formed from the equatorial division are called *spermatids.* Each contains half the original chromosome number, or 23 chromosomes, and is said to be haploid. Each primary spermatocyte therefore produces four spermatids by meiosis (reduction division and equatorial division). Spermatids lie close to the lumen of the seminiferous tubule.

The final stage of spermatogenesis involves the maturation of spermatids into spermatozoa. This maturation process is called *spermiogenesis.* During this process, each spermatid embeds in a Sertoli cell and develops a head with an acrosome and a flagellum (tail). Sertoli cells nourish the developing spermatids. The end product of the maturation process is mature spermatozoa. Since there is no cell division involved in spermiogenesis, each spermatid develops into a single spermatozoan.

duction division. It is the first of two nuclear divisions that takes place as part of meiosis. In the first nuclear division, DNA is replicated, and the 46 chromosomes formed move toward the equatorial plane of the nucleus. Here, they line up by homologous pairs so that there are 23 pairs of chromosomes in the center of the nucleus. The four chromatids of each of the homologous pairs of chromosomes twist around each other. Such a group of four chromatids is called a *tetrad.* Within a tetrad, portions of one chromatid are exchanged with portions of another. This process, called *crossing-over,* permits an exchange of genes among chromatids (Figure 20-2). In other words, crossing-over results in the recombination of genes. This means that the spermatozoa produced by any one male are most unlikely to carry the same sets of genetic information. This genetic recombination accounts for the great variation among humans. Follow-

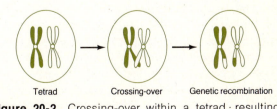

Tetrad   Crossing-over   Genetic recombination

Figure 20-2 Crossing-over within a tetrad, resulting in genetic recombination.

In summary, then, a single primary spermatocyte develops into four spermatozoa by the processes of meiosis and spermiogenesis. Spermatozoa enter the lumen of the seminiferous tubule and migrate to the epididymis, where they are temporarily stored. After spermatozoa have been in the epididymis for about 10 days, they complete their maturation and become capable of fertilizing an ovum. Most spermatozoa are probably stored in the ductus deferens.

Oogenesis

The formation of a haploid ovum by meiosis in the ovary is referred to as *oogenesis*. With some exceptions, oogenesis occurs in essentially the same manner as spermatogenesis. These exceptions will be noted in the discussion that follows. Like spermatogenesis, oogenesis involves mitosis, meiosis, and maturation.

The female germ cells, called the *oogonia,* first appear in the interior stroma of the ovary during early embryonic development. These germ cells migrate into the ovary from their origin in the blood islands of the embryonic yolk sac. During fetal development, the numbers of oogonial cells reach several million, but then their number begins to decrease rapidly as most of these germinal cells die. At the time the female child is born, there are only about 2 million oogonial cells in the two ovaries. During postnatal growth, the total continues to diminish so that at puberty about 400,000 are still present. The normal female usually releases only one mature egg in each menstrual cycle (about every 28 days). This corresponds to 13 ova per year for up to about 40 years. This means that less than 500 eggs are liberated during her reproductive years. Although the number of viable egg cells decreases during the reproductive years of a female, at the time of menopause, there are still a significant number present. Toward the end of her reproductive life, a woman releases eggs that are 30 or more years older than those she released early in her reproductive life. Some investigators find significance in the correlation between the increase in certain congenital defects in the children of older mothers and the aging of the ova.

During the early embryonic development of the ovaries, the oogonial cells become enveloped by a layer of squamous epithelial cells to form a new structure called a *primary follicle.* Toward the end of the third month of embryonic development, the oogonial cells within the follicle begin to increase in size and now become known as *primary oocytes.* At this time, the primary oocytes begin the meiotic process. However, the process is only initiated; the first step or prophase stage is entered into, but is not completed until some time after puberty. The primary oocytes remain in this arrested phase of meiosis within the follicle until the follicle is about to rupture and release the potential ovum. Release of the oocyte from the follicle is primarily under the control of LH.

Starting at puberty, several follicles respond each month to the rising level of FSH. As the cycle proceeds and LH is secreted from the anterior pituitary, one of the follicles reaches a stage in which the diploid ovum, now a *primary oocyte,* undergoes its reduction division (Figure 20-3). In this division, the first of the two nuclear divisions of meiosis, two cells of unequal size, both with 23 chromosomes of two chromatids each are produced. The smaller cell is called the *first polar body* and is essentially a packet of discarded nuclear material. The larger cell, which receives most of the cytoplasm, is known as the *secondary oocyte.*

At this stage in the ovarian cycle, ovulation takes place. Since the secondary oocyte with its polar body and surrounding supporting cells is discharged at the time of ovulation, the "ovum" discharged is not yet fully mature. The discharged secondary oocyte enters the uterine tube. The second meiotic division does not begin unless the ovum is fertilized. If this division (the equatorial division) occurs, the secondary oocyte also produces two cells of unequal size, each of which

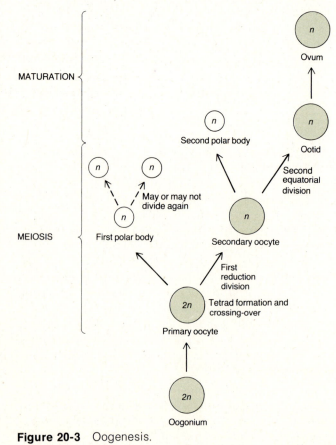

Figure 20-3 Oogenesis.

is haploid. The larger of the cells is called the *ootid* and the smaller is known as the *second polar body*. Quickly, the ootid completes fusion with the spermatozoan to become a zygote—a one-celled fertilized egg.

The first polar body may undergo a second division. If it does, meiosis of the original primary oocyte results in a single haploid ovum and three polar bodies. Eventually, the polar bodies disintegrate. Thus, in the female, one oogonium produces a single ovum that requires years to complete, whereas in the male, each spermatocyte produces four spermatozoa and takes only weeks.

Movement of Sperm

During sexual intercourse, ejaculation introduces millions of sperm into the vagina. Within 3 minutes, sperm have reached the entrance to the cervix. Sperm swim at a speed of about 0.5 cm per minute. So, in order to cover distances of up to 10 cm in 3 minutes, they must be aided in their transport by other forces. Cilia lining the route beat to set up fluid currents, and muscular contractions of the cervix occur. Both seem to aid the movement of the sperm into the uterine cavity. Inside the uterus, rhythmic contractions of the muscular walls greatly aid the sperm as they swim toward the uterine tubes. Sperm can be detected entering the uterine tubes about 30 minutes following ejaculation into the vagina. There is some evidence that suggests the hormone oxytocin may be released from the posterior pituitary during orgasm and/or that prostaglandins in the seminal fluid function to cause the uterine contractions.

The slightly acidic pH of the vagina greatly hinders the motility of the sperm. The cervical mucus, however, is alkaline and helps to neutralize the vaginal acidity. At the time of ovulation, the cervical mucus reaches its highest level of alkalinity and thus supports the sperm movement from the vagina into the uterus. Even so, the loss of active sperm is considerable. The exact manner by which the sperm and egg find each other is unknown. It is most likely that the relatively large size of the egg and the great number of sperm make the probability of encounter high. Actually, several sperm attach to the ovum, but usually only one penetrates the egg to cause fertilization.

PREGNANCY

Pregnancy is a sequence of events including fertilization, implantation, embryonic growth, and in normally, fetal growth that terminates in birth.

Fertilization

The term **fertilization** is applied to the union of the sperm nucleus and the nucleus of the ovum. It normally occurs in the uterine tube when the ovum is about one-third of the way down the tube, usually within 24 hours after ovulation (Figure 20-4a). Peristaltic contractions and the action of cilia transport the ovum through the uterine tube to meet the sperm.

Sperm must remain in the female genital tract for 4 to 6 hours before they are capable of fertilizing an ovum. During this time, a series of enzymes is released or exposed on the surface of the acrosome. One enzyme has been identified as hyaluronidase. Hyaluronidase apparently dissolves parts of the membrane covering the ovum. Normally, only one spermatozoan fertilizes an ovum, because once union is achieved, the egg membrane and its surrounding coat undergo rapid chemical changes to produce a fertilization membrane that is impermeable to the entrance of other spermatozoa. When the spermatozoan has entered the ovum, the tail is shed and the nucleus in the head develops into a structure called the *male pronucleus*. The nucleus of the ovum also develops into a *female pronucleus*. After the pronuclei are formed, they fuse to produce a *segmentation nucleus* —a process termed fertilization. The segmentation nucleus contains 23 chromosomes from the male pronucleus and 23 chromosomes from the female pronucelus. Thus, the fusion of the two haploid pronuclei restores the diploid number. The fertilized ovum, consisting of a segmentation nucleus, cytoplasm, and enveloping membrane, is referred to as a *zygote*.

Immediately after fertilization, rapid cell division of the zygote begins (Figure 20-4b). This division of the zygote is called *cleavage*. The progressively smaller cells produced are called *blastomeres*. Successive cleavages produce a solid mass of cells, the *morula*, which is only slightly larger in size than the original zygote.

Implantation

As the morula descends through the uterine tube, it continues to divide and form itself into a hollow ball of cells. At this stage of development, the mass is referred to as a *blastocyst* (Figure 20-4c). The blastocyst consists of an outer covering of cells called the *trophoblast* and an *inner cell mass*, and the internal cavity which is the *blastocoel*. The trophoblast ultimately will form the membranes composing the fetal portion of the placenta, and the inner cell mass will develop into the embryo. About the third or fourth day after fertilization, the blastocyst enters the uterine cavity.

The attachment of the blastocyst to the endometrium occurs 7 to 8 days following fertilization, and the process is called *implantation* (Figure 20-4d). At this time, the endometrium of the uterine wall is in its postovulatory phase. During implantation, the cells of the trophoblast secrete an enzyme that enables the blastocyst to literally "eat a hole" in this uterine lining and become completely buried in the endometrium, usually on the posterior wall of the fundus, or body, of the uterus. The portion of the endometrium to which the blastocyst adheres and in which it becomes implanted is the decidua functionalis layer. Implantation enables the blastocyst to absorb nutrients from the glands and blood vessels of the endometrium for its subsequent growth and development.

Embryonic Period

The first 2 months of development are considered the **embryonic period.** During this period, the developing human is called an **embryo.** After the second month, it is called a **fetus.** By the end of the embryonic period, the rudiments of all the principal adult organs are present, the embryonic membranes are developed, and the placenta is functioning.

Beginnings of Organ Systems

Following implantation, the inner cell mass of the blastocyst begins to differentiate into the *three primary germ layers:* the *ectoderm, endoderm,* and *mesoderm.* The primary germ layers are the embryonic tissues from which all tissues and organs of the body will develop. The fetal membranes, structures that lie outside the embryo and protect and nourish it, also develop from these three germ layers.

In the human being, the germ layers are formed so quickly that it is difficult to determine the exact sequence of events. Before implantation, a layer of *ectoderm* (the trophoblast) already has formed around the blastocoel (Figure 20-5a). The trophoblast will become part of the chorion—one of the fetal membranes. Within 8 days after implantation, the inner cell mass moves downward creating a space called the amnionic cavity between the inner cell mass and the trophoblast (Figure 20-5b). The bottom layer of the inner cell mass develops into an *endodermal* germ layer.

About the twelfth day after fertilization, the striking

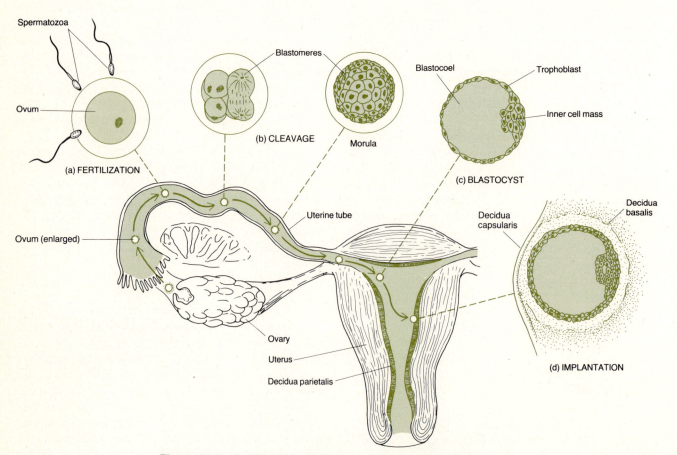

Figure 20-4 Fertilization, cleavage, and implantation of an ovum.

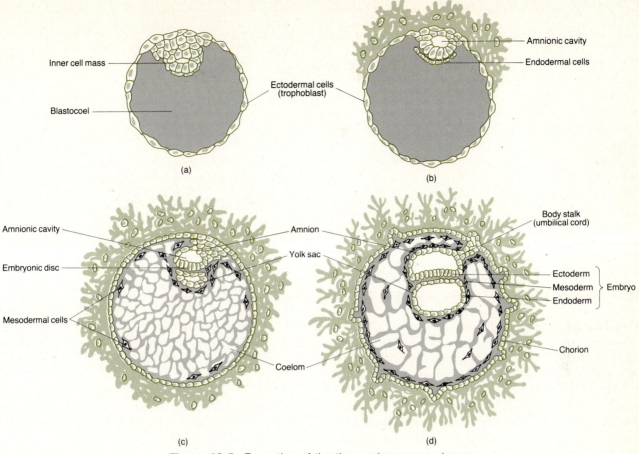

Figure 20-5 layer images labeled:

(a) Inner cell mass · Blastocoel · Ectodermal cells (trophoblast)

(b) Amnionic cavity · Endodermal cells

(c) Amnionic cavity · Embryonic disc · Mesodermal cells · Amnion · Yolk sac · Coelom

(d) Amnion · Yolk sac · Body stalk (umbilical cord) · Ectoderm · Mesoderm · Endoderm } Embryo · Chorion

Figure 20-5 Formation of the three primary germ layers.

changes shown in Figure 20-5c appear. A layer of cells from the inner cell mass has grown around the top of the amnionic cavity. These cells will become the *amnion,* another fetal membrane. The cells below the cavity are called the *embryonic disc;* these cells will form the embryo. The embryonic disc contains scattered ectodermal, mesodermal, and endodermal cells in addition to the endodermal layer observed in Figure 20-5b. Notice in Figure 20-5c that the cells of the endodermal layer have been dividing rapidly, so groups of them now extend downward in a circle. This circle is the *yolk sac,* another fetal membrane. The *mesodermal cells* have also been dividing, and many have left the area of the embryonic disc and can be seen around the structures that are becoming fetal membranes.

About the fourteenth day, the scattered cells in the embryonic disc separate into three distinct layers: the upper ectoderm, the middle mesoderm, and the lower endoderm (Figure 20-5d). At this time, the two ends of the embryonic disc draw together, squeezing off the yolk sac. The resulting cavity inside the disc is the endoderm-lined *primitive gut.* The mesoderm within the disc soon splits into two layers, and the space between the layers becomes the *coelom,* or body cavity.

As the embryo develops, the endoderm becomes the epithelium lining the digestive tract and a number of other organs. The mesoderm forms the peritoneum, muscle, bone, and other connective tissue, and the ectoderm develops into the skin and nervous system. Exhibit 20-1 provides more details about the fates of these primary germ layers.

Embryonic Membranes

During the embryonic period, the *embryonic membranes* form. These membranes lie outside the embryo and will protect and nourish the fetus. The membranes are the yolk sac, the amnion, the chorion, and the allantois (Figure 20-6).

The *yolk sac* is an endoderm-lined membrane that encloses the yolk. In many species the yolk provides the primary or exclusive nutrient for the embryo, and consequently, the ova of these animals contain a great deal of yolk. However, since the human embryo receives its nourishment from the endometrium, the human yolk sac never develops further, and during an early stage of development it becomes a nonfunctional part of the umbilical cord.

The *amnion* is a thin, protective membrane that initially overlies the embryonic disc. As the embryo grows, the amnion entirely surrounds the embryo and

Exhibit 20-1 STRUCTURES PRODUCED BY THE THREE PRIMARY GERM LAYERS

Endoderm	Mesoderm	Ectoderm
Epithelium of digestive tract and its glands	Skeletal, smooth, cardiac muscle	Epidermis of skin
Epithelium of urinary bladder and gallbladder	Cartilage, bone, other connective tissues	Hair, nails, skin glands
Epithelium of pharynx, auditory tube, tonsils, larynx, trachea, bronchi, lungs	Blood, bone marrow, lymphoid tissue	Lens of eye
	Endothelium of blood vessels and lymphatics	Receptor cells of sense organs
Epithelium of thyroid, parathyroid, thymus glands	Mesothelium of coelomic and joint cavities	Epithelium of mouth, nostrils, sinuses, oral glands, anal canal
Epithelium of vagina, vestibule, urethra, associated glands	Epithelium of kidneys and ureters	Enamel of teeth
Adenohypophysis	Epithelium of gonads and associated ducts	Entire nervous tissue, except adenohypophysis
	Epithelium of adrenal cortex	
	Stroma of most soft organs, except those of central nervous system	

becomes filled with a fluid called *amnionic fluid*. Amnionic fluid serves as a shock absorber for the fetus. The amnion usually ruptures just before birth, and it and its fluid constitute the so-called bag of water.

The *chorion* derives from the trophoblast of the blastocyst and its associated mesoderm. It surrounds the embryo and, later, the fetus. Eventually the chorion becomes the principal part of the placenta, the structure through which materials are exchanged between the mother and fetus. The amnion also surrounds the fetus and eventually fuses to the inner layer of the chorion.

The *allantois* is a small vascularized membrane. Later its blood vessels serve as connections in the placenta between the mother and fetus. This connection is the umbilical cord.

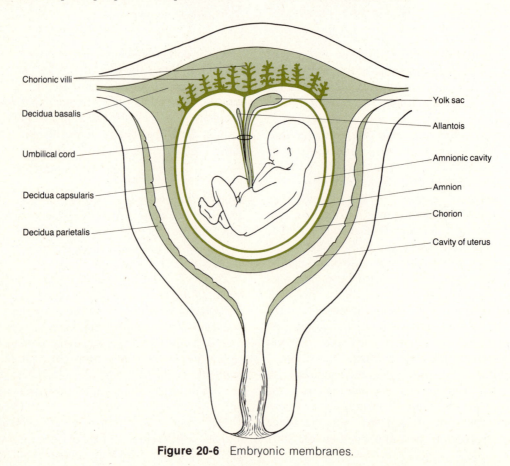

Figure 20-6 Embryonic membranes.

Placenta and Umbilical Cord

Development of the placenta, the third major event of the embryonic period, is accomplished by the third month of pregnancy. The **placenta** has the shape of a flat cake when fully developed and is formed by the chorion of the embryo and a portion of the endometrium of the mother (Figure 20-7). It provides an exchange of nutrients and wastes between fetus and mother and secretes the hormones necessary to maintain pregnancy.

If implantation occurs, a portion of the endometrium becomes modified and is known as the **decidua.** The decidua includes all but the deepest layer of the endometrium and is shed when the fetus is delivered. Different regions of the decidua are named on the basis of their positions relative to the site of the implanted ovum. The *decidua parietalis* is the portion of the modified endometrium that lines the entire pregnant uterus, except for the area where the placenta is forming (Figure 20-6). The *decidua capsularis* is the portion of the endometrium that overlies the developing embryo (Figure 20-4d and Figure 20-6). The *decidua basalis* is the portion of the endometrium between the chorion and the muscularis of the uterus (Figure 20-4d and Figure 20-6). The decidua basalis becomes the maternal part of the placenta.

During embryonic life, fingerlike projections of the chorion, called *chorionic villi,* grow into the decidua basalis of the endometrium. These will contain fetal blood vessels of the allantois. They continue growing until they are bathed in the maternal blood in the sinuses of the decidua basalis. Thus, maternal and fetal blood vessels are brought into close proximity. It should be noted, however, that maternal and fetal blood normally do not mix. Oxygen and nutrients from the mother's blood diffuse across the walls and into the capillaries of the villi. From the capillaries the nutrients circulate into the umbilical vein. Wastes leave the fetus through the umbilical arteries, pass into the capillaries of the villi, and diffuse into the maternal blood. The **umbilical cord** consists of an outer layer of amnion containing the umbilical arteries and umbilical vein, supported internally by mucous connective tissue from the allantois called Wharton's jelly. At delivery, the placenta detaches from the uterus and is referred to as the "afterbirth." At this time, the umbilical cord is severed, leaving the baby on its own. The scar that marks the site of the entry of the fetal umbilical cord into the abdomen is called the *umbilicus.*

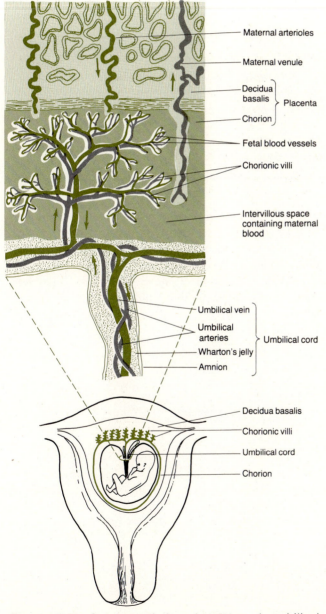

Maternal arterioles

Maternal venule

Decidua basalis
Chorion } Placenta

Fetal blood vessels

Chorionic villi

Intervillous space containing maternal blood

Umbilical vein
Umbilical arteries } Umbilical cord
Wharton's jelly
Amnion

Decidua basalis

Chorionic villi

Umbilical cord

Chorion

Figure 20-7 Structure of the placenta and umbilical cord.

Fetal Growth

During the **fetal period,** organs established by the primary germ layers grow rapidly. The organism takes on a human appearance. A summary of changes associated with the fetal period is presented in Exhibit 20-2.

Hormones of Pregnancy

Following fertilization, the corpus luteum is maintained until about the fourth month of pregnancy. For most of this time, it continues to secrete estrogens and progesterone. Both these hormones maintain the lining of the uterus during pregnancy and prepare the mammary glands to secrete milk. The amounts of estrogens and progesterone secreted by the corpus

Exhibit 20-2 CHANGES ASSOCIATED WITH FETAL GROWTH

End of Month	Approximate Size and Weight	Representative Changes
1	0.6 cm ($\frac{3}{16}$ inch)	Eyes, nose, and ears not yet visible. Backbone and vertebral canal form. Small buds that will develop into arms and legs form. Heart forms and starts beating. Body systems begin to form.
2	3 cm ($1\frac{1}{4}$ inches) 1 g ($\frac{1}{30}$ oz)	Eyes far apart, eyelids fused, nose flat. Ossification begins. Limbs become distinct as arms and legs. Digits are well formed. Major blood vessels form. Many internal organs continue to develop.
3	7.5 cm (3 inches) 28 g (1 oz)	Eyes almost fully developed but eyelids still fused, nose develops bridge, and external ears are present. Ossification continues. Appendages are fully formed and nails develop. Heartbeat can be detected. Body systems continue to develop.
4	18 cm ($6\frac{1}{2}$–7 inches) 113 g (4 oz)	Head large in proportion to rest of body. Face takes on human features and hair appears on head. Skin bright pink. Many bones ossified, and joints begin to form. Continued development of body systems.
5	25–30 cm (10–12 inches) 227–454 g ($\frac{1}{2}$–1 lb)	Head less disproportionate to rest of body. Fine hair (laguno hair) covers body. Skin still bright pink. Rapid development of body systems.
6	27–35 cm (11–14 inches) 567–681 g ($1\frac{1}{4}$–$1\frac{1}{2}$ lb)	Head becomes less disproportionate to rest of body. Eyelids separate and eyelashes form. Skin wrinkled and pink.
7	325–425 cm (13–17 inches) 1,135–1,362 g ($2\frac{1}{2}$–3 lb)	Head and body become more proportionate. Skin wrinkled and pink. Seven-month fetus (premature baby) is capable of survival.
8	41–45 cm ($16\frac{1}{2}$–18 inches) 2,043–2,270 g ($4\frac{1}{2}$–5 lb)	Subcutaneous fat deposited. Skin less wrinkled. Testes descend into scrotum. Bones of head are soft. Chances of survival much greater at end of eighth month.
9	50 cm (20 inches) 3,178–3,405 g (7–$7\frac{1}{2}$ lb)	Additional subcutaneous fat accumulates. Laguno hair shed. Nails extend to tips of fingers and maybe even beyond.

luteum, however, are only slightly higher than those produced after ovulation in a normal menstrual cycle. The high levels of estrogens and progesterone needed to maintain pregnancy and develop the breasts for lactation are provided by the placenta.

During pregnancy, the chorion of the placenta secretes a hormone called *human chorionic gonadotropin*, or *HCG*. This hormone is excreted in the urine of pregnant women from about the middle of the first month of pregnancy, reaching its peak of excretion during the third month. The HCG level decreases sharply during the fourth and fifth months and then levels off until childbirth. Excretion of HCG in the urine serves as the basis for some pregnancy tests. The primary role of HCG seems to be to maintain the activity of the corpus luteum, especially with regard to continuous progesterone secretion—an activity necessary for the continued attachment of the fetus to the lining of the uterus (Figure 20-8).

The placenta begins to secrete estrogens and progesterone no later than the sixtieth day of pregnancy (Figure 20-9). They are secreted in increasing quantities until the time of birth. Once the placenta is established, the secretion of HCG is cut back drastically at about the fourth month. The corpus luteum then disintegrates—it is no longer needed

because the placenta supplies the levels of estrogens and progesterone needed to maintain the pregnancy. The fetal hormones thus take over the management of the mother's body in preparation for parturition (birth) and lactation. Following delivery, estrogens and progesterone in the blood decrease to normal levels.

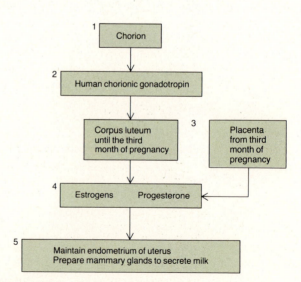

Figure 20-8 Hormones of pregnancy.

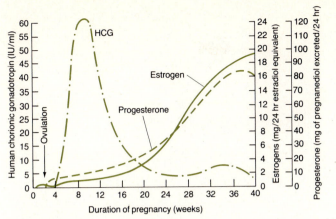

Figure 20-9 Rates of secretion of estrogens, progesterone, and human chorionic gonadotropin at different stages of pregnancy.

Parturition and Labor

The time during which the embryo or fetus is carried in the uterus is called **gestation,** and in the human this period lasts about 266 days. It begins at the time of fertilization and ends at the time of delivery. Pregnancy, however, is timed somewhat differently. In medical practice, it is common to date the onset of pregnancy from the time of the last menstrual period. This adds about 14 days to the period of gestation. The most common duration for a pregnancy is therefore 280 days after the last menstrual period.

The term **parturition** refers to birth. Parturition is preceded by a sequence of events commonly called **labor.** The onset of labor stems from a complex interaction of many factors, especially hormones. Just prior to birth, the muscles of the uterus contract rhythmically and forcefully. Both placental and ovarian hormones play a dominant role in the contractions. Recall that estrogens stimulate uterine contractions, whereas progesterone inhibits them. Until the effects of progesterone are effectively diminished, labor cannot take place. Reference to Figure 20-9 will show that after about 26 weeks, the level of estrogens rises to a greater value than that of progesterone. At about the time of delivery the level of progesterone begins to fall. The ratio of estrogens to progesterone increases to the point where the action of estrogens prevails and helps to cause uterine contractions. Coupled with the estrogenic stimulation is the action of *oxytocin,* a hormone released from the posterior pituitary. This hormone appears in increased amounts during labor and aids in causing uterine contractions. Some recent animal experiments indicate that a neurogenic reflex may cause the increase in oxytocin levels. Stretching of the cervix (as might occur when the baby's head moves into the cervical canal) triggers nerve impulses that reach the hypothalamus and then the posterior pituitary.

Uterine contractions occur in waves, quite similar to peristaltic waves, that start at the top of the uterus and move downward. These waves expel the fetus. *True labor* begins when pains occur at regular intervals. The pains correspond to uterine contractions. As the interval between contractions shortens, the contractions intensify. Another sign of true labor is localization of pain in the back, which is intensified by walking. The final indication of true labor is the "show" and dilation of the cervix. The "show" is a discharge of a blood-containing mucus that accumulates in the cervical canal during pregnancy. Cervical dilation will be discussed shortly. In *false labor,* by contrast, pain is felt in the abdomen at long, irregular intervals. The pain does not intensify and is not altered significantly by walking. Also, in false labor there is no "show" and no cervical dilation.

The first stage of labor, called the *stage of dilation,* is the period of time from the onset of labor to the complete dilation of the cervix (Figure 20-10). During this stage there are regular contractions of the uterus, a rupturing of the amnionic sac, and complete opening (10 cm) of the cervix. The next stage of labor, the *stage of expulsion,* is the period of time from complete cervical dilation to delivery. In the final stage, the *placental stage,* the placenta or "afterbirth" is expelled. A few minutes after delivery, powerful uterine contractions expel the placenta. These contractions also constrict blood vessels that were torn during delivery, reducing the possibility of hemorrhage.

Prenatal Photography

Great strides in prenatal care have been achieved because of a photographic process that has enabled doctors to photograph a fetus before it is born. It is basically the same method used by the navy for tracking submarines. For medical diagnosis, it involves projecting ultra high frequency sound waves into body tissues and charting the echoes that bounce back. The procedure is painless. The only discomfort an expectant mother experiences is when a cold gel is applied to her abdomen, and this is done simply to enhance conduction of sound waves. The echo returning from the interfaces between tissues of different densities is used to form a picture. The picture, called a *β-scan,* appears on an oscilloscope. The differing echo strengths can be seen as the picture's varying lines form. Visible in most sonar scan pictures are the placenta, the amnionic fluid, and a general outline of the fetus' trunk and limbs. The head appears as a solid line.

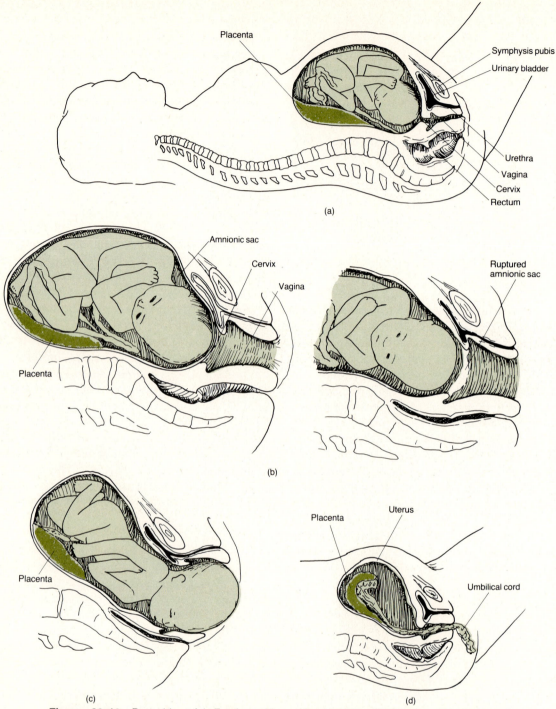

Figure 20-10 Parturition. (a) Fetal position prior to birth. (b) Dilation. Protrusion of amnionic sac through partly dilated cervix. Amnionic sac ruptured and complete dilation of cervix. (c) Stage of expulsion. (d) Placental stage.

The use of diagnostic sonar in obstetrics and gynecology is relatively new. It is preferred over x-ray for use in obstetrics because the sound waves are apparently harmless to the fetus. The value of such a development is considerable. If an initial β-scan showed an enlarged head, for example, other tests could then be used to look at the brain for signs of hydrocephalus. In the future, doctors hope that limb deformities might also be detected by this technique.

Presently, diagnostic sonar is used on pregnant women to determine approximate fetal age, fetal growth, fetal abnormalities such as an ectopic pregnancy or placental bleeding, and fetal position—normal or breech. It also shows if a multiple birth is expected. In addition to the routine monitoring of fetal age and development, sonar is particularly useful in helping determine the proper care of women with high-risk pregnancies.

Sonar is also used in some cancer diagnoses and has been used by cardiologists for heart examinations.

Lactation

The term **lactation** refers to the secretion of milk by the mammary glands. During pregnancy, the mammary glands develop a rich system of ducts and alveoli under the influence of the estrogens and progesterone. At the time of delivery, the mammary glands secrete only a fluid called *colostrum,* an essentially "fat free" kind of milk. True lactation is suppressed by the inhibitory actions of estrogens and progesterone. The high levels of these hormones prevent the release of the hormone *prolactin* from the anterior pituitary. When the levels of estrogens and progesterone in the mother's blood decrease at birth, the hormone prolactin is released to stimulate the glandular tissue of the mammary glands (Figure 20-11). The glands respond to the stimulus by secreting large quantities of fat, the sugar lactose, and a protein called casein. Within 3 to 4 days, the mammary glands are able to produce large volumes of milk.

Although milk can be produced in large volumes, it does not necessarily pour out freely from the nipples. The milk is obtained or released from the breasts by the suckling action of the newborn child. This suckling action triggers an unusual combined neurogenic and hormonal reflex called the *milk "letdown"* phenomenon. The suckling action by the infant on the breast sends afferent nerve impulses via the spinal cord to the hypothalamus. Oxytocin is released from the posterior pituitary on receipt of nerve stimuli from the hypothalamus. The oxytocin is distributed by the blood and, in the breast, the hormone causes the myoepithelial cells around the alveoli to *contract.* The contents of the alveoli are thus ejected or "let down" into the ducts and made available to the suckling baby. The entire reflex phenomenon requires less than one minute. Sometimes a mother can trigger the "letdown" reflex merely by thinking about nursing her baby. Other new and anxious mothers discover they have difficulty in nursing their child. Since it is known that many forces can inhibit the "letdown" reflex, it is likely that anxiety, fear, apprehension—any one of a gamut of psychogenic stimuli—can block the secretion of oxytocin and milk delivery. A quiet, undisturbed, and peaceful environment where mother and child can "get to know each other" very often solves this type of problem. Lactation often delays the onset of the normal menstrual cycles for periods up to months following delivery. The continued suckling of the child maintains the elevated level of prolactin production by the anterior pituitary. Presumably, this synthetic pathway has priority over those producing FSH and LH, so the hormones for initiating the female cycle are insufficient to start the cycle. Breast feeding, therefore, has often been used as a contraceptive technique. However, the effectiveness decreases with time following parturition.

INHERITANCE

Inheritance is the passage of parental traits from one generation to another. It is the process by which we

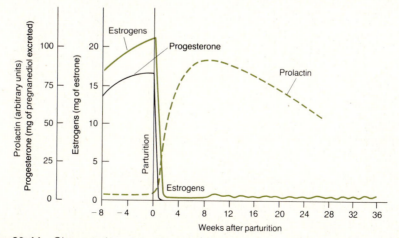

Figure 20-11 Changes in rates of secretion of estrogens, progesterone, and prolactin at parturition and during the succeeding weeks. Note the rapid increase in prolactin secretion immediately after parturition.

acquire our peculiar characteristics from our parents and the way by which we pass on our characteristics to our children. The branch of biology that deals with inheritance is called **genetics**.

Genotype and Phenotype

Recall that the nuclei of almost all human cells, except gametes, contain 23 pairs of chromosomes, the diploid number. One chromosome from each pair originally came from the mother, and the other came from the father. The two chromosomes that make up the pair are called *homologous chromosomes*. Homologues contain genes that control the same traits. For instance, if a certain chromosome contains a gene for height, its homologue, or mate, will also contain a gene for height.

To explain the relationship of genes to heredity, we shall look at the disorder called PKU (phenylketonuria) (Figure 20-12). People with PKU are unable to manufacture the enzyme phenylalanine hydroxylase, and as a consequence phenylalanine metabolism is detoured into an unusual pathway. The end product of this pathway, phenylpyruvic acid (a phenylketone), accumulates in the blood and is excreted in the urine. In young children, circulating phenylpyruvate can produce severe mental retardation. It is believed that PKU is brought about by an abnormal gene, which can be symbolized as p. By convention, the normal gene is then symbolized as P. The chromosome that is concerned with directions for phenylalanine hy-

droxylase production will have either p or P on it. Its homologue will also have p or P. Thus, every individual will have one of the following genetic makeups, or **genotypes**: PP, Pp, or pp. Although people with genotypes of Pp have the abnormal gene, it is only those with the genotype pp who will suffer from the disorder. The reason is that the normal gene dominates and inhibits the abnormal one. A gene that dominates is called the *dominant gene,* and the trait expressed is said to be the *dominant trait.* The gene that is inhibited is called the *recessive gene,* and if that trait is expressed, it is called a *recessive trait.*

An individual who has the same genes on homologous chromosomes (for example, PP or pp) is said to be *homozygous* for the trait. If, however, the genes on homologous chromosomes are different (for example, Pp), the individual is said to be *heterozygous* for the trait. The word **phenotype** refers to how the genetic makeup is expressed in the body. A person with Pp has a different genotype from one with PP, but both will have the same phenotype since the dominant trait is expressed in both instances. In this case, it is the normal production of phenylalanine hydroxylase.

To determine how gametes containing haploid chromosomes unite to form diploid fertilized eggs, special charts called *Punnett squares* are used. Usually, the male gametes (sperm cells) are placed at the side of the chart, and the female gametes (ova) are placed at the top of the chart (Figure 20-13). The four spaces on the chart represent the possible combinations of male and female gametes that could form fertilized eggs. Possible combinations are determined simply by "dropping" each female gamete into the two spaces under it, then moving each male gamete across to the two spaces in line with it.

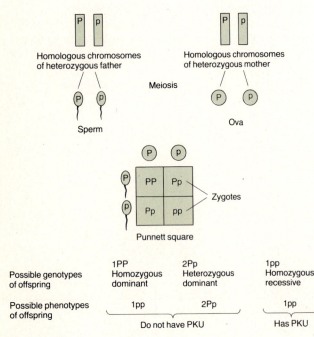

Figure 20-12 Inheritance of PKU.

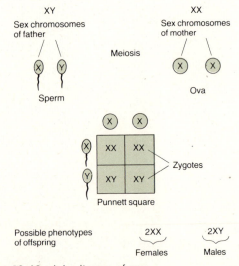

Figure 20-13 Inheritance of sex.

Several dominant and recessive traits are listed in Exhibit 20-3. Normal traits do not always dominate over abnormal ones, but genes for severe disorders are more frequently recessive than they are dominant. People who have severe disorders very often do not live long enough to pass the abnormal gene to the next generation. In this way, expression of the gene tends to be selectively lost from the population. Huntington's chorea is one example of a major disorder caused by a dominant gene. This disorder is characterized by degeneration of nervous tissue, usually leading to mental disturbance and death. The first signs of Huntington's chorea do not occur until adulthood, very often after the person has already produced offspring.

Inheritance of Sex

Microscopic examination of the chromosomes in cells reveals that one pair differs in males and in females (see Figure 7-1). In females, the pair consists of two rod-shaped chromosomes designated as X chromosomes. One X chromosome is present in males, but its mate is a hook-shaped structure called a Y chromosome. The XX pair in the female and the XY pair in the male are called the *sex chromosomes*. All other chromosomes are called *autosomes*.

The sex chromosomes are responsible for the sex of the individual (see Figure 20-13). When a spermato-

Exhibit 20-4 PROCEDURES FOR INCREASING PROBABILITY OF SEX OF OFFSPRING

For Female Offspring	For Male Offspring
1. Last intercourse should be timed to occur 2–3 days prior to ovulation.	1. Intercourse should be timed to occur immediately following ovulation.
2. Intercourse should be preceded by an acid douche (two tablespoons of white vinegar to a quart of water).	2. Intercourse should be preceded by an alkaline douche (two tablespoons of baking soda to a quart of water).
3. Female should try to avoid orgasm.	3. Female orgasm not necessary but desirable.
4. Shallow penetration of penis at time of ejaculation is best.	4. Deep penetration of penis at time of ejaculation is best.
5. Low sperm count is beneficial. Repeated intercourse prior to the last one would lower the count.	5. Intercourse should be avoided for 3–4 days prior to the critical one on day of ovulation.
6. Male partner should avoid coffee, tea, or other caffeinated beverages.	6. Male partner should drink 2–3 cups of a caffeinated beverage 15–30 minutes prior to intercourse.

cyte undergoes meiosis to reduce its chromosome number, one daughter cell will contain the X chromosome and the other will contain the Y chromosome. Oocytes have no Y chromosomes and produce only X-containing ova. If the ovum is subsequently fertilized by an X-bearing sperm, the offspring normally will be female (XX). Fertilization by a Y sperm normally produces a male (XY).

Choosing the Sex of Offspring

In recent years, considerable publicity has been received by a simple and popular set of procedures that increases the probability of parents selecting the sex of their offspring. Exhibit 20-4 lists the conditions that enhance this probability.

Color Blindness and X-Linked Inheritance

The sex chromosomes also are responsible for the transmission of a number of *nonsexual* traits. Genes for these traits appear on X chromosomes, but many of these genes are absent from Y chromosomes. This produces a pattern of heredity that is different from the pattern described earlier. Consider color blindness as an example. The gene for color blindness is a recessive one, and is designated as c. Normal vision, designated as C, dominates. The C/c genes are located on the X

Exhibit 20-3 HEREDITARY TRAITS IN HUMAN BEINGS

Dominant	Recessive
Curly hair	Straight hair
Dark brown hair	All other colors
Coarse body hair	Fine body hair
Normal skin pigmentation	Albinism
Brown eyes	Blue or gray eyes
Near or farsightedness	Normal vision
Normal hearing	Deafness
Normal color vision	Color blindness
Broad lips	Thin lips
Large eyes	Small eyes
Short stature	Tall stature
Polydactylism (extra digits)	Normal digits
Brachdactylism (short digits)	Normal digits
Syndactylism (webbed digits)	Normal digits
Hypertension	Normal blood pressure
Diabetes insipidus	Normal excretion
Huntington's chorea	Normal nervous system
Normal mentality	Schizophrenia
Migraine headaches	Normal
Normal resistance to disease	Susceptability to disease
Enlarged spleen	Normal spleen
Enlarged colon	Normal colon
A or B blood factor	O blood factor
Rh blood factor	No Rh blood factor

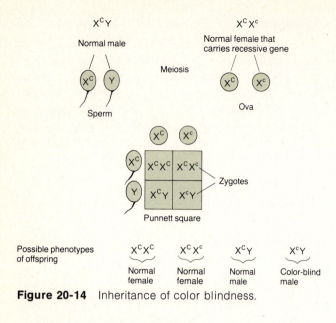

Figure 20-14 Inheritance of color blindness.

chromosome. The Y chromosome, however, does not contain the segment of DNA that programs this aspect of vision. Thus the ability to see colors depends entirely on the X chromosomes. The genetic possibilities are:

X^CX^C	Normal female
X^CX^c	Normal female that carries the recessive gene
X^cX^c	Color-blind female
X^CY	Normal male
X^cY	Color-blind male

As you can see, only females who have two X^c chromosomes are color-blind. In X^CX^c females, the trait is inhibited by the normal, dominant gene. Males, on the other hand, do not have a second X chromosome that would inhibit the trait. Therefore, all males with an X^c chromosome will be color-blind. The inheritance of color blindness is illustrated in Figure 20-14.

Traits that are inherited in the manner we have just described are called **X-linked traits.** Another example of an X-linked trait is *hemophilia,* a condition in which the blood fails to clot or clots very slowly after a surface or internal injury. It is a much more serious defect than color blindness because people with severe hemophilia can bleed to death from even a small cut. Like the trait for color blindness, hemophilia is caused by a recessive gene. If H represents normal clotting, and h represents abnormal clotting, then X^hX^h females will have the disorder. Males with X^HY will be normal, and males with X^hY will be hemophiliac. Actually, clotting time varies somewhat among hemophiliacs, so the condition may be affected by other genes as well.

A few other X-linked traits in human beings are nonfunctional sweat glands, certain forms of diabetes, some types of deafness, uncontrollable rolling of the eyeballs, absence of central incisors, night blindness, one form of cataract, white forelocks, juvenile glaucoma, and juvenile muscular dystrophy.

Amniocentesis

Amniocentesis is a technique of withdrawing some of the amnionic fluid that bathes the developing fetus. Though this technique has been available for nearly a century, only in the last decade has the procedure been used to diagnose possible genetic disorders. Sixteen to twenty weeks after conception, a small amount of amnionic fluid can be removed from the uterus with a hypodermic needle after the embryo has been located by the ultrasound β-scan procedure. Cells contained in the fluid are examined for biochemical defects and/or for abnormalities in chromosome number or structure. There are over 50 biochemical, inheritable disorders and close to 300 chromosomal disorders that can be detected through amniocentesis. Some disorders that can be detected include Down's syndrome, hemophilia, certain muscular dystrophies, Tay-Sachs disease, myelocytic leukemia, Klinefelter's and Turner's syndromes, sickle-cell anemia, thalassemia, and cystic fibrosis. When *both* parents are known or suspected to be genetic carriers of any one of these disorders, amniocentesis is advised.

Amniocentesis has been of great value in the new science of **genetic counseling.** Doctors who have specialized in studying genetic diseases or congenital anomalies can give advice as to whether a pregnancy should proceed or be aborted.

Figure 20-15 Down's syndrome. Appearance of chromosomes in an individual with Down's syndrome.

On example of a chromosome disorder that may be diagnosed through amniocentesis is *Down's syndrome*, or mongolism (Figure 20-15). This disorder is characterized by mental retardation; retarded physical development (short stature and stubby fingers); distinctive facial structures (large tongue, broad skull, slanting eyes, and round head); and malformation of the heart, ears, and feet. Sexual maturity is rarely attained.

Individuals with the disorder have 47 chromosomes instead of the normal 46. The extra chromosome is responsible for the syndrome. All the chromosomes of a person with Down's syndrome are in pairs except for the twenty-first pair. Here the chromosomes are present in triplicate.

Attention will now be turned to a discussion of blood and lymph. Here you will see the importance of these two fluids in maintaining homeostasis.

CHAPTER SUMMARY OUTLINE

FORMATION OF GAMETES

Chromosome Number

1. Ova and sperm are collectively called gametes.
2. Chromosome number is the number of chromosomes contained in each nucleated cell that is not a gamete.
3. The human chromosome number is 23 pairs—a total of 46 chromosomes (diploid).
4. Meiosis occurs in the process of producing sex cells. It causes a developing sperm or ovum to halve its duplicate set of chromosomes so that the mature gamete has only 23 chromosomes (haploid).

Spermatogenesis

1. Spermatogenesis occurs in the testes. It is the formation of haploid spermatozoa by meiosis.
2. In spermatogenesis the process called crossing-over permits an exchange of genes among the four chromatids or tetrads.
3. The final stage of spermatogenesis involves the maturation of spermatids into spermatozoa by a process called spermiogenesis.
4. In summary, a single primary spermatocyte develops into four spermatozoa by the process of meiosis and spermiogenesis.

Oogenesis

1. Oogenesis occurs in the ovaries. It is the formation of haploid ova by meiosis.
2. A primary oocyte undergoes reduction division, producing two cells of unequal size. The smaller cell is called the first polar body, and the larger cell is known as the secondary oocyte.
3. The secondary oocyte is discharged at the time of ovulation as the ovum.

4. If the ovum is fertilized, the second meiotic division occurs, producing two cells of unequal size. The smaller cell is called the second polar body, and the larger cell is called the ootid.
5. The ootid completes fusion with the spermatozoan to become a zygote—a one-celled fertilized egg.
6. Thus, in the female, one oogonium produces a single ovum.

Movement of Sperm

1. Many forces contribute to the transportation of sperm from the vagina, where they are deposited during ejaculation, to the uterine tubes where fertilization normally takes place.
2. Sperm swim very well, but the movement of the sperm into the uterine cavity is aided by cilia and muscular contractions of the uterus.
3. It is most likely that the relatively large size of the egg and the great number of sperm make the probability of encounter (fertilization) as high as it is.

PREGNANCY

1. Pregnancy is a sequence of events that includes fertilization, implantation, embryonic growth, fetal growth, and birth.

Fertilization

1. Fertilization is the union of the sperm nucleus and the nucleus of the ovum, and normally occurs in the uterine tube.
2. The fusion of the two haploid pronuclei restores the diploid number.
3. The fertilized ovum, consisting of a segmentation nucleus, cytoplasm, and enveloping membrane, is called a zygote.

Implantation

1. Immediately after fertilization, rapid cell division of the zygote begins until a blastocyst forms.
2. The attachment of the blastocyst to the endometrium occurs 7 to 8 days following fertilization and the process is called implantation.
3. Implantation enables the blastocyst to absorb nutrients from the glands and blood vessels of the endometrium for its subsequent growth and development.

Embryonic Period

1. The first 2 months of development are considered the embryonic period.
2. The three primary germ layers—ectoderm, mesoderm, and endoderm—are the embryonic tissues from which all tissues and organs of the body will develop.
3. Embryonic membranes include the yolk sac, the amnion, the chorion, and the allantois.
4. Fetal and maternal materials are exchanged through the placenta.

Hormones of Pregnancy

1. During pregnancy, the chorion of the placenta secretes a hormone called human chorionic gonadotropin (HCG). This hormone acts on the corpus luteum so that it provides a continuous secretion of progesterone.
2. The placenta begins to secrete estrogens and progesterone no later than the sixtieth day of pregnancy.

Parturition and Labor

1. The time that the embryo or fetus is carried in the uterus is called gestation; in the human this time is 266 days.
2. The term parturition refers to birth, and parturition is preceded by a sequence of events called labor.
3. The stages of labor are: the stage of dilation, the stage of expulsion, and the placental stage.

Prenatal Photography

1. Prenatal photography, called a β-scan, employs ultra high frequency sound waves and is used on pregnant women to determine: fetal age, fetal growth, fetal abnormalities, and fetal position.
2. The same technique is also used in cancer diagnosis and by cardiologists for heart examinations.

Lactation

1. Lactation is the secretion of milk by the mammary glands, and is influenced by estrogens, progesterone, prolactin, and oxytocin.
2. The milk is obtained or released from the breasts by the suckling action of the newborn child.
3. This suckling action triggers an unusual combined neurogenic and hormonal reflex called the milk "letdown" phenomenon.

INHERITANCE

1. Inheritance is the passage of hereditary traits from one generation to another, and the branch of biology that deals with inheritance is called genetics.
2. The genetic makeup of an organism is called its genotype. The traits expressed are called its phenotype.
3. Dominant genes control a particular trait; expression of recessive genes is inhibited by dominant genes.

Inheritance of Sex

1. The last, or twenty-third pair of chromosomes are called the sex chromosomes, and in this pair the Y chromosome of the male is responsible for the sex of the individual.

Color Blindness and X-Linked Inheritance

1. Color blindness and hemophilia primarily affect males because there are no counterbalancing dominant genes on the Y chromosomes.
2. Traits that are inherited in this manner are called X-linked traits; in addition to color blindness and hemophilia, other disorders that follow this pattern include certain forms of diabetes, night blindness, juvenile glaucoma, and juvenile muscular dystrophy.

Amniocentesis

1. Amniocentesis is a technique of withdrawing some of the amnionic fluid for diagnosis of possible genetic disorders.
2. Some disorders that can be detected include Down's syndrome, hemophilia, certain muscular dystrophies, Tay-Sachs disease, Klinefelter's and Turner's syndromes, sickle-cell anemia, thalassemia, and cystic fibrosis.

REVIEW QUESTIONS

1. Define chromosome number, gametes, homologous chromosomes, meiosis, diploid number, and haploid number.
2. Compare the events associated with spermatogenesis and oogenesis.
3. Describe the sequence of events that must take place following ejaculation, in order for fertilization to occur.
4. Define fertilization. Where does it normally occur? How is a morula formed?
5. What is implantation? How does the fertilized ovum implant itself? Why is implantation vital to the blastocyst?
6. Define the embryonic period. Describe some body structures formed by the ectoderm, mesoderm, and endoderm.
7. Explain the importance of the placenta and umbilical cord to fetal growth.
8. Outline some of the major developmental changes during fetal growth.
9. Compare the sources and functions of estrogens, progesterone, and the human chorionic gonadotropic hormone. How does the placenta serve as an endocrine organ during pregnancy?
10. Define parturition and gestation. Distinguish between false and true labor.
11. Describe what happens during the stage of dilation, the stage of expulsion, and the placental stage of delivery.
12. What is prenatal photography? What is its value?
13. What is lactation? How is the female prepared for it?
14. What is the milk "letdown" phenomenon?
15. Define inheritance. What is genetics?
16. Explain how the sex of the new individual is determined.
17. Define the following terms: genotype, phenotype, dominant, recessive, homozygous, heterozygous.
18. What is a Punnett square?
19. List several dominant and recessive traits inherited in human beings.
20. Set up Punnett squares to show the inheritance of the following traits: sex, color blindness, and hemophilia.
21. Describe the technique of amniocentesis. Of what value is this procedure?
22. List some disorders that are detectable by this procedure.

Chapter 21

BLOOD AND LYMPH

STUDENT OBJECTIVES

After reading this chapter, you should be able to

- Contrast the general roles of blood, interstitial fluid, and lymph in maintaining homeostasis
- Define the principal physical characteristics of blood and its functions in the body
- Identify the plasma and formed element constituents of blood
- Compare the origins of the formed elements in blood
- Describe the structure of erythrocytes and their function in the transport of oxygen and carbon dioxide
- Define erythropoiesis and describe erythrocyte production and destruction, and iron metabolism
- Describe the importance of a reticulocyte count in the diagnosis of abnormal rates of erythrocyte production
- List the structural features and types of leucocytes
- Explain the significance of a differential count
- Discuss the role of leucocytes in phagocytosis and antibody production
- Explain the details of the immune response
- Differentiate between innate immunity and acquired immunity
- Discuss lymphoid tissue and the areas of the body where it is located
- Contrast the two types of lymphocytes— T cells and B cells—as to structure
- Define immunoglobulins, describe their structure, and differentiate between the five different classes of immunoglobulins
- Explain various antibody reactions, and define agglutination, neutralization, the complement system, and opsonization
- Describe the responses of the B cell and the T cell to antigen
- Discuss cooperation between T cells and B cells in the immune response
- Explain immune tolerance and autoimmunity, and describe the mechanism of autoimmunity
- List the components of plasma and explain their importance
- Discuss the structure of thrombocytes and explain their role in blood clotting
- Describe the stages involved in blood clotting and the various coagulation factors
- Name the factors that promote and inhibit blood clotting
- Contrast a thrombus and an embolus
- Define clotting time, bleeding time, and prothrombin time
- Explain ABO and Rh blood grouping
- Define the antigen-antibody reaction of the Rh blood grouping system
- Define erythroblastosis fetalis as a harmful antigen-antibody reaction
- Compare the location, composition, and function of interstitial fluid and lymph
- Contrast the causes of hemorrhagic, hemolytic, aplastic, and sickle-cell anemia
- Compare the clinical symptoms of polycythemia, infectious mononucleosis, and leukemia

During the process of evolution on the earth, organisms appeared that demonstrated increased abilities to control their immediate environment. Cells within these organisms became capable of performing certain specialized tasks with increasing efficiency. As the specialization and cellular efficiency increased, those cells began to lose some other—often vital—functions. The cells became less able to survive independently and so came to rely on other cells for certain of their needs. A specialized cell might be less capable of protecting itself from extreme temperatures, toxic substances, or changes in pH. It might lose its mobility and therefore be unable to pursue nutrients, and if it became part of a tissue, it might not be able to escape the effects of its own wastes. Therefore, as cells became more specialized, they also became more dependent on other cells to improve and maintain the environment in a manner most suitable for their mutual survival. As the organisms moved toward the great complexities of the higher forms of life, many specialized cells organized themselves into tissues, organs, and organ systems, with the overall goal of cooperating to maintain homeostasis for the survival of the whole. The fluid that surrounded the cells and became crucial to the cells' existence is the *interstitial fluid*. The regulation of this fluid became the ultimate and prime function of each cell in order to attain homeostasis.

In the human organism, the cardiovascular and lymphatic systems reach into all parts of the body to interact directly with interstitial fluid and deliver needed materials and remove unneeded wastes. The **cardiovascular system** includes the heart, blood vessels, and blood, whereas the **lymphatic system** includes the lymph glands, lymph vessels, and lymph fluid.

Blood picks up oxygen from the lungs, nutrients from the digestive tract, hormones from the endocrine glands, and enzymes from still other parts of the body. The blood then transports these substances to all the tissues of the body, where they diffuse into the interstitial fluid. From the interstitial fluid, the substances are taken up by the cells and exchanged for wastes. Since the blood exchanges materials with the interstitial fluid, it can also distribute disease-causing organisms throughout the body. To protect itself from spreading diseases, however, the body employs the lymphatic system, a collection of vessels containing a fluid called lymph. The lymph picks up material, including wastes, from the interstitial fluid, cleanses it of bacteria, and returns the material and wastes to the blood. The blood carries the wastes to the lungs, kidneys, and sweat glands, where they are eliminated from the body. The blood also flows to the liver, where certain poisons are detoxified.

Blood within the blood vessels, lymph within the lymph vessels, and the interstitial fluid constitute the total internal environment of the human body. Its constancy is vital to survival and the mechanisms that operate to keep it constant are vital processes of homeostasis.

BLOOD

At some time in our lives, each of us has seen blood flowing from a cut or scrape in our skin. This remarkable red body fluid performs so many important

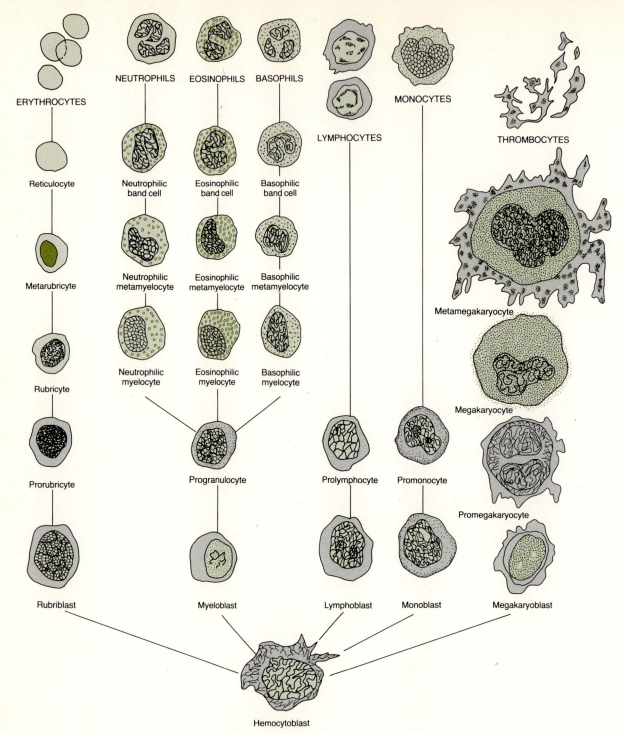

ERYTHROCYTES

Reticulocyte

Metarubricyte

Rubricyte

Prorubricyte

Rubriblast

NEUTROPHILS EOSINOPHILS BASOPHILS

Neutrophilic
band cell

Eosinophilic
band cell

Basophilic
band cell

Neutrophilic
metamyelocyte

Eosinophilic
metamyelocyte

Basophilic
metamyelocyte

Neutrophilic
myelocyte

Eosinophilic
myelocyte

Basophilic
myelocyte

Progranulocyte

Myeloblast

LYMPHOCYTES

Prolymphocyte

Lymphoblast

MONOCYTES

Promonocyte

Monoblast

THROMBOCYTES

Metamegakaryocyte

Megakaryocyte

Promegakaryocyte

Megakaryoblast

Hemocytoblast

Figure 21-1 Origin, development, and structure of blood cells.

functions that loss of a large volume could be fatal. Normally, it is contained within a closed system of blood vessels called the circulatory system, and it is kept flowing in a continual unidirectional movement, mostly by the pumping action of the heart.

Blood is a viscous fluid, four to five times thicker than water, and thus, is prone to flow more slowly than water. Blood is slightly heavier than water, is

maintained within the body at a temperature of 38°C, has a pH between 7.35 and 7.45, and survives in a solution consisting mostly of 0.85 percent NaCl. Blood constitutes about 8 percent of the total body weight. Thus, the total blood volume of the standardized 70-kg male is about 5.6 liters.

Despite its rather simple physical appearance, blood is an exceedingly complex tissue that transports a number of substances to different parts of the body that have an important role in homeostasis. For example, blood transports:

1. *Oxygen* from the lungs to all cells of the body.
2. *Carbon dioxide* from the cells of the body to the lungs.
3. *Nutrients* from the digestive organs to the cells of the body.
4. *Waste products* from the cells of the body to the kidneys.
5. *Hormones* from endocrine glands to cells of the body.
6. *Enzymes* from secretory cells to various other cells of the body.
7. *Buffering systems and molecules* that serve to maintain the body pH.
8. *Heat* from the deeper parts of the body to the surface when necessary.
9. *Clotting molecules* produced in the liver to damaged regions of the circulation to prevent body fluid loss.
10. *Antibody molecules and cells* to areas of infection where foreign agents are neutralized and removed.

Microscopically, blood is composed of two parts: the *formed elements* or blood cells and *plasma,* the liquid phase in which the cells are suspended. The formed elements include the *erythrocytes* or red blood cells, the *leucocytes* or white blood cells, and the *platelets* or thrombocytes.

Formed Elements

A common classification of the **formed elements** of blood is the following (Figure 21-1):

I. Erythrocytes, or red blood cells
II. Leucocytes, or white blood cells
   A. Granular leucocytes (granulocytes)
      1. Neutrophils
      2. Eosinophils
      3. Basophils
   B. Agranular leucocytes (agranulocytes)
      1. Lymphocytes
      2. Monocytes
III. Thrombocytes, or platelets

If blood is drawn from a person and set aside in a test tube, within minutes its consistency changes from a fluid to a gel. The blood is said to have *coagulated* or *clotted.* After additional time, a clear yellow liquid called *serum* separates from the clot. Serum has the identical composition of plasma except that it lacks *fibrinogen,* a large protein molecule normally found in plasma, and small amounts of some additional clotting factors needed to trigger the clotting process.

If the blood is collected and prevented from clotting by the addition of an appropriate anticoagulant, it can be centrifuged into its component parts. (Figure 21-2). The special tubes used are called hematocrit tubes and are graduated so that after centrifugation the percent of the formed elements can be easily read. The packed cell volume (PCV), or the *hematocrit,* expresses the volume of the packed erythrocytes as a percent of the total blood volume. The normal values are 40 to 50 percent in the adult male, 35 to 45 percent in the adult female, and about 35 percent in a child up to 10 years of age. The value in a newborn child is about 45 to 60 percent. The upper portion of the tube is filled with a clear, straw-colored, viscous fluid—the plasma. Separating the plasma and red blood cells is a thin white or grayish layer called the *buffy coat.* This layer consists of the leucocytes and platelets, and together they make up about 1 percent of the blood volume.

Origin of Blood Cells

The process by which blood cells are formed is called **hemopoiesis** or **hematopoiesis.** During embryonic and fetal life, there is no single center for blood cell production. The yolk sac, liver, spleen, thymus gland, lymph nodes, and bone marrow participate at various times in producing the formed elements. In the adult,

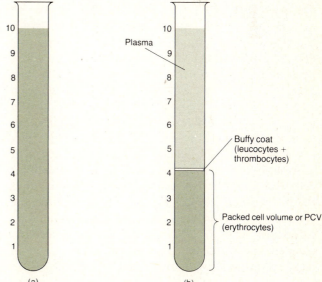

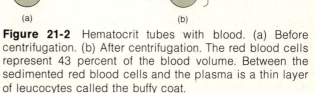

Figure 21-2 Hematocrit tubes with blood. (a) Before centrifugation. (b) After centrifugation. The red blood cells represent 43 percent of the blood volume. Between the sedimented red blood cells and the plasma is a thin layer of leucocytes called the buffy coat.

however, the hemopoietic tissue is divided into two types: the *myeloid tissue* or bone marrow and the *lymphoid tissue* located mostly in the lymph nodes. The red blood cells, platelets, and white blood cells, with the important exception of the lymphocytes, are formed in the myeloid or bone marrow tissue. It is significant to note that the lymphocytes which are produced in the lymphoid tissue after a person is born had their origin in the bone marrow tissues. During embryonic development, some of the stem cells (*hemocytoblasts*) that will become lymphocyte-producing cells leave the bone marrow and migrate to the thymus gland, where they are modified in some manner before they settle in the lymphoid tissues. These lymphocytes thereafter become known as the *T cells* —T for thymus. Other stem cells leave the bone marrow and take an alternate and as yet unknown route to the lymphoid tissues. On their journey, they also are processed in some manner so that they give rise to what is known as *B cells* — B for the term *bursa of Fabricius*. Research has shown that in birds, the second type of lymphocyte travels an embryonic route to the lymph nodes via this bursa, where they are modified and become the B cells. But this bursa is not found in mammals, and the structure in which the modification of the stem cells occurs in the human embryo is yet unknown, although the liver and/or the spleen are considered the most likely candidates. More will be said later about the defensive roles of these two types of lymphocytes.

Current opinion favors the idea that all of the formed elements originate from some single type of precursor cell called the *hemocytoblast*. According to this idea, these cells arise directly from undifferentiated embryonic mesenchymal cells and the direction of development of any hemocytoblast cell is determined by its microenvironment. For example, Figure 21-1 shows that hemocytoblasts can develop into: (1) rubriblasts, which later develop into mature red cells; (2) myeloblasts, which later develop into mature neutrophils, eosinophils, and basophils; and (3) megakaryoblasts, which later mature and fragment to form the circulating platelets. Other hemocytoblast cells end up in the lymphoid tissue to produce either the T or B cells, depending on their route from the bone marrow to the lymphoid tissue. The origin of the monocyte type of white blood cell is somewhat clouded, but it probably also arises from a hemocytoblast cell in the bone marrow.

Erythrocytes

Microscopically, the **erythrocytes** or **red blood cells** appear as biconcave discs with a mean diameter of 8.1 μm, a maximum thickness of 2.7 μm, and a mini-

mum thickness in the center of the disc of about 1.0 μm. Red blood cells lack a nucleus and cannot reproduce or carry on any extensive metabolic activities. Water takes up about 70 percent of the cellular volume and the pigment *hemoglobin* (Hb) occupies about 25 percent of the volume, while other constituents such as proteins and lipids, including cholesterol, occupy the remainder of the volume.

The main function of the erythrocytes is to transport hemoglobin. Although hemoglobin functions primarily to carry oxygen and carbon dioxide, it also plays an important part in regulating the acid-base balance in the body.

Hemoglobin

The hemoglobin molecule is a conjugated protein with a molecular weight of about 64,500. It is made up of two component parts: (1) an iron-containing pigment called *heme* (Figure 21-3) bonded to (2) a polypeptide molecule called *globin* to form a conjugated protein subunit. The functional hemoglobin molecule of a red blood cell is assembled from four such subunits. Two of the subunits contain a globin molecule designated as an α (alpha) chain, and the other two subunits contain a globin molecule designated as a β (beta) chain. The four subunits are fitted together in an interlocking fashion that separates the identical chains but promotes contact between the unlike chains. The flat (planar) heme portions are equidistant from each other and form different angles in relationship to each other (Figure 21-4). Each is located at the bottom of a pocket, accessible from the surface of the molecule. In these surface locations, the iron-containing pigment portion is readily accessible for interaction with oxygen molecules.

Each functional hemoglobin molecule contains four iron-containing atoms in the *ferrous* (Fe^{2+}) or divalent state, and the hemoglobin is able to combine with four molecules of oxygen (O_2). When oxygen com-

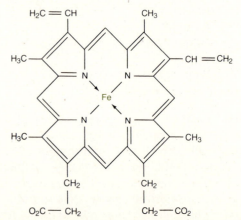

Figure 21-3 Structure of the heme portion of hemoglobin.

bines with the first iron atom, subsequent combinations of oxygen with the remaining three iron atoms are greatly enhanced. Thus, as blood passes through the lungs, the hemoglobin of the red blood cells loads up with O_2, which it transports to all the other tissues of the body. In the body tissues, the oxygen is released for use by the cells of the body.

The ability of hemoglobin to combine reversibly with oxygen is the feature that makes this molecule so remarkable and thus provides it with a significant role in homeostasis. Its ability to bind and release oxygen can be altered by changing the temperature, the ionic composition of the plasma, the pH, or the level of carbon dioxide (CO_2). The most important and variable regulators of the oxygen-hemoglobin binding capability are the pH and partial pressure of CO_2 (pCO_2). The manner in which the changes in pH and pCO_2 affect the percent of hemoglobin saturation is shown in Figure 21-5. The pH and pCO_2 are interrelated by the fact that as CO_2 levels increase, the concentration of H^+ also increases. This is indicated as follows:

$$CO_2 + H_2O \rightleftharpoons H_2CO_3 \rightleftharpoons H^+ + HCO_3^-$$

The interrelated actions of pH and pCO_2 on the oxygen-hemoglobin binding capability are known as the *Bohr effect*. The graphs in Figure 21-5 show that, as the levels of CO_2 or H^+ increase, the ability for hemoglobin to cling to O_2 decreases. Thus, as the red blood cells arrive at the peripheral tissues with their load of oxygen, they encounter increasing levels of CO_2 that have been produced by the metabolizing cells. As a consequence, the ability to cling to the O_2 drops. The sigmoid curves in Figure 21-5 also show that, as the partial pressure of oxygen around the hemoglobin decreases, the saturation of the hemoglobin also decreases. This means that when blood flows through tissues where the level of oxygen is low (as in metabolizing cells), the hemoglobin releases its oxygen very easily. Conversely, in tissues where the oxygen level is high (as in the lungs) the hemoglobin quickly loads up with oxygen.

In the lungs, when the hemoglobin molecule combines with the oxygen molecule it also gives up a H^+. The hydrogen ions that are released combine with bicarbonate ions in the plasma to form carbonic acid, which then releases the CO_2 to be exhaled. This occurs as follows:

$$\underset{\substack{\text{Reduced}\\\text{hemoglobin}}}{HHb} + \underset{\text{Oxygen}}{O_2} \longrightarrow \underset{\substack{\text{Oxygenated}\\\text{hemoglobin}}}{HbO_2} + \underset{\substack{\text{Hydrogen}\\\text{ion}}}{H^+}$$

$$\underset{\substack{\text{Hydrogen}\\\text{ion}}}{H^+} + \underset{\substack{\text{Bicarbonate}\\\text{ion}}}{HCO^-} \longrightarrow \underset{\substack{\text{Carbonic}\\\text{acid}}}{H_2CO_3} \longrightarrow \underset{\substack{\text{Carbon}\\\text{dioxide}\\\text{(exhaled)}}}{CO_2 \uparrow} + \underset{\text{Water}}{H_2O}$$

In the peripheral tissues, much CO_2 is produced from cellular metabolism. The CO_2 forms carbonic acid, which dissociates in solution to H^+ and HCO_3^-. Thus, the low pO_2 and elevated H^+ cause hemoglobin to release the oxygen and pick up the H^+. About two-thirds of the CO_2 is removed from the tissues in the form of dissolved bicarbonate.

Red blood cells contain the enzyme *carbonic anhydrase*, which functions in the following reaction:

$$CO_2 + H_2O \rightleftharpoons H_2CO_3$$

In the peripheral tissues where CO_2 is produced, carbonic anhydrase promotes the reaction going to the right. In the low CO_2 environment of the lungs, the enzyme moves the reaction toward the left.

Carbon dioxide is also carried directly on the hemoglobin molecule in the form of a *carbamino group*. Certain of the amino acids in the α and β chains of the hemoglobin molecule possess extra amino groups (NH_2). These combine with CO_2 to form the carbamino group as follows:

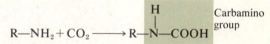

$$R-NH_2 + CO_2 \longrightarrow R-\overset{\overset{\displaystyle H}{|}}{N}-COOH \quad \begin{array}{l}\text{Carbamino}\\\text{group}\end{array}$$

When hemoglobin is loaded with oxygen, its ability to form carbamino complexes is low. After releasing the O_2, its ability to complex with CO_2 increases. Thus, the hemoglobin loads itself with CO_2 in the tissues after releasing the oxygen, and in the lungs, where the pO_2 is higher, the hemoglobin loads up with oxygen. Loading up with oxygen alters the hemoglobin molecule so that it releases the CO_2 present in the carbamino complex.

Shape of Red Blood Cells

The biconcave, discoid shape of nonnucleated red blood cells provides them with a very large ratio of surface area to volume. The large surface area enables a fast exchange of gases from the interior to the exterior and vice versa. The gas molecules localized in the interior are never far from the cell's surface. Therefore, diffusion can be completed quickly through the membrane. The biconcavity also permits the cell to adjust to osmotic irregularities that it may encounter by permitting volume changes that exert little or no tension on the membrane.

Life Span of Red Blood Cells

The concentration of red blood cells is so closely regulated that any departure from normal values is used as a criterion for some disorders. The normal constant value for red blood cell (rbc) concentration reflects the fact that the rates of production and destruction of the cells must be well balanced. A healthy male has about 5.4 million red blood cells per cubic millimeter

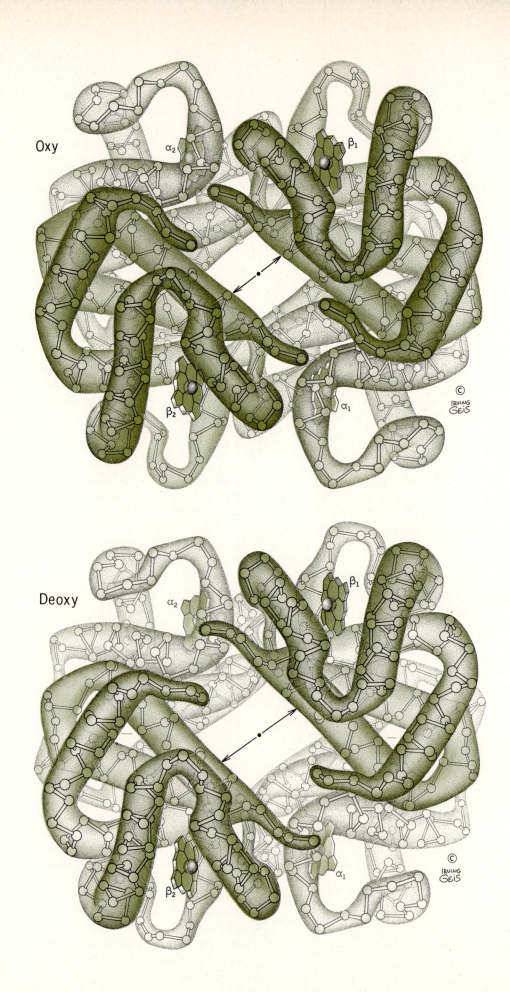

Oxy

β₁

α₂

β₂

α₁

© IRVING GEIS

Deoxy

β₁

α₂

β₂

α₁

© IRVING GEIS

404

Figure 21-4 Two states of the hemoglobin molecule, oxy (top) and deoxy (bottom). Hemoglobin, with four polypeptide chains, consists of two identical α-chains and two identical β-chains (α_1, α_2, and β_1, β_2). Note the change in subunit relationships on going from the oxy to the deoxy state. When oxygen is taken up, the two β-subunits are close together. When oxygen is released, the β-subunits move further apart as indicated by arrows in the center of the molecule. (From R. E. Dickerson and I. Geis, *Proteins: Structure, Function, and Evolution,* 2nd Edition to be published in 1982. Benjamin/Cummings, Menlo Park, Calif. Illustrations copyright by Irving Geis.)

of blood. A healthy female has about 4.8 million red blood cells per cubic millimeter. The cumulative effects of wear and tear eventually reaches a critical level for each cell at which point it is destroyed and removed from the circulation by a phagocytic cell of the reticuloendothelial system. Figure 21-6 shows that the life spans for red blood cells follow a normal distribution, with a mean life span of about 127 days.

Erythropoiesis

To maintain normal quantities of erythrocytes, the body must produce new mature cells at the astonishing rate of 2 million per second. In the adult, rbc production takes place in the myeloid tissue located in the bone marrow of the cranial bones, ribs, sternum, bodies of the vertebrae, and proximal epiphyses of the humerus and femur. The process by which erythrocytes are formed is called *erythropoiesis.*

Erythropoiesis starts with the transformation of a hemocytoblast into a rubriblast. Subsequently, other intermediate cells are formed until the final stage of red blood cell formation is reached (see Figure 21-1). Hemoglobin synthesis and the loss of the nucleus characterize the developmental sequence of erythropoiesis. The *rubriblast* is the first differentiated stage

in the sequence. It gives rise to a *prorubricyte.* The prorubricyte then develops into a *rubricyte,* the first cell in the sequence that begins to synthesize hemoglobin. The rubricyte next develops into a *metarubricyte.* In the metarubricyte, hemoglobin synthesis is maximum and the nucleus is lost by extrusion. In the next stage, the metarubricyte develops into a *reticulocyte* which, in turn, becomes an *erythrocyte,* or mature red blood cell. Once the erythrocyte is formed, it circulates through the blood vessels until its life span comes to an end. Then, hemoglobin molecules are split apart; the iron is reused and the rest of the molecule is converted into other substances for reuse or elimination.

Normally, erythropoiesis and red cell destruction are such finely tuned homeostatic mechanisms that they proceed at the same pace. But if the body suddenly needs more erythrocytes, or if erythropoiesis is not keeping up with red blood cell destruction, a homeostatic mechanism steps up erythrocyte production. The stimulus for the homeostatic mechanism is oxygen deficiency (hypoxia) within the cells of the kidney. This is not surprising because the chief function of the erythrocytes is to deliver oxygen. As soon as the kidney cells become oxygen deficient, they release an

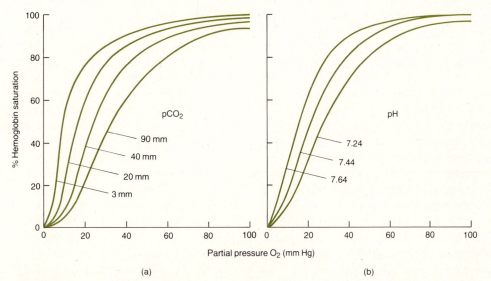

(a)   (b)

Figure 21-5 Relationship of (a) pCO_2 and (b) pH to oxygen-hemoglobin binding capacity.

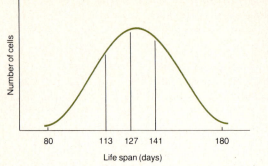

Figure 21-6 Distribution of life spans of red blood cells.

enzyme called the *renal erythropoietic factor (REF)* into the blood. The REF attacks a particular plasma globulin and produces an active glycoprotein hormone called *erythropoietin*. This hormone circulates through the blood to the red bone marrow, where it stimulates both the proliferation of stem cells to become hemocytoblasts and the proliferation of hemocytoblasts. It also accelerates the maturation process and the synthesis of new hemoglobin. Even with the acceleration of red cell production, it still takes about five days before peak production rate is achieved. The accelerated rate of production continues as long as the hypoxia exists and erythropoietin is released from the plasma.

Cellular hypoxia can occur if inhaled air is deficient in oxygen. This commonly happens at high altitudes where the air is "thin." Oxygen deficiency also occurs in anemia. The term *anemia* means that the number of functional red blood cells and/or their hemoglobin content is below normal. Consequently, the erythrocytes are unable to transport enough oxygen from the lungs to the cells. Anemia has many causes. The most common arise from a deficiency of iron, certain amino acids, or vitamin B_{12}. Iron is needed for the oxygen-carrying part of the hemoglobin molecule, and the amino acids are needed for the protein, or globin, part of hemoglobin. Vitamin B_{12} does not actually become part of the new blood cell. Its function is to assist the red bone marrow to produce erythrocytes. Vitamin B_{12} which is obtained from meat — especially liver — cannot be absorbed through the walls of the digestive tract without the help of another substance. This substance, called the *intrinsic factor*, is produced by the mucous cells that line the stomach. The intrinsic factor combines with vitamin B_{12} to protect it from the digestive process. Vitamin B_{12} is absorbed in the small intestine (ileum) and stored in the liver. The inability to produce intrinsic factor is the cause of a disorder called *pernicious anemia*. An anemia that arises simply from an inadequate diet is called *nutritional anemia*.

A diagnostic test that informs the physician about the rate of erythropoiesis is the *reticulocyte count*. A number of reticulocytes are normally released into the bloodstream before they become mature red blood cells (see Figure 21-1). If the number of reticulocytes in a sample of blood is less than 0.5 percent of the number of mature red blood cells in the sample, erythropoiesis is occurring too slowly. A low reticulocyte count might confirm a diagnosis of nutritional or pernicious anemia. Or, it might indicate a kidney disease that prevents the kidney cells from initiating the production of erythropoietin. If the reticulocyte number is more than 1.5 percent of the mature red blood cells, erythropoiesis is abnormally rapid. This is most likely a normal physiological response to any one of a number of problems, including most types of anemia and oxygen deficiency. Uncontrolled red blood cell production caused by a cancer in the bone marrow would be a pathology requiring medical intervention. If the individual has been suffering from a nutritional or pernicious anemia, the high count may indicate that tratement has been effective, and the bone marrow is making up for lost time.

Iron Metabolism

Iron is required by our bodies not only for the synthesis of hemoglobin, but also for the formation of myoglobin, the "hemoglobin of muscle." Iron is also needed for the production of other iron-containing substances such as the cytochromes, which are important electron carriers in energy production, and for a variety of enzymes such as catalase and peroxidase.

The human body contains a total of about 4 g of iron of which 65 to 70 percent is located in hemoglobin, with most of the remainder stored in the liver, spleen, and bone marrow.

The regulation of the iron content of the body is carried out by the mucosal cells of the digestive tract. Though ferrous (Fe^{2+}) iron can be absorbed from the digestive tract anywhere between the stomach and and colon, most of it is absorbed in the upper end of the small intestine — particularly the duodenum. The cells lining the lumen of the small intestine transport the iron from the cavity of the digestive tract across their cell membranes into their interior by a mechanism not yet completely understood. Once absorbed into the cell, the iron seems to have two possible fates, depending on the content of iron already in the body.

(1) If there is a need for iron, the iron moves from the luminal side of the cell to the basal side, which is adjacent to a capillary. From the cell, the iron is moved into the plasma, where two ferric (Fe^{3+}) iron atoms are bound to a β globulin molecule called *transferrin*. This bond is weak, so as the blood flows to various parts of

the body, the iron can be removed easily. The requirement for iron by most cells is negligible. The major exception, of course, is the bone marrow cells. Excess iron is stored primarily in the liver, where the iron is removed from the transferrin molecule by the hepatocytes (liver cells) and bound to a protein molecule called *apoferritin*. The combined form is called *ferritin*. Apoferritin is a large molecule (molecular weight = 460,000) that is able to bind iron in varying amounts, depending on the amount delivered to the cells. It is possible to saturate the apoferritin stores of the body. After this, additional iron is stored as *hemosiderin*—an extremely insoluble form of iron. Hemosiderin can be seen "precipitated out" inside many cells when they are examined by the light microscope after simple staining procedures. When the need for iron by body cells increases, the iron stored in the liver as ferritin is easily mobilized, whereas that in the form of hemosiderin is exceedingly difficult to recover.

Iron is also recycled from the hemoglobin of red cells that are destroyed by the reticuloendothelial cells of the body. About 0.5 mg of iron is lost daily from our bodies in the feces. Females lose additional iron in the menstrual flow, so their averaged daily loss might be three times that of the male. Consequently, the need for replacement iron by the female is three times that of the male. The maximum absorption rate varies depending on the body stores. However, even in times of great need, the body can only take up a few milligrams in one day.

(2) If the body stores are filled, then the transferrin molecule becomes unable to unload its two ferric iron atoms. As a consequence, the iron absorbed by the columnar epithelial cells of the intestine enter into their second, or alternative, fateful pathway. The cells retain the iron by synthesizing more of their apoferritin and thus bind the iron to form ferritin. The mucosal cells have a normal life span of about four days before they are lost into the gut lumen and replaced. When they are shed, they carry the stored iron with them. The regulation of the iron content of the body is entirely done by control of the absorption rate—by a *mucosal block*. There is no regulating process that increases the excretion rate of stored iron.

Leucocytes

Unlike red blood cells, **leucocytes,** or **white blood cells,** have nuclei and do not contain hemoglobin. In addition, they are far less numerous, averaging between 5,000 and 9,000 cells per cubic millimeter of blood. Red blood cells, therefore, outnumber white blood cells about 700 to 1. There are five kinds of leucocytes, divided into two major groups (see Figure 21-1). The first group contains the *granular leucocytes*. These

develop from red bone marrow. They have granules in the cytoplasm and possess lobed nuclei. Three kinds of granular leucocytes exist. These are the *neutrophils*, the *eosinophils*, and the *basophils*. The second principal group of leucocytes is called the *agranular leucocytes*. They develop from lymphoid and myeloid tissue. When they are placed under a light microscope, no cytoplasmic granules can be seen. Their nuclei are more or less spherical. The two kinds of agranular leucocytes are *lymphocytes* and *monocytes*.

The general function of the leucocytes is to combat inflammation and infection. Some leucocytes are actively *phagocytotic*—they can ingest bacteria and dispose of dead matter. Most leucocytes also possess, to some degree, the ability to move through minute pores between the cells that form the walls of the capillaries, the smallest type of blood vessel. This flowing movement called *diapedesis,* starts when a part of the cell flows into an armlike projection that threads its way into and then through a small pore. The rest of the cell cytoplasm flows slowly through the pore into the extended arm on the other side of the capillary wall. In this way, the entire cell moves through the pore from one side of the capillary wall to the other.

Leucocytes are guided to infection sites by a process called *chemotaxis*. A variety of substances released either by invading microorganisms or by killed tissue cells serves to guide the leucocytes toward the source of the chemotaxic agent. Diffusion of the substance sets up a gradient of concentration that the leucocytes follow. Chemotaxis can have a positive (attractive) effect or a negative (repellent) effect.

Inflammation

If a tissue of the body becomes injured or infected, an *inflammation*, or inflammatory response, is the body's defense. Key to the inflammatory response is the release of various chemicals from the injured tissue—especially one called histamine. Histamine causes blood vessels in the region of the injury to dilate, thus increasing local blood flow. It also increases the permeability of the capillaries so that the outward filtration of fluid from the capillaries into the tissues increases. As a consequence of the increased blood flow, the tissue becomes redder and warmer. As a result of the increased tissue fluid, the injured tissue, becomes swollen—a condition called *edema*. The tissue fluid, which becomes loaded with extra clotting proteins from the plasma, begins to coagulate and prevents the normal flow of tissue fluid. As a result the spread of bacteria or their toxins is greatly slowed and is more or less confined to the area of tissue injury.

The rapidity of the inflammatory response is proportional to the extent of tissue destruction. Therefore, a staphylococcal infection, which produces great tissue destruction, is normally quickly confined by the inflammatory response. Streptococcal infections, which are less destructive, elicit a much slower inflammatory response. As a consequence, the confinement is less likely to be successful, and the bacterial invasion can continue to spread throughout the body.

With tissue destruction and the release of chemical substances, the attraction for leucocytes to the site of injury is increased. By the process of diapedesis the neutrophils escape from the capillaries, and by the process of chemotaxis they are guided to the site of injury. As the leucocytes engulf the bacteria, some *suppuration,* or pus formation, may occur. Essentially, pus consists of dead and living bacteria, white blood cells, cellular debris, and body fluid. If the leucocytes destroy the invading bacteria effectively, the affected area returns to normal and the process of repair takes place. If the leucocytes are unsuccessful, then suppuration increases and the infection continues to spread.

Differential Count

The diagnosis of an injury or infection often can be aided by a *differential count.* In this procedure, the *percentages* of the five types of leucocytes in the circulation are estimated. Blood from a normal healthy person would give a differential count as follows:

Neutrophils	60–70%
Eosinophils	2–4%
Basophils	0.5–1%
Lymphocytes	20–25%
Monocytes	3–8%
Total	100%

Interpretation of the results of a differential count requires particular attention to the neutrophils. The neutrophils are the most active cell type in response to tissue destruction. Their major role is phagocytosis. They also release the enzyme lysozyme, which destroys certain bacteria. Usually, a high neutrophil count indicates that the damage is caused by invading bacteria. An increase in the number of monocytes generally indicates a chronic (of long duration) infection such as tuberculosis. It is hypothesized that monocytes take longer to reach the site of infection than neutrophils, but they arrive in larger numbers and destroy more microbes. Monocytes, like neutrophils, are phagocytic. They clean up debris and cellular material following an infection. High eosinophil counts typically indicate allergic or parasitic conditions, since eosinophils are believed to combat allergens, the causative agents of allergies. Eosinophils produce antihistamines. Basophils are also believed to be involved in allergic reactions. They are believed to carry histamine.

The term *leucocytosis* refers to an increase in the number of white blood cells. If the increase exceeds 10,000, a pathological condition is usually indicated. A decrease below normal in the number of white blood cells is termed *leucopenia.*

Some leucocytes afford protection in another extremely important way. These are the lymphocytes, which are key elements in the immune response of the body.

THE IMMUNE RESPONSE

Our bodies provide protection for our well-being and survival by employing a variety of mechanisms. Most of these mechanisms are present from the time of our birth and constitute what is called **innate immunity.** Included under this umbrella of protective mechanisms are the leucocytes, described in the preceding section. The fixed and wandering cells of the reticulo-endothelial system serve as macrophage cells that also phagocytose bacteria and foreign materials. Enzymes, acids, and a variety of specialized chemical substances, such as lysozyme, that cause bacteria to fall apart, also protect our bodies from invasion.

The ultimate response that aims at and singles out specific invaders is referred to as **acquired immunity.** Beginning shortly after birth, this type of immunity "learns" how to combat certain types of invaders by mobilizing specific members of its defense systems and by causing them to increase in number so that the defenders overpower the invaders. Often the first encounter with an invader is a long drawn out battle that severely taxes the resources of the host. If the defense is successful and the host survives, subsequent attacks by the same invader are usually easily repulsed. This is so because the acquired immune system "remembers" the invader and next time is able to muster its defenses quickly and in sufficient numbers to prevent the invader from gaining a foothold.

Acquired immunity operates by means of **antibodies.** The antibody defense system comes in two forms—*molecules* and *cells.* The origin of the lymphocytes or antibody-producing cells was described earlier.

The antibody is said to be *specific* because it attacks only the foreign material that triggered its production. The foreign material, called the *antigen,* contains some molecular species, usually protein or glycoprotein, that is not normally present in the host organism. Therefore, bacterial cell membranes or toxins produced by bacteria are considered antigenic in the human body because they possess molecular species not normally present there. For the same reason, viruses, parasites, and tissue transplants also elicit antibody responses.

Research has elaborated certain characteristics of antigenic molecules. For example, they need to be over a certain size and usually have a molecular weight over 10,000. A good antigen (one that elicits a high level of antibody production) has a molecular structure of many regularly repeating chemical groups over its length. For this reason, most antigens are protein-linked molecules. Certain more complex polysaccharides also have antigenic properties.

Units with molecular weights less than 10,000 can bond to larger antigenic molecules and still be a target for antibody production. When bonded to a larger molecule, the smaller unit is called a *hapten*. Antibodies that are formed when the combined complex is injected into an animal are directed against the hapten as well as the larger molecule. If the haptenic unit is later reintroduced into the animal unbonded, the antibodies will destroy the hapten. Sometimes drugs or chemicals that are inhaled or absorbed through the skin behave as haptens and elicit antibody production.

Lymphoid Tissue

The **lymphoid tissue** guards the main entrances into the body. Composed mainly of lymph nodes, the lymphoid tissue is located in the throat in the form of tonsils and adenoids, and here they intercept airborne invaders. The intestinal tract is also well guarded by lymphoid tissue so that invaders that are swallowed can be quickly removed. In addition, lymph nodes are located strategically in the peripheral tissues where they can capture foreign materials that penetrate the skin. Finally, should antigenic material penetrate the other body defenses and reach the bloodstream, the lymphoid tissue of the spleen, liver, and bone marrow will remove and destroy the foreign material as the blood circulates through these organs.

Lymphoid tissue is composed primarily of a meshwork of interlocking reticular cells and fibers. Clinging to the interstices of the tissues are large numbers of lymphocyte cells and other cells in various stages of differentiation, such as plasma cells, lymphoblasts, monocyte-macrophages, and a few eosinophils and mast cells.

The two types of lymphocytes (T cells and B cells) are indistinguishable under the light microscope. However, the scanning electron microscope indicates that most of the B cells have a complex villous surface, whereas most of the T cells have a smoother, much less villous surface topography (Figure 21-7). Other

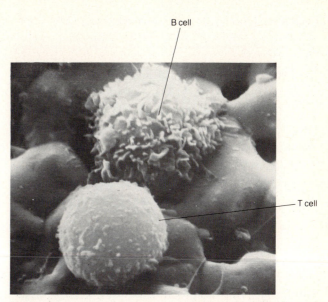

Figure 21-7 Scanning electron micrograph of B cells and T cells. (Courtesy of Dr. Aaron Polliack, Hadassah University Hospital.)

techniques have demonstrated additional surface differences on the lymphocytes. For example, the B cells are distinguished by the presence of characteristic surface *immunoglobulins*—antibody molecules that are fixed to the cell membrane. Most of the B type lymphocytes also have receptor sites for accepting antigen-antibody complexes formed in the body fluids. Some of the B cells also have a receptor site for a component part of *complement*—an activity system that is responsible for the lysis of foreign cells, the destruction of viruses, and the release of histamine from mast cells.

Distribution of the T and B type lymphocytes in various tissues in humans has been determined by the use of the aforementioned markers, and Exhibit 21-1 shows this distribution.

Structure of Immunoglobulins

Immunoglobulins are the special protein molecules in body fluids that function as antibodies—they are able to combine specifically with the foreign compound, the antigen, that triggered their formation. Other than a few naturally occurring antibodies that are genetically programmed, antibodies in the body are the result of an immune response to foreign material that has gained entrance into the body.

Immunoglobulins comprise about 20 percent of the total plasma proteins, and most fall into the category of gamma globulins, although a few important antibodies fall into the beta globulin group. The activity of the immunoglobulins is better understood if their structure is described.

Exhibit 21-1 LYMPHOCYTE DISTRIBUTION IN VARIOUS TISSUES IN HUMANS

Location	APPROXIMATE PERCENT	
	B Cells	T Cells
Peripheral blood	15–30	55–75
Bone marrow	>75	<25
Lymph	<25	>75
Lymph node	25	75
Spleen	50	50
Tonsil	50	50
Thymus	<25	>75

Analyses of immunoglobulins show that they are a very heterogeneous group of molecules. They are composed of protein (82 to 96 percent) and carbohydrate (4 to 18 percent), with the protein portion manifesting the biological responses usually associated with the reactions of the whole molecule.

Figure 21-8 illustrates a simplified structure of the basic antibody molecule. Each antibody molecule consists of four polypeptide chains—two designated as *heavy (H) chains* and two designated as *light (L) chains*. The L chain contains approximately one-half the number of amino acids present in the H chain and is about one-half the size and molecular weight. The four chains are linked together by disulfide (S—S) bonds to form the Y-shaped structure. The portions at the ends of the four polypeptide chains represent the *variable (V) regions* of the chains. Here, the sequence of the amino acids in the polypeptide chains differs for each kind of antibody. It is the V region that endows each immunoglobulin with its specificity. The remaining portion of each of the four polypeptides is called the *constant (C) region* because the sequence of amino acids in this region is the same for each class of antibody molecule. Scientists in the field of immunochemistry have further subdivided the four chains into what they call *domains*. For example, examination of the heavy chain as drawn in Figure 21-8 shows that the variable (V) region is also called the V_H (variable-heavy) domain. Adjacent to the V_H domain and extending to the point of junction of the L and H chains is the domain called C_{H_1} (constant-heavy; number one). Continuing along the heavy chain are two additional domains named C_{H_2} (constant-heavy; two) and C_{H_3} (constant-heavy; three). In a similar fashion the light (L) chain is divided into two domains called the V_L (variable-light) and C_L (constant-light) domains. It is in the variable regions of the antibody molecule that the binding sites for the specific antigen are located.

Dissection of the antibody molecule by an enzymatic attack with papain produces two antigen-binding fragments designated F_{ab} and one fragment designated F_c because it is easily crystallized.

Classes of Immunoglobulins

The immunoglobulins have been subdivided into five classes on the basis of the amino acid sequences in the constant (C) region of the heavy chains. Immunoglobulins are abbreviated Ig, and the five immuno-

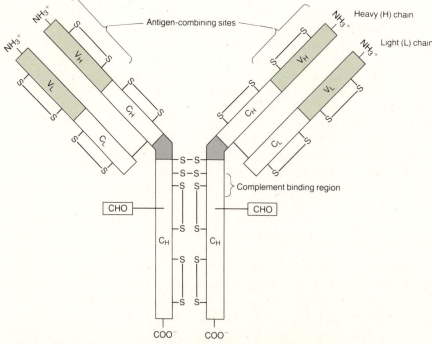

Figure 21-8 Structure of an antibody molecule. Shown here is IgG.

globulin classes are referred to as IgG, IgM, IgA, IgD, and IgE. Since it has also been demonstrated that the constant portion of the light chains divides them into two groups, the κ (kappa) and λ (lambda), the combinations of heavy and light chains produce a considerable number of subclasses. In addition, critical examination of the constant-heavy portions of the polypeptides has revealed additional differences, so the IgG class of immunoglobulins now consists of four forms: IgG_1, IgG_2, IgG_3, IgG_4. More recent evidence shows that IgA can be divided into two subclasses: IgA and IgA_2; and that the IgM consists of at least two subclasses: IgM and IgM_2. The purpose of indicating the number of classes and subclasses is to show the great number of possible antibody forms that the body can produce, each with its own specific target and mode of inactivating the antigenic material.

IgG and IgM Classes. Of the major classes of immunoglobulins, the IgG group is the largest. In serum of normal adults, 75 percent of the total immunoglobulins are of the IgG class. IgM makes up about 10 percent of the total. It normally exists as a pentamer, that is, five molecules linked together, with a molecular weight about 900,000 (Figure 21-9). The IgM antibody is the first of the immunoglobulins to appear in plasma in response to most antigens. It is also present on the surface of the B lymphocytes. As the immune response progresses, the level of IgM antibodies in the plasma decreases and the IgG antibodies appear in increasing numbers. These two classes of immunoglobulins (IgM and IgG) are responsible for the reactions usually associated with antibody responses, such as precipitation, agglutination, hemolysis, and complement fixation.

IgG shows much greater binding power (avidity) than IgM in attacking foreign material. Despite this difference, IgM is very efficient when it attacks a bacterial cell. When this pentameric antibody attaches to the antigenic receptor site on the bacterium, it activates the complement system that is present in the circulation. This system can destroy cells by chewing a hole in the bacterial cell's membrane. When the membrane of a bacterium destroyed by complement action is examined with a scanning electron microscope, it looks like a battlefield with many craters. IgM attached to foreign cells also promotes a process called *opsonization*. It seems as though macrophage cells prefer to attack foreign cells that have been roughened by a coating of antibody molecules. One might liken this process to putting a special sauce on your meal to improve its taste.

When foreign material enters the body, the IgM response is rapid but not very specific. The arrival of IgG, though somewhat delayed, brings into action

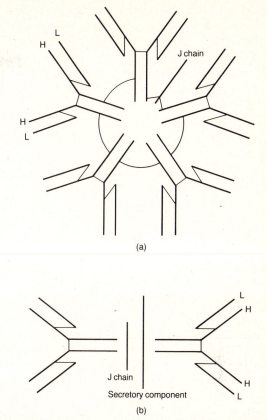

Figure 21-9 Structure of IgM and IgA. (a) Pentamer structure of IgM and (b) dimer structure of IgA. Polypeptide chains are shown as thick lines; disulfide bonds linking different polypeptide chains are shown as thin lines.

very specific antibodies that affix themselves to the antigen. These antibodies persist for weeks and provide protection long after all traces of the foreign material have been removed. IgG, because it is much smaller (molecular weight = 150,000) than IgM, is able to cross the placental barrier and enter the fetal circulation. These long-lived immunoglobulins obtained from the mother persist in the fetus to protect the child even after it is born. It is weeks before the newborn can begin to manufacture its own antibodies. During this time, the newborn can receive additional IgGs from maternal milk—a good reason to encourage breast feeding for at least 10 weeks.

IgA Class. The IgA immunoglobulins make up about 15 percent of the total serum antibodies. They consist of two molecules of antibody linked together to form a *dimer* plus an additional J chain and secretory component (see Figure 21-9). IgA antibodies are found in saliva, tears, vaginal and nasal secretions, bronchial secretions, prostatic fluid, and secretions produced by the intestinal mucosa. The belief, at this time, is that these antibodies function *not* to destroy foreign invaders but, by combining with them, to prevent them from gaining entrance to the body and the IgM-IgG systems.

IgD Class. The IgD antibody, like the IgG, is a monomer but it has a slightly heavier molecular weight (180,000). This immunoglobulin is present in minute amounts (0.2 percent of the total serum immunoglobulin). The function of IgD is still unknown though it has been shown to attack various materials such as insulin, penicillin, thyroid antigens, and some milk products. It has also been shown to be present on human B type lymphocytes.

IgE Class. One of the more important breakthroughs in the study of allergic diseases came about when it was discovered that IgE molecules were the antibodies involved in the allergic response. The IgE molecule is a monomer with a molecular weight of about 196,000. There is so little (0.004 percent) of this class of immunoglobulin in serum that it was not detected until very recently. When the IgE antibodies are synthesized, they attach to the surface of mast cells, where they await the arrival of the *allergen* (the antigen that causes the allergic response). The allergen is clasped by two adjacent IgE molecules (Figure 21-10). This creates a bridge between the two antibodies in a way that causes distortion of the plasma membrane. This triggers a cascading series of events that eventually leads to the release of mast cell granules. The granules dissolve in the tissue fluids and release *histamine, serotonin,* and several other substances. The histamine and serotonin act on ad-jacent smooth muscle cells to produce the *anaphylactic* response. This includes bronchospasm (constriction of bronchioles), which can cause severe obstruction of the airways, and a marked vasodilation of arterioles and capillary leakage that can produce shock (a precipitous drop in blood pressure).

Antibody Reactions

The mechanism of action of the IgE immunoglobulins describes one way in which antibodies operate within our bodies. Another way requires the bivalent antibody (bivalent because each immunoglobulin has two identical combining sites) to *precipitate* the antigen molecules by formation of a growing lattice of antigen-antibody reaction (Figure 21-11). If the antigen is multivalent (many combining sites), then any bivalent antibody can link two antigen molecules together. If the proportions of antigen to antibody are optimal, the lattice can grow and fall out of solution as a precipitate. A similar reaction is that of *agglutination*. If the antigenic combining sites are fixed to cell surfaces, then the bivalent antibodies will link together the surfaces of two adjacent cells. *Neutralization* reactions are those in which the toxic portion of a poison is blocked by the attached antibody molecule, hence preventing the destructive effect of the poison.

The *complement system* consists of at least 15 chemically and immunologically distinct proteins that are normally nonreactive and dissolved in the plasma. These 15 components, which are identified by numbers C_1, C_2, C_3, etc., are activated in a sequence that

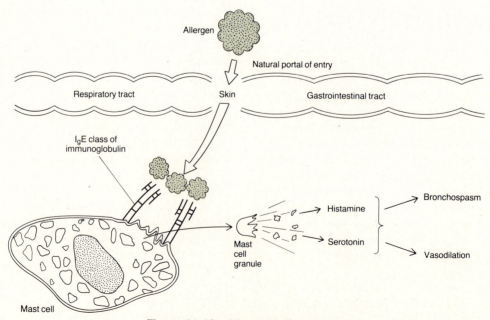

Figure 21-10 Allergen-IgE response.

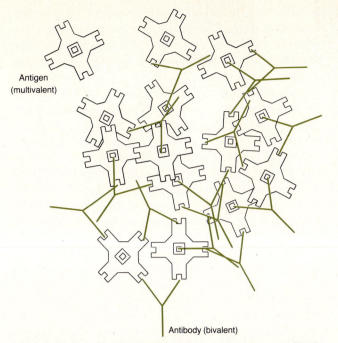

Figure 21-11 The lattice hypothesis of antigen-antibody reactions.

be triggered by antigen-antibody complexes that contain either IgG or IgM immunoglobulins. Activation begins when the C_1 component of the complement system binds to a site in the F_c region of the bound IgG or IgM antibody. Component C_1, when bound, becomes enzymatically active. It brings components C_4 and C_2 together to form another enzymatically active complex, C_{42}. Figure 21-12 presents a schematic diagram that illustrates the activation of the complement system by two routes—the *classical* and the *alternative*. Although the alternative route employs some still unknown factors, both routes utilize C_3 and C_5 components. Both routes process the remaining C_6–C_9 components in identical fashion. When component 5 is acted upon, it is split into two parts— C_{5a} and C_{5b} (Figure 21-12). The C_{5b} binds to the cell membrane. Components C_6 and C_7 bind to the C_{5b} already on the membrane. The complex of C_{567} then receives C_8 to form a tetramolecular complex, C_{5678}, which in turn binds up to six C_9 molecules. This large complex initiates lysis and destroys the cell.

Response of the B Cells to Antigen

The B cell produces and releases the immunoglobulins into the circulation. Each person (or animal) is able to produce antibodies to *any* antigen introduced into the body. The explanation for this astounding ability rests in the fact that each of us possesses a vast number of lymphocytes, each of which is able to make

produces complex proteins of various combinations. Some of the protein combinations possess enzymatic activity that is responsible for the lytic and cytotoxic effects on bacteria.

The entire sequence of complement activation can

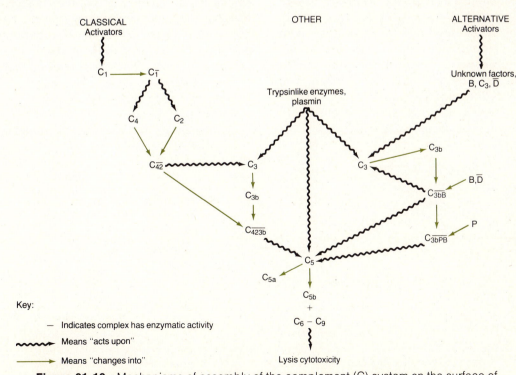

Figure 21-12 Mechanisms of assembly of the complement (C) system on the surface of a complement activator.

a unique antibody. The *clonal theory* of antibody formation proposes that the antigenic molecules are discovered by a particular lymphocyte cell that is one of very few able to make antibodies to that particular antigen. Once the encounter occurs, the lymphocyte cell proliferates to produce a great number of identical cells also able to form that specific antibody. The colony of lymphocytes, all "offspring" of the same parent cell, is termed a *clone*. Thus, the clone is formed in response to the introduction of a specific antigen (Figure 21-13). How can such a great population of lymphocytes come into existence—each committed to produce a different antibody? Two theories try to answer this question. The *germ line theory* states that all the information needed to code for all the antibodies an animal could possibly need is present in the genetic information inherited from its parents. Each lymphocyte that is developed utilizes, or turns on, a different antibody gene. The *somatic mutation theory* proposes that what is inherited is a core of antibody information that is somehow enlarged and modified during development to produce the wide diversity of antibody-forming cells that are present in the mature animal.

During its patrol of the body, when a B cell encounters and binds to its antigen, the cell undergoes repeated divisions to produce a clone of cells. Within this population of "identical" cells, some mature to become antibody factories that release the immunoglobulins into the blood. When they are fully mature, they become identified as *plasma cells*—cells that are capable of releasing about 2,000 identical antibody molecules per second until they die, generally within 2 or 3 days after reaching maturity. During the developmental changes, the plasma cells switch from producing the "emergency" IgM type antibodies to producing the IgG type (Figure 21-14).

Other cells within the clone never produce antibodies but function as *memory cells,* which carry the

program for the specific IgG antibody. As a consequence of the initial challenge by an antigen there are now many more cells identical to the parent cell, each of which is able to respond in the same way to the antigen as the original B cell (that is, by proliferation and clone formation). Consequently, if the antigen appears a second time, it will encounter one of the correct lymphocytes sooner, and since they are programed for the IgG antibody, the immune response will begin sooner, accelerate faster, be more specific, and produce greater numbers of antibodies. Figure 21-15 shows a comparison of the primary and secondary (or anamnestic) response. Immunity can persist for years because memory cells survive for months or years and also because the foreign principle is sometimes reintroduced in minute doses that are sufficient to constantly trigger low-level immune responses. In this way, the memory cells are periodically replenished.

Response of the T Cells to Antigen

The T cells provide a defense against microorganisms such as fungi, bacteria that reside *inside* host cells, and viruses. T cells are responsible for the immunologic rejection of allografts (transplants between members of the same species) and for antitumor immunity. T cells are coated with IgM receptor molecules that enable them to recognize specific antigens. When singled out, a foreign cell is quickly and firmly held by the T cell by means of these antigen-antibody linkages. Large areas of contact are established, and then, over the next several hours, the membrane permeability of the foreign cell gradually changes until the cell is completely lysed. It appears that, once the process of *cytotoxic* destruction begins, it can continue without the attached lymphocyte.

Most of the protection provided by the T cells is the result of their ability to release minute quantities of a

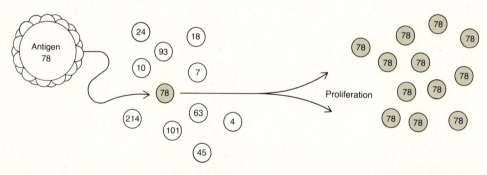

Assorted lymphocytes   Clone of identical lymphocytes

Figure 21-13 Clonal selection of antibody formation. The body contains small numbers of assorted lymphocytes capable of synthesizing different antibodies (numbers). When a specific antigen makes contact with its corresponding lymphocyte, that lymphocyte is stimulated to proliferate. The result is a clone of identical cells.

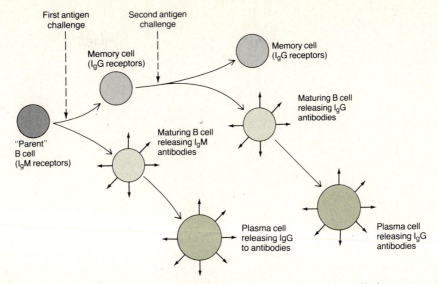

Figure 21-14 Formation of antibody-producing cells and memory cells in response to antigen.

large variety of very powerful proteins called *lympho-kines*. These soluble substances trigger the multiple responses associated with a cellular type of immune reaction. For example, the *macrophage (monocyte) chemotactic factor* released by the activated lymphocytes selectively attracts macrophages or monocytes to the reaction site. Another lymphokine, named the *migration inhibitory factor*, prevents macrophages from "migrating" away from the site of infection. *Mitogenic factor*, another lymphokine, causes uncommitted lymphocytes to dedifferentiate and divide more rapidly. Coupled to this last response is the action of yet another lymphokine, called *transfer factor*. The dedifferentiated lymphocyte can be made to take on the same capabilities as the original activated lymphocyte by assimilating transfer factor. Present research indicates that transfer factor is an informational polynucleotide that can enter and transform unsensitized lymphocytes.

Cooperation Between T Cells and B Cells in the Immune Response

That there is cooperation and interaction between the two types of lymphocytes is not questioned. But how they support each other is not yet completely settled. One of the best current models suggests that the following sequence of events occurs in many immune responses (Figure 21-16). The detection of the antigen occurs first by the T cells, which react quickly to bind the antigens with their surface receptor molecules. The T cells begin to proliferate by rapidly dividing (clone formation) and by producing the monomeric immunoglobulin that is localized on the membrane surface. The next step involves the attachment of the antigen molecules to these surface antibodies. After the combination occurs, the complexes of antigen-antibody (ag-ab) are released from the T cell and are picked up by macrophage cells. The ag-ab complex lands on the

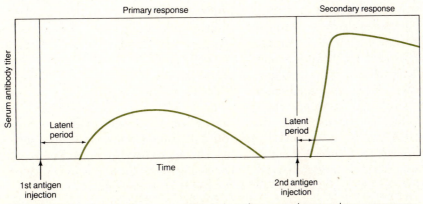

Figure 21-15 Level of serum antibody in primary and secondary responses.

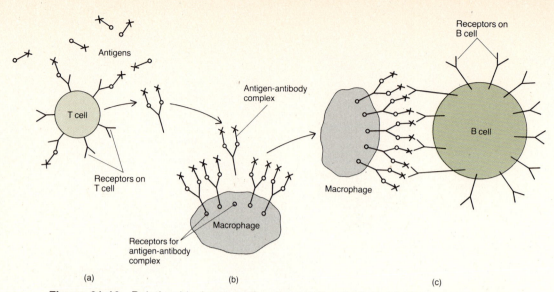

Figure 21-16 Relationship between T cells and B cells in the immune response. (a) Antigens attach to surface receptors on T cells. (b) Antigen-antibody complexes released from T cells are picked up by macrophages. (c) Macrophage presents antigens to B cell.

surface of the macrophage by the insertion of the F_c portion of the antibody molecule into a receptor slot. The macrophage eventually becomes covered with ag-ab complexes that protrude from the surface with the bound antigens facing away from the cell.

The protrusion of the antigens seems to be very important for the next step, because the macrophage now presents the antigens to the specific B cell. The B cell, which has 50,000 to 100,000 receptors for that antigen, apparently requires that the antigens be presented in a very ordered manner. When the antigens have been accepted, the B cell then is triggered into proliferation and maturation.

Although some immune responses are more T cell mediated and others more B cell mediated, most immune responses are a mixture of both. The cooperative nature is such that the B cell response is almost entirely dependent upon T cell assistance. T cells are essential for almost every kind of immune response.

Immune Tolerance

Immune tolerance basically means that a host animal will accept materials into itself *without responding* by producing antibodies. A more mechanistic definition would describe immune tolerance as a condition in which the responsive cell clones have been eliminated or inactivated by prior contact with antigen. As a result, there is no immune response when the antigen is introduced into the host.

Tolerance occurs separately in T and B lymphocytes. T cells can be suppressed or *tolerized* more quickly and easily than B cells, but only by prior *low*

doses of antigen. This kind of T cell tolerance will last as long as the low level of antigen persists. B cells also can be suppressed or tolerized, but to achieve B cell tolerance the prior dose of antigen must be high. The high level of antigen actually tolerizes both the T cells and B cells. Therefore, to achieve tolerance, the antigen can be presented in very low levels (*low-zone tolerance*) or in very high levels (*high-zone tolerance*). In either situation, antibody production is so suppressed that, if a later intermediate level or a normally immunogenic dose is administered, no antibodies are produced.

How can the suppression be produced by either very low or very high levels? The answer seems to lie in the triggering mechanism located on the lymphocyte membrane. Recall that antigenic molecules are *multivalent*—that is, they have functional groups that are repeated at intervals down their length. When the antigenic molecule is grasped by the correct lymphocyte, many of the repeating functional groups on the antigen bind to a separate receptor molecule on the lymphocyte (Figure 21-17). The antigen thus cross-links a series of neighboring antibody molecules. Apparently the amount of cross-linking required to activate the lymphocyte is critical—too little or too much "turns off" or paralyzes the cell's immune response.

In terms of producing immune tolerance, it seems that in low-zone tolerance *insufficient cross-linking* occurs on the T cells, and therefore they are turned off and cannot respond. In high-zone tolerance, the excess of antigenic molecules produces *too much cross-linking* in both T cells *and* B cells, so both cell systems are paralyzed.

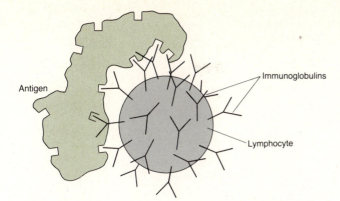

Antigen

Immunoglobulins

Lymphocyte

Figure 21-17 Suppression of antibody production. Immunoglobulins on the lymphocyte are cross-linked by binding to determinant or functional groups on the antigen. Distortion may then result in the membrane that triggers a sequence of intracellular reactions. Too many or too few cross-links will "turn off" the intracellular reactions.

Autoimmunity

When an animal produces antibodies that attack its *own tissues*, the condition of **autoimmunity** exists. The likelihood of autoimmunity increases with age and wear and tear of body tissues. Tissue destruction can result from the action of bacteria, viruses, or toxins. One result is the production of many new haptenlike antigenic determinants that are able to recombine with other large protein molecules. In these new combinations, the haptens can elicit immune responses that will produce antibodies to attack the hapten whether it is on its new carrier protein or still in its original locus in healthy tissue. It is this attack on the normal tissue by the antibodies that produces the tissue damage. As a consequence of the attack on the normal tissue, additional haptenlike antigens are produced, which in turn perpetuate a constant low level of circulating antibody.

Autoimmune diseases are divided into two groups — those in which one organ is affected and those in which the destruction is widespread into many regions of the body. *Thyroiditis* is an example of the first group. Antibodies are formed against the thyroid hormone, which circulates in the plasma at very low concentrations. The antibodies, in attacking the hormone, also progressively destroy the thyroid gland.

Systemic lupus erythematosus (SLE) is an example of widespread destruction caused by a number of autoantibodies. Skin, joints, lung, spleen, stomach, intestine, liver, and mainly kidneys are attacked by the autoantibodies.

The Mechanism of Autoimmunity

The haptenlike antigens are normally bound to their tissues of origin, but when cleaved from their normal location, they bind to new carrier molecules. In the healthy state, the original carrier for the molecule plus the hapten is considered "self." Tolerance of self suggests that the T cells are paralyzed by the normally very low levels of antigen presented to the T cells. However, when the hapten is attached to a new carrier, the carrier inserts itself onto a *different set of nonparalyzed T cells*. These loaded T cells deliver the antigens to the specific B cells, which then produce the antibodies directed against the hapten portion of the molecule.

One other way that autoimmunity can occur is the result of *cross-reactivity*. Consider two different antigenic molecules that happen to contain somewhere in their structure the same haptenic group. One antigenic molecule is part of some tissue and is tolerated because of the low level of antigen, and those T cells directed against it are paralyzed. If the other antigen is introduced, it activates the appropriate and different T cells, which in turn present the antigenic molecule with the same haptenic group to the B cells, which produce the antibodies. These antibodies are directed against the common haptenic groups and thus destroy both antigenic carriers. Unfortunately, one is part of normal tissue.

Plasma

When the formed elements are removed from blood, a straw-colored liquid called **plasma** is left. Exhibit 21-2 outlines the chemical composition of plasma. However, a few of the solutes should be pointed out. About 7 to 9 percent of the solutes are proteins. Some of these proteins are also found elsewhere in the body, but when they occur in blood they are called *plasma proteins*. Albumins, which constitute the majority of plasma proteins, are responsible for the viscosity or thickness of blood. Along with the electrolytes, albumins also regulate blood volume by preventing all the water in the blood from diffusing into the interstitial fluid. Recall that water moves by osmosis from an area of low solute (high water) concentration to an area of high solute (low water) concentration. Globulins, which are antibody proteins released by plasma cells, form a small component of the plasma proteins. Gamma globulin is especially well known because it is able to form an antigen-antibody complex with the proteins of the hepatitis and measles viruses and the tetanus bacterium. Fibrinogen, a third plasma protein, takes part in the blood clotting mechanism, along with the platelets.

Before examining the clotting mechanism, we shall first discuss the structure of thrombocytes, or platelets, the third type of formed element in blood.

Thrombocytes

If a hemocytoblast does not become an erythrocyte or granular leucocyte, it may develop into still another kind of cell, called a megakaryoblast, that is transformed into a megakaryocyte. Megakaryocytes are large cells whose cytoplasm breaks up into fragments. Each fragment becomes enclosed by a piece of the cell membrane and is called a **thrombocyte** or **platelet** (see Figure 21-1). Platelets are disc-shaped cells without a nucleus. They average from 2 to 4 μm in diameter. Between 250,000 and 500,000 platelets appear in each cubic millimeter of blood.

The platelets prevent fluid loss by initiating a chain of reactions that results in blood clotting. Like the

Exhibit 21-2 CHEMICAL COMPOSITION AND DESCRIPTION OF SUBSTANCES IN PLASMA

Constituent	Description
Water	Constitutes about 92 percent of plasma and is liquid portion of blood. Ninety percent of water is derived from absorption from digestive tract; 10 percent comes from cellular respiration. Water acts as solvent and suspending medium for solid components of blood and absorbs heat.
Solutes	
1. Proteins	Constitute 7 to 9 percent of solutes in plasma.
Albumins	Constitute 55 to 64 percent of plasma proteins and are smallest plasma proteins. Produced by liver and provide blood with viscosity, a factor related to maintenance and regulation of blood pressure. Also exert considerable osmotic pressure to maintain water balance between blood and tissues and regulate blood volume.
Globulins	Constitute about 15 percent of plasma proteins. Protein group to which antibodies produced by leucocytes belong. Gamma globulins attack measles and hepatitis viruses, tetanus bacteria, and possibly poliomyelitis virus.
Fibrinogen	Represents small fraction of plasma proteins (4 percent). Produced by liver and plays essential role in clotting.
2. Nonprotein nitrogen (NPN) substances	Contain nitrogen but are not proteins. These substances include urea, uric acid, creatine, creatinine, ammonium salts. Represent breakdown products of protein metabolism and are carried by blood to organs of excretion.

Exhibit 21-2 (Continued)

Constituent	Description
3. Food substances	Once foods are broken down in digestive tract, products of digestion are passed into blood for distribution to all body cells. These products include amino acids (from proteins), glucose (from carbohydrates), and fats (from lipids).
4. Regulatory substances	Enzymes are produced by body cells and catalyze chemical reactions. Hormones, produced by endocrine glands, regulate growth and development in body.
5. Respiratory gases	Oxygen and carbon dioxide are carried by blood. These gases are more closely associated with hemoglobin of red blood cells than with plasma itself.
6. Electrolytes	A number of ions constitute inorganic salts of plasma. Cations include Na^+, K^+, Ca^{2+}, Mg^{2+}. Anions include Cl^-, PO_4^{3-}, SO_4^{2-}, HOC_3^-. Salts help maintain osmotic pressure, normal pH, physiological balance between tissues and blood.

leucocytes, platelets have a short life, probably only 1 week, because they are used up in clotting and are too simple to carry on much metabolic activity.

BLOOD CLOTTING

Several aspects of blood clotting we shall now consider are clot formation, clot retraction and fibrinolysis, clot prevention, and clotting tests.

Clot Formation

Under normal circumstances, blood maintains its liquid state as long as it remains in the vessels. If, however, it is drawn from the body, it first becomes very thick and then forms a soft jelly. The gel eventually separates from the liquid component. The straw-colored liquid component, called *serum*, is plasma minus its clotting proteins. The gel, called a **clot**, consists of a network of insoluble fibers in which the cellular components of blood are trapped.

The clotting or coagulation of blood is an essential part of the hemostatic response—it prevents the loss of large volumes of blood. Clotting depends on a very delicate balance of a number of interrelated factors. Disruption of this balance can have fatal conse-

Exhibit 21-3 COAGULATION FACTORS*

Factor and Synonym	Comments
I: Fibrinogen	Produced by liver. Important factor in stage 3 of clotting, in which it is converted to fibrin. Plasma minus fibrinogen is called serum.
II: Prothrombin	Produced by liver. Its synthesis requires vitamin K. Important in stage 2 of clotting, in which it is converted to thrombin.
III: Thromboplastin	In extrinsic pathway it is known as extrinsic thromboplastin and is formed from tissue thromboplastin. In intrinsic pathway it is called intrinsic thromboplastin and is formed from platelet disintegration. Formation of thromboplastin signifies end of stage 1 of clotting.
IV: Calcium ions	Apparently involved in all three stages of clotting. Removal of calcium or its binding in plasma prevents coagulation.
V: Proaccelerin or labile factor	Produced in liver. Required for stages 1 and 2 of both extrinsic and intrinsic pathways.
VI:	No longer used in coagulation theory. Number has not been reassigned.
VII: Serum prothrombin conversion accelerator (SPCA) or stable factor	Synthesized in liver. Formation requires vitamin K. Required in stage 1 of extrinsic pathway.
VIII: Antihemophilic factor	Synthesized by liver. Required for stage 1 of intrinsic pathway. Deficiency causes classic hemophilia or hemophilia A—an inherited disorder, primarily of males, in which bleeding may occur spontaneously or after only minor trauma.
IX: Christmas factor or plasma thromboplastin component (PTC)	Synthesized by liver. Formation requires vitamin K. Required for stage 1 of intrinsic pathway. Deficiency causes disorder called hemophilia B, a disease similar to hemophilia A.
X: Stuart factor or Stuart-Prower factor	Synthesized by liver. Formation dependent on vitamin K. Required for stages 1 and 2 of extrinsic and intrinsic pathways. Deficiency results in nosebleeds, bleeding into a joint, or bleeding into soft tissues.
XI: Plasma thromboplastin antecedent (PTA)	Synthesized by liver. Required for stage 1 of intrinsic pathway. Deficiency results in hemophilia C, a mild hemophilia of both males and females.

Exhibit 21-3 (Continued)

Factor and Synonym	Comments
XII: Hageman factor	Required for stage 1 of intrinsic pathway. Known to be activated by contact with glass and may assume role in initiating coagulation outside body.
XIII: Fibrin stabilizing factor (FSF)	Required for stage 3 of clotting.
Pf$_1$: Platelet factor 1 or platelet accelerator	Essentially same as plasma coagulation factor V.
Pf$_2$: Platelet factor 2 or thrombin accelerator	Accelerates formation of thrombin in stage 1 of intrinsic pathway and conversion of fibrinogen to fibrin.
Pf$_3$: Platelet factor 3 or platelet thromboplastic factor	In stage 1 of intrinsic pathway.
Pf$_4$: Platelet factor 4	Binds heparin, an anticoagulant, during clotting.

* Roman numerals I to XIII are plasma coagulation factors. Pf$_1$ to Pf$_4$ are platelet coagulation factors.

quences. For example, if the blood clots too easily, the result can be thrombosis—clotting within an unbroken blood vessel. Or, if the blood takes too long to clot, the result can be a hemorrhage—a large loss of blood from the circulation.

When a blood vessel is injured, two kinds of reactions occur to stop the bleeding. One reaction is closure of the blood vessel. The trauma to a blood vessel triggers reflexes that cause contraction of the smooth muscle in the wall of the blood vessel, resulting in a spastic closure of the vessel. This closure, which occurs within seconds after the blood vessel is damaged, helps greatly to prevent a large blood loss.

The second reaction that occurs to stop bleeding is the formation of a clot. The clot plugs the rupture in the blood vessel. The various chemicals involved in clotting are known as *coagulation factors,* and since most of them are found in plasma, they are called *plasma coagulation factors. Platelet coagulation factors* are released by platelets.

Another coagulation factor is released by body tissues when they are damaged. Each plasma coagulation factor has been assigned a Roman numeral as well as a name. The coagulation factors are summarized in Exhibit 21-3.

The clotting of blood is a very complex process, but it can be viewed more simply as a sequence of three general stages. Stage 1 is concerned with the release or formation of a substance called thromboplastin. Stage 2 involves the conversion of prothrombin, a plasma protein, into thrombin, an enzyme. This stage requires the presence of thromboplastin and several other plasma coagulation factors. In stage 3, thrombin catalyzes the conversion of fibrinogen, another plasma protein, into fibrin. Fibrin is the clot. A summary of these stages is as follows:

Stage 1   Thromboplastin (formation or release)
Stage 2   Prothrombin ⟶ Thrombin
Stage 3   Fibrinogen ⟶ Fibrin

On the basis of whether thromboplastin is released by damaged tissues or formed by platelet disintegration, blood clotting may proceed along one of two routes, either the extrinsic pathway or the intrinsic pathway.

Extrinsic Pathway

The extrinsic pathway of blood clotting begins when a blood vessel is ruptured (Figure 21-18). The damaged tissues surrounding the blood vessel or the ruptured area of the blood vessel itself releases a lipoprotein called *tissue thromboplastin*. Tissue thromboplastin, in reaction with plasma coagulation factors IV, V, VII, and X forms *extrinsic thromboplastin*. This is stage 1 of the extrinsic pathway.

In stage 2, prothrombin is converted to thrombin. This conversion requires extrinsic thromboplastin and several plasma coagulating factors such as IV, V, VII, and X.

In stage 3, fibrinogen, which is soluble, is converted to fibrin, which is insoluble, by the thrombin formed in stage 2. This reaction requires plasma coagulation factors IV and XIII.

Intrinsic Pathway

The intrinsic pathway of blood clotting begins with the roughened surface created by a ruptured blood vessel Figure 21-18). Under normal conditions, the negatively charged platelet membrane and negatively charged endothelial lining of a blood vessel repel each other. In this way, platelets do not adhere to the endothelial lining. However, a ruptured blood vessel results in the loss of the negative charge by the endothelial lining. Thus, the platelets adhere to the ruptured area. This clumping together of platelets causes them to disintegrate and release platelet coagulation factors into the plasma.

In stage 1 of the intrinsic pathway, four platelet coagulation factors (Pf_1, Pf_2, Pf_3, and Pf_4) in reaction with plasma coagulation factors IV, V, VIII, IX, X, XI, and XII form *intrinsic thromboplastin*.

In stage 2, prothrombin is converted to thrombin by intrinsic thromboplastin and several plasma coagulating factors including IV, V, VII, and X.

Stage 3 of the intrinsic pathway is the same as that of the extrinsic pathway. It involves the conversion of fibrinogen to fibrin in the presence of thrombin and plasma coagulation factors IV and XIII.

In addition to helping to convert fibrinogen to fibrin, thrombin serves another very important function. It causes more platelets to adhere to each other, disintegrate, and release even more platelet coagulation factors. This cyclic feature of the intrinsic pathway ensures continual platelet disintegration until the clot is formed.

Once the clot is formed, it plugs the ruptured area of the blood vessel and prevents hemorrhage (bleeding). After this is accomplished, permanent repair of the blood vessel can take place. In time, fibroblasts form connective tissue in the ruptured area and newly produced endothelial cells repair the lining.

Clot Retraction and Fibrinolysis

Normal coagulation involves two additional events that occur after the formation of the clot: clot retraction and fibrinolysis. **Clot retraction** or **syneresis** is the consolidation or tightening of the fibrin clot. The fibrin threads attached to the damaged surfaces of the blood vessel gradually contract. As the clot retracts, the ruptured area of the blood vessel becomes smaller. Thus, the risk of hemorrhage is further decreased. During retraction, some serum escapes between the fibrin threads, but the formed elements in blood remain trapped within the fibrin threads. Normal clot retraction depends on the presence of an adequate number of platelets. It is speculated that platelets in the clot disintegrate and, through the intrinsic pathway, form more fibrin.

The second event following clot formation is **fibrinolysis**, the dissolution of the blood clot. Once the blood vessel is repaired, the clot dissolves. At the beginning of the coagulation pathway, certain lysing enzymes are released by damaged tissues. These enzymes activate an inactive plasma enzyme called *plasminogen* into an active form called *plasmin*. Plasmin dissolves the fibrin clot.

Clot formation is a vital mechanism that controls excessive loss of blood from the body. In order to form clots, the body needs calcium and vitamin K. Vitamin K is not involved in the actual clot formation, but it is required for the synthesis of prothrombin, which occurs in the liver. The vitamin is normally

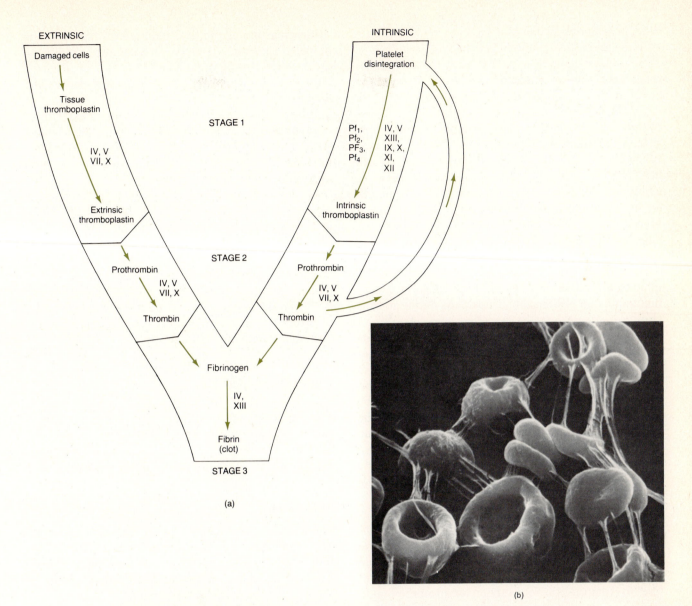

Figure 21-18 Blood clotting. (a) Extrinsic and intrinsic pathways. (b) Scanning electron micrograph of fibrin threads and red blood cells at a magnification of 5,000×. (Courtesy of Fisher Scientific Company and S.T.E.M. Laboratories, Inc., Copyright 1975.)

produced by bacteria that live in the intestine. It is also fat-soluble, which means that it can be absorbed through the walls of the intestine and into the blood only if it is attached to fat. People suffering from disorders that prevent absorption of fat often suffer uncontrolled bleeding. Clotting may also be encouraged by application of a thrombin or fibrin spray, a rough surface such as gauze, or heat.

Clot Prevention

Unwanted clotting may be brought on by the formation of cholesterol-containing masses called *plaques* on the walls of the blood vessels. The plaques supply a rough surface that is perfect for the adhesion of platelets and therefore the sites of clotting. Clotting within an unbroken blood vessel is referred to as *thrombosis*. The clot itself is called a *thrombus*. A thrombus may dissolve, or it may remain intact and interfere with circulation. In the latter case, it may damage tissues by cutting off part of their oxygen supply. If the thrombus becomes dislodged and is carried with the blood to a smaller vessel, the clot can block the circulation to a portion of a vital organ. A blood clot, bubble of air, or piece of debris that is transported by the bloodstream is called an *embolus*.

When an embolus becomes lodged in a vessel and cuts off circulation, the condition is called an *embolism.*

No matter how healthy the body is, occasional rough spots appear on uncut vessel walls. In fact, it is believed that blood clotting is a continuous process inside blood vessels and that it is continually combated by clot preventing and clot dissolving mechanisms. This is another excellent example of homeostasis. Blood contains antithrombic substances—substances that prevent thrombin formation. One of these is *heparin,* which inhibits the conversion of prothrombin to thrombin and prevents most thrombus formation. Heparin is used in open-heart surgery to prevent clotting. Should the thrombus form after all, the enzyme plasmin dissolves the clot—a process called *fibrinolysis.* During coagulation, platelet factor IV inhibits the action of heparin. Thus, heparin cannot function as an antithrombic substance during normal clotting.

In general, any chemical substance that prevents clotting is an **anticoagulant.** Examples of anticoagulants are heparin, dicumarol, and the sodium salts of citrate and oxalate. Heparin is a quick-acting anticoagulant that blocks the clotting mechanism in five places. It is extracted from donated human blood. The pharmaceutical preparation *dicumarol* may be given to patients who are prone to thrombosis. Dicumarol acts as an antagonist to vitamin K and thus lowers the level of prothrombin. Dicumarol is slower acting than heparin, and it is used primarily as a preventative. The *citrates* and *oxalates* are used by laboratories and blood banks to prevent donated blood from clotting. These substances react with calcium to form insoluble calcium compounds. In this way, the ionized calcium (Ca^{2+}) in blood is tied up and is no longer free to catalyze the conversion of prothrombin to thrombin.

Clotting Tests

The time required for blood to coagulate, usually from 5 to 15 minutes, is known as **clotting time.** This time is used as an index of a person's blood-clotting properties. One method for determining clotting time involves taking a sample of blood from a vein and placing 1 ml into each of three Pyrex tubes. The tubes are then submerged in a water bath at 37°C and examined every 30 seconds for the formation of a clot. The clotting process is initiated when the platelets break up upon contact with the glass tubing. When the clot adheres to the walls of the tube, the end point is reached, and the time is recorded. Blood taken from individuals with hemophilia clots very slowly or not at all.

Bleeding time is the time required for the cessation of bleeding from a small skin puncture, usually on the ear lobe. As the droplets of blood escape, they are blotted gently with filter paper. When the paper is no longer stained, the bleeding has stopped. Normally, bleeding time varies from 1 to 4 minutes. Unlike coagulation time, which involves only the biochemical steps initiated by the breakdown of platelets, bleeding time includes the response of the injured blood vessels and all the steps of clot formation.

Prothrombin time is a test used to determine the amount of prothrombin in the blood. First, the blood is treated with oxalate to tie up the calcium. This makes the blood incoagulable. Then, calcium, thromboplastin, plasma coagulation factor V, and plasma coagulation factor VII are mixed with the sample of blood. The length of time required for the blood to clot is the prothrombin time and depends on the amount of prothrombin in the sample of blood. Normal prothrombin time is about 12 seconds.

BLOOD GROUPING

The surfaces of erythrocytes contain genetically determined antigens called **agglutinogens.** These proteins are responsible for the two major blood group classifications: the ABO group and Rh system.

ABO Group

The **ABO blood grouping** is based upon two agglutinogens, which are symbolized as A and B (Figure 21-19). Individuals whose erythrocytes manufacture only agglutinogen A are said to belong to blood group A. Those whose erythrocytes manufacture only agglutinogen B belong to group B. Individuals whose erythrocytes manufacture both A and B are in group AB. Other individuals, whose erythrocytes manufacture neither, belong to group O.

Agglutinogen on Erythrocyte Membrane	Blood Group
A	A
B	B
AB	AB
Neither A nor B	O

The percentages of individuals possessing these four blood types are not equally distributed. The incidence of the various groups in the Caucasian population in the United States is as follows: blood group A, 41 percent; group B, 10 percent; group AB, 4 percent; and group O, 45 percent. Among blacks the incidence

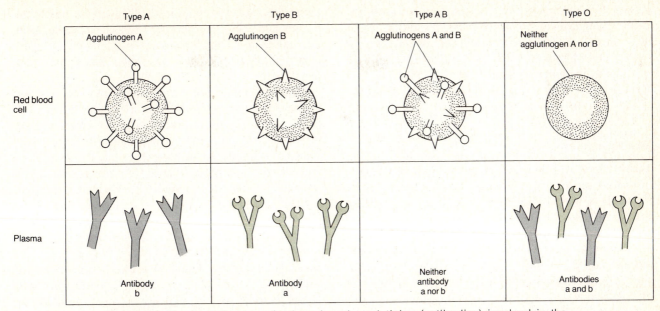

Figure 21-19 Agglutinogens (antigens) and agglutinins (antibodies) involved in the ABO blood grouping system.

of the groups is as follows: group A, 27 percent; group B, 20 percent; group AB, 7 percent; and group O, 46 percent.

The blood plasma of many people contains genetically determined antibodies referred to as *agglutinins*. These are *antibody a* (anti-A), which attacks agglutinogen A, and *antibody b* (anti-B), which attacks agglutinogen B. The antibodies formed by each of the four blood types are shown in Figure 21-19. Individuals do not have antibodies that attack the antigens of their own erythrocytes. For instance, a person with blood group A does not have antibody a. But all people have antibodies against the agglutinogens that they themselves do not synthesize. For example, suppose group A blood is accidentally given to a person who does not have A agglutinogens. The individual's immune system recognizes that the A protein is foreign and therefore treats it as an antigen. Antibodies directed against foreign erythrocytes attack them and cause them to *agglutinate* (clump). This is another example of an antigen-antibody response.

When blood is given to a patient, care must be taken to ensure that the individual's antibodies will not agglutinate the donated erythrocytes and cause clumping. The destruction of the donated cells will not only undo the work of the transfusion, but the clumps can block vessels and cause serious problems that may lead to death.

Cross-matching

When blood transfusions are given, blood of the same group should always be used if it is available, and if not, in an emergency, blood from a different group can be used but only if it is compatible. *Cross-matching* is a test that examines the compatibility of the two bloods. The red cells from the donor and recipient are isolated from their plasma, washed, and resuspended in a suitable solution. Serum is obtained from the blood samples (after plasma has clotted the serum is expressed) and the following combinations of cells and serum are prepared. In the *major cross-matching* test, the donor's cells are mixed with the recipient's serum, and agglutination of the red cells is sought. In the *minor cross-matching* test the recipient's cells are challenged with the donor's serum and checked for possible agglutination. If there is no agglutination in either cross-matching test, then it can be assumed that the transfusion will be acceptable.

Identification and Interactions of the Four Blood Groups

The blood group of a person is routinely determined by obtaining two drops of blood from the patient and diluting each drop with a drop of saline previously placed on either end of a microscope slide. A drop of specially prepared anti-A serum is mixed with one sample of the diluted blood, and a drop of anti-B serum is mixed with the other sample of diluted blood. After a few minutes, the slide is carefully examined for agglutination in either or both of the test samples.

Since the agglutinins seek out and react only with their specific antigen, clumping or agglutination of the cells will occur only if antigens are present on the cells and the specific antigen is detected by the known antiserum. For example, if the anti-A serum causes the cells to agglutinate, then it can be concluded the A agglutinogen is present on the cells. Similarly, if anti-B causes cell clumping, then the B agglutinogen can be said to be present. The possible reactions shown in Exhibit 21-4 allow easy determination of the blood group.

The Rh System

When blood is transfused, the technician must make sure that the donor and recipient blood groups are correctly matched—not only for ABO group type, but also for Rh type.

The **Rh system** is so named because the Rh antigen was identified originally on the red blood cells of the rhesus monkey. Like the ABO grouping, the Rh system is based upon agglutinogens that are present on the surfaces of erythrocytes. Individuals whose erythrocytes contain the Rh agglutinogens are designated as Rh positive (+). Those who lack Rh agglutinogens are designated as Rh negative (−). It is estimated that 85 percent of Caucasians and 88 percent of blacks in the United States are Rh positive. Therefore, 15 percent of Caucasians and 12 percent of blacks are Rh negative.

Under normal circumstances, human plasma does not contain anti-Rh antibodies. However, if an Rh negative person receives Rh positive blood, anti-Rh antibodies will appear in the plasma. If a second transfusion of Rh positive blood is given later, the previously formed anti-Rh antibodies will attack the donated red blood cells and a severe reaction may occur. One of the most common problems with Rh incompatibility arises from pregnancy. During delivery, some of the fetus' blood is apt to leak from the afterbirth into the mother's bloodstream. If the fetus is Rh positive and the mother is Rh negative, the mother will be immunized and will make anti-Rh antibodies. If the woman becomes pregnant again, her previously made anti-Rh antibodies can enter the bloodstream of the fetus. If the baby is Rh negative, no problem will occur since the fetus does not have the Rh antigen. If the fetus is Rh positive, an antigen-antibody response called *hemolysis* may occur in its blood.

Hemolysis means a disruption of the membrane of the erythrocytes so that there is a release of hemoglobin into the blood. The hemolysis brought on by fetal-maternal incompatibility is called *erythroblastosis fetalis.*

When a baby is born with erythroblastosis, an immediate exchange transfusion is begun. The newborn's blood is slowly removed and simultaneously replaced with nonantigenic Rh negative blood. This process attempts to remove the anti-Rh antibodies that entered the infant's blood from the mother. Also, replacing the infant's blood with Rh negative blood reduces any attack by the antibodies aimed at the transfused blood cells since they lack the Rh antigen.

Present treatment of the Rh incompatibility between mother and fetus employs the technique of *passive immunity.* The most likely times for Rh antigens to enter the mother's circulation are just before, during, or just after the birth of the baby. At this time, it is possible to passively immunize the mother by injecting her with immunoglobulins directed against the Rh antigen. These immunoglobulins are obtained from an Rh negative person who has been injected with the antigen and has produced the immunoglobulins. When introduced into the mother's circulation, these antibodies immediately act to destroy the Rh antigenic material *before* the mother's immune system can detect them and respond. Thus, the mother makes no anti-Rh antibodies. There are commercial preparations of anti-Rh immunoglobulins (RhoGAM) that are available for use by physicians.

Other Blood Cell Antigens

Red blood cell membranes—in fact, all cell membranes—contain a great assortment of antigens. Some physiologists liken the membrane surface to a mosaic pattern that shows antigenic determinants repeating over the surface. These other antigenic determinants seldom create incompatibility problems in transfusion, so their value is usually restricted to medical-legal purposes. Some of the other known blood group antigens are the M, N, P, L, S, Kell, Duffy, Diego, Lewis, Lutheran, and Kidd.

INTERSTITIAL FLUID AND LYMPH

For all practical purposes, interstitial fluid and lymph are the same. The major difference between the two is location. When the fluid bathes the cells, it is called **interstitial fluid,** or **tissue fluid.** When it flows through the lymphatic vessels, it is called **lymph** (Figure 21-20). Both fluids are similar in composition to plasma. The principal chemical difference is that they contain less

Exhibit 21-4 BLOOD GROUP INTERACTIONS AND IDENTIFICATION

REACTION TO SERUM		BLOOD GROUP
Anti-A	**Anti-B**	
−	−	O
+	−	A
−	+	B
+	+	AB

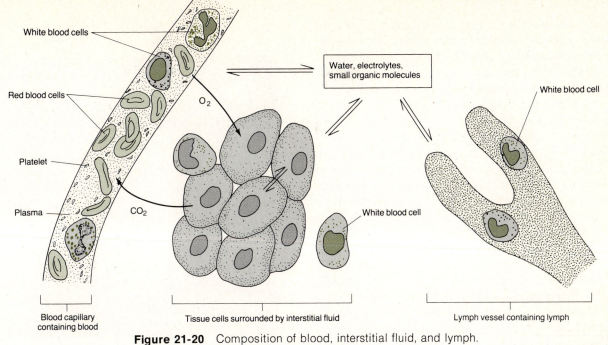

White blood cells

Red blood cells

Platelet

Plasma

O_2

CO_2

Water, electrolytes, small organic molecules

White blood cell

White blood cell

Blood capillary containing blood

Tissue cells surrounded by interstitial fluid

Lymph vessel containing lymph

Figure 21-20   Composition of blood, interstitial fluid, and lymph.

protein because the plasma protein molecules are not easily filtered through the cells that form the capillary walls. Keep in mind that whole blood does not flow into the tissue spaces; it remains within closed vessels. Certain dissolved constituents of the plasma leave the circulation with water as it leaks through the capillary walls. This fluid becomes the interstitial fluid. The transfer of materials between blood and interstitial fluid occurs by osmosis, diffusion, and filtration across the cells that make up the capillary walls. Both interstitial fluid and lymph contain variable numbers of leucocytes. Leucocytes can enter the tissue fluid by diapedesis, and the lymphoid tissue produces the nongranular type leucocytes. However, interstitial fluid and lymph both lack erythrocytes and platelets.

Other substances, especially organic molecules, in interstitial fluid and lymph vary in kinds and amounts in relation to the location of the sample analyzed. The lymph vessels of the digestive tract, for example, contain a great deal of lipid that has been absorbed from food.

HOMEOSTATIC IMBALANCES OF BLOOD

The various blood disorders that can arise affect differing portions of the blood. For example, a patient with anemia may have an abnormally low number of red blood cells, whereas a patient with mononucleosis will have an abnormally high number of white blood cells. There are numerous blood disorders, and all may have wide-ranging effects.

Anemia

Anemia is a sign rather than a diagnosis. The many kinds of anemia are all characterized by abnormally low erythrocyte numbers and/or hemoglobin content. The lowered hemoglobin content of the patient usually produces a paleness, or pallor. These conditions lead to fatigue and intolerance to cold. Both conditions are related to lack of oxygen needed for energy and heat production. However, diagnosis cannot be made, nor can treatment begin, until the cause of the anemia is discovered.

Hemorrhagic (Blood Loss) Anemia

An excessive loss of erythrocytes through bleeding is called **hemorrhagic anemia**. Common causes are large wounds, stomach ulcers, and excessive menstrual bleeding. If bleeding is extraordinarily heavy, the anemia is termed *acute*. Excessive loss of blood fluid may endanger the individual's life. Slow, prolonged bleeding is more apt to produce a *chronic* anemia, and its chief symptom is fatigue.

Hemolytic Anemia

The term **hemolytic anemia** comes from the word *hemolysis*—the rupturing of erythrocyte cell membranes. When the cell is destroyed, its hemoglobin pours into the plasma. Distortions in the shapes of erythrocytes are characteristic signs that the cells are progressing toward hemolysis. There may also be a sharp increase in the number of reticulocytes due to the stimulation of erythropoietin. This premature destruction of red cells may result from a number of causes such as: defects in the synthesis of hemoglobin, abnormalities of red cell enzymes, or defects of the red cell membrane. Parasites, toxins, and antibodies from incompatible blood (for example, erythroblastosis fetalis), may cause hemolytic anemia. **Thalassemia,** one of the more common hereditary hemolytic anemias, includes a group of hereditary anemias that result from the quantitative reduction in the synthesis of one or more of the normal globin polypeptide chains. It occurs mainly in the area around the Mediterranean Sea. Treatment generally consists of blood transfusions, and prevention and early management of infection.

Aplastic Anemia

Destruction or inhibition of the red bone marrow results in **aplastic anemia.** Typically, the marrow is replaced by fatty tissue, fibrous tissue, or tumor cells. It is well known that excessive doses or long-term exposure to high-energy radiation will produce injury to most body cells. Especially vulnerable are those tissues in which rapid cell division is a recognized function. Therefore, the germinal cells in the testes, hematopoietic cells in bone marrow, intestinal cells, and germinal cells in the epidermis are easily suppressed by radiation. The suppression of the bone marrow cells by radiation, or by drugs, chemicals, or toxins results in aplastic anemia.

Bone marrow transplants have been successful in some patients with aplastic anemias. However, the transplanted tissue must be from an identical twin or a well-matched individual in order to prevent rejection by the host's immune system. The transplantation operation is usually preceded by several days of pretreatment with immunosuppressive drugs and then followed with a posttreatment regimen of immunosuppressive drugs that are usually administered with decreasing frequency for about 100 days.

Sickle-cell Anemia

Sickle-cell anemia is an inherited disease that occurs in about 1 to 2 percent of black populations. Its symptoms include a severe anemia and infarctions of vital organs, such as the lungs and kidneys, that can lead to death, often in the teen years. Red blood cells, when reaching regions of low oxygen tension, deform and become elongated and sickle shaped (see Figure 21-21). About 8 percent of the Negroes in the United States demonstrate sickling but do not have the disease. These individuals are said to possess the trait for the disease and are heterozygous for the abnormal gene, whereas those demonstrating the disease symptoms are homozygous for the abnormal gene. Sickled

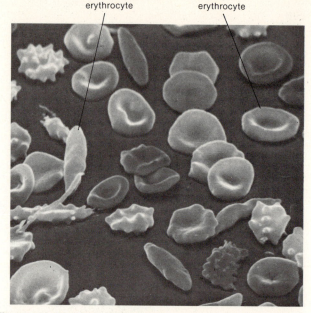

Sickled erythrocyte

Normal erythrocyte

Figure 21-21 Scanning electron micrograph of erythrocytes in sickle-cell anemia at a magnification of 5,000×. (Courtesy of Fisher Scientific Company and S.T.E.M. Laboratories, Inc., Copyright 1975.)

cells rupture easily, and even though erythropoiesis is stimulated by the loss of cells, it cannot keep pace with the destruction. The individual consequently suffers from a hemolytic anemia that reduces the amount of oxygen supplied to the tissues. Prolonged oxygen deficiency aggravates the sickling phenomenon and can cause extensive tissue damage. Because of their shape, the cells tend to aggregate and form thrombi that lodge in blood vessels. Blood flow into the tissue "downstream" is hindered or completely blocked, resulting in tissue infarction. There is evidence that the sickle-cell tendency prevents the cells from rupturing during a malaria crisis. Support for the selective advantage given to those with the sickling trait comes from analysis of the distribution of peoples with these abnormal genes. Populations, or descendants of populations, that live in the malaria belts around the world have the highest incidence of the genes for sickle-cell anemia. This includes the peoples of Mediterranean Europe and subtropical Africa and Asia.

The abnormal gene is expressed in the protein portion of the hemoglobin molecule. The structure of hemoglobin consists of four polypeptide chains—two alpha chains and two beta chains. The apparently very minor change in the number 6 amino acid of the beta chains (it has a total of 146 amino acids) from glutamic acid to valine promotes the linear aggregation of neighboring hemoglobin molecules inside the cells. This generates the internal stresses for the distortion and folding of the cell.

Polycythemia

The term **polycythemia** refers to a condition characterized by an abnormal increase in the number of red blood cells. Increases of 2 to 3 million cells per cubic millimeter are considered polycythemic. The disorder is harmful because the thickness of the blood (viscosity) is greatly increased, and viscosity contributes to a tendency for thrombosis and hemorrhage. It also causes a rise in blood pressure. The tendency for thrombosis to develop results from too many red blood cells piling up as they try to enter smaller vessels. The tendency to hemorrhage results from widespread hyperemia (unusually large amount of blood in an organ part).

There are two basic types of polycythemia: *primary* and *secondary*. The primary type is characterized by an overactivity of the red bone marrow and by an enlarged liver and spleen. Its cause is unknown. The secondary type results from oxygen deficiency in the arteries of people suffering from chronic cardiac or pulmonary disease. Other causes include very high altitudes and liver cancer, which can interfere with the normal function of the liver in disposing of millions of old red blood cells daily.

A clinical test that is important in the diagnosis of polycythemia and anemia is the hematocrit, which indicates the percentage of red blood cells in the blood. Whereas the average hematocrit is 45 percent, anemic blood may have a hematocrit of 15 percent, and polycythemic blood may have a hematocrit of 65 percent.

Disorders Involving White Blood Cells

Two well-known disorders affecting white blood cells are infectious mononucleosis and leukemia.

Infectious Mononucleosis

Infectious mononucleosis is a contagious disease of viral origin that occurs mainly in children and young adults. The trademark of the disease is an elevated white count with an abnormally high percentage of mononuclear cells—primarily lymphocytes. As mentioned earlier in the chapter, an increase in the number of monocytes usually indicates a chronic infection. The various signs and symptoms include slight fever, sore throat, brilliant red throat and soft palate, stiff neck, cough, and malaise. The spleen may enlarge, and secondary complications involving the liver, heart, kidneys, and nervous system may develop.

There is no cure for mononucleosis, and treatment consists of watching for and treating any complications. Usually, the disease runs its course in a few weeks and the individual suffers no permanent ill effects.

Leukemia

Also called "cancer of the blood," **leukemia** is an uncontrolled, greatly accelerated production of white cells. Many of the cells fail to reach maturity. As with most cancers, the symptoms and the cause of death result less from the cancer cells themselves than from their interference with normal body processes. The accumulation of cells leads to abnormalities in organ functions. For example, the anemia and bleeding problems commonly seen in leukemia result from the "crowding out" of normal bone marrow cells. This interferes with the normal production of red blood cells and platelets. The most common cause of death from leukemia is internal hemorrhage, especially cerebral hemorrhage that destroys the vital centers in the brain. The second most frequent cause of death is uncontrolled infection due to a lack of mature or normal white blood cells.

Therapy may temporarily stop the pathologic process. The abnormal accumulation of leucocytes may be reduced or even eliminated by x-rays and anti-leukemic drugs. Partial and complete remissions have been reported, and the progress in treatment and understanding of the disease has been very encouraging. Individuals who have had remissions for over 15 years are considered cured.

CHAPTER SUMMARY OUTLINE

BLOOD AND LYMPH

1. Blood picks up oxygen from the lungs, nutrients from the digestive tract, hormones from the endocrine glands, and enzymes, and transports these substances to all the tissues of the body, where they diffuse into the interstitial fluid.
2. Interstitial fluid is the fluid that surrounds all cells. From the interstitial fluid, the substances are taken up by the cells and exchanged for wastes.

3. Lymph picks up material, including wastes, from the interstitial fluid, cleanses it of bacteria, and returns the material and wastes to the blood.

BLOOD

1. The principal function of blood is transportation of O_2, CO_2, nutrients, wastes, hormones, and enzymes. It regulates pH, body temperature, and water content of cells and protects against disease. Blood consists of plasma and formed elements.

Formed Elements

1. Formed elements are erythrocytes, leucocytes, and thrombocytes. Wandering epithelial cells become hemocytoblasts. Hemocytoblasts in red bone marrow develop into erythrocytes, granular leucocytes, and thrombocytes. Hemocytoblasts entrapped in lymphatic tissue develop into agranular leucocytes.
2. Erythrocytes, or red blood cells, are biconcave discs without nuclei and contain hemoglobin. Erythrocyte formation is called erythropoiesis and occurs in adult red marrow of certain bones. A reticulocyte count is a diagnostic test that indicates the rate of erythropoiesis.
3. Leucocytes, or white blood cells, are nucleated cells. Two principal types are granular (neutrophils, eosinophils, basophils) and agranular (lymphocytes and monocytes).
4. One function of leucocytes, especially neutrophils and monocytes, is to combat inflammation and infection through phagocytosis.

THE IMMUNE RESPONSE

1. Our bodies provide protection for our well-being and survival by employing a variety of mechanisms.
2. Two types of mechanisms involved in this protection are innate immunity and acquired immunity.
3. Innate immunity is present from the time of our birth, while acquired immunity begins weeks after birth and develops as time progresses. It aims at and singles out specific invaders, and operates by means of antibodies.

Lymphoid Tissue

1. Lymphoid tissue is distributed strategically throughout the body, so that it guards the main entrances into the body.

2. Locations include the throat and pharynx, in the form of tonsils and adenoids; the intestinal tract; peripheral tissue; and the lymphoid tissue of the spleen, liver, and bone marrow.
3. Acquired immunity operates by means of antibodies. The antibody defense system comes in two forms—molecules and cells. The antibody is said to be specific because it attacks only that foreign material which triggered its production. The foreign material is called the antigen.
4. The two main types of lymphocytes active in the antigen-antibody response are the T cells and the B cells. B cells possess antibody molecules called immunoglobulins and an activity system called complement.

Structure and Classes of Immunoglobulins

1. Immunoglobulins comprise about 20 percent of the total plasma proteins and most fall into the category of gamma globulins.
2. Immunoglobulins are a heterogeneous group of molecules composed of protein (82 to 96 percent) and carbohydrate (4 to 18 percent) with the protein portion containing their biological activities.
3. Each antibody molecule consists of four polypeptide chains—two designated as heavy (H) chains and two designated as light (L) chains.
4. The variable (V) region contains the sequence of amino acids in the polypeptide chains that differs for each kind of antibody, and it is this region that endows each immunoglobulin with its specificity.
5. Immunoglobulins are subdivided into five classes on the basis of the amino acid sequences in the constant region of the heavy chains.
6. The five immunoglobulin classes are referred to as IgG, IgM, IgA, IgD, and IgE, with the IgG group the largest, making up 75 percent of the total immunoglobulins in the serum of normal adults.

Antibody Reactions

1. The mechanism of action of the IgE immunoglobulins describes one way in which antibodies operate within our bodies. Other methods involve precipitation, agglutination, and neutralization.
2. The complement system consists of at least 15 chemically and immunologically distinct proteins that are normally nonreactive and dissolved in the plasma.
3. The entire sequence of complement activation can be triggered by antigen-antibody complexes that contain either IgG or IgM immunoglobulins.

Response of the B Cells to Antigen

1. The B cells produce and release the immunoglobulins into the circulation.
2. When a B cell encounters and binds to its antigen, it produces a clone of cells that includes plasma cells and memory cells.

Response of the T Cells to Antigen

1. The T cells provide a defense against microorganisms such as fungi, bacteria that reside inside host cells, and viruses. T cells are responsible for the immunologic rejection of allografts and for antitumor immunity.
2. Most of the protection provided by the T cells is the result of their ability to release minute quantities of a large variety of very powerful proteins called lymphokines.

Immune Tolerance and Autoimmunity

1. Immune tolerance means that a host animal will accept materials into itself without producing antibodies. As a result there is no immune response when the antigen is introduced into the host.
2. When an animal produces antibodies that attack its own tissues, the condition is called autoimmunity.
3. The likelihood of autoimmunity increases with age and wear and tear of body tissues. Tissue destruction can result from the action of bacteria, viruses, or toxins.
4. Autoimmune diseases are divided into two groups — those in which one organ is affected, and those in which the destruction is widespread.

Plasma

1. The liquid portion of blood, called plasma, consists of 92 percent water and 7 to 9 percent solutes. Principal solutes include proteins (albumins, globulins, fibrinogen), foods, enzymes and hormones, gases, and electrolytes.
2. Thrombocytes, or platelets, are disc-shaped structures without nuclei. They are formed from megakaryocytes and are involved in clotting.

CLOTTING

1. A clot is a network of insoluble protein (fibrin) fibers in which formed elements of blood are trapped.
2. The chemicals involved in clotting are known as coagulation factors.

3. There are two kinds of coagulation factors: plasma and platelet coagulation factors.
4. Blood clotting occurs by either of two pathways: the intrinsic or the extrinsic.
5. Clotting in a blood vessel is called thrombosis. A thrombus that moves from its site of origin is called an embolus.
6. Clinically important clotting tests are: clotting time (time required for blood to coagulate), bleeding time (time required for the cessation of bleeding from a small skin puncture), and prothrombin time (time required for the blood to coagulate, which depends on the amount of prothrombin in the blood sample).

BLOOD GROUPING

1. ABO and Rh systems are based on antigen-antibody responses.
2. In the ABO system, agglutinogens (antigens) A and B determine blood type. Plasma contains agglutinins (antibodies) that clump agglutinogens which are foreign to the individual.
3. In the Rh system, individuals whose erythrocyte membranes have Rh agglutinogens are classified as Rh positive (Rh+). Those which lack the antigen are Rh negative (Rh−).
4. A harmful antigen-antibody reaction involving the Rh system is the fetal-maternal incompatibility called erythroblastosis fetalis.

INTERSTITIAL FLUID AND LYMPH

1. Interstitial fluid bathes body cells, whereas lymph is found in lymphatic vessels.
2. These fluids are similar in chemical composition. They differ chemically from plasma in that both contain less protein, a variable number of leucocytes, and no platelets or erythrocytes.

HOMEOSTATIC IMBALANCES OF BLOOD

Anemia

1. Anemia is a sign rather than a diagnosis. The many kinds of anemia are all characterized by abnormally low erythrocyte numbers and/or hemoglobin content.
2. Hemorrhagic anemia is an excessive loss of erythrocytes through bleeding.
3. Hemolytic anemia occurs when the erythrocyte

cell membrane ruptures, releasing the cells' hemoglobin into the plasma. Thalassemia is one of the more common hereditary hemolytic anemias.

4. Aplastic anemia results from destruction or inhibition of the red bone marrow by radiation, drugs, chemicals, or toxins.

5. Sickle-cell anemia is an inherited disease that occurs in about 1 to 2 percent of black populations. Its symptoms include a severe anemia and infarctions of vital organs such as the lungs and kidneys. Severe infarctions can lead to death.

Polycythemia

1. Polycythemia is a condition characterized by an abnormal increase in the number of red blood cells, and can be either primary or secondary.

2. A clinical test that is important in the diagnosis of polycythemia and anemia is the hematocrit.

Disorders Involving White Blood Cells

1. Two well-known disorders that affect white blood cells are infectious mononucleosis and leukemia.

2. Infectious mononucleosis is a contagious disease of viral origin that occurs mainly in children and young adults. Its trademark is an elevated white count with an abnormally high percentage of mononuclear cells, especially lymphocytes.

3. Leukemia, or cancer of the blood, is an uncontrolled, greatly accelerated production of white cells, many of which fail to reach maturity. The accumulation of cells leads to abnormalities in organ functions.

REVIEW QUESTIONS

1. How are blood, interstitial fluid, and lymph related to homeostasis?
2. Distinguish between the cardiovascular system and lymphatic system.
3. Define the principal physical characteristics of blood. List the functions of blood and their relationship to other systems of the body.
4. What are formed elements? Where are the different formed elements produced?
5. Describe the microscopic appearance of erythrocytes. What is the essential function of erythrocytes?
6. Explain the role of heme and globin as related to the function of erythrocytes.
7. What is the Bohr effect?
8. Define erythropoiesis. Relate erythropoiesis to the homeostasis of the red blood cell count. What factors accelerate and decelerate erythropoiesis?
9. What is a reticulocyte count? What is its diagnostic significance?
10. Explain iron metabolism and indicate why iron is so necessary, especially in females.
11. Describe the classification of leucocytes. What are their functions?
12. What is the importance of diapedesis, chemotaxis, and phagocytosis in fighting bacterial invasion?
13. Explain inflammation.
14. What is a differential count? What is its significance?
15. Distinguish between leucocytosis and leucopenia.

16. What is your understanding of the body's immune response?
17. Contrast innate immunity with acquired immunity.
18. Define antigen, antibody, and hapten.
19. List different parts of the body that contain lymphoid tissue.
20. What is the structure of immunoglobulins? What particular part of their structure gives each immunoglobulin its specificity?
21. On what basis are immunoglobulins divided into five classes?
22. List the classes of immunoglobulins and indicate which are the largest and most helpful.
23. Define precipitation, agglutination, neutralization, opsonization, allergen, and the anaphylactic response.
24. How does the complement system operate?
25. What is the response of the B cells to antigen? Contrast the germ line theory with the somatic mutation theory.
26. Compare plasma cells with memory cells as to synthesis and function.
27. What is the response of the T cells to antigen? What are lymphokines?
28. Name some lymphokines and describe their functions in protection.
29. Explain the cooperation between T cells and B cells in the immune response.
30. What is immune tolerance?
31. What is autoimmunity? How does it work?

32. Name some autoimmune disorders.
33. What are the major chemicals in plasma? What do they do?
34. Describe thrombocytes. Of what value are they?
35. Briefly describe the process of clot formation. List the various coagulation factors.
36. What are the pathways involved in blood clotting?
37. What are syneresis and fibrinolysis?
38. Why does blood usually not remain clotted in vessels?
39. Define the following: thrombus, embolus, anticoagulant, clotting time, bleeding time, and prothrombin time.
40. What is the basis for ABO blood grouping? What are agglutinogens and agglutinins?
41. Define cross-matching.
42. What is the basis for the Rh system? How does erythroblastosis fetalis occur? How is it treated?
43. Compare interstitial fluid and lymph with regard to location, chemical composition, and function.
44. What is always characteristic in anemia?
45. List the different kinds of anemia with their characteristic symptoms and effects on the body. Which ones are the most serious?
46. What is polycythemia? Differentiate between primary and secondary polycythemia.
47. Why is polycythemia considered so dangerous?
48. What is a hematocrit? Of what value is it?
49. Contrast infectious mononucleosis and leukemia as to cause, symptoms, and possible serious complications.

Chapter 22

CARDIOVASCULAR PHYSIOLOGY

STUDENT OBJECTIVES

After reading this chapter, you should be able to

- Distinguish between the structure and location of fibrous and serous pericardium
- Identify the blood vessels, chambers, and valves of the heart
- Describe the initiation and conduction of nerve impulses through the electrical conduction system of the heart
- Label and explain the deflection waves of a normal electrocardiogram
- Describe the route of blood in coronary circulation
- Define systole and diastole
- Describe the pressure changes associated with blood flow through the heart
- Relate the events of the cardiac cycle to time
- Describe the sounds of the heart and their clinical significance
- Explain the variations in coronary arterial flow and explain cardiac tissue metabolism
- Define cardiac output and explain how it is determined
- Define Starling's law of the heart
- Contrast the effects of sympathetic and parasympathetic stimulation of the heart
- Define the role of pressoreceptors and chemoreceptors in controlling heart rate
- Describe the homeostatic mechanisms that compensate for circulatory shock
- Contrast the structures and functions of arteries, capillaries, and veins
- Explain the hemodynamics of circulation
- List the factors that assist the return of venous blood to the heart and explain capillary circulation
- Define pulse and identify the arteries where pulse may be felt
- Compare the several abnormal pulse rates
- Define blood pressure
- Describe one clinical method for recording systolic and diastolic pressure
- List the six risk factors involved in heart attacks
- Explain why inadequate blood supply, anatomical disorders, and malfunctions of conduction are primary reasons for heart trouble
- Compare angina pectoris and myocardial infarction as abnormalities of coronary circulation
- Describe patent ductus arteriosus, septal defects, and valvular stenosis as congenital heart defects
- Define tetralogy of Fallot
- Define atrioventricular block, atrial flutter, atrial fibrillation, and ventricular fibrillation as abnormalities of the conduction system of the heart
- Define aneurysm and compare the different types
- List the causes and symptoms of atherosclerosis and explain its possible dangers
- Describe the diagnosis of atherosclerosis by the use of angiography
- Discuss hypertension, its different types, and its effects on the body's various organs

The cardiovascular system delivers blood to all parts of the body. The blood in turn carries oxygen, metabolic fuels, vitamins, hormones, salts, and heat to every living cell and simultaneously collects wastes such as carbon dioxide, urea, water, and extra heat from the cells.

The needs of the tissues and cells of the body vary according to their changing levels of activity. Muscle cells, for example, require a great increase in fuel and oxygen during periods of exercise. The cardiovascular system is designed to accommodate the varying needs of all body tissues by delivering volumes of blood to the tissues that are always adequate for their survival.

In order to accomplish this delicate balancing of distribution, the cardiovascular system has two component parts. (1) A variable pump called the *heart* operates under nervous and hormonal control to generate blood pressure. (2) The blood vessels are hollow tubes that are classified into four basic types, each specially designed to fulfill a special job. The *arteries, arterioles, capillaries,* and *veins* and their special features and roles will be described later in the chapter. First, let us examine this incredible pump called the heart.

THE HEART

The **heart** is a hollow, muscular organ that pumps the blood into the blood vessels. It lies obliquely between the lungs in the mediastinal space with about two-thirds of its mass to the left of the midline of the body (Figure 22-1). The heart is shaped like a blunt cone about the size of a clenched fist—an average size is 12 cm long by 9 cm wide at its broadest point and 6 cm thick.

Parietal Pericardium (Pericardial Sac)

The heart is enclosed and held in place by the **parietal pericardium** or **pericardial sac,** an ingenious structure designed to confine the heart to its position in the mediastinum, yet allow it sufficient freedom of movement so that it can contract vigorously and rapidly when the need arises.

The parietal pericardium consists of two layers referred to as the fibrous layer and the serous layer (Figure 22-2). The *fibrous layer* or *fibrous pericardium* is the outer layer and consists of a very heavy fibrous connective tissue. The fibrous pericardium prevents overdistension of the heart, provides a tough protective membrane around the heart, and anchors the heart in the mediastinum. The inner layer of the parietal pericardium is referred to as the *serous layer* or *serous pericardium*. This is a thinner, more delicate membrane. The *epicardium* or *visceral pericardium* is the thin, transparent outer layer of the wall of the heart that is composed of fibrous tissue and mesothelium. Between the serous pericardium and the epicardium is a potential space called the *pericardial cavity*. The cavity contains a watery fluid, known as pericardial fluid, which prevents friction between the membranes as the heart moves.

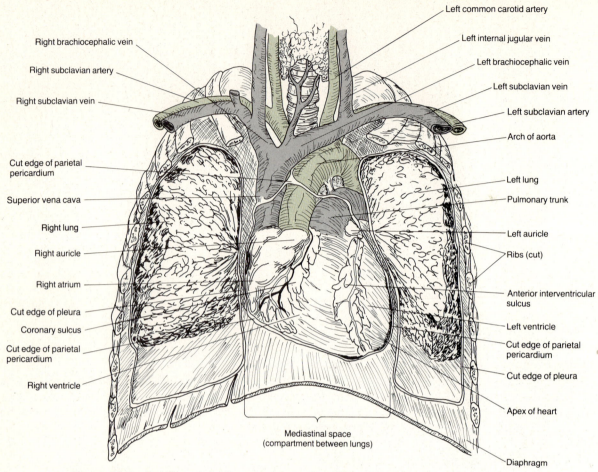

Right brachiocephalic vein

Right subclavian artery

Right subclavian vein

Cut edge of parietal pericardium

Superior vena cava

Right lung

Right auricle

Right atrium

Cut edge of pleura

Coronary sulcus

Cut edge of parietal pericardium

Right ventricle

Left common carotid artery

Left internal jugular vein

Left brachiocephalic vein

Left subclavian vein

Left subclavian artery

Arch of aorta

Left lung

Pulmonary trunk

Left auricle

Ribs (cut)

Anterior interventricular sulcus

Left ventricle

Cut edge of parietal pericardium

Cut edge of pleura

Apex of heart

Mediastinal space (compartment between lungs)

Diaphragm

Figure 22-1 Position of the heart and associated blood vessels in the thoracic cavity.

Chambers

The interior of the heart is divided into four chambers that receive the circulating blood (Figure 22-3). The two upper chambers are called the *right* and *left atria*. The atria are separated by a partition called the *interatrial septum*. The two lower chambers, the *right and left ventricles*, are separated by a partition called the *interventricular septum*. The right atrium receives blood from all parts of the body except the lungs. It receives the blood through three veins. One of these veins is the *superior vena cava*, which brings blood from the upper portion of the body. Another of the veins is the *inferior vena cava*, which brings blood from the lower portions of the body. The third vein is the *coronary sinus*, which drains blood from the vessels supplying the walls of the heart. Blood flows from the right atrium into the right ventricle, which pumps it into the *pulmonary trunk*. The pulmonary trunk then divides into a *right* and *left pulmonary artery*, each of which carries blood to the lungs. In the lungs, the blood releases its carbon dioxide and takes

on oxygen. It returns to the heart via four *pulmonary veins* that empty into the left atrium. The blood then flows into the left ventricle and exits from the heart through the *ascending aorta*. From here aortic blood is distributed to all parts of the body.

An examination of the sectioned heart in Figure 22-3 shows that the sizes of the four chambers vary according to their functions. The right atrium, which must collect blood coming from almost all parts of the body, is slightly larger than the left atrium, which receives only the blood coming from the lungs. The thickness of the walls of the chambers varies too. The atria are relatively thin-walled because they need only enough cardiac muscle tissue to squeeze the fluid into the ventricles. The right ventricle has a much thicker layer of myocardium since it must force the blood through the lungs and then back to the left atrium. The left venticle has the thickest walls, since it must pump blood through literally miles of vessels in the head, trunk, and extremities. As each chamber of the heart contracts, it pushes a portion of blood into a ventricle or out of the heart into an artery. As the walls of the

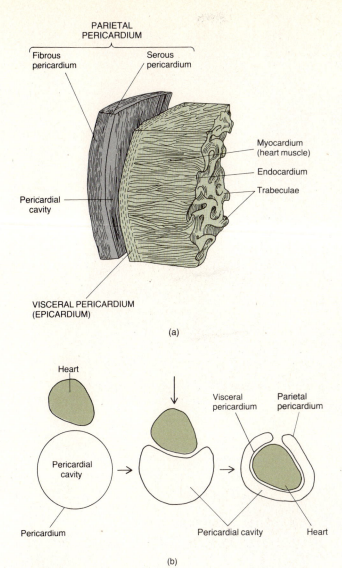

PARIETAL
PERICARDIUM

Fibrous
pericardium

Serous
pericardium

Myocardium
(heart muscle)

Endocardium

Trabeculae

Pericardial
cavity

VISCERAL PERICARDIUM
(EPICARDIUM)

(a)

Heart

Visceral
pericardium

Parietal
pericardium

Pericardial
cavity

Pericardium

Pericardial cavity

Heart

(b)

Figure 22-2 Parietal pericardium. (a) Structure. (b) Development.

chambers relax, valves prevent the blood from flowing back.

Valves

Atrioventricular valves lie between the atria and their ventricles (Figure 22-3). The atrioventricular valve between the right atrium and right ventricle is called the *tricuspid valve* because it consists of three flaps, or cusps. Other names for this valve are the right atrioventricular or right AV valve. As the atrium relaxes and the ventricle contracts, any blood that is driven back toward the atrium drives the cusps upward until their edges meet and close the opening. The atrioventricular valve between the left atrium and left ventricle is called the *bicuspid* or *mitral valve.*

It has two cusps that work in the same way as the cusps of the tricuspid valve. The bicuspid valve is also known as the left atrioventricular or left AV valve.

Each of the arteries that leaves the heart has a valve that prevents blood from flowing back into the heart. These are the *semilunar valves* — semilunar meaning half-moon or crescent-shaped. The *pulmonary semilunar valve* lies in the opening where the pulmonary artery leaves the right ventricle. The *aortic semilunar valve* is situated at the opening between the left ventricle and the aorta (Figure 22-3). Both valves consist of three semilunar cusps. Like the atrioventricular valves, the semilunar valves permit the blood to flow in only one direction. In this case, the flow is from the ventricles into the arteries.

Physiological Properties

The heart demonstrates three properties related to homeostasis that are developed to a very high degree of proficiency: (1) autorhythmicity, (2) conductibility, and (3) contractibility.

Autorhythmicity

A heart removed from an animal can be maintained and kept beating for several hours even though it is completely devoid of any nerve or vascular connections. Beating continues at a definite rate due to the activity of very specialized cells, called P cells, located in the *sino-atrial,* or *S-A node* (Figure 22-4). These P cells are located in the posterior wall of the right atrium close to the opening of the superior vena cava, and they are capable of generating spontaneous rhythmic electrical potentials. With a microscope, the P cells can be identified by their oval or rounded shapes and their smooth surface.

Conductibility

The electrical activity generated by the sino-atrial node is distributed to all parts of the heart by tissues specialized to conduct impulses rapidly. These tissues include the interatrial and internodal conduction pathways, the atrioventricular or A-V node, the bundle of His, the right bundle branch, the left bundle branch, and the peripheral Purkinje network (Figure 22-4). The P cells of the S-A node ordinarily function as the pacemaker for the rate of cardiac contraction. The electrical activity generated within the S-A node is relayed through the atria by preferential conduction pathways composed of selected muscle cells. The A-V nodal cells located near the inferior portion of the interatrial septum are also stimulated by the spreading impulses. The impulses reaching the A-V node are conducted slowly through the A-V nodal tissue to the

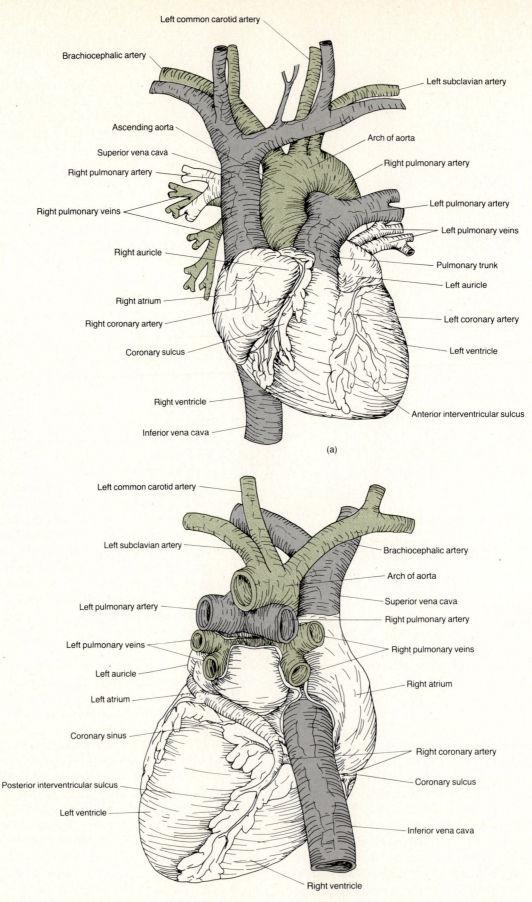

Left common carotid artery

Brachiocephalic artery

Ascending aorta

Superior vena cava

Right pulmonary artery

Right pulmonary veins

Right auricle

Right atrium

Right coronary artery

Coronary sulcus

Right ventricle

Inferior vena cava

Left subclavian artery

Arch of aorta

Right pulmonary artery

Left pulmonary artery

Left pulmonary veins

Pulmonary trunk

Left auricle

Left coronary artery

Left ventricle

Anterior interventricular sulcus

(a)

Left common carotid artery

Left subclavian artery

Left pulmonary artery

Left pulmonary veins

Left auricle

Left atrium

Coronary sinus

Posterior interventricular sulcus

Left ventricle

Brachiocephalic artery

Arch of aorta

Superior vena cava

Right pulmonary artery

Right pulmonary veins

Right atrium

Right coronary artery

Coronary sulcus

Inferior vena cava

Right ventricle

(b)

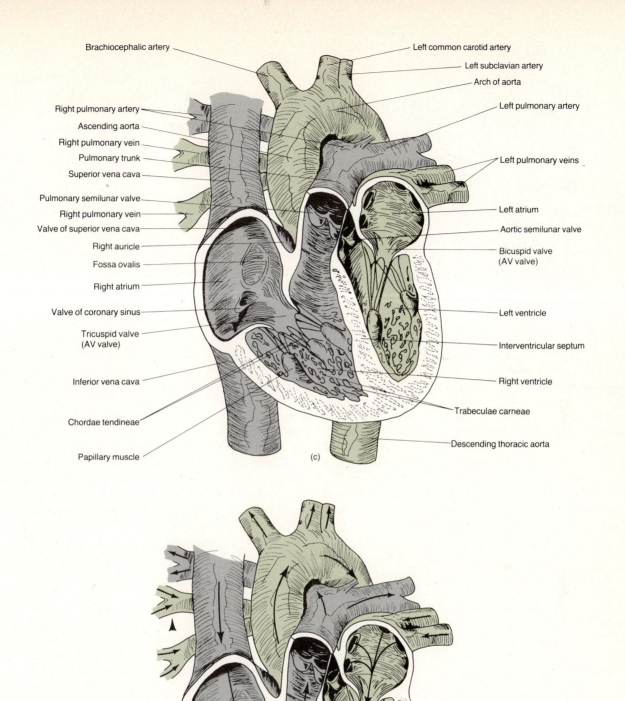

Brachiocephalic artery

Right pulmonary artery

Ascending aorta

Right pulmonary vein

Pulmonary trunk

Superior vena cava

Pulmonary semilunar valve

Right pulmonary vein

Valve of superior vena cava

Right auricle

Fossa ovalis

Right atrium

Valve of coronary sinus

Tricuspid valve
(AV valve)

Inferior vena cava

Chordae tendineae

Papillary muscle

Left common carotid artery

Left subclavian artery

Arch of aorta

Left pulmonary artery

Left pulmonary veins

Left atrium

Aortic semilunar valve

Bicuspid valve
(AV valve)

Left ventricle

Interventricular septum

Right ventricle

Trabeculae carneae

Descending thoracic aorta

(c)

(d)

Figure 22-3 Structure of the heart. (a) Anterior external view. (b) Posterior external view. (c) Anterior internal view. (d) Path of blood through the heart.

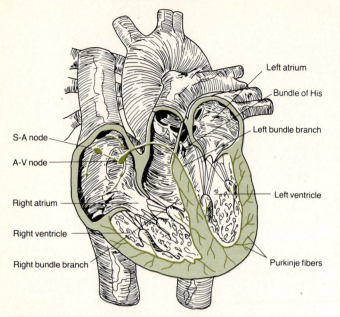

Figure 22-4 The conduction system of the heart.

bundle of His, with which it is continuous and which is located at the top of the interventricular septum. The tissue of the A-V node also contains P cells capable of spontaneous activity, but these cells are fewer in number here and deeper in the nodal tissue. Their automaticity is easily overcome by that of the S-A node. The bundle of His projects a single right bundle branch down the right side of the interventricular septum just under the endocardium. The bundle fibers then branch repeatedly and become continuous with the Purkinje network. On the left side, the bundle of His sends a number of branches subendocardially down the left side of the interventricular septum to distribute electrical impulses to the Purkinje network. The Purkinje networks course under the endocardium and send many branches deep into the myocardium, where the Purkinje cells merge and become indistinguishable from the contracting muscle cells.

Contractibility

A typical histologic section of heart muscle is shown in Figure 22-5a. The cross striations show that cardiac muscle is like skeletal muscle in appearance, but since cardiac muscle behaves involuntarily, it is also unlike skeletal muscle.

The muscle fibers are not arranged in parallel fashion as in skeletal muscle, but rather they form branching networks in which the cells appear to split lengthwise and fuse with neighboring cells. For many years, therefore, it was thought that there were no individual cardiac cells, but that the heart tissue was composed of two huge protoplasmic networks each

called a *syncytium*. The muscular walls of the atria and interatrial septum composed one network and the muscular walls and septum of the ventricles composed the other. It is true that when a single fiber of either network is stimulated, all the fibers in the network respond. Thus, each network operates as a *functional* unit. When cardiac tissue was studied with the electron microscope, the individual cell boundaries became apparent in the junctional complex called the *intercalated disc* (Figure 22-5b). The region of contact shows that the adjoining membranes interdigitate rather elaborately in end-to-end connections and much less so in side-to-side connections. In this region, there are also other junctional complexes (see Chapter 10), such as the zonula occludens, gap junctions, and desmosomes. These are thought to bind the cardiac muscle cells together. Also, there is considerable evidence now that the gap junctions alongside the intercalated discs are regions of low electrical resistance that probably permit an easy spread of excitation to adjacent cells. This would explain why the heart functions as if it were a syncytium—and indeed, some physiologists use the term *functional syncytium* to emphasize this unique behavior.

Cardiac muscle cells have the same basic arrangement of myofilaments, sarcoplasmic reticulum, and T tubules that are found in skeletal muscle cells. The sarcoplasmic reticulum is less abundant and simpler than in skeletal muscle, and the T tubules in cardiac muscle cells are fewer but larger in diameter.

The Electrocardiogram

As the impulses generated by the sino-atrial node spread over the atrial muscle cells, the muscle cells contract. The impulses that are transmitted via the conducting system into the ventricular muscle cells cause these muscle cells to contract.

The order in which the ventricular muscle cells contract has been carefully mapped. It shows two constant features. First, the wave of excitation travels down the *right* bundle branch a little faster than down the left bundle branches. Therefore the cells of the right ventice are depolarized slightly before those of the left ventricle. However, because the left ventricular muscle mass is much greater than that of the right, the total depolarization or total amount of *negative voltage* generated on the surface of the left ventricular cells is much greater than that of the right.

Second, the depolarization wave going through the heart wall moves from the inside cells—those just under the endocardium—through the musculature to the outermost cells under the pericardium. Therefore,

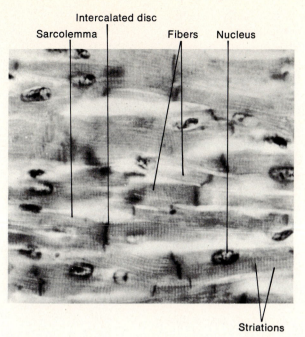

Intercalated disc

Sarcolemma   Fibers   Nucleus

Striations

(a)

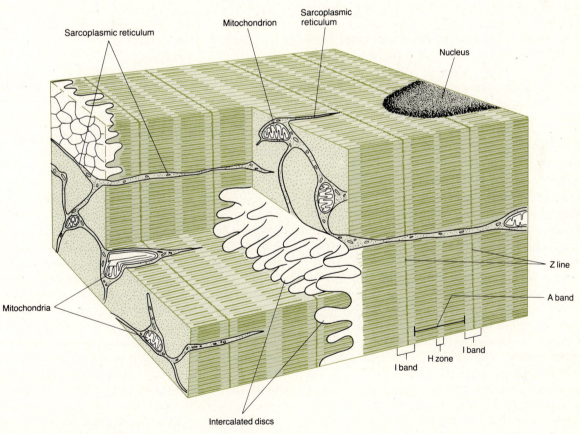

Sarcoplasmic reticulum   Mitochondrion   Sarcoplasmic reticulum   Nucleus

Sarcoplasmic reticulum

Z line

A band

Mitochondria

I band   H zone   I band

Intercalated discs

(b)

Figure 22-5 Histology of cardiac muscle tissue. (a) Photomicrograph in longitudinal section. (Courtesy of Edward J. Reith, from *Atlas of Descriptive Histology,* by Edward J. Reith and Michael H. Ross, New York: Harper & Row, 1970.) (b) Diagram showing cellular junction between muscle cells and the internal arrangement of myofibrils and cellular organelles. Diagram based on electron microscopic studies.

the location of the negativity is constantly changing as it migrates through the ventricular tissue. Other portions of the cardiac tissue that have not been excited or have already repolarized following the excitation will be positively charged on the surface of the cells.

Three hundred years ago, Sir Isaac Newton, in conceiving the concept of center of gravity, proved that it is possible to treat the effect of all the positive charges acting at many different points in the same object as if they were gathered at the center of their effect. The negative charges can also be treated as having a center of charge. During the period of electrical activity in the heart, the positions of the center of negativity and positivity are constantly moving and do not coincide. Only when the heart is at rest do the centers coincide. When the centers of charge do not coincide, an *electrical dipole* exists. This means that the centers of opposite charges are separated by some distance during cardiac excitation, and that a line joining these two centers is constantly changing in length and direction.

This changing electrical activity of the heart can be recorded by electrodes placed on the surface of the body. Routinely, electrodes are attached to the right arm, the left arm, and the left leg (Figure 22-6). Instantaneous differences in potential measured between pairs of electrodes are called *leads*. The so-called *three standard leads* measure the voltage difference between:

Lead I (RL): Right arm (R) and left arm (L)
Lead II (RF): Right arm (R) and left leg (F)
Lead III (LF): Left arm (L) and left leg (F)

The instantaneous changes in voltage differences are plotted on a strip chart that is called the **electrocardiogram (ECG)**. The three leads produce graphs with the same wave components, but their shapes, heights, and directions differ. A typical record obtained from lead II is shown in Figure 22-7. Three clearly recognizable waves accompany each cardiac cycle. The first, called the *P wave*, is generated by the wave of excitation spreading over both atria. The first region to depolarize in the atria is the area inside and around the sino-atrial node. This area acts as a center of negativity, while a center of positivity is in the yet unexcited portion of the atria. This period of charge separation is recorded on the electrocardiogram as a *positive* curve traced *above* the baseline that generates about 0.2 mV and lasts about 0.1 second. During the period of repolarization the first portion to return to its original state is the first area that was depolarized, the S-A node. Thus, this area now acts as a center of positivity, whereas a center

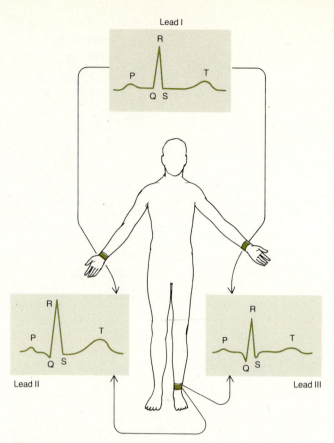

Figure 22-6 The standard limb leads of the electrocardiogram and the usual normal records from these leads.

of negativity exists in some further area of the atria. This period of charge separation is recorded on the electrocardiogram as a *negative* curve traced *below* the baseline. However, this second wave representing the repolarization of the atria occurs during the period when the ventricles are depolarizing and is thus overwhelmed by the large burst of ventricular electrical activity. The *QRS complex* records the depolarization of the Purkinje cells and ventricular muscle. The Q portion is a downward deflection initially produced by the Purkinje cells, followed by the steeply rising Q-R curve produced by the increasing number of responding ventricular cells. The R-S portion of the curve results from the declining number of responding ventricular cells. The S-T interval represents a brief time during which the entire ventricular muscle is depolarized. The centers of negativity and positivity coincide and there is no dipole. Therefore, the voltage recorded on the electrocardiogram during this time is close to zero. The third recognizable deflection is a dome-shaped deflection called the *T wave*. Its shape and height are the most variable of the electrocardiogram waves, and these variations are good indicators of a problem in ventricular conduction.

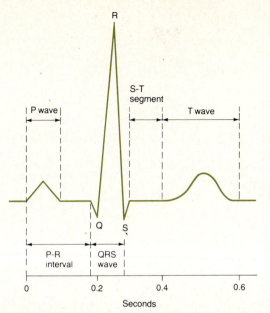

Figure 22-7 *Recordings of a normal electrocardiogram.*

During the initial stages of ventricular excitation, the interior of the heart is depolarized first and thus is negatively charged while the outside of the heart is positively charged. This charge separation is recorded on the electrocardiogram as the positive QRS curve that is traced above the baseline in the same fashion as that drawn for the depolarization curve of the atria. One would reasonably expect the pattern of ventricular repolarization to follow the pattern of depolarization, that is, to begin where it depolarized first as it did in the atria. But this is *not* the case for the ventricle. The interventricular septum and endocardium remain depolarized and contracted for a longer period of time than the outer regions of the heart. Thus, in the ventricles, it is the outer surface that returns to a positively charged state first while the interior surface is still negatively charged. As repolarization continues inward, the curve traced on the electrocardiogram, instead of being negatively drawn as might have been expected, is the positively drawn T wave.

In reading an electrocardiogram, it is important to note the size of the deflection waves and certain time intervals (Figure 22-7). Since the P wave represents atrial contraction, enlargement of the P wave indicates an increase in the mass of the atria, as might occur in mitral stenosis. In this condition, which is described later, the left atrium enlarges because the mitral valve (between the left atrium and left ventricle) narrows. This causes blood to back up into the atrium, ultimately resulting in growth of the atrial wall.

The *P–R interval* is measured from the beginning of the P wave to the beginning of the Q wave. It repre-

sents the conduction time from the beginning of the atrial excitation to the beginning of ventricular excitation. The P–R interval normally lasts 0.16 to 0.20 second and indicates the time required for an impulse to travel through the atria and atrioventricular node to the remaining conducting tissues. The lengthening of this interval, as in arteriosclerotic heart disease and rheumatic fever (described later), occurs because the atrial or A-V nodal tissue is scarred or inflamed. Thus, the impulse travels at a slower rate and the interval is lengthened.

In the QRS wave, an enlarged Q wave may indicate an old myocardial infarction and an enlarged R wave generally indicates enlarged ventricles. The S–T segment is elevated in acute myocardial infarction and depressed when the heart muscle receives insufficient oxygen.

The T wave represents ventricular repolarization. It is flat when the heart muscle is receiving insufficient oxygen, as in arteriosclerotic heart disease. It may be elevated when the body's potassium level is increased.

The ECG is invaluable in diagnosing abnormal cardiac rhythms and conduction patterns, detecting the presence of fetal life, determining multiple pregnancies, and following the course of recovery from a heart attack.

Blood Supply of the Heart

Cardiac tissue, like any other, has its own system of blood vessels that delivers nutrients and oxygen and removes wastes and carbon dioxide in order to maintain homeostasis. Its blood supply originates directly from the aorta via two arteries—the *left coronary artery* and the *right coronary artery*. Figure 22-8 shows how these arteries originate from within the cup portion of the semilunar valves immediately beyond the left ventricle. This immediate origin indicates the prime importance of the heart, for it is the organ that receives the first blood ejected from the heart under the highest possible pressure.

In general, the left coronary artery distributes its blood to the left ventricle and the anterior portion of the septum, while the right coronary artery supplies the right ventricle and the posterior portion of the septum. There are numerous *anastomoses* between the two branching systems so that blood from one system can flow easily into the other via connecting vessels. These interconnections provide a safety factor should a branch artery of one system become occluded (a myocardial infarction). At rest, the heart receives about 5 percent of the total cardiac output. The branching system of the blood supply to the heart

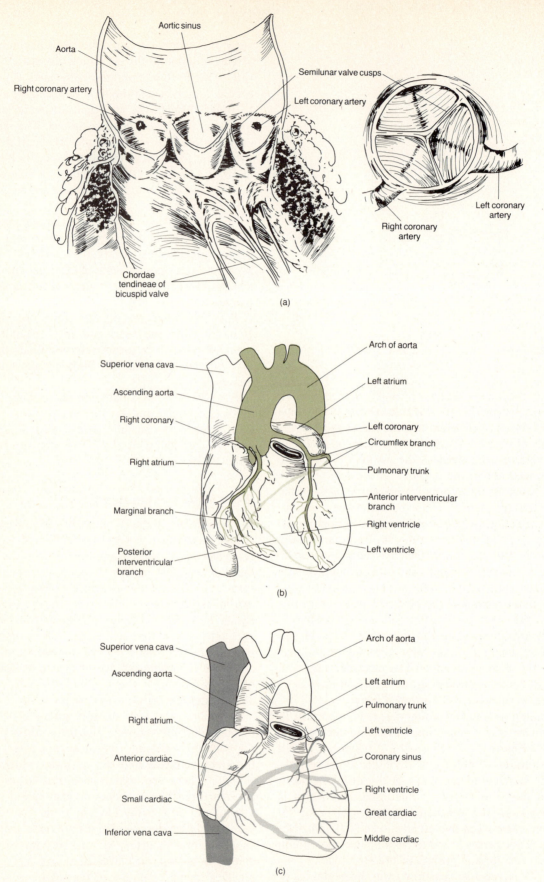

Figure 22-8 Blood supply of the heart. (a) Origin of the coronary arteries seen in sectional view (left) and from above (right). (b) Arterial distribution. (c) Venous drainage.

is so great that every cardiac muscle cell has very close access to a capillary. The drainage from the heart is via a venous system that returns 90 percent of the blood directly into the heart. Most of the blood from the myocardium is collected into a large venous channel—the *coronary sinus*—that empties into the right atrium. Other small veins, the *anterior cardiac veins*, which collect some of the blood from the right side of the myocardium, empty independently into the right atrium. In addition to the venous drainage there are *Thebesian vessels* that deliver some of the coronary outflow directly to the ventricular chambers. The function of these vessels is debatable, however. Some physiologists think they return venous blood to the ventricular chambers. Others feel that, in the left ventricle at least, these vessels carry oxygenated blood from the ventricular chamber backward into the muscular walls.

The Cardiac Cycle

In a normal heartbeat, the two atria contract simultaneously while the two ventricles are relaxing. Then, when the two ventricles are contracting, the two atria are relaxing. The term **systole** refers to the phase of contraction, and the term **diastole** refers to the phase of relaxation. A **cardiac cycle,** or complete heartbeat, consists of the systole and diastole of the atria, plus the systole and diastole of the ventricles.

An examination of the cardiac cycle can be started near the end of atrial diastole (Figure 22-9a). At this time, both atrioventricular valves are open and the semilunar valves are closed. Throughout most of atrial diastole, the ventricles receive blood from the atria. The atria are constantly filling with blood. The right atrium receives deoxygenated blood from the superior and inferior vena cava and coronary sinus. The left atrium receives oxygenated blood from the pulmonary veins. It is estimated that 75 percent of ventricular filling occurs during atrial diastole.

When the S-A node fires, it marks the end of atrial diastole and causes the atria to depolarize and contract. This is the P wave on the ECG, and it signals the start of *atrial systole*. Contraction of the atria forces more blood into the ventricles. Deoxygenated blood from the right atrium passes into the right ventricle through the open tricuspid valve. Oxygenated blood passes from the left atrium into the left ventricle through the open mitral valve. While the atria are contracting, the ventricles, which are in diastole, are stretched further in order to accommodate the "extra" blood. During this phase of *ventricular diastole,* the semilunar valves to the aorta and pulmonary trunk are closed.

When atrial systole and ventricular diastole are completed, the events are reversed. The atria go into diastole, and the ventricles go into systole. During *atrial diastole,* as noted earlier, deoxygenated blood from the various parts of the body enters the right atrium. Simultaneously, oxygenated blood from the lungs enters the left atrium. During the first part of atrial diastole, the atrioventricular valves are closed, since the ventricles are in systole. In *ventricular systole,* the ventricles contract and force blood into their respective vessels. Ventricular systole is initiated by the spread of the QRS wave through the ventricles. The right ventricle pumps deoxygenated blood through the semilunar valve into the pulmonary trunk. The blood is conveyed to the lungs, where it is oxygenated. The left ventricle forces oxygenated blood through the open semilunar valve into the aorta. At the end of the ventricular systole, the semilunar valves close, and both atria and ventricles relax.

In a complete cardiac cycle, the atria are in systole 0.1 second and diastole 0.7 second. By contrast, the ventricles are in systole 0.3 second and diastole 0.5 second.

Two phenomena control the movement of blood through the heart: the opening and closing of the valves, and the contraction and relaxation of the myocardium. Both these activities occur without direct stimulation from the nervous system. The valves are controlled by pressure changes that occur within each heart chamber. The contraction of the cardiac muscle is initiated by its own conduction system.

Pressure Changes in the Cycle

The pressure developed in a heart chamber is related primarily to the size of the chamber and the volume of blood it contains. The greater the volume of blood pushed into the chamber, the higher the pressure. Figure 22-9b illustrates the pressure changes occurring on the left side of the heart. Although the pressures in the right side of the heart are somewhat lower, the pattern is similar. The general principle operating to move blood through the vessels is that blood flows from an area of higher pressure to an area of lower pressure.

The pressure within the atria is referred to as *intraatrial pressure*. When the atria are in diastole, the pressure within them steadily increases. This happens because blood flows into the atria continuously from the veins. When the ventricles go into diastole, the atrioventricular valves open and blood drains into the ventricles, which now have a lower pressure. When atrial pressure builds to point *a* in Figure 22-9b, the atria contract and force more blood into the ventricles. The ventricles are stretched further and their pres-

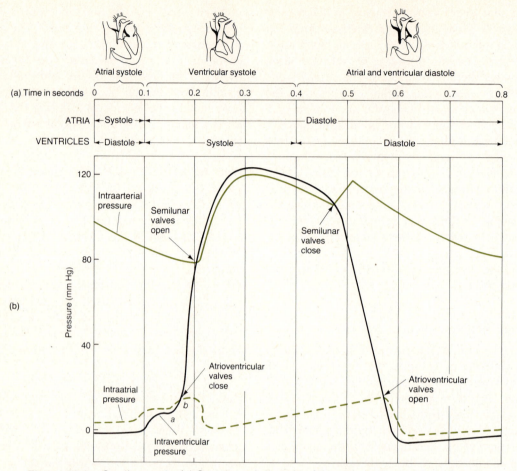

Figure 22-9 Cardiac cycle. (a) Systole and diastole of the atria and ventricles related to time. (b) Pressure changes and valvular openings and closings.

sure, called *intraventricular pressure,* increases. The ventricles start to contract and the intraventricular pressure increases further. At this point, point *b* in Figure 22-9b, a backflow of blood closes the atrioventricular valves. The semilunar valves remain closed. At this instant, no blood is moving out into the aorta or pulmonary artery. The ventricles are completely closed chambers.

The continual contraction of the ventricular myocardium causes the intraventricular pressure to rise until it is greater than the pressure in the aorta and pulmonary artery (the *intraarterial pressure*). The instant intraventricular pressure rises above intraarterial pressure, the semilunar valves burst open, and blood is ejected from the ventricles into the great vessels with great pressure.

Once the ventricles eject their blood, intraventricular pressure decreases and falls below that in the great vessels. As a result, the semilunar valves are pushed closed by blood attempting to return from the vessels into the ventricles. Once again, the ventricles are closed chambers. As the ventricles relax, intraventricular pressure decreases until it becomes less than

intraatrial pressure. At that point, intraatrial pressure forces the tricuspid and mitral valves open, blood fills the ventricles, and another cycle begins.

Timing of the Cycle

We can now relate the events of the cardiac cycle to time. If we assume that the average heart beats 72 times per minute, then each beat with its short pause requires about 0.8 second (Figure 22-9a). During the first 0.1 second, the atria are contracting and the ventricles are relaxing. The atrioventricular valves are open and the semilunar valves are closed. For the next 0.3 second, the atria are relaxing and the ventricles are contracting. During the first part of this period, all valves are closed. During the second part, the semilunar valves are open. The last 0.4 second of the cycle is the relaxation, or quiescent, period. All chambers are in diastole. For the first part of the quiescent period, all valves are closed. During the latter part of the relaxation period, the atrioventricular valves open, and blood starts draining into the ventricles. When the heart beats at a faster rate than normal, it is the quiescent period that is shortened.

Sounds of the Heart

The first sound, which is described as a lubb (o͞o) sound, is a comparatively long, vibrating sound. The lubb is the sound of the atrioventricular valves closing soon after ventricular systole begins. The second sound, which is heard as a short, sharp snap, is described as a dupp (ŭ) sound. Dupp is the sound of the semilunar valves closing toward the end of ventricular systole. A pause about two times longer comes between the second sound and the first sound of the next cycle. Thus, the sound of the cardiac cycle is heard as a lubb, dupp, pause; lubb, dupp, pause; lubb, dupp, pause.

Heart sounds provide valuable information about the valves of the heart. If the sounds are muffled, they are referred to as murmurs. Murmurs are frequently the noise made by a little blood bubbling back into an atrium because of the failure of one of the atrioventricular valves to close properly. However, murmurs do not always indicate a valve problem, and many have no clinical significance.

Although the heart sounds are produced in part by the closure of the valves, they are not necessarily best heard over these valves. Instead, each sound tends to be clearest in the location where the vascular chamber distal to the valve in terms of blood flow lies closest to the surface of the body. These locations are illustrated in Figure 22-10.

Variations in Coronary Arterial Flow

The blood flow pattern through the left and right coronary arteries is shown in Figure 22-11. The irregular pattern reflects the great changes in resistance that occur during the cardiac cycle. Ventricular contraction increases the pressure on the small blood vessels nourishing the heart to the point where flow is temporarily halted. The graph for flow through the left coronary artery illustrates this phenomenon. As the ventricle begins to contract, flow through the coronary artery drops precipitously and may even reverse if the "squeeze" on the blood vessels is great enough. The ejection of blood increases the blood pressure in the aorta so that flow into the coronary vessel increases momentarily to an intermediate level, only to fall again as the ventricular "squeeze" continues to increase to its maximum. When the ventricle begins to relax (diastole), the greatly elevated aortic blood pressure quickly forces blood into the coronary arteries, which are now being released from the ventricular squeeze. Consequently, the flow during the onset of ventricular relaxation reaches the maximum level and then slowly declines. It is therefore important to realize that because of this varying flow pattern, most of the coronary blood flow takes place during ventricular relaxation (the period of diastole). Measurements indicate that between 70 and 90 percent of the coronary blood flow occurs at this time. This also

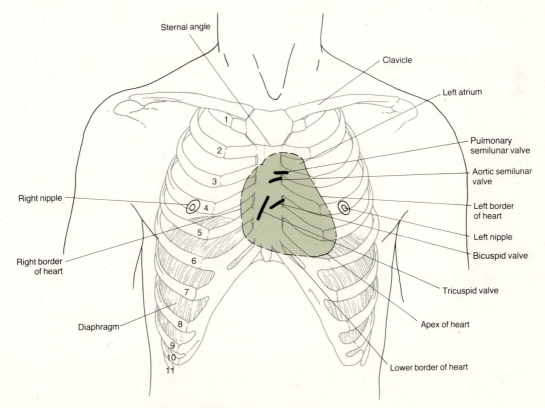

Figure 22-10 Surface projection of the heart.

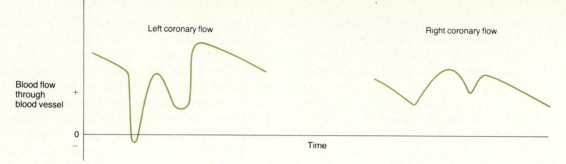

Figure 22-11 Curves showing variation in rate of flow in the left and right coronary arteries.

points out the great importance of the *diastolic* pressure in the aorta. It is probably much more important than the systolic pressure in maintaining an adequate coronary flow.

During periods of increased activity (for example, during exercise) the oxygen requirement of the heart tissue also increases. This requirement is met by an automatic increase in the blood flow through the coronary vessels. It would be expected that coronary flow in a more rapidly beating heart would decrease, since the heart would spend a lesser percentage of its time in diastole and consequently have less time for blood flow into the contracted cardiac musculature. However, this effect is apparently more than compensated by a marked dilation of the coronary vessels. Very small drops in the level of oxygen (hypoxia) produce marked local vasodilation. Thus, the increased metabolic need for oxygen is automatically regulated by the tissues that need it.

About 35 percent of the oxygen consumed by the cardiac tissue is used to metabolize *carbohydrates* for energy production. Most of the remaining oxygen is used to oxidize *fatty acids* for energy. This apportionment of oxygen utilization varies greatly from that of other tissues. The brain depends entirely upon glucose

for energy, and skeletal muscle obtains about 60 percent of its energy from carbohydrate. Cardiac muscle has a preference for fatty acids that is very different from other tissues. Also unusual is the ability of the heart to utilize the carbohydrate *lactate*. This metabolite is produced in large amounts by exercising skeletal muscle. Thus, during a time of great muscular exercise when heart activity also increases, the by-product of the skeletal muscle metabolism, lactate, is used by heart muscle for energy production. Exhibit 22-1 shows the proportions of various nutrients utilized by the heart for its energy production.

Cardiac Output

Although the heart possesses properties that enable it to beat independently, its operation is regulated by events occurring in the rest of the body. All body cells must receive a certain amount of oxygenated blood each minute in order to maintain health and life. When cells are very active, as during exercise, they need even more blood. During rest periods, cellular need is reduced, and the heart cuts back on its output.

The amount of blood ejected from the left ventricle into the aorta per minute is called the *cardiac output,* or minute volume. Cardiac output is determined by two factors: (1) the amount of blood that is pumped by the left ventricle during each beat, and (2) the number of heartbeats per minute. The amount of blood ejected by a ventricle during each systole is called the *stroke volume*. In a resting adult, stroke volume averages 70 ml, and heart rate is about 72 beats per minute. The average cardiac volume, then, in a resting adult is:

$$\text{Cardiac output} = \text{stroke volume} \times \text{ventricular systole/minute}$$
$$= 70 \text{ ml} \times 72/\text{minute}$$
$$= 5{,}040 \text{ ml/minute}$$

In general, any factor that increases the heart rate or its stroke volume tends to increase cardiac output. Factors that decrease the heart rate or its stroke vol-

Exhibit 22-1 RELATIVE CONTRIBUTION OF CARBOHYDRATES AND NON-CARBOHYDRATES TO TOTAL MYOCARDIAL OXYGEN USAGE

Carbohydrate		Noncarbohydrate	
Glucose	17.90%	Fatty acids	67.0%
Pyruvate	0.54%	Amino acids	5.6%
Lactate	16.46%	Ketones	4.3%
Total	34.90%*	Total	76.9%*

Source: courtesy of R. J. Bing, "Myocardial Metabolism," *Circulation 12*:635–647 (1955.

* The total is more than 100 percent because some of the substrates removed from the blood, in amounts measured by the arterial-venous differences, were stored by the heart or not completely metabolized.

ume tend to decrease cardiac output. If stroke volume falls dangerously low, the body can compensate to some extent by increasing the heartbeat and vice versa.

Stroke Volume

Stroke volume is determined by the force of the ventricular contraction. The more strongly the cardiac fibers contract, the more blood they eject. The strength of contraction is directly related to the amount of venous blood that is returned to the heart.

In Chapter 10 it was stated that, within limits, the contraction is more forceful when skeletal muscle fibers are stretched. The same principle applies to cardiac muscle. During exercise, for example, a large amount of blood enters the heart, and the increased diastolic filling stretches the fibers of the right ventricle. This increased length of the cardiac muscle fibers intensifies the force of the ventricular contraction—that is, the force of the beat. The result is that the increased incoming volume of blood is handled by an increased output through a more forceful ventricular contraction. As the increased amount of blood returns from the lungs to the left side of the heart, left ventricular stroke volume also increases. Thus, during exercise, cardiac output is increased. This phenomenon, by which the length of the cardiac muscle fiber determines the force of contraction, is referred to as *Starling's law of the heart.*

In normal situations, the amount of blood returning to the heart regulates the stroke volume and thereby affects cardiac output. However, during certain pathological conditions, stroke volume may fall to a dangerously low level. For instance, if the ventricular myocardium is weak, or if it is damaged by an infarction, it cannot contract strongly. If blood volume is reduced by excessive bleeding, stroke volume falls because the cardiac fibers are not sufficiently stretched. In these cases, the body attempts to maintain a safe cardiac output by increasing the rate of contraction. The rate of the heartbeat is controlled by the endocrine and autonomic nervous systems.

Regulation of Heart Rate

Left to its own devices, the S-A node, which functions as the pacemaker for the heart, would set a steady heart rate that never varied, regardless of the needs of the body. Consequently, a number of reflexes exist to quicken the heartbeat when tissues need more oxygen and to slow down the heart during periods of relative inactivity. This is an excellent example of homeostasis. The reflex arcs start in receptors located in blood vessels, pass to cardiac centers in the brain, and finally travel over autonomic nerves to the heart.

Autonomic Nervous Control

The S-A node receives nerves from both the parasympathetic and sympathetic divisions of the autonomic nervous system. The parasympathetic neurons that influence the heart originate in the *cardioinhibitory* center of the medulla and travel down the vagus nerve as far as the neck region. Here, they separate from the main trunk and combine with sympathetic nerves on their way to the heart. The parasympathetic nerves continue to the atria, where they synapse in a ganglion to the postganglionic cells. Some of the postganglionic cells travel short distances to the S-A and A-V nodes; the others terminate in the atrial and ventricular myocardium.

Stimulation of the parasympathetic vagal fibers has both *negative chronotropic* and *negative inotropic* effects. The term *chronotropic* (time or rate altering) refers to changes in heart rate, whereas *inotropic* (influencing muscle contraction) refers to changes in the strength of cardiac muscle contraction. A negative description implies a decrease in effect, whereas a positive description means an increased effect. Thus, vagal fibers have a tonic effect that continually acts to decrease heart rate and the overall strength of cardiac contraction.

The sympathetic neurons that influence the heart originate in the *cardioacceleratory* center of the medulla. They travel in a tract down the spinal cord and emerge at different levels to synapse in ganglia with postganglionic fibers. These postganglionic fibers are carried to the heart via cardioaccelerator nerves. Sympathetic stimulation counteracts parasympathetic stimulation. It quickens the heartbeat and increases the strength of contraction (that is, *positive* chronotropic and *positive* inotropic effects).

Cardiac Reflexes

The most important automatic control system of the heart is mediated by the *carotid sinus* and *aortic arch reflexes.* These reflexes monitor the blood pressure in two very crucial regions—the brain and the heart. Located in the bifurcation region of the common carotid artery (where it divides into the external and internal carotid arteries) is a slight enlargement called the carotid sinus. The outer covering of the carotid sinus contains stretch receptors similar to the proprioceptors of skeletal muscles. These carotid sinus receptors are called *baroreceptors* or *pressure receptors* (Figure 22-12). When the blood pressure in the common carotid artery increases, the walls distend and stretch these receptors. The receptors, in turn, fire impulses more frequently in the *sinus nerve,* which then conveys the action potentials to the cardioinhibitory center in the medulla oblongata. Similar baroreceptors are located in the wall of the aortic arch.

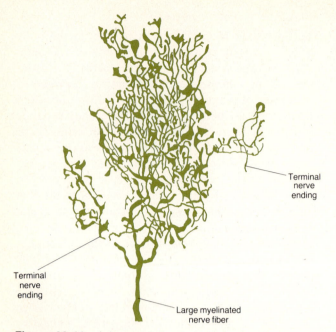

Figure 22-12 A baroreceptor from the wall of the carotid sinus of an adult male.

These baroreceptors send their impulses to the cardio-inhibitory center via an afferent branch of the vagus nerve known as the *aortic* or *cardiac depressor nerve.*

Figure 22-13 illustrates the nerve connections between the baroreceptors of the carotid sinus and aortic arch with the control centers in the medulla. By the use of microelectrode stimulation and destruction, four centers controlling the heart activity and blood vessel diameter have been mapped. The two vasomotor (blood vessel regulator) centers have been shown to be reciprocally innervated. The pressor center, if stimulated, causes blood vessel constriction and simultaneously inhibits its partner, the depressor center. Stimulation of the depressor center inhibits the pressor center and causes blood vessels to dilate. The cardioregulatory centers are reciprocally innervated in a similar manner. Stimulation of the cardioacceleratory center turns off the cardioinhibitory center while it increases cardiac activity.

To illustrate the action of the cardiac and associated vasomotor reflex actions, let us consider what occurs when a person hemorrhages severely. With the loss in blood volume, the filling of the arteries is lessened, and thus blood pressure falls. The fall in blood pressure is detected quickly by the baroreceptor mechanisms in the carotid sinus and aortic arch. The reduced stretch on these receptors leads to a drop in the activity of the nerves connecting the receptors to the regulatory centers. In the medulla, the cardioacceleratory center first inhibits the cardioinhibitory center, which acts via the vagus to inhibit cardiac activity. Thus, removal of the tonic inhibition allows

the heart rate to accelerate. The pressor center inhibits the depressor center while it also increases its activity to the blood vessels, causing them to constrict. The pressor center also stimulates the adrenal medulla. This gland responds by releasing the hormones epinephrine and norepinephrine, which act to produce cardiac acceleration, increased strength of contraction, and a generalized vasoconstriction. The combined effect of the enhanced cardiac activity and vasoconstriction is to raise the blood pressure to within a normal range and thus maintain the vital areas of the body supplied with nutrients and oxygen.

Another cardiac reflex that is initiated by increased blood pressure on the *right* side of the heart is known as the *Bainbridge reflex.* This reflex seems to be complicated, because a rise in blood pressure in the right atrium can cause *either* an increase *or* a decrease in heart rate. Stretch receptors in the walls of the atria and adjoining large veins are connected by large afferent nerve fibers that carry impulses via the vagus nerve to the regulatory centers in the medulla. Current research suggests that this reflex keeps the heart beating within a range where it functions most efficiently. For instance, if an initial heart rate is low (for example, 50 beats/minute), a rapid venous return causes the blood pressure in the right atrium to rise. This in turn produces an *increase* in the heart rate. On the other hand, if the heart rate is initially fast (for example, 120 beats/minute), then a rapid venous return causes the heart rate to *decrease.* The optimum heart rate for maximum cardiac output is about 80 beats/minute.

Other Factors that Influence Heart Activity

Many other factors can influence heart activity. The regulation of deep-body temperature depends upon adjustments of blood flow in the peripheral parts of the circulation. The temperature regulating center in the hypothalamus is continually relaying impulses to the vasomotor centers. When the peripheral blood flow increases, skin temperature rises and heat is lost from the body.

Respiration can also influence blood pressure regulation. In situations of lowered oxygenation of the blood, not only do breathing rate and depth increase, but a marked pressor effect is observed. It is most likely that the respiratory centers in the medulla and the vasomotor centers are interconnected.

Strong emotions such as fear, anger, and anxiety, along with a multitude of physiological stressors, *increase* rate through the general stress response. For some reason, mental states such as depression and grief tend to stimulate the cardioinhibitory center so that the heart rate *decreases.* The exact neurophysiology involved is not clear, but the hypothalamus,

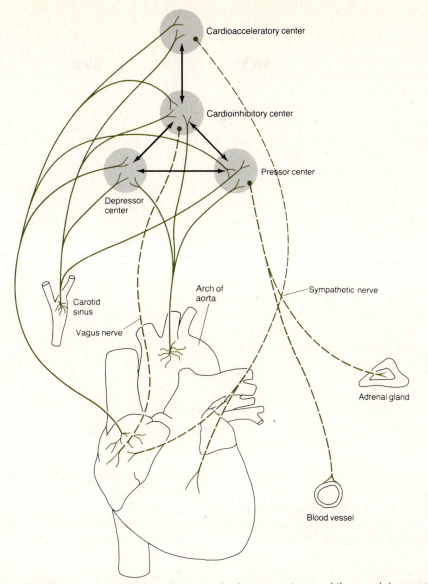

Figure 22-13 Nerve connections between the baroreceptors and the regulatory centers (solid lines). Nerve connections between regulatory centers and effector organs (broken lines).

which is the area concerned with emotions, probably sends impulses to the regulatory centers in the medulla.

Females generally have a faster heart rate than males. Also, heart rate is fastest at birth and then gradually declines throughout a lifetime. Muscular exercise also accelerates heart rate—usually in proportion to the work done.

CIRCULATORY SHOCK AND HOMEOSTASIS

When cardiac output or blood volume is reduced to the point where body tissues do not receive an adequate blood supply, **circulatory shock** results. The principal cause of circulatory shock is loss of blood volume, or decreased cardiac output. Blood volume can be lost through hemorrhage or through the release of histamine due to damage to body tissues (trauma). The characteristic symptoms of circulatory shock are a pale, clammy skin; cyanosis of the ears and fingers; a feeble though rapid pulse; shallow and rapid breathing; lowered body temperature; and some degree of mental confusion or unconsciousness.

If the shock is relatively mild, certain homeostatic mechanisms of the circulatory system become operative and compensate so that no serious damage results. This is called the compensatory stage. During the compensatory stage, lowered blood pressure is compensated by constriction of blood vessels and water retention. These are accomplished by the secretion of renin by the kidneys (Chapter 26), aldosterone by the adrenal cortex, epinephrine from the adrenal medulla, and ADH by the posterior pituitary (Chapter 17). The result is that, even though some blood is lost from the circulation, the volume of blood returning to the heart is normal and cardiac output remains essentially unchanged. Although veins and many arterioles are constricted during compensation, there is no constriction of arterioles supplying the heart and brain. As a consequence, blood flow to the heart and brain is normal, or nearly so. Compensation is an effective homeostatic mechanism until about 900 ml of blood is lost.

If the shock is severe, death may occur. If the return of venous blood is greatly diminished due to excessive blood loss, the compensatory mechanisms are insufficient. As a result, cardiac output is reduced. When the cardiac output decreases, the heart fails to pump enough blood to supply its own coronary vessels, and the heart muscle becomes progressively weakened. In addition, prolonged vasoconstriction ultimately leads to tissue hypoxia, and vital organs such as the brain, kidneys, and liver are damaged. Essentially, the initial shock promotes more shock, and a *circulatory shock cycle* is established (Figure 22-14). When the shock reaches a certain level of severity, damage to the circulatory organs is so extensive that death ensues.

BLOOD VESSELS

The vascular system consists of thousands of miles of branching hollow tubes whose function is to deliver blood to all parts of the body in proportion to their need. This function provides a key element in maintaining homeostasis. The hollow tubes are blood

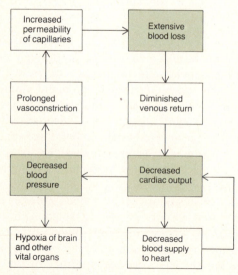

Figure 22-14 The circulatory shock cycle.

vessels, and they are generally classified into four general types on the basis of different anatomic features and functions. First, there are the **arteries,** large enough to be seen easily by the unaided eye; strong, elastic vessels that distribute blood under high pressure to the tissues of the body. Second, there are the **arterioles,** at the threshold of visibility to the unaided eye (0.2 mm). They have muscular walls, are innervated, and function to reduce the high pressure in the tissues and regulate the flow into the capillary beds. The third type of vessels is the **capillaries.** These vessels form an enormous network of microscopic, branching and interconnecting, thin-walled, permeable tubes in the tissues. They exchange the fluid of the blood, its nutrients and gases, with the interstitial fluid that bathes the cells. The fourth general type of blood vessel is the **veins.** These vessels range in size from microscopic to nearly an inch in diameter. Veins are thin-walled, passive vessels that collect the blood from the capillary beds and return it to the heart.

The blood flows within the vessels in two recirculating systems—the pulmonary and systemic systems. Each system can be traced from the ventricles of the heart. The **pulmonary system** arises from the right ventricle, which pumps the blood to the lungs via arteries and arterioles. Gases exchange in the lungs through the capillary walls, and the blood is returned to the left atrium of the heart via the pulmonary veins. The blood flows into the left ventricle, which then pumps it into the **systemic circulation.** Blood is delivered to every tissue of the body by means of the blood vessels and then returned to the right atrium of the heart. Each system pumps the same volume of blood every minute, and the coordination of the two systems is exact.

Hemodynamics of Circulation

Hemodynamics is the study of the factors that influence the flow characteristics of blood as it moves through the body. Before reading this section, you may find it helpful to review the properties of liquids discussed in Chapter 3.

Blood Distribution

At any given moment, blood in the vascular system is distributed so that about 84 percent of the volume is located within the systemic circulation, about 8.8 percent is within the pulmonary circulation, and about 7.2 percent is within the chambers of the heart (Exhibit 22-2). These percentages, while not invariable, are normally controlled so that they change very little. The arteries contain the most constant volume (about 14 percent), since they are the least dilatable. The low-pressure vessels (the capillaries and veins) are the

Exhibit 22-2 ESTIMATED DISTRIBUTION OF BLOOD IN VASCULAR SYSTEM OF HYPOTHETICAL ADULT MAN*

Region	Volume	
	(Milliliters)	(Percent)
Heart (diastole)	360	7.2
Pulmonary		
Arteries	130 ⎫	2.6 ⎫
Capillaries	110 ⎬ 440	2.2 ⎬ 8.8
Veins	200 ⎭	4.0 ⎭
Systemic		
Aorta and large arteries	300 ⎫	6.0 ⎫
Small arteries	400 ⎪	8.0 ⎪
Capillaries	300 ⎬ 4,200	6.0 ⎬ 84.0
Small veins	2,300 ⎪	46.0 ⎪
Large veins	900 ⎭	18.0 ⎭
	5,000	100

Source: From William R. Milnor, "Cardiovascular System," in Vernon B. Mountcastle, ed., *Medical Physiology,* 14th ed. (St. Louis: C. V. Mosby Co., 1980).
* Age, 40 years; weight, 75 kg; surface area, 1.85 m².

most variable in volume, since they are the most dilatable. For example, when a transfusion is given to a healthy person, only about 1 percent of the blood ends up in the arterial volume, but 99 percent ends up in the low-pressure vessels. Thus, the veins and capillanics serve as resorvoirs for the cardiovascular system.

Pressure Gradients

Blood flows through the system of closed vessels because of pressure gradients (differences). Blood flows from regions of higher pressure to regions of lower pressure (Figure 22-15). The mean blood pressure in the aorta is about 100 mm Hg, and as the figure shows, the pressure decreases continually from the aorta through the blood vessels and back to the heart. The graph also shows the dampening of the fluctuations in blood pressure into the arterioles so that the pressure of the blood entering the capillaries is uniform at an average value of about 35 mm Hg. Blood flowing from the capillaries into the veins is at an average pressure of about 15 mm Hg. The mean capillary pressure is, therefore, about 25 mm Hg. In other specialized situations, such as the glomerular capillaries of the kidney, the mean pressure is probably about 75 mm Hg; and the mean pressure in the pulmonary capillaries is about 10 mm Hg.

Figure 22-15 also shows that the greatest drop in pressure occurs in the arterioles—from 95 mm Hg to 35 mm Hg—illustrating the very important pressure reducing function of the arterioles.

Velocity

The average speed or mean velocity at which blood flows is determined primarily by the geometry of the vessels. Common sense tells us that the same volume of blood must flow through each division of the vascular system in the same time interval. That is, if 100 ml flows through the aorta in one second, then a total of 100 ml must flow through the large arteries of the body in one second, a total of 100 ml must be flowing through the arterioles in one second, a total of 100 ml must be flowing through the sum total of all the capillaries in the body in one second, and a total of 100 ml of blood must return to the heart in one second.

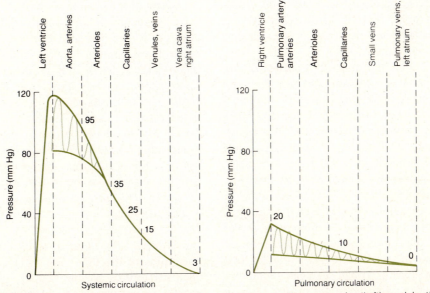

Figure 22-15 Distribution of intravascular pressures in the systemic (left) and in the pulmonary (right) circuits.

The *total cross-sectional area* of an arterial bed increases with each branching. This means, in the simplest example, that a blood vessel that is 1 cm in radius has a cross-sectional area equal to πr^2 or $3.14 \times (1)^2 \cong 3$ cm². When this blood vessel divides into two branches, the sum of the cross-sectional areas of the two branches will be greater than 3 cm². The sum of the cross-sectional areas of the branches usually runs between 1.2 and 1.7 times greater than the trunk. In this example, the sum of the areas would be between 3.5 and 5 cm². Multiple branching in some locations can produce very rapid increases in the cross-sectional area.

Velocity can be explained by a very simple equation:

$$v = \frac{V}{A}$$

v = mean velocity (cm/second)
V = volume flow/per second (ml/second)
A = *total* cross-sectional area (cm²)

Consider the human aorta to be 2 cm in diameter (radius = 1 cm). The total cross-sectional area of the aorta is $3.14 \times 1^2 \cong 3$ cm². The volume flow per second in the aorta is about 83 ml/second (cardiac output = 5,000 ml/minute, or 5,000/60 = 83 ml/second). The mean velocity *(v)* of blood flow in the aorta is 83/3 = 28 cm/second. During exercise, cardiac output can increase to values of 20,000 ml/minute or 333 ml/second. The diameter of the aorta, and thus the cross-sectional area, cannot change, so the mean velocity of volume flow increases greatly.

$$v = \frac{333}{3} = 111 \text{ cm/second}$$

If the same volume of blood per second flows through all divisions of the circulation while the total cross-sectional area increases, then our formula predicts that the blood flow velocity will decrease as the vascular tree branches more and more.

It has been estimated that total cross-sectional area of the capillary beds is 800 times greater than the cross-sectional area of the aorta. Therefore, the mean velocity of blood flow in a capillary is 1/800 that in the aorta, or about 28 cm/second/800 = 0.035 cm/second = 0.35 mm/second. This means it will take a red blood cell about 3 seconds to travel 1 mm in a capillary vessel. It has been calculated that, at this rate, it would take over $1\frac{2}{3}$ years for 1 mm³ of blood to flow through a single capillary! The reason, of course, that 83 ml or 333 ml of blood can flow through the capillary beds in one minute is because there are an incredible number of short capillaries in the bed connected in parallel array. It has been calculated that in the mesentery of the dog, there are over 1.2 billion (1.2×10^9) capillaries.

Relationship Between Flow, Pressure, and Resistance

The most important relationship in circulatory physiology is expressed as:

$$\text{Flow} = \frac{\text{pressure difference}}{\text{resistance}} \quad \text{or} \quad F = \frac{\Delta P}{R}$$

The Greek letter delta, Δ, indicates difference, or change.

The blood flow (in milliliters per second) through a blood vessel is determined directly by the net pressure pushing it and inversely by the resistance it encounters in flowing through the vessel.

Pressure is produced in the vascular system by the pumping action of the heart. The heart pushes blood forcibly and abruptly into the large, elastic arteries, which are those nearest to the heart. These vessels then stretch about 10 percent larger to accommodate the blood. The aorta stretches because the pressure on the inside of the vessel wall exceeds the pressure on the outside of the wall. The blood moves down the aorta into branches because the pressure is greater in the aorta than in the vessels downstream. Flow is always *from* the region of higher pressure and *toward* the region of lower pressure. The strength of the pushing force is determined by the *difference in pressure* that exists between two points on the pathway.

When the ejected blood is forced into the filled aorta, it must push the blood already there further downstream. The stretched aorta recoils during the period of ventricular diastole and it also pushes blood further downstream. Force is needed to push the blood because of a *resistance* to flow that is ultimately a measure of the inner slipperiness, or *viscosity,* of the blood. The plasma molecules in contact with the vessel wall are impeded by friction. A cylindrical shell or lamina of the plasma in contact with the wall cannot flow. A second infinitesimally thin lamina of plasma adjacent to the first also is retarded from flowing because it encounters friction from the first stationary lamina. How much it is retarded depends upon the inner slipperiness of the molecular layers. The retardation decreases as the distance from the wall or first lamina increases. Thus, the least retarded, or the fastest flowing, blood is that in the very center of the vessel, the central axial stream (Figure 22-16). It follows, therefore, that the vessels with the largest diameters will have the fastest flowing blood because they offer the least retardation or resistance to flow.

The resistance encountered by the blood is related first to its viscosity, second to the length of blood vessel wall with which it is in contact, and third to the radius of the tube. The radius factor is related to viscosity, since as the vessel increases in size, a greater proportion of the blood is unhindered by the friction of the first stationary lamina.

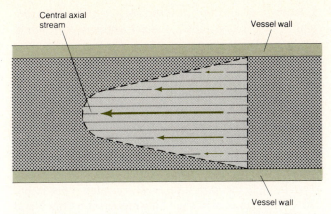

Figure 22-16 Velocity of blood flow in vessels. Flow is fastest in the central axial stream and progressively decreases to zero at the vessel wall.

The equation that expresses the relationship of all the resistance components is known as Poiseuille's law (pronounced pwah-zoo 'iz).

$$R = \frac{8}{\pi} \times \eta \times \frac{l}{r^4}$$

R = the resistance

$\frac{8}{\pi}$ = a constant that comes from the mathematical development of the law

η (pronounced eta) = the viscosity of the fluid

l = the length of the vessel

r = the radius of the vessel

The important fact to remember is that R, the resistance, is inversely proportional to the *fourth power of the radius*. Thus, *small* changes in the radius can *greatly* alter the resistance.

To appreciate how much the flow changes when the different components of the resistance are changed, first rewrite the most important relationship in circulatory physiology:

$$F = \frac{\Delta P}{R}$$

For R substitute from the Poiseuille equation.

$$F = \frac{\Delta P}{\frac{8}{\pi} \cdot \eta \cdot \frac{l}{r^4}} = \frac{\pi \cdot \Delta P \cdot r^4}{8 \cdot \eta \cdot l}$$

Consider now the effect of changing the η, or viscosity. Viscosity is a *number* that compares the flow properties of any substances to those of water. Plasma has a normal viscosity of about 1.8, which indicates that it is a "stickier" fluid than water. Blood has a normal viscosity between 3 and 4.

If all other factors on the right-hand side of the above equation remain the same but the viscosity doubles, the total value of the right-hand side of the equation will be halved. This shows that the flow rate will decrease to one-half of its original value. Therefore, as viscosity increases, flow rate decreases and vice versa.

When an illness causes the blood viscosity to increase, flow rates tend to decrease. Usually some other factor adjusts to provide an adequate flow to the tissues, and most often the adjustment is an increase in the pumping pressure. This means that the heart labors more to overcome the viscous resistance.

Changes in the length (*l*) of the blood vessel are not a usual problem in the vascular system. Additional capillary beds do open in certain tissues (for example, muscle) when their level of activity increases. This might be considered as lengthening the vessels.

If the value of *l* doubles, again it can be seen that the flow rate is halved. Thus, lengthening the vessel decreases the flow rate.

To illustrate the effect of a change in the radius of the vessel, consider a blood vessel initially with a radius of 1 cm which dilates so that the radius doubles to 2 cm.

By inserting first the value, 1 cm, for r into the Poiseuille equation, we obtain:

$$F = \frac{\pi}{8} \cdot \frac{\Delta P}{\eta} \cdot \frac{(1)^4}{l} = \left(\frac{\pi}{8} \cdot \frac{\Delta P}{\eta} \cdot \frac{1}{l} \right) \text{ ml/minute}$$

Repeating the calculation for flow rate, using 2 cm for the radius, we get:

$$F = \frac{\pi}{8} \cdot \frac{\Delta P}{\eta} \cdot \frac{(2)^4}{l} = \left(\frac{\pi}{8} \cdot \frac{\Delta P}{\eta} \cdot \frac{16}{l} \right) \text{ ml/minute}$$

Comparison of the flow values shows that the second situation has a flow rate 16 times greater than the original. In other words, doubling the radius increases the flow rate 16 fold!

Similarly, if the radius were increased to 3 cm, then the flow rate would become $(3)^4$ or 81 times greater than the initial rate.

Very slight changes in the radius of a blood vessel can produce great changes in the flow rate through that vessel.

Recall that the arterioles are muscular vessels richly innervated by the autonomic nervous system. These blood vessels, by their muscular action, can constrict or dilate and thus regulate the inflow to the downstream capillary beds. This varying impediment to flow by the arterioles is called the *peripheral resistance*. Thus, under appropriate nervous impulses, arterioles in one region of the body (for example, the digestive tract after a meal) may dilate to increase blood flow, while in another region (such as the skeletal muscles), they constrict to reduce the blood flow. During exercise, the reverse actions would occur.

The center for this regulation is in the *vasomotor*

center, located in the medulla. The term *vas* relates to vessel, whereas *motor* means mover or movement. Functionally, the vasomotor center consists of a vasoconstrictor and a vasodilator center.

Vasoconstrictor nerves conduct impulses from the medulla to the spinal cord and through sympathetic nerves to the arterioles. The impulses bring about a reduction in the diameter of blood vessels by stimulating the smooth muscle to contract. Normally, the nerves continually send impulses to the smooth muscle of arterioles. As a result, the arterioles are kept in a state of tonic contraction. However, if there is an increase in blood pressure, the vasoconstrictor nerves are inhibited. The arteriole diameter increases, and blood pressure falls due to the reduction in peripheral resistance.

Vasodilator nerves, on the other hand, carry impulses that bring about an increase in the diameter of vessels. When there is an increase in blood pressure, the vasodilator nerves are stimulated, arteriole diameter increases, and blood pressure falls due to the decrease of peripheral resistance. Essentially, the vasomotor center, in conjunction with arterioles, adjusts the pressure with which blood flows in response to varying needs of the body.

Production of the Pressure Gradient

Blood always flows from an area of greater pressure to an area of lesser pressure. What causes this continual decrease in pressure that directs the flow of the blood? Although resistance increases pressure at the point where the resistance occurs, the overall effect is a slow loss of pressure. The resistance of the walls of the arteries against the blood produces friction. And any time friction occurs, some of the kinetic energy of the flowing fluid is converted to heat energy and escapes. Thus, as the blood passes through the arteries to the capillaries and on to the veins, it continually loses pressure as it rubs against the walls of the vessels.

Factors that Aid Venous Return

The establishment of a pressure gradient is the primary reason why blood flows. But a number of other factors help the blood to return through the veins. These include an increased rate of flow, contractions of the skeletal muscles, valves, and respiratory movements.

Rate of Flow

The formula for pressure shows that the less resistance offered the blood, the faster it flows. Thus,

$$Flow = \frac{pressure\ difference}{resistance}$$

A larger vessel offers less resistance than a smaller one. Thus, blood flows fastest through the aorta, which has the largest diameter and therefore the least resistance of any vessel in the body. The blood flows most slowly through the capillaries, which have minute diameters and offer a great deal of resistance. The slow flow through the capillaries allows time for the exchange of materials and is another example of how body structure is admirably suited to function. But as the blood reaches the smaller veins and finally the very large veins, it picks up more and more speed even though the pressure is decreasing. This is because the diameter of the veins increases as the blood nears the heart. The larger diameters of the veins offer much less resistance, which more than compensates for the progressive loss in blood pressure.

Muscles and Valves

The contraction of skeletal muscles is very important in returning blood to the heart. As the muscles contract, they tighten around the walls of the veins running through them. This provides a presssure that squirts the blood forward from one part of a vein to another. The presence of one-way valves in the veins ensures that blood is moved toward the heart. This action is called milking. Individuals who are immobilized through injury or disease cannot take advantage of these skeletal muscular contractions. As a result, the return of venous blood to the heart is slower, and the heart has to work harder.

Respiratory Movements

Another very important factor in maintaining venous circulation is breathing. During each inspiration, pressure in the thoracic cavity decreases, and this draws blood toward the chest. Each expiration would drive blood back away from the chest, but the valves in the veins prevent the backward flow. Veins in many places of the body, especially the extremities, contain valves that offer little resistance to blood flowing toward the heart. Any pressure on the valves from blood moving backward closes the valves and stops the backflow.

The Micro or Capillary Circulation

The French phrase *raison d'être* most adequately expresses the importance of the capillary circulation. Its reason for being, for the circulatory system of course, is to nourish the tissue cells, and it is at the capillary level that the exchange of materials between blood and interstitial fluid occurs. The roles of the other parts of the cardiovascular system, the controlling mechanisms, both nervous and chemical, are supportive and directed toward the capillaries' fulfilling their job.

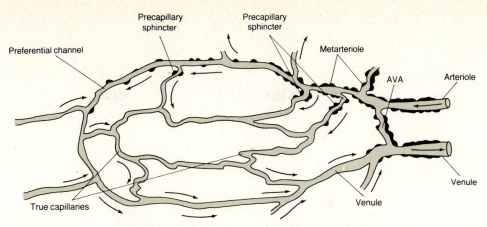

Figure 22-17 Architecture of capillary circulation.

Architecture of the Capillary Circulation

In most tissues, the blood normally flows through only a small portion of the total available capillaries. When a tissue becomes active and its metabolic needs increase, the entire bed of capillaries fills with blood. In muscle at rest, it has been estimated that only about one-hundredth of the capillaries are functioning.

The general arrangement of capillaries in the tissues is shown in Figure 22-17. The many flow patterns are evident, and it is easy to see how the blood volume in the capillary bed can vary so greatly. The *true capillaries* are not on the direct flow route from arteriole to venule (small vein). The direct flow route is via a *preferential channel*, that is, a select capillary with a slightly thicker supporting tissue coat. It runs directly from the arteriole to the venule. The true capillaries emerge from vessels termed *metarterioles*, which branch from their parent arteriole at almost right angles. They are identified by the irregular distribution of smooth muscle cells that gradually decrease in number along the length of the vessel. At the point where each capillary leaves the metarteriole, one or two of the last smooth muscle cells encircle the capillary and constitute what is called the *precapillary sphincter*. When this sphincter closes, blood flow from that metarteriole into the true capillary bed stops. These smooth muscle cells, as well as those in the metarterioles and arterioles, are controlled by sympathetic nerves. There appears to be a cycling activity that causes the metarterioles and precapillary sphincters to alternately constrict and relax. During periods of greater activity, the relaxation phase is longer, whereas when the tissue is "resting," the constrictor phase is prolonged. In addition, change in local conditions—oxygen lack or the accumulation of the wastes, CO_2, and lactic acid—can cause the sphincters and metarterioles to relax. In this way, local controls can also operate to regulate the local blood flow.

In addition to a preferential channel *through* the capillary bed, there exists a path that *avoids* the capillary bed entirely. The *arteriovenous anastomosis* (AVA) is a short muscular vessel that connects very small arteries to small veins or arterioles to venules. These AVAs are found in most tissues and organs and generally tend to operate fully open or completely closed.

Since they are devoid of smooth muscle or any supporting tissue, the capillaries are passive tubes that fill to varying degrees depending upon the volume of blood permitted to enter by the controlling sphincters. For this reason, when the smooth muscle cells constrict around the metarteriole or precapillary sphincter, the capillaries collapse and seem to disappear as the blood drains out into the venules.

Fluid Exchange Across Capillary Walls

The composition and volume of the interstitial fluid are maintained fairly constant by the continual exchange of materials with the blood. Most of the exchange occurs by simple diffusion. Water and dissolved substances move easily through the capillary wall. Larger molecules, such as plasma proteins and polysaccharides, may cross the endothelial cell membranes by pinocytosis. Vesicles form on one surface membrane and release their contents on the opposite side of the cell.

Another process that contributes to the exchange is a bulk flow circulation of fluid. Fluid *flows out of* capillary pores from the end of the capillary nearest to the metarteriole, and interstitial fluid *flows into* the capillary at the venule end of the capillary. The factors involved in these microcirculating movements of fluid are illustrated in Figure 22-18.

At the metarteriolar end of the capillary, the following forces are acting: (1) The hydrostatic pressure due to the blood in the capillary at this point is about 35 mm Hg. This force pushes fluid out through the capillary pores into the interstitium. (2) A negative *interstitial fluid pressure* of about 6 mm Hg exists which also tends to suck fluid out of the capillaries. This negative pressure can sometimes be observed

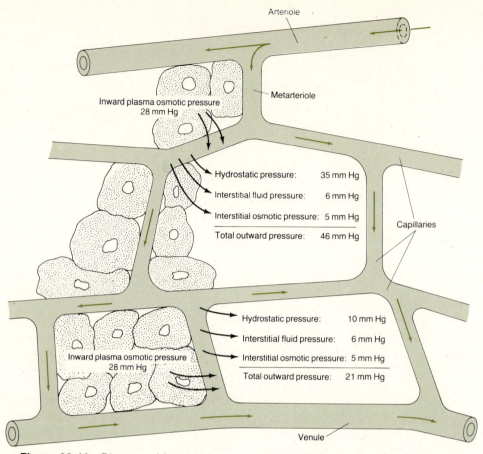

Figure 22-18 Diagram of factors involved in microcirculation of fluid over tissue cells.

when a filled hypodermic syringe is introduced subcutaneously. The plunger will be drawn slowly into the barrel of the syringe. (3) The presence of small amounts of protein in the interstitium creates an osmotic force that also pulls water from the capillary. This *interstitial osmotic pressure* has a value of about 5.0 mm Hg. (4) The proteins in plasma create an osmotic pressure within the capillary that pulls water into the capillary. The strength of this inward *plasma osmotic pressure* is about 28 mm Hg.

Therefore, at the **metarteriolar end** of the capillary, forces operating to move fluid out are:

1. Hydrostatic pressure	35.0 mm Hg
2. Interstitial fluid pressure	6.0 mm Hg
3. Interstitial osmotic pressure	5.0 mm Hg
Total outward pressure	46.0 mm Hg

The force that pulls fluid into the capillary is the inward plasma osmotic pressure, which is 28 mm Hg.

Therefore, the net pressure operating at the "front end" of the capillary is about 46− 28, or 18 mm Hg. This force moves fluid *out* of the capillary.

At the venule end of the capillary, a similar examination of the operating forces shows a different value

for the hydrostatic pressure. At this end of the capillary its value drops to about 10 mm Hg.

Thus, the outward operating forces at the **venule end** include:

1. Hydrostatic pressure	10.0 mm Hg
2. Interstitial fluid pressure	6.0 mm Hg
3. Interstitial osmotic pressure	5.0 mm Hg
Total outward pressure	21.0 mm Hg

The inward plasma osmotic pressure is 28 mm Hg.

The net pressure operating at the **venule end** of the capillary is about 28− 21, or 7 mm Hg. This force draws fluid *back into* the capillary.

This reabsorbing action recovers most, but not all, of the fluid that was filtered out at the "front end" of the capillary. The remainder of the fluid is collected by the lymphatic vessels of the tissue.

CHECKING CIRCULATION

The pulse can be used to measure the rate of the heartbeat. The blood pressure measures the pressure that the blood exerts on the walls of the blood vessels. First, let us discuss some aspects of the pulse.

Pulse

The alternate expansion and elastic recoil of an artery with each systole of the left ventricle is called the **pulse.** Pulse is strongest in the arteries closest to the heart. It becomes weaker as it passes over the arterial system, and it disappears altogether in the arterioles. The pulse may be felt in any artery that lies near the surface of the body and over a bone or other firm tissue. The radial artery at the wrist is most commonly used for this purpose. Other arteries that may be used for determining pulse are: (1) the temporal artery, which is above and toward the outside of the eye; (2) the facial artery, which is at the lower jawbone on a line with the corners of the mouth; (3) the common carotid artery, which is in the neck and can be located beside and somewhat behind the trachea; (4) the brachial artery along the inner side of the biceps brachii muscle; (5) the femoral artery near the pelvic bone; (6) the popliteal artery behind the knee; (7) the posterior tibial artery behind the medial malleolus of the fibia; and (8) the dorsalis pedis artery over the instep of the foot.

The pulse rate is the same as the heart rate and averages between 70 and 80 per minute in the resting state. The term *tachycardia* (*tachy* = fast) is applied to a very rapid heart rate or pulse rate—usually over 100 per minute. The term *bradycardia* (*brady* = slow) indicates a very slow heart rate or pulse rate—usually 60 per minute or less. In addition to the rate of the pulse, other factors should be noted. For example, the intervals between beats should be equal in length. If a pulse is missed at regular or irregular intervals, the pulse is said to be irregular. Also, all pulse beats should be of equal strength. Irregularities in strength may indicate a lack of muscle tone in the heart or arteries.

Blood Pressure

Although the term **blood pressure** may be defined as the pressure exerted by the blood on the walls of any blood vessel, in clinical settings it refers to the pressure only in the large arteries. Blood pressure is usually taken in the left brachial artery, and it is measured by a *sphygmomanometer* (*sphygmo* = pulse). A commonly used kind of sphygmomanometer consists of a rubber cuff attached by a rubber tube to a compressible hand pump or bulb. Another tube attaches to the cuff and to a column of mercury that is marked off in millimeters. This column measures the pressure. The cuff is wrapped around the arm over the brachial artery and inflated by squeezing the bulb. This causes a pressure on the outside of the artery. The bulb is squeezed until the pressure in the cuff exceeds the pressure in the artery. At this point, the walls of the brachial artery are compressed tightly against each other, and no blood can flow through it. Compression of the artery may be evidenced in two ways. First, if a stethoscope is placed below the cuff over the artery, no pulse can be heard. Second, no pulse can be felt if the fingers are placed over the radial artery at the wrist.

Next, the cuff is deflated gradually until the pressure in the cuff is just lower than the maximal pressure in the brachial artery. At the maximum pressure, the arterial pressure forces open the artery and a spurt of blood passes through. The pulse may be heard through the stethoscope as a repetitive thumping sound. As the cuff pressure is further reduced, the sound suddenly becomes more faint or muffled. Finally, the sound becomes muffled and disappears altogether. At the time the first sound is heard, a reading on the mercury column is made. This sound corresponds to the peak or *systolic blood pressure*. This pressure is the force with which blood is pushing against arterial walls during ventricular contraction. The pressure recorded on the mercury column when the sounds suddenly become faint and muffled is called *diastolic blood pressure*. It measures the force of blood in arteries during ventricular relaxation. Whereas systolic pressure typically indicates the force of the left ventricular contraction, the diastolic pressure typically provides information about the resistance of blood vessels.

The average blood pressure of a young adult male is about 120 mm Hg systolic and about 80 mm Hg diastolic. For convenience, these pressures are indicated as 120/80. In young adult females, the pressures are 8 to 10 mm Hg less. The difference between the systolic and diastolic pressures is called *pulse pressure*. This pressure, which averages 40 mm Hg, provides information about the condition of the arteries. The higher the systolic pressure and the lower the diastolic pressure, the greater the pulse pressure. The normal ratio of systolic pressure to diastolic pressure to pulse pressure is about 3:2:1.

HOMEOSTATIC IMBALANCES OF THE CARDIOVASCULAR SYSTEM

Risk Factors in Heart Attacks

It is estimated that one in every five persons who reaches the age of 60 will have a heart attack. It is also estimated that one in every four persons between 30 and 60 has the potential to be stricken. Heart disease is epidemic in this country, despite the fact that some of the causes can be foreseen and prevented.

The Framingham, Massachusetts, Heart Study, which began in 1950, is the longest and most famous study ever made of the susceptibility of a community to heart disease. Approximately 13,000 people in the town have participated in the investigation by receiving examinations every 2 years since the study began. The results of this research indicate that people who develop combinations of certain risk factors eventually have heart attacks. These factors are: (1) high blood cholesterol level, (2) high blood pressure (hypertension), (3) cigarette smoking, (4) overweight, (5) lack of exercise, and (6) diabetes mellitus.

The first five risk factors all contribute to increasing the work load of the heart. High blood cholesterol level and hypertension increase the heart's work load and increase the probability of a stroke. Cigarette smoking increases the work load through the effects of nicotine, which stimulates the adrenal gland to oversecrete aldosterone, epinephrine, and norepinephrine —powerful vasoconstrictors. Overweight people develop miles of extra capillaries to nourish their fat tissue. This means that the heart has to work harder to pump the blood through more vessels. Lack of exercise means that venous return gets less help from contracting skeletal muscles. In addition, regular exercise strengthens the smooth muscle of blood vessels and enables them to assist general circulation more efficiently. Exercise also increases cardiac efficiency and output. Recall that in diabetes mellitus, fat metabolism dominates glucose metabolism. As a result, cholesterol levels get progressively higher and result in plaque formation, a situation that can contribute to high blood pressure.

People with three or more risk factors form an especially high-risk group. The incidence of serious heart attacks in this high-risk group is far greater than in groups that have no risk factor or only one. The people who are most apt to develop atherosclerosis and who, consequently, run the highest risk of all have three risk factors: high cholesterol, hypertension, and cigarette smoking. Other researchers list emotional stress as an important risk factor. However, there is still controversy among medical people as to the relative importance of this factor.

The risk factors make a person more susceptible to heart trouble because they strain the heart or increase the likelihood that its oxygen supply will be shut off at some time. Generally, the immediate cause of the heart trouble is one of the following: (1) failure of the heart's blood supply, (2) faulty heart architecture, or (3) failure of the heart's conductivity. Of these three reasons, the first two are far more common than the third.

Disorders Associated with Failure of Blood Supply

The majority of heart problems result from some malfunction in the coronary circulation. If a reduced oxygen supply weakens the cells, but does not actually kill them, the condition is called **ischemia. Angina pectoris** is ischemia of the myocardium. The name comes from the area in which the pain is felt. Pain impulses originating from most visceral muscles are referred to an area on the surface of the body. Angina pectoris occurs when the coronary circulation is somewhat reduced. Stress, which produces constriction of vessel walls, is a common cause. Equally common is strenuous exercise after a heavy meal. When food is taken into the stomach, the body increases blood flow to the digestive tract. The digestive glands can then receive enough oxygen for their increased activities, and the digested food can be quickly absorbed into the bloodstream. As a consequence, some blood is diverted away from other organs, including the heart. Exercise, however, increases heart muscle activity and thus increases its need for oxygen. Thus, doing heavy work while food is in the stomach can lead to oxygen deficiency in the myocardium. Angina pectoris weakens the heart muscle, but it does not produce a full-scale heart attack. The simple remedy of taking nitroglycerin, a drug that dilates vessels and thereby increases the area of blood flow, brings the coronary circulation back to normal and stops the pain of angina. Because repeated attacks of angina can weaken the heart and lead to serious heart trouble, angina patients are told to avoid activities and stresses that bring on the attacks.

A much more serious problem is **myocardial infarction,** commonly called a **coronary** or **heart attack.** *Infarction* means death of an area of tissue because of a drastically reduced or completely interrupted blood supply. Myocardial infarction results from a thrombus or embolus in one of the coronary arteries. The tissue on the far side of the obstruction dies, and the heart muscle loses some of its strength. The consequences depend partly on the size and location of the infarcted area, which is replaced by scar tissue.

Angina pectoris and myocardial infarction result from insufficient oxygen supply to the myocardium. Coronary artery disease claims about one in twelve of all Americans who die between the ages of 25 and 34. It claims almost one in four of all those who die between 35 and 44. It has been reported that 50 to 65 percent of all sudden deaths are due to coronary heart disease.

At least half of the deaths from myocardial infarction occur before the patient reaches the hospital. These early deaths could result from an irregular heart rhythm, which is called an **arrhythmia.** Sometimes this progresses to the stage called **cardiac arrest** or ventricular fibrillation, in which the heart stops functioning. An arrhythmia is an abnormal, irregular rhythm change of the heart, caused by disturbances in the conduction system. It can result in cardiac arrest because the heart is not capable of meeting the oxygen demands of the body. Serious arrhythmias can be controlled, and the normal heart rhythm can be reestablished, if they are detected and treated early enough. Coronary care units have reduced hospital mortality rates from acute myocardial infarctions by about 30 to 20 percent or less by preventing or controlling serious arrhythmias.

Disorders Associated with Faulty Architecture

Less than 1 percent of all new babies have a **congenital,** or **inborn,** heart defect. Even so, the total number in this country each year is estimated to be 30,000 to 40,000. Some of these infants may be able to live quite healthy and long lives without any need for repairing their hearts. But sometimes an inborn heart defect is so severe that an infant lives only a few hours. One of the more common of these defects is **patent ductus arteriosus.** This condition is serious because the blood vessel that connects the aorta with the pulmonary artery during fetal development remains open instead of closing completely after birth. This results in aortic blood flowing into the lower-pressure pulmonary trunk, increasing the pulmonary trunk blood pressure. This increases considerably the work of both ventricles and overworks the heart.

Another common group of congenital problems are the septal defects. A **septal defect** is an opening in the septum that separates the interior of the heart into a left and right side. **Atrial septal defect** is a hole caused by the failure of the fetal foramen ovale to seal the two atria from one another. Because pressure in the right atrium is low, atrial septal defect generally allows a sizable volume of blood to flow from the left atrium into the right atrium. This results in an overload of the pulmonary circulation, producing fatigue, increased respiratory infections, and growth failure. If it occurs early in life, it may deprive the systemic circulation of a considerable portion of the blood destined for the organs and tissues of the body. **Ventricular septal defect** is caused by an abnormal development of the interventricular septum. Deoxygenated blood mixes with the oxygenated blood that is pumped into the systemic circulation. Consequently, the victim suffers *cyanosis,* a blue or dark purple discoloration of the skin. Cyanosis results from increased amounts of deoxygenated blood in the skin circulation. Septal openings can now be sutured or covered with synthetic patches.

A third defect is **valvular stenosis.** In this disorder there is a narrowing, or *stenosis,* of one of the valve leaflets inside the heart. Narrowing occurs most commonly in the mitral valve, usually from rheumatic heart disease. The aortic valve may be injured from sclerosis or rheumatic fever. The seriousness of all types of stenoses stems from the fact that they all place a severe work load on the heart by making it work harder to push the blood through the abnormally narrow valve openings. As a result of mitral stenosis, blood pressure is increased, and angina pectoris and heart failure may accompany the progress of this disorder. The majority of stenosed valves are totally replaced with artificial valves.

Another congenital defect is *tetralogy of Fallot,* which is a combination of four heart defects causing a "blue baby." These are: (1) a ventricular septal opening, (2) an aorta that emerges from both ventricles instead of solely from the left ventricle, and as a result of these, (3) a stenosed pulmonary semilunar valve and (4) an enlarged right ventricle. Because of the ventricular septal defect, both oxygenated and deoxygenated blood are mixed in the ventricles. However, the tissues of the body are much more starved for oxygen than are those of a child with simple ventricular septal defect. Because the aorta also emerges from the right ventricle and the pulmonary artery is stenosed, very little blood ever gets to the lungs and the pulmonary circulation is bypassed almost completely. Today, it is possible to surgically correct cases of tetralogy of Fallot when the patient is of proper age and condition. In open-heart surgery, the narrowed pulmonary valve is cut open and the septal defect is sealed with a Dacron patch.

Disorders Associated with Faulty Conduction

As noted earlier, the term arrhythmia refers to any variation in the rate, rhythm, or synchrony of the heart. An arrhythmia can arise when electrical impulses through the heart are blocked at a critical point in the conduction system. One such arrhythmia is called a **heart block.** Perhaps the most common blockage is in the atrioventricular node, which conducts impulses from the atria to the ventricles. This disturbance is called **atrioventricular (AV) block.** It can result from a myocardial infarction, arteriosclerosis, rheumatic heart disease, diphtheria, or syphilis. In a first-degree AV block, which can be detected by an electrocardiograph, the transmission of impulses from the atria to the ventricles is delayed. Here, the P–R interval is

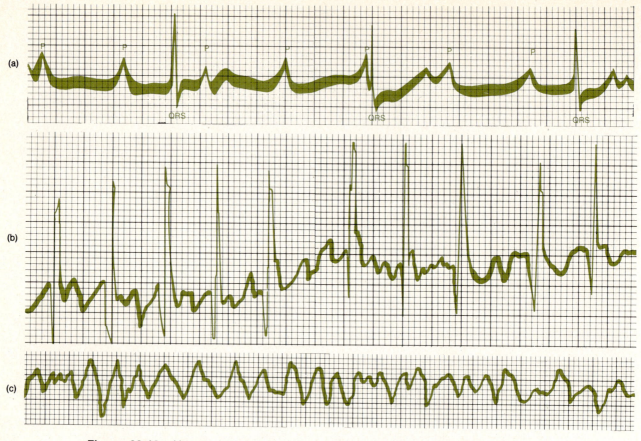

Figure 22-19 Abnormal electrocardiograms. (a) Complete heart block. There is no fixed ratio between atrial contractions (P waves) and ventricular contractions (QRS waves). (b) Atrial fibrillation. There is no regular atrial contraction and, therefore, no P wave. Since the ventricles contract irregularly and independently, the QRS wave appears at irregular intervals. (c) Ventricular fibrillation. In general, there is no rhythm of any kind.

greater than it should be. In a second-degree AV block, every second impulse fails to reach the ventricles, so the ventricular rate is about one-half that of the atrial rate. When ventricular contraction does not occur (dropped beat), oxygenated blood is not pumped efficiently to all parts of the body. The patient may feel faint and dizzy or may collapse if there are many dropped ventricular beats. In a third-degree or complete AV block, impulses reach the ventricle at irregular intervals, and some never reach it at all. The result is that atrial and ventricular rates are out of synchronization (Figure 22-19a). The ventricles may go into systole at any time. This can occur when the atria are in systole or just before the atria go into systole. Or the ventricles may take a rest for a few cardiac cycles.

With complete AV block, many patients experience vertigo, unconsciousness, or convulsions. These symptoms result from a decreased cardiac output with consequent diminished cerebral blood flow and cerebral hypoxia, or lack of sufficient oxygen. Among the causes of AV block are excessive stimulation by the vagus nerves that depress conductivity of the

junctional fibers, destruction of the AV bundle as a result of coronary infarction, arteriosclerosis, myocarditis, or depression caused by various drugs. Other heart blocks include *intraatrial (IA) block, interventricular (IV) block,* and *bundle branch block (BBB).* In the last condition, the ventricles do not contract together because of the delay in the impulse in the blocked branch.

Flutter and Fibrillation

Two rhythms that indicate heart trouble are atrial flutter and fibrillation. In **atrial flutter** the atrial rhythm averages between 240 and 360 beats per minute. The condition is essentially very rapid atrial contractions accompanied by a second-degree AV block. It is typically indicative of severe damage to heart muscle. Atrial flutter usually becomes fibrillation after a few days or weeks. **Atrial fibrillation** is an asynchronous contraction of the atrial muscles that causes the atria to contract irregularly and still faster. An electrocardiogram of atrial fibrillation is shown in Figure 22-19b. Atrial flutter and fibrillation occur in myocardial infarction, acute and chronic rheumatic heart disease, and hyperthyroidism. Atrial fibrillation results in complete uncoordination of atrial contraction so that atrial pumping ceases altogether. When the muscle fibrillates, the muscle fibers of the atrium quiver individually instead of contracting together. The quivering cancels the pumping function of the atrium. In a strong heart, atrial fibrillation reduces the pumping effectiveness of the heart by 25 to 30 percent.

Ventricular fibrillation is another kind of rhythm that indicates heart trouble. It is characterized by asynchronous, irregular, haphazard, ventricular muscle contractions. The rate may be rapid or slow. The impulse travels to the different parts of the ventricles at different rates. Thus, part of the ventricle may be contracting while other parts are still unstimulated. Ventricular contractions, therefore, become ineffective. Circulatory failure results and death will occur within 3 to 4 minutes unless the arrhythmia is reversed quickly (Figure 22-19c). Ventricular fibrillation may also be caused by coronary occlusion. It sometimes occurs during surgical procedures on the heart or pericardium. It may be the cause of death in electrocution.

Aneurysm

A blood-filled sac formed by an outpouching in an arterial or venous wall is called an **aneurysm.** Aneurysms may occur in any major blood vessel of the body and include the following types:

1. Berry, which is a small aneurysm frequently in a cerebral artery. If it ruptures, it may cause a hemorrhage below the dura mater (Figure 22-20a). Hemorrhaging is one cause of stroke.
2. Ventricular, which is a focal dilation of a ventricle of the heart (Figure 22-20b).
3. Aortic, which is a focal dilation of the aorta (Figure 22-20c).

Atherosclerosis

Atherosclerosis is the major form of arteriosclerosis, which includes many diseases of the arterial wall. Atherosclerosis is a lipid-related arterial lesion, and is responsible for the most important and prevalent of all clinical complications. In this disorder, the inner layer of an artery becomes thickened with soft fatty deposits called *atheromatous plaques* (Figure 22-21a). An *atheroma* is an abnormal mass of fatty or lipid material existing as a discrete deposit in an arterial wall. Atheromas involve the abdominal aorta and major leg arteries more extensively than the thoracic aorta. They may also be found in coronary, cerebral, and main-stem peripheral arteries.

The atheroma looks like a pearly gray or yellow mound of tissue on the inside of the blood vessel wall. It usually consists of a core of lipid (mainly cholesterol) covered by a cap of fibrous (scar) tissue. As the atheromas increase in size, they may impede or cut off blood flow in affected arteries. This damages the downstream tissues supplied by these arteries.

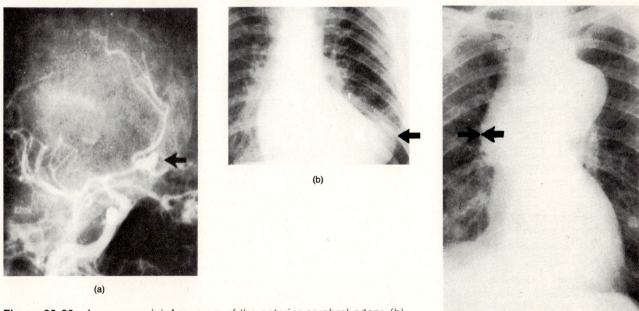

(a)

(b)

(c)

Figure 22-20 Aneurysms. (a) Aneurysm of the anterior cerebral artery. (b) Ventricular aneurysm. (c) Aortic aneurysm. (Courtesy of Lester W. Paul and John H. Juhl, *The Essentials of Roentgen Interpretation,* 3d ed., New York: Harper & Row, 1972.)

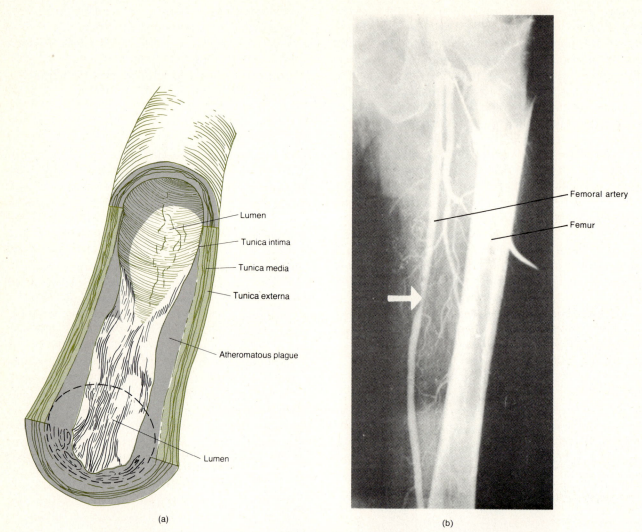

(a)

(b)

Figure 22-21 Atherosclerosis. (a) Atheromatous plaque formation. The broken circular line indicates the approximate size of a normal lumen. (b) Femoral arteriogram showing an atheromatous plaque (arrow) in the middle third of the thigh. (Courtesy of Lester W. Paul and John H. Juhl, *The Essentials of Roentgen Interpretation,* 3d ed., New York: Harper & Row, 1972.)

An additional danger is that the lipid core of the plaques may be washed into the bloodstream. There, it can become an embolus and obstruct small arteries, capillaries, and veins quite a distance from the original site of formation. A third possibility is that the plaque will provide a rough surface for clot formation.

The exact degree of arterial stenosis necessary to produce symptoms varies with the site of stenosis and with the anatomic location of the artery. When the insidious process is gradual, a collateral circulation can develop around the stenotic block. These collaterals, if adequate, prevent ischemic necrosis. But when they do not, ischemia results. This may progress to gangrene if multiple episodes involve more than one artery, or if the occlusion reduces available collaterals to a given area.

Atherosclerosis is generally a slow, progressive disease. It may start in childhood, and its development may produce absolutely no symptoms for 20 to 40 years or longer. Even when it reaches the advanced stages, the individual may feel no symptoms, and the condition may be discovered only at postmortem examination. Diagnosis during life is performed by injecting radiopaque substances into the blood and then taking x-rays of the arteries. This technique is called *angiography* or *arteriography.* The film is called an *arteriogram* (Figure 22-21b).

Animal experiments have given us considerable scientific information about the plaques. They begin as yellowish fatty streaks of lipids that appear under the tunica intima. It is possible to produce the streaks in many animals by feeding them a diet that is high in fat and cholesterol. This raises the blood lipid levels — a condition called *hyperlipidemia* (*hyper* = over or above; *lipo* = fat). Hyperlipidemia is an important factor in increasing the risk of atherosclerosis. Patients with high blood levels of cholesterol should be identified and treated with appropriate diet and drug therapy.

Hypertension

Hypertension, or high blood pressure, is the commonest of the diseases affecting the heart and blood vessels. Statistics from a recent National Health Survey indicate that hypertension afflicts at least 17 million American adults and perhaps as many as 22 million.

Primary hypertension, or essential hypertension, is a persistently elevated blood pressure that cannot be attributed to any particular organic cause. Specifically, the diastolic pressure continually exceeds 95 mm Hg. Approximately 85 percent of all hypertension cases fit this definition. The other 15 percent is called **secondary hypertension.** Secondary hypertension is caused by disorders such as arteriosclerosis, kidney disease, and adrenal hypersecretion. Arteriosclerosis increases blood pressure by reducing the elasticity of the arterial walls and by narrowing the space through which the blood can flow. Both kidney diseases and obstruction of blood flow to the kidney may cause the kidney to release an enzyme called renin into the blood. This enzyme catalyzes the formation of angiotensin from a plasma protein. Angiotensin is a powerful blood vessel constrictor. It is the most potent agent known for raising blood pressure. Aldosterone is the adrenal cortex hormone that promotes the retention of salt and water by the kidneys. Thus it tends to increase plasma volume. Pheochromocytoma is a benign tumor of the adrenal medulla. It produces and releases into the blood large quantities of norepinephrine and epinephrine. These hormones also raise blood pressure by stimulating the heart and constricting blood vessels.

High blood pressure is of considerable concern because of the harm it can do to certain body organs such as the heart, brain, and kidneys if it remains uncontrolled for long periods. The heart is most commonly affected by high blood pressure. When pressure is high, the heart uses more energy in pumping. As a result of the increased effort, the heart muscle thickens and the heart becomes enlarged. The heart also needs more oxygen. If it cannot meet the demands put on it, angina pectoris or even myocardial infarction may occur. Continued high blood pressure may produce a cerebral vascular accident, or stroke. In this case, severe strain has been imposed on the cerebral arteries that supply the brain. These arteries are usually less protected by the surrounding tissues than are the major arteries in other parts of the body. As a result, one or more of these weakened cerebral arteries may finally rupture, and a brain hemorrhage follows.

The kidney is another prime target of hypertension. The principal site of damage is in the arterioles that supply this vital organ. The continual high blood pressure pushing against the walls of the arterioles causes them to thicken, thus narrowing the lumen. The blood supply to the kidney is thereby gradually reduced. In response, the kidney may secrete renin, which raises the blood pressure even higher and complicates the problem. The reduced blood flow to the kidney cells may eventually lead to the death of the cells.

At present, the causes of primary hypertension are unknown. Medical science cannot cure it. However, almost all cases of hypertension, whether mild or very severe, can be controlled by a variety of effective drugs that reduce elevated blood pressure.

We shall now turn our attention to another body system that assumes a key function in maintaining homeostasis, the respiratory system. The essential role of this system is to raise the oxygen content of the blood while simultaneously removing carbon dioxide. Thus, the cardiovascular and respiratory systems coordinate their functions in the maintenance of homeostasis.

CHAPTER SUMMARY OUTLINE

THE HEART

Parietal Pericardium (Pericardial Sac)

1. The pericardium, consisting of an outer fibrous layer and an inner serous layer, encloses the heart.

2. Between the outer fibrous layer and the inner serous layer is a space called the pericardial cavity.
3. The pericardial cavity contains fluid that prevents friction between the membranes.

Chambers

1. The chambers of the heart include two upper atria and two lower ventricles.
2. The blood flows through the heart from the superior and inferior venae cavae and the coronary sinus to the right atrium. It then goes through the tricuspid valve to the right ventricle and through the pulmonary trunk to the lungs. It returns to the heart through the pulmonary veins into the left atrium, through the bicuspid valve to the left ventricle, and out through the aorta.

Valves

1. Atrioventricular valves lie between the atria and their ventricles.
2. The atrioventricular valves are the tricuspid valve on the right side of the heart and the bicuspid on the left.
3. The two arteries that leave the heart both have a semilunar valve.
4. All heart valves prevent the backflow of blood.

Physiological Properties

1. The heart demonstrates three properties related to homeostasis that are developed to a very high degree of proficiency: (1) autorhythmicity, (2) conductibility, (3) contractibility.
2. P cells are capable of generating spontaneous rhythmic electrical potentials.
3. The electrical activity generated by the sino-atrial node is distributed to all parts of the heart by tissues specialized to conduct impulses rapidly.
4. These tissues include the interatrial and internodal conduction pathways, the A-V node, the bundle of His, the right bundle branch, the left bundle branch, and the Purkinje network.

The Electrocardiogram

1. The record of electrical changes during each cardiac cycle is referred to as an electrocardiogram (ECG).
2. A normal ECG consists of a P wave (spread of impulse from S-A node over atria), QRS wave (spread of impulse through ventricles), and T wave (ventricular repolarization). The P–R interval represents the conduction time from the beginning of atrial excitation to the beginning of ventricular excitation.
3. The ECG is invaluable in diagnosing abnormal cardiac rhythms and conduction patterns, detecting the presence of fetal life, determining multiple pregnancies, and following the course of recovery from a heart attack.

Blood Supply of the Heart

1. The blood supply of the heart itself originates directly from the aorta via the left and right coronary arteries.
2. Deoxygenated blood returns to the right atrium via the coronary sinus.

The Cardiac Cycle

1. This cycle consists of the systole (contraction) and diastole (relaxation) of both atria, plus the systole and diastole of both ventricles, followed by a short pause.
2. Two phenomena control the movement of blood through the heart. These are the opening and closing of the valves, and the contraction and relaxation of the myocardium.
3. With an average heartbeat of 72/minute, a complete cardiac cycle requires 0.8 second.
4. Blood flows through the heart from an area of higher pressure to an area of lower pressure.
5. The first sound (lubb) represents the closing of the atrioventricular valves. The second sound (dupp) represents the closing of semilunar valves.
6. Most of the coronary arterial flow takes place during ventricular relaxation with 70 to 90 percent of the flow occurring at this time.
7. About 35 percent of the oxygen consumed by the cardiac tissue is used to metabolize carbohydrates for energy production. Most of the remaining oxygen is used to oxidize fatty acids for energy.

Cardiac Output

1. Cardiac output is the amount of blood ejected from the left ventricle into the aorta per minute. It is calculated as follows: cardiac output = stroke volume × ventricular systole/minute.
2. Stroke volume is the amount of blood ejected by a ventricle during each systole. According to Starling's law of the heart, the length of the cardiac muscle fiber determines the force of contraction.

Regulation of Heart Rate

1. The S-A node, the pacemaker, receives nerves from both the parasympathetic and sympathetic divisions of the autonomic nervous system.
2. The parasympathetic neurons originate in the cardioinhibitory center of the medulla and slow down the pacemaker.

3. The sympathetic neurons originate in the cardio-acceleratory center of the medulla and accelerate the pacemaker.
4. Pressure receptors (baroreceptors) and hormones control heart rate.
5. Other influences on heart rate include strong emotions, depression, sex, age, and muscular exercise.

CIRCULATORY SHOCK AND HOMEOSTASIS

1. Circulatory shock results when cardiac output or blood volume are reduced to the point where body tissues become hypoxic.
2. Mild shock is compensated by vasoconstriction and water retention.
3. In severe shock, venous return is diminished and cardiac output decreases. The heart becomes hypoxic, prolonged vasoconstriction leads to hypoxia of other organs, and the shock cycle is intensified.

BLOOD VESSELS

1. Arteries distribute blood under high pressure to the tissues of the body. Arterioles reduce the high pressure in the tissues and regulate the flow into the capillary beds. Capillaries exchange the fluid of the blood, its nutrients and gases, with the interstitial fluid that bathes the cells. Veins collect the blood from the capillary beds and return it to the heart.
2. The blood flows within the vessels in two recirculating systems—the pulmonary and systemic systems.

Hemodynamics of Circulation

1. Hemodynamics is the study of all the factors that influence the flow characteristics of blood as it moves through the body.
2. Blood distribution in the vascular system is divided so that about 84 percent of the volume is in the systemic circulation, about 8.8 percent is within the pulmonary circulation, and about 7.2 percent is within the chambers of the heart.
3. Blood always flows from regions of greater pressure to regions of lesser pressure.
4. The average speed or mean velocity at which blood flows is determined primarily by the geometry of the vessels. Blood flow is alo related to the total cross-sectional area of the blood vessels. Blood flows most rapidly where the cross-sectional area is least.

5. The most important relationship i~ circulatory physiology is expressed in the following equation.

$$\text{Flow} = \frac{\text{pressure difference}}{\text{resistance}}$$

Factors that Aid Venous Return

1. The return of venous blood to the heart is assisted by an increased rate of flow, contractions of the skeletal muscles, valves, and respiration.

The Micro or Capillary Circulation

1. The function of the capillary circulation is to nourish the tissue cells, and it is at the capillary level that the exchange of materials between blood and interstitial fluid occurs.
2. The composition and volume of the interstitial fluid are maintained fairly constant by the continual exchange of materials with the blood. Most of this exchange occurs by simple diffusion.

CHECKING CIRCULATION

1. The pulse tells us the rate of the heartbeat. The blood pressure measures the pressure the blood is under.

Pulse

1. Pulse is the alternate expansion and elastic recoil of an artery with each systole of the left ventricle. It may be felt in any artery that lies near the surface or over a hard tissue.
2. A normal pulse rate averages between 70 and 80 per minute in the resting state.
3. Tachycardia indicates a very rapid heart rate or pulse rate, whereas bradycardia indicates a very slow heart rate or pulse rate.

Blood Pressure

1. Blood pressure is the pressure exerted by the blood on the walls of any blood vessel; however, in clinical settings, it refers to the pressure only in the large arteries. It is measured by the use of a sphygmomanometer.
2. Systolic blood pressure is the force of blood recorded during ventricular contraction. Diastolic blood pressure is the force of blood recorded during ventricular relaxation. The average blood pressure is 120/80 mm Hg.
3. Pulse pressure is the difference between systolic and diastolic pressure. It averages 40 mm Hg and

provides information about the condition of the arteries.

HOMEOSTATIC IMBALANCES OF THE CARDIOVASCULAR SYSTEM

Risk Factors in Heart Attacks

1. Research indicates that people who develop combinations of certain risk factors eventually have heart attacks.
2. These factors are: high blood cholesterol level, high blood pressure, cigarette smoking, overweight, lack of exercise, and diabetes mellitus.
3. People with three or more risk factors form an especially high-risk group.

Disorders Associated with Failure of Blood Supply

1. Ischemia is a reduced oxygen supply that weakens but does not kill cells.
2. Angina pectoris is ischemia of the myocardium or cardiac muscle, and it occurs when coronary circulation is reduced.
3. Myocardial infarction is a much more serious problem, since infarction means death of an area of tissue because of a drastically reduced or completely interrupted blood supply.

Disorders Associated with Faulty Architecture

1. The total number of babies born each year with a congenital or inborn heart defect is between 30,000 and 40,000.
2. Patent ductus arteriosus is a serious defect that occurs when the blood vessel that connects the aorta with the pulmonary artery during fetal growth remains open instead of closing completely after birth.
3. Septal defects are openings in the septa or partitions between either the atria or the ventricles.
4. Valvular stenosis is a narrowing of one of the valve leaflets inside the heart placing a severe work load on the heart.
5. Tetralogy of Fallot is a combination of four heart defects causing a "blue baby."

Disorders Associated with Faulty Conduction

1. Arrhythmia refers to any variation in the rate, rhythm, or synchrony of the heart, and it can arise when electrical impulses through the heart are blocked.

2. The most common blockage is in the atrioventricular node which conducts impulses from the atria to the ventricles and is called atrioventricular (AV) block.
3. Other heart blocks include intraatrial (IA) block, interventricular (IV) block, and bundle branch block (BBB).

Flutter and Fibrillation

1. Two rhythms that indicate heart trouble are atrial flutter and fibrillation.
2. In atrial flutter the atrial rhythm is very rapid, averaging between 240 and 360 beats per minute, and is indicative of severe damage to heart muscle.
3. Atrial fibrillation results in complete uncoordination of atrial contraction so that atrial pumping ceases altogether, reducing the pumping effectiveness of the heart by 25 to 30 percent.
4. Ventricular fibrillation is characterized by asynchronous, irregular, haphazard, ventricular muscle contractions that produce circulatory failure, resulting in death within 3 to 4 minutes unless the arrhythmia is reversed quickly.

Aneurysm

1. A blood-filled sac formed by an outpouching in an arterial or venous wall is called an aneurysm.
2. Different types of aneurysms include the berry, ventricular, and aortic.

Atherosclerosis

1. Atherosclerosis is a form of arteriosclerosis, and is associated with the lipid-related arterial lesion responsible for the most important of all clinical complications.
2. In atherosclerosis the inner layer of an artery becomes thickened with atheromatous plaques that may impede or cut off blood completely.
3. Atherosclerosis is a slow, progressive disease, but it can be diagnosed during life by a technique called angiography or arteriography.

Hypertension

1. Hypertension, or high blood pressure, is the commonest of the diseases affecting the heart and blood vessels, with at least 17 million American adults being afflicted.
2. Primary hypertension is a persistently elevated blood pressure that cannot be attributed to any particular organic cause. It accounts for approximately 85 percent of all hypertension.

3. Secondary hypertension is caused by disorders such as arteriosclerosis, kidney disease, and adrenal hypersecretion.
4. High blood pressure that remains uncontrolled for long periods can harm the heart, brain, and kidneys.
5. At present, the causes of primary hypertension are unknown, but all cases of hypertension can be controlled by drugs.

REVIEW QUESTIONS

1. Describe the location of the heart in the mediastinum. Distinguish the subdivisions of the pericardium. What is the function of this structure?
2. Define atria and ventricles. What vessels enter or exit the atria and ventricles?
3. Discuss the principal valves in the heart and how they operate.
4. The heart demonstrates three properties related to homeostasis that are developed to a very high degree of proficiency. Name these three properties and explain each of them.
5. Describe the path of a nerve impulse through the heart's conducting system. Define syncytium and intercalated discs.
6. Define and label the deflection waves of a normal electrocardiogram. Why is the ECG an important diagnostic tool?
7. Describe the route of blood in the coronary circulation.
8. Define systole and diastole and their relationship to the cardiac cycle.
9. Distinguish the principal events that occur during atrial systole, ventricular diastole, atrial diastole, and ventricular systole.
10. Describe the pressure changes associated with the movement of blood through the heart.
11. By means of a diagram, relate the events of the cardiac cycle to time. What is the quiescent period?
12. Describe the first and second heart sounds and indicate their clinical significance. Define a murmur.
13. What are the variations in coronary arterial flow? Explain cardiac tissue metabolism.
14. What is cardiac output? How is it calculated? What factors alter cardiac output?
15. Define Starling's law of the heart. Why is it important?
16. Compare the effects of parasympathetic and sympathetic stimulation of the heart.
17. What is a baroreceptor? Describe hormonal effects on the heart. Outline the operation of the carotid sinus reflex, the aortic arch reflex, and the Bainbridge reflex.
18. Explain some other factors that influence heart activity.
19. Define circulatory shock. What are its symptoms?
20. How is the body's homeostasis restored during mild shock?
21. Describe the effects of a severe shock on circulatory organs by drawing a shock cycle.
22. Differentiate between arteries, arterioles, capillaries, and veins as to location and function.
23. Compare the pulmonary circulation system with the systemic circulation system.
24. What is the distribution of blood (in percent) at any given moment?
25. Correlate pressure gradients with blood flow and give exact values in millimeters of mercury.
26. Describe blood flow in terms of its relationship to flow, pressure, and resistance.
27. Identify the factors that assist the return of venous blood to the heart.
28. What is capillary circulation? What are some of the factors that make capillary circulation so efficient?
29. Define pulse. Where may pulse be felt?
30. Contrast the following: tachycardia, bradycardia, and irregular pulse.
31. What is blood pressure? Describe how systolic and diastolic blood pressure may be recorded by means of a sphygmomanometer.
32. Compare the clinical significance of systolic and diastolic pressure. How are these pressures written?
33. Define pulse pressure. What does this pressure indicate?
34. What are the six risk factors in the probability of developing a heart attack?
35. List the three more immediate causes of heart trouble.
36. Define ischemia, infarction, and arrhythmia, and compare angina pectoris and myocardial infarction as abnormalities of coronary circulation.
37. Compare the abnormalities of patent ductus arteriosus, septal defects, and valvular stenosis.
38. What are the four heart defects that produce tetralogy of Fallot?

39. Define arrhythmia and compare AV block, IA block, IV block and BBB.
40. Contrast atrial flutter, atrial fibrillation, and ventricular fibrillation as to symptoms and effects.
41. Define aneurysm and compare the different types.
42. What is atherosclerosis? How is it produced, and why is it considered such a serious disorder?
43. Describe the technique of detecting atherosclerosis.
44. Differentiate between primary and secondary hypertension.
45. Explain how prolonged hypertension can damage other organs of the body, giving specific examples.

Chapter 23

RESPIRATORY PHYSIOLOGY

STUDENT OBJECTIVES

After reading this chapter, you should be able to

- Explain pulmonary circulation
- Compare blood pressures and resistance in the pulmonary circulation
- Define the distribution of blood flow in the lungs and explain the constriction of pulmonary vessels
- Identify the organs of the respiratory system
- Contrast the functions of the external and internal nose in filtering, warming, and moistening air
- Differentiate the three regions of the pharynx and describe their roles in respiration
- Identify the anatomical features of the larynx related to respiration and voice production
- Describe the tubes that form the bronchial tree with regard to structure and location
- Contrast tracheostomy and intubation as alternative methods for clearing air passageways
- Identify the gross anatomical features of the lungs and their coverings
- Describe the structure of a lobule of the lung
- Describe the role of alveoli in the diffusion of respiratory gases
- Explain the functions of the diaphragm and intercostal muscles in breathing
- Define the dynamics of air flow between the air within the alveoli and the air in the atmosphere
- Describe surface tension and its relationship to breathing
- Compare the volumes and capacities of air exchanged in respiration
- Describe the areas of the nervous system that control respiration
- Compare the roles of the Hering-Breuer reflex and the pneumotaxic center in controlling respiration
- Describe the mechanisms of external and internal respiration based on differences in partial pressure of O_2 and CO_2
- Describe how O_2 and CO_2 are carried by the blood
- Define the mechanisms that modify the rate of normal breathing by describing the effects of chemical stimuli and pressure
- Describe several modified respiratory movements
- List the basic steps involved in heart-lung resuscitation including the Heimlich maneuver.
- Describe the effects of pollutants on the epithelium of the respiratory system
- Describe the administration of medication by nebulization
- Define hay fever, bronchial asthma, emphysema, pneumonia, tuberculosis, and hyaline membrane disease

Respiration can be defined as the exchange of gases between a living cell and its environment. If an organism consists of only one cell, then it exchanges oxygen and carbon dioxide directly with its external environment. In an organism as complex as the human body, only a very small percent of the total cells can exchange gases directly with the external environment. Therefore, in order to satisfy the total cellular requirement for oxygen and carbon dioxide removal, the human body employs two systems: (1) the cardiovascular system, which transports the gases to and from the tissues, and (2) the respiratory system, which performs the task of raising the oxygen content of the blood while simultaneously removing the excess of carbon dioxide it carries.

PULMONARY CIRCULATION

The pulmonary circulation is that portion of the cardiovascular system that delivers blood from the right ventricle of the heart to the left atrium. In this journey, the blood passes through the lungs, where it exchanges carbon dioxide for oxygen. Blood leaving the right ventricle enters the main pulmonary trunk, which shortly branches into the right and left pulmonary arteries. These arteries then lead to their respective lungs, where they branch rapidly and successively into ever smaller vessels. The few arterioles open into the capillary beds, consisting of thin-walled vessels through which the exchange of gases occurs. Blood flow through the capillary beds is so rich that some microscopists describe it as a lake of blood flowing between two thin parallel sheets of epithelial tissue that is frequently interrupted where the two sheets are pinched together. This thin layer of blood flows over and around tiny air sacs called *alveoli,* which are constantly being refilled with "fresh air" from the atmosphere (Figure 23-1a). This air, which has a higher content of oxygen and a lower content of carbon dioxide than the blood, makes it possible for the oxygen and carbon dioxide gases to exchange. Figure 23-1b illustrates the red blood cells flowing in the capillaries and the large volumes of air surrounding the blood vessels. To appreciate the thinness of the barrier separating the red blood cells from the air, examine the separate structures shown in Figure 23-1c. Blood flows from the capillaries into small venules, which unite to form larger and larger veins. Eventually the blood is returned to the left atrium of the heart by four pulmonary veins, two returning from each lung. It takes an average time of about 4.5 seconds for the blood to complete the pulmonary circuit, and only about 1 second of that time is spent in the capillary beds.

Blood Pressures and Resistance

The blood pressure in the main pulmonary trunk is about 25 mm Hg systolic and 8 mm Hg diastolic. In contrast to the thick walls of the aorta, the walls of

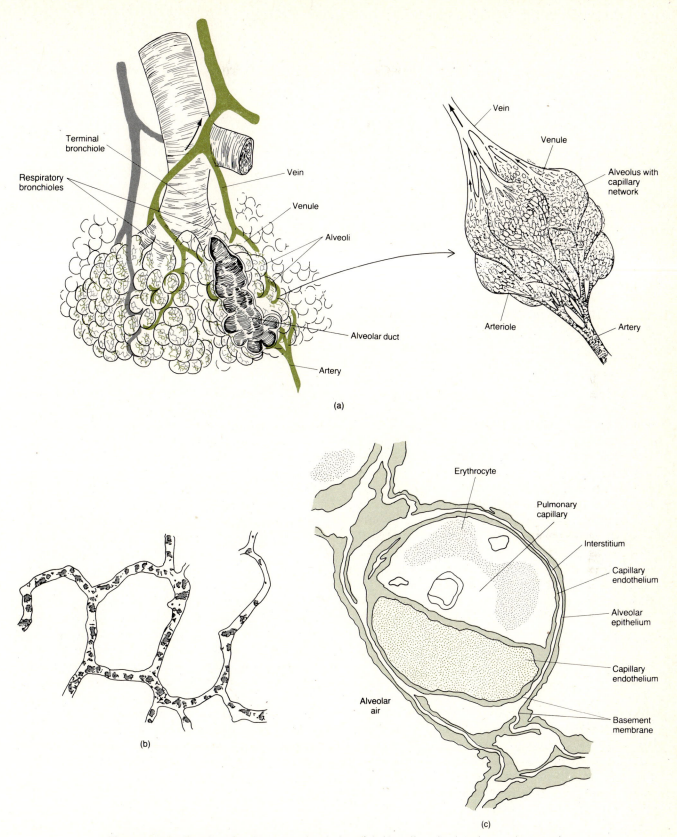

Figure 23-1 Details of pulmonary circulation. (a) Alveoli and related vascular supply. (b) Capillaries in alveolar walls. (c) Pulmonary capillary in the alveolar wall.

the pulmonary arteries are thin and contain relatively little smooth muscle. The thinner walls are consistent with the lower pressures in these arteries. The pulmonary capillary pressure is about 9 to 6 mm Hg.

The resistance to flow through the pulmonary vessels can be estimated by using the formula:

$$R = \frac{\Delta P}{F}$$

The values for ΔP can be approximated by subtracting the outflow pressures (in veins) from the mean inflow pressures (in arteries) as the blood flows through the lungs. Therefore, $\Delta P = 16 - 6 = 10$ mm Hg. The value for F is equivalent to the cardiac output, which at rest is about 6 liters/minute.

$$\text{Pulmonary } R = \frac{10}{6} = 1.7 \text{ mm Hg/liter/minute}$$

In comparison, the systemic circulation has a pressure drop of almost 100 mm Hg with the same value for flow rate. Thus the resistance in the systemic system is at least 10 times that of the pulmonary.

$$\text{Systemic } R = \frac{100}{6} = \text{about 17 mm Hg/liter/minute}$$

The lower resistance of the pulmonary circulation is due, in large part, to the general lack of muscular arterioles such as exist in the systemic circulation. If the pressure in the pulmonary artery increases for any reason, then the resistance in the pulmonary capillaries decreases even further. This can take place because of two mechanisms. First, additional capillaries open. As the blood pressure rises, these previously closed vessels begin conducting blood and the vascular resistance drops. Second, the vessels already conducting blood distend even more. With only air surrounding the capillaries, they easily increase their radius and thus further lower the resistance to blood. This second mechanism is most responsible for the resistance drop as vascular pressures increase.

Distribution of Blood Flow in the Lungs

Blood flow in the lungs is not uniform. When you stand upright, the superior or apical parts of your lungs receive less blood flow, whereas the lower basal regions receive a very large portion of the pulmonary flow. However, when you lie on your back, the blood flow pattern changes so that delivery is equalized between the superior and inferior regions. Now the blood flow is greater in the posterior regions of the lung and less in the anterior.

The explanation for the unequal blood flow is related to the pulmonary blood pressure variation that

exists in the lung. In the superior parts of the lung, the arterial pressure is usually just sufficient to overcome the atmospheric pressure in the air sacs, and the capillaries stay open. If the blood pressure were to drop below this marginal value, then the capillaries in this part of the lung would collapse and flow would cease. This portion of the lung is then useless for gaseous exchange. In the middle regions, the arterial pressure is greater than the atmospheric air pressure, and blood flows into the capillaries. However, the venous pressure is often less than the atmospheric pressure, and the venous ends of the capillaries may be shut, thus restricting flow.

In the inferior regions, the blood pressure in the arteries, capillaries, and veins is always greater than the atmospheric air pressure, and therefore flow is always present. Figure 23-2 illustrates the situations possible in the three cases. However, it is important to realize that the three regions are used only for illustration. The blood pressure changes gradually and continuously from the highest value at the bottom to the lowest at the top.

Active Constriction of Pulmonary Vessels

It has been shown that a marked constriction of the blood vessels occurs if the partial pressure of oxygen in the inhaled air drops below 70 mm Hg in any portion of the lung. Current thought is that cells surrounding the blood vessels suffer hypoxia at the lowered oxygen levels and release some chemical that acts on the blood vessels, causing them to constrict. In this way, blood can be shunted to regions where the oxygen is in higher concentrations. There may be partial obstruction in a bronchial tube that acts to reduce the amount of oxygen reaching a section of the lung. In high altitudes, where the partial pressure of oxygen is lower, there may be a generalized vasoconstriction that not only hinders proper oxygenation of the blood but also may increase pulmonary arterial pressure and add to the work load of the heart.

This same mechanism can be seen dramatically operating at the time of birth. Prior to birth, the lungs are not oxygenated and the perivascular cells may produce the substance that causes a great vasoconstriction. Thus, the blood flow to the lungs is greatly impeded, and only about 15 percent of the cardiac output transverses the lungs. With its first breath, the newborn inhales sufficient oxygen to cause the blood vessels to relax, and blood flow increases dramatically.

THE RESPIRATORY ORGANS

Air that will eventually participate in the oxygenation

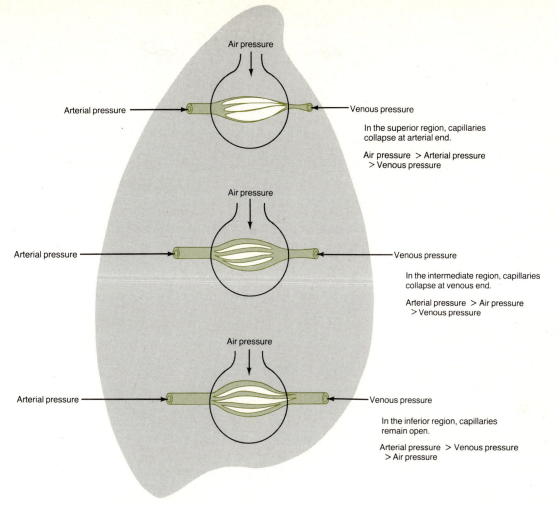

Air pressure

Arterial pressure → ← Venous pressure

In the superior region, capillaries collapse at arterial end.

Air pressure > Arterial pressure
> Venous pressure

Air pressure

Arterial pressure → ← Venous pressure

In the intermediate region, capillaries collapse at venous end.

Arterial pressure > Air pressure
> Venous pressure

Air pressure

Arterial pressure → ← Venous pressure

In the inferior region, capillaries remain open.

Arterial pressure > Venous pressure
> Air pressure

Figure 23-2 Flow distribution at different levels of the lung, based on pressures acting on the capillaries.

of the blood enters the respiratory system via the nose or mouth and then proceeds through a branching system of smaller and smaller tubes (Figure 23-3).

Nose

The interior structures of the **nose** are specialized for three functions. First, incoming air is warmed, moistened, and filtered. Second, olfactory stimuli are received; and third, large, hollow resonating-chambers are provided for speech sounds. These three functions are accomplished in the following manner. When air enters the external nares (nostrils), it passes first through the vestibule. The *vestibule* is lined by skin containing coarse hairs that filter out large dust particles. The air then passes into the rest of the cavity. Three shelves formed by projections of the *superior, middle,* and *inferior conchae* extend out of the lateral wall of the cavity. The conchae, almost reaching the septum, subdivide each nasal cavity into a series of grovelike passageways called *meati*. These are called the *superior, middle,* and *inferior*

meati. Mucous membrane lines the cavity and its shelves. The olfactory receptors lie in the membrane lining the upper portion of the cavity, also called the olfactory region. Below the olfactory region, the membrane contains pseudostratified ciliated columnar cells with many goblet cells and capillaries. As the air whirls around the conchae and meati, it is warmed by the capillaries. Mucus secreted by the goblet cells moistens the air and traps dust particles. Drainage from the lacrimal ducts, and perhaps secretion from the paranasal sinuses, also help to moisten the air. The cilia move the resulting mucus-dust packages along to the throat so that they can be eliminated from the body. As the air passes through the top of the cavity, chemicals in the air may stimulate the olfactory receptors.

Pharynx

The **pharynx,** or throat, is a tube about 13 cm (5 inches) long that starts at the internal nares (openings from the nasal cavity into the nasopharynx) and runs

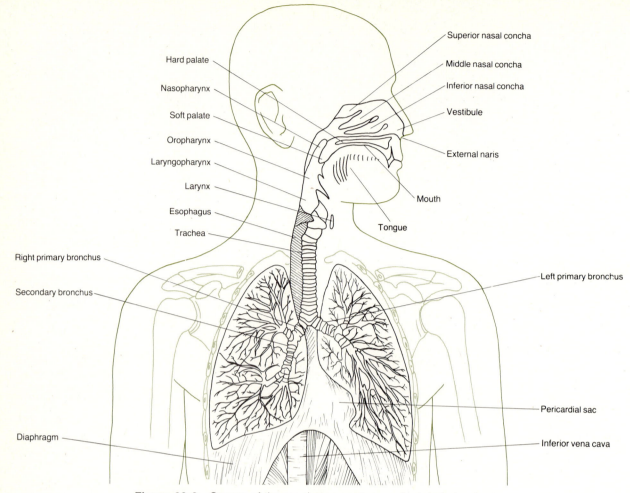

Labels on figure:
- Hard palate
- Nasopharynx
- Soft palate
- Oropharynx
- Laryngopharynx
- Larynx
- Esophagus
- Trachea
- Right primary bronchus
- Secondary bronchus
- Diaphragm
- Superior nasal concha
- Middle nasal concha
- Inferior nasal concha
- Vestibule
- External naris
- Mouth
- Tongue
- Left primary bronchus
- Pericardial sac
- Inferior vena cava

Figure 23-3 Organs of the respiratory system and related structures.

partway down the neck (Figure 23-4). The functions of the pharynx are limited to serving as a passageway for air and food and providing a resonating chamber for speech sounds.

The uppermost portion of the pharynx is called the *nasopharynx*. Its posterior wall contains the pharyngeal tonsils or adenoids. The nasopharynx has a lining of pseudostratified ciliated epithelium. Cilia in the walls of the nasopharynx move the mucus down toward the mouth. The nasopharynx also exchanges small amounts of air with the Eustachian tubes so that the air pressure inside the middle ear equals the pressure of the atmospheric air flowing through the nose and pharynx.

The second portion of the pharynx, the *oropharynx*, is both respiratory and digestive in function since it is a common passageway for both air and food. Two pairs of tonsils, the palatine tonsils and the lingual tonsils, lie at the base of the tongue.

The lowest portion of the pharynx is called the *laryngopharynx*. The laryngopharynx extends downward from the hyoid bone and empties into the esophagus (food tube) posteriorly and into the larynx (voice box) anteriorly. Like the oropharynx, the laryngopharynx has both respiratory and digestive functions.

Larynx

The **larynx,** or voice box, is a short passageway that connects the pharynx with the trachea. Like other respiratory passageways, the larynx is lined with a ciliated mucous membrane. Dust not removed in the upper passages can be trapped by the mucus and moved toward the throat, where the mucus can be swallowed or eliminated.

The mucous membrane of the larynx is arranged into two pairs of folds, an upper pair called the *ventricular folds* or *false vocal cords*, and a lower pair called simply the *vocal folds* or *true vocal cords*

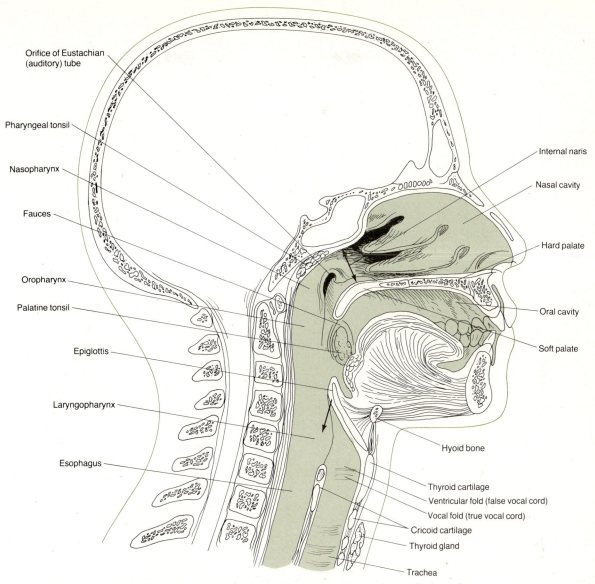

Figure 23-4 Head, neck, and upper chest seen in sagittal section.

Labels (clockwise from top left):
Orifice of Eustachian (auditory) tube
Pharyngeal tonsil
Nasopharynx
Fauces
Oropharynx
Palatine tonsil
Epiglottis
Laryngopharynx
Esophagus
Internal naris
Nasal cavity
Hard palate
Oral cavity
Soft palate
Hyoid bone
Thyroid cartilage
Ventricular fold (false vocal cord)
Vocal fold (true vocal cord)
Cricoid cartilage
Thyroid gland
Trachea

(Figure 23-5). If air is directed against the vocal folds, they vibrate and set up sound waves in the column of air in the pharynx, nose, and mouth. The greater the pressure of air, the louder the sound.

Pitch is controlled by the tension on the true vocal cords. If the cords are pulled taut, they vibrate more rapidly and a higher pitch results. Lower sounds are produced by decreasing the muscular tension on the cords. Vocal cords are usually thicker and longer in males than they are in females, and they vibrate more slowly. This is why men have a lower range of pitch than women.

Sound originates from the vibration of the true vocal cords, but other structures are necessary for convert-ing the sound into recognizable speech. For instance, the pharynx, mouth, nasal cavities, and paranasal sinuses act as resonating chambers that give the voice its human and individual quality. By constricting and relaxing the muscles in the walls of the pharynx, we produce the vowel sounds. Muscles of the face, tongue, and lips help us to enunciate words.

Laryngitis is an inflammation of the larynx that is most often caused by a respiratory infection or by irritants, such as cigarette smoke. Inflammation of the vocal folds themselves themselves causes hoarseness or loss of voice by interfering with the contraction of the cords or by causing them to swell to the point where they cannot vibrate freely. Many long-term

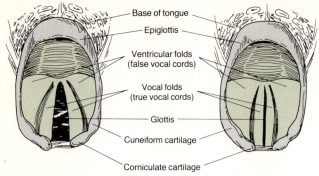

Figure 23-5 Larynx viewed from above.

smokers acquire a permanent hoarseness from the damage done by chronic inflammation.

Trachea

The **trachea,** or windpipe, is a tubular passageway for air, about 12 cm (4.5 inches) and 2.5 cm (1 inch) in diameter. It is located in front of the esophagus and extends from the larynx to the fifth thoracic vertebra, where it divides into right and left primary bronchi (Figure 23-6a).

The trachea is lined with pseudostratified ciliated columnar epithelium that provides the same protection against dust as the membrane lining the larynx. The walls of the trachea are composed of muscle and elastic connective tissue. They are encircled by a series of horizontal incomplete rings of cartilage that look like a stack of letter Cs. The open parts of the Cs face the esophagus and permit it to expand into the trachea during swallowing. The solid parts of the Cs provide a rigid support so that the tracheal walls do not collapse inward and obstruct the air passageway.

Occasionally the respiratory passageways are unable to protect themselves from obstruction. For instance, the rings of cartilage may be accidentally crushed, or the mucous membrane may become inflamed and swell so much that it closes off the air space. Inflamed membranes also secrete a great deal of mucus that may clog the lower respiratory passageways. Or a large object may be inhaled (aspirated) while the glottis is open. In any case, the passageways must be cleared quickly. If the obstruction is above the level of the chest, a *tracheostomy* may be performed. The first step in a tracheostomy is to make an incision through the neck and into the part of the trachea below the obstructed area. A tube is then inserted through the incision, and the patient breathes through the tube. Another method that may be employed is *intubation*. A tube is inserted into the mouth and passed through the larynx and trachea. The firm walls of the tube push back any flexible obstruction, and the inside of the tube provides a passageway for air. If mucus is clogging the airways, it can be suctioned up through the tube.

Bronchi

The trachea terminates in the chest by dividing into a *right primary bronchus,* which goes to the right lung, and a *left primary bronchus,* which goes to the left lung (Figure 23-6a).

Upon entering the lungs, the primary bronchi divide to form smaller bronchi, the *secondary* or *lobar bronchi,* one for each lobe of the lung. (The right lung has three lobes, the left lung has two). The secondary bronchi continue to branch, forming still smaller bronchi, called *tertiary* or *segmental bronchi,* which divide into *bronchioles*. Bronchioles, in turn, branch into even smaller tubes, called the *terminal bronchioles*. The branching of the trachea resembles a tree trunk with its branches and is commonly referred to as the *bronchial tree*. As the branching becomes more extensive in the bronchial tree, rings of cartilage are replaced by plates of cartilage in the smaller branches, and even they finally disappear in the bronchioles. As the cartilage decreases, the amount of smooth muscle increases. The fact that the walls of the bronchioles contain a great deal of muscle but no cartilage is clinically significant. During an asthma attack the muscles constrict. Because there is no supporting cartilage, the constrictions tend to close off the air passageways.

Bronchography is a technique for examining the bronchial tree. In this procedure, the patient inhales air that contains a safe dosage of radiopaque material, which emits rays that penetrate the chest walls and expose a film. The developed film, a *bronchogram,* provides a picture of the tree (Figure 23-6b).

Lungs

The **lungs** are paired, cone-shaped organs lying in the thoracic cavity (Figure 23-7a). They are separated from each other by the heart and other structures in the mediastinum.

Each lung is divided into lobes by one or more fissures. Each lobe receives its own secondary or lobar bronchus. The right primary bronchus gives rise to three lobar bronchi called the *superior, middle,* and *inferior lobar* or *secondary bronchi*. The left primary bronchus gives rise to a *superior* and an *inferior lobar* or *secondary bronchus*. Within the substance

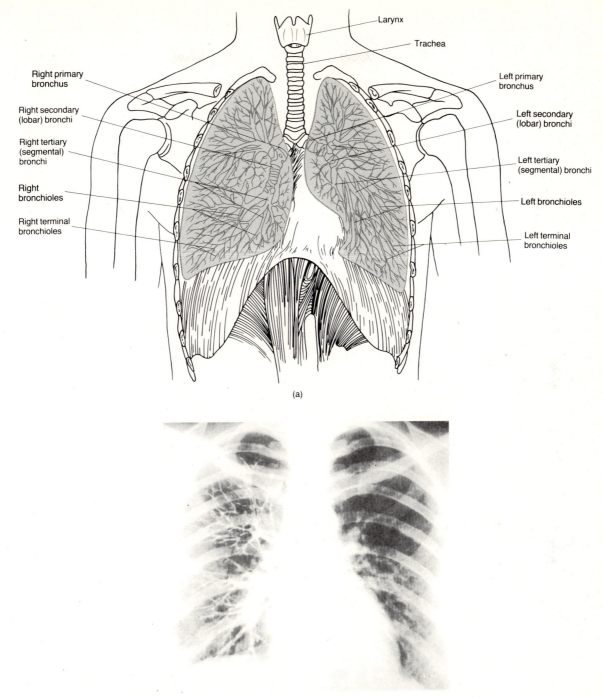

Figure 23-6 Air passageways to the lungs. (a) Bronchial tree. (b) Bronchogram. (Courtesy of Lester W. Paul and John H. Juhl, *The Essentials of Roentgen Interpretation*, 3d ed., New York: Harper & Row, 1972.)

of the lung, the lobar bronchi give rise to branches that are relatively constant in both origin and distribution. Such branches are called *tertiary* or *segmental bronchi*, and the segment of lung tissue that

each supplies is referred to as a *bronchopulmonary segment*.

Each bronchopulmonary segment of the lungs is broken up into many small compartments called

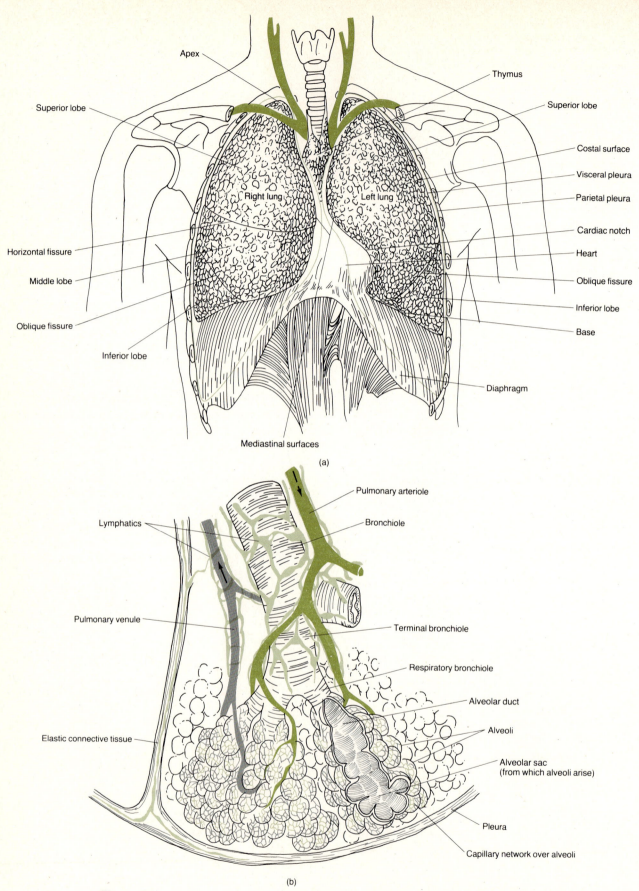

Figure 23-7 Lungs. (a) Coverings and external anatomy. (b) A lobule of the lung.

lobules (Figure 23-7b). Every lobule is wrapped in elastic connective tissue and contains a lymphatic vessel, an arteriole, a venule, and a branch from a terminal bronchiole. Terminal bronchioles subdivide into microscopic branches called *respiratory bronchioles*. In the respiratory bronchioles, the epithelial lining changes from cuboidal to squamous as they become more distal. In addition, respiratory bronchioles contain some alveoli which appear as cup-shaped outpouchings of their walls. Respiratory bronchioles, in turn, subdivide into several (2 to 11) *alveolar ducts*. Around the circumference of the alveolar ducts are numerous alveoli and alveolar sacs. An *alveolus* is a cup-shaped outpouching lined by squamous epithelium and supported by a thin elastic membrane. An *alveolar sac* or air sac is a collection or cluster of alveoli that share a common opening. Over the alveoli, the arteriole and venule disperse into a network of capillaries. The exchange of gases between the lungs and blood takes place by diffusion across the alveoli and the walls of the capillaries. It has been estimated that each lung contains 150 million alveoli—providing an immense surface area for the exchange of gases.

Pleural Membranes

Each lung is enclosed by a double layer of serous membrane called the **pleura** (Figure 23-8). It essentially is one continuous, moist, slick-surfaced membrane that folds back upon itself in the region of the root of the lung. Each layer completely encloses its lung. The inner or *visceral pleura* adheres firmly to

the surface of the lung tissue and is inserted into the fissures that separate the lobes. The outer or *parietal layer* of the serous membrane is anchored firmly so that it lines the interior of each cavity. It lines the rib cage, diaphragm, and organs of the central mediastinal space.

The two parts of the pleura are held together by a thin film of serous fluid—much in the manner of two flat, moistened pieces of glass. Thus, as the chest expands, the membranes hold the lung tissue to the walls and diaphragm, causing the lung to stretch. At the same time, the slippery film holding the membranes together permits the parts to slip and slide laterally as dimensions change. The small volume occupied by the thin film is a potential space, the *pleural cavity*, that can be demonstrated if air is introduced between the membranes. When this occurs, as in a puncture wound, the elastic recoil of the lung tissue is sufficient to cause the membranes to separate so that the lung collapses.

A negative or subatmospheric pressure exists between the parietal and visceral pleura, and this also serves to draw and hold the membranes together. When the chest increases in size during inhalation, the pressure in the intrapleural space becomes even more negative. At the moment of rest, just before inspiration begins, the intrapleural pressure is about −5 cm H_2O. When inspiration reaches its peak during a quiet breathing maneuver, the intrapleural pressure reaches a value between −8 and −9 cm H_2O.

This subatmospheric pressure is maintained by the constant activity of the pleural membranes. Any gas or fluid that appears in the intrapleural space is quickly reabsorbed and thus removed.

A negative intrapleural pressure is also produced by the force created by the reabsorptive action of the pulmonary capillaries. The hydrostatic pressure in the pulmonary capillaries is only about 7 mm Hg. This is considerably less than the average found in systemic capillaries, which is about 20 mm Hg (see Chapter 22, Microcirculation). This lower hydrostatic pressure in the pulmonary capillaries permits a much greater proportion of the reabsorptive osmotic pressure of the pulmonary capillaries to be realized. As a consequence, a negative intrapleural fluid pressure is created.

MECHANICS OF BREATHING

The delivery of air from the environment into the alveoli of the lungs and its return to the environment is a task accomplished with the aid of muscles. The

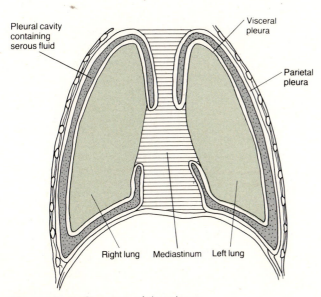

Figure 23-8 Structure of the pleura.

muscles used in breathing include the diaphragm and the intercostal muscles.

Diaphragm

The **diaphragm** forms the floor of the thoracic cavity, completely sealing this compartment from the abdominal cavity below. This thin but strong sheet of muscle is attached circumferentially to fixed structures, such as the vertebrae and sternum, and to movable structures, such as the lower portions of the rib cage. In its relaxed position, the diaphragm is cone-shaped, following the general contour of the base of the rib cage. When the diaphragm contracts, the domed portion lowers and the muscle flattens out across the bottom of the thoracic cavity. The result of the contraction is an increase in the vertical length of the thorax. As with any muscle, the strength of contraction and thus the distance of descent are related to the number of motor units that are activated. The diaphragm is innervated by the *phrenic nerves:* nerves that arise from the cervical plexuses (C1–C4) and travel to the diaphragm. As the diaphragm contracts and flattens, the viscera in the abdomen are displaced to accommodate the lowering of the diaphragm. Therefore, there is a concurrent relaxation of the abdominal muscles during inhalation. Advanced pregnancy, excessive obesity, or confining abdominal clothing can interfere with the abdominal relaxation and prevent a complete descent of the diaphragm.

In most people, the diaphragm probably contributes more to the inspiratory (inhalation) movement than the enlargement of the chest by the intercostal muscles.

Intercostal Muscles

The **intercostal muscles** arise from one rib and insert upon the next rib. They are angled so that the *external intercostals* slope downward and forward and the *internal intercostals* slope obliquely upward and backward.

When the external intercostals receive stimuli via the intercostal nerves, they contract and pull the ribs upward and outward so that the thorax increases in dimensions both laterally and anteroposteriorly. The movement of the ribs can be illustrated in an exaggerated way by standing upright with the fingers of both hands interlocked and the arms hanging down against the abdomen. Raising the arms slowly while slightly flexing the elbows until the arms are horizontal illustrates the rib movement during inhalation. Slowly reversing the movements illustrates the rib movements during exhalation.

The internal intercostal muscles are employed dur-

ing a forced exhalation or beyond a normal quiet breathing pattern. They cause the ribs to be depressed further so as to decrease the thoracic volume. It is this set of muscles that tense to keep the ribs from lifting and extending during straining efforts.

Dynamics of Air Flow

The air in the alveoli is continuous with the atmosphere, and therefore its pressure equals that of the atmosphere. However, in order for air to move through the respiratory passages, a pressure gradient (difference) must develop. During inspiration, this is achieved by increasing the volume of the lungs.

The pressure exerted by a gas in a closed container is inversely proportional to the volume of the container. That is, when the size of the container is increased, the pressure of the gas inside decreases, and conversely, when the size of the container is decreased, the pressure inside increases (see Figure 3-11).

The piston arrangement of a simple bicycle pump also illustrates this concept. The alveolar pressure fluctuation normally is not very great; it alternates between -1 cm H_2O and $+1$ cm H_2O (Figure 23-9). Yet this is a sufficient gradient to provide for the normal movement of air into and out of the bronchial tree.

In cases where there may be an obstruction in the air passages, the pressures can build to very high values. During an extreme cough maneuver, pressure in the alveoli can reach a level of 100 mm Hg.

There is an analogous location of the "resistance" structures in both the cardiovascular and respiratory systems. Recall that in the cardiovascular system, it is the arterioles, the intermediate-sized blood vessels, that are the pressure regulating vessels. Similarly, in the respiratory system, it is the intermediate-sized bronchi that contribute most of the resistance to air

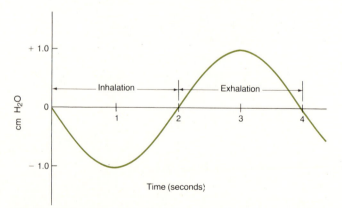

Figure 23-9 Alveolar pressure. Alveolar pressure normally oscillates between -1 cm H_2O and $+1$ cm H_2O.

flow. The thousands of very small bronchioles—those less than 2 mm in diameter—are responsible for less than 20 percent of the total resistance. The smooth muscle of the bronchi also contracts reflexively to narrow the air passages. Irritants in the inhaled air (cigarette smoke and other noxious fumes) stimulate receptors in the large bronchi which trigger the smooth muscle contraction. The sympathetic nerves of the autonomic nervous system, when stimulated, cause bronchodilation, whereas the parasympathetic nerves cause bronchoconstriction. The effect of histamine is well known to allergy sufferers. This chemical, released from surface cells during antigen-antibody reactions, causes the smooth muscles in the walls of the alveolar ducts to constrict. Great distress results because the alveoli are not ventilated and movement of air requires great effort.

Surface Tension

The term **compliance** is used to describe the ease with which a tissue stretches. In the lungs, elastic and collagen fibers are present in the walls of the small blood vessels, bronchi, and alveoli. The elastic fibers permit an "easy" stretch or yield to the point where the collagen fibers, which are tougher and do not yield, restrain the further stretch. The recoil of the lung tissue is due to an actual shortening of the stretched elastic fibers and a geometrical rearrangement of all the fibers.

A second very important component affecting the compliance of the lung tissue is the **surface tension** of a special fluid secreted onto the inner surface of the alveoli. Surface tension is a force generated by the molecules on the *surface* of a liquid. The molecules on the surface are exposed to the gases in the air, but rather than react with them, they react more strongly with each other so that the intermolecular tightness increases. This causes the surface to produce the smallest exposed area possible. Hence, one can almost envision a taut "skin" stretched over the surface of liquids. Because water has a high surface tension, raindrops are nearly spherical. This shape represents the smallest surface area possible for any given volume of matter. In still water, the tight skin across the surface makes it possible for certain species of insects, such as the water strider, to walk on water.

A bubble assumes a spherical shape because it also tries to obtain the smallest surface area. A bubble, however, is hollow, and the surface molecules often generate sufficient force to collapse the bubble. The force of the surface tension is measured in *dynes* and is considered to exist between a *line* of molecules 1 cm in length. The units of surface tension are thus in dynes per centimeter. Consider again the hollow

bubble. The surface tension, in pulling the skin tightly around the enclosed gas, produces a pressure on the gas. The pressure generated, P, is equal to four times the surface tension (T) divided by the radius (r).

$$P = \frac{4T}{r}$$

If the liquid surface under examination has only one surface exposed, as is the case in the alveolus, the formula is modified to:

$$P = \frac{2T}{r}$$

This relationship has important application to lung pathologies. Examine Figure 23-10. In part c of this figure note that the smaller bubble, because it has a smaller r, produces a greater pressure on the enclosed gas than the larger bubble. This means that the smaller bubble can collapse by forcing its gas into the larger bubble, where the pressure is less. As a consequence, the larger bubble gets even bigger and may burst.

There are cells in the alveolar walls that secrete *surfactant*, a material that greatly *lowers* the surface tension of the fluid moistening the inside of the alveoli. Although the exact composition of this secretion is not known, one important component, *dipalmitoyl lecithin*, has been identified. This surfactant can be produced quickly by the cells to replenish that which is continually lost to the air entering the alveoli.

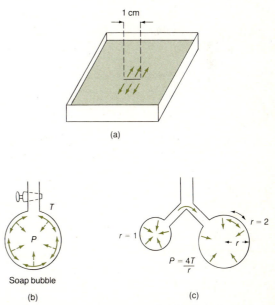

Figure 23-10 Surface tension. (a) Surface tension is the force in dynes acting across an imaginary line 1 cm long on a liquid surface. (b) Surface forces in a soap bubble tend to reduce the surface area and generate a pressure within the bubble. (c) Because the smaller bubble generates a larger pressure, it blows up the large bubble.

The importance of the surfactant is threefold. First, because it decreases the surface tension within the effective alveolar wall, it increases compliance. This means the alveoli can be expanded with much less work than would be the case without surfactant. Second, the decreased surface tension reduces the pressure on the enclosed gases and thus reduces the tendency for the smaller alveoli to blow up the larger ones. In some pathologies, the loss of surfactant in one area of the lungs does cause smaller alveoli to collapse and blow up larger alveoli until they burst. Third, the surfactant prevents water from accumulating in the alveoli. Consider the lack of surfactant. The surface tension would be higher, and one effect of this elevated surface tension is to pull water from the lung capillaries into the alveoli. The presence of surfactant lowers the surface tension and thus counters the leakage of fluid into the alveoli.

Air Volumes Exchanged in Respiration: Spirometry

In clinical practice, the word *respiration* is used to mean one inspiration plus one expiration. The average healthy adult has 14 to 18 respirations a minute (that is, the individual inspires 14 to 18 times and expires 14 to 18 times). During each respiration, the lungs exchange given volumes of air with the atmosphere. A lower than normal exchange volume is usually a sign of pulmonary malfunction. The apparatus commonly used to measure the amount of air exchanged during breathing is referred to as a *respirometer* or *spirometer* (Figure 23-11).

A respirometer consists of a drum inverted over a chamber of water. The drum is counterbalanced by a weight. The drum contains a mixture of gases, usually oxygen or air. A tube connects the air-filled chamber with the subject's mouth. During inspiration, air is removed from the chamber, the drum sinks, and an upward deflection is recorded by the stylus on the graph paper on the kymograph (rotating drum). During expiration, air is added to the chamber, the drum rises, and a downward deflection is recorded on the kymograph. Such a record of respirations is called a *spirogram* (Figure 23-12). Spirometric studies provide records of lung capacities and rates and depths of ventilation. Spirometry is usually indicated for individuals who exhibit labored breathing. It is also used in the diagnosis of respiratory disorders such as emphysema and bronchial asthma. These disorders will be described later.

During normal quiet breathing, about 500 ml of air moves into the respiratory passageways with each inspiration, and the same amount moves out with each expiration. This volume of air inspired (or expired) is called *tidal volume*. Actually, only about 350 ml of the tidal volume reaches the alveoli. The other 150 ml remains in the dead spaces of the nose, pharynx, larynx, trachea, and bronchi and is known as *dead air*.

By taking a very deep breath, we can inspire a good deal more than 500 ml. This excess inhaled air, called the *inspiratory reserve volume,* averages 3,100 ml above the 500 ml of tidal volume. Thus, the respiratory system can pull in as much as 3,600 ml of air. If we inhale normally and then exhale as forcibly as possible, we should be able to push out 1,200 ml of air in addition to the 500 ml tidal volume. This extra 1,200 ml is called the *expiratory reserve volume.* Even after the expiratory reserve volume is expelled, a good deal of air still remains in the lungs because the lower intrathoracic pressure keeps the alveoli slightly inflated. This air, the *residual volume,* amounts to about 1,200 ml. Opening the thoracic cavity allows the intrathoracic pressure to equal the atmospheric pressure, forcing out the residual volume. The air still remaining is called the *minimal volume.*

The presence of minimal volume can be demonstrated by placing a piece of lung in water and seeing that it floats. Minimal volume provides a medical and legal tool for determining whether a baby was born dead or died after birth. Fetal lungs contain no air. If a baby is born dead, no minimal volume will be observed, but if the child died after taking its first breath, a minimal volume will be detected.

Lung capacity can be calculated by combining various lung volumes. *Inspiratory capacity,* the total inspiratory ability of the lungs, is the sum of tidal volume plus inspiratory reserve volume (3,600 ml). *Functional residual capacity* is the sum of residual volume plus expiratory reserve volume (2,400 ml). *Vital capacity* is the sum of inspiratory reserve volume, tidal volume, and expiratory reserve volume (4,800 ml). Finally, *total lung capacity* is the sum of all volumes (6,000 ml).

Minute Volume of Respiration

The total air taken in during 1 minute is called the **minute volume of respiration.** It is calculated by multiplying the tidal volume by the normal breathing rate per minute:

$$\underset{\text{of respiration}}{\text{Minute volume}} = \underset{\substack{\text{(tidal}\\\text{volume)}}}{500} \times \underset{\substack{\text{(average}\\\text{rate per}\\\text{minute)}}}{16} = 8{,}000 \text{ ml/minute}$$

The measurement of respiratory volumes and capacities is an essential tool for determining how well the lungs are functioning. For instance, during the early stages of emphysema, many of the alveoli lose

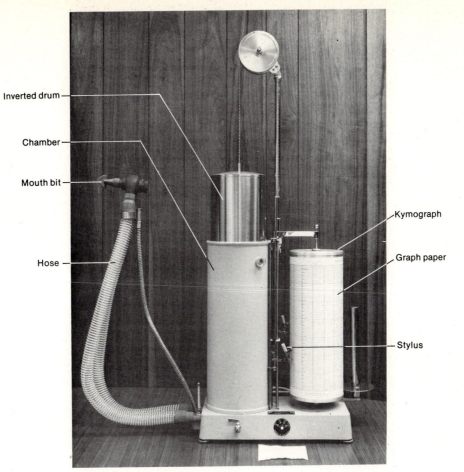

Inverted drum

Chamber

Mouth bit

Hose

Kymograph

Graph paper

Stylus

(a)

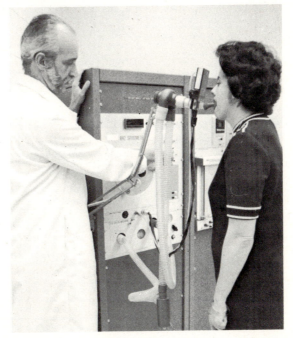

(b)

Figure 23-11 Respirometers. (a) Collins respirometer. This type of respirometer is the one commonly used in college biology laboratories. (Courtesy of Charles I. Foster, Vice President, Warren E. Collins, Inc., Braintree, Mass.) (b) Ohio 842 respirometer. This instrument is a highly sophisticated respirometer that utilizes a computerized mechanism for recording results. (Courtesy of Lenny Patti.)

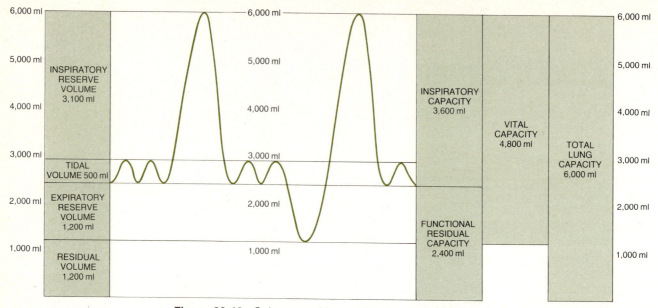

Figure 23-12 Spirogram of lung volumes and capacities.

their elasticity. During expiration, they fail to snap inward, and consequently they fail to force out a normal amount of air. Thus, the residual volume is increased at the expense of the expiratory reserve volume. Pulmonary infections can cause inflammation and an accumulation of fluid in the air spaces of the lungs. The fluid reduces the amount of space available for air and consequently decreases the vital capacity.

CONTROL OF BREATHING

The muscles of respiration are not inherently active. Like all skeletal muscles, they require nervous stimulation to contract. The mechanisms that regulate the oscillatory "in-and-out" breathing movements that are essential to homeostasis are located within the central nervous system.

Location of the Respiratory Centers

The search for areas within the central nervous system that regulate breathing patterns has gone on for almost 200 years. **Respiratory centers** are groups of neurons and synapses that are able to receive, evaluate, and emit signals to the respiratory muscles. The signals sent out can be very simple or extremely complex, since they are designed to meet the varying and sometimes demanding needs of the body. These centers have been localized in the medulla oblongata and pons by the use of three basic techniques. The first involves removing portions of the brain while simultaneously watching for changes in the breathing pattern. A second and much more precise technique involves electrically stimulating very small, discrete areas of the brain stem and noting any resultant respiratory changes. In the third approach, attempts are made to record from groups of nerve cells any spontaneous activity that correlates with breathing movements. The end result of much research using all three methods is the discovery of three major centers that control breathing, the *medullary respiratory center,* the *apneustic center* situated in the middle and lower pons, and the *pneumotaxic center* located in the upper third of the pons.

The Medullary Respiratory Center

The medullary respiratory center consists of a collection of nerve cells in the brain stem that issues the signals that travel to the diaphragm and intercostal muscles. The center consists of two divisions—one inspiratory and the other expiratory. Both divisions are represented in a *bilateral* pattern with neurons on either side of a midline. The cells that fire during inspiratory movements are located deep in the floor of the fourth ventricle in the posterior one-third. If these cells are electrically stimulated, an inspiratory effort results. Cells located more posterior to the inspiratory region fire during expiratory efforts and constitute the expiratory center. If these cells are stimulated, an exhalation results. There are synaptic connections between the inspiratory and expiratory neurons on the same side of the medulla as well as connections between the same types of neurons on the opposite side. This means that, if the inspiratory neurons on the right side are stimulated, the inspiratory neurons on the left side are also activated. Simultaneously, the connections to the expiratory centers convey impulses that inhibit the expiratory neurons. In this manner, the two divisions are reciprocally activated and inhibited and thus produce the oscillating breathing pattern—"in-and-out."

At present, the most likely explanation for the alternating activity of these centers is proposed by Pitts and his co-workers. The inspiratory neurons are inherently active—much like the P cells of the S-A node. When these cells fire, they send a train of action potentials (1) down the intercostal nerves to the intercostal muscles and (2) down the phrenic nerve to the diaphragm. This initiates the inhalation process as the ribs are lifted and the diaphragm is lowered. A *third* pathway of impulses is delivered to the pneumotaxic center in the upper pons. The arrival of the train of impulses at this center activates the pneumotaxic neurons, which respond by sending impulses to the expiratory neurons in the medulla. The expiratory neurons, when activated, *inhibit* the inspiratory neurons. The inspiratory neurons, having been inhibited, cease sending impulses to the intercostal muscles and diaphragm, and thus, the *passive* act of exhalation occurs. Also, as a consequence of inhibiting the inspiratory center, the volley of impulses traveling to the pneumotaxic center stops. After the last impulse has reached the pneumotaxic center, the center loses its source of stimulation and stops firing impulses to the expiratory neurons in the medulla. These cells, therefore, cease their inhibitory activity on the inspiratory neurons, which are inherently active, so the latter resume their firing to the intercostals, diaphragm, and pneumotaxic center. The time for the "delay circuit" via the pneumotaxic center determines the length of time for the inhalation maneuver. Generally, during quiet breathing, the inspiratory effort lasts 2 seconds while expiration takes about 3 seconds.

Other investigators postulate that the entire reciprocating "on-off" firing of the inspiratory and expiratory neurons is contained within the medullary centers themselves. There may be some "delay circuit" entirely within this area that inhibits the inherently active inspiratory neurons at regular intervals.

The Pontine Respiratory Centers

While these centers are known to exist and can be made to operate in experimental situations, their normal functional significance is still not completely understood. It is known that the pneumotaxic center in the upper pons normally has a strong inhibitory influence on the apneustic center in the lower pons. The effect is to suppress or shut off the activity of the apneustic neurons. If released, the apneustic neurons drive or further prolong the normal firing period of the inspiratory neurons. Thus, the end result is a prolonged or gasping type of inspiratory maneuver—often called *apneustic breathing*.

Afferent impulses from vagal fibers originating from stretch receptors in the lung also inhibit the neurons in the apneustic center. This effect is described next.

The Inflation Reflex (Hering-Breuer Reflex)

Located in the walls of the bronchi and bronchioles are sensory receptors connected to afferent vagal fibers. As the lungs inflate, the frequency of impulses in the nerve fibers increases while more and more receptors are also recruited into the firing process. Some of these impulses reach the pons and act directly on the cells of the apneustic center to inhibit its driving action on the inspiratory neurons. Other vagal afferents act directly on the inspiratory neurons to inhibit their activity. The result is that inflation of the lungs slows and ends. Deflation then occurs passively during the time the inspiratory neurons are inhibited. As the deflation progresses, the degree of stretch experienced by the receptors is reduced and their rate of firing decreases. Thus, the vagal afferents release the inspiratory neurons so they again initiate an inspiratory effort.

The Hering-Breuer inflation reflex regulates the extent of lung inflation so that a minimal amount of work achieves the greatest amount of alveolar ventilation. Various physiologists have discovered that, if a person breathes more deeply (that is, increases tidal air), the work done increases in order to overcome the increasing elastic resistance of the lung. Therefore, the smaller the tidal air volume, the less work required. However, moving sufficient air into and out of the lung every minute requires an increase in the number of breaths every minute. This process also requires muscular work, and the work required to move the air increases with the rate of breathing. It turns out that the least amount of work is a compromise between tidal volume and respiratory rate. A tidal volume of about 400 ml and a respiratory rate of about 15 times per minute is the least expensive energetically. Therefore, it appears that, by cutting off the inspiratory inflation of the lungs at the right level (about 400 ml), the Hering-Breuer reflex achieves the most efficient rate of breathing.

Exchange of Respiratory Gases

As soon as the lungs fill with air, oxygen moves from the alveoli to the blood, through the interstitial fluid, and finally into the cells. Carbon dioxide moves in the opposite direction—from the cells, through interstitial fluid to the blood, and to the alveoli. You can gain a better understanding of how respiratory gases are exchanged in the body by reviewing the gas laws already discussed in Chapter 3.

External Respiration

External respiration is the exchange of oxygen and carbon dioxide between the alveolar sacs and the blood (Figure 23-13a). During inspiration, atmospheric air containing oxygen is brought into the alveolar sacs. Meanwhile, venous blood, which is low in oxygen and high in carbon dioxide, is pumped through the pulmonary artery into the capillaries overlying the

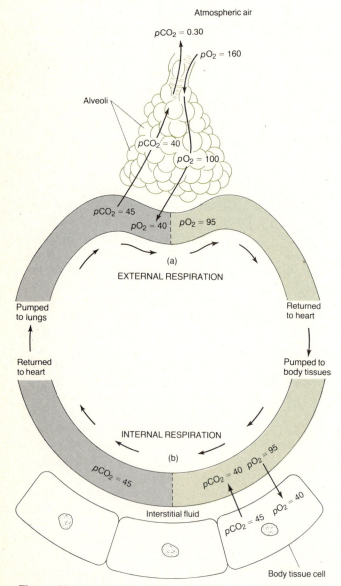

Atmospheric air
$pCO_2 = 0.30$
$pO_2 = 160$
Alveoli
$pCO_2 = 40$
$pO_2 = 100$
$pCO_2 = 45$
$pO_2 = 40$ $pO_2 = 95$

(a)
EXTERNAL RESPIRATION

Pumped to lungs

Returned to heart

Returned to heart

Pumped to body tissues

INTERNAL RESPIRATION

(b)

$pCO_2 = 45$

$pCO_2 = 40$ $pO_2 = 95$

$pCO_2 = 40$ $pO_2 = 40$

Interstitial fluid

$pCO_2 = 45$ $pO_2 = 40$

Body tissue cell

Figure 23-13 Partial pressures involved in respiration. (a) External. (b) Internal. All pressures are in millimeters of mercury (mm Hg).

alveoli. Exhibit 23-1 shows that the oxygen in the alveolar air has a partial pressure of 100 mm Hg, and the oxygen in venous (deoxygenated) blood has a partial pressure of only 40 mm Hg. Oxygen moves down its partial pressure gradient from the alveolar sacs to the blood until the blood's pO_2 reaches 100 mm Hg, the pO_2 of arterial blood. While the blood is being oxygenated, carbon dioxide is also moving down its partial pressure gradient. Thus, carbon dioxide diffuses from the venous blood, where its partial pressure is 45 mm Hg, to the alveolar sacs, where its partial pressure is 40 mm Hg.

External respiration is aided by several anatomic adaptations. The total thickness of the alveolar-capillary membranes is only 0.004 mm. Thicker membranes would inhibit diffusion. The blood and air are also given maximum surface exposure to each other. The total surface area of the alveoli is about 540 ft², many more times the total surface area of the skin. Lying over the alveoli are countless capillaries—so many, that up to 900 ml of blood can participate in gas exchange at peak activity. Finally, the capillaries are so narrow that the red blood cells must flow through them in single file. This feature gives each red blood cell maximum exposure to the available oxygen.

The efficiency of external respiration depends on several factors. One of the most important is the alveolar pO_2 relative to the blood pO_2. As long as alveolar pO_2 is higher than venous blood pO_2, oxygen diffuses from the alveolar sacs into the blood. However, as a person ascends in altitude, the atmospheric pO_2 decreases. The alveolar pO_2 correspondingly decreases, and less oxygen diffuses into the blood. The common symptoms of altitude sickness—shortness of breath, nausea, and dizziness—are attributable to the lower concentrations of oxygen in the blood. Another factor that affects external respiration is the total surface area available for O_2-CO_2 exchange. Any pulmonary disorder that decreases the functional surface area formed by the alveolar-capillary membranes decreases the efficiency of external respiration. A third factor that influences external respiration is the minute volume of respiration (tidal volume times rate of respiration per minute). Certain drugs, such as morphine, slow the respiration rate, thereby decreasing the amounts of oxygen and carbon dioxide that can be exchanged between the alveoli and the blood.

Internal Respiration

As soon as the task of external respiration is completed, the blood moves through the pulmonary veins to the heart, where it is pumped out to the body tissues.

Exhibit 23-1 PARTIAL PRESSURES OF OXYGEN
AND CARBON DIOXIDE*

	Atmospheric Air	Alveolar Air	Venous Blood	Arterial Blood	Body Tissues
pO_2	160	100	40	95	40
pCO_2	0.30	40	45	40	45

* All pressures (mm Hg) are approximate under normal conditions.

In the capillaries of the body tissues a second exchange, called **internal respiration,** takes place. This is the exchange of O_2 and CO_2 between the blood and the body tissues (Figure 23-13b). Exhibit 23-1 shows that the pO_2 of body tissues (40 mm Hg) is much lower than the pO_2 of the arterial blood (95 mm Hg). On the other hand, the pCO_2 of body tissues (45 mm Hg) is higher than the pCO_2 of the arterial blood (40 mm Hg). Therefore, the two gases move down their concentration gradients. Oxygen diffuses from the blood through the interstitial fluid into the body tissues, and carbon dioxide diffuses from the body tissues through the interstitial fluid into the blood until the blood attains partial pressures that are typical of venous blood. The blood now returns to the lungs before it can exchange more gas with body tissues.

Transport of Respiratory Gases

When oxygen and carbon dioxide enter the blood, certain physical and chemical changes occur to aid the gaseous exchange involved in maintaining homeostasis.

Oxygen

When oxygen enters the blood, it initially dissolves in the plasma (Figure 23-14). When 0.5 ml of oxygen has dissolved in 100 ml of blood, the pO_2 in the blood approximately equals the pO_2 inside the alveolar sac. Because the partial pressures are equal, oxygen normally could not continue to move into the blood. However, the cells of the body need much more oxygen than this to survive. In fact, they need 20 ml of oxygen/100 ml of blood. To get around this problem, most of the oxygen quickly leaves the plasma and combines with the hemoglobin of red blood cells. Oxygen that has become attached to hemoglobin is no longer a free gas. It cannot behave as a gas and consequently cannot affect partial pressure. In this way, the pO_2 of the blood is lowered again, and more oxygen from the alveolar sacs can diffuse into the plasma. Most of these molecules are captured by hemoglobin until all the hemoglobin molecules are bound to oxygen. At this point, the pO_2 of the alveolar sacs and the plasma is equalized, and the blood leaves the lungs.

The chemical union between oxygen and hemoglobin is symbolized as follows:

$$\text{Hb} + \text{O}_2 \underset{\substack{\text{low } pO_2 \\ \text{high H}^+ \\ \text{high temp.}}}{\overset{\substack{\text{high } pO_2 \\ \text{low } pCO_2}}{\rightleftharpoons}} \text{HbO}_2$$

Reduced hemoglobin (uncombined hemoglobin) Oxygen Oxyhemoglobin (combined hemoglobin)

The reduced hemoglobin represents hemoglobin that has not combined yet with oxygen. Oxyhemoglobin is the compound formed by the union of oxygen and hemoglobin. This is a reversible reaction; so when the pO_2 is high, as it is in the lungs, there are more oxygen molecules available to make contact with the hemoglobin, and thus more molecules of oxyhemoglobin can be formed.

When the blood reaches the tissue cells of the body, three factors operate to separate the oxygen from the hemoglobin. The first factor is simply the low pO_2 in the cells. When an oxygen molecule separates from the oxyhemoglobin molecule, it moves into the tissue cells before it has a chance to recombine with the hemoglobin. All the free oxygen gas also moves into the tissue, so no other oxygen molecules are available to take its place.

The second factor is acidity. Oxyhemoglobin quickly splits apart when it is placed in an acid solution. Acid comes from the large quantity of carbon dioxide moving from tissue cells through the interstitial fluid into the blood. As the carbon dioxide is taken up by the blood, much of it is temporarily converted into carbonic acid. Thus, a high pCO_2 encourages the oxyhemoglobin to release its oxygen. Lactic acid, the reaction product of contracting muscles, also increases the blood acidity. Thus, active muscles are able to obtain oxygen more quickly than resting ones.

The third factor that encourages the blood to give up its oxygen is an increase in temperature. Heat energy is a by-product of the metabolic reactions of all cells, and contracting muscle cells release an especially large amount of heat. Splitting of the oxyhemoglobin molecule is another example of how homeostatic mechanisms adjust body activities to cellular needs. Active cells require more oxygen, and active cells liberate more acid and heat. The acid and

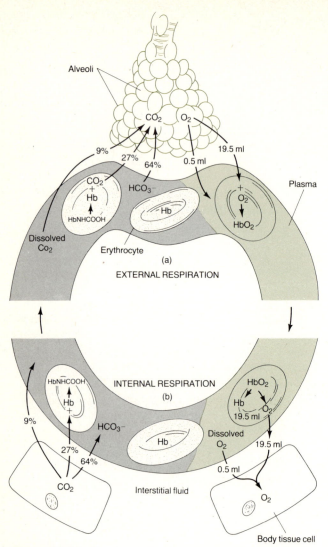

Figure 23-14 Carriage of respiratory gases in respiration. (a) External. (b) Internal.

heat, in turn, stimulate the oxyhemoglobin to release its oxygen.

Carbon Dioxide

Venous blood contains about 56 ml of carbon dioxide/100 ml. Like oxygen, carbon dioxide is carried by the blood in several forms (Figure 23-14). The smallest percentage, about 9 percent, is dissolved in plasma. A somewhat higher percentage, about 27 percent, combines with the protein part of the hemoglobin molecule to form carbaminohemoglobin. This reaction may be represented as follows:

$$Hb \quad + \quad CO_2 \quad \rightleftharpoons \quad HbNHCOOH$$

Hemoglobin Carbon dioxide Carbaminohemoglobin

However, the greatest percentage of carbon dioxide, about 64 percent, is converted into the bicarbonate ion (HCO_3^-) in the following way:

$$CO_2 + H_2O \rightleftharpoons H_2CO_3 \rightleftharpoons H^+ + HCO_3^-$$

Carbon   Water   Carbonic   Hydrogen   Bicarbonate
dioxide   acid   ion   ion

As the carbon dioxide diffuses from tissues into blood plasma and then into red blood cells, an enzyme called carbonic anhydrase stimulates the major portion of the gas to combine with water to form *carbonic acid*. Inside red blood cells, carbonic acid dissociates into H^+ and HCO_3^- ions. The H^+ ions provide the acid stimulus for the release of oxygen from oxyhemoglobin. Some of the HCO_3^- ions remain within the cell and combine with potassium, the chief positive ion of intracellular fluid, to form potassium bicarbonate ($KHCO_3$). The majority of the bicarbonate ions, however, diffuse into the plasma and combine with sodium, the principal positive ion of extracellular fluid, to form sodium bicarbonate ($NaHCO_3$).

When the blood reaches the lungs, the above events reverse. The high pO_2 causes oxygen to replace the carbon dioxide on the hemoglobin molecule. Bicarbonate breaks apart and releases CO_2. Finally, the CO_2 that has been traveling dissolved in the plasma and the CO_2 that has been released from the reversed reactions diffuse down the partial pressure gradient and into the alveolar sacs.

Mechanisms that Modify Rhythmicity

The oxygen consumption and carbon dioxide production of the body fluctuate widely, but the respiratory system is able to respond in accordance with these changes so that demands are always adequately met. This is another excellent example of homeostasis. In addition, many changes in our normal breathing occur during speaking, singing, blowing wind instruments, sniffing, coughing, sneezing, sighing, yawning, swallowing, hiccuping, snoring, and vomiting. These changes are unrelated to the need for alveolar ventilation but serve to illustrate that other nervous activities can alter a normal breathing pattern. Various emotional states (fear, anxiety, surprise, etc.) also alter breathing patterns.

Response to Changes in Arterial pO_2

It might be reasonable to think that ventilation (breathing) would increase when the body cells needed more oxygen and formed more carbon dioxide. Measurements by many investigators assure us that ventilation increases under these conditions and in proportion to the need. This argues strongly for the existence of some sensitive devices strategically located to monitor these changes in the body fluids. Indeed, it might also be reasonable to assume that oxygen, with its life-sustaining capability, would be the variable most closely monitored. Wrong! The body's ability to measure O_2 lack is relatively poor. Figure 23-15 shows that the respiratory minute volume changes very little even when the percent of

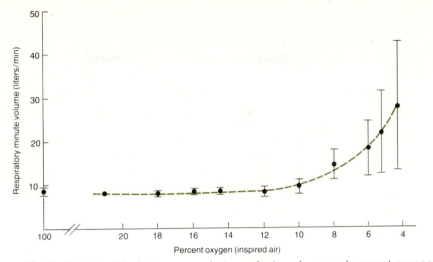

Figure 23-15 Relationship between respiratory minute volume and percent oxygen in inspired air. Cross lines represent one standard deviation from the mean.

oxygen in the air breathed drops from a normal 20 percent to 10 percent. When the percent of oxygen in the air drops below this level to 8, 6, or 4 percent we note a quick rise in ventilation. This type of result suggests that there are O_2 measuring devices, which are not very sensitive to small drops in O_2 concentration. They only respond when the level becomes life threatening.

Such receptor devices have been found. They contain chemosensitive cells. They are located within the carotid bodies near the bifurcation of the common carotid arteries and in the aortic bodies near the arch of the aorta. The chemosensitive cells are able to detect changes in arterial levels of carbon dioxide and hydrogen ions as well as oxygen. The carotid bodies are small, oval nodules about 4 to 5 mm long, located in the space between the internal and external carotid arteries. The aortic bodies are clustered in the region between the arch of the aorta and the dorsal surface of the pulmonary artery. The afferent nerve fibers from the carotid body join with those from the carotid sinus to form the carotid sinus nerve, which joins the glossopharyngeal nerve. The afferent fibers from aortic bodies run into the vagus nerve.

Investigators have recorded the action potentials occurring in the carotid sinus nerve of animals and have shown that the impulse frequency increases very slowly when the arterial pO_2 is lowered from 500 mm Hg to about 100 mm Hg. When the pO_2 is lowered further, the rate of firing increases very quickly. The precise stimulus for the chemoreceptors seems to be an inadequate supply of oxygen to meet their own needs. This can be a result of a very low arterial pO_2 or a normal arterial pO_2 but a very poor blood flow into the bodies.

The focus of our attention is on the respiratory consequences of stimulating the carotid and aortic bodies. However, stimulation of the chemoreceptors produces an array of responses that includes:

1. Increase in depth and frequency of breathing (minute volume increase) (Figure 23-16).
2. Vasoconstriction of blood vessels in the limbs.
3. Tachycardia (increased heart rate and pulse rate) when the *aortic body* is stimulated (Figure 23-17).
4. Bradycardia (decreased heart rate and pulse rate) when the *carotid body* is stimulated (Figure 23-17).
5. Overall increase in systemic arterial blood pressure.

However, the main function of the chemoreceptors is to regulate breathing. Oxygen lack must be relatively severe before the majority of the chemosensitive cells begin firing to increase ventilation (see Figure 23-15). Small decreases in arterial pO_2 are ineffective in triggering an increase in ventilation.

Response to Changes in Arterial pCO$_2$

The response to breathing air containing increased concentrations of carbon dioxide is compared to the effects of oxygen lack and decreasing pH in Figure 23-18. The graph shows that when a healthy man breathes increasing levels of CO_2, the ventilation response increases almost linearly. Shown again is the flat response to breathing air with lowered O_2 content. The O_2 lack does not trigger a respiratory response until it is a life-threatening situation.

It is very clear that the primary effect of CO_2 buildup in the body fluids is *not* on the carotid and aortic bodies. Research has provided much evidence to support the idea that there are *centrally* located chemoreceptors. For a long time it was believed that the medullary respiratory centers were themselves

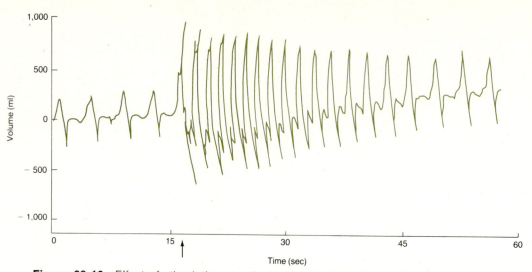

Figure 23-16 Effect of stimulating carotid bodies. At the point of stimulation (arrow), the response shows an increase in tidal volume, frequency of breathing, and thus minute volume of respiration.

exquisitely sensitive to slight changes in CO_2. More recent evidence indicates that the central chemoreceptors are anatomically separated from the medullary respiratory centers. The CO_2-sensitive receptors are located on the ventrolateral surfaces of the upper medulla near the origin of cranial nerves ix and x. When simulated cerebrospinal fluid containing high pCO_2 or (H^+) is applied to these regions, an increase in breathing results.

Since these chemoreceptors are located on the surface of the medulla, they would normally be bathed in cerebrospinal fluid (CSF). The existence of the blood-brain barrier does not permit an elevation of H^+ or HCO_3^- in the blood to reach the chemoreceptor centers. However, CO_2 can diffuse easily from the blood across the blood-brain barrier into the CSF. So it is now thought that when CO_2 from the blood crosses into the CSF, it forms carbonic acid (H_2CO_3), which dissociates to form H^+ and HCO_3^-. The H^+ ions so formed are the direct stimulators of the

chemoreceptors on the surface of the medulla. Thus, the stimulation of the central chemoreceptors when CO_2 rises in arterial blood is due to the resultant elevation of hydrogen ions in the CSF. The stimulation of the central chemoreceptors, in turn, is conveyed to the respiratory centers, which respond by increasing the rate and depth of breathing.

The location of the central chemoreceptors on the medullary surface, where they are bathed in CSF, has a certain logic. In this location, the presence of increased levels of H^+ ions is more readily detected, since the buffering capacity of CSF is considerably less than that of blood. This means that the appearance of even a few additional H^+ ions will be easily detected. Also, this location for the chemoreceptors suggests that the pH of the CSF is of critical concern for proper functioning of the central nervous system.

The control of breathing during daily minute-to-minute activity is primarily regulated by the pCO_2 in arterial blood. In fact, this system is so well designed

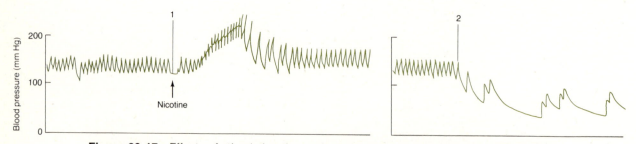

Figure 23-17 Effects of stimulating the aortic and carotid bodies. Recordings show the opposite effects produced by stimulating first the aortic bodies and then, 75 seconds later, the carotid bodies. Nicotine is introduced, as the stimulator, into the aorta. At point 1, the nicotine is injected and it stimulates aortic bodies, producing an increase in carotid artery blood pressure and heart rate. At point 2, the nicotine stimulus reaches the carotid bodies and causes a drop in heart rate and blood pressure.

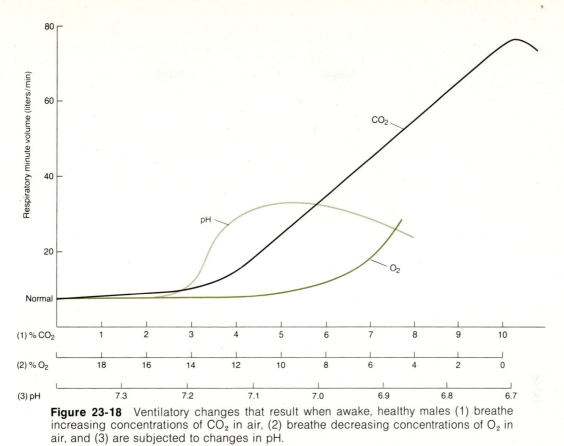

Figure 23-18 Ventilatory changes that result when awake, healthy males (1) breathe increasing concentrations of CO_2 in air, (2) breathe decreasing concentrations of O_2 in air, and (3) are subjected to changes in pH.

that the pCO_2 of arterial blood is normally held to within 3 mm Hg.

The peripheral chemoreceptors in the aortic and carotid bodies can also respond to elevated arterial pCO_2, whereas the central chemoreceptors are not affected by decreased pO_2. The contribution of the peripheral chemoreceptors to elevations in CO_2 during normal situations is probably very little. However, the CO_2-sensitive cells in the peripheral chemoreceptors are apparently responsible for adjusting to very sudden increases in arterial pCO_2. Also, it seems that the CO_2-sensitive cells in both central and peripheral chemoreceptors are more susceptible to drugs such as morphine and barbiturates. When they are depressed, breathing slows greatly, and then as hypoxia increases, the *oxygen-sensitive* cells of the peripheral chemoreceptors operate to drive respiration.

Response to Changes in Arterial pH

When the pH of arterial blood drops, respiration increases. In most situations, it is difficult to distinguish between the effects of a drop in pH and a rise in the pCO_2 of arterial blood. In experimental situations using animals, it has been possible to keep the pCO_2 constant, while the pH was reduced. In such experiments, it has been shown that the increased H^+ concentration (drop in pH) directly stimulates respiration. The mechanism for such stimulation operates via the peripheral chemoreceptors.

Response to Changes in Arterial Blood Pressure

Although the pressoreceptors in the carotid sinus and aortic arch are concerned mainly with the control of circulation, they assume some function in the control of respiration. For example, a sudden rise in blood pressure decreases the rate of respiration, and a drop in blood pressure brings about an increase in the respiratory rate.

Some of the other factors that control respiration are:

1. A sudden cold stimulus, such as plunging into cold water, causes a temporary cessation of breathing, called *apnea*.
2. A sudden, severe pain brings about apnea, but a prolonged pain triggers the general stress syndrome and increases respiration rate.
3. Stretching of the anal sphincter muscle increases the respiratory rate. Medical personnel sometimes employ this technique to stimulate respiration during emergencies.
4. Irritation of the pharynx or larynx by touch or by chemicals brings about an immediate cessation of breathing followed by coughing.

Respiration Control During Exercise

The breathing pattern that occurs during exercise is shown in Figure 23-19. It shows that with the onset of exercise there is an *abrupt* increase in pulmonary ventilation and a similar *abrupt* but *larger drop* at the conclusion of the exercise. In addition, as the metabolic rate of the individual increases during the exercise, the pulmonary ventilation increases in direct proportion (Figure 23-20a).

Attempts to discover the mechanism that controls respiration during exercise have been mostly unsatisfactory. Measurements of exercising healthy men show that as the ventilation increases as much as 15 times, the arterial pCO_2 and pO_2 change very little (Figure 23-20b and c). It has become apparent that other mechanisms must function during times of strenuous exercise.

Some physiologists believe that when the motor cortex begins sending signals to the skeletal muscles, it also sends a proportionate number of signals to the respiratory and vasomotor centers. In this fashion, the brain arranges for the appropriate blood distribution and ventilation to take place in anticipation of the ensuing exercise. While this is an attractive idea, there is no experimental evidence to confirm it.

Other physiologists believe that muscle and joint proprioceptors signal to the respiratory center the increased amount of limb movements. This idea has some validity. Passive movements of limbs, such as performed by a physical therapist on a recumbent patient, cause a reflex increase in the patient's breathing. While such signals do exist, they cannot account for the direct relationship between increased metabolism and ventilation. The muscle and joint receptors relay information that movement has occurred, but they do not convey any information about the strength of the contraction—which is a major factor determining energetic cost.

If the exercise is prolonged, body temperature rises. It is a well-known physiological response that ventilation increases when body temperature rises. It may

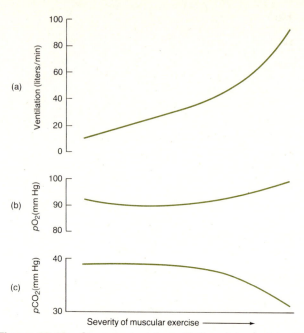

Figure 23-20 Changes in (a) ventilation, (b) arterial pO_2, and (c) arterial pCO_2 as the severity of muscular exercise increases. The severity is measured by metabolic rate—O_2 consumption.

be that during the time of elevated body temperature, some ventilation may be due to the temperature rise.

At this time, it seems that more than one factor contributes to the hyperapnea of muscular exercise. No one satisfactory explanation is available.

MODIFIED RESPIRATORY MOVEMENTS

Respirations provide human beings with methods for expressing emotions, such as laughing, yawning, sighing, and sobbing. Moreover, respiratory air can expel foreign matter from the upper air passages through actions such as sneezing and coughing. Some of the modified respiratory movements that express emotion or clear the air passageways are listed in Exhibit 23-2. These movements are reflexes, but some of them also can be initiated voluntarily.

HEART AND LUNG RESUSCITATION

A serious decrease in respiration or heart rate presents an urgent crisis, because the body's cells cannot survive long if they are starved of oxygenated blood. The disruption of homeostasis would be total. In fact, if oxygen is withheld from the cells of the brain for 4 to 6 minutes, brain damage or death will result. Heart-lung resuscitation is the artificial reestablishment of

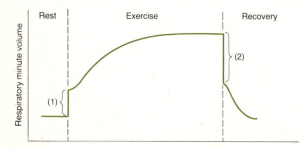

Figure 23-19 Breathing pattern during exercise showing (1) an abrupt increase at the onset and (2) an even larger abrupt decrease at the end.

Exhibit 23-2 MODIFIED RESPIRATORY MOVEMENTS

Movement	Comment
Coughing	Preceded by a long-drawn and deep inspiration that is followed by a complete closure of the glottis—resulting strong expiration suddenly pushes glottis open and sends a blast of air through the upper respiratory passages. Stimulus for this reflex act could be a foreign body lodged in the larynx, trachea, or epiglottis.
Sneezing	Spasmodic contraction of muscles of expiration forcefully expels air through the nose and mouth. Stimulus may be an irritation of the nasal mucosa.
Sighing	A deep and long-drawn inspiration immediately followed by a shorter but forceful expiration.
Yawning	A deep inspiration through the widely opened mouth producing an exaggerated depression of the lower jaw. May be stimulated by drowsiness or fatigue, but precise stimulus-receptor cause is unknown.
Sobbing	Starts with a series of convulsive inspirations. Glottis closes earlier than normal after each inspiration so only a little air enters the lungs with each inspiration. Immediately followed by a single prolonged expiration.
Crying	An inspiration followed by many short convulsive expirations. Glottis remains open during the entire time, and the vocal cords vibrate. Accompanied by characteristic facial expressions.
Laughing	Involves the same basic movements as crying, but the rhythm of the movements and the facial expressions usually differ from those of crying. Laughing and crying are sometimes indistinguishable.
Hiccuping	Spasmodic contraction of the diaphragm followed by spasmodic closure of the glottis. Produces a sharp inspiratory sound. Stimulus is usually irritation of the sensory nerve endings of the digestive tract.

normal or near normal respiration and circulation. The two simplest techniques for heart-lung resuscitation are exhaled air ventilation and external cardiac compression. Both techniques can be administered by a layman at the site of the emergency, and both are highly successful. They can be used for any sort of heart or respiratory failure, whether the cause be drowning, strangulation, carbon monoxide or insecticide poisoning, overdose of a drug or anesthesia, electrocution, or myocardial infarction. However, the success of heart-lung resuscitation is directly related to the speed and efficiency with which it is applied. Delay may be fatal.

Exhaled Air Ventilation

A technique for reestablishing respiration in order to restore homeostasis is **exhaled air ventilation.** The first and most important step is immediate opening of the airway. This is accomplished easily and quickly by tilting the victim's head backward as far as it will go without being forced. The tilted position opens the upper air passageways to their maximum size (Figure 23-21a). If the patient does not resume spontaneous breathing after the head has been tilted backward, artificial ventilation must be given immediately either mouth to mouth or mouth to nose. In the more usual mouth-to-mouth method, the nostrils are pinched together with the thumb and index finger of the hand. The rescuer then opens his mouth widely, takes a deep breath, makes a tight seal with his mouth around the patient's mouth, and blows in about twice the amount the patient normally breathes. He then removes his mouth and allows the patient to exhale passively. This cycle is repeated approximately 12 times per minute for adults. Atmospheric air contains about 21 percent O_2 and a trace of CO_2. Exhaled air still contains about 16 percent O_2 and 5 percent CO_2. This is more than adequate to maintain a victim's blood pO_2 and pCO_2 at normal levels if air is given at the prescribed rate and amount.

If the rescuer observes the following three signs, he knows that adequate ventilation is occurring:

1. The chest rises and falls with every breath.
2. He feels the resistance of the lungs as they expand.
3. He hears the air escape during exhalation.

The three most common errors in mouth-to-mouth resuscitation are *inadequate extension* of the victim's head, *inadequate opening* of the rescuer's mouth, and an *inadequate seal* around the patient's mouth or nose. If the rescuer is sure he has not made these errors and he is still unable to inflate the lungs, a foreign object is probably lodged in the respiratory passages. The rescuer should sweep his fingers through the patient's mouth to remove such material. An adult victim with this problem should next be rolled quickly onto his side. Firm blows should be delivered over his spine between the shoulder blades in an attempt to dislodge the obstruction. Then exhaled air ventilation should be resumed quickly. A child with an obstructive foreign object should be picked up quickly and inverted over the rescuer's forearm while firm blows are delivered over the spine between the shoulder blades. Then ventilation can be resumed.

If the patient is an infant or small child, the rescuer should make the following adjustments in his technique. The neck of an infant is so pliable that forceful

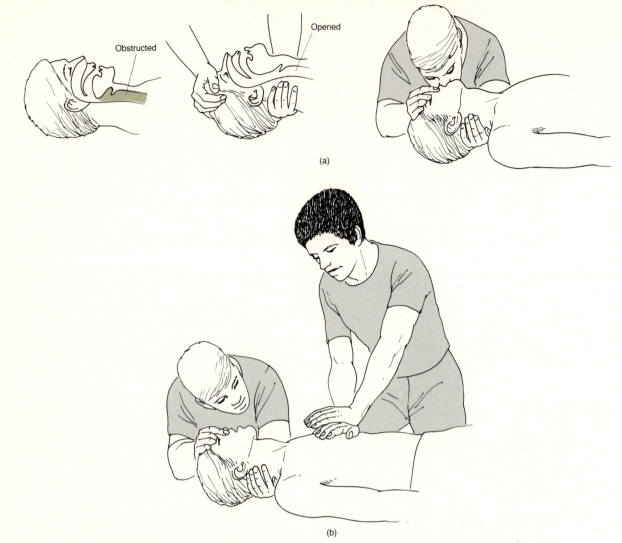

Figure 23-21 Heart-lung resuscitation. (a) Exhaled air ventilation. Shown on the left is the procedure for immediate opening of the airway. Shown on the right is the procedure for mouth-to-mouth respiration. (b) External cardiac compression technique in conjunction with exhaled air ventilation.

backward tilting of the head may obstruct breathing passages, so the tilted position should not be exaggerated. The rescuer should also remember that the lungs of a small child do not have a large capacity. To avoid overinflating the child's lungs, he should cover both the victim's mouth and nose with his mouth and blow gently, using less volume, at a rate of 20 to 30 times a minute.

External Cardiac Compression

External cardiac compression or **closed-chest cardiac compression** (CCCC), consists of the application of rhythmic pressure over the sternum (Figure 23-21b). The rescuer places the heels of his hands on the lower half of the sternum and presses down firmly and smoothly at least 60 times a minute. This action com-

presses the heart and produces an artificial circulation because the heart lies almost in the middle of the chest between the lower portion of the sternum and the spine. When properly done, external cardiac compression can produce systolic blood pressure peaks of over 100 mm Hg. It can also bring carotid arterial blood flow up to 35 percent of normal.

Complications that can occur from the use of cardiac compression include fracture of the ribs and sternum, laceration of the liver, and the formation of fat emboli. They can be minimized by adhering to the following precautions:

1. Never compress over the xiphoid process at the tip of the sternum. This bony prominence extends down over the abdomen, and pressure on it may cause laceration of the liver, which can be fatal.

2. Never let your fingers touch the patient's ribs when you compress. Keep your fingers off the patient, and place the heel of your hand in the middle of the patient's chest over the lower half of his sternum.

3. Never compress the abdomen and chest simultaneously since this action traps the liver and may rupture it.

4. Never use sudden or jerking movements to compress the chest. Compression should be smooth, regular, and uninterrupted, with 50 percent of the cycle compression and 50 percent relaxation.

Compression of the sternum produces some artificial ventilation, but not enough for adequate oxygenation of the blood. Therefore, exhaled air ventilation must always be used with it. This combination constitutes **heart-lung resuscitation,** also called **cardiopulmonary resuscitation** (CPR). When there are two rescuers, the most physiologically sound and practical technique is to have one rescuer apply cardiac compression at a rate of at least 60 compressions per minute. The other rescuer should exhale into the patient's mouth between every fifth and sixth compression. The sequential steps in emergency heart-lung resuscitation must be continued uniformly and without interruption until the patient recovers or is pronounced dead.

Food Choking

Food choking (café coronary), the sixth leading cause of accidental death, is often mistaken for myocardial infarction. However, it can be recognized easily. The victim cannot speak or breathe; he may become panic-stricken and run from the room. He becomes pale, then deeply cyanotic, and collapses. Without intervention, death occurs in 4 to 5 minutes.

The food or other obstructing object causing asphyxiation may lodge in the back of the throat or enter the trachea to occlude the airway. Tracheostomy, even when a physician performs it, can be hazardous in a nonclinical setting. An instrument for removing food from the back of the throat has been developed, but is seldom at hand at the time of emergency. However, there is a first aid procedure that does not require special instruments and can be performed by any informed layman. This procedure is called the **Heimlich maneuver.**

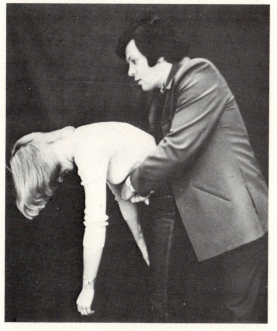
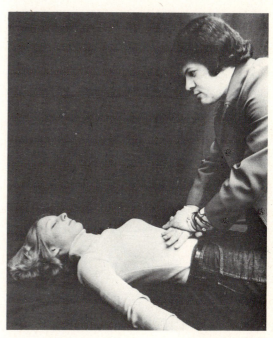

(a)   (b)

Figure 23-22 Heimlich maneuver. The principle employed in this antichoke first aid method is the application of force to compress the air in the lungs to expel objects in the air passageways. (a) If the victim is upright, the rescuer stands behind the person and places both arms around the waist just above the belt line. Then he grasps his right wrist with his left hand and allows the victim's head, arms, and upper torso to fall forward. Next, he compresses the victim's abdomen rapidly and strongly. If another person is present, he or she can assist by removing the ejected object from the victim's mouth. (b) If the victim is recumbent, the rescuer may rapidly and strongly force the victim's diaphragm upward by applying pressure as shown. (Courtesy of Donald Castellaro and Deborah Massimi.)

Food choking occurs during inspiration, and this causes the bolus of food to be sucked against the opening into the larynx. Therefore, at the time of the accident, the lungs are expanded. Even during normal expiration, however, some tidal air (500 ml) and the entire expiratory reserve volume (1,200 ml) are present in the lungs.

Pressing one's fist upward into the epigastrium elevates the diaphragm suddenly—compressing the lungs in the rib cage and increasing the air pressure in the tracheobronchial tree. This increased air pressure is conveyed through the trachea and ejects the food (or object) blocking the airway. The action can be simulated by inserting a cork in a compressible plastic bottle and then squeezing it suddenly. The cork flies out because of the increased pressure. Figure 23-22 shows how to administer the Heimlich maneuver.

SMOKING AND THE RESPIRATORY SYSTEM

As part of ordinary breathing, many irritating substances are inhaled. Almost all pollutants, including inhaled smoke, have an irritating effect on the bronchial tubes and lungs and may be regarded as stresses or irritating stimuli that disrupt homeostasis.

Close examination of the epithelium of a bronchial tube reveals that it consists of three kinds of cells

(Figure 23-23). The uppermost cells are columnar cells that contain the cilia on their surfaces. At intervals between the ciliated columnar cells are the mucus-secreting goblet cells. The bottom of the epithelium normally contains two rows of basal cells above the basement membrane. The bronchial epithelium is important clinically because researchers have learned that one of the most common types of lung cancer, *bronchogenic carcinoma*, starts in the walls of the bronchi.

The stress of constant irritation by inhaled smoke and pollutants causes an enlargement of the goblet cells of the bronchial epithelium. They respond by secreting excessive amounts of mucus. The basal cells respond to stress by undergoing cell division so fast that they push into the area occupied by the goblet and columnar cells. As many as 20 rows of basal cells may be produced. Many researchers believe that if the stress is removed at this point, the epithelium can return to normal.

If the stress persists, more and more mucus is secreted and the cilia become less effective. As a result, mucus is not carried toward the throat; instead, it remains trapped in the bronchial tubes. The individual then develops a "smoker's cough." Moreover, the constant irritation from the pollutant slowly destroys the alveoli, which are replaced with thick, inelastic connective tissue. Mucus that has accumulated becomes trapped in the air sacs. Millions of the sacs

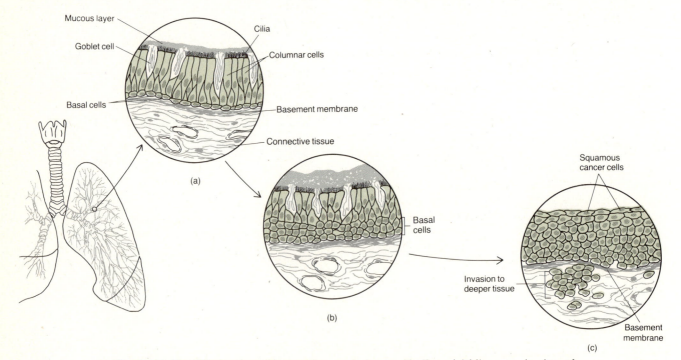

Figure 23-23 Effects of smoking on the respiratory epithelium. (a) Microscopic view of the normal epithelium of a bronchial tube. (b) Initial response of the bronchial epithelium to irritation by pollutants. (c) Advanced response of the bronchial epithelium.

rupture. This results in a loss of diffusion surface for the exchange of oxygen and carbon dioxide. The individual has now developed emphysema. Even if the stress is removed at this point, there is little chance for improvement. Alveolar tissue that has been destroyed cannot be repaired. Removal of the stress only stops further destruction of lung tissue.

If the stress continues, the emphysema gets progressively worse, and the basal cells of the bronchial tubes continue to divide and break through the basement membrane. At this point, the stage is set for bronchogenic carcinoma. Columnar and goblet cells disappear and may be replaced with squamous cancer cells. If this happens, the malignant growth spreads throughout the lung and may block a bronchial tube. If the obstruction occurs in a large bronchial tube, very little oxygen enters the lung, and disease-producing bacteria thrive on the mucoid secretions. In the end, the patient may develop emphysema, carcinoma, and a host of infectious diseases. Treatment involves surgical removal of the diseased lung. However, metastasis of the growth through the lymphatic or blood system may result in new growths in other parts of the body such as the brain and liver.

Other factors may be associated with lung cancer. For instance, breast, stomach, and prostate malignancies can metastasize to the lungs. People who apparently have not been exposed to pollutants do occasionally develop bronchogenic carcinoma. However, the occurrence of bronchogenic carcinoma is probably over 20 times higher in heavy cigarette smokers than it is in nonsmokers.

THERAPY BY NEBULIZATION

Many of the previously mentioned respiratory disorders are treated by means of a comparatively new method called **nebulization.** This procedure is the administering of medication, in the form of droplets that are suspended in air, to selected areas of the respiratory tract. The patient inhales the medication as a fine mist. Droplet size is directly related to the number of droplets suspended in the mist. Smaller droplets (approximately 2 μm in diameter) can be suspended in greater numbers than can large droplets and will reach the alveolar ducts and sacs. The larger droplets (approximately 7 to 16 μm in diameter) will be deposited mostly in the bronchi and bronchioles. Droplets of 40 μm and larger will be deposited in the upper respiratory tract—the mouth, pharynx, trachea, and main bronchi. Nebulization therapy can be used with many different types of drugs, such as chemicals that relax the smooth muscle of the respiratory passage-ways, chemicals that reduce the thickness of mucus, and antibiotics.

HOMEOSTATIC IMBALANCES OF THE RESPIRATORY SYSTEM

Hay Fever

An allergic reaction to the proteins contained in foreign substances such as plant pollens, dust, and certain foods is called **hay fever.** Allergic reactions are a special type of antigen-antibody response that initiate either a localized or a systemic inflammatory response. In hay fever, the response is localized in the respiratory membranes. The membranes become inflamed, and a watery fluid drains from the eyes and nose.

Bronchial Asthma

Bronchial asthma is a reaction, usually allergic, characterized by attacks of wheezing and difficult breathing. Attacks are brought on by spasms of the smooth muscles that lie in the walls of the smaller bronchi and bronchioles, causing the passageways to close partially. The patient has trouble exhaling, and the alveoli may remain somewhat inflated during expiration. Usually the mucous membranes that line the respiratory passageways become irritated and secrete excessive amounts of mucus that may clog the bronchi and bronchioles and worsen the attack. About three out of four asthma victims are allergic to something they eat or to substances they breathe in, such as pollens, animal dander, house dust, or smog. Others are usually sensitive to the proteins of relatively harmless bacteria that inhabit the sinuses, nose, and throat. Asthma might also have a psychosomatic origin.

Emphysema

In **emphysema,** the alveolar walls lose their elasticity and remain filled with air during expiration. The name means "blown up" or "full of air." Reduced forced expiratory volume is the first symptom. Later, alveoli in other areas of the lungs are damaged. The lungs become permanently inflated because they have lost their elasticity. To adjust to the increased lung size, the size of the chest cage increases. The patient has to work hard to exhale. Oxygen diffusion does not occur as easily across the damaged alveoli, blood pO_2 is somewhat lowered, and any mild exercise that raises the oxygen requirements of the body tissues leaves the patient breathless. As the disease progresses, the alveoli degenerate and are replaced with thick, fibrous connective tissue. Even carbon dioxide does not diffuse easily through this fibrous tissue. If the blood cannot buffer all the carbonic acid that accumulates, the blood pH drops, and unusually high amounts of carbon dioxide dissolve in the plasma. High carbon dioxide levels are toxic to the brain cells. Consequently, the inspiratory center becomes less active and the respiration rate slows, further aggravating the problem. The capillaries that lie around the deteriorating alveoli are compressed and damaged and may no longer be able to receive blood. As a result, pressure increases in the pulmonary artery, and the right atrium overworks as it attempts to force blood through the remaining capillaries.

Emphysema is generally caused by a long-term irritation. Air pollution, occupational exposure to industrial dusts, and cigarette smoke are the most common irritants. Chronic bronchial asthma also may produce alveolar damage. Cases of emphysema are becoming more and more frequent in the United States. The irony is that the disease can be prevented and the progressive deterioration can be stopped by eliminating the harmful stimuli.

Pneumonia

The term **pneumonia** means an acute infection or inflammation of the alveoli. In this disease, the alveolar sacs fill with fluid and dead white blood cells, reducing the amount of air space in the lungs. (Remember that one of the cardinal signs of inflammation is edema.) Oxygen has difficulty diffusing through the inflamed alveoli, and the blood pO_2 may be drastically reduced. Blood pCO_2 usually remains normal because carbon dioxide always diffuses through the alveoli more easily than oxygen does. If all the alveoli of a lobe are inflamed, the pneumonia is called *lobar pneumonia*. If only parts of the lobe are involved, it is called *lobular* or *segmental pneumonia*. If both the alveoli and the bronchial tubes are included, it is called *bronchopneumonia*.

The most common cause of pneumonia is the pneumococcus bacterium, but other bacteria or a fungus may be a source of the trouble. Viral pneumonia is caused by any of several viruses, including the influenza virus.

Tuberculosis

The bacterium called *Mycobacterium tuberculosis* produces an inflammation called **tuberculosis.** This disease is still one of the most serious of present-day illnesses and ranks as the number one killer in the communicable disease category. Tuberculosis most often affects the lungs and the pleura. The bacteria destroys parts of the lung tissue, and the tissue is replaced by fibrous connective tissue. Because the connective tissue is inelastic and relatively thick, the affected areas of the lungs do not snap back during expiration, and larger volumes of air are retained. Gases no longer diffuse easily through the fibrous tissue.

Tuberculosis bacteria are airborne and can infect a person when they are inhaled. Although they can withstand exposure to many disinfectants, they die quickly in sunlight. This is why tuberculosis is sometimes associated with crowded, poorly lit housing conditions. Many drugs are successful in treating tuberculosis. Rest, sunlight, and good diet are vital parts of treatment.

Hyaline Membrane Disease (HMD)

Sometimes called glassy-lung disease or infant respiratory distress syndrome (RDS), **hyaline membrane disease (HMD)** is responsible for approximately 20,000 newborn infant deaths per year. Before birth, the respiratory passages are filled with fluid. Part of this fluid is amniotic fluid inhaled during respiratory movements in utero. The remainder is produced by the submucosal glands and the goblet cells of the respiratory epithelium.

At birth, this fluid-filled airway must become an air-filled airway, and the collapsed primitive alveoli (terminal sacs) must expand and function in gas exchange. The success of this transition depends largely on the pulmonary surfactant—a mixture of lipoproteins that lowers surface tension in the fluid layer lining the primitive alveoli once air enters the lungs. Surfactant is present in the fetus's lungs as early as the twenty-third week. By 28 to 32 weeks, however, the amount of surfactant is great enough to prevent alveolar collapse during breathing. Surfactant is produced continuously by alveolar cells. The presence of surfactant can be detected by amniocentesis.

Although in a normal, full-term infant the second and subsequent breaths require less respiratory effort than the first, breathing is not completely normal until about 40 minutes after birth. The entire lung is not inflated fully with the first one or two breaths. In fact, for the first 7 to 10 days, small areas of the lungs may remain uninflated.

In the newborn whose lungs are deficient in surfactant, the effort required for the first breath is essentially the same as that required in normal newborns. However, the surface tension of the alveolar fluid is 7 to 14 times higher than the surface tension of alveolar fluid with a monomolecular layer of surfactant.

Consequently, during expiration after the first inspiration, the surface tension of the alveoli increases as the alveoli deflate. The alveoli collapse almost to their original uninflated state.

Idiopathic RDS usually appears within a few hours after birth. Affected infants show difficult and labored breathing with withdrawal of the intercostal and subcostal spaces. Death may occur soon after the onset of respiratory difficulty or may be delayed for a few days, although many infants survive. At autopsy, the lungs are underinflated and areas of atelectasis are prominent. (In fact, the lungs are so airless they sink in water.) If the infant survives for at least a few hours after developing respiratory distress, the alveoli are often filled with a fluid of high-protein content that resembles a hyaline (or glassy) membrane. RDS occurs frequently in premature infants and also in infants of diabetic mothers, particularly if the diabetes is untreated or poorly controlled.

A new treatment currently being developed, called PEEP—positive end expiratory pressure—could reverse the mortality rate from 90 percent deaths to 90 percent survival. This treatment consists of passing a tube through the air passage to the top of the lungs to provide needed oxygen-rich air at continuous pressures of up to 14 mm Hg. Continuous pressure keeps the baby's alveoli open and available for gas exchange.

In the next chapter we shall examine how the body utilizes nutrients in order to maintain homeostasis. We shall study the chemistry of the molecules that serve as nutrients, the digestive process, and the mechanisms by which nutrients are metabolized.

CHAPTER SUMMARY OUTLINE

RESPIRATORY PHYSIOLOGY

1. Respiration can be defined as the exchange of gases between a living cell and its environment.
2. In order to satisfy the total cellular requirement for oxygen and carbon dioxide removal, the human body employs both the cardiovascular system and the respiratory system.

PULMONARY CIRCULATION

1. The pulmonary circulation is that portion of the cardiovascular system that delivers blood from the right ventricle of the heart into the left atrium.
2. It takes an average time of about 4.5 seconds for the blood to complete the pulmonary circuit.

Blood Pressures and Resistance

1. The blood pressures in the main pulmonary trunk are about 25 mm Hg systolic and 8 mm Hg diastolic. The pulmonary capillary pressure is about 9 to 6 mm Hg.
2. The resistance to flow through the pulmonary vessels can be estimated by using the formula:
$$R = \Delta P/F.$$
3. Blood flow in the lungs is not uniform, and the unequal blood flow is related to the pulmonary blood pressure variation that exists in the lung.
4. A marked constriction of the blood vessels occurs if the partial pressure of oxygen in the inhaled air drops below 70 mm Hg in any portion of the lung.

THE RESPIRATORY ORGANS

1. The interior structures of the nose are specialized for warming, moistening, and filtering air, for olfaction, and for speech.
2. The functions of the pharynx are limited to serving as a passageway for air and food, and providing a resonating chamber for speech sounds.
3. The larynx is a passageway that connects the pharynx with the trachea, and contains true vocal cords that produce sound.
4. The trachea is located in front of the esophagus and extends from the larynx to the fifth thoracic vertebra, where it divides into right and left primary bronchi.
5. Two methods of overcoming obstruction of the respiratory passageway are tracheostomy and intubation.
6. The bronchial tree consists of primary bronchi, secondary bronchi, bronchioles, and terminal bronchioles. Bronchography is a technique for examining the bronchial tree.
7. Each lung is divided into lobes. Each lobe consists of lobules, which contain lymphatics, arterioles, venules, terminal bronchioles, respiratory bronchioles, alveolar ducts, alveolar sacs, and alveoli.
8. The exchange of gases between the lungs and blood takes place by diffusion across the alveoli and the walls of the capillaries.
9. Each lung is enclosed by a double layer of serous membrane called the pleura.

MECHANICS OF BREATHING

1. The delivery of air from the environment into the alveoli of the lungs and its return to the environment is a task accomplished with the aid of muscles.
2. The muscles used in breathing include the diaphragm and the intercostal muscles.
3. The diaphragm contributes more to the inspiratory (inhalation) movement than does the enlargement of the chest by the intercostal muscles.
4. The external intercostals contract and pull the ribs upward and outward so that the thorax increases in dimensions both laterally and anteroposteriorly. The internal intercostal muscles tense to keep the ribs from lifting and extending during straining efforts.

Dynamics of Air Flow

1. In order for air to move through the respiratory passages, a pressure gradient must develop. During inspiration this is achieved by increasing the volume of the lungs.
2. The term compliance is used to describe the ease with which a tissue stretches. The first important component affecting the compliance of lung tissue is the elastic and collagenous fibers that it contains. The second component is the surface tension of a special fluid secreted onto the inner surface of the alveoli.

Air Volumes Exchanged in Respiration

1. The apparatus commonly used to measure the amount of air exchanged during breathing is referred to as a respirometer or spirometer.
2. Among the air volumes exchanged in ventilation are tidal volume, inspiratory reserve, expiratory reserve, residual volume, and minimal volume.
3. The minute volume of respiration is the total air taken in during 1 minute (tidal volume times 16 respirations per minute).

CONTROL OF BREATHING

Location of the Respiratory Centers

1. Nervous control is regulated by the respiratory centers in the medulla oblongata and pons.
2. The three major centers that control breathing are: the medullary respiratory center, the apneustic center, and the pneumotaxic center.
3. The inflation reflex (Hering-Breuer reflex) serves to regulate the extent of lung inflation so that a

minimal amount of work achieves the greatest amount of alveolar ventilation.

External Respiration

1. External respiration is the exchange of oxygen and carbon dioxide between the alveolar sacs and the blood.
2. In internal and external respiration O_2 and CO_2 move from areas of their higher partial pressure to areas of their lower partial pressure.
3. External respiration is aided by a thin alveolar-capillary membrane, a large alveolar surface area (about 540 ft^2), a rich blood supply, and narrow capillaries.

Internal Respiration

1. Internal respiration takes place in the capillaries of the body tissues and is the exchange of O_2 and CO_2 between the blood and body tissues.

Transport of Respiratory Gases

1. In each 100 ml of oxygenated blood, there is 20 ml of O_2 — 0.5 ml is dissolved in plasma, and 19.5 ml is carried by hemoglobin as oxyhemoglobin (HbO_2).
2. In each 100 ml of deoxygenated blood there is 56 ml of CO_2. About 9 percent of CO_2 is dissolved in plasma, about 27 percent combines with hemoglobin as carbaminohemoglobin ($HbNHCOOH$), and about 64 percent is converted to the bicarbonate ion (HCO_3^-).

Mechanisms that Modify Rhythmicity

1. Chemical control is regulated by chemical stimuli (CO_2, O_2 and H^+ ions) in the blood that stimulate chemoreceptors in the carotid sinuses, the aorta, and the medulla itself.
2. Pressoreceptors in the carotid and aortic bodies also influence rate of respiration.

HEART AND LUNG RESUSCITATION

1. Heart-lung resuscitation is the artificial reestablishment of normal or near normal respiration and circulation. The two simplest techniques for heart-lung resuscitation are exhaled air ventilation and external cardiac compression.
2. Exhaled air ventilation is used to reestablish respirations, and the first and most important step is immediate opening of the airway.
3. External cardiac compression is used to reestablish circulation, and consists of the application of rhythmic pressure over the sternum.

4. The Heimlich maneuver is a new and effective first aid procedure in case of food choking.

SMOKING AND THE RESPIRATORY SYSTEM

1. Pollutants, including inhaled smoke, act as stresses on the epithelium of the bronchi and lungs. Constant irritation results in excessive secretion of mucus and rapid division of bronchial basal cells.
2. Additional irritation may cause retention of mucus in bronchioles, loss of elasticity of alveoli, and less surface area for gaseous exchange.
3. In the final stages, bronchial epithelial cells may be replaced by cancer cells. The growth may block a bronchial tube and spread throughout the lung and other body tissues.

NEBULIZATION

1. Nebulization is a comparatively new method of introducing different types of drugs, chemicals, and antibiotics directly into the respiratory tracts.

HOMEOSTATIC IMBALANCES OF THE RESPIRATORY SYSTEM

1. Hay fever is an allergic reaction of respiratory membranes.
2. Bronchial asthma occurs when spasms of smooth muscle in bronchial tubes result in partial closure of air passageways, inflammation, inflated alveoli, and excess mucus production.
3. Emphysema is characterized by deterioration of alveoli leading to loss of their elasticity. Symptoms are reduced expiratory volume, inflated lungs, and enlarged chest.
4. Pneumonia is an acute inflammation or infection of alveoli.
5. Tuberculosis is an inflammation of pleura and lungs produced by a specific bacterium.
6. Hyaline membrane disease is an infant disorder in which surfactant is lacking and alveolar ducts and alveoli have a glassy appearance.
7. Sudden infant death syndrome has recently been linked to laryngospasm, possibly triggered by a viral infection of the upper respiratory tract.

REVIEW QUESTIONS

1. Define respiration. List the two systems involved in respiration and give their functions.
2. What is pulmonary circulation?
3. What is the correlation between blood pressure and resistance in the pulmonary circulation?
4. Explain the distribution of blood flow in the lungs, and describe constriction of pulmonary vessels.
5. What organs make up the respiratory system?
6. Describe the structures of the external and internal nose and describe their functions in filtering, warming, and moistening air.
7. What is the pharynx? Differentiate the three regions of the pharynx and indicate their roles in respiration.
8. Describe the structures of the larynx and explain how they function in respiration and voice production.
9. Describe the location and structure of the trachea. Compare the technique of tracheostomy with that of intubation.
10. What is the bronchial tree? Describe its structure. What is a bronchogram?
11. Where are the lungs located? Distinguish the parietal pleura from the visceral pleura.
12. What is a lobule of the lung? Describe its composition and function in respiration.
13. Define each of the following parts of a lung: respiratory bronchioles, alveolar ducts, alveoli, and alveolar sacs.
14. Indicate several ways in which the respiratory organs are structurally adapted to carry on their respiratory functions.
15. What are the functions of the diaphragm and intercostal muscles in breathing?
16. Describe surface tension and its relationship to breathing. Define compliance and surfactant.
17. What is a respirometer? Define the various lung volumes and capacities. How is the minute volume of respiration calculated?
18. Which areas of the nervous system control respirations? Compare the roles of the medullary respiratory center, the apneustic center, and the pneumotaxic center.
19. Compare the Hering-Breuer reflex and the pneumotaxic center in controlling respiration.
20. Describe the mechanisms of external and internal respirations based on differences in partial pressure of O_2 and CO_2.

21. What are the anatomic adaptations that aid external respirations?
22. What are the partial pressures of oxygen and carbon dioxide in the atmosphere, alveolar air, arterial blood, body tissues, and venous blood?
23. Construct a diagram to illustrate how and why the respiratory gases move during external and internal respiration.
24. How are oxygen and carbon dioxide carried by the blood?
25. What are the three factors that conspire to split the oxyhemoglobin molecule apart?
26. Explain how the following chemical stimuli affect respiration: pCO_2, pO_2, and H^+ ions.
27. How do pressoreceptors affect the control of respiration?
28. How does the control of respiration demonstrate the principle of homeostasis?
29. What are some modified respiratory movements?
30. What is the objective of heart-lung resuscitation? Compare the procedures of exhaled air ventilation with external cardiac compression.
31. What precautions must be taken in each of these procedures?
32. Describe the steps involved in the Heimlich maneuver.
33. Describe in proper sequence the effects of pollutants on the epithelium of the respiratory system.
34. What is therapy by nebulization? How is it useful in most disorders?
35. For each of the following disorders, list the principal clinical symptoms: hay fever, bronchial asthma, emphysema, pneumonia, tuberculosis, hyaline membrane disease, and sudden infant death syndrome.

Chapter 24

PHYSIOLOGY OF DIGESTION

STUDENT OBJECTIVES

After reading this chapter, you should be able to

- Define the terms hunger, appetite, and satiety, and describe the area of the brain that is responsible for their control
- Explain each of the three important classes of food molecules
- Define digestion as a chemical and mechanical process
- Identify the organs of the alimentary canal and the accessory organs of digestion
- Describe the role of the tongue in digestion
- Identify the location of the salivary glands and define the function of saliva in digestion
- Define the function of salivary amylase and describe the mechanisms that regulate the secretion of saliva.
- Discuss the sequence of events involved in swallowing
- Describe the structural features of the stomach and the relationship between these features and digestion
- Compare mechanical and chemical digestion in the stomach
- Describe the factors that control the secretion of gastric juice
- Describe the three separate phases that regulate digestion in the stomach
- Describe the relationship of the pancreas to digestion
- Define the role of the liver and gallbladder in digestion
- Describe the structural features of the

- small intestine that adapt it for digestion and absorption
- Describe the mechanisms involved in the hormonal control of digestion in the stomach and small intestine
- Describe the digestive activities of the small intestine that reduce carbohydrates, proteins, and fats to their final products
- Describe the mechanical movements of the small intestine
- Define absorption and compare the fates of absorbed nutrients
- List the structural features of the large intestine that adapt it for absorption, formation of feces, and elimination
- Describe the mechanical movements of the large intestine
- Describe the processes involved in the formation of feces
- Discuss the mechanisms involved in defecation
- List the causes and symptoms of dental caries and periodontal disease
- Contrast the location and effects of gastric and duodenal ulcers
- Explain peritonitis and describe cirrhosis as a disorder of an accessory organ of digestion
- Describe the location of tumors of the gastrointestinal tract

When the human body reaches its adult size, its weight stabilizes at a particular weight around which some slight changes occur very gradually. It is apparent, therefore, that a balance has developed between the total energy taken in by the body and the utilization of that energy by cells of the body. If this were not the case, our bodies would continue to gain weight and enlarge as we grew older—much like a tree which grows larger as it ages. When we encounter an extremely obese person—one who has gained an excessive amount of weight—we usually consider this an abnormal situation that has resulted from a pathological condition.

The normal steady weight condition is the result of two opposing sets of control systems—one regulating the amount of food taken into the body and the other regulating the utilization or disposal of the extracted energy.

Consider the following. If a person were to eat a predetermined *excessive* amount of food at each meal, the person would certainly increase in weight. However, certain normal physiological responses begin to operate in order to maintain a consistent body weight. One response would be an increase in the basal metabolic rate that would burn up the accumulating energy faster. In a short time, the increased metabolic rate would catch up to and match the rate of food intake. At that point, the energy uptake and consumption would again be in balance although, in the interim, there would have been a slight gain in weight.

Many animal experiments have been carried out that clearly demonstrate that food consumption is regulated in some way so that consumption always matches the energy requirement of the animal. When mature, fully grown animals are given freedom to eat when they wish and as much as they wish, they will eat only as much food as is necessary to maintain their adult weight. This control is also able to adjust for the caloric value of the food. For example, if the caloric value of the food is increased, then the animal eats less so that the caloric intake just remains equal to the energy needs and no more.

HUNGER AND APPETITE

The terms **hunger** and **appetite** are used to describe the conscious desire for food. Often the words are loosely used and interchangeable. However, a distinction should be made. Hunger should be used to refer to the rather indiscriminate need for food that is derived from the energy requirements of the body, whereas appetite should refer to the more selective desire for a particular food that may or may not be related to the nutritional needs of the body.

The term **satiety** is used to describe the onset of hunger satisfaction or the feeling of satisfaction after energy needs have been met. All of these food-regulating sensations—hunger, appetite, and satiety—are housed within nuclei of the hypothalamus.

The distinction of hunger and appetite can be illustrated in the following way. You eat a wholesome meal because you are *hungry* and need to satisfy a manifest energy requirement. You may consume dessert following the meal because of your *appetite* for that high-caloric, tasty food.

Two regions of the hypothalamus have been pinpointed to control hunger and satiety. The region of the ventromedial nucleus has been designated as the "satiety center." In experiments with rats, if this minute area of neural tissue is destroyed, the animals continue to eat as long as food is available. Consequently, they become very obese. The second region of the hypothalamus, termed the "feeding center," is in the lateral hypothalamus nucleus. If this region is destroyed, the rats completely stop eating and starve to death. Further experiments have shown that the medial feeding center inhibits the lateral feeding center.

Electrical stimulation of these regions with microelectrodes has produced the expected results. Electrodes implanted in the lateral hypothalamic nuclei are used to stimulate the feeding center. Such stimulation produces hyperphagia (excessive eating) in mice, rats, cats, and monkeys.

Further research has shown that these centers do not by themselves regulate eating behavior but function to turn on or off other centers in the brain stem involved with the capture, eating, and uptake of food.

Other so-called *peripheral mechanisms* also are involved in the phenomena of hunger and satiety. Figure 24-1 outlines some of the mechanisms involved in the satisfying of the hunger sensation. Gastrointestinal factors might include the stimulation of sensory nerves in the gastrointestinal tract. Hunger contractions may be related to hunger sensations as well as a fall in blood sugar level (hypoglycemia). A distended stomach or elevated blood sugar usually accompanies a feeling of satiety. It has been suggested that a rise in temperature, such as the specific dynamic action (SDA) that accompanies the digestion of food, could be a factor in limiting food consumption.

FOOD MOLECULES

Before examining the mechanisms by which food molecules are digested, absorbed, and utilized by the body, we shall first examine the structure of three important classes of food molecules. These are the carbohydrates, lipids, and proteins.

Carbohydrates

A large and diverse group of organic compounds found in the body is the **carbohydrates,** also known as sugars and starches. The carbohydrates perform a number of major functions in living systems. A few even form structural units. For instance, one type of sugar is a building block of genes, the molecules that carry hereditary information. Some carbohydrates are converted to proteins and to fats or fatlike substances, which are used to build structures and provide an emergency source of energy. Other carbohydrates function as food reserves. One example is glycogen, which is stored in the liver and skeletal muscles. The principal function of carbohydrates, however, is to provide the most readily available source of energy to sustain life.

Carbon, hydrogen, and oxygen are the elements found in carbohydrates. The ratio of hydrogen to oxygen atoms is always 2 to 1. This ratio can be seen in the formulas for carbohydrates such as ribose ($C_5H_{10}O_5$), glucose ($C_6H_{12}O_6$), and sucrose ($C_{12}H_{22}O_{11}$). Although there are exceptions, the general formula for carbohydrates is $(CH_2O)_n$, where n symbolizes three or more CH_2O units. Carbohydrates can be divided into three major groups: monosaccharides, disaccharides, and polysaccharides.

1. **Monosaccharides.** Simple sugars are called *monosaccharides*. These compounds contain from three to seven carbon atoms and cannot be broken down into simpler sugar molecules. Simple sugars with three carbons in the molecule are called trioses. The number of carbon atoms in the molecule is indicated by the prefix *tri*. There are also tetroses (four-carbon sugars), pentoses (five-carbon sugars), hexoses (six-carbon sugars), and heptoses (seven-carbon sugars). Pentoses and hexoses are exceedingly important to the human organism. The pentose called deoxyribose is a component of genes. The hexose called glucose is the main energy-supplying molecule of the body.

2. **Disaccharides.** A second group of carbohydrates, called the *disaccharides,* consists of two monosaccharides joined chemically. In the process of disaccharide formation, two monosaccharides combine to form a disaccharide molecule and a molecule of water is lost. This reaction is known as *dehydration synthesis* (*dehydration* = loss of water). The following reaction shows disaccharide formation. Molecules of the monosaccharides glucose

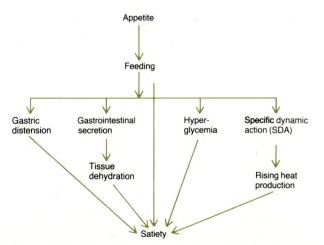

Figure 24-1   Mechanisms involved in satisfying hunger.

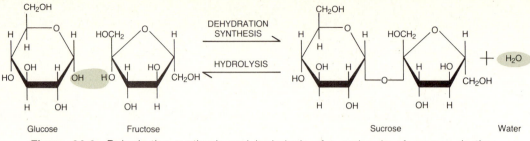

Figure 24-2 Dehydration synthesis and hydrolysis of a molecule of sucrose. In the dehydration synthesis (read from left to right), the two smaller molecules, glucose and fructose, are joined to form a larger molecule of sucrose. Note the loss of a water molecule. In hydrolysis (read from right to left), the larger sucrose molecule is broken down into the two smaller molecules, glucose and fructose. Here, a molecule of water is added to sucrose for the reaction to occur.

and fructose combine to form a molecule of the disaccharide sucrose (cane sugar):

$$C_6H_{12}O_6 + C_6H_{12}O_6 \rightarrow C_{12}H_{22}O_{11} + H_2O$$

Glucose	Fructose	Sucrose	Water
(monosaccharide)	(monosaccharide)	(disaccharide)	

You may be puzzled to see that glucose and fructose have the same chemical formulas. Actually, they are different monosaccharides since the relative positions of the oxygens and carbons vary in the two different molecules (Figure 24-2). The formula for sucrose is $C_{12}H_{22}O_{11}$ and not $C_{12}H_{24}O_{12}$ since a molecule of H_2O is lost in the process of disaccharide formation. In every dehydration synthesis, a molecule of water is lost. Along with this water loss, there is the synthesis of two small molecules, such as glucose and fructose, into one large, more complex molecule, such as sucrose (Figure 24-2). Similarly, the dehydration synthesis of two monosaccharides such as glucose and galactose forms the disaccharide lactose (milk sugar).

Disaccharides can also be broken down into smaller, simpler molecules by adding water. This reverse chemical reaction is called *digestion* or *hydrolysis*, which means to split by using water. A molecule of sucrose, for example, may be digested into its components of glucose and fructose by the addition of water. The mechanism of this reaction also is represented in Figure 24-2.

3. **Polysaccharides.** The third major group of carbohydrates, the *polysaccharides*, consists of eight or more monosaccharides joined together through dehydration synthesis. Polysaccharides have the formula $(C_6H_{10}O_5)_n$. Like disaccharides, polysaccharides can be broken into their constituent sugars through hydrolysis reactions. Unlike monosaccharides or disaccharides, however, they usually lack the characteristic sweetness of sugars like fructose or sucrose and are usually not soluble in water. One of the chief polysaccharides is glycogen.

Lipids

A second group of organic compounds that is vital to the human organism is the **lipids.** Like carbohydrates, lipids are composed of carbon, hydrogen, and oxygen, but they do not have a 2:1 ratio of hydrogen to oxygen. Most lipids are insoluble in water, but they readily dissolve in solvents such as alcohol, chloroform, and ether. Since lipids are a large and diverse group of compounds, only one kind, the *fats,* is discussed here. Pertinent information regarding other lipids is provided in Exhibit 24-1.

A molecule of fat consists of two basic components. The first component is called *glycerol;* the second is a group of compounds called *fatty acids* (Figure 24-3). A single molecule of fat is formed when a molecule of glycerol combines with three molecules of fatty acids. This reaction, like the one described for disaccharide formation, is a dehydration synthesis reaction. During hydrolysis, a single molecule of fat is broken down into fatty acids and glycerol.

Fats represent the body's most highly concentrated source of energy. In fact, they provide twice as many calories (a form of energy) per weight as either carbohydrates or proteins. In general, however, fats are about 10 to 12 percent less efficient as body fuels than are carbohydrates. A great amount of the fat calorie is wasted and thus not available for the body to use.

Proteins

A third principal group of organic compounds is **proteins.** These compounds are much more complex in structure than the carbohydrates or lipids. They are also responsible for much of the structure of body cells and are related to many physiological activities. For example, proteins in the form of enzymes speed up many essential biochemical reactions. Other proteins assume a necessary role in muscular contraction. Antibodies are proteins that provide the human organism

Exhibit 24-1 RELATIONSHIPS OF REPRESENTATIVE LIPIDS TO THE HUMAN ORGANISM

Lipids	Relationship	Lipids	Relationship
Fats	Protection, insulation, source of energy.	Hemoglobin	Oxygen-transporting pigment in red blood cells.
Phospholipids		Bile pigments	Bilirubin, a reddish pigment, and biliverdin, a greenish pigment, are both formed from hemoglobin and are responsible for the brown color of feces.
Lecithin	Major lipid component of cell membranes; constituent of plasma.		
Cephalin and sphingomyelin	Found in high concentrations in nerves and brain tissue.	Cytochromes	Coenzymes involved in the respiration of all cells.
Steroids			
Cholesterol	Constituent of all animal cells, blood, and nervous tissue; suspected relationship to heart disease and "hardening of the arteries."	**Other lipid substances**	
		Carotenes	Pigment in egg yolk, carrots, and tomatoes; vitamin A is formed from carotenes; retinene, also formed from vitamin A, is a photoreceptor in the retina of the eye.
Bile salts	Substances that emulsify or suspend fats before their digestion.		
Vitamin D	Produced in skin on exposure to ultraviolet radiation; necessary for bone growth and development.	Vitamin E	"Antisterility" vitamin in rats; necessary for the synthesis of connective tissue in wound healing in humans.
Estrogens	Sex hormones produced in large quantities by females.	Vitamin K	Vitamin that promotes blood clotting and prevents excessive bleeding.
Androgens	Sex hormones produced in large quantities by males.	Prostaglandins	Membrane-associated lipids that stimulate smooth muscle to contract, raise and lower blood pressure, and regulate metabolism.
Porphyrins (lipid portions of organic molecules)			

with defenses against invading microbes. And some hormones that regulate body functions are also proteins. A classification of proteins on the basis of function is shown in Exhibit 24-2.

Chemically, proteins always contain carbon, hydrogen, oxygen, and nitrogen. Many proteins also contain sulfur and phosphorus. Just as monosaccharides are the building units of sugars and fatty acids and glycerol are the building units of fats, *amino acids* are the building blocks of proteins. In protein

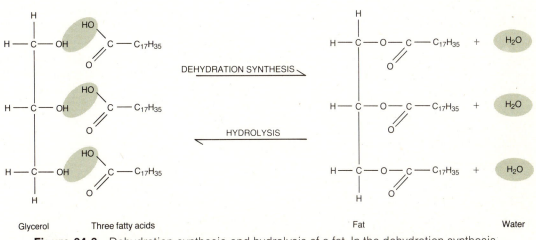

Glycerol   Three fatty acids   Fat   Water

Figure 24-3 Dehydration synthesis and hydrolysis of a fat. In the dehydration synthesis (read from left to right), one molecule of glycerol combines with three fatty acid molecules, and there is a loss of three molecules of water. In hydrolysis (read from right to left), a molecule of fat is broken down into a single molecule of glycerol and three fatty acid molecules upon the addition of three molecules of water. The fatty acid shown is stearic acid, a component of corn oil, coconut oil, beef fat, and pork fat.

Exhibit 24-2 CLASSIFICATION OF PROTEINS
BY FUNCTION

Nature of Protein	Description
Structural	Proteins that form the structural framework of various parts of the body. Examples: keratin in the skin, hair, and fingernails and collagen in connective tissue.
Regulatory	Proteins that function as hormones and regulate various physiological processes. Examples: insulin, which regulates blood sugar, and ADH (anti-diuretic hormone) which regulates the volume and concentration of urine excreted.
Contractile	Proteins that serve as contractile elements in muscle tissue. Examples: myosin and actin.
Immunological	Proteins that serve as antibodies to protect the body against invading microbes. Example: gamma globulin.
Transport	Proteins that transport vital substances throughout the body. Example: hemoglobin, which transports oxygen.
Catalytic	Proteins that act as enzymes and function by controlling biochemical reactions. Examples: salivary amylase, pepsin, and lactase.

formation, amino acids combine to form more complex molecules, while water molecules are lost. The process is a dehydration synthesis reaction, and the bonds formed between amino acids are called *peptide bonds* (Figure 24-4).

When two amino acids combine, a *dipeptide* results. Adding another amino acid to a dipeptide produces a *tripeptide*. Further additions of amino acids result in the formation of *polypeptides*, which are large protein molecules. At least 20 different amino acids are found in proteins. A great variety of proteins is possible because each variation in the number or sequence of amino acids can produce a different protein. The situation is similar to using an alphabet of 20 letters to form words. Each letter could be compared to a different amino acid, and each word would be a different protein.

Food is vital for life because it is the ultimate source of energy that drives the chemical reactions that occur in every cell—those that synthesize new enzymes, cell structure, bone, and all the other components of the body. Energy is needed for muscle contraction, the conduction of nerve impulses, and the secretory and absorptive activities of many cells. The food as it is consumed, however, is not in a state suitable for use as an energy source by any cell. The food must be digested into molecule-sized pieces so that it can be transported through the cell membranes. This process of digestion involves both mechanical and chemical breakdown components.

CHEMICAL AND MECHANICAL DIGESTION

Chemical digestion is a series of catabolic reactions that break down the large carbohydrate, lipid, and protein molecules that we eat into monosaccharides, some glycerol and fatty acids, and amino acids, respectively. These products of digestion are small enough to pass through the walls of the digestive organs, into the blood and lymph capillaries, and eventually into the cells of the body. **Mechanical digestion** consists of various movements that aid chemical digestion. Food must be pulverized by the teeth before it can be swallowed. Then the smooth muscles of the stomach and small intestine churn the food so it is thoroughly mixed with the enzymes that are responsible for most of the chemical breakdown of the food.

The digestive system prepares food for consumption by the cells. It does this through five basic activities:

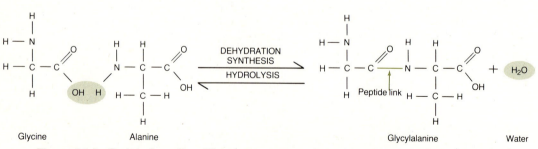

Glycine   Alanine   Glycylalanine   Water

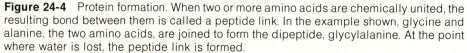

Figure 24-4 Protein formation. When two or more amino acids are chemically united, the resulting bond between them is called a peptide link. In the example shown, glycine and alanine, the two amino acids, are joined to form the dipeptide, glycylalanine. At the point where water is lost, the peptide link is formed.

1. *Ingestion,* or eating, which is taking food into the body.
2. *Movement* of food along the digestive tract.
3. Mechanical and chemical *digestion.*
4. *Absorption* of the digested food from the digestive tract into the circulatory and lymphatic systems for distribution to the cells.
5. *Defecation,* the elimination of indigestible substances from the body.

GENERAL ORGANIZATION OF THE GASTROINTESTINAL TRACT

The organs of digestion are traditionally divided into two main groups. First is the **gastrointestinal tract (GI)** or **alimentary canal,** a continuous tube running through the ventral body cavity and extending from the mouth to the anus (Figure 24-5). The length of a tract taken from a cadaver is about 9 m (30 ft). In a living person, it is somewhat shorter because the muscles lying in its walls are in a state of tone. Organs composing the gastrointestinal or GI tract include the mouth, pharynx, esophagus, stomach, small intestine, and large intestine. The GI tract contains the food while it is being eaten, digested, and prepared for elimination. Muscular contractions in the walls of the GI tract break down the food physically by churning it. Secretions produced by cells along the GI tract break down the food chemically.

The second group of organs composing the digestive system are the **accessory organs**—the teeth, tongue, salivary glands, liver, gallbladder, pancreas, and appendix. Teeth are cemented to bone, protrude into the GI tract, and aid in the physical breakdown of food. The other accessory organs, except for the tongue, lie

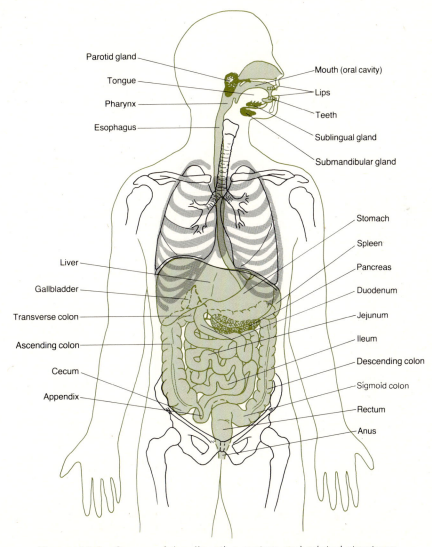

Figure 24-5   Organs of the digestive system and related structures.

totally outside the tract and produce or store secretions that aid in the chemical breakdown of the food. These secretions are released into the tract through ducts.

Tongue

The **tongue** is composed of skeletal muscles covered with mucous membrane (Figure 24-6). The lingual frenulum is a fold of mucous membrane in the midline of the undersurface of the tongue that limits the movement of the tongue posteriorly. If the frenulum is too short, tongue movements are restricted, speech is faulty, and the person is said to be "tongue-tied." These functional problems can be corrected very easily by cutting the lingual frenulum.

The upper surface and sides of the tongue are covered with *papillae*. Taste buds are located in some papillae. *Filiform papillae* are conical projections distributed in parallel rows over the anterior two-thirds of the tongue and contain no taste buds. *Fungiform papillae* are mushroomlike elevations distributed among the filiform papillae and are more numerous near the tip of the tongue. They appear as red dots on the surface of the tongue, and most of them contain taste buds. *Circumvallate papillae* are

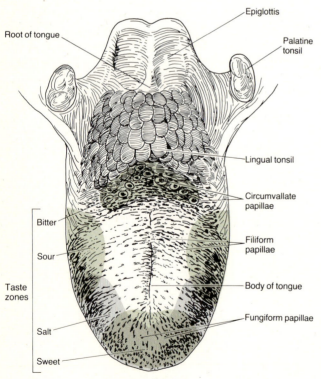

Figure 24-6 Dorsum of the tongue showing the locations of the papillae and the four taste zones.

arranged in the form of an inverted V on the posterior surface of the tongue, and all of them contain taste buds. Note the relative positions of the taste zones of the tongue in Figure 24-6.

Salivary Glands

Saliva is a fluid that is continuously secreted by glands lying in or near the mouth. Ordinarily, just enough saliva is secreted to keep the mucous membranes of the mouth moist. But when food enters the mouth, secretion increases so the saliva can lubricate, dissolve, and chemically break down the food. The mucous membrane lining the mouth contains many small glands, the *buccal glands,* that secrete small amounts of saliva. However, the major portion of saliva is secreted by the **salivary glands,** which lie outside the mouth and deliver their contents via ducts that empty into the oral cavity. There are three pairs of salivary glands: the parotid, submandibular, and sublingual glands (Figure 24-7). The *parotid glands* are located under and in front of the ears between the skin and masseter muscle. Each empties into the oral cavity via Stensen's duct, which opens into the mouth opposite the upper second molar tooth. The *submandibular glands* are found beneath the base of the tongue in the posterior part of the floor of the mouth, and their ducts (Wharton's ducts) empty into the oral cavity just behind the central incisors. The *sublingual glands* are anterior to the submandibular glands, and their ducts open into the floor of the mouth in the oral cavity proper. The parotid glands are compound tubuloacinar glands, whereas the submandibular and sublinguals are compound acinar glands.

Saliva

The fluids secreted by the buccal glands and the three pairs of salivary glands constitute **saliva.** Amounts of saliva secreted daily vary considerably but range from 1,000 to 1,500 ml. Chemically, saliva is 99.5 percent water and 0.5 percent solutes. Among the solutes are salts, such as chlorides, bicarbonates, and phosphates of sodium and potassium. Some dissolved gases and various organic substances including urea and uric acid, serum albumin and globulin, mucin, the bacteriolytic enzyme lysozyme, and the digestive enzyme amylase are also present.

The water in saliva provides a medium for dissolving foods so they can be tasted and in which digestive reactions can take place. The chlorides in the saliva activate the amylase. The bicarbonates and phosphates buffer chemicals that enter the mouth and keep the saliva at a slightly acidic pH of 6.35 to 6.85. Urea and uric acid are found in saliva because

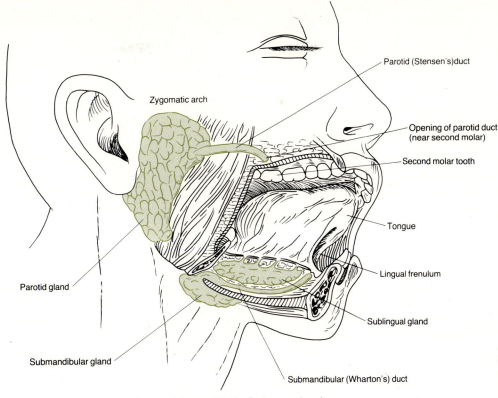

Figure 24-7 Salivary glands.

the saliva producing glands, like the sweat glands of the skin, help the body to get rid of wastes. Mucin is a protein that forms mucus when it is dissolved in water. Mucus lubricates the food so it can be easily turned in the mouth, and the enzyme salivary amylase breaks the long-chain polysaccharides into shorter polysaccharides called *dextrins*. Given sufficient time, salivary amylase also can break down the dextrins into disaccharides, but food usually is swallowed too quickly for more than 3 to 5 percent of the carbohydrates to be reduced to disaccharides in the mouth. However, salivary amylase in the bolus continues to act on polysaccharides for another 15 to 30 minutes in the stomach before the stomach acids eventually inactivate it.

Control of Salivary Secretion

Normally, moderate amounts of saliva are continuously secreted, but the rate may vary considerably from moment to moment. This continuous or resting level of secretion keeps the mucous membranes moist and lubricates the tongue and lips during speech. The saliva is then swallowed and the water reabsorbed to prevent the loss of fluid.

The secretions of the salivary glands, unlike the other digestive juices, are controlled entirely by the autonomic nervous system. To date, no hormone that regulates salivation has been found. The production of

saliva—salivation—has been shown to be triggered in one of three ways: (1) by the placing of food or some substance in the mouth: (2) by the thought, smell, or sight of food; (3) by the presence of food, particularly irritating types, in the stomach and small intestine.

The presence of food in the mouth triggers tactile and taste responses that are relayed to the salivary center in the medulla via the trigeminal (V), facial (VII), glossopharyngeal (IX), and vagus (X) cranial nerves. Parasympathetic efferent fibers evoke a vasodilation of vessels to the parotid gland, promoting copious amounts of serous fluid from the parotids and a large volume of mucus fluid from the submaxillary glands. Sympathetic efferents have an overall vasoconstrictor action on the blood vessels and promote secretion of small amounts of mucus type saliva.

The stimulation of salivation by the sight or smell of food falls into the category of a *conditioned reflex*. A conditioned reflex is learned by associating an accompanying stimulus with the original or primary stimulus. Experiments of this type were pioneered by the famed Russian physiologist, Pavlov. Each time he fed his dogs, Pavlov rang a bell. After a period of training, he could provoke salivation in the dogs by simply ringing the bell. Similarly, we learn to associate certain smells, visions, and even memories of meals with the taste and texture they elicited in our mouths.

Salivation has also been initiated by placing food

directly into the stomach through a gastric fistula (an opening into the stomach through the wall of the abdomen). This reflex may operate normally to wash down or dilute any remaining food or irritants present in the upper part of the digestive tract.

Deglutition

Food in the mouth is chewed, or **masticated,** by the tearing and grinding action of the teeth and mixed with saliva. As a result, the food is reduced to a soft, pulpy consistency. Exhibit 24-3 summarizes the digestion that occurs in the mouth.

Exhibit 24-3 DIGESTION IN THE MOUTH

Structure	Activity	Result
Cheeks	Keep food between teeth during mastication.	Foods are uniformly chewed.
Lips	Keep food between teeth during mastication.	Foods are uniformly chewed.
Tongue		
Extrinsic muscles	Move tongue from side to side and in and out.	Maneuver food for mastication and deglutition (swallowing).
Intrinsic muscles	Alter shape of tongue.	Deglutition.
Taste buds	Serve as receptors for food stimulus.	Nerve impulses from taste buds to brain to salivary glands stimulate secretion of saliva.
Buccal glands	Secrete saliva.	Lining of mouth and pharynx moistened and lubricated.
Salivary glands	Secrete saliva.	Same as above. Saliva softens, moistens, and dissolves food, coats food with mucin, cleanses mouth and teeth. Salivary amylase reduces polysaccharides to dextrins and the disaccharide maltose.
Teeth	Cut, tear, and pulverize food.	Solid foods are reduced to smaller particles for swallowing.

The mass of food in the mouth is rolled and pressed into a slightly elongated ball referred to as a *bolus.* The process of swallowing, or **deglutition,** starts by centering the bolus on the upper side of the tongue. Then the tip of the tongue rises and presses against the palate (Figure 24-8). The bolus slides to the back of the mouth and is pulled through the fauces by muscles that lie in the pharynx. During this period, the respiratory passageways close, and breathing is temporarily interrupted. The soft palate and uvula move upward to close off the nasopharynx, and the larynx is pulled forward and upward under the tongue. As the larynx rises, it meets the epiglottis, which seals off the glottis. The movement of the larynx also pulls the vocal cords together, further sealing off the respiratory tract, and widens the opening between the pharynx and esophagus. The bolus passes through the pharynx and enters the esophagus in 1 second. The respiratory passageways then reopen, and breathing resumes.

The **esophagus,** the third organ involved in deglutition, is a muscular, collapsible tube that lies behind the trachea (see Figures 24-5 and 24-8). It does not produce digestive enzymes and does not carry on absorption. It does, however, secrete mucus and transport food to the stomach.

Food is pushed through the esophagus by muscular movements called **peristalsis** (Figure 24-9). Peristalsis is a function of the tunica muscularis, which performs a sequential series of muscular movements. In the section of the esophagus just above and around the top of the bolus, the circular muscle fibers contract. The contraction constricts the esophageal wall and squeezes the bolus downward. Meanwhile, longitudinal fibers around the bottom of the bolus and just below it also contract. Contraction of the longitudinal fibers shortens this lower section, pushing its walls outward so it can receive the bolus. The contractions are repeated in a wave that moves down the esophagus, pushing the food toward the stomach. Passage of the bolus is further facilitated by glands secreting mucus. The passage of solid or semisolid food from the mouth to the stomach takes about 4 to 8 seconds. Very soft foods and liquids pass through in about 1 second.

Just above the level of the diaphragm, the esophagus is very slightly narrowed. This narrowing has been attributed to a physiological sphincter in the inferior part of the esophagus known as the *lower esophageal* or *gastroesophageal sphincter.* A *sphincter* is an opening that has a thick circle of muscle around it. The lower esophageal sphincter relaxes during swallowing and thus aids the passage of food from the esophagus into the stomach. The movement of the diaphragm against the stomach during ventilation presses on the

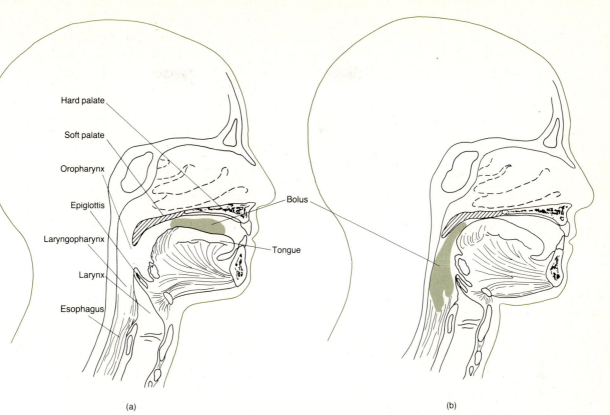

Figure 24-8 Deglutition. (a) Position of structures prior to swallowing. (b) During swallowing, the tongue rises against the palate, the nose is closed off, the larynx rises, the epiglottis seals off the larynx, and the bolus is passed into the esophagus.

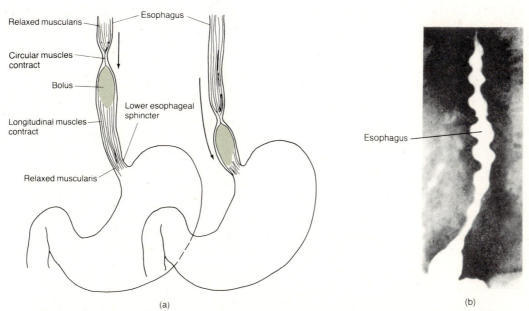

Figure 24-9 Peristalsis. (a) Diagrammatic representation. (b) Anteroposterior projection of peristalsis made during fluoroscopic examination while a patient was swallowing a barium "meal." (Courtesy of Lester W. Paul and John H. Juhl, *The Essentials of Roentgen Interpretation*, 3d ed., New York: Harper & Row, 1972.)

stomach and helps to prevent the regurgitation of gastric contents from the stomach to the esophagus.

Exhibit 24-4 contains a summary of the digestion-related activities of the pharynx and esophagus.

Stomach

The **stomach** is a J-shaped enlargement of the GI tract directly under the diaphragm (see Figure 24-5). The superior portion of the stomach is a continuation of the esophagus, and the inferior portion empties into the duodenum, the first part of the small intestine. Within each individual, the position and size of the stomach vary continually. For instance, the diaphragm pushes the stomach downward with each inspiration and pulls it upward with each expiration. When the stomach is empty, it is about the size of a large sausage, but the stomach can enlarge greatly to accommodate large amounts of food.

Anatomy

The stomach is divided into four areas: the cardia, fundus, body, and pylorus (Figure 24-10). The *cardia* surrounds the lower esophageal sphincter, and the rounded portion above and to the left of the cardia is the *fundus.* Below the fundus, the large central portion of the stomach is called the *body,* and the more narrow, inferior region is the pylorus. The concave medial border of the stomach is called the *lesser curvature,* and the convex lateral border is referred to as the *greater curvature.* The pylorus communicates with the duodenum of the small intestine via a sphincter called the *pyloric valve.*

Two abnormalities of the pyloric valve sometimes are found in infants. One of these abnormalities, *pylorospasm,* is characterized by failure of the muscle fibers encircling the opening to relax normally. As a result, ingested food does not pass easily from the stomach to the small intestine. The stomach becomes overly full, and the infant vomits frequently to relieve the pressure. Pylorospasm is treated by adrenergic drugs that relax the muscle fibers of the valve. The other abnormality, called *pyloric stenosis,* is a narrowing of the pyloric valve caused by a tumorlike mass that apparently is formed by enlargement of the circular muscle fibers. The mass obstructs the passage of food and must be surgically corrected.

The wall of the stomach is composed of the same four basic layers as the rest of the alimentary canal, with certain modifications. When the stomach is empty, the mucosa lies in large folds that can be seen with the naked eye. These folds are called *rugae.* As the stomach fills and distends, the rugae gradually smooth out and disappear. Microscopic inspection of the mucosa reveals a layer of simple columnar epithelium containing many narrow openings, or pits, that extend into the lamina propria. These pits are called *gastric glands,* and they are lined with three kinds of secreting cells; zymogenic, parietal, and mucous (Figure 24-10b). The *zymogenic,* or *chief cells,* secrete the principal gastric precursor or proenzyme called pepsinogen. Hydrochloric acid, which activates one of the digestive enzymes, is produced by the *parietal cells.* The *mucous cells* secrete mucus and the intrinsic factor, a substance involved in the absorption of vitamin B_{12}. Secretions of the gastric glands are collectively called *gastric juice.*

Digestion in the Stomach

Several minutes after food enters the stomach, gentle, rippling, peristaltic movements called *mixing waves* pass over the stomach. These waves occur about every 15 to 25 seconds and serve to macerate food, mix it with the secretions of the digestive glands, and reduce it to a thin liquid called *chyme.* Relatively few mixing waves are observed in the fundus, which is primarily a storage area. Foods may remain in the fundus for an hour or more without becoming mixed with gastric juice. During this time, salivary digestion continues.

The principal chemical activity of the stomach is to begin the digestion of proteins. In the adult, this is achieved primarily by the enzyme *pepsin.* Pepsin breaks some of the peptide bonds between the amino acids making up proteins. Thus a protein chain of many amino acids is broken down into fragments containing 4 to 12 amino acids—longer fragments are called *proteoses,* and the shorter ones are called *peptones.* Pepsin is most effective in the very acidic environment of the stomach, which has a pH of 1. It becomes inactive in an alkaline environment. What keeps pepsin from digesting the cells of the stomach along with the food? First of all, pepsin is secreted in

Exhibit 24-4 DIGESTIVE ACTIVITIES OF THE PHARYNX AND ESOPHAGUS

Structure	Activity	Result
Pharynx	Deglutition	Food is passed from the pharynx into the esophagus. Air passageways are closed off, and opening to the esophagus is widened.
Esophagus	Peristalsis	Bolus is forced down the esophagus into the stomach.
	Secretion of mucus	Bolus passes smoothly down esophagus.

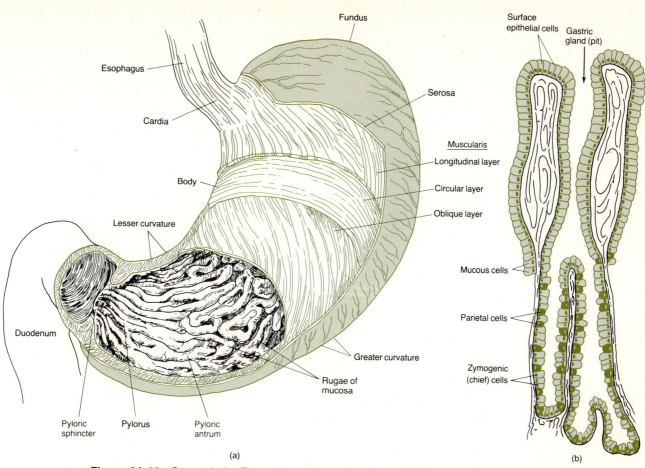

Figure 24-10 Stomach. (a) External and internal anatomy. (b) Diagram of gastric glands from the fundic wall.

an inactive form called *pepsinogen,* so it cannot digest the proteins in the zymogenic cells that produce it. When pepsinogen comes in contact with the hydrochloric acid secreted by the parietal cells, it is converted to active pepsin. The cells of the stomach are protected by mucus from the digestive action of the active enzyme, pepsin. The mucus coats the mucosa and forms a barrier between the acid rich gastric juice and the cells.

Another enzyme of the stomach is *gastric lipase.* In the infant stomach, gastric lipase splits the butter-fat molecules found in milk. This enzyme operates best at a pH of 5 to 6 and has a limited role in the adult stomach. Adults rely exclusively on an enzyme found in the small intestine to digest fats.

As digestion proceeds in the stomach, more vigorous peristaltic waves begin at about the middle of the stomach, pass downward, reach the pyloric valve, and sometimes go into the duodenum. The actual movement of chyme from the stomach into the duodenum depends on a pressure gradient between the two organs. When the pressure in the stomach (intragastric pressure) is greater than that in the duodenum (intraduodenal pressure), chyme is forced into the duodenum. Peristaltic waves are largely responsible for

increased intragastric pressure. It is estimated that 2 to 5 ml of chyme are passed into the duodenum with each peristaltic wave. When intraduodenal pressure exceeds intragastric pressure, the pyloric valve closes and prevents the regurgitation of chyme from the duodenum to the stomach. The stomach empties all its contents into the duodenum after 2 to 6 hours following ingestion. Food rich in carbohydrate leaves the stomach in the shortest time. Protein foods are somewhat slower, and emptying is slowest after a meal containing large amounts of fat. The stomach wall is impermeable to the passage of most materials into the blood, so most substances are not absorbed until they reach the small intestine. However, the stomach does participate in the absorption of some water and salts, certain drugs, and alcohol.

Production of Hydrochloric Acid

The mechanism of hydrochloric acid synthesis and release into the gastric gland is not completely understood. One currently accepted idea is based on the active transport of H^+ ions into the lumen of the gland. Figure 24-11 illustrates the reactions believed to be involved. The parietal cell obtains *carbon dioxide* from either its own metabolism or the blood. A

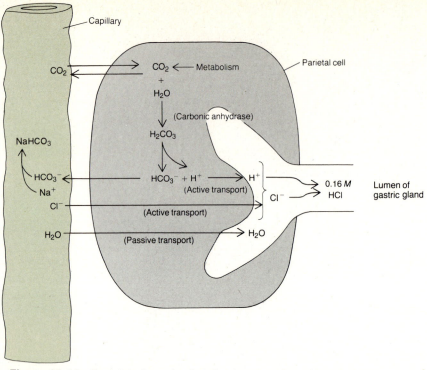

Figure 24-11 Postulated mechanism for the secretion of hydrochloric acid.

reaction of carbon dioxide with water is greatly accelerated by means of the enzyme *carbonic anhydrase* to form *carbonic acid*. This weak acid dissociates into hydrogen and bicarbonate ions. An active transport system then pumps the hydrogen ions into a canaliculus of the parietal cell. The bicarbonate ion diffuses out of the cell into the blood plasma in an exchange for a chloride ion. The chloride is then also pumped into a canaliculus in a manner that is coupled to that for the hydrogen ion. As the concentration of H^+ and Cl^- build up in the canaliculi and lumen of the gastric gland, water passively diffuses from the plasma through the cell into the lumen to complete the formation of the hydrochloric acid solution.

Regulation of Gastric Digestion

Digestion in the stomach is regulated by both hormonal and neural reflexes, and can be best described as occurring in three separate phases: *cephalic, gastric,* and *intestinal.* Each phase is identified by the place where the reflex is initiated.

Cephalic Phase. Gastric secretion is initiated prior to the entry of food into the stomach. The term *cephalic* indicates that the reflex stimulation starts in the head (*kephalē* = head). A variety of stimuli such as the sight, smell, or even thought of food can trigger afferent impulses to the vagus nucleus. From there, efferent stimuli emerge and travel via the vagal fibers to the secretory portions of the gastric mucosa. This phase produces an average volume of between 50 to 150 ml of gastric juice during a 20-minute interval

(Exhibit 24-5). Nerve impulses reaching the gastric mucosa via the vagus cause secretion by two mechanisms. The release of acetylcholine from the nerve terminals acts directly on the parietal cells to promote acid secretion. The acetylcholine causes the release of *gastrin,* a hormone from cells in the pyloric antrum region of the stomach. Sectioning the vagus nerve (termed a vagotomy) completely abolishes the cephalic phase.

Gastric Phase. The presence of food in the stomach activates the gastric phase of digestion. Distension of the stomach walls and the chemical action of the food (especially certain amino acids, alcohol, caffeine, and peptide molecules) initiate secretion from the stomach glands. Both mechanisms trigger the release of the hormone gastrin from specialized cells of the mucosa in the pyloric antrum. Nerve endings pick up the distension and chemical stimuli and then carry action potentials to the gastrin-producing cells. The release of acetylcholine activates the gastrin cells. Since this

Exhibit 24-5 VOLUME OF DIFFERENT PHASES OF GASTRIC SECRETION IN HUMANS

Period or Phase	Amount
Interdigestive	30–60 ml/hour
Cephalic	50–150 ml/20 minutes
Gastric	225–350 ml/5 hours
Intestinal	200–300 ml/5 hours

Source: Courtesy of A. C. Ivy, M. I. Grossman, and H. Bachrach, *Peptic Ulcer,* New York: Blakiston, 1950.

phase of digestion also employs a nervous reflex, drugs such as atropine will block the effect of acetylcholine on the gastrin-producing cells. The hormone gastrin is taken up by the bloodstream and delivered to the secretory portions of the stomach—mainly the fundic and body regions. The hormone acts primarily upon the parietal HCl-secreting cells, which can increase their acid secretion eightfold, and to a lesser degree upon the enzyme-producing chief cells, which may double their rate of secretion. As long as food is present in the stomach—and this may be for hours after a large meal—the gastrin stimulating reflex will continue to act. The amino acid composition of gastrin is shown in Figure 24-12a. The gastric phase is also mediated by local reflexes that operate within Auerbach's plexus, an autonomic nerve supply of the muscle layer of the gastrointestinal tract. Sensory endings detect the food in the stomach, and the secretory cells are stimulated. Also there are vagal reflexes that sense the food in the stomach and then relay impulses via afferent vagal fibers to the brain stem centers, which then return efferent signals via the vagus back to the secretory cells in the stomach. The combined effects of the gastrin and vagal stimulation of the secretory cells is greater than the sum of their separate effects. That is, the two mechanisms act upon each other in some manner to enhance their actions. The gastric phase of digestion, since it can last hours, produces about 2,000 ml of gastric juice daily and is responsible for about two-thirds of the total gastric secretion.

Intestinal Phase. Gastric secretion also occurs when food is placed directly into the duodenum. A variety of partially digested foods, milk, alcohol, and even water can evoke the response. Though the exact mechanism

for this reflex is not known, most workers believe that it must operate via some "intestinal gastrin" hormone. Completely denervated pouches prepared from portions of the stomach will secrete in response to food placed in the upper part of the small intestine. In a contrary fashion, it has also been shown that when acid, fat, or hypertonic solutions are placed in the duodenum, gastric secretion is inhibited. Some investigators believe the mechanism operates by the action of two hormones, secretin and cholecystokinin. Their primary function is to stimulate the pancreas and gallbladder, respectively, but these workers believe that secretin also inhibits the action of gastrin on the secretory cells of the gastric mucosa. The structure of cholecystokinin partially resembles that of gastrin, so it competes with gastrin for the receptor sites on the target cells. Since its effect is very weakly stimulatory, the net result of displacing the gastrin molecule is a reduction in gastric secretions.

Interdigestive Phase. When the digestive tract is empty and in between digestive episodes, the stomach is said to be in the interdigestive phase. During this time period, the stomach continues to secrete small volumes of gastric juice, but a juice that is nonparietal —that is, one almost devoid of HCl. The juice contains mostly mucus and small amounts of pepsin. However, emotional stress can often provoke the secretion of hydrochloric acid during the interdigestion phase through a mechanism similar to that employed during the cephalic phase of gastric secretion. It is thought that this excessive acid squirted through the pylorus eats through the mucus coating of the duodenum faster than it can be restored and thus produces a painful peptic duodenal ulcer.

Exhibit 24-6 summarizes the chief activities of the stomach. The next step in the breakdown of food is digestion in the small intestine. However, chemical digestion in the small intestine is dependent not only on its own secretions but on those from three organs that lie outside the alimentary canal. These organs—pancreas, liver, and gallbladder—will be discussed before we continue with the alimentary canal.

Glu- Gly- Pro- Trp- Leu- Glu- Glu- Glu- Glu- Glu- Ala- Tyr- Gly- Trp- Met-

Asp- Phe- NH$_2$

|
HSO$_3$

(a)

Lys- (Ala, Gly, Pro, Ser)- Arg- Val- (Ile, Met, Ser)- Lys- Asn- (Asn, Gln, His, Leu$_2$,

Pro, Ser$_2$)- Arg- Ile- (Asp, Ser)- Arg- Asp- Tyr- Met- Gly- Trp- Met- Asp- Phe- NH$_2$

|
HSO$_3$

(b)

His- Ser- Asp- Gly- Thr- Phe- Thr- Ser- Glu- Leu- Ser- Arg- Leu- Arg- Asp- Ser-

Ala- Arg- Leu- Gln- Arg- Leu- Leu- Gln- Gly- Leu- Val- NH$_2$

(c)

Figure 24-12 Comparison of the amino acid sequence in (a) gastrin, (b) cholecystokinin, and (c) secretin.

Pancreatic Secretion

The **pancreas** is a soft, yellowish, oblong gland about 13 cm (6 inches) long and 2.5 cm (1 inch) thick. It lies behind and below the greater curvature of the stomach and can grossly be divided into a head, body, and tail region (Figure 24-13). Microscopically, the greater portion of the cells of the pancreas are arranged into compound tubulo-acinar or alveolar glands that are interconnected by a duct network. The ducts collect the secretions of the acinar cells and deliver the pancreatic juice finally into a single, large main tube

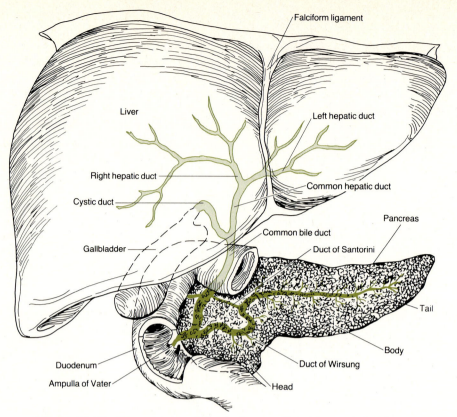

Figure 24-13 Pancreas in relation to liver, gallbladder, and duodenum.

called the *pancreatic duct* or *duct of Wirsung*. In most people, the pancreatic duct unites with the common bile duct from the liver and gallbladder and enters the duodenum as a single duct called the *ampulla of Vater*. Scattered among the alveoli are the *islets of Langerhans*, clusters of endocrine cells, that empty their secretions of insulin and glucagon into the adjacent blood vessels. There are about a million islets in the pancreas, and they receive a very rich blood supply.

Each day, the pancreas empties 1,200 to 1,500 ml (1.0 to 1.5 qt) of a clear, colorless liquid called *pancreatic juice* into the duodenum. Pancreatic juice consists mostly of water, some salts, sodium bicarbonate,

Exhibit 24-6 GASTRIC DIGESTION

Structure	Activity	Result
Mucosa		
Rugae	Provide large surface area for stretching of stomach.	Allow for distension of stomach.
Mucous cells	Secrete mucus.	Prevents digestion of stomach wall.
	Secrete intrinsic factor.	Required for erythrocyte formation.
Zymogenic cells	Secrete pepsinogen.	Its active form (pepsin) digests proteins into proteoses and peptones.
Parietal cells	Secrete hydrochloric acid.	In its presence, pepsinogen is converted into pepsin.
Muscularis	Mixing waves.	Macerate food, mix it with gastric juice, and reduce food to chyme.
	Peristaltic waves.	Force chyme through pyloric valve into duodenum.
Lower esophageal sphincter	Regulates passage of bolus from esophagus into stomach.	Prevents backflow of food from stomach to esophagus.
Pyloric valve	Opens to permit passage of chyme into duodenum.	Prevents backflow of food from duodenum to stomach.

and enzymes. The sodium bicarbonate gives pancreatic juice a slightly alkaline pH (7.1 to 8.2) that stops the action of pepsin from the stomach and creates the proper environment for the enzymes in the small intestine. The enzymes in pancreatic juice include *pancreatic amylase;* a carbohydrate-digesting enzyme, *trypsin; chymotrypsin* and *carboxypeptidase,* protein-digesting enzymes; *pancreatic lipase,* which breaks neutral fats into glycerol and fatty acids; and the enzymes *ribonuclease* and *deoxyribonuclease,* which hydrolyze RNA and DNA, respectively.

The exocrine pancreas protects itself from autodigestion in two ways. First, it synthesizes the proteolytic enzymes in the form of inactive proenzymes. That is, the enzyme trypsin is produced within the cells as *trypsinogen,* a polypeptide chain possessing 229 amino acid residues. When the terminal six amino acids are removed, the active enzyme trypsin is released to act upon protein and produce the short-chain polypeptides known as proteoses and peptides. Trypsinogen is activated by an enzyme in the mucosa called *enterokinase.* Similarly, *chymotrypsinogen* is a polypeptide with 246 amino acids that is activated by an attack with trypsin. Fifteen terminal amino acids are removed to release the active chymotrypsin molecule, which then attacks proteins at different linkages from the active trypsin enzyme so that their combined effect is to digest protein molecules at many locations to produce short-chain residues. *Procarboxypeptidase* is also activated by trypsin to the active carboxypeptidase form that attacks proteins starting at the C terminal end.

A second method preventing autodigestion within the pancreas is the production of *trypsin inhibitor,* a small polypeptide of molecular weight around 5,500. This small molecule combines with trypsinogen to prevent its activation while it is stored within the cells, acini, or ducts of the pancreas. If a duct becomes blocked or damaged, secretions are backed up, and their concentrations increase to the point where they overpower the trypsin inhibitor. This permits trypsinogen activation and the subsequent activation of the other proteolytic enzymes, with the rapid end result being digestion of the pancreas.

Regulation of Pancreatic Secretion

The pancreas is controlled by both neural and hormonal mechanisms. The neural mechanisms, though not insignificant, are not as active as the hormonal ones.

The neural mechanism is mediated almost entirely by branches of the vagus nerve, which terminate not only upon the acinar cells and smooth muscles of the ducts, but also upon the islet cells. Increase in the tone of the vagus results in secretion of enzymes into the pancreatic juice. This secretion is included in the cephalic phases of salivary and gastric digestion.

The hormonal mechanisms are somewhat more elaborate. Two hormones are involved, and the end product of the actions are very different. *Secretin* is a very small molecule that contains only 27 amino acids (see Figure 24-12c). When chyme enters the duodenum or jejunum, the mucosal cells release the hormone secretin into the bloodstream. Upon reaching the pancreas, the hormone causes production of a watery secretion that is rich in *bicarbonate ion* but contains virtually no enzymes. The most adequate stimulus for secretin release from the intestinal wall is the acidity (low pH) of the chyme entering the duodenum. The end result of this secretory activity is the neutralization of the stomach acid still mixed with the partially digested food. The following chemical reaction occurs:

$$\underset{\substack{\text{Hydrochloric} \\ \text{acid}}}{\text{HCl}} + \underset{\substack{\text{Sodium} \\ \text{bicarbonate}}}{\text{NaHCO}_3} \rightarrow \underset{\substack{\text{Sodium} \\ \text{chloride}}}{\text{NaCl}} + \underset{\substack{\text{Carbonic} \\ \text{acid}}}{\text{H}_2\text{CO}_3}$$

The neutralization of the hydrochloric acid and the consequent elevation of pH to values near 8 provide a favorable environment for the pancreatic enzymes whose optimal activity occurs at slightly alkaline pH values.

Cholecystokinin is the second hormone released from the intestinal mucosa. This hormone contains 33 amino acids and is somewhat similar to gastrin in portions of its amino acid sequencing (see Figure 24-12b). When cholecystokinin reaches the pancreas, it promotes the pancreas to release quantities of enzymes into the pancreatic juice. The most adequate stimulus for cholecystokinin apparently is distension of the intestinal wall by the entrance of food.

Liver

The **liver** performs so many vital functions that we cannot live long without it:

1. The liver manufactures the anticoagulant heparin and most of the other plasma proteins.
2. The reticuloendothelial cells of the liver phagocytose worn-out red blood cells, and some bacteria.
3. Liver cells contain enzymes that either break down poisons or transform them into less harmful compounds. When amino acids are burned for energy, they leave behind toxic nitrogenous wastes that are converted to urea by the liver cells. Moderate amounts of urea are harmless to the body and are easily excreted by the kidneys and sweat glands.

4. Newly absorbed nutrients are collected in the liver. It can change any excess monosaccharides into glycogen or fat, both of which can be stored, or it can transform glycogen, fat, and protein into glucose, depending on the body's needs.

5. The liver stores glycogen, copper, iron, and vitamins A, D, E, and K. It also stores some poisons that cannot be broken down and excreted. (This is why high levels of DDT are found in the livers of animals, including humans, who have eaten sprayed fruits and vegetables.)

6. Finally, the liver manufactures bile, which is used in the small intestine for the digestion and absorption of fats.

The liver is the largest single organ in the body, weighing about 1.4 kg (4 lb) in the average adult. It is located under the diaphragm and occupies most of the right hypochondrium and part of the epigastrium of the abdomen. The liver is covered largely by peritoneum and completely by a dense connective tissue layer that lies beneath the peritoneum. Anatomically, the liver is divided into two principal lobes—the **right lobe** and the **left lobe** (Figure 24-14). The right, or main lobe, also has associated with it an inferior **quadrate lobe** and a posterior **caudate lobe.**

Microscopically the liver can be divided into functional units called *lobules.* A lobule consists of cords of *hepatic* (liver) *cells* arranged in a radial pattern around a central vein. Between the cords are endothelial-lined spaces called *sinusoids* through which blood passes. The sinusoids are also partly lined with phagocytic cells, termed *Kupffer cells,* that remove worn-out white and red blood cells (Figure 24-14c).

The liver receives a double supply of blood. From the hepatic artery, it obtains oxygenated blood, and from the hepatic portal vein it receives deoxygenated blood containing nutrients just absorbed from the intestinal tract. Branches of both the hepatic artery and the hepatic portal vein carry the blood into the sinusoids of the lobules, where oxygen, most of the nutrients, and certain poisons are extracted by the hepatic cells. Nutrients are stored or used to synthesize new materials, and the poisons are stored or detoxified. Products manufactured by the hepatic cells and nutrients needed by other cells are secreted into the blood. The blood then drains into the central vein and eventually is collected into a hepatic vein.

Bile is manufactured by the hepatic cells and secreted into *bile capillaries* or *canaliculi* so that it flows between the hepatocytes of a cord to empty into a small bile duct. These small ducts eventually merge to form the larger *right* and *left hepatic ducts,* which unite to leave the liver as the *common hepatic duct.*

Further on, the common hepatic duct joins the *cystic duct* from the gallbladder, and the two tubes become the *common bile duct,* which empties into the duodenum. The *sphincter of Oddi* is a valve in the common bile duct. When the small intestine is empty, the sphincter closes and the bile is forced up the cystic duct to the gallbladder, where it is stored.

Gallbladder

The **gallbladder** is a sac attached to the underside of the liver (see Figure 24-14b). Its inner walls consist of a mucous membrane arranged in rugae resembling those of the stomach. When the gallbladder fills with bile, the rugae allow it to expand to the size and shape of a pear. Bile is continually collected into the gallbladder where it is concentrated at least fivefold by the absorption of water, sodium, chloride, and other electrolytes.

The arrival of the partially digested food into the duodenum from the stomach, particularly when rich in fat and protein products, triggers the release of cholecystokinin from the intestinal mucosal cells into the blood. Upon reaching the gallbladder, cholecystokinin causes rhythmic contraction of the smooth muscle fibers in the muscular coat. Cholecystokinin also causes the relaxation of the sphincter of Oddi. The actions of cholecystokinin, combined with peristaltic waves that pass over the duodenum, permit the bile to squirt into the lumen of the gut.

Small Intestine

Continued digestion and the major portion of absorption occurs within a long tube called the **small intestine.** The small intestine begins at the pyloric valve of the stomach, coils through the central and lower part of the abdominal cavity, and eventually empties into the large intestine (see Figure 24-5). In a living human, it averages about 2.5 cm (1 inch) in diameter and 6.35 m (21 ft) in length.

Anatomy

The small intestine is divided into three segments: duodenum, jejunum, and ileum. The **duodenum,** the broadest part of the small intestine, originates at the pyloric valve of the stomach and extends about 25 cm (10 to 12 inches) until it merges with the jejunum. The **jejunum** is about 2.5 m (8 ft) long and gradually merges into the final portion of the small intestine, the **ileum.** This measures 3.6 m (12 ft) long and joins the large intestine at the *ileocecal valve.*

The wall of the small intestine is composed of the same four tunics or coats that make up most of the GI

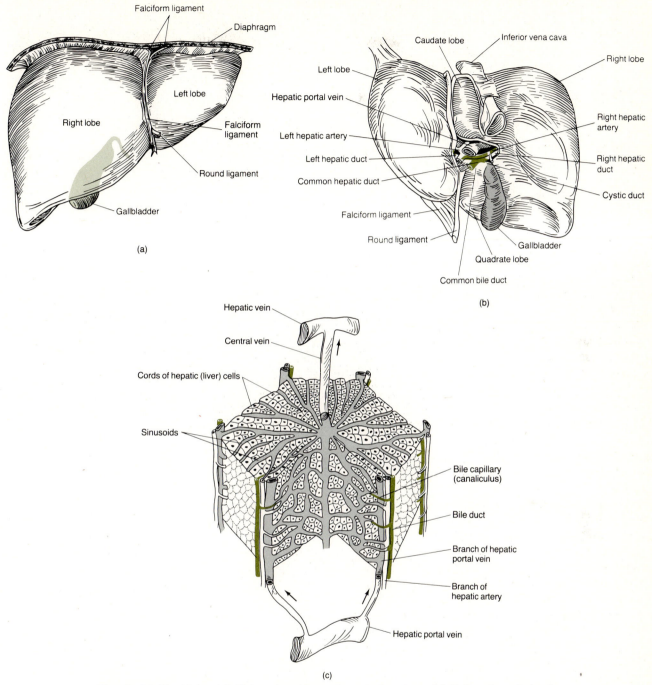

Figure 24-14 Liver. (a) External anatomy in anterior view. (b) External anatomy in posteroinferior view. (c) Diagrammatic representation of the microscopic appearance of a lobule.

tract. However, both the mucosa and submucosa are modified to allow the small intestine to complete the processes of digestion and absorption. The mucosa contains many pits lined with glandular epithelium. These pits—the intestinal glands, or crypts of Lieberkühn—secrete the intestinal juice, which is almost pure extracellular fluid, devoid of enzymes (Figure 24-15). The submucosa of the duodenum contains *Brunner's glands*, which secrete mucus to protect the walls of the small intestine from the action of the gastric enzymes still present in the partially digested food.

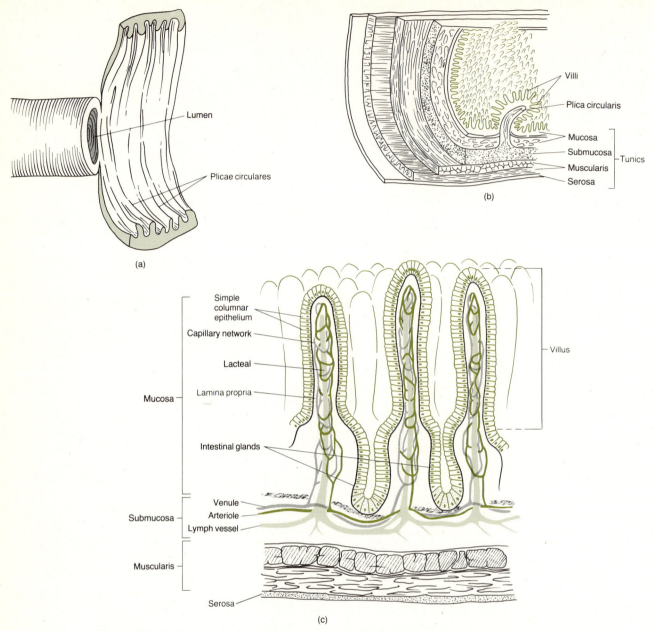

Figure 24-15 Small intestine. (a) Section of small intestine cut open to expose plicae circulares. (b) Villi in relation to the tunics of the small intestine. (c) Enlarged aspect of several villi.

Since almost all the absorption of nutrients occurs in the small intestine, its walls need to be specially equipped to do this job. The epithelium covering and lining the mucosa consists of simple columnar epithelium. Some of the epithelial cells have been transformed to goblet cells, which secrete additional mucus. The rest contain microvilli—fingerlike projections of the plasma membrane. The final steps of digestion occur on the microvilli of the epithelial cells. Enzymes, such as sucrase, maltase, lactase, and intestinal lipase, are fixed on the outer surface of the cell membrane. Therefore, the final digestive steps and the absorption of the end products both occur at the same place and at the same time. The tremendous surface area created by the microvilli enables both processes to occur rapidly.

The mucosa lies in a series of villi, projections 0.5 to 1.5 mm high, giving the intestinal mucosa its velvety appearance. The enormous number of villi (4 to 5 million) also serve to vastly increase the surface area of epithelium available for absorption. Each villus has a core of lamina propria, the connective tissue layer of the mucosa. Embedded in this connective tissue are an artery, a venule, a capillary, and a *lacteal* (lymphatic vessel). Nutrients that diffuse through the adjacent epithelial cells are able to pass through the walls of the capillary and lacteal and enter the blood.

In addition to the microvilli and villi, a third set of projections called plicae circulares further increase the surface area for absorption. The plicae are permanent deep folds in the mucosa and submucosa. Some of the folds extend all the way around the intestine, and others extend only part way around.

There is an abundance of lymphatic tissue in the walls of the small intestine. Single lymph nodules, called *solitary lymph nodules*, are most numerous in the lower part of the ileum. Aggregated lymph nodules, referred to as *Peyer's patches*, are also most numerous in the ileum.

Chemical Digestion

The digestion of carbohydrates, proteins, and lipids in the small intestine requires the combined actions of the secretions from the pancreas, liver, and intestinal glands.

Secretions. Each day the liver secretes about 800 to 1,000 ml (almost 1 qt) of the yellow, brownish, or olive-green liquid called *bile*. Bile consists of water, bile salts, bile acids, a number of lipids, and two pigments called biliverdin and bilirubin. Bile is partially an excretory product and partially a digestive secretion. When worn-out red blood cells are removed from the circulation and broken down by the liver cells, iron, globin, and bilirubin are released. The iron and globin are recycled, but the bilirubin is excreted into the bile ducts. Bilirubin eventually is broken down in the intestines, and its breakdown products give feces their color. If the liver is unable to export its bile because the bile ducts are obstructed, large amounts of bilirubin back up into the liver sinusoids. The pigment thus enters the blood to circulate through the bloodstream and collect in other tissues, giving the skin and eyes a yellow color. This condition is called *jaundice*. The most important secretory products of the liver are substances known as bile salts, which aid in the digestion of fats and are required for their absorption.

The intestinal juice, or *succus entericus,* is a clear yellow fluid secreted in amounts of about 2 to 3 liters (2 to 3 qt) per day. It has a pH of 7.6, which is slightly alkaline, and contains water, mucus, and a low concentration of enzymes from other sources that complete the digestion of carbohydrates and proteins.

The hormone control of digestion is summarized in Exhibit 24-7.

The Digestive Process. When chyme reaches the small intestine, the carbohydrates and proteins have been digested considerably but are not yet ready for absorption. Lipid digestion has not even begun. Digestion in the small intestine continues as follows:

1. **Carbohydrates.** In the mouth, the carbohydrates are broken down into dextrins containing several monosaccharide units (Figure 24-16). Even though the action of salivary amylase may continue in the stomach, very few of the carbohydrates are reduced to disaccharides by the time chyme leaves the stomach. *Pancreatic amylase,* an enzyme in pancreatic juice, breaks dextrins into the disaccharides maltose, sucrose, and lactose. Next, three enzymes secreted by the intestinal mucosa digest the disaccharides into monosaccharides. *Maltase* splits maltose into two molecules of glucose, *sucrase* breaks sucrose into a molecule of glucose and a molecule of fructose, and *lactase* digests lactose into a molecule of glucose and a molecule of galactose. This completes the digestion of carbohydrates.

2. **Proteins.** Protein digestion starts in the stomach, where most of the proteins are fragmented into short chains of amino acids called peptones and proteoses (Figure 24-17). Three enzymes found in pancreatic juice continue the digestion. *Trypsin*

Exhibit 24-7   SUMMARY OF THE HORMONAL CONTROL OF DIGESTION IN THE STOMACH AND SMALL INTESTINE

Hormone	Where Produced	Stimulant	Action
Gastrin	Pyloric mucosa	Partially digested proteins	Causes gastric glands to secrete gastric juice.
Gastrinlike hormone	Intestinal mucosa	Partially digested proteins	Same as above.
Secretin	Intestinal mucosa	Acidity of chyme	Stimulates secretion of pancreatic juice rich in carbonate and promotes production of bile by liver.
Enterocrinin	Intestinal mucosa	Acidity of chyme	Stimulates secretion of succus entericus.
Cholecystokinin	Intestinal mucosa	Combination of acid and fat	Stimulates secretion of pancreatic juice rich in enzymes and causes ejection of bile from gallbladder and opening of sphincter of Oddi.

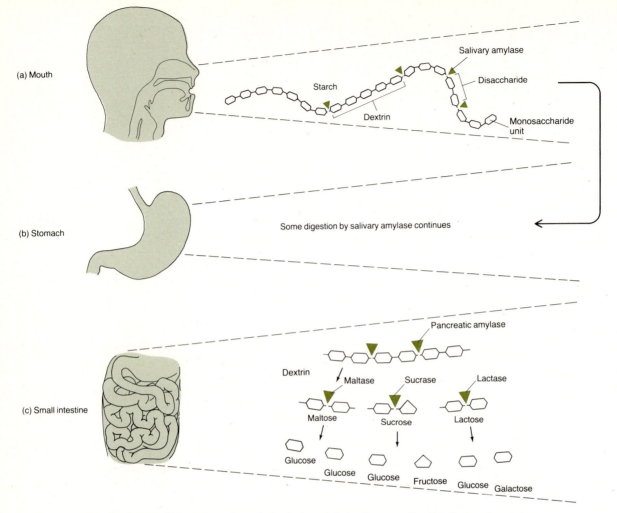

Figure 24-16 Digestion of carbohydrates. (a) In the mouth. (b) Some breakdown continues in the stomach. (c) In the small intestine.

digests any intact proteins into peptones and proteoses, breaks the peptones and proteoses into dipeptides (containing only two amino acids), and breaks some of the dipeptides into single amino acids. *Chymotrypsin* duplicates trypsin's activities. *Carboxypeptidase,* the third enzyme, reduces whole or partially digested proteins to amino acids. Protein digestion is completed by several intracellular intestinal enzymes that include amino polypeptidase and a variety of dipeptidases. They convert all the remaining dipeptides into single amino acids.

3. **Lipids.** In an adult, almost all lipid digestion occurs in the small intestine. The first step in the process is the *emulsification* of fats, which is a function of bile. Neutral fats, or just fats, are the most abundant lipids in the diet. They are also called triglycerides because they consist of a molecule of glycerol bonded to three molecules of fatty acid (Figure 24-18). Bile salts break the globules of fat into tiny droplets (emulsification) so the fat-splitting enzyme can get at the lipid molecules more easily. In the second step, *pancreatic lipase* hydrolyzes each fat molecule into glycerol and fatty acids, the end products of fat digestion.

Mechanical Digestion

In the small intestine, three distinct types of movement occur as a result of contractions of the longitudinal and circular muscles. These movements are *rhythmic segmentation, pendular movements,* and *propulsive peristalsis.* Rhythmic segmentation and pendular movements are strictly localized contractions occurring in areas containing food. The two movements mix the chyme with the digestive juices and bring every particle of food into contact with the mucosa for absorption. They do not push the intestinal contents along the tract. Rhythmic segmentation starts with the contractions of some of the circular muscle fibers in a portion of the intestine, an action

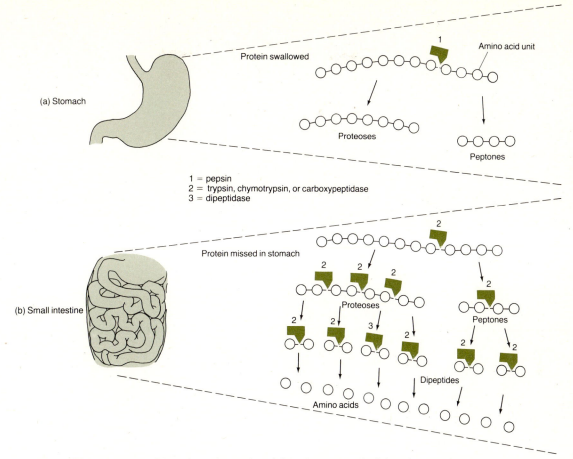

1 = pepsin
2 = trypsin, chymotrypsin, or carboxypeptidase
3 = dipeptidase

Figure 24-17 Digestion of proteins. (a) In the stomach. (b) In the small intestine.

that constricts the intestine into segments. Next, muscle fibers that encircle the middle of each segment also contract, dividing each segment into two smaller segments. Finally, the fibers that contracted first relax, and each small segment unites with an adjoining small segment so that large segments are re-formed. This sequence of events is repeated at a rate of 12 to 16 times a minute, spilling the chyme back and forth. Pendular movements consist of alternating contractions and relaxations of the longitudinal muscles. The contractions cause a portion of the intestine to shorten and lengthen, which also sends the chyme spilling back and forth.

The third kind of movement, propulsive peristalsis, propels the chyme onward through the intestinal tract. Peristaltic movement in the intestine is similar to that in the esophagus. In the intestine, these waves may be as slow as 5 cm (2 inches)/minute or as fast as 50 cm (20 inches)/second.

Absorption

All the chemical and mechanical phases of digestion that start in the mouth and continue into the small intestine are directed toward changing foods into forms that can diffuse through the epithelial cells lining the mucosa into the underlying blood and lymph vessels. The diffusible forms are monosaccharides (glucose, fructose, and galactose), amino acids, fatty acids, glycerol, and glycerides. Passage of these digested nutrients from the alimentary canal into the blood or lymph is called **absorption.**

About 90 percent of all absorption takes place throughout the length of the small intestine. The other 10 percent occurs in the stomach and large intestine. Absorption of materials in the small intestine occurs specifically through the villi (see Figure 24-15c) and depends on diffusion, facilitated diffusion, osmosis, and active transport. Monosaccharides and amino acids are actively absorbed into the blood capillaries of the villi and are transported in the bloodstream to the liver via the hepatic portal system. Fatty acids, glycerol, and glycerides do not enter the bloodstream immediately. They cluster together and become surrounded by bile salts to form water-soluble particles called *micelles.* The micelles attach themselves to the surface of the epithelial cells, where the triglycerides and fatty acids are absorbed through the cell membrane. The bile salts are released and

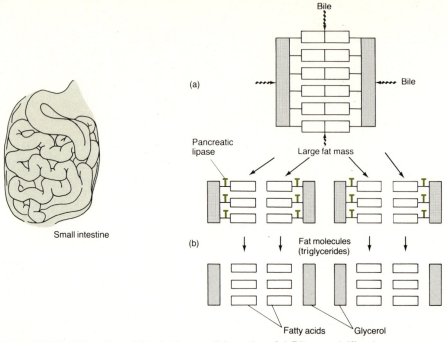

Figure 24-18 Digestion of fats in the small intestine. (a) Bile emulsifies large masses of fat into smaller fragments so that pancreatic lipase can break down the fat molecules. (b) The enzyme pancreatic lipase breaks fats into fatty acids, glycerol, and glycerides.

return to the chyme in the intestinal tract to absorb more fats. The bile salts are reused until all the fats have been removed or moved further through the intestinal tract. Then the bile salts, themselves, are absorbed in the distal end of the ileum. Eventually, they are returned to the liver and gallbladder. In the course of a day the same salts may be reused several times. Only about 5 percent of the bile salts are lost in each cycle via the feces. The absorbed fats in the intestinal epithelial cells enter the smooth endoplasmic reticulum, where the triglycerides are resynthesized. The triglycerides, along with small quantities of phospholipids and cholesterol, are then organized into protein-coated lipid droplets called *chylomicrons*. The protein coat keeps the chylomicrons suspended and prevents them from sticking to each other or to the walls of the lymphatics or blood vessels. Small chylomicrons leave the intestinal cells and enter the blood capillaries in the villus. Larger chylomicrons enter the lacteal in the villus and are transported by way of lymphatic vessels to the thoracic duct, and there they enter the cardiovascular system at the left subclavian vein. Finally, they arrive at the liver through the hepatic artery. Most of the products of carbohydrate, protein, and lipid digestion are processed by the liver before they are delivered to the other cells of the body. Large amounts of water, electrolytes, mineral salts, and some vitamins also are absorbed in the small intestine.

In summary, then, the principal chemical activity of the small intestine is to complete the digestion of all

foods into forms that are usable by body cells (Exhibit 24-8). Any undigested materials that are left behind are processed in the large intestine.

Large Intestine

The overall functions of the large intestine are the completion of absorption, the manufacture of some vitamins, the formation of feces, and the expulsion of feces from the body.

Anatomy

The **large intestine** is about 1.5 m (5 ft) long, averages 6.5 cm (2.5 inches) in diameter, and extends from the ileum to the anus. Structurally, the large intestine is divided into four principal regions: the cecum, colon, rectum, and anal canal (Figure 24-19).

The opening from the ileum into the large intestine is guarded by a fold of mucous membrane called the *ileocecal valve*. This structure allows materials from the small intestine to pass into the large intestine. Hanging below the ileocecal valve is the *cecum*, a blind pouch about 6 cm (2 to 3 inches) long. Extending from the cecum is a twisted, coiled tube, measuring about 8 cm (3 inches) in length on the average, that is called the *vermiform appendix* (*vermis* = worm). Inflammation of the appendix is called *appendicitis*.

The open end of the cecum merges with a long tube called the *colon*. Based on location, the colon is divided into ascending, transverse, descending, and

Exhibit 24-8 DIGESTION AND ABSORPTION IN THE SMALL INTESTINE

Structure	Description	Function
Pancreas (pancreatic juice)		
Trypsin	Protein-digesting enzyme activated by the intestinal enzyme enterokinase	Digests intact proteins into proteoses and peptones. Digests partially digested proteins into dipeptides plus some amino acids.
Chymotrypsin	Protein-digesting enzyme activated by trypsin	Same as above.
Carboxypeptidase	Protein-digesting enzyme activated by trypsin	Reduces proteins to amino acids.
Pancreatic amylase	Enzyme that digests carbohydrates	Converts dextrins into disaccharide maltose.
Pancreatic lipase	Fat-splitting enzyme	Converts neutral fat (triglyceride) into fatty acids, glycerol, and glycerides.
Liver (bile)	Bile salts	Emulsifies neutral fats in preparation for digestion by pancreatic lipase. Bile salts also allow products of neutral fat digestion to be absorbed.
Small intestine		
Mucosa and submucosa		
Intestinal glands	Secrete succus entericus	
Maltase		Converts disaccharide maltose into monosaccharide glucose.
Sucrase		Converts disaccharide sucrose into monosaccharides glucose and fructose.
Lactase		Converts disaccharide lactose into monosaccharides glucose and galactose.
Dipeptidases		Change dipeptides into amino acids.
Microvilli	Projections of plasma membranes of intestinal epithelial cells	Increase cellular surface area for absorption.
Villi	Fingerlike projections of mucous membrane	Serve as sites for absorption of digested foods and increase absorptive area.
Plicae circulares	Circular folds of mucosa and submucosa	Increase surface area for digestion and absorption.
Goblet cells	Secrete mucus	Lubricates foods and protects mucosa.
Intestinal glands	Secrete intestinal digestive juices	Digest carbohydrates and proteins.
Brunner's glands	Secrete mucus	Lubricates foods and protects mucosa.
Solitary lymph nodules	Lymphatic tissue associated with small intestine	Filter lymph.
Peyer's patches	Collection of large numbers of lymph nodules in distal end of ileum	Filter lymph. Defense against bacteria.
Muscularis		
Rhythmic segmentation	Alternating contractions of circular fibers produce segmentation and resegmentation of portions of small intestine	Mixes chyme with digestive juices and brings food into contact with mucosa for absorption.
Pendular movement	Contractions of longitudinal muscle pull portions of intestine forward and backward	Same as for rhythmic segmentation.
Peristalsis	Waves of contraction and relaxation of circular and longitudinal muscle passing length of small intestine	Moves chyme forward.

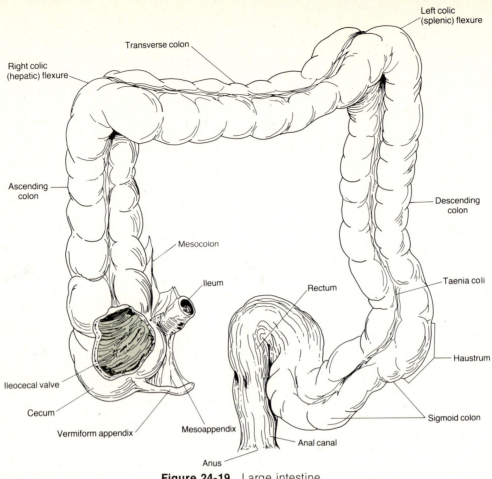

Figure 24-19 Large intestine.

Labels on figure:
- Transverse colon
- Right colic (hepatic) flexure
- Left colic (splenic) flexure
- Ascending colon
- Descending colon
- Mesocolon
- Ileum
- Rectum
- Taenia coli
- Ileocecal valve
- Cecum
- Vermiform appendix
- Mesoappendix
- Anal canal
- Anus
- Haustrum
- Sigmoid colon

sigmoid portions. The *ascending colon* ascends on the right side of the abdomen, reaches the undersurface of the liver, and turns abruptly to the left. The colon continues across the abdomen to the left side as the *transverse colon*. It curves beneath the lower end of the spleen on the left side and passes downward to the level of the iliac crest as the *descending colon*. The *sigmoid colon* is the S-shaped portion that begins at the iliac crest, projects inward to the midline, and terminates as the rectum at about the level of the third sacral vertebra.

The *rectum,* the last 20 cm (7 to 8 inches) of gastrointestinal tract, lies anterior to the sacrum and coccyx. The terminal 2 to 3 cm of the rectum is referred to as the *anal canal* (Figure 24-20). Internally, the mucous membrane of the anal canal is arranged in longitudinal folds called *anal columns* that contain a network of arteries and veins. Inflammation and enlargement of the anal veins is known as *hemorrhoids* or *piles*. The opening of the anal canal to the exterior is called the *anus*. It is guarded by an internal sphincter of smooth muscle and an external sphincter muscle. Normally the anus is closed except during the elimination of the wastes of digestion.

The wall of the large intestine differs from that of the small intestine in some respects. No villi or permanent circular folds are found in the mucosa. The mucosa does, however, contain simple columnar epithelium with numerous goblet cells that secrete mucus that lubricates the colonic contents as they pass through the colon. Solitary lymph nodes also are found in the mucosa. The submucosa of the large intestine is similar to that found in the rest of the alimentary canal. The muscularis consists of an external layer of longitudinal muscles and an internal layer of circular muscles. Unlike other parts of the digestive tract, however, the longitudinal muscles do not form a continuous sheet around the wall but are broken up into three flat bands called *taeniae coli* (see Figure 24-19). Each band runs the length of the large intestine. Tonic contractions of the bands gather the colon into a series of pouches called *haustra,* which give the colon its puckered appearance. The serosa of the large intestine is part of the visceral peritoneum.

Activities of the Large Intestine

The principal activities of the large intestine are mechanical movements, absorption, and the formation and elimination of feces.

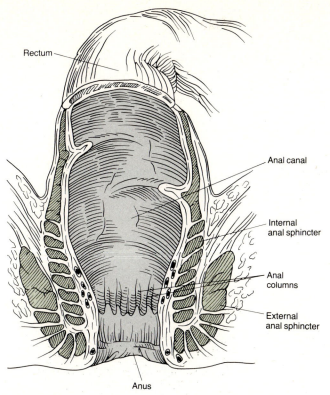

Rectum

Anal canal

Internal anal sphincter

Anal columns

External anal sphincter

Anus

Figure 24-20   Anal canal seen in longitudinal section.

Movements. Movements of the colon begin when substances enter through the ileocecal valve. Since chyme moves through the small intestine at a fairly constant rate, the time required for a meal to pass into the colon is determined by gastric evacuation time. As food passes through the ileocecal valve, it fills the cecum and accumulates in the ascending colon.

One movement characteristic of the large intestine is *haustral churning.* In this process, the haustra remain relaxed and distended while they fill up. When the distension reaches a certain point, the walls contract and squeeze the contents into the next haustrum. Peristalsis also occurs, although at a slower rate than in other portions of the tract (3 to 12 contractions per minute). A final type of movement is *mass peristalsis,* a very strong peristaltic wave that drives the colonic contents into the rectum. Food in the stomach initiates this reflex action in the colon. Thus, mass peristalsis usually takes place three or four times a day, during or immediately after a meal.

Absorption and Formation of Feces. By the time the intestinal contents arrive at the large intestine, digestion and absorption are almost complete.

By the time the chyme has remained in the large intestine for about 3 to 10 hours, it has become solid or semisolid as a result of water absorption and is now known as *feces.* Chemically, feces consist of water, inorganic salts, mucus, epithelial cells from the mucosa of the alimentary canal, bacteria, products of bacterial decomposition, and undigested parts of food not attacked by bacteria.

The glands of the large intestine secrete mucus but not enzymes. The mucus serves as a lubricant to aid the movement of the colonic materials and acts as a protective covering for the mucosa.

Chyme is prepared for elimination in the large intestine by the action of bacteria. These bacteria ferment any remaining carbohydrates and release hydrogen, carbon dioxide, and methane gas. They also convert remaining proteins to amino acids and break down the amino acids into simpler substances, such as indole, skatole, hydrogen sulfide, and fatty acids. Some of the indole and skatole is carried off in the feces and contributes to its odor, and the rest is absorbed. Bacteria also decompose bilirubin, the breakdown product of red blood cells that is excreted in bile, to simpler pigments that give feces its color. Intestinal bacteria also aid in the synthesis of several vitamins needed for normal metabolism, including some B vitamins (riboflavin, nicotinic acid, biotin, and folic acid) and vitamin K.

Although most water absorption occurs in the small intestine, the large intestine absorbs enough to make it an important organ in maintaining the water balance of the body. Water absorption in the large intestine is greatest in the cecum and ascending colon. The large intestine also absorbs some inorganic solutes and some products of bacterial action, including vitamins and large amounts of indole and skatole. The indole and skatole are transported to the liver, where they are converted to less toxic compounds that are excreted in the urine.

Defecation. Mass peristaltic movement pushes fecal material from the sigmoid colon into the rectum. The resulting distension of the rectal walls stimulates pressure-sensitive receptors, initiating a reflex for *defecation,* which is emptying of the rectum. Contraction of the longitudinal rectal muscles shortens the rectum, thereby increasing the pressure inside it. The pressure forces the sphincters open, and the feces are expelled through the anus. Voluntary contractions of the diaphragm and abdominal muscles aid defecation by increasing the pressure inside the abdomen, which pushes the walls of the sigmoid colon and rectum inward. If defecation does not occur, the feces remain in the rectum until the next wave of peristalsis again stimulates the pressoreceptors, creating an awareness of the desire to defecate.

Activities of the large intestine are summarized in Exhibit 24-9.

Exhibit 24-9 SUMMARY OF DIGESTIVE ACTIVITIES OF THE LARGE INTESTINE

Structure	Action	Function
Mucosa	Secretes mucus	Lubricates the colon and protects the mucosa.
	Absorbs water and other soluble compounds	Water balance is maintained and feces become solidified. Vitamins and minerals are obtained, and toxic substances are sent to the liver to be detoxified.
	Bacterial action	Undigested carbohydrates, proteins, and amino acids are broken down so they can be expelled in feces or absorbed and detoxified by liver. Certain vitamins are synthesized.
Muscularis	Haustral churning	Haustra fill and contract, moving contents from haustra to haustra.
	Peristalsis	Contractions of circular and longitudinal muscles continually move contents along length of colon.
	Mass peristalsis	Strong peristaltic wave forces contents into sigmoid colon and rectum.
	Defecation	Contractions in sigmoid colon and rectum rid the body of feces.

HOMEOSTATIC IMBALANCES OF THE DIGESTIVE SYSTEM

Now that we have discussed the physiology of the digestive system, we shall look at some disorders that are related to it.

Dental Caries

Dental caries, or tooth decay, involves a gradual demineralization (softening) of the enamel and dentin. If this condition remains untreated, various microorganisms may invade the pulp, causing inflammation and infection with subsequent death (necrosis) of the dental pulp and abscess of the alveolar bone surrounding the root's apex. Such teeth are treated by root canal therapy.

The process of dental caries is initiated when bacteria act on carbohydrates deposited on the tooth, giving off acids that demineralize the enamel. Microbes that digest carbohydrates include two bacteria, *Lactobacillus acidophilus* and *Streptococcus mutans*. Research suggests that the streptococci break down carbohydrates into *dental plaque,* a polysaccharide that adheres to the tooth surface. When other bacteria digest the plaque, acid is produced. Saliva cannot reach the tooth surface to buffer the acid because the plaque covers the teeth.

Certain measures can be taken to prevent dental caries. First, the diet of the mother during pregnancy is very important in forestalling tooth decay of the newborn. Simple, balanced meals are the best diet during pregnancy. Supplementation with multivitamins, with emphasis on vitamin D, and the minerals calcium and phosphorus is customary because they are responsible for the normal development of bones and teeth.

Other preventive measures have centered around fluoride treatment because teeth are less susceptible to acids when they are permeated with fluoride. Fluoride may be incorporated in the drinking water or applied topically to erupted teeth. Maximum benefit occurs when fluoride is present in drinking water during the period when teeth are being calcified. Naturally occurring excessive fluoride may cause a light brown to brownish-black discoloration of the enamel of the permanent teeth called mottling.

Brushing the teeth immediately after eating removes the plaque from flat surfaces before the bacteria have a chance to go to work. Dentists also suggest that the plaque between the teeth be removed every 24 hours with dental floss, or flushing with a water irrigation device.

Periodontal Diseases

Periodontal disease is a collective term for a variety of conditions characterized by inflammation and/or degeneration of the gingivae, alveolar bone, periodontal ligament, and cementum. The initial symptoms are enlargement and inflammation of the soft tissue and bleeding gums. Without treatment, the soft tissue may deteriorate and the alveolar bone may be resorbed, causing loosening of the teeth and receding of the gums.

Periodontal diseases are frequently caused by local irritants, such as bacteria, impacted food, and cigarette smoke, or by a poor "bite." The latter may put a strain on the tissues supporting the teeth. Methods of prevention and treatment include good mouth care to remove plaque and other sources of irritation. Periodontal diseases may also be caused by allergies, vitamin deficiencies, and a number of systemic disorders, especially those that affect bone, connective tissue, or circulation. In these cases, the systemic disorder must be treated as well.

Peptic Ulcers

An **ulcer** is a craterlike lesion in a membrane. Ulcers that develop in areas of the alimentary canal exposed to acid gastric juice are called **peptic ulcers.** Peptic ulcers occasionally develop in the lower end of the esophagus. However, most of them occur on the lesser curvature of the stomach, in which case they are called **gastric ulcers,** or in the first part of the duodenum, where they are called **duodenal ulcers.**

The cause of ulcers is obscure. However, hypersecretion of an acidic gastric juice seems to be the immediate cause in the production of duodenal ulcers and in the reactivation of healed ulcers. Hypersecretion of acidic gastric juice is not implicated as much in gastric ulcer patients because the stomach walls are highly adapted to resist gastric juice through their secretion of mucus. A possible cause of gastric ulcers is hyposecretion of mucus. Hypersecretion of pepsin also may contribute to ulcer formation.

Among the factors believed to stimulate an increase in acid secretion are emotions, certain foods or medications, alcohol, coffee or aspirin, and overstimulation of the vagus nerve. Normally, the mucous membrane lining the stomach and duodenal walls resists the secretions of hydrochloric acid and pepsin, and no ulcer develops. In some people, however, this resistance breaks down, and an ulcer develops.

The danger inherent in ulcers is the eventual erosion into the muscular portion of the wall of the stomach or duodenum. At this level, damage to blood vessels could produce a fatal hemorrhage. If an ulcer continues and eventually erodes all the way through the wall, the condition is called *perforation.* Perforation allows bacteria and partially digested food in the stomach or intestine to pass into the peritoneal cavity, producing peritonitis.

Peritonitis

Peritonitis is an acute inflammation of the serous membrane lining the abdominal cavity and covering the abdominal viscera. One possible cause is contamination of the peritoneum by pathologic bacteria from the external environment. This contamination could result from accidental or surgical wounds in the abdominal wall or from perforation or rupture of organs exposed to the outside environment. Another possible cause is perforation of the walls of organs that contain bacteria or chemicals that are normally beneficial to the organ but are toxic to the peritoneum. For example, the large intestine contains colonies of bacteria that live on undigested nutrients and break them down so they can be eliminated more easily. But, if the bacteria enter the peritoneal cavity, they attack the cells of the peritoneum for food and produce acute infection. As another example, the normal bacteria of the female reproductive tract protect the tract by giving off acid wastes that produce an acid environment unfavorable to many yeasts, protozoa, and bacteria which might otherwise attack the tract. However, these acid-producing bacteria are harmful to the peritoneum. A third cause may be chemical irritation. The peritoneum does not have any natural barriers that keep it from being irritated or digested by chemical substances such as bile and digestive enzymes. However, it does contain a great deal of lymphatic tissue and can fight infection extremely well. The danger stems from the fact that the peritoneum is in contact with most of the abdominal organs. If the infection gets out of hand, it may destroy vital organs and bring on death. For these reasons, perforation of the alimentary canal from an ulcer or perforation of the uterus from an incompetent abortion are considered serious. A patient who is going to have extensive surgery on the colon may be given high doses of antibiotics for several days preceding surgery to kill intestinal bacteria and reduce the risk of peritoneal contamination.

Cirrhosis

Cirrhosis is a chronic disease of the liver in which the parenchymal (functional) liver cells are replaced by fibrous connective tissue, a process called *stromal repair.* Often, there is a lot of replacement with adipose connective tissue as well. The liver has a high ability for parenchymal regeneration, so stromal repair occurs whenever any parenchymal cell is killed or when damage to the cells occurs continuously over a long time. These conditions could be caused by *hepatitis* (inflammation of the liver), certain chemicals that may destroy liver cells, parasites that sometimes infect the liver, and alcoholism.

Tumors

Both **benign** and **malignant tumors** occur in all parts of the gastrointestinal tract. The benign growths are much more common, but the malignant tumors are responsible for 30 percent of all deaths from cancer in the United States. To achieve relative early diagnosis, complete, periodic routine examinations are necessary. Cancers of the mouth usually are detected through routine dental checkups.

A regular physical checkup should include a rectal examination. Fifty percent of all rectal carcinomas are within reach of the finger, and 75 percent of all colonic carcinomas can be seen with the sigmoidoscope. Both the fiber optic sigmoidoscope and the more recent fiber optic endoscope are flexible tubular instruments composed of a light and many tiny glass fibers. They allow visualization, magnification, biopsy, electrosurgery, and even photography of the entire length of the gastrointestinal tract. The greatest contribution of colonoscopy may be its ability to allow identification and removal of malignant polyps of the colon (gastric polypectomy) before invasion of the bowel wall or lymphatic metastasis occurs. It has proved to be a safe and effective method of treatment that avoids the significant patient risk, discomfort, and inconvenience of major surgery and is more economical as well. It may prove to be the most important advance toward lowering the death rate from cancer of the colon that has appeared in the last quarter century. Unfortunately, this type of cancer has shown a considerable increase in incidence over the last 20 years.

A test in a routine examination for intestinal disorders is the filling of the gastrointestinal tract with barium solution, which is either swallowed or given in an enema. Barium, a mineral, shows up on x-rays the same way that calcium appears in bones. Tumors as well as ulcers can be diagnosed this way. The only definitive treatment of gastrointestinal carcinomas is surgery.

CHAPTER SUMMARY OUTLINE

PHYSIOLOGY OF DIGESTION

1. The normal steady weight condition is the result of two opposing sets of control systems—one regulating the amount of food taken into the body and the other regulating the utilization or disposal of the extracted energy.

HUNGER AND APPETITE

1. Hunger refers to the rather indiscriminate need for food that is derived from the energy requirements of the body.
2. Appetite refers to the more selective desire for a particular food that may or may not be related to the nutritional needs of the body.
3. Satiety describes the onset of hunger satisfaction or the feeling of satisfaction after energy needs have been met.
4. All of these food-regulating sensations are housed within nuclei of the hypothalamus.

FOOD MOLECULES

1. The three important classes of food molecules are the carbohydrates, lipids, and proteins.

2. Carbohydrates (also known as sugars and starches) form a large and diverse group of organic compounds in the body. Carbohydrates can be divided into three major groups: monosaccharides, disaccharides, and polysaccharides.
3. A second group of organic compounds that is vital to the human organism are the lipids. The lipids are also a large and diverse group of compounds and include the fats, which represent the body's most highly concentrated source of energy.
4. The third principal group of organic compounds is proteins. These compounds are much more complex in structure than the carbohydrates or lipids. They are also responsible for much of the structure of body cells and are related to many physiological activities.

CHEMICAL AND MECHANICAL DIGESTION

1. Chemical digestion is a series of catabolic reactions that break down the large carbohydrate, lipid, and protein molecules that are used by body cells.
2. Mechanical digestion consists of movements that aid chemical digestion.

GENERAL ORGANIZATION OF THE GASTROINTESTINAL TRACT

1. The organs of digestion are traditionally divided into two main groups.
2. First is the gastrointestinal (GI) tract, or alimentary canal, a continuous tube running through the ventral body cavity from the mouth to the anus.
3. The second group of organs composing the digestive system are the accessory organs: teeth, tongue, salivary glands, liver, gallbladder, pancreas, and appendix.

Tongue

1. The tongue, together with its associated muscles, forms the floor of the oral cavity. It is composed of skeletal muscle covered with mucous membrane.
2. The upper surface and sides of the tongue are covered with papillae. Some papillae contain taste buds.

Salivary Glands

1. The major portion of saliva is secreted by the salivary glands, which lie outside the mouth and pour their contents into ducts that empty into the oral cavity.
2. There are three pairs of salivary glands: the parotid, submandibular (submaxillary), and sublingual glands.
3. The salivary glands produce saliva that lubricates food and starts the chemical digestion of carbohydrates.

Deglutition

1. Both pharynx and esophagus assume a role in deglutition, or swallowing.
2. When a bolus is swallowed, the respiratory tract is sealed off and the bolus moves into the esophagus.
3. Peristaltic movements of the esophagus pass the bolus into the stomach.

Stomach

1. The stomach begins at the bottom of the esophagus and ends at the pyloric valve.
2. Adaptations of the stomach for digestion include rugae that permit distension; glands that produce mucus, hydrochloric acid, and enzymes that break down food molecules; and a three-layered muscularis for efficient mechanical movement.
3. Digestion in the stomach is regulated by both hormonal and neural reflexes, and can be best described as occurring in three separate phases: cephalic, gastric, and intestinal.
4. Proteins are chemically digested into peptones and proteoses through the action of pepsin in the stomach.

Pancreatic Secretion

1. Pancreatic acinar cells produce enzymes that enter the duodeun via the pancreatic duct. Pancreatic enzymes digest proteins, carbohydrates, and fats.
2. The pancreas is controlled by both neural and hormonal mechanisms.

Liver

1. Cells of the liver produce bile, which is needed to emulsify fats.
2. The sinusoids of the liver are partly lined with phagocytic cells, called Kupffer cells, that remove worn-out white and red blood cells.

Gallbladder

1. Bile is stored in the gallbladder and passed into the duodenum via the common bile duct.

Small Intestine

1. The small intestine extends from the pyloric valve to the ileocecal valve.
2. It is highly adapted for digestion and absorption. Its glands produce enzymes and mucus, and its wall contains microvilli, villi, and plicae circulares.
3. The enzymes of the small intestine digest carbohydrates, proteins, and fats into the end products of digestion: monosaccharides, amino acids, fatty acids, and glycerol.
4. The entrance of chyme into the small intestine stimulates the secretion of several hormones that coordinate the secretion and release of bile, pancreatic juice, and intestinal juice, and inhibit gastric activity.
5. Mechanical digestion in the small intestine involves rhythmic segmentation, pendular movements, and propulsive peristalsis.
6. Absorption is the passage of the end products of digestion from the alimentary canal into the blood or lymph.
7. Absorption in the small intestine occurs through the villi. Monosaccharides and amino acids pass into the blood capillaries, small aggregations (chylo-

microns) of fatty acids and glycerol pass into the blood capillaries, and large chylomicrons enter the lacteal.

Large Intestine

1. The overall functions of the large intestine are the completion of absorption, the manufacture of some vitamins, the formation of feces, and the expulsion of feces from the body.
2. Mechanical movements of the large intestine include haustral churning and mass peristalsis.
3. The elimination of feces from the large intestine is called defecation. Defecation is a reflex action aided by voluntary contractions of the diaphragm and abdominal muscles.

HOMEOSTATIC IMBALANCES OF THE DIGESTIVE TRACT

Dental Caries

1. Dental caries or tooth decay involves a gradual demineralization (softening) of the enamel and dentin.
2. The process of dental caries is initiated when bacteria act on carbohydrates deposited on the tooth. This produces acids that demineralize the enamel.

Periodontal Diseases

1. Periodontal disease is a collective term for a variety of conditions characterized by inflammation and/or degeneration of the gingivae, alveolar bone, periodontal ligament, and cementum.
2. Periodontal diseases are frequently caused by local irritants, such as bacteria, impacted food, and cigarette smoke, or by a poor "bite."

Peptic Ulcers

1. An ulcer is a craterlike lesion in a membrane. Ulcers that develop in areas of the alimentary canal exposed to acidic gastric juice are called peptic ulcers.
2. Hypersecretion of acidic gastric juice seems to be the immediate cause in the production of duodenal ulcers. A possible cause of gastric ulcers is hyposecretion of mucus, or hypersecretion of pepsin.
3. If an ulcer erodes all the way through the wall of the stomach or duodenum, the condition is called perforation, and it can produce fatal hemorrhage.

Peritonitis

1. Peritonitis is an acute inflammation of the serous membrane lining the abdominal cavity and covering the abdominal viscera.
2. The main danger of peritonitis is that the peritoneum is in contact with most of the abdominal organs, and if the infection gets out of hand, it may destroy vital organs and bring on death.

Cirrhosis

1. Cirrhosis is a chronic disease of the liver in which the parenchymal (functional) liver cells are replaced by fibrous connective tissue, a process called stromal repair.
2. These conditions could be caused by hepatitis, chemicals that destroy liver cells, parasites that infect the liver, and alcoholism.

Tumors

1. Both benign and malignant tumors occur in all parts of the gastrointestinal tract.
2. Fifty percent of all rectal carcinomas are within reach of the finger, and 75 percent of all colonic carcinomas can be seen with the sigmoidoscope.
3. The sigmoidoscope examination called colonoscopy has proved to be a safe and effective method of treatment that avoids the significant patient risk, discomfort, and inconvenience of major surgery.

REVIEW QUESTIONS

1. Define the terms hunger, appetite, and satiety. Which specific areas of the brain are responsible for these food-regulating sensations?
2. Explain each of the three important classes of food molecules and give specific examples of each.
3. Differentiate between the three major groups of carbohydrates and give examples of each.
4. Classify the various proteins according to their functions.
5. Define digestion. Distinguish between chemical and mechanical digestion.
6. In what respect is digestion an important component of your homeostatic mechanism?
7. Identify the organs of the alimentary canal in

sequence. How does the alimentary canal differ from the accessory organs of digestion?

8. What is the role of the tongue in digestion? Make a simple diagram of the tongue. Indicate the location of the papillae and the four taste zones.

9. Describe the location of the salivary glands and their ducts. What are buccal glands?

10. Describe the composition of saliva and the role of each of its components in digestion. What is the pH of saliva?

11. Briefly explain the mechanisms that control saliva secretion.

12. By means of a labeled diagram, outline the action of salivary amylase in the mouth.

13. What is a bolus? How is it formed?

14. Define deglutition. List the sequence of events involved in passing a bolus from the mouth to the stomach.

15. Describe the esophagus, the process called peristalsis, and the lower esophageal sphincter.

16. Describe the location of the stomach. List and briefly explain the anatomical features of the stomach.

17. Distinguish between pyloric stenosis and pylorospasm.

18. What is the importance of rugae, zymogenic cells, parietal cells, and mucous cells in the stomach?

19. What is chyme? Why are protein-digesting enzymes secreted in an inactive form?

20. By means of a labeled diagram, outline the action of pepsin in the stomach.

21. Describe the action of gastric lipase in the infant stomach.

22. Explain the mechanism of hydrochloric acid synthesis and release into the gastric gland.

23. What two systems of the body regulate digestion in the stomach?

24. List the three separate phases of digestion in the stomach, and summarize the activities of each of them.

25. Explain the interdigestive phase of digestion. How can a peptic or duodenal ulcer be produced?

26. What factors control the secretion of gastric juice?

27. Where is the pancreas located? Describe the duct system connecting the pancreas to the duodenum.

28. Contrast the functions of the acinar cells with those of the cells of the islets of Langerhans.

29. Describe the enzymes that are found in pancreatic juice and their particular functions.

30. How is pancreatic juice regulated?

31. What are the main functions of the liver? Where is it located?

32. Draw a labeled diagram of a liver lobule. What are Kupffer cells?

33. How is blood carried to and from the liver?

34. Once bile has been formed by the liver, how is it collected and transported to the gallbladder for storage?

35. Where is the gallbladder located? How is it connected to the duodenum?

36. What are the subdivisions of the small intestine? How are the coats of the small intestine adapted for digestion and absorption?

37. By means of a labeled diagram, outline the chemical digestion that occurs in the small intestine.

38. List the hormones, and their actions, that control digestion in the stomach and small intestine.

39. Describe the movements in the small intestine.

40. Why is the small intestine considered the principal area of the digestive tract?

41. Define absorption. How are the end products of carbohydrate and protein digestion absorbed? How are the end products of fat digestion absorbed?

42. Suppose you have just eaten a roast beef sandwich with butter. Describe or diagram the chemical changes that occur in the sandwich as it passes through the mouth, stomach, and small intestine. Name the enzymes involved and the glands that secrete them. Include the role of bile. Remember that roast beef is a protein, bread is a carbohydrate, and butter is a fat.

43. What routes are taken by absorbed nutrients to reach the liver?

44. What are the principal subdivisions of the large intestine? How does the wall, and especially the muscularis, differ from that of the rest of the digestive tract?

45. Describe the mechanical movements that occur in the large intestine.

46. Explain the activities of the large intestine that change its contents into feces.

47. Define defecation. How does it occur?

48. Describe the process that results in dental caries.

49. What are two bacteria that digest carbohydrates?

50. Define periodontal disease. What are some of its causes?

51. Describe various ways in which an ulcer can form. What is the main danger of ulcers?

52. What is peritonitis? Why is it considered dangerous?

53. Describe cirrhosis and some of its possible causes.

54. Discuss gastrointestinal tumors and new methods of diagnosis.

Chapter 25

METABOLIC PHYSIOLOGY

STUDENT OBJECTIVES

After reading this chapter, you should be able to

- Define a nutrient and list the functions of the six classes of nutrients
- Define metabolism and contrast the physiological effects of catabolism and anabolism
- Correlate ATP and cellular energy
- Describe the fate of glucose as it is catabolized via glycolysis, the Krebs cycle, the phosphogluconate pathway, and the electron transport system
- Describe fat storage in adipose tissue
- Explain the catabolism of triglycerides
- Describe the catabolism of fatty acids via beta oxidation and ketogenesis
- Describe the biosynthesis of saturated fatty acids and the biosynthesis of triglycerides
- Describe the mechanism involved in protein metabolism
- Contrast essential and nonessential amino acids, and describe the oxidation of amino acids
- Define transamination and deamination reactions
- Describe the urea cycle
- Describe the hormonal control of metabolism by contrasting the roles of insulin, glucagon, HGH, ACTH, TSH, epinephrine, and sex hormones.
- Compare the sources, functions, and importance of minerals in metabolism
- Define a vitamin and differentiate between fat-soluble and water-soluble vitamins
- Compare the sources, function, deficiency symptoms, and disorders of the principal vitamins
- Explain how heat is measured and produced by the body
- Describe how the caloric value of foods is determined
- Explain how the basal metabolic rate is measured
- Describe the loss and conduction of body heat
- Explain how normal body temperature is maintained
- Describe several body temperature abnormalities
- Explain fever with regard to causes and its effects on heat-promoting and heat-losing mechanisms of the body
- Describe how fever is beneficial and how it can cause death

Nutrients are chemical substances in food that provide energy, form new body components, or assist body processes. There are six classes of nutrients: carbohydrates, lipids, proteins, minerals, vitamins, and water. Carbohydrates, proteins, and lipids are the raw materials for reactions in cells. The cells either break them down to release energy or use them to build new structures and new regulatory substances such as hormones and enzymes. Some minerals and many vitamins are used by enzyme systems that catalyze the reactions undergone by carbohydrates, proteins, and lipids. Water has four major functions. It acts as a reactant in hydrolysis reactions, as a solvent and suspending medium, as a lubricant, and as a coolant.

METABOLISM

In its broadest sense, **metabolism** refers to all the chemical activities of the body. Since chemical reactions either release or require energy, the body's metabolism may be thought of as an energy balancing act. Metabolism has two phases: catabolism, an energy-generating process, and anabolism, an energy-requiring process.

Catabolism

Catabolism is the term used for processes that are degradative. When large, complex molecules (for example, carbohydrates, proteins, and fats) are taken into the body cells, they are systematically broken down to produce ever smaller and simpler molecules such as carbon dioxide, ammonia, and urea. Catabolism is an energy-releasing process. The bond energies that hold the atoms of a complex molecule together are gradually released as the molecule is disassembled stepwise. A considerable portion of the released energy is captured and stored in the form of a high-energy molecule known as *adenosine triphosphate (ATP)*.

Multienzyme Systems

The stepwise degradation of the nutrient molecules is carried out by a series of enzyme molecules. Each enzyme in the series acts sequentially to further disassemble the nutrient molecule. Such a series of enzymes is known as a **multienzyme system.** The end product of the first enzyme becomes the substrate for the second enzyme in the sequence, the end product of the second enzyme becomes the substrate for the third enzyme in the sequence, and so on. In this manner, a multienzyme system, containing anywhere from 2 to 20 or more enzymes, can completely degrade a complex molecule into very small and simple component parts.

Multienzyme systems are present inside cells in any of three increasing levels of organization. In the first level, the enzymes may be free in solution within the cytoplasm as separate and free molecules. Such a system is called a soluble or dissociated multienzyme system. In such a low level of organization, the multienzyme system depends upon the diffusion characteristics of the substrate of each enzymatic step. The substrate must diffuse some distance and locate the next enzyme in the sequence (Figure 25-1a). In the second level, the enzymes may aggregate into a large multienzyme cluster. In this arrangement, called a multienzyme complex, the distance between one enzyme and the next is minimal. Thus, the substrate goes through its series of reactions without ever leaving the complex (Figure 25-1b). In the third level of organization, the enzymes may be associated or incorporated into an even larger structure. Supramolecular structures, such as ribosomes and membranes, array the enzymes in specific sequences for an efficient, speedy process. A membrane-bound enzyme system is shown in Figure 25-1c.

Anabolism

Anabolism is the term used to describe processes that build up or synthesize new, more complex molecules from simpler molecules. Nucleic acids, proteins, and polysaccharides are built from nucleotides, amino acids, and monosaccharides, respectively. The synthesis of these bio-organic molecules requires the expenditure of cellular energy furnished by the ATP produced from the cell's catabolic activities.

One example of an anabolic process is the formation of peptide bonds between amino acids, making long chains of amino acids called *polypeptides*. Large polypeptides are the protein molecules that function as structural components, enzymes, or antibodies. Fats also participate in the body's anabolism. For

instance, fats can be built up into the lipids that form the middle layer of the cell membrane. Fats are also incorporated into the structure of steroid molecules.

Catabolism and anabolism are performed simultaneously in living cells, but each is regulated independently of the other. The degradation of complex molecules (catabolism) and the synthesis of new molecules (anabolism) proceed stepwise through a series of intermediate stages. Thus, the term *intermediary metabolism* is used to describe the study of these biochemical pathways. The sequences of compounds resulting from the actions of the enzymes are referred to as *metabolites*.

Catabolism in cells can be envisioned as taking place in three stages (Figure 25-2). In stage I, the large polymeric molecules are degraded to their unit monomeric pieces. For example, polysaccharides are reduced to the simpler unit sugars such as the hexose sugar, glucose; lipids are reduced to glycerol and fatty acid molecules; and proteins are degraded to yield their component array of amino acids. In stage II, the molecules are further degraded to simpler and smaller molecules such as acetic acid, which is always found in the cell combined with coenzyme A (in this form it is called acetyl coenzyme A or acetyl CoA). This key compound then funnels the acetic acid into stage III of catabolism. Here, the final degradation occurs. The end products are the wastes carbon dioxide and water and large amounts of energy in the form of ATP.

Anabolism can also be considered in three stages

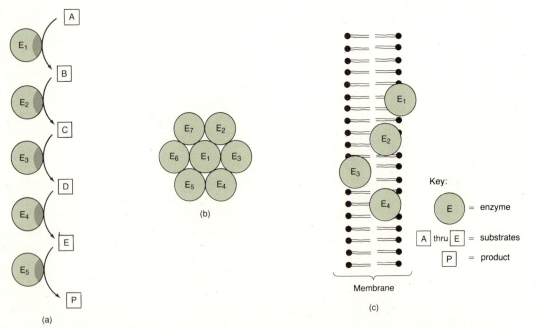

Figure 25-1 Kinds of multienzyme systems. (a) Soluble or dissociated multienzyme system. The substrate must diffuse to locate the next enzyme in sequence. (b) Multienzyme complex. The substrates do not diffuse away from the complex. (c) Membrane-bound enzyme system.

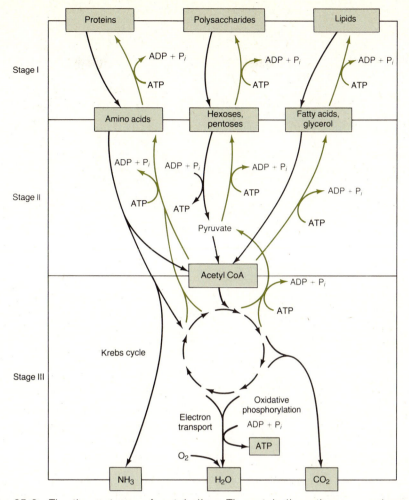

Figure 25-2 The three stages of metabolism. The catabolic pathways are drawn with downward black arrows and demonstrate that the waste end products of catabolism formed in stage III include ammonia, water, and carbon dioxide. The important useful product of catabolism is the high-energy molecule, ATP. The anabolic pathways are drawn with upward colored arrows and demonstrate that simple molecules available from stage III, coupled to the energy available from ATP, can be converted to a variety of more complex molecules useful for cellular functions.

by simply reversing those of catabolism. In stage I of anabolism, certain small molecules are selected from the many intermediates present in stage III breakdown. These intermediates are built up to form more complex stage II level molecules. These molecules are then assembled into the long chains, forming the new polymeric forms of stage I.

Adenosine Triphosphate (ATP)

Every living cell, in order to stay alive and thus be a functioning entity, requires energy. Energy is obtained from the environment surrounding each cell. In the human body, the environment of every cell is the interstitial fluid, and this fluid is regulated to provide a constant supply of organic nutrient molecules for use by cells. These molecules are catabolized by multi-enzyme systems within the cell. Energy released by these degradation processes is captured by the cell and stored in the form of a molecule known as *adenosine triphosphate (ATP)*. The structure of this molecule is illustrated in Figure 25-3. It consists of the organic base, *adenine*, a five-carbon sugar (a pentose), called *ribose*, and three phosphoric acid residues linked together to form the *triphosphate* portion of the ATP molecule. The three phosphates can be identified, as shown, by the symbols α (alpha), β (beta), and γ (gamma).

The most important feature of the ATP molecule is its ability to shed the γ-phosphate group and in the

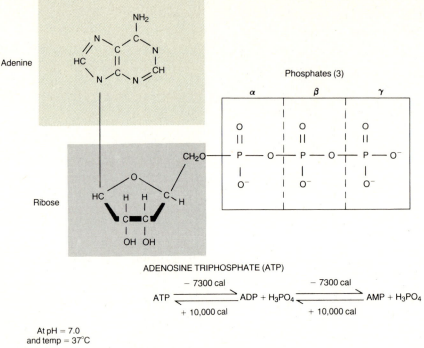

Figure 25-3 Structure of ATP.

process release 7,300 cal/mole. When hydrolyzed, the ATP molecule forms a molecule of adenosine diphosphate (ADP) and H_3PO_4 which is identified as inorganic phosphate (P_i). The released energy is coupled to an energy-requiring reaction and thus fuels an otherwise unlikely process.

In some instances, the ATP molecule is hydrolyzed so that two phosphates are liberated in the form of pyrophosphate (PP_i), and 14,600 cal/mole is available to drive an energy-requiring process. Such a reaction is employed in muscle cells that are driven during vigorous and prolonged exercise.

The resynthesis of ATP is an absolute requirement, not only for muscle cells, but for all cells. The process of energy capture in the form of ATP involves multienzyme systems. The first system contains 11 enzymes and degrades glucose to lactic acid. This sequence is referred to as *glycolysis* (*glykys* = sweet) because it describes the breakdown or *lysis* (dissolution) of sugar. This process can occur in the absence of oxygen and is therefore said to be *anaerobic*. It is a preparatory pathway that sends its end product into a second multienzyme system that further degrades the product of the first system. The second system is dependent upon the availability of oxygen and is therefore said to be *aerobic* because pairs of hydrogen atoms removed from the degraded glucose are transported by a third series of carrier molecules to oxygen to form water.

CARBOHYDRATE METABOLISM

During digestion, complex carbohydrates are hydrolyzed to become the simple sugars—glucose, fructose, and galactose—which are then absorbed into the capillaries of the villi of the small intestine and carried through the hepatic portal vein to the liver. The liver is the only organ that converts fructose and galactose into glucose (Figure 25-4). Thus, the description of carbohydrate metabolism that occurs in all other cells is really a description of the metabolism of glucose.

Since glucose is the body's most direct source of energy, the fate of any absorbed glucose molecule depends on the energy needs of the body cells. If the cells require immediate energy, they transport glucose from the blood plasma into the cell. The liver, in turn, releases stored glucose into the blood plasma and restores the blood sugar level to its normal constant value. The glucose not needed for immediate use is handled in several ways. First, the liver can convert excess glucose to glycogen (*glycogenesis*) and store it in this compact form (Figure 25-5). Skeletal muscles can also store excess glucose as glycogen, but unlike the liver, they can never release glucose back into the blood plasma. Second, if the glycogen storage areas are filled, the liver cells can transform the glucose to fat (*glycogenolysis*), which can then be stored in adipose

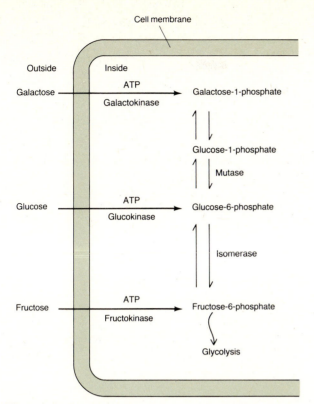

Figure 25-4 Interconversion of the major monosaccharides in liver cells.

phorylated compounds. In this fashion, the added phosphate prevents the phosphorylated molecule from escaping the cell. It attaches the enzyme for the next enzymatic attack and it serves very importantly in storing energy that will be transferred into the formation of ATP.

The pathway of glycolysis can be separated into two stages. The first stage starts with glucose—a six-carbon molecule—and ends at the point where the six-carbon molecule is split into three-carbon molecules of glyceraldehyde-3-phosphate. During this first stage, the glucose is activated or prepared for the next stage. This activation requires the cell to expend two ATP molecules and so is an energy-consuming stage. The second stage operates with the three-carbon molecules, and during this part of glycolysis, four molecules of ATP are formed. The net yield of

tissue. Later, if the cells' need for energy continues, the stored glycogen and fat can be converted back to glucose for cellular use. Third, if the level of blood glucose reaches very high concentrations and there are no storage areas available, it will be excreted in the urine. Normally, this happens only when a meal contains mostly carbohydrates and no fat. Without the inhibiting effects of fats, the stomach empties its contents quickly and the carbohydrates are digested very rapidly. As a result, large numbers of monosaccharides suddenly flood into the bloodstream. Since the liver is unable to process all of them simultaneously, the blood glucose level rises, causing hyperglycemia, and the glucose is excreted in the urine.

Glycolysis

The sequence of enzymatic steps that degrades glucose to lactic acid is known as **glycolysis** or the **Embden-Meyerhof** pathway, named in honor of two pioneers who did the basic research on this pathway during the 1920s and 1930s.

The entire sequence of glycolysis is outlined in Figure 25-6. The enzymes of glycolysis are not believed to be arranged in a multienzyme cluster but are freely dispersed in the cytoplasm of the cell. All of the intermediate products of glycolysis are phos-

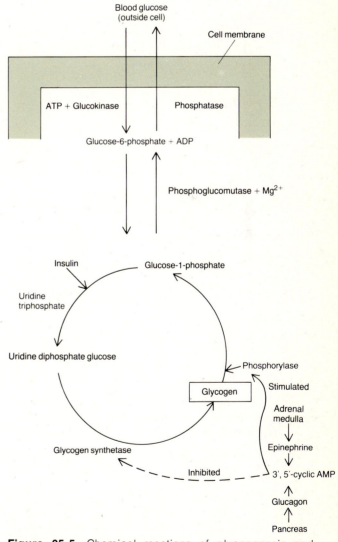

Figure 25-5 Chemical reactions of glycogenesis and glycogenolysis in liver cells.

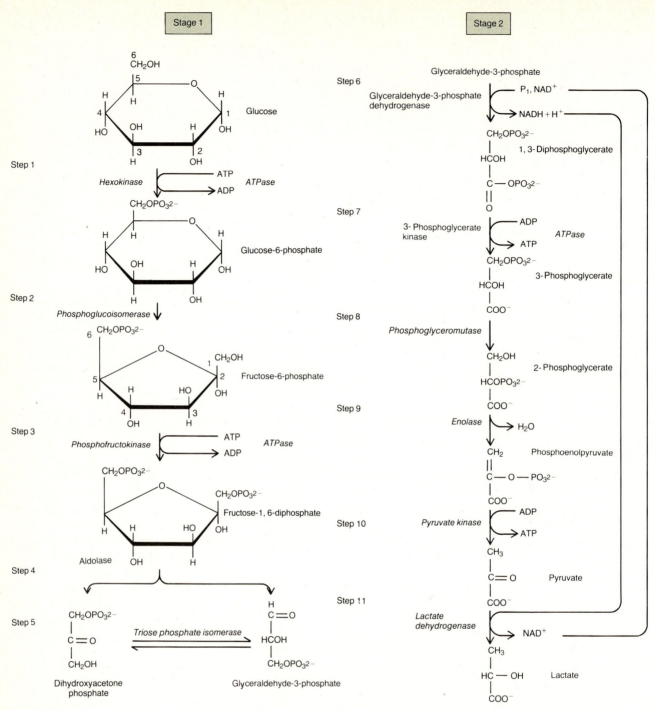

Figure 25-6 Overview of the steps in glycolysis.

ATP molecules from glycolysis is only two ATP molecules for each glucose molecule that is metabolized by glycolysis to form two molecules of pyruvate (pyruvic acid).

Overview of the Steps in Glycolysis

The biochemical steps of glycolysis are indicated in Figure 25-6. There is very little free glucose in cells. Intracellular glucose exists as a phosphorylated molecule, glucose-6-phosphate (the 6 is a position number that indicates the point in the glucose molecule where phosphate is attached). The process of glucose phosphorylation is powered by an ATP molecule as follows:

$$ATP + Glucose \rightarrow ADP + Glucose\text{-}6\text{-}phosphate$$

The next step in the sequence is the conversion of glucose-6-phosphate to fructose-6-phosphate. This is an enzymatic isomerization reaction that does not add to or remove any atoms from the molecule but rearranges the internal bonding of the atoms. Thus:

$$\text{Glucose-6-phosphate} \rightleftharpoons \text{Fructose-6-phosphate}$$

In the third step, also powered by ATP, a second phosphate is added to the fructose-6-phosphate on the number 1 carbon as follows:

$$\text{ATP} + \text{Fructose-6-phosphate} \rightarrow \text{ADP} + \text{Fructose-1,6-diphosphate}$$

During the fourth step, the six-carbon fructose-1,6-diphosphate is broken into two three-carbon (triose) pieces. Each piece has one of the phosphates. The triose phosphates are called dihydroxyacetone phosphate and glyceraldehyde-3-phosphate, and their structures can be compared by referring to Figure 25-6.

The fifth step is crucial for understanding the numbers of ATP and pyruvate produced by the glycolytic process. Of the two triose phosphates present at this point, only the glyceraldehyde-3-phosphate can be further metabolized. Fortunately, an enzyme is present in all cells that converts the dihydroxyacetone phosphate into glyceraldehyde-3-phosphate. This is an enzymatic isomerization reaction that does not add or remove atoms but only rearranges the bonding of the atoms thus:

$$\text{Dihydroxyacetone phosphate} \rightleftharpoons \text{Glyceraldehyde-3-phosphate}$$

At this point, the first stage of glycolysis is complete and the original glucose molecule is now in the form of *two* identical glyceraldehyde-3-phosphate molecules. Both of these molecules proceed through the succeeding steps.

The sixth step is one of the most important steps in the glycolytic pathway. In this step, the glyceraldehyde-3-phosphate is provided with a second phosphate group. Simultaneously, it loses two hydrogen atoms which associate with the coenzyme *NAD+ (nicotinamide adenine dinucleotide)*. The NAD+ is reduced in the process, and the reduced form is designated as NADH + H+. The phosphate that is employed by the enzyme in this reaction is *not* high-energy phosphate, but is inorganic phosphate (H_3PO_4), abbreviated as P_i. Thus:

$$\text{Glyceraldehyde-3-phosphate} + \text{NAD}^+ + P_i \rightleftharpoons \text{1,3-Diphosphoglycerate} + \text{NADH} + \text{H}^+$$

In the seventh step, ATP molecules are produced. The enzyme for this step removes the phosphate from position 1 and transfers it to an ADP molecule, putting a third phosphate on the nucleotide.

$$\text{1,3-Diphosphoglycerate} + \text{ADP} \rightleftharpoons \text{3-Phosphoglycerate} + \text{ATP}$$

The eighth step is a simple one. The phosphate in the 3 position is moved to the number 2 position. The reaction is:

$$\text{3-Phosphoglycerate} \rightleftharpoons \text{2-Phosphoglycerate}$$

The ninth step is removal of a molecule of water (H_2O) from the 2-phosphoglycerate. The resultant molecule is called phosphoenolpyruvate:

$$\text{2-Phosphoglycerate} \rightleftharpoons \text{Phosphoenolpyruvate} + \text{H}_2\text{O}$$

In the tenth step, another ATP molecule is produced. The enzyme for this step removes the phosphate from the phosphoenolpyruvate and adds it to ADP:

$$\text{Phosphoenolpyruvate} + \text{ADP} \rightarrow \text{Pyruvate} + \text{ATP}$$

The eleventh step occurs only when *anaerobic* conditions exist within the cell. When oxygen is in short supply to the cell, an enzyme is used that adds two hydrogen atoms to pyruvate to form lactate. The hydrogens are provided by the reduced nucleotide NADH + H+:

$$\text{Pyruvate} + \text{NADH} + \text{H}^+ \rightleftharpoons \text{Lactate} + \text{NAD}^+$$

If the cells finds it necessary to perform this last step, as is often the case during a quick and vigorous burst of muscular activity, lactate formed within the cell diffuses out into the body fluids and plasma. Fortunately, the liver captures lactate molecules and uses them to synthesize new glucose molecules.

A careful tally of the glycolytic process shows that for *one* glucose molecule entering the pathway and reaching the tenth step, *two* pyruvate molecules, *four* ATP molecules, and two NADH + H+ molecules have been produced. Stage-two steps were performed twice because two glyceraldehyde-3-phosphate molecules travelled through the enzymatic reactions. However, there is a net gain of only *two* ATP molecules because two were consumed in stage one of the glycolytic process.

Respiration

Respiration is the process in which cells extract energy by using molecular oxygen to oxidize the fuel molecules. Glucose is metabolized to pyruvate whether or not oxygen is available to the cell. When oxygen is not available to the cell, pyruvate is reduced to lactate. However, when oxygen is available to the cell, the pyruvate molecules enter the mitochondria of the cell, where they are oxidized first to acetate, with the loss of a CO_2 molecule and two hydrogen atoms to NAD+. The acetate is never really set free by the catalyzing complex of three enzymes. The enzyme complex temporarily retains the acetate and binds it to the key coenzyme known as *coenzyme A*. The bonded acetate is then released from the enzyme in *acetyl coenzyme A*, or *acetyl CoA*. In this combined form, the acetate moiety has been energized up to a level that now enables it to proceed through a second multienzyme complex known by three different names: the *Krebs cycle, citric acid cycle,* or *tricarboxylic acid (TCA) cycle* (Figure 25-7).

The oxidation of pyruvate to acetyl CoA is an irreversible process that is summarized as follows:

$$\text{Pyruvate} + \text{NAD}^+ + \text{CoA} \rightarrow \text{Acetyl CoA} + \text{NADH} + \text{H}^+ + \text{CO}_2$$

Overview of the Steps in the Krebs Cycle

The acetyl CoA combines with a molecule of a four-carbon compound called *oxaloacetate* to form a six-carbon molecule called *citrate* and coenzyme A. Thus,

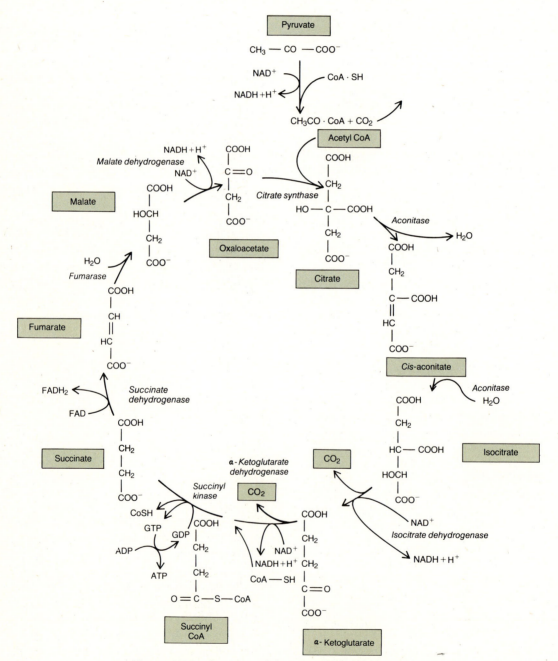

Figure 25-7 Overview of the steps in the Krebs cycle.

the first step of the Krebs cycle introduces the "active acetate" to oxaloacetate and the first intermediate of the Krebs cycle, citrate is produced.

During the succeeding sequence of steps, the citrate is gradually degraded in a way that produces two molecules of CO_2 and regenerates the oxaloacetate. Thus, there is no loss of oxaloacetate and the cycle is able to repeat the process indefinitely.

In the first step of the cycle, citrate is converted into *isocitrate*. The enzyme catalyzing the reaction combines with the citrate and for a very brief time removes a water molecule. During this brief instant, the citrate is changed to a molecule called *cis-aconitate*. Then, the atoms of the water molecule are restored to this enzyme-bound intermediate, but in different positions. Thus, the same total atoms are present, but arranged in a new way.

$$\text{Citrate} \rightleftharpoons [\textit{cis-}\text{aconitate}] \rightleftharpoons \text{Isocitrate}$$

The second step is the oxidation of isocitrate to α-ketoglutarate. The enzyme catalyzing this reaction removes CO_2 from the isocitrate (called a decarboxylation reaction) and also transfers two hydrogen atoms from the isocitrate to the coenzyme NAD^+. The reaction is as follows:

$$\text{Isocitrate} + NAD^+ \rightleftharpoons \alpha\text{-Ketoglutarate} + CO_2 + NADH + H^+.$$

The third step is the oxidation of α-ketoglutarate to succinyl CoA. This enzyme-catalyzed reaction is irreversible, and is very similar to the conversion of pyruvate to acetyl CoA. The α-ketoglutarate molecule is *decarboxylated* (loses CO_2) and *dehydrogenated* (loses hydrogens) and *energized* by combining with coenzyme A. The multiplicity of events is carried out by a complex of molecules that includes three enzymes and several coenzymes. The reaction is described as:

$$\alpha\text{-Ketoglutarate} + NAD^+ + CoA \rightarrow \text{Succinyl CoA} + NADH + H^+ + CO_2$$

The crucial fourth step is a *substrate-level phosphorylation* reaction. The succinyl CoA utilizes its energy to phosphorylate a molecule of guanosine diphosphate (GDP) to guanosine triphosphate by using inorganic phosphate (P_i). This molecule of GTP, in turn, catalyzes the phosphorylation of ADP to ATP. Thus, the sequence of reactions is as follows:

$$\text{Succinyl CoA} + P_i + \text{GDP} \rightleftharpoons \text{Succinate} + \text{GTP} + \text{CoA}$$

then:

$$\text{GTP} + \text{ADP} \rightleftharpoons \text{GDP} + \text{ATP}$$

In the fifth step, an enzyme is employed that depends upon a covalently bound coenzyme called *flavin*

adenine dinucleotide (FAD), which accepts two hydrogen atoms. The dehydrogenation reaction removes two hydrogen atoms from succinate to yield a molecule of fumarate:

$$\text{Succinate} + \text{Enzyme FAD} \rightleftharpoons \text{Fumarate} + \text{Enzyme } FADH_2$$

The sixth step of the cycle is the addition of a molecule of water to the fumarate to produce a molecule of malate:

$$\text{Fumarate} + H_2O \rightleftharpoons \text{Malate}$$

The seventh, and last, step of the cycle is another dehydrogenation reaction that utilizes an enzyme which removes two hydrogen atoms from malate and transfers them to the coenzyme NAD^+. The reaction is as follows:

$$\text{Malate} + NAD^+ \rightleftharpoons \text{Oxaloacetate} + NADH + H^+$$

The oxaloacetate is regenerated and can function to combine with another "active acetate" to produce citrate.

The overall result of one revolution of the Krebs cycle can be summed as follows. From each acetyl group introduced to the cycle, the two carbons are released as two carbon dioxide molecules. Four pairs of hydrogen atoms are released in the dehydrogenation reactions. Three of the four pairs are in the $NADH + H^+$ combination and one pair of hydrogens is combined in the $FADH_2$. One ATP molecule is produced by a substrate-level phosphorylation reaction.

The Phosphogluconate Pathway

Many cells possess in varying degrees an additional pathway for the metabolism of glucose. This enzyme sequence has been named the *phosphogluconate pathway* but is also referred to as the *hexose monophosphate shunt (HMPS)*. Varying proportions of the glucose molecules presented to the cell travel this alternate route. It is more actively utilized in cells that are involved in the synthesis of fatty acids and steroids—cells of the liver, adrenal cortex, and mammary tissue. The purposes of this alternate metabolic route are threefold: (1) It is used primarily as a means of generating the reduced form of the high-energy coenzyme $NADPH + H^+$, *nicotinamide adenine dinucleotide phosphate*. This reduced molecule plays an essential role in the synthesis of fatty acids. (2) This pathway also generates five-carbon sugars. Of particular importance is the pentose, ribose-5-phosphate, that is used in DNA synthesis. (3) This pathway is used for the complete oxidation of five-carbon sugars by converting them back into hexose sugars which then reenter the glycolytic sequence. The pathway, shown in Figure 25-8, shows the possible interconversions of various three-, four-, five-, six-, and seven-carbon sugars. These different-sized carbon molecules can be selected by synthetic pathways as needed—much like precut lumber used in building a new structure.

The Electron Transport System

The pairs of hydrogen atoms liberated from glycolysis and the Krebs cycle are potential sources of large amounts of energy. When released from the enzymatic reactions that produced them, the hydrogen atoms are bonded to or associated closely with either coenzyme NAD^+ or FAD. The **electron transport system** consists of a series of electron carrier enzymes that are incorporated into the membranes of the mitochondrial cristae. The system operates by transferring first the hydrogen atoms, and then just their electrons, through the carrier system finally to reduce molecular oxygen. The important consequence of this transport process is that portions of the energy carried by the hydrogen atoms and electrons are gradually released. Some of the energy released is given off as heat, but much of the energy is conserved in the formation of ATP. The capture of released energy into ATP formation is termed *oxidative phosphorylation*. The exact description of how this energy is captured and used to make ATP in the mitochondrial membrane is not completely understood. If the capture mechanism is inactivated—as can be done by certain inhibitors—the energy is still released by the transport system, but it is *all* liberated as heat. The process of oxidative phosphorylation is coupled into the electron transport system, and a portion of the energy released is captured and used to drive the reaction:

$$ADP + P_i \rightleftharpoons ATP$$

The sequence of carriers in the electron transport is shown in Figure 25-9. The first member of the sequence is $NADH + H^+$. The reduced form of this enzyme is produced during glycolysis and the Krebs cycle. The hydrogen atoms are transferred from the $NADH + H^+$ to the coenzyme FAD. This converts, or reduces, FAD to $FADH_2$, and the $NADH + H^+$, having given up its two hydrogen atoms, is reoxidized to NAD^+.

An examination of Figure 25-9 shows the cyclic behavior of each member in the system. The $NADH_2$ is oxidized while the flavoprotein is reduced. During this first transfer of hydrogen atoms, sufficient energy is made available and captured to drive the $ADP + P_i$ reaction to form ATP. That coupling step is shown also in Figure 25-9.

The next transfer of hydrogens is from $FADH_2$ to *coenzyme Q*. The reaction oxidizes the flavoprotein back to FAD and the coenzyme Q is reduced to $CoQ \cdot H_2$.

The next member of the transport system is the first of a series of cytochrome molecules—a type of iron-containing pigment molecule. The first member of the cytochrome sequence is called *cytochrome b*. However, when the transfer between $CoQ \cdot H_2$ and cytochrome b occurs, the hydrogen atom is ionized to form H^+ and an electron, and only the electron is transferred to the cytochrome b. Within the cytochrome b molecule, the iron atom Fe^{3+} is reduced to Fe^{2+}. The H^+ portion of the hydrogen atom is released briefly from the complex of carriers to be used eventually at the end of the transport system. Since we are transferring two hydrogen atoms from $CoQ \cdot H_2$ to cytochrome b, and the iron atom can only accept one electron, it is necessary to have two cytochrome b molecules available at this point. The next carrier in the series is *cytochrome c*. Two oxidized cytochrome c molecules (Fe^{3+}) are employed to accept the two electrons held, at this point, by the two cytochrome b molecules. This transfer can be written thus:

$$2 \text{ cyt b } (Fe^{2+}) + 2 \text{ cyt c } (Fe^{3+}) \rightarrow 2 \text{ cyt b } (Fe^{3+}) + 2 \text{ cyt c } (Fe^{2+})$$

Again, reference to Figure 25-9 shows the cyclic nature and the regeneration of the oxidized (Fe^{3+}) form of the cytochromes. During this transfer of electrons, sufficient energy is released and captured to drive the phosphorylation of ADP to ATP (the second ATP molecule produced by the movement of the hydrogen atom or its electron through the transport system).

The next carrier in the series is *cytochrome a*. Two molecules of oxidized cytochrome a (Fe^{3+}) accept the two electrons from the two reduced cytochrome c molecules. This transfer can be written thus:

$$2 \text{ cyt c } (Fe^{2+}) + 2 \text{ cyt a } (Fe^{3+}) \rightarrow 2 \text{ cyt c } (Fe^{3+}) + 2 \text{ cyt a } (Fe^{2+})$$

During the next transfer, when the electrons move from cytochrome a (Fe^{2+}) to the next carrier *cytochrome a_3* (Fe^{3+}), sufficient energy is again made available for the phosphorylation of ADP to ATP (the third ATP generated by the electron transport system).

Often, the term *cytochrome oxidase* was used to describe the enzymelike property of the last carrier. In fact, the molecules of cytochrome a and cytochrome a_3 are both part of the same large protein molecule. Now, the term cytochrome oxidase refers to the large complex that contains both cytochromes. The electrons are finally transferred to one-half of an oxygen molecule ($\frac{1}{2}$ of $O_2 = O$) to form O^{2-}.

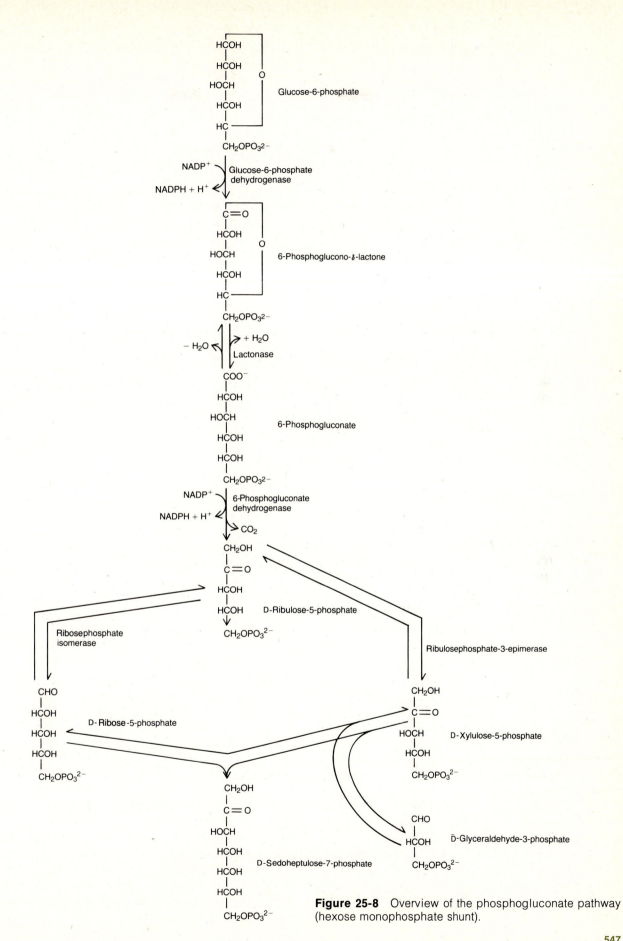

Figure 25-8 Overview of the phosphogluconate pathway (hexose monophosphate shunt).

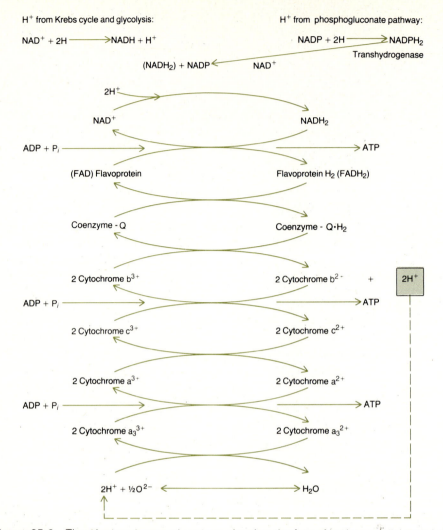

Figure 25-9 The electron transport system, showing the fate of hydrogens released via the Krebs cycle and glycolysis.

This O^{2-} reacts with H_3O to form two OH^- ions: $O^{2-} + H_2O \rightarrow 2OH^-$.

The two hydroxyl ions (OH^-) then react with the two H^+ ions liberated from the $CoQ \cdot H_2$ to 2 cytochrome b^{3+} step. The end result is the formation of two molecules of water. Thus:

$$2OH^- + 2H^+ \rightarrow 2H_2O$$

The oxygen used in this last step is the oxygen that is inhaled from the atmosphere, carried by the hemoglobin in the red blood cells, and released into the tissue fluid bathing the cells and that diffuses into the cell and mitochondrion. This is certainly an example of complex but coordinated processes that support and depend upon each other. The electron transport system is an evolutionary marvel in its efficient transferral of electrons down an energy pathway to meet an oxygen atom. The oxygen atom, in turn, is de-

livered precisely when needed by the flowing blood, regulated and modified by pressure and CO_2 reflexes.

An ATP Accounting

It is important to realize the amount of energy captured when one glucose molecule is completely metabolized to CO_2 and H_2O.

First, glycolysis produces two pyruvate molecules, two NADH + H$^+$ molecules, and a net gain of *two ATP molecules*.

Second, the Krebs cycle accepts the two pyruvate molecules and revolves twice to produce *two ATP molecules* by means of substrate-level phosphorylation. In addition, the conversion of the two pyruvate molecules to acetyl CoAs and the two revolutions of the Krebs cycle produces a total of eight molecules of NADH + H$^+$ and two molecules of FADH$_2$.

Third, the reduced coenzymes transfer their hydrogen into the electron transport system to yield additional ATP molecules.

Two NADH + H⁺ molecules from glycolysis and eight from the Krebs cycle and pyruvate conversion mean that ten NADH + H⁺ molecules will produce *30 ATP* molecules when they travel the electron transport system. In addition, each FADH₂ will produce two ATP molecules when their hydrogens complete their journey through the electron transport system. Thus, the two FADH₂ molecules yield *four ATP* molecules.

The total number of ATP molecules that can be produced from one glucose molecule when it is metabolized completely to CO_2 and H_2O is 38.

Therefore, $38 \times 7{,}700$ or 292,600 cal of energy is stored in the form of ATP. The total energy content of one gram-mol of glucose, as determined by bomb calorimetry, has been determined to be 686,000 cal. Thus 292,600/686,000 or about 43 percent of the energy of glucose is captured in ATP molecules, and the remaining portion, about 57 percent, is given off as heat.

LIPID METABOLISM

About 50 percent of body fat is located in the subcutaneous tissue, about 12 percent is packed around the kidneys, 10 to 15 percent is stored in the omenta, 20 percent is in genital areas, and 5 to 8 percent is between muscles. Fat is also located behind the eyes, in the sulci of the heart, and in the folds of the large intestine. Fat molecules are constantly being turned over, that is, they are continually released from storage, transported through the blood, and redeposited in other adipose tissue cells. Some investigators estimate that as much as one-half of the total body fat turns over daily.

Fat stored in the body's depots constitutes the largest reserve of energy. The body can store much more fat than it can glycogen—hundreds of pounds of it in very obese persons. Moreover, the energy yield per molecule of fat is more than twice that of carbohydrate. Nevertheless, fats are only the body's second favorite source of energy because they are more difficult to catabolize than carbohydrates.

Catabolism of Triglycerides

Triglycerides are the lipid-type molecules that are used mainly for energy storage. Other lipid-type molecules—phospholipids and cholesterol—are used intracellularly for the production of hormones and other functional molecules.

The triglyceride molecule consists of a molecule of

glycerol and three fatty-acid molecules. They are linked together by an ester bond to form the triglyceride. The structure of a triglyceride is illustrated in Figure 25-10.

The first step in the breakdown of triglycerides stored in fat or adipose tissue is their separation into one molecule of glycerol and three molecules of fatty acids by the action of hormonally controlled enzymes called *lipases*. The glycerol is converted by many other tissues immediately into glyceraldehyde, which then enters the phosphogluconate pathway for glucose metabolism. The fatty acids released from fat depots are transported, bound to plasma albumin, to many other tissues that can metabolize fatty acids. The long-chain fatty-acid molecules are oxidized completely to CO_2 and H_2O by all tissues except nervous tissue. There are times when even the brain can utilize certain

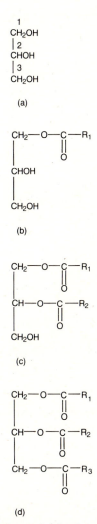

Figure 25-10 Structure of (a) glycerol, (b) a monoglyceride (1-monoacylglycerol), (c) a diglyceride (1,2-diacylglycerol), and (d) a triglyceride (triacylglycerol).

fatty-acid breakdown products such as β-hydroxy-butyrate. Heart tissue is somewhat unusual in that it obtains most of its energy from fatty-acid metabolism.

Beta Oxidation of Fatty Acids

The metabolic degradation of fatty acids is carried out in the matrix of the mitochondria. There, a series of soluble enzymes catalyze a process referred to as **beta oxidation.** In this process, the long-chain molecules are reduced to two-carbon units of acetic acid bound to coenzyme A, producing *acetyl CoA*. Since by far the majority of fatty acids contain an even number of carbons, the number of acetyl CoA molecules produced is easily calculated by dividing the number of carbons in the fatty acid by two. For example, the fatty acid *palmitic acid* contains 16 carbon atoms and, upon complete hydrolysis, will yield 8 acetyl CoA molecules.

The first step of fatty-acid oxidation requires that the fatty acid be transported from the cytoplasm of the cell into the mitochondrial matrix. This is a rather complex reaction that requires the free fatty acid in the cytoplasm to be activated:

$$R^* \cdot COOH + ATP + CoA \rightarrow R \cdot COCoA + AMP + PP_i$$

Free fatty acid	Fatty acyl CoA

Once activated, the fatty acid is then transferred to the transport molecule, *carnitine,* located in the inner mitochondrial membrane. The reaction is catalyzed by an enzyme to produce the *acyl carnitine complex,* which then crosses the membrane. The formation of the transport complex is shown in Figure 25-11.

A second enzyme located on the inner surface of the inner mitochondrial membrane causes the release of the fatty acyl group from carnitine and its transfer to intramitochondrial coenzyme A. Thus:

$$\text{Acyl carnitine} + CoA \rightarrow \text{Acyl CoA} + \text{Carnitine}$$

This fatty acyl CoA complex presents the fatty acid to the enzymes of the beta oxidation pathway.

First Dehydrogenation Step

The intramitochondrial fatty acyl CoA undergoes an enzymatic attack that removes two hydrogen atoms, one each from the α and β carbons (Figure 25-12). Starting from the activated end of the fatty acid, the carbons are numbered 1, 2, 3, 4. . . . Carbon number 2

---

* R is a shorthand symbol used to represent the remaining large portion of a molecule. Generally, it assumes the student already knows the entire molecular structure. In this instance for palmitic acid which has 16 carbons R stands for:

$$CH_3-CH_2-CH_2-CH_2-CH_2-CH_2-CH_2-CH_2-CH_2-$$
$$CH_2-CH_2-CH_2-CH_2-CH_2-CH_2-$$

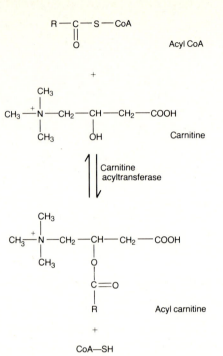

Figure 25-11 Transfer of an acyl group to carnitine.

is also designated the α (alpha) carbon, the third carbon is designated the β (beta) carbon, and so forth: γ (gamma), δ (delta), etc.

Dehydrogenation of the α and β carbons is carried out by an enzyme that contains a tightly bound FAD coenzyme molecule. Thus the reaction can be outlined as follows:

$$R-CH_2-CH_2-\underset{O}{\overset{\|}{C}}-sCoA + Enz \cdot FAD \rightarrow$$

Fatty acyl CoA

$$R-\underset{H}{\overset{H}{\underset{|}{C}}}=\underset{}{\overset{}{C}}-\underset{O}{\overset{\|}{C}}-CoA + Enz \cdot FADH_2$$

The $FADH_2$ transfers its hydrogens to the CoQ molecule of the electron transport system so that the hydrogens can travel the rest of the electron transport system to form H_2O and *two ATP molecules.*

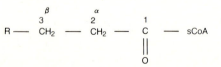

Figure 25-12 The numbering and naming of the carbons in the activated fatty acid. The -sCoA linkage shows that the bonding of the fatty acid to the coenzyme is via a sulfhydral (SH) oxidation reaction. The sulfur atom is shown in the linkage.

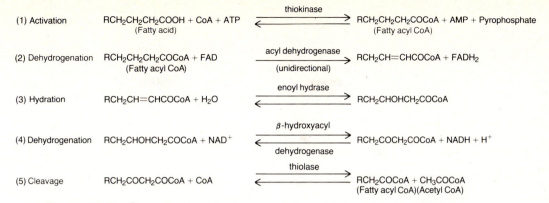

Figure 25-13 Steps involved in beta oxidation of fatty acids to yield acetyl coenzyme A.

Hydration Step

The next step is the addition of water across the double bond between the α and β carbons (OH^- is added to the β carbon and H^+ to the α carbon).

$$R—\overset{\underset{\textstyle H}{|}}{\underset{\underset{\textstyle H}{|}}{C}}=\overset{H}{C}—\overset{\underset{\textstyle O}{\|}}{C}—sCoA + H_2O \rightarrow R—\overset{\underset{\textstyle OH}{|}}{\underset{\underset{\textstyle H}{|}}{C}}—CH_2—\overset{\underset{\textstyle O}{\|}}{C}—sCoA$$

Second Dehydrogenation Step

The second dehydrogenation step removes two hydrogen atoms from the β carbon. The hydrogens removed are accepted by NAD^+ and passed down the electron transport system to form H_2O and *three ATP molecules.*

$$R—\overset{\underset{\textstyle OH}{|}}{\underset{\underset{\textstyle H}{|}}{C}}—CH_2—\overset{\underset{\textstyle O}{\|}}{C}—sCoA + NAD^+ \rightarrow$$

$$R—\overset{\underset{\textstyle O}{\|}}{C}—CH_2—\overset{\underset{\textstyle O}{\|}}{C}—sCoA + NADH + H^+$$

Cleavage Step

The cleavage reaction can now occur because of the resultant weakened bond between α and β carbons. The two terminal carbons still bound to coenzyme A are set free from the remainder of the fatty acyl portion. The two-carbon portion released is acetyl CoA. The remaining fatty acyl portion reacts with a molecule of free CoA to yield a fatty acyl CoA molecule two carbons shorter than before.

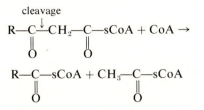

The complete steps of beta oxidation are summarized in Figure 25-13.

An Accounting

The preceding five steps outline the sequence of enzymatic steps necessary to cleave the two terminal carbons from a fatty acid to produce one acetyl CoA molecule, one $FADH_2$, and one $NADH + H^+$. This process repeats itself, this time beginning with a fatty acyl CoA that is two carbons shorter.

Consider the energetic product of palmitic acid (C_{16}). Since this fatty acid contains 16 carbons, it will yield eight acetyl CoA molecules after complete hydrolysis. The five steps of beta oxidation need be carried out only seven times.

Thus, eight acetyl CoAs produced can enter the Krebs cycle for complete oxidation, where they will produce:

$$8 \times 3\ NADH + H^+ = 24\ NADH^+ + H$$
$$8 \times FADH_2 \qquad = 8\ FADH_2$$
$$8 \times ATP \qquad = 8\ ATP$$

Recall that the steps of beta oxidation also produce reduced coenzymes. Each operation of the beta oxidation sequence yields one $FADH_2$ and one $NADH + H^+$. Thus, seven operations of beta oxidation will produce:

$$7 \times NADH + H^+ = 7\ NADH^+ + H$$
$$7 \times FADH_2 \qquad = 7\ FADH_2$$

Addition of reduced coenzymes and ATP shows that hydrolysis of palmitic acid (C_{16}) yields:

$$31\ NADH + H^+ + 15\ FADH_2 = 8\ ATP$$

The reduced coenzymes, in turn, channel the hydrogen atoms into the electron transport system. Each $NADH + H^+$ can produce three molecules, whereas each $FADH_2$ yields two ATP molecules.

$$31\ NADH + H^+\ produce\ 3 \times 31 = 93\ ATP$$
$$15\ FADH_2\ produce\ 2 \times 15 \qquad = 30\ ATP$$
$$8\ ATP\ from\ the\ Krebs\ cycle \qquad = 8\ ATP$$
$$Total\ production \qquad = 131\ ATP$$

However, the initial activation process that enabled the fatty acid to enter the mitochondrial matrix required: ATP $\rightarrow$ AMP $+$ PP$_i$. To repay the energy spent in this reaction requires that two ATP molecules be "paid back."

Thus, the net energy capture from the complete oxidation of a 16-carbon fatty acid (palmitic acid) is $131 - 2$ or 129 ATP molecules. The efficiency of this capture can be estimated by comparing the total energy content of one mole of palmitic acid with the amount present in the ATPs. When one gram molecular weight of palmitic acid is burned in a bomb calorimeter, it yields 2,340 kcal/mole.

$$\text{Efficiency} = \frac{129 \times 7.7}{2,340} \times 100 = \frac{993}{2,340} \times 100 = 42 \text{ percent}$$

Ketone (Acetone) Bodies

The liver has enzymes capable of coping with excessive amounts of acetyl CoA that result from excessive fatty-acid breakdown such as can occur immediately following an excessively fatty meal, during a fast, or in diabetes mellitus. The liver is able to condense two molecules of acetyl CoA into *acetoacetyl CoA* and, from this molecule, to derive a series of molecules known collectively as *ketone (acetone) bodies*:

Acetyl CoA + Acetyl CoA $\rightleftharpoons$ Acetoacetyl CoA + CoA

Acetoacetyl CoA can shed the coenzyme A to yield free acetoacetate.

Acetoacetyl CoA + H$_2$O $\rightleftharpoons$ Acetoacetate + CoA

The acetoacetate can be enzymatically reduced by the addition of two hydrogen atoms obtained from NADH + H$^+$ and be converted to β-hydroxybutyrate.

Acetoacetate + NADH + H$^+$ $\rightleftharpoons$
β-Hydroxybutyrate + NAD$^+$

A small portion of the acetoacetate can be converted to acetone by a decarboxylation (CO$_2$ loss) reaction.

Acetoacetate $\rightarrow$ Ketone (acetone) + CO$_2$ $\uparrow$

Figure 25-14 summarizes the reactions.

As acetoacetate and β-hydroxybutyrate accumulate within the cell, a gradient for outward diffusion results, and these ketone bodies then appear in the blood. Once in the blood, they are carried to the peripheral tissues and diffuse into the cells, where the above reactions are reversed and acetyl CoA molecules are re-formed. Thus, these ketone bodies can be utilized for energy production by processing them in the Krebs cycle.

When the levels of ketone bodies in the plasma and tissue fluids become high, the condition *ketosis* is said to exist.

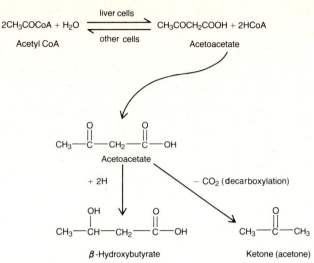

Figure 25-14 Synthesis of ketone (acetone) bodies.

Biosynthesis of Saturated Fatty Acids

The storage of energy in the form of carbohydrate is limited to that which can be retained as glycogen in the liver and skeletal muscle. Excess carbohydrate ingested by the body is metabolized to pyruvate through the glycolytic pathway, converted to acetyl CoA, and then, through a somewhat reverse process of beta oxidation, polymerized to form fatty acids. These fatty acids are conjugated in threes to glycerol to form the triglycerides. The biosynthetic pathway is very active in the liver, adipose tissue, and the mammary glands. In this compact form of triglycerides, energy is stored most efficiently. On a weight or volume basis fats contain more energy than polysaccharides or protein.

The biosynthesis of saturated fatty acids, however, is not simply a reversal of the process of beta oxidation. The biosynthesis of fatty acids takes place in the cellular cytoplasm, as opposed to the matrix of the mitochondria where beta oxidation occurred. Unexpected also is the need for citrate and CO$_2$. A simplified explanation of the synthetic process is outlined below.

Source of Carbon for Fatty Acids

The carbons ultimately making up the saturated fatty acids are provided by acetyl CoA molecules, formed in the mitochondria from either glycolysis, fatty-acid degradation, or transformation of certain amino acids.

Acetyl CoA is unable to escape the mitochondrion and so combines with oxalacetate to form citrate and CoA. Citrate is able to leave the mitochondrion and, once in the cytoplasm, again is converted to acetyl CoA and oxalacetate with the aid of energy from ATP.

Thus:

$$\text{Citrate} + \text{ATP} + \text{CoA} \rightarrow \text{Acetyl CoA} + \text{ADP} + \\ P_i + \text{Oxalacetate}$$

An alternative way in which acetyl CoA leaves the mitochondrion is via the *carnitine* carrier system (see Figure 25-11). This is the same carrier that transports fatty acids into the mitochondrial matrix. The carnitine carrier accepts the acetate portion of the acetyl CoA on the inside of the inner membrane, transports it through the membrane, and releases the acetate to another CoA on the cytoplasm side and thus re-forms acetyl CoA.

Formation of Malonyl CoA

The first step of biosynthesis requires the production of an activated three-carbon molecule named *malonyl CoA* from acetyl CoA. The synthesis of this molecule requires CO_2 (in the form of H_2CO_3) and ATP.

$$\text{Acetyl CoA} + H_2CO_3 + \text{ATP} \rightleftharpoons \text{Malonyl CoA} + \text{ADP} + P_i$$

Assembly of the Fatty Acid

The polymerization reactions that follow are initiated with an acetyl CoA molecule that behaves as a "primer." Malonyl CoA undergoes a condensation reaction with the acetyl CoA molecule while both molecules are anchored to a very special *acyl carrier protein (ACP)*. During the condensation reaction, CO_2 is released (the CO_2 originally present in H_2CO_3) and the complex anchored to ACP becomes a four-carbon molecule of acetoacetate.

The acetoacetate, still anchored to the ACP, undergoes a reduction reaction that utilizes $NADPH + H^+$ that was formed as a by-product of the phosphogluconate pathway.

$$\text{Acetoacetyl ACP} + NADPH + H^+ \rightleftharpoons \\ \beta\text{-Hydroxybutyryl-ACP} + NADP^+$$

The result of the reduction reaction is that acetoacetate is converted to β-hydroxybutyrate.

The following step involves the removal of a molecule of water from the β-hydroxybutyrate to produce a molecule of crotonate.

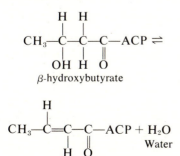

β-hydroxybutyrate

Crotonate

The next step requires $NADPH + H^+$ for the reduction of crotonate to butyrate.

Crotonate

Butyrate

The creation of butyryl ACP completes the first revolution of a cyclic process that will repeat itself a total of seven times. Each addition will initially require a malonyl CoA molecule, and as this condenses with the growing fatty acid, CO_2 is released that can be reused as H_2CO_3 to react with acetyl CoA to form more malonyl CoA. An overall accounting for the synthesis of palmitic acid starting from acetyl CoA can be summarized as:

$$8 \text{ Acetyl CoA} + 14(NADPH + H^+) + 7\text{ATP} + H_2O \rightarrow \\ \text{Palmitic acid } (C_{16}) + 8\text{CoA} + 14NADP^+ + 7\text{ADP} + 7P_i$$

Formation of Longer Fatty Acids from Palmitic Acid

The normal end product of fatty-acid biosynthesis is a saturated fatty acid with a chain length of 16 carbons (palmitic acid). One of the key enzymes in the biosynthetic pathway can only accommodate chain lengths up to 14 carbons, and thus the last addition of carbons up to 16 is the last one possible. However, longer fatty acids are found within cells and as structural components. The longer fatty acids are formed either within mitochondria by condensing with additional acetyl CoA molecules using $NADPH + H^+$ or in the endoplasmic reticulum by condensing with malonyl CoA molecules directly (not bound to ACP).

Biosynthesis of Triglycerides

The fat stored in liver and in the fat depots is in the form of triglycerides. Triglycerides are formed from three activated fatty acids and a molecule of glycerol-3-phosphate.

The usual source of glycerol-3-phosphate is the reduction of dihydroxyacetone phosphate—an intermediate of the glycolytic pathway.

$$\text{Dihydroxyacetone phosphate} + NADH + H^+ \rightleftharpoons \\ \text{Glycerol-3-phosphate} + NAD^+$$

Three fatty acyl CoA molecules are then added to the glycerol-3-phosphate in sequence starting from the number 1 position to form a complete triglyceride. The triglycerides usually stored in fat depots are of the simple type (the fatty acids in the triglyceride are of a single kind) See Exhibit 25-1.

PROTEIN METABOLISM

Proteins function in a variety of ways in the human body. The largest group of proteins—almost 2,000 different kinds—function as enzymes. Some proteins are used as a source of nutrition or as a source of amino acids for a growing embryo and newborn; for example, ferritin stores iron in the spleen, in the liver, and in the casein of milk. Other proteins function as transport molecules. Serum albumins bond to fatty acids in plasma, hemoglobin carries oxygen molecules, and plasma globulins carry iron in the blood. Molecules that function as antibodies or in blood clotting, and thus have protective functions, are also protein. Certain hormones such as insulin, human growth hormone, and ACTH function to regulate body activities. The contractile machinery of the muscles is protein in nature—the actin and myosin molecules. A large number of structual components are, or include, protein. Collagen, elastin, α-keratin, and mucoproteins are specialized proteins that are designed for specific purposes. Other proteins are found in cellular membranes and on cell membrane surfaces.

The amazing feature of proteins is that all of the astronomical number of different proteins that exist are synthesized from the same 20 amino acids (Figure 25-15). The individual amino acids have virtually no biological action. Only when they are linked into chains that assume a three-dimensional, stabilized, coiled structure do they have a specific function.

Essential and Nonessential Amino Acids

Of the 20 naturally occurring amino acids, 10 are considered *essential*. They are designated essential because the human body is unable to synthesize them

Exhibit 25-1 OCCURRENCE OF SIMPLE AND MIXED TRIACYLGLYCERIDES IN DEPOT FAT OF THE RAT (S = SATURATED; U = UNSATURATED)

Type	Symbol	Percent
Simple	SSS	0.3
	UUU	61.8
Mixed	SSU	4.1
	SUS	1.6
	SUU	19.5
	USU	12.8

from other molecules present in human cells. Enzymes that would be needed for the synthesis of these 10 amino acids are not part of the profile of human cells. So, these essential amino acids are synthesized for the human organism by plants or bacteria. Therefore, foods containing these amino acids are "essential" for human growth and survival and must be part of our diet.

Oxidation of Amino Acids

After proteins are taken into the digestive tract, they are broken down into their constituent amino acids before they can be absorbed into the bloodstream and transported to the cells of the body. Amino acids can be presented to the cells from an exogenous source (proteins that were just eaten), but they can also be obtained from endogenous sources (proteins already in use in the body). Experiments using radioactively tagged protein have shown that in the standard 70-k man, about 400 g of protein is turned over every day. Of this total, about 100 g is completely oxidized or converted to glucose. This portion must be replaced by the ingestion of new protein. The remaining 300 g is reincorporated into functional proteins. Therefore, this portion is not lost but is reused. Between 6 and 20 g of nitrogen derived from the oxidative degradation of proteins is excreted daily in the urine, mostly in the form of urea. If a person were to fast or not to eat any protein, nitrogen would still be excreted in the urine. As much as 5 g per day could be lost in this way—indicative that as much as 30 g of protein is being broken down daily.

Pathways of Amino Acid Oxidation

Each of the 20 naturally occurring amino acids has its own pathway of oxidative catabolism. Figure 25-16 shows that the multienzyme pathways converge to enter the general metabolic mill in six locations. The largest number of amino acids enters the scheme as acetyl CoA either directly or stepwise to pyruvate or acetoacetyl CoA first, then acetyl CoA. Five are ultimately converted to α-ketoglutarate; three are converted to succinyl CoA; and two are converted to oxaloacetate. Phenylalanine and tyrosine are unusual in that when they are degraded, they each produce two pieces. One enters the mill as acetyl CoA and the other enters as fumarate. In the human body, the liver is the major organ of protein catabolism, with the kidney also able to carry out these reactions at a somewhat lower level.

The first step in the oxidative degradation requires that the α-amino nitrogen of the amino acid be removed. It is this nitrogen that is incorporated into urea and excreted in the urine.

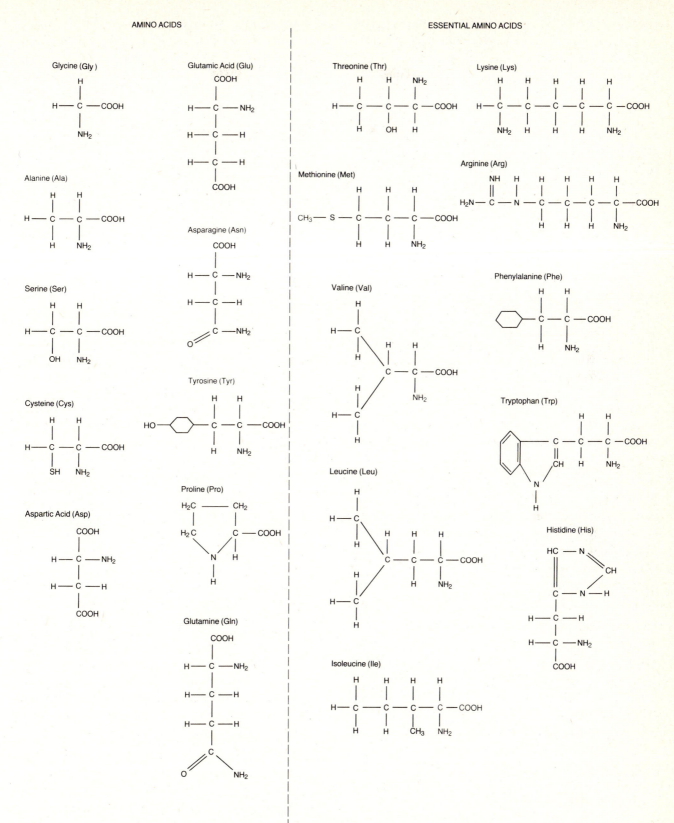

Figure 25-15 The amino acids, including the 10 essential amino acids, which cannot be synthesized at all or in sufficient quantity in the body. The three-letter symbols for each amino acid are shown in parentheses.

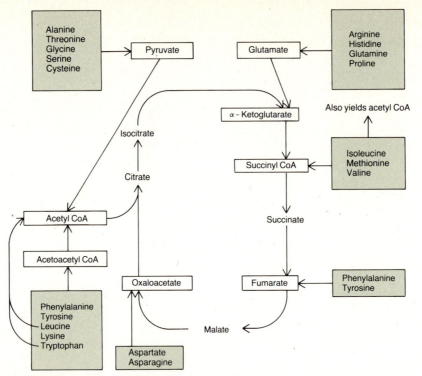

Figure 25-16 Pathways by which the carbon skeletons of amino acids enter into the tricarboxylic acid cycle.

Transamination and Deamination Reactions

Transamination reactions are the most common ones employed in the initial steps of amino acid catabolism. They require a transaminase enzyme and an α-keto acid as an acceptor for the amino group that will be removed from the amino acid. The most commonly employed α-keto acid is a familiar one found in the Krebs cycle: α-ketoglutarate. The transamination reaction actually is one where the two molecules involved trade functional groups. The amino acid gives up its amino group (NH_2) and receives an oxygen (O^{2-}) from the α-keto acid, which in turn gives up its oxygen atom and receives an amino group. An example of a transamination reaction is shown in Figure 25-17. Since α-ketoglutarate is the usual amino group acceptor, then most amino groups will end up as part of the molecule glutamate. The glutamate molecule represents a final, common end product that can be *deaminated* to yield ammonia (NH_3).

Deamination reactions are also confined mainly to liver cells. In this reaction, the amino group and two hydrogen atoms are removed from the glutamate molecule to re-form the α-ketoglutarate,

$$\text{Glutamate} + NAD^+ + H_2O \rightleftharpoons \alpha\text{-Ketoglutarate} \\ + NH_3 + NADH + H^+$$

Ammonia, if allowed to escape into the circulation, can be very toxic to the brain. The normal fate of ammonia produced in the liver is conversion into a nontoxic substance, *urea*. The urea is then carried by the blood from the liver to the kidneys and is excreted in the urine.

The Urea Cycle

The *urea cycle*, or *ornithine cycle*, as it is also known, is an unusual multienzyme system. Part of the pathway requires enzymes located within the mitochondrion and part of the cycle employs enzymes found in the cytoplasm of the liver cell. Figure 25-18 outlines the steps of the cycle.

Step One

The cycle is best examined starting within the mitochondrion, where $2ATP + CO_2 + NH_3 + H_2O$ react to form carbamoyl phosphate + $2ADP + P_i$.

Step Two

The carbamoyl phosphoric acid, which has incorporated the toxic ammonia, is an activated molecule that reacts with a molecule of *ornithine*. Ornithine is constantly being transported into the mitochondrion from the cytoplasm, and it is converted to a molecule of citrulline (Figure 25-19).

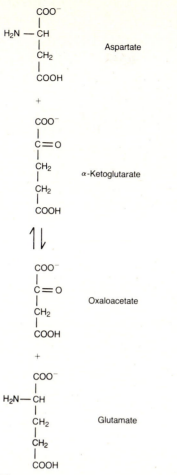

Figure 25-17 The aspartate transaminase reaction. The specific amino group acceptor of this and most other transaminases of animal tissue is α-ketoglutanate, thus yielding glutamate as the product in which α amino groups from other amino acids are collected.

Step Three

Citrulline leaves the mitochondrion for the cytoplasm, where it receives a second amino group from the amino acid *aspartate*. It is converted to a molecule of argininosuccinate (Figure 25-20).

$$\text{Citrulline} + \text{Asparate} + \text{ATP} \rightleftharpoons \text{Argininosuccinate} + \text{AMP} + \text{PP}_i$$

Step Four

In this reaction, argininosuccinate is split into two molecules, arginine and fumarate. The fumarate becomes part of the pool of Krebs cycle intermediates (Figure 25-21).

$$\text{Argininosuccinate} \rightarrow \text{Arginine} + \text{Fumarate}$$

Step Five

The arginine reacts with water and is hydrolyzed to form ornithine and urea. The ornithine is transported into the mitochondrion to maintain the operation of the cycle. The urea so formed diffuses out of the liver cells into the blood and is carried to the kidneys, where it is removed and excreted in the urine.

$$\text{Arginine} + H_2O \rightarrow \text{Ornithine} + \text{Urea}$$

Genetic Defects

Defects have been found in each of the enzymes of the urea cycle. In every instance, the person suffering the defect has shown intolerance to proteins. A high level of ammonia in the blood plasma during early development hinders proper formation of the nervous system which, in turn, produces mental retardation and other disorders of the central nervous system. In the nerve cells, the appearance of ammonia removes the essential Krebs cycle intermediate α-ketoglutarate.

$$NH_3 + \alpha\text{-Ketoglutarate} + \text{NADH} + H^+ \rightarrow \text{Glutamate} + \text{NAD}^+ + H_2O$$

This reaction is essentially a one-way reaction to the right, forming glutamate. The removal of the α-ketoglutarate severely hinders the respiration of the nerve cells.

CONTROL OF METABOLISM

The metabolism of carbohydrates, lipids, and proteins is controlled by a number of hormones. Two major hormones, insulin and glucagon, are secreted in response to blood glucose level. Under normal circumstances, the blood sugar level ranges from 80 to 120 mg/100 ml blood. After a meal, the level may rise to 120 through 130 mg/100 ml. The increase stimulates the pancreatic beta cells to secrete insulin, which pushes the blood glucose level down to normal in a few hours. Insulin lowers blood sugar by stimulating the liver cells to convert glucose to glycogen and fat, stimulating all other cells to catabolize glucose, and inhibiting the catabolism of fats and proteins. When the body fails to produce enough insulin, the blood sugar is somewhat lowered as the glucose is excreted in the urine. Thus sugar in the urine is a sign of untreated diabetes mellitus. The body also excretes glucose if a large amount of carbohydrate is absorbed all at once. In this case, the condition is called *alimentary glycosuria*. Alimentary glycosuria is temporary and is not considered pathological.

After a period of fasting, the blood glucose level falls below normal. Low blood glucose (hypoglycemia) stimulates the pancreatic alpha cells to release glucagon, which stimulates the liver to change glycogen to glucose (glycogenolysis) and to catabolize fats.

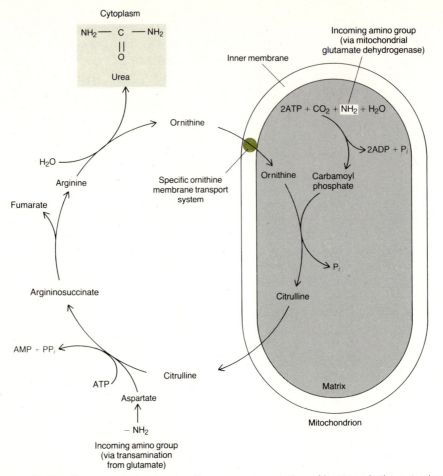

Figure 25-18   The urea cycle, showing the compartmentation of its steps in the cytoplasm and mitochondrion of a liver cell.

The anterior pituitary gland also assumes a role in the regulation of metabolism through the actions of HGH, ACTH, and TSH. HGH, the human growth hormone, encourages new tissue to be laid down by stimulating protein anabolism. HGH also stimulates the body cells to catabolize fats instead of carbohydrates. This action causes the blood sugar level to increase. Adrenocorticotropic hormone (ACTH) stimulates the adrenal cortex to increase its secretion of glucocorticoid hormones. The chief effect of the glucocorticoids is to increase the amount of energy available to cells during extreme stress. Glucocorticoids stimulate the liver to accelerate the breakdown of proteins into amino acids, which the liver converts to glucose by gluconeogenesis. Since the resulting glucose is released into the bloodstream, glucocorticoids are hyperglycemic. The third anterior pituitary hormone, thyroid stimulating hormone (TSH), stimulates the thyroid to release thyroxine, a hormone that encourages glucose catabolism and is therefore hypoglycemic.

The adrenal medulla, the ovaries, and the testes also produce hormones that affect metabolism. Epinephrine and norepinephrine, the adrenal medulla hormones, accelerate the conversion of liver glycogen to glucose and are therefore hyperglycemic. Testosterone, the male hormone, and progesterone, the female hormone, encourage protein anabolism. Progesterone is secreted in copious amounts during pregnancy, when uterine and mammary tissues are growing.

Hormones are the primary regulators of metabolism. However, hormonal control is ineffective without the proper minerals and vitamins. Some minerals and many vitamins are components of the enzyme systems that catalyze the metabolic reactions.

MINERALS

Minerals are inorganic substances. They may appear in combination with each other or in combination with organic compounds. Minerals constitute about 4

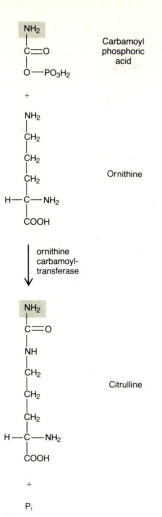

Figure 25-19 Conversion of ornithine to citrulline. The amino groups introduced by carbamoyl phosphoric acid are in the boxes.

percent of the total body weight, and they are concentrated most heavily in the skeleton. Minerals known to perform functions essential to life include calcium, phosphorus, sodium, chlorine, potassium, magnesium, iron, sulfur, iodine, manganese, cobalt, copper, and zinc. Other minerals—aluminum, silicon, arsenic, nickel—are present in the body, but their functions have not yet been determined.

Calcium and phosphorus form part of the structure of bone. But since minerals do not form long-chain compounds, they are otherwise poor building materials. Their chief role is to help regulate body processes. Calcium, iron, magnesium, and manganese are constituents of some coenzymes. Magnesium also serves as a catalyst for the conversion of ADP to ATP. Without these minerals, metabolism would stop and the body would die. Minerals such as sodium and phosphorus work in buffer systems. Sodium helps

regulate the osmosis of water and, along with other ions, is involved in the generation of nerve impulses. Exhibit 25-2 describes the functions of some minerals vital to the body. Note that the body generally uses the ions of the minerals rather than the nonionized forms. Some minerals, such as chlorine, are toxic or even fatal to the body if ingested in the nonionized form.

VITAMINS

Organic nutrients required in minute amounts to maintain growth and normal metabolism are called **vitamins**. Unlike carbohydrates, fats, or proteins, vita-

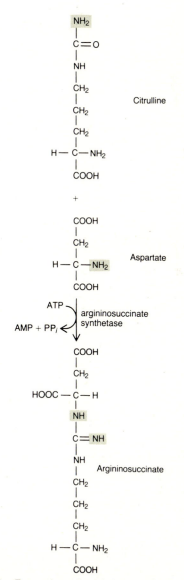

Figure 25-20 Formation of argininosuccinate. The amino groups that end up in urea are in the boxes.

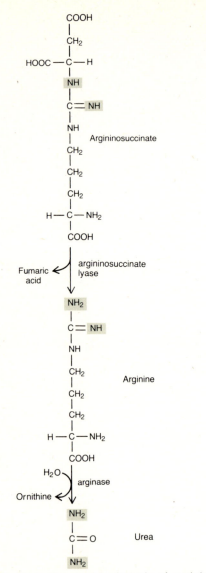

Figure 25-21 Formation and hydrolysis of arginine. The amino groups that end up in urea are in the boxes.

mins do not provide energy or serve as building materials. The essential function of vitamins is the regulation of physiological processes. Accordingly,

some vitamins act as enzymes, others are precursors (forerunners) of enzymes, and still others serve as coenzymes.

Most vitamins cannot be synthesized by the body. One source of vitamins is ingested foods—for example, vitamin C in citrus fruits. Another source is vitamin pills. Other vitamins, such as vitamin K, are produced by bacteria in the gastrointestinal tract. The body can assemble some vitamins if the raw materials called *provitamins* are provided. Vitamin A is produced by the body from the provitamin carotene, a chemical present in spinach, carrots, liver, and milk. No single food contains all the required vitamins— one of the best reasons for eating a balanced diet. The term *avitaminosis* refers to a deficiency of any vitamin in the diet.

On the basis of solubility, vitamins are divided into two principal groups: fat-soluble and water-soluble. *Fat-soluble* vitamins are absorbed along with digested dietary fats by the small intestine as micelles. In fact, they cannot be absorbed unless they are ingested with some fat. Fat-soluble vitamins are generally stored in cells, particularly liver cells, so reserves can be built up. Examples of fat-soluble vitamins are vitamins A, D, E, and K. *Water-soluble* vitamins, by contrast, are absorbed along with water in the gastrointestinal tract and dissolve in the body fluids. Excess quantities of these vitamins are excreted in the urine. Thus, the body does not store water-soluble vitamins well. Examples of water-soluble vitamins are the B vitamins and vitamin C. Exhibit 25-3 lists the principal vitamins, their sources, functions, and related disorders.

METABOLISM AND HEAT

The relationship of foods to body heat, mechanisms of heat gain and loss, and the regulation of body temperature will now be considered.

Measuring Heat

Recall that *heat* is a form of energy that can be measured as *temperature* and expressed in units called

Exhibit 25-2 MINERALS VITAL TO THE BODY

Mineral	Comments	Importance
Calcium	Most abundant cation in body. Appears in combination with phosphorus in ratio of 2:1.5. About 99 percent is stored in bone and teeth. Remainder stored in muscle, other soft tissues, and blood plasma. Blood calcium level controlled by thyrocalcitonin and parathyroid hormone. Most is excreted in feces and small amount in urine. Recommended daily intake for children is 1.2–1.4 g and 800 mg for adults. Good sources are milk, egg yolk, shellfish, green leafy vegetables.	Formation of bones and teeth, blood clotting, and normal muscle and nerve activity.

Exhibit 25-2 *(Continued)*

Mineral	Comments	Importance
Phosphorus	About 80 percent found in bones and teeth. Remainder distributed in muscle, brain cells, blood. More functions than any other mineral. Most excreted in urine, small amount eliminated in feces. Recommended daily intake is 1.2–1.4 g for children and 1,200 mg for adults. Good sources are dairy products, meat, fish, poultry, nuts.	Formation of bones and teeth. Constitutes a major buffer system of blood. Plays important role in muscle contraction and nerve activity. Component of many enzymes. Involved in transfer and storage of energy (ATP). Component of DNA and RNA.
Iron	About 66 percent found in hemoglobin of blood. Remainder distributed in skeletal muscles, liver, spleen, enzymes. Normal losses of iron occur by shedding of hair, epithelial cells, and mucosal cells and in sweat, urine, feces, and bile. Recommended daily intake for children is 7–12 mg and for adults 10–15 mg. Sources are meat, liver, shellfish, egg yolk, beans, legumes, dried fruits, nuts, cereals.	As component of hemoglobin, carries O_2 to body cells. Component of coenzymes involved in formation of ATP from catabolism.
Iodine	Essential component of thyroxine. Excreted in urine. Estimated daily requirement: 0.15–0.30 mg. Adequate sources are iodized salt, seafoods, cod-liver oil, and vegetables grown in iodine-rich soils.	Required by thyroid gland to synthesize thyroxine and triiodothyronine, hormones that regulate metabolic rate.
Copper	Some stored in liver and spleen. Most excreted in feces. Daily requirement about 2 mg. Good sources include eggs, whole-wheat flour, beans, beets, liver, fish, spinach, asparagus.	Required with iron for synthesis of hemoglobin. Component of enzyme necessary for melanin pigment formation.
Sodium	Most found in extracellular fluids, some in bones. Excreted in urine and perspiration. Recommended daily intake of NaCl (table salt) is 5 g.	As most abundant cation in extracellular fluid, strongly affects distribution of water through osmosis. Part of bicarbonate buffer system.
Potassium	Principal cation in intracellular fluid. Most is excreted in urine. Recommended daily intake not known. Normal food intake supplies required amounts.	Functions in transmission of nerve impulses and muscle contraction.
Chlorine	Found in extracellular and intracellular fluids. Principal anion of extracellular fluid. Most excreted in urine. Normal intake of NaCl supplies required amounts.	Assumes role in acid-base balance of blood, water balance, and formation of HCl in stomach.
Magnesium	Component of soft tissues and bone. Excreted in urine and feces. Suggested minimum daily intake: 250–350 mg. Widespread in various foods.	Required for normal functioning of muscle and nervous tissue. Participates in bone formation. Constituent of many coenzymes.
Sulfur	Constituent of many proteins (such as insulin) and some vitamins (thiamine and biotin). Excreted in urine. Good sources include beef, liver, lamb, fish, poultry, eggs, cheese, beans	As component of hormones and vitamins, regulates various body activities.
Zinc	Important component of certain enzymes. Widespread in many foods.	Necessary for normal growth.
Fluorine	Component of bones, teeth, other tissues.	Appears to improve tooth structure.
Manganese	Distribution throughout body similar to that of copper. Human daily requirement about 2.5 mg.	Activates several enzymes. Needed for hemoglobin synthesis. Required for growth, reproduction, lactation.
Cobalt	Constituent of vitamin B_{12}.	As part of B_{12} required for stimulation of erythropoeisis.

calories. One *calorie* (cal) is the amount of heat energy required to raise the temperature of 1 g of water 1°C. Since the calorie is a small unit, we use the *Calorie*, or *kilocalorie*, which is equal to 1,000 cal and is defined as the amount of heat required to raise the temperature of 1,000 g of water 1°C. The Calorie is the unit used in determining the heating value of foods and in measuring body metabolism rates.

Exhibit 25-3 THE PRINCIPAL VITAMINS

Vitamin	Comment and Source	Function	Related Symptoms and Deficiency Disorders
Fat-soluble			
A	Formed from provitamin carotene (and other provitamins) in intestinal tract. Requires bile salts and fat for absorption. Stored in liver. Sources of carotene and other provitamins include yellow and green vegetables; sources of vitamin A include fish, liver oils, milk, butter.	Maintains general health and vigor of epithelial cells.	Deficiency results in atrophy and keratinization of epithelium, leading to dry skin and hair, increased incidence of ear, sinus, respiratory, urinary, and digestive infections, inability to gain weight, drying of cornea with ulceration (xerophthalmia), nervous disorders, and skin sores.
		Essential for formation of rhodopsin, light-sensitive chemical in rods of retina.	Night blindness or decreased ability for dark adaptation.
		Growth of bones and teeth by apparently helping to regulate activity of osteoblasts and osteoclasts.	Slow and faulty development of bones and teeth.
D	In presence of sunlight, provitamin D_3 (derivative of cholesterol) converted to vitamin D in the skin. Dietary vitamin D requires moderate amounts of bile salts and fat for absorption. Stored in tissues to slight extent. Most excreted via bile. Sources include liver oils of bony fish, egg yolk, fortified milk.	Essential for absorption and utilization of calcium and phosphorus from gastrointestinal tract. May work with parathyroid hormone that controls calcium metabolism.	Defective utilization of calcium by bones leads to rickets in children and osteomalacia in adults. Possible loss of muscle tone.
E (tocopherols)	Stored in liver, adipose tissue, and muscles. Requires bile salts and fat for absorption. Sources include fresh nuts and wheat germ, seed oils, green leafy vegetables.	Believed to inhibit catabolism of certain fatty acids that help form cell structures, especially membranes. Involved in formation of DNA, RNA, and red blood cells. Believed to help protect liver from toxic chemicals such as carbon tetrachloride.	Catabolism of certain fatty acids on exposure to oxygen (fatty acids in membranes of red blood cells may cause hemolytic anemia). Deficiency also causes muscular dystrophy in monkeys and sterility in rats.
K	Produced in considerable quantities by large intestinal bacteria. Requires bile salts and fat for absorption. Stored in liver and spleen. Other sources include spinach, cauliflower, cabbage, liver.	Coenzyme essential for synthesis of prothrombin by liver and thus for normal blood clotting. Also known as antihemorrhagic vitamin.	Prolonged clotting time results in excessive bleeding.

Exhibit 25-3 *(Continued)*

Vitamin	Comment and Source	Function	Related Symptoms and Deficiency Disorders
Water-soluble			
B$_1$ (thiamine)	Rapidly destroyed by heat. Not stored in body. Excessive intake eliminated in urine. Sources include whole-grain cereals, eggs, pork, nuts, liver, yeast.	Acts as coenzyme for 24 different enzymes involved in carbohydrate metabolism of pyruvic acid to CO_2 and H_2O. Essential for synthesis of acetylcholine.	Improper carbohydrate metabolism leads to buildup of pyruvic and lactic acids and insufficient energy for muscle and nerve cells. Deficiency leads to two syndromes: (1) *Beriberi*—partial paralysis of smooth muscle of GI tract, causing digestive disturbances, skeletal muscle paralysis, atrophy of limbs. (2) *Polyneuritis*—reflexes related to kinesthesia are impaired, impairment of sense of touch, decreased intestinal motility, stunted growth in children, poor appetite.
B$_2$ (riboflavin)	Not stored in large amounts in tissues. Most is excreted in urine. Small amounts supplied by bacteria of GI tract. Other sources include yeast, liver, beef, veal, lamb, eggs, whole-wheat products, asparagus, peas, beets, peanuts.	Component of certain coenzymes concerned with carbohydrate and protein metabolism, especially in cells of eye, integument, mucosa of intestine, blood.	Deficiency may lead to improper utilization of oxygen resulting in blurred vision, cataracts, and corneal ulcerations. Also dermatitis and cracking of skin, lesions of intestinal mucosa, and development of one type of anemia.
Niacin (nicotin-amide)	Derived from amino acid tryptophan. Sources include yeast, meats, liver, fish, whole-grain breads and cereals, peas, beans, nuts.	Essential component of coenzyme concerned with energy-releasing reactions. In lipid metabolism, inhibits production of cholesterol and assists in fat breakdown.	Principal deficiency is *pellagra*, characterized by dermatitis, diarrhea, and psychological disturbances.
B$_6$ (pyridoxine)	Formed by bacteria of GI tract. Stored in liver, muscle, brain. Other sources include salmon, yeast, tomatoes, yellow corn, spinach, whole-grain cereals, liver, yogurt.	May function as coenzyme in fat metabolism. Essential coenzyme for normal amino acid metabolism. Assists production of circulating antibodies.	Most common deficiency symptom is dermatitis of eyes, nose, and mouth. Other symptoms are retarded growth and nausea.
B$_{12}$ (cyanoco-balamin)	Only B vitamin not found in vegetables; only vitamin containing cobalt. Absorption from GI tract dependent on HCl and intrinsic factor secreted by gastric mucosa. Sources include liver, kidney, milk, eggs, cheese, meat.	Coenzyme necessary for red blood cell formation, formation of amino acid methionine, entrance of some amino acids into Krebs cycle, and manufacture of choline (chemical similar in function to acetylcholine).	Pernicious anemia and malfunction of nervous system due to degeneration of axons of spinal cord.

Exhibit 25-3 *(Continued)*

Vitamin	Comment and Source	Function	Related Symptoms and Deficiency Disorders
Pantothenic acid	Stored primarily in liver and kidneys. Some produced by bacteria of GI tract. Other sources include kidney, liver, yeast, green vegetables, cereal.	Constituent of co-enzyme A essential for transfer of pyruvic acid into Krebs cycle, conversion of lipids and amino acids into glucose, and synthesis of cholesterol and steroid hormones.	Experimental deficiency tests indicate fatigue, muscle spasms, neuromuscular degeneration, insufficient production of adrenal steroid hormones.
Folic acid	Synthesized by bacteria of GI tract. Other sources include green leafy vegetables and liver.	Component of enzyme systems synthesizing purines and pyrimidines built into DNA and RNA. Essential for normal production of red and white blood cells.	Production of abnormally large red blood cells—macrocytic anemia.
Biotin	Synthesized by bacteria of GI tract. Other sources include yeast, liver, egg yolk, kidneys.	Essential coenzyme for conversion of pyruvate to oxaloacetate and synthesis of fatty acids and purines.	Mental depression, muscular pain, dermatitis, fatigue, nausea.
C (ascorbic acid)	Rapidly destroyed by heat. Some stored in glandular tissue and plasma. Sources include citrus fruits, tomatoes, green vegetables.	Exact role not understood. Promotes many metabolic reactions, particularly protein metabolism, including laying down of collagen in formation of connective tissue. As coenzyme may combine with poisons, rendering them harmless until excreted. Works with antibodies.	Scurvy: many symptoms related to poor connective tissue growth and repair, including tender swollen gums, loosening of teeth (alveolar processes also deteriorate), poor would healing, bleeding (vessel walls fragile because of connective tissue degeneration), and retardation of growth. Anemia and low resistance to infection of scurvy.

Calories

The apparatus used to determine the caloric value of foods is called a *calorimeter* (see Figure 2-8). A weighed sample of a dehydrated food is burned completely in an insulated metal container. The energy released by the burning food is absorbed by the container and then transferred to a known volume of water that surrounds the container. The change in the water's temperature is directly related to the number of calories released by the food. Knowing the caloric value of foods is important—if we know the amount of energy the body uses for various activities, we can adjust our food intake. In this way, we can control body weight by taking in only enough calories to sustain our activities.

Production of Body Heat

Most of the heat produced by the body comes from oxidation of the food we eat. The rate at which this heat is produced—the **metabolic rate**—is also measured in Calories. Among the factors that affect your metabolic rate are the following:

1. **Exercise.** During strenuous exercise, your metabolic rate may increase as much as 40 times your normal rate.

2. **Nervous system.** In a stress situation, your sympathetic nervous system is stimulated and the nerves release norepinephrine, which increases the metabolic rate of body cells. Strong sympathetic stimulation may increase the metabolic rate by 160 times, but only for a few minutes.

3. **Hormones.** In addition to norepinephrine, two other hormones affect metabolic rate: epinephrine produced by the adrenal glands and thyroid hormones produced by the thyroid gland. Epinephrine is secreted in stress situations. Increased secretions of thyroid hormones increase the metabolic rate.

4. **Body temperature.** The higher your body temperature, the higher your metabolic rate. Each 1°C rise in temperature increases the rate of biochemical reactions by about 10 percent. In fact, the metabolic rate may be doubled if you have a high fever.

Since many factors affect metabolic rate, it is difficult to measure and compare your metabolic rate with someone else's. Therefore, in standardized tests used world wide certain conditions must be controlled. The subject should not exercise for 30 minutes to an hour before the measurement is taken. Also, the individual must be completely at rest, but awake, and air temperature should be comfortable. Finally, the person should have fasted for at least 12 hours and body temperature should be normal. All these conditions produce the **basal state**—a state in which factors that could affect metabolic rate have been greatly reduced or eliminated. **Basal metabolic rate (BMR)** is a measure of the rate at which your body breaks down foods (and therefore releases heat) under the basal state conditions. It is also a measure of how much thyroxine your thyroid gland is producing, since thyroxine regulates the rate of food breakdown and is not a controllable factor under basal conditions.

Metabolic Rate

The most convenient way to measure basal metabolic rate is to do it indirectly by measuring oxygen consumption. If a given amount of food releases a given amount of heat energy when it is oxidized, it must combine with a given amount of oxygen. Thus, by measuring the amount of oxygen needed for the metabolism of foods, we can determine how many Calories are produced. The amount of heat energy released when 1 liter of oxygen combines with carbohydrates is 5.05 Cal; with fats, the heat released is 4.70 Cal; with proteins, the heat released is 4.60 Cal. The average of the three values is 4.825 Cal. Therefore, every time a liter of oxygen is consumed, 4.825 Cal is produced.

The instrument used to determine metabolic rate indirectly is called a *respirometer* or *spirometer* (Figure 25-22). After pinching the nose closed, the subject places the mouthpiece in the mouth and breathes in and out of an inverted bell that rides up and down in a water bath. The respirometer bell contains O_2 that is breathed back and forth into the lungs. The O_2 is gradually taken into the blood and in its place CO_2 is exhaled. The exhaled air passes through a valve into a container of soda lime, which absorbs the CO_2, and then passes back into the respirometer bell. As oxygen is consumed, the level of the respirometer bell falls. The top of the bell is connected by a cord running from a pulley to a writing lever, which records the level of the gas container on a rotating kymograph. As the level of the respirometer bell falls, the writing level falls. From the record on the kymograph, the amount of oxygen consumed in a given period of time can be measured.

Basal metabolic rate is usually expressed in Calories per square meter of body surface area per hour. Suppose you use 1.8 liters of oxygen in 6 minutes as recorded on a respirometer. This means that your oxygen consumption in an hour would be 18 liters (1.8 × 10). Your basal metabolic rate would be 18 × 4.825 Cal or 86.85 Cal/hour. To express the Calories per square meter of body surface, a standardized

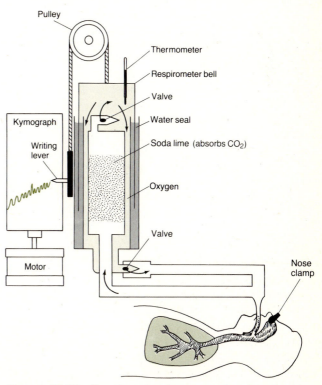

Figure 25-22 Respirometer for measuring basal metabolic rate.

chart is used. Such a chart shows square meters of body surface relative to height in centimeters and weight in kilograms. If you weigh 75 kg and are 190 cm tall, your body surface area is 2 m³. Your basal metabolic rate is equal to 86.85 Cal divided by 2, or 43.43 Cal/m²/hour.

The normal basal metabolic rates for various age groups by sex are also listed in standardized charts. Suppose a 27-year-old male has a recorded basal metabolic rate of 45.7 Cal/m²/hour. The chart would show that his basal metabolic rate should be about 40.3 — obviously above "normal." What does this mean? The recorded basal metabolic rate is 5.4 Cal/m²/hour, or about 14 percent, above normal. Since a basal metabolic rate between +15 and −15 is considered normal, this male is judged to be normal. Values above or below 15 may indicate an excess or deficiency of thyroid hormones. When the thyroid is secreting extreme quantities of thyroid hormones, the basal metabolic rate can go as high as +100. If, on the other hand, the thyroid is secreting very little of its thyroid hormones, the basal metabolic rate may be as low as −50.

The BMR is no longer widely used to measure thyroid function because of its unreliability. It is often elevated in nonthyroid diseases such as leukemia, emphysema, and congestive heart failure, for instance, and depressed in nonthyroid diseases such as vitamin B_1 deficiency and Addison's disease.

Another test for measuring thyroid function is the *protein-bound iodine (PBI) test*. Recall that thyroxine is transported in the blood in combination with a plasma protein called thyroxine-binding globulin (TBG). The PBI test measures the iodine bound to TBG, which is virtually all the iodine in the body that is found in thyroxine. Thus, the PBI is a fairly good measure of the amount of circulating thyroxine.

The *serum thyroxine (T_4) test* for measuring thyroid function directly measures the level of thyroxine in the blood. It is becoming the most common test for thyroid function. Other tests of thyroid function are the T_3 *resin uptake test*, which indirectly measures the amount of triiodothyronine (T_3) in the blood, and the *radioactive iodine (I^{131}) uptake test*, which measures the amount of iodine used by the thyroid gland to synthesize the thyroid hormones.

Loss of Body Heat

Body heat is produced by the oxidation of foods we eat. This heat must be removed continually or body temperature would rise steadily. The principal routes of heat loss include radiation, conduction, convection, and evaporation.

Radiation

Radiation is the transfer of heat from one object to another without physical contact. Your body loses heat by the radiation of heat waves to cooler objects nearby such as ceilings, floors, and walls. If these objects are at a higher temperature, you absorb heat by radiation. Incidentally, the air temperature has no relationship to the radiation of heat to and from objects. Skiers can remove their shirts in bright sunshine, even though the air temperature is very low, because the radiant heat from the sun is adequate to warm them. In a room at 70°F, about 60 percent of heat loss is by radiation.

Conduction

Another method of heat transfer is *conduction*. In this process, body heat is transferred to a substance or object in contact with the body, such as chairs, clothing, and jewelry.

Convection

When cool air makes contact with the body, it becomes warmed and is carried away by *convection* currents. Then more cool air makes contact with the body and is carried away. The faster the air moves, the faster the rate of convection. About 15 percent of body heat is lost to the air by convection, and about 3 percent is conducted to nearby cooler objects such as clothing or jewelry.

Evaporation

Evaporation means the conversion of a liquid such as water to a vapor. Water has a high *heat of evaporation*, the amount of heat necessary to evaporate 1 g of water at 30°C. Because of water's high heat of evaporation, every gram of water evaporating from the skin takes with it a great deal of heat — about 0.58 Cal/g of water. Under normal conditions, about 22 percent of heat loss occurs through evaporation. The evaporation of only 150 ml of water per hour is enough to remove all the heat produced by the body under basal conditions. Under extreme conditions, about 4 liters (1 gal) of sweat is produced each hour, and this volume can remove 2,000 Cal of heat from the body. This is approximately 32 times the basal level of heat production. The rate of evaporation is inversely related to relative humidity, the ratio of the actual amount of moisture in the air to the greatest amount it can hold at a given temperature. The higher the relative humidity, the lower the rate of evaporation.

Temperature Regulation

If the amount of heat production equals the amount of heat loss, you maintain a constant body temperature of 98.6°F. If your heat-producing mechanisms generate more heat than is lost by your heat-losing mechanisms, your body temperature rises. If your heat-losing mechanisms are giving off more heat than is generated by heat-producing mechanisms, your temperature falls.

Body temperature is regulated by systems that attempt to keep heat production and heat loss in balance. One such regulatory system is found in the hypothalamus in a group of neurons in the anterior portion referred to as the *preoptic area.* If blood temperature increases, the neurons of the preoptic area fire impulses more rapidly. If something causes the temperature of the blood to decrease, these neurons fire impulses more slowly. The preoptic area is adjusted to maintain normal body temperature and thus serves as your thermostat.

Impulses from the preoptic area are sent to other portions of the hypothalamus to control either heat production or heat loss. These other portions are known as the heat-losing center and the heat-promoting center. The **heat-losing center,** when stimulated by the preoptic area, sets into operation a series of responses that raise body temperature. The heat-losing center is mainly parasympathetic in function; the heat-promoting center is primarily sympathetic.

Suppose the environmental temperature is low or blood temperature falls below normal. Both stresses stimulate the preoptic area. The preoptic area, in turn, activates the heat-promoting center. In response, the **heat-promoting center** discharges impulses that automatically set into operation a number of responses designed to increase body heat and bring body temperature back to normal. One such response is vasoconstriction. Impulses from the heat-promoting center stimulate sympathetic nerves that cause blood vessels of the skin to constrict. The net effect of vasoconstriction is to decrease the flow of warm blood from the internal organs to the skin, thus decreasing the transfer of heat from the internal organs to the skin. This reduction in heat loss helps raise the internal body temperature.

Another response triggered by the heat-promoting center is the sympathetic stimulation of metabolism. The heat-promoting center stimulates sympathetic nerves leading to the adrenal medulla. This stimulation causes the medulla to secrete epinephrine and norepinephrine into the blood. The hormones, in turn, bring about an increase in cellular metabolism, a reaction that also increases heat production.

Heat production is also increased by responses of skeletal muscles. For example, stimulation of the heat-promoting center causes stimulation of parts of the brain that increase muscle tone and hence heat

production. In fact, shivering is caused by a high degree of muscle tone. As the muscle tone increases, the stretching of the agonist muscle initiates the stretch reflex and the muscle contracts. This contraction causes the antagonist muscle to stretch, and it too develops a stretch reflex. The repetitive cycle — called shivering — increases the rate of heat production by several hundred percent.

A final body response that increases heat production is increased production of thyroxine. A low environmental temperature causes the secretion of the releasing factor TRF produced by the preoptic area of the hypothalamus. The TRF, in turn, stimulates the anterior pituitary to secrete TSH, which causes the thyroid to release thyroxine into the blood. Since increased levels of thyroxine increase the metabolic rate, body temperature is increased.

Suppose some stress raises body temperature above normal. The stress stimulates the preoptic area, which in turn stimulates the heat-losing center and inhibits the heat-promoting center. Instead of blood vessels in the skin constricting, they dilate. The skin becomes warm and the excess heat is lost to the environment. At the same time, the metabolic rate is decreased, muscle tone decreases, and there is no shivering. These responses reverse the heat-promoting effects and bring body temperature down to normal.

When the body is subjected to high environmental temperatures or strenuous exercise, the high temperature of the blood signals the hypothalamus, which activates the heat-losing center. In response, impulses are sent out to the sweat glands of the skin and they produce more perspiration. As the perspiration evaporates from the surface of the skin, the skin is cooled and body temperature drops to normal.

Temperature Abnormalities

Most of us have probably had a **fever** — an abnormally high body temperature. The most frequent cause of fever is infection from bacteria (and their toxins) and viruses. Other conditions that result in fever are heart attacks, tumors, tissue destruction by x-rays or trauma, and reactions to vaccines. The mechanism of fever production is not completely understood. However, it is believed that foreign proteins (antigens) affect the hypothalamus by somehow setting the thermostat at a higher temperature. It has been shown that 0.001 g of protein from the bacterium that causes typhoid fever can set the thermostat as high as 110°F, and the body temperature will continue to be regulated at this temperature until the protein is eliminated.

Suppose that disease-producing microbes set the thermostat at 103°F. Now the heat-promoting mechanisms (vasoconstriction, increased metabolism, shivering) are operating at full force. Thus, even though body temperature is climbing higher than normal, say 101°F, the skin remains cold and shivering occurs. This condition, called a **chill,** is a definite sign that body temperature is rising. After several hours, body temperature reaches the setting of the thermostat and the chills disappear. But the body will continue to regulate temperature at 103°F until the stress is removed. When the stress is removed, the thermostat is reset at normal (98.6°F). Since body temperature remains high in the beginning, the heat-losing mechanisms (vasodilation and sweating) go into operation to de-crease body temperature. The skin becomes warm and the person begins to sweat. This phase of the fever, called the *crisis,* indicates that the body temperature is falling.

Up to a point, fever is beneficial. The high body temperature is believed to inhibit the growth of some bacteria and viruses. Moreover, heat speeds up the rate of chemical reactions. This increase may help body cells to repair themselves more quickly during a disease. As a rule, death results if body temperature rises to 112 to 114°F (44 to 46°C). On the other end of the scale, death usually results when body temperature falls to 70 to 75°F (21 to 24°C)

CHAPTER SUMMARY OUTLINE

METABOLISM

1. Nutrients are chemical substances in food that provide energy, form new body components, and assist body processes.
2. There are six classes of nutrients: carbohydrates, lipids, proteins, minerals, vitamins, and water.
3. Metabolism has two phases: catabolism and anabolism.
4. Catabolism is the term for processes that are degradative and energy-releasing. Digestion is a catabolic process because the breaking of bonds releases energy.
5. Anabolism consists of a series of synthetic reactions whereby small molecules are built up into larger ones that form the body's structural and functional components.
6. Much of the energy released by degradation processes is captured by the cell and stored in the form of a chemical molecule known as ATP.

CARBOHYDRATE METABOLISM

1. Carbohydrate metabolism is primarily concerned with glucose.
2. Glucose is broken down via glycolysis, the Krebs cycle, the phosphogluconate pathway, and the electron transport system to produce energy in the form of ATP.
3. About 43 percent of the energy of glucose is captured in ATP molecules, and the remaining 57 percent is given off as heat.

LIPID METABOLISM

1. Fats are stored in adipose tissue, mostly in subcutaneous tissue, and constitute the largest reserve of energy.
2. Triglycerides are the lipid-type molecules that are used mainly for energy storage.
3. In fat catabolism, glycerol is converted in the liver into glucose, and fatty acids undergo beta oxidation and transformation into ketone bodies. Ketone bodies are transformed by nonliver cells to acetyl CoA, which enters the Krebs cycle. The presence of excess ketone bodies is called ketosis.

PROTEIN METABOLISM

1. Amino acids are built into proteins that serve as cell structures, enzymes, antibodies, and glandular secretions.
2. Protein catabolism involves transamination reactions and the deamination of amino acids. Both are confined mainly to liver cells.
3. Amino acids may then be converted to keto acids, pyruvate, and acids of the Krebs cycle.

CONTROL OF METABOLISM

1. The metabolism of carbohydrates, lipids, and proteins is controlled by a number of hormones.
2. Two major hormones, insulin and glucagon, are secreted in response to blood glucose level.
3. The anterior pituitary gland helps regulate me-

tabolism through the actions of HGH, ACTH, and TSH.
4. The adrenal medulla, the ovaries, and the testes produce hormones that affect metabolism.

MINERALS

1. Minerals are inorganic substances that help regulate body processes.
2. Calcium and phosphorus are necessary for growth of bones and teeth. Iron and copper are used in the synthesis of hemoglobin. Iodine is necessary for thyroxine and triiodothyronine synthesis. Sodium is used in water balance, buffers, and is involved in the generation of nerve impulses.
3. Potassium is necessary for nerve impulse transmission. Chlorine is required for acid-base balance. Magnesium is used for proper muscle and nerve functioning. Sulfur is a component of hormones.

VITAMINS

1. Vitamins are organic nutrients that regulate metabolism. Many function in enzyme systems.
2. Fat-soluble vitamins are absorbed with fats and include A, D, E, and K.
3. Water-soluble vitamins are absorbed with water and include the B vitamins and vitamin C.
4. Representative physiological functions of the vitamins are: A—healthy epithelium and vision; D—proper utilization of calcium; K—blood clotting; B_1, B_2, niacin, B_6—regulation of energy metabolism; B_{12} and folic acid—blood cell formation.

METABOLISM AND HEAT

1. The Calorie is the unit of heat used in determining the caloric value of foods and in measuring body metabolism rates.

2. The apparatus used to determine the caloric value of foods is called a calorimeter.
3. Most body heat is a result of oxidation of the food we eat. The rate at which this heat is produced is known as the metabolic rate.
4. Metabolic rate is affected by exercise, the nervous system, hormones, and body temperature.
5. Measurement of the metabolic rate under basal conditions is called the basal metabolic rate (BMR).
6. The instrument used to determine metabolic rate indirectly is called a respirometer.
7. The principal methods of heat loss are radiation, conduction, convection, and evaporation.
8. A normal body temperature is determined by a delicate balance between heat-production and heat-loss mechanisms.
9. Temperature abnormalities are possible, and some of them may be beneficial, while others may be harmful.
10. Fever is an abnormally high body temperature, and the most frequent cause of fever is infection from bacteria and viruses.
11. The mechanism of fever production is not completely understood, but it is believed that foreign proteins affect the hypothalamus by setting the thermostat at a higher temperature.
12. A chill indicates that body temperature is rising, while sweating indicates that body temperature is falling.
13. Fever may be beneficial since a high body temperature inhibits growth of some bacteria and viruses, and it speeds up the rate of chemical reactions, helping the body cells repair themselves quickly.
14. Death results if body temperature rises to 112 to 114°F (44 to 46°C) or falls to 70 to 75°F (21 to 24°C).

REVIEW QUESTIONS

1. Define a nutrient. List the six classes of nutrients and indicate a function of each.
2. What is metabolism? Distinguish between catabolism and anabolism and give examples of each.
3. What is ATP? Explain how it is formed and its main function.
4. What is the body's most direct source of energy?
5. Explain what happens to glucose during glycolysis and the Krebs cycle, the phosphogluconate pathway, and the electron transport system.

6. Indicate areas where fat is stored in the body.
7. Explain the catabolism of triglycerides.
8. Define beta oxidation. What are ketone bodies? What is ketosis?
9. Describe the biosynthesis of saturated fatty acids and the biosynthesis of triglycerides.
10. List some of the many types of protein and give their specific functions.
11. Explain the difference between essential and nonessential amino acids.

12. Briefly describe the mechanism involved in protein synthesis.
13. Define transamination and deamination reactions.
14. Describe some of the pathways of amino acid oxidation.
15. Explain the conversions that occur between amino acids and keto acids. What is the normal fate of ammonia?
16. Describe the urea cycle.
17. Indicate the role of the following hormones in the control of metabolism: insulin, glucagon, HGH, ACTH, TSH, epinephrine, sex hormones.
18. What is a mineral? Briefly describe the functions of the following minerals: calcium, phosphorus, iron, iodine, copper, sodium, potassium, chlorine, magnesium, sulfur, zinc, fluorine, manganese, and cobalt.
19. Define a vitamin. Explain how we obtain vitamins. Distinguish between a fat-soluble and a water-soluble vitamin.
20. What are the functions of vitamin A? Relate its functions to health of the epithelium, night blindness, and growth of bones and teeth.
21. How is sunlight related to vitamin D? What are the functions of vitamin D?
22. What is believed to be the principal physiological activity of vitamin E?
23. How does vitamin K function in blood clotting?
24. Relate the roles of vitamin B_1 to beriberi and polyneuritis.
25. How does vitamin B_2 function in the body?

26. Why does a niacin deficiency cause pellagra?
27. Relate the role of vitamin B_6 to dermatitis.
28. How does vitamin B_{12} function in red blood cell formation?
29. What are the principal physiological effects of pantothenic acid?
30. What relationship exists between folic acid and macrocytic anemia?
31. What are the functions of biotin in the body?
32. Relate vitamin C deficiency to the symptoms of scurvy.
33. Define a calorie. Describe the mechanism of a calorimeter.
34. What is metabolic rate? List some factors that affect your metabolic rate.
35. Define BMR. What is its relationship to your thyroid gland?
36. Name the principal routes or methods of heat loss.
37. How is a normal body temperature determined?
38. Describe some temperature abnormalities.
39. What are the most frequent causes of fever? What are its other possible causes?
40. What is your understanding of the mechanism of fever production?
41. Describe what happens to the physiology of the body in a chill, and explain crisis as a phase of fever.
42. How is fever beneficial?
43. What extremes in body temperature could produce death?

Chapter 26

RENAL PHYSIOLOGY

STUDENT OBJECTIVES

After reading this chapter, you should be able to

- Identify the gross anatomical features of the kidneys
- Describe the blood supply to the kidneys
- Define the structural adaptations of a nephron for urine formation
- Explain the blood supply to the kidney tubules
- Describe the nerve supply to the kidneys
- Describe the process of urine formation
- Explain in detail the processes of glomerular filtration, tubular reabsorption, and tubular secretion
- Compare the chemical composition of plasma, glomerular filtrate, and urine
- Define the forces that support and oppose the filtration of blood in the kidneys
- Discuss renal suppression as a disorder resulting from a decreased filtration pressure
- Describe the physiological role of tubular reabsorption
- Compare the obligatory and facultative reabsorption of water
- Discuss the secretion of PAH
- Describe urine concentration and dilution and explain the countercurrent mechanisms
- Discuss the regulation of sodium and potassium excretion
- Explain the contribution of the kidney to acid-base balance
- Compare the lungs, integument, and alimentary canal as organs of excretion that help maintain body pH
- Describe the effects of blood pressure, blood concentration, temperature, diuretics, and emotions on urine production
- List the physical characteristics of urine
- List the normal chemical constituents of urine
- Define albuminuria, glycosuria, hematuria, pyuria, ketosis, casts, and calculi
- Describe the structure and physiology of the ureters and urinary bladder
- Describe the physiology of micturition
- Compare the causes of incontinence, retention and suppression
- Describe the structure and physiology of the urethra
- Discuss the causes of ptosis, kidney stones, gout, glomerulonephritis, pyelitis, and cystitis
- Discuss the operational principles of hemodialysis and peritoneal dialysis

The metabolism of nutrients by body cells results in the production of wastes such as carbon dioxide, excess water, and heat. Oxidative degradation of protein produces toxic nitrogenous wastes such as ammonia and urea. In addition, the concentrations of essential ions such as sodium, chloride, sulfate, phosphate, and hydrogen must be kept within strict homeostatic limits. All toxic materials must be eliminated and essential materials must be either retained or excreted, depending upon their plasma concentration.

The primary function of the **urinary system** is to aid the body to maintain homeostasis by controlling the concentration and volume of blood. It does so by removing and restoring selected amounts of water and solutes. It also excretes selected amounts of various wastes. Two kidneys, two ureters, one urinary bladder, and a single urethra make up the system (Figure 26-1). The kidneys regulate the concentration and volume of the blood and remove wastes from the blood in the form of urine. Urine is excreted from each kidney through its ureter and is stored in the urinary bladder until it is expelled from the body through the urethra. Other systems that aid in waste elimination are the respiratory, integumentary, and digestive systems.

KIDNEYS

The **kidneys** are paired, slightly flattened organs that resemble kidney beans in shape. They are surrounded and packed in a mass of fat and loose areolar tissue. They are pressed against the posterior abdominal wall behind the parietal peritoneal membrane and are therefore said to be *retroperitoneal*. The kidneys are high in the abdominal cavity, each against the poste-

rior part of the diaphragm and positioned between the levels of the last thoracic and first three lumbar vertebrae. The right kidney is slightly lower than the left because of the large area occupied by the liver.

The concave medial border of each kidney faces the vertebral column. The renal artery and nerves enter and the renal vein, lymph vessels, and ureter exit the kidney through the *hilum*—a notch occupying the middle third of the medial border of the organ. The hilum is the entrance into a C-shaped cavity in the kidney called the *renal sinus*.

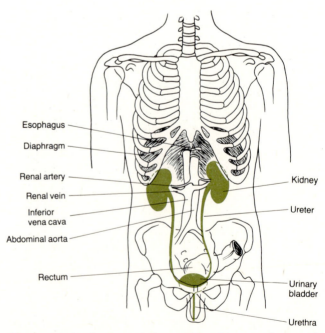

Esophagus
Diaphragm
Renal artery
Renal vein
Inferior vena cava
Abdominal aorta
Rectum
Kidney
Ureter
Urinary bladder
Urethra

Figure 26-1 Organs of the male urinary system and related structures.

Internal Anatomy

An analysis of the internal anatomy of the kidney is crucial to understanding the functional components of this organ. The relationships of vessels to the kidney tubules, the blood distribution patterns, and the pressure regulation of the blood are all interrelated. An examination of Figure 26-2 first reveals an outer reddish area called the *cortex* and an inner reddish brown region called the *medulla*. The outer two-thirds of the cortex is designated the *cortical region*, the inner one-third is termed the *juxtamedullary region*. Within the medulla are 8 to 18 striated, triangular structures termed *renal* or *medullary pyramids*. The striated appearance results from the general parallel arrangement of collecting ducts and blood vessels packed into these structures. The broad base of the pyramids is directed toward the cortex, and the pointed tips, called *renal papillae*, are directed toward the center of the kidney.

In the renal sinus of the kidney is a large cavity called the *renal pelvis*. The edge of the renal pelvis is interrupted by cuplike extensions called the *major* and *minor calyces*. There are usually only 2 or 3 major calyces and 7 to 13 minor calyces. Each minor calyx collects urine from collecting ducts of the pyramids and empties it into a major calyx. From the major calyces, the urine collects in the renal pelvis, which funnels the urine out of the kidney and into the narrow ureter.

Blood Supply

The human kidneys contribute less than 0.5 percent to the weight of the body, yet they use 8 percent of the total oxygen consumed by the body at rest, and they receive about 25 percent of the cardiac output (1,200 ml/minute).

Blood is delivered to each kidney by a short *renal artery* (Figure 26-3), which arises from the abdominal aorta, thus assuring the kidneys of a large volume of blood delivered at a high pressure. As the renal artery enters the kidney, it immediately divides into several branches called *interlobar arteries*. These vessels pass between the renal pyramids and extend to the junction of the cortical and medullary regions, where they divide and arch over the bases of the pyramids as halves of incomplete arches. Their arching pattern has earned them the name of *arciform arteries*. From these arciform arteries, a series of *interlobular arteries* arise at right angles and course radially toward the outermost regions of the cortex — some reaching and nourishing the capsule enclosing the kidney.

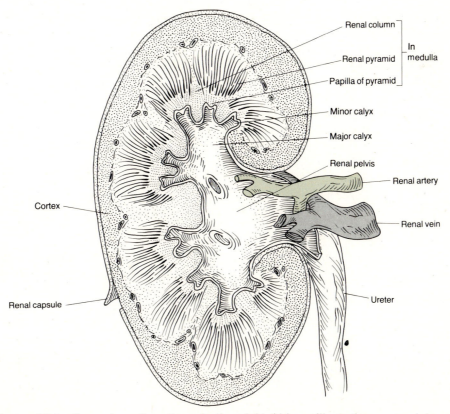

Figure 26-2 Coronal section through the right kidney illustrating gross internal anatomy.

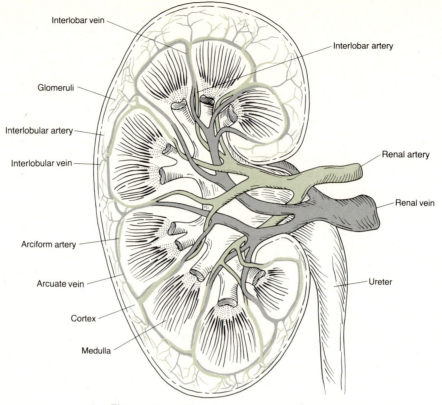

Figure 26-3 Blood supply of the kidney.

The interlobular arteries, in turn, give rise, at approximately right angles, to short branches called the *afferent arterioles* (Figure 26-4). Each afferent arteriole branches abruptly into a tuft of capillaries termed the *glomerular capillaries* or sometimes simply the *glomerulus*. The glomerular capillaries converge to form a single *efferent arteriole* leading away from the glomerulus. The efferent arteriole is somewhat narrower in diameter than the afferent arteriole. This arrangement aids in maintaining the blood pressure in the glomerulus at a high level. The afferent-efferent arrangement of vessels is unusual because blood usually flows out of capillaries into venules and not into other arterioles. Each efferent arteriole divides to form a second network of capillaries called the *peritubular capillaries*—also known as the *vasa recta*—which surround the convoluted tubules. The peritubular capillaries eventually reunite to form *interlobular veins*. Blood from the interlobular veins is collected by the *arcuate veins* that arch over the bases of the renal pyramids. The arcuate veins are arranged as complete arches which anastomose freely with other arcuate veins. *Interlobar veins* are given off in a pattern similar to that seen for interlobar arteries. These veins drain blood from the arcuate veins and, in the region of the sinus, converge to form the single *renal vein,* which emerges from the kidney and bridges the short gap to the inferior vena cava.

Nephron

Each human kidney contains between 1 and 1.25 million long, coiled tubules termed **nephrons** (Figure 26-4). Each nephron is subdivided into functional portions and is best examined beginning with its blind-ended origin, a double-walled globe called *Bowman's capsule.* The hollow tube continues as the *proximal convoluted tubule* (*PCT*), a region called the *loop of Henle,* and then a final portion called the *distal convoluted tubule* (*DCT*). Many distal convolutions deliver their fluid into collecting ducts and these, in turn, join to form the *papillary ducts* or *ducts of Bellini,* which empty their contents into a renal calyx.

Nephrons are classified into two kinds. *Cortical nephrons,* which make up about seven-eighths of the total number, have their glomerulus in the outer two-thirds of the cortical zone, and the remainder of the tubule rarely penetrates into the medulla. The remaining one-eighth of the total nephrons are *juxtamedullary nephrons.* These have their glomerulus in the inner one-third of the cortical region, which is next to the medulla. These tubules possess a long loop of Henle that penetrates deeply into the medulla (Figure 26-4).

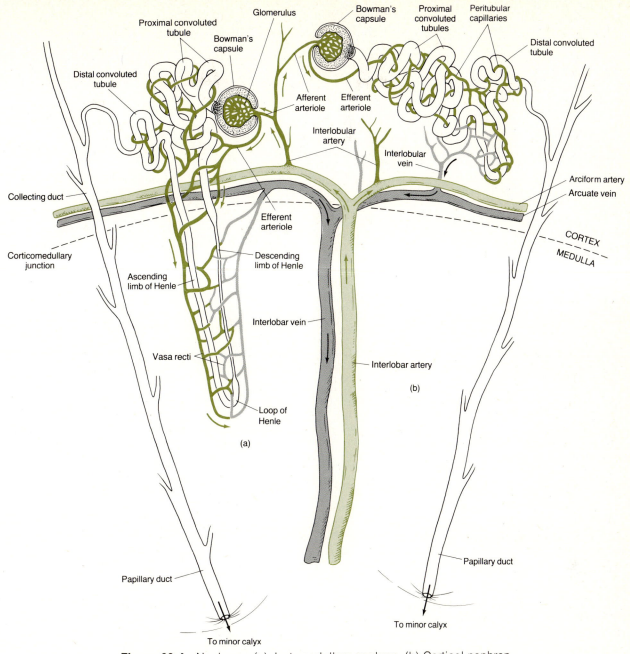

Figure 26-4 Nephrons. (a) Juxtamedullary nephron. (b) Cortical nephron.

Bowman's Capsule

The expanded blind end of the nephron that is Bowman's capsule has two component layers. The outer or *parietal layer* is composed of simple squamous epithelium and is separated from the inner or *visceral layer* by the interior cavity.

A very simple analogy would be to push a finger into the end of a long balloon. The intruding portion corresponds to the visceral layer, and the outer layer is analogous to the parietal layer. The inner visceral layer consists of specialized cells called *podocytes*. These cells wrap around the glomerular capillaries that intrude into the blind end of the tubule. Collectively,

Bowman's capsule and the enclosed glomerulus constitute a *renal corpuscle*. In the human kidney, renal corpuscles have a diameter of about 0.1 mm and, if filled with blood, can be seen by the unaided eye.

The visceral layer of Bowman's capsule and the endothelium of the glomerulus form an *endothelial-capsular membrane*. This membrane consists of the following parts, listed in the order in which substances are filtered from the blood in the glomerular capillaries into the cavity of Bowman's capsule (Figure 26-5).

1. Endothelium of the glomerular capillaries. This single layer of endothelial cells has open pores averaging 500 to 1,000 Å in diameter.

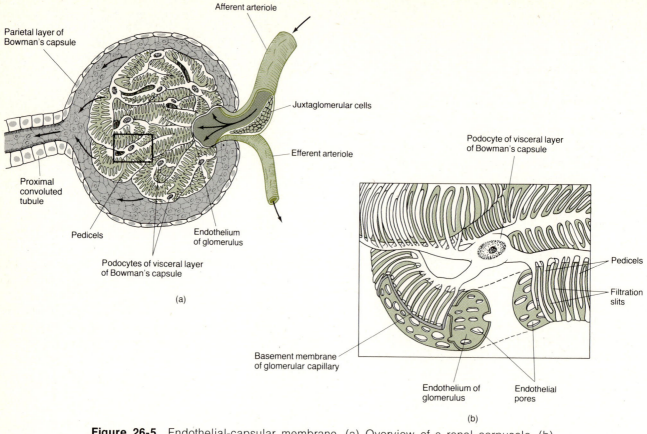

Figure 26-5 Endothelial-capsular membrane. (a) Overview of a renal corpuscle. (b) Enlarged aspect of a portion of the endothelial-capsular membrane.

2. Basement membrane. This layer lies beneath the endothelium and does not contain any pores. It consists of a mat of fibrils enmeshed in a mucopolysaccharide matrix. It serves as the filtering surface.
3. Epithelium of the visceral layer of Bowman's capsule. These epithelial cells, because of their peculiar shape, are called *podocytes*. The podocytes contain footlike structures called *pedicels*. The pedicels are arranged parallel to the circumference of the glomerulus and cover the basement membrane except for spaces between them called *filtration slits* or *slit pores*.

The endothelial-capsular membrane filters water and solutes in the blood. Large molecules, such as proteins, and the formed elements in blood do not normally pass through it. The substances that are filtered pass into the space between the visceral and parietal layers of Bowman's capsule. The filtered fluid moves into the renal tubule.

Proximal Convoluted Tubule (PCT)

Bowman's capsule funnels into the narrower proximal convoluted tubule. Convoluted means the tubule is coiled, and proximal means this portion is nearest to the origin of the tubule (closest to Bowman's capsule). The renal corpuscle seems to be engulfed in a tangled mass of loops of proximal convoluted tubule. The entire length of the proximal tubule is composed of a single layer of cuboidal cells with microvilli on the apical surface that faces the fluid-filled cavity. These cytoplasmic extensions, like those of the small intestine, increase the cellular surface area for absorption and secretion. The basal surface of the cells is lined with mitochondria separated by infoldings of the cellular membrane. The length of the proximal tubules range from 12 to 24 mm, and their width is about 50 to 65 μm.

Loop of Henle

The next section of the tubule begins with the short section of the descending thick limb of the loop of Henle. This thick portion continues as the thin portion of the descending limb, which makes an abrupt hairpin turn and reverses its direction as the thin portion of the ascending limb. This changes to a thick ascending portion before it becomes the distal convoluted tubule. The length of the loop of Henle varies depending upon the location of the renal corpuscle from which the tubule arises. Cortical nephrons have relatively short loops that are often contained entirely within the cortex and have very little or no thin portion. Juxtamedullary nephrons have relatively long loops, some of which reach as far as the tips of the papillae. The thick portions of the loop of Henle consist of simple cuboidal epithelium, whereas the thin portion consists of simple squamous epithelium. The hairpin configuration of the loop of Henle plays a very important role in a countercurrent multiplication process that occurs in the long juxtamedullary loops. The lengths of the thin portions of the loop of Henle range from 0 to 14 mm, and their width is between 14 and 22 μm. The ascending thick portion is about 6 to 18 mm long.

Distal Convoluted Tubule (DCT)

The distal convoluted tubule arises as the continuation of the thick ascending portion of the loop of Henle. It travels a fairly straight course back to its own renal corpuscle and then makes several convolutions. One region of the distal convoluted tubule comes into a very close relationship with the afferent arteriole where it just enters its renal corpuscle. At this point, the cells of the distal tubule change from cuboidal to tall columnar cells that are crowded very closely together. This plaquelike region is referred to as the *macula densa*. At the point of contact between tubule and blood vessel, the afferent arteriole also shows some modifications. The smooth muscle cells of the tunica media become modified. Their nuclei become rounded (instead of elongated) and their cytoplasm contains granules (instead of myofibrils). Such modified cells are called *juxtaglomerular cells*. Together with the macula densa of the distal tubule, this combined relationship constitutes the *juxtaglomerular apparatus* (Figure 26-6).

The cuboidal cells of the distal convoluted tubule, unlike the cells of the proximal tubule, have very few microvilli, if any at all. They do show the same type of striations along the basal border where the cell membrane invaginations separate the densely packed mitochondria. The overall length and width of the distal tubule is shorter, only about 2 to 9 mm, and narrower, 20 to 50 μm, and its convolutions are not as extensive.

Collecting Ducts

The collecting ducts arise in the outer edges of the cortical region when two or more distal tubules join to form a larger tubule, the collecting duct. Additional terminal portions of distal tubules contribute to the growth of the collecting duct as it approaches the medulla. Here, several collecting ducts fuse to form a papillary duct, which then empties into a calyx. Cells of the collecting ducts are cuboidal; those of the papillary ducts are columnar.

Blood Supply to the Tubules

The blood supply to the tubules is via a secondary set of capillaries. Blood first flows through the capillaries of the glomerulus, converges into the efferent arterioles, and then disperses into the second capillary network. Figure 26-4 also shows the differences in the blood supply of the cortical and juxtamedullary nephrons.

The efferent arteriole in cortical nephrons is smaller than the afferent, and immediately upon emerging from the glomerulus, it branches rapidly to form a large anastomosing network of capillaries that covers the proximal and distal tubule convolutions. The capillary bed arising from one glomerulus connects freely with other adjacent capillary networks.

The blood supply to juxtamedullary nephrons is more complicated. The efferent arteriole, which has an equal or greater diameter than the afferent arteriole, sends one or two major branches to distribute over the tubular convolutions in a pattern similar to that of the cortical supply. In addition, the efferent arteriole runs an extended course parallel to the long loop of Henle giving off a series of looping branches. It finally forms a hairpin turn at the tip of the papilla and returns to connect with the looping branches again. Here, it forms a collecting venule that connects with the drainage from the convoluted tubule network and delivers the blood into an interlobar vein. It is important to recognize that the pattern of blood flow around the long loops of Henle sets up a second countercurrent arrangement—the first being the long loop of Henle. These countercurrent mechanisms will be described later.

Nerve Supply to the Kidneys

The rich nerve supply to the kidneys is entirely sympathetic in nature. No parasympathetic fibers have been found. The sympathetic nerves originate from the renal plexus, accompany the renal arteries and their branches, and generally terminate in the smooth muscle of the afferent and efferent arterioles. Some fibers terminate between the cells of the proximal convoluted tubule. The nerves are generally vasoconstrictor in function and are thus able to regulate the blood pressure in the glomerular capillaries by regulating the diameters of the afferent and efferent arterioles.

URINE PRODUCTION

Although each kidney contains about 1 million tubules, it is much simpler to discuss kidney function as though there were only one tubule. Therefore, our

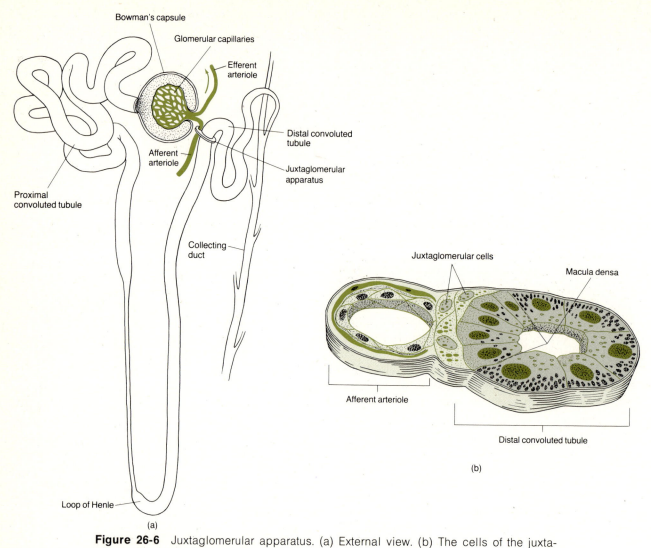

Figure 26-6 Juxtaglomerular apparatus. (a) External view. (b) The cells of the juxtaglomerular apparatus.

Glomerular Filtration

The first step in the production of urine is called **glomerular filtration.** Filtration—the forcing of fluids and dissolved substances through a membrane by an outside pressure—occurs in the renal corpuscle of the kidneys across the endothelial-capsular membrane.

simpler concept of the kidney has only one glomerulus, one Bowman's capsule, one proximal convoluted tubule, one loop of Henle, and one distal tubule. Since each tubule functions in much the same manner as all the others, this approximation is not too inaccurate. The value given for glomerular filtration rate represents, in reality, the volume of fluid filtered through 1 million glomeruli, but is best and most easily represented diagrammatically and mathematically as though there were only one very large one.

When blood enters the glomerulus, the blood pressure forces water and dissolved blood components (plasma) through the walls of the capillaries, basement membrane, and on through the adjoining visceral wall of Bowman's capsule (Figure 26-7a). The resulting fluid is the *filtrate*. In a healthy person, the filtrate consists of all the materials present in the blood except for the formed elements and most proteins, which are too large to pass through the endothelial-capsular barrier. Exhibit 26-1 compares the constituents of plasma, glomerular filtrate, and urine during a 24-hour period. Although the values shown are typical, they vary considerably according to diet. The chemicals listed under plasma are those present in glomerular blood plasma before filtration. The chemicals listed under filtrate immediately after Bowman's capsule are those that pass from the glomerular blood plasma through the endothelial-capsular membrane before

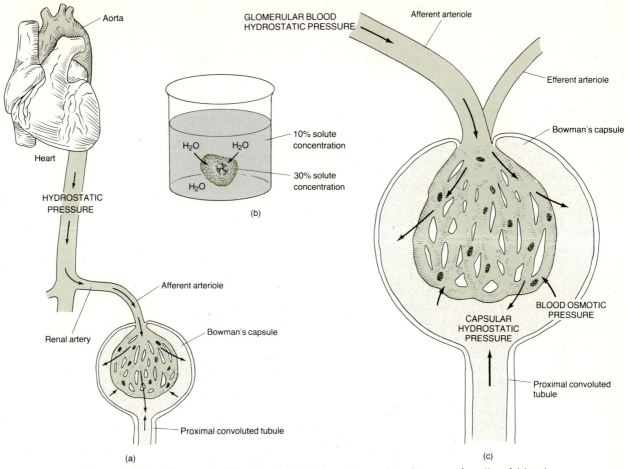

Figure 26-7 Glomerular filtration. (a) Heart action and resistance of walls of blood vessels provide the hydrostatic pressure for filtration. The hydrostatic pressure is counteracted by the walls of the Bowman's capsule and the already present filtrate. (b) The development of osmotic pressure. When water moves into the cell, it swells as its osmotic pressure increases. (c) Application of hydrostatic pressure and osmotic pressure to glomerular filtration.

reabsorption. The chemicals listed as being reabsorbed from the filtrate are the ones that have been filtered.

Renal corpuscles are especially structured for filtering blood. First, each capsule contains a tremendous length of highly coiled glomerular capillaries, presenting a vast surface area for filtration. Second, the endothelial-capsular membrane is structurally adapted for filtration. Although the endothelial pores generally do not restrict the passage of substances, the basement membrane permits the passage of smaller molecules. Thus, water, glucose, vitamins, amino acids, small proteins, nitrogenous wastes, and ions pass into the Bowman's capsule. Large proteins and the formed elements in blood do not normally pass through the basement membrane. The filtration slits permit only the occasional passage of very small plasma proteins such as albumins. Third, the blood flow pattern to cortical nephrons is such that the efferent arteriole is smaller in diameter than the afferent arteriole, so great resistance to the outflow of blood from the glomerulus is established. Consequently, blood pressure is higher in the glomerular capillaries than in other capillaries. Glomerular blood pressure averages 75 mm Hg, whereas the blood pressure of other capillaries averages only 30 mm Hg. Fourth, the endothelial-capsular membrane separating the blood from the space in Bowman's capsule is very thin (0.1 μm).

The filtering of the blood depends on a number of opposing pressures. The chief one is the *glomerular blood hydrostatic pressure. Hydrostatic (hydro = water) pressure* is the force that a fluid under pressure exerts against the walls of its container. Glomerular blood hydrostatic pressure means the blood pressure in the glomerulus. This pressure tends to move fluid out of the glomeruli at a force averaging 75 mm Hg.

Exhibit 26-1 CHEMICALS IN PLASMA, FILTRATE, AND URINE DURING 24-HOUR PERIOD*

Chemical	Plasma	Filtrate Immediately After Bowman's Capsule	Reabsorbed from Filtrate	Urine
Water	180,000 (180 liters)	180,000 (180 liters)	178,000–179,000 (178–179 liters)	1,000–2,000 (1–2 liters)
Proteins	7,000–9,000	10–20	10–20	0
Chloride (Cl⁻)	630	630	625	5
Sodium (Na⁺)	540	540	537	3
Bicarbonate (HCO₃⁻)	300	300	299.7	0.3
Glucose	180	180	180	0
Urea	53	53	28	25
Potassium (K⁺)	28	28	24	4
Uric acid	8.5	8.5	7.7	0.8
Creatinine	1.4	1.4	0	1.4

* All values, except for water, are expressed in grams/24 hours. The chemicals are arranged in sequence from highest to lowest concentration in plasma.

However, glomerular blood hydrostatic pressure is opposed by two other forces. The first of these, *capsular hydrostatic pressure,* develops in the following way. When the filtrate is forced into the space between the walls of Bowman's capsule, it meets with two forms of resistance: the walls of the capsule and the fluid that has partly filled the renal tubule. As a result, some filtrate is pushed back into the capillary. The amount of "push" is the capsular hydrostatic pressure. It usually measures about 20 mm Hg (Figure 26-7c).

The second force opposing filtration into the Bowman's capsule is the *blood osmotic pressure. Osmotic pressure* is the pressure that develops because of water movement into a contained solution. Suppose we place a cell in a hypotonic solution. As water moves from the solution into the cell, the volume inside the cell increases, forcing the cell membrane outward (Figure 26-7b). Osmotic pressure always develops in the solution with the higher concentration of solutes since water will distribute itself to the same concentration in both. Hydrostatic pressure develops because of a force outside a solution. Osmotic pressure develops because of the concentration of the solution itself. Since the blood contains a much higher concentration of proteins than the filtrate does, water moves out of the filtrate and into the blood vessel. This blood osmotic pressure is normally about 30 mm Hg.

To determine how much filtration finally occurs, we have to subtract the forces that oppose filtration from the glomerular blood hydrostatic pressure. The net result is called the *effective filtration pressure,* which is abbreviated P_{eff}:

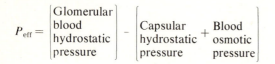

$$P_{eff} = \begin{vmatrix} \text{Glomerular} \\ \text{blood} \\ \text{hydrostatic} \\ \text{pressure} \end{vmatrix} - \begin{vmatrix} \text{Capsular} \\ \text{hydrostatic} \\ \text{pressure} \end{vmatrix} + \begin{vmatrix} \text{Blood} \\ \text{osmotic} \\ \text{pressure} \end{vmatrix}$$

By substituting the values just discussed, we can calculate a normal P_{eff} as follows:

$$P_{eff} = (75 \text{ mm Hg}) - (20 \text{ mm Hg} + 30 \text{ mm Hg})$$
$$= (75 \text{ mm Hg}) - (50 \text{ mm Hg})$$
$$= 25 \text{ mm Hg}$$

This means that a pressure of 25 mm Hg causes a normal amount of plasma to filter from the glomerulus into the Bowman's capsule. This is about 125 ml of filtrate per minute.

Certain conditions may alter any of these three pressures and thus the P_{eff}. In some forms of kidney disease, the glomerular capillaries become so permeable that the plasma proteins are able to pass from the blood into the filtrate. As a result, the capsular filtrate exerts an osmotic pressure that draws water out of the blood. Thus, if a capsular osmotic pressure develops, the P_{eff} will increase. At the same time, blood osmotic pressure decreases, further increasing the P_{eff}.

The P_{eff} is also affected by changes in the general arterial blood pressure. Severe hemorrhaging produces a drop in general blood pressure that also decreases the glomerular blood hydrostatic pressure. If the blood pressure falls to the point where the hydrostatic pressure in the glomeruli reaches 50 mm Hg, no filtration occurs because the glomerular blood hydrostatic pressure in the glomeruli reaches 50 mm Hg, no filtration called *renal suppression.*

A final factor that may affect the P_{eff} is the regulation of the size of the afferent and efferent arterioles. In this case, glomerular blood hydrostatic pressure is regulated separately from the general blood pressure. Sympathetic impulses and small doses of epinephrine cause constriction of both afferent and efferent arterioles. However, intense sympathetic impulses and large doses of epinephrine cause greater constriction of afferent than efferent arterioles. This intense stimulation results in a decrease in glomerular hydrostatic pressure even though blood pressure in other parts of the body may be normal or even higher than normal. Intense sympathetic stimulation is most likely to occur during the alarm reaction of the general adaptation syndrome. Blood may also be shunted away from the kidneys during hemorrhage.

Measurement of Glomerular Filtration Rate (GFR)

The glomerular filtration rate (GFR) is a *measure of the volume* of fluid that enters Bowman's capsule every minute. Knowledge of the glomerular filtration rate is central to a complete and correct evaluation of kidney function. For example, the difference between the rate at which a substance *(x)* is filtered through the glomerulus, F_x, and the rate at which it is excreted by the kidney, E_x, is a measurement of the rate at which the substance is removed (reabsorbed) or added (secreted) to the urine:

$$E_x = F_x - Reab \quad \text{or} \quad E_x = F_x + Sec_x$$

The filtration rate, F, of a substance is a measurement of the amount of the substance filtered from the blood plasma into Bowman's capsule in one minute. This is determined by multiplying the GFR by the concentration of the substance in the glomerular filtrate. Since the glomerular filtrate has essentially the identical concentration of dissolved substances as plasma, it is common practice to obtain the concentration of a substance in the glomerular filtrate by measuring its concentration in a sample of blood plasma. Thus, the filtered rate of a substance is expressed by the following equation.*

$$F_x = GFR_x \cdot [P]_x$$

Similarly, the amount of substance excreted per minute can also be easily determined. It is the product of the rate of urine production, V, and the concentration of the substance in the urine $[U]$.

$$\text{Excretion} = [U] \cdot V$$

In order to determine the rate of glomerular filtration, it is necessary to use some nontoxic substance that can be injected into the circulation, that is freely filterable through the glomerular capillaries, and that is not reabsorbed from or secreted into the kidney tubule. Under these conditions, the rate of excretion must be equal to the rate of filtration.

Consider the following example. Inulin, which is a carbohydrate polymer of fructose obtained from dahlia

---

* Square brackets placed around a symbol indicate concentration. $[P]_x$ means the plasma concentration of x.

tubers, is a nontoxic substance that is filterable through the glomerulus and is neither reabsorbed nor secreted by the tubular cells. Sufficient inulin is injected intravenously to raise the plasma concentration $[P]_{inulin}$ to 2 mg/ml. The urinary bladder is emptied immediately, and 15 minutes later, the urine is collected. Its volume is measured to be 30 ml, and the concentration of inulin in urine is determined to be 120 mg/ml.

$$\text{Rate of excretion} = \frac{30 \text{ ml}}{15 \text{ minutes}} \times \frac{120 \text{ mg}}{\text{ml}} = 240 \text{ mg/minute}$$

Since the filtration rate of inulin must equal the rate of its excretion, then:

$$F_{inulin} = 240 \text{ mg/minute}$$
$$GFR \times [P]_{inulin} = 240 \text{ mg/minute}$$
$$GFR = \frac{240 \text{ mg}}{1 \text{ minute}} \times \frac{\text{ml}}{2 \text{ mg}} = 120 \text{ ml/minute}$$

Thus, the rate of glomerular fluid filtration is determined by the formula:

$$GFR = \frac{[U]_{inulin} \times V}{[P]_{inulin}} = 120 \text{ ml/minute}$$

In the normal adult, GFR is about 125 ml/minute—about 180 liters a day.

Concept of Clearance

The concept of **clearance** originated in 1929. It was an early attempt to express, in quantitative terms, how well the kidney could rid the blood of urea. By measuring the amount of urea excreted in the urine and knowing its concentration in plasma, one could calculate how many milliliters of plasma were completely "cleared" of urea.

However, we know that as plasma flows through the kidneys much but not all of the urea is removed from the plasma. No single milliliter is ever completely cleared of urea. Thus, the term clearance describes an imaginary or *virtual* number of milliliters that must have to be completely rid of urea to account for the amount excreted. In fact, some urea is removed from each milliliter of plasma that perfuses the kidneys.

Calculations of clearance values, therefore, give the same answer, if 100 ml of plasma are rid of all their urea, or if 200 ml are rid of exactly one-half of their urea, or if 400 ml are rid of exactly one-quarter of their urea.

Measurement of GFR was a measurement of the clearance of inulin. Because inulin is neither reabsorbed nor secreted, when the determination of "how many milliliters of filtrate must have been completely rid of inulin every minute to account for the amount excreted every minute" is done, the answer is the volume of fluid actually filtered into Bowman's capsule every minute—the GFR.

The clearance value (milliliters/minute) of any substance can be determined. Its significance depends upon the understanding of how the kidney handles the substance. In humans, no substance other than inulin has been found that will serve for the accurate determination of GFR.

Tubular Reabsorption

A second process that is involved in urine production is **reabsorption.** As the filtrate flows through the renal tubules, water is reabsorbed into the blood at a rate of about 123 to 124 ml/minute, or about 178 to 179 liters a day. Thus only about 1 percent of the water actually leaves the body—about 1 to 2 ml/minute, or about 1 to 2 liters a day. The movement of water and/or solutes from the filtrate back into the blood of the peritubular capillaries or vasa recta is called *tubular reabsorption.* Tubular reabsorption is carried out by the epithelial cells of the renal tubule. It is a very discriminating process. Only specific amounts of certain substances are reabsorbed, depending on the body's needs at the time. The maximum amount of a substance that can be reabsorbed under any condition is called the *transport maximum (T_m)* for that substance. Substances that are reabsorbed include water, glucose, and amino acids. Also reabsorbed are ions such as Na^+, K^+, Ca^{2+}, Cl^-, HCO_3^-, and HPO_3^-. Tubular reabsorption permits the body to recover necessary amounts of vital nutrients and by limited reabsorption of nonessential or waste materials rid the body of unwanted materials. Exhibit 26-1 compares the values for the chemicals in the filtrate immediately after Bowman's capsule with those reabsorbed from the filtrate.

Reabsorption of Glucose

The removal of glucose from the tubule and its placement back into the plasma is achieved by the process of *active transport.* The rate of glucose filtration, or any other substance, can be calculated simply as the product of GFR and its plasma concentration $[P]_{glucose}$.

Filtration rate = 120 ml/minute × 80 mg/100 ml

Glucose filtration rate = $\dfrac{120 \text{ ml}}{1 \text{ minute}} \times \dfrac{80 \text{ mg}}{100 \text{ ml}} = 96$ mg/minute

For convenience, consider the filtration rate about 100 mg/minute. Of this glucose, none is excreted. An examination of the urine for glucose may turn up a trace, but the value for excretion is considered normally to be zero. This means that all of the glucose carried into the kidney tubule by filtration is selectively removed by the tubular active transport systems.

A quantification of the rate of glucose reabsorption can be carried out as follows. The infusion of glucose intravenously will raise the plasma glucose concentration to higher levels. As a result, the filtration rate for glucose also increases—since it is proportional to $[P]_{glucose}$. The graph in Figure 26-8 shows that as the plasma glucose level increases, the filtered load increases in a linear fashion. This increased filtered load presents the proximal tubular cells with greater amounts of glucose, and they respond by increasing tubular reabsorption. However, as the filtered load continues to increase, the ability of the proximal cells to reabsorb the glucose is exceeded. At this point, the T_m for glucose reabsorption is passed. Some of the glucose remains in the filtrate as it continues past the proximal tubular cells, and this glucose is excreted in the urine. The graph shows that, as the plasma glucose concentration approaches 300 mg percent, glucose begins to appear in the urine. The proximal tubular cells are reabsorbing at a maximal and constant rate. As the plasma glucose increases—and thus also the filtered load—the excess glucose is lost and the rate of excretion also increases linearly. When glucose appears in the urine, the condition is called *glycosuria.*

Reabsorption of Ions and Water

The removal of ions and water from the proximal tubule is an *isosmotic* process. This means that the fluid that is removed from the tubule and the fluid that remains in the tubule always have the same osmotic pressure, which is also equal in value to that of the plasma.

The ions that are responsible for the major portion of the osmotic pressure (90 to 95 percent) include

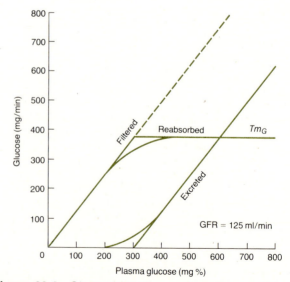

Figure 26-8 Glucose filtration, reabsorption, and excretion.

sodium, chloride, and bicarbonate. Since the osmotic pressures never change, it is obvious that the rates of salt and water reabsorption must be equal. The cells of the proximal tubule neither concentrate nor dilute the tubular fluid.

The key process to the reabsorption occurring in the proximal tubule is the *active* transport of sodium and potassium ions from the tubular fluid through the cell and into the peritubular capillaries. This movement begins to set up an electrochemical gradient as the positive ions leave the tubular fluid and are deposited in the capillaries. As a consequence, the negatively charged chloride and bicarbonate ions follow the positively charged sodium and potassium ions into the plasma by *passive* diffusion.

As the ions move into the plasma—by either an active or a passive process—an osmotic gradient is generated across the tubule wall with the consequence that water flows passively into the capillaries. Thus, the active transport of sodium ions—and to a much lesser degree, potassium (Exhibit 26-2)—produces circumstances that cause the passive reabsorption of chloride, bicarbonate, and water in an isosmotic fashion. Between 65 and 90 percent of the water presented to the proximal tubular cells is reabsorbed in this portion of the tubule. This is often referred to as *obligatory water reabsorption*. The remainder of the fluid proceeds through the rest of the tubule and collecting ducts. In these parts, additional varying portions of the remaining water are reabsorbed. The passive reabsorption of water from the distal tubule and collecting ducts is controlled by the changes in permeability of these tubes. Antidiuretic hormone (ADH), when released from the posterior pituitary, causes an increase in permeability of the tubules and ducts. This permits a passive removal of water into the long capillary loops that parallel the collecting ducts. This variable reabsorption is termed *facultative (optional) water reabsorption* and is responsible for the recovery

of an additional 12 to 20 percent of the water in the filtrate.

Tubular Secretion

A third process involved in urine production is **tubular secretion**. Whereas tubular reabsorption removes substances from the filtrate into the blood, tubular secretion adds materials to the filtrate from the blood. These secreted substances include K^+, H^+, ammonia, creatine, and drugs such as penicillin and para-aminohippuric acid (PAH). Materials moved from the blood are added to the filtrate by active and passive processes.

Secretion of Para-aminohippurate (PAH)

The quantification of the rate of secretion of any substance is illustrated by the secretion of PAH. Figure 26-9 shows the curves for excretion, secretion, and filtration.

Para-aminohippurate is not only freely filtered through the glomerulus, but it is also secreted by the cells of the kidney tubule. Therefore, the rate of excretion of PAH in the urine is the sum of the rate of filtration and rate of secretion:

$$\text{Excretion rate}_{PAH} = \text{Filtration rate} + \text{Secretion rate}_{PAH}$$
$$[U]_{PAH} \times V = (\text{GFR} \times [P]_{PAH}) + \text{Secretion rate}_{PAH}$$

Rewriting this equation isolates secretion rate on the left-hand side of the equation, and shows it is equal to excretion rate minus the filtration rate.

$$\text{Secretion rate}_{PAH} = ([U]_{PAH} \times V) - (\text{GFR} \times [P]_{PAH})$$

In order to obtain the data needed for calculation of secretion rate of PAH, the following general procedures need to be carried out. First, to obtain GFR, a clearance of inulin must be done. Therefore, inulin is infused into the subject at a rate sufficient to keep the plasma level around 15 mg percent. Concurrently, PAH is also gradually infused so that plasma level

Exhibit 26-2 FILTRATION, REABSORPTION, AND EXCRETION OF IONS AND WATER BY HUMANS

	Plasma Concentration (meq/liter)	Rate of Glomerular Filtration (liters/24 hr)	Gibbs Donnan Factor	Quantity Filtered (meq/24 hr)	Quantity Excreted (meq/24 hr)	Quantity Reabsorbed (meq/24 hr)	Percent Reabsorbed
Sodium	140	180	0.95	23,940	103	23,837	99.6
Chloride	105	180	1.05	19,845	103	19,742	99.5
Bicarbonate	27	180	1.05	5,103	2	5,101	99.9+
Potassium	4	180	0.95	684	51	633	92.6
	(liters/liter)			(liters/24 hr)	(liters/24 hr)	(liters/24 hr)	
Water	0.94	180	—	169.2	1.5	167.7	99.1

Source: Reproduced with permission from Pitts, R.F.: *Physiology of the Kidney and Body Fluids*, 3rd edition. Copyright © 1974 by Year Book Medical Publishers, Inc., Chicago.

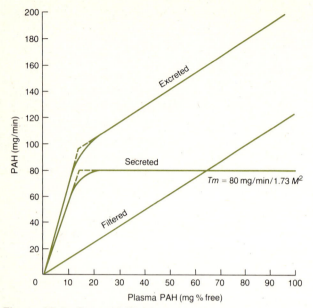

Figure 26-9 Rates of filtration, secretion, and excretion of PAH as functions of plasma concentration in humans.

slowly approaches 100 mg percent. Then urine and plasma samples are taken at frequent levels and analyzed, and the necessary calculations for inulin clearance (thus GFR) and rate of PAH filtration and excretion are carried out.

As the plasma concentration of PAH increases up to about 10 mg percent (Figure 26-9) the rate of filtration (GFR × $[P]_{PAH}$) and rate of excretion ($[U]_{PAH}$ × V) increase proportionately. Therefore, since the rate of secretion is the difference between the rates of excretion and filtration, it also increases in proportion to the plasma concentration.

However, when the level of plasma PAH exceeds 10 mg percent, the rate of secretion begins to plateau and remain constant. This occurs because the tubular cells have reached their maximum secretory level. Any further increase in the plasma PAH level does not increase their secretory contribution of PAH into the tubule. The rate of excretion also changes to a new, slower rate that is now the same as the rate of filtration. This is so because as plasma PAH continues to increase, the filtered load also increases and adds more PAH into the tubule. An examination of the graph shows that the maximum rate of PAH secretion *(Tm)* by the human kidney is about 80 mg/minute.

Measurement of Renal Blood Flow

All the plasma that enters the kidney reaches the glomerular capillaries. There, some of it (about 125 ml/minute) is filtered into Bowman's capsule. The rest of the plasma flows out of the glomerular capillaries into the efferent arteriole and then into the peritubular

capillaries. PAH is introduced into the tubule by filtration through the glomerulus and also by secretion from the peritubular capillaries. If the peritubular capillaries carry less than 10 mg percent PAH (say 5 mg percent), then the tubular cells are able to secrete all the PAH carried by the capillaries into the tubule. Therefore, all of the PAH that was present in the plasma entering the kidney would end up in the tubule, and the plasma leaving the kidney would have none.

If the clearance of PAH is measured with plasma levels set below 10 mg percent, then the value obtained describes the volume of plasma from which all the excreted PAH was obtained. That volume was the total plasma volume delivered to the kidney. Therefore, the clearance value for PAH is, in fact, a measure of *renal plasma flow* (RPF).

The average clearance value for PAH in normal humans is about 660 ml/minute. *Renal blood flow* can easily be calculated if the hematocrit value is known. For example, given a hematocrit value of 45 percent and the clearance of PAH value of 660 ml/minute, the renal blood flow is equal to 660/(1 − 0.45), or 1,200 ml/minute.

The normal cardiac output of a person at rest is about 5.0 liters/minute. Under such conditions, the kidney receives almost 25 percent of the cardiac output. The ratio of renal blood flow to cardiac output is called the *renal fraction*. The ratio of glomerular filtration rate to renal plasma flow is called the *filtration fraction*. Interestingly, the filtration fraction is also the ratio of the clearance of inulin (which measures GFR) to the clearance of PAH (which measures RPF).

URINE CONCENTRATION AND DILUTION

The various metabolic wastes in urine must be eliminated from the body regardless of its state of hydration. During dehydration, for example, little urine is excreted and it is concentrated. Such urine is said to be *hypertonic* or *hyperosmotic* to the blood plasma. When there are large amounts of water in the body, more urine is excreted but it is diluted. Such urine is said to be *hypotonic* or *hyposmotic* to blood plasma. The ability of the kidneys to produce either hyperosmotic or hyposmotic urine is a result of the operation of two **countercurrent mechanisms:** a countercurrent *multiplication* system and a countercurrent *exchange* system.

Principle of Countercurrent Multiplication

The principle of countercurrent multiplication is shown in Figure 26-10. A sodium chloride solution flows into the system, down the left side to the bottom

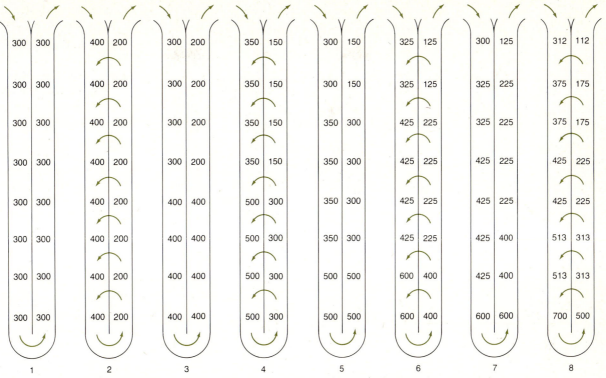

Figure 26-10   The principle of countercurrent multiplication of concentration.

of the tube, and then it flows up the right side and out of the system. A dividing wall separates the fluid compartments in which the solution flows in opposite directions (countercurrent). In this example, it is necessary to assume that the dividing wall is able to transport sodium ions from the ascending fluid through the wall into the descending fluid. By employing a stop-and-look technique, we shall examine the concentrating (multiplication) effects.

In step 1, the salt solution flows in to fill both sides of the system with 300 mOsm/liter solution, and the flow is stopped. In step 2, the separating wall begins to actively transport, or pump, sodium ions from the right side into the left side. This transport process occurs down the entire length of the tube, and the pumping mechanism is capable of generating a gradient across the wall of 200 mOsm/liter. The figures in the step 2 diagram show that the salt concentration in the left side is increased to 400 mOsm/liter, whereas that on the right side is decreased to 200 mOsm/liter. The gradient difference of 200 mOsm/liter is achieved because 100 mOsm/liter was removed from the right, or ascending, side and delivered into the left, or descending side. In step 3, flow is permitted to resume for a brief time. Thus, more 300 mOsm/liter solution enters the descending side. This pushes the 400 mOsm/liter solution further down the tubule, around the hairpin turn, and up the other side. Some of the 200 mOsm/liter fluid

flows out of the top away from the system. Again the flow is stopped and the dividing wall is permitted to transport sodium ions from the ascending limb into the descending limb until the gradient difference directly across the wall is 200 mOsm/liter. Step 4 shows the end result of such transport activity. Steps 5 through 8 repeat the start-stop-and-look technique. In this fashion, it is possible to produce gradients *across the wall* of *200 mOsm/liter* while the continual delivery of sodium into the descending limb produces a lengthwise gradient of sodium down the tube, which reaches 700 mOsm/liter in this example.

It should be noted that the fluid that leaves the system from the top of the ascending limb is always hypotonic to that flowing in by 200 mOsm/liter. The sodium chloride extracted from the ascending solution is added to the fluid entering the descending limb. The solution in the descending limb becomes progressively more concentrated—reaching its peak value at the hairpin turn.

If the stop-and-look periods are progressively shortened, then the flow eventually becomes continuous, as it really is in the long loops of Henle.

One additional condition needs to be added to the model. The walls of the ascending limb must be impermeable to water; otherwise the movement of the salt would set up an osmotic gradient that would cause water to flow with it into the descending limb. This condition is crucial to the multiplier process.

Operation of Countercurrent Multiplication in the Kidney

The long loops of Henle extend into the medulla several millimeters and are arranged in a countercurrent pattern (Figure 26-11). Fluid entering the descending loop of Henle is isotonic at 300 mOsm/liter. The outline of the ascending limb is heavier to indicate the impermeable nature of its walls. Sodium ions are pumped from the ascending limb into the surrounding interstitium. So, as fluid flows up the ascending limb, it becomes hypotonic and the surrounding space becomes increasingly concentrated. The descending limb is permeable to both water and sodium, so as the fluid flows down the descending limb, water is drawn into the concentrated salt region of the interstitium, while salt can also diffuse into the descending fluid. Therefore, the concentration of salt in the descending limb increases as the fluid flows toward the hairpin turn. When the fluid makes its turn and begins ascending, it starts to pump sodium into the interstitium. The concentration of the fluid in the tubule now decreases

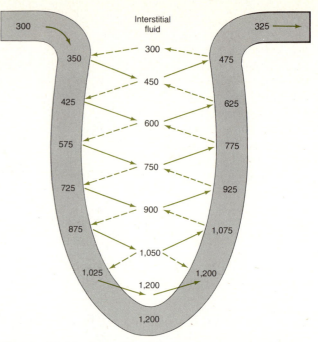

Figure 26-12 Operation of countercurrent exchange mechanism across capillaries to reduce the rate of dissipation of the osmolar gradient between the cortex and the medulla.

progressively as it flows upward so that, when the fluid leaves the thin-walled part of the loop of Henle and enters the thicker cuboidal portion of the distal convoluted tubule, the concentration of the fluid is only 100 mOsm/liter.

Countercurrent Exchange in the Kidney

The long capillary loops of the peritubular capillaries extend deep into the medulla of the kidney parallel to the loops of Henle and collecting ducts. Figure 26-12 illustrates a capillary loop and its changing salt concentrations. As the blood enters the loop, it is isotonic with a salt concentration of 300 mOsm/liter. As the blood flows down the capillary, its permeable walls permit an equilibration of concentration to begin. However, the flow rate is sufficiently swift that a complete equilibration at any point is not possible. Thus, the salt concentration inside the capillary increases, but it never reaches the concentration of the interstitium immediately outside. At the tip of the loop, the interior of the capillary comes closest to achieving a complete equilibration. As the blood flows up the loop, the salt concentration of the plasma inside is now higher than that of the interstitium. However, again the blood flow rate through the capillary is too swift to permit equilibration. When the blood reaches the upper portion of the loop, the nature of the blood vessel changes to a more venous one and the perme-

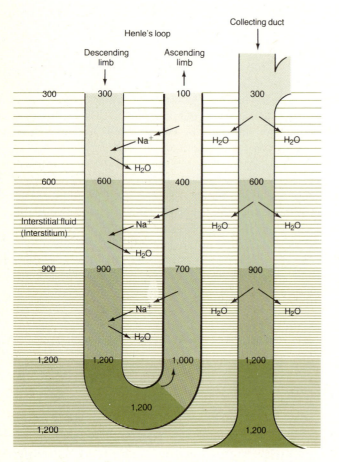

Figure 26-11 The operation of countercurrent multiplication of concentration in the formation of hypertonic urine.

ability of the vessel wall drops. Thus the blood in the veins is slightly hypertonic. In this fashion, the blood flow carries away some of the salt of the interstitium and prevents an unlimited accumulation of sodium. There is, therefore, a very precise balance set up between the two countercurrent systems. The multiplier system sets up a longitudinal salt gradient in the interstitium, and the capillary exchange system prevents a continual buildup of salt. In other words, a steady-state situation exists, and the salt concentration of the interstitium is maintained constant by the continual removal of salt and water by the blood vessels at a rate equal to that at which salt is deposited in the interstitium by the ascending limb of the loop of Henle.

Regulation of Urine Volume and Concentration

The end result of the countercurrent multiplier is that a salt gradient exists in the interstitium of the kidney medulla. This gradient reaches a high value of 1,200 mOsm/liter at the tips of the renal papillae and decreases to values of 300 mOsm/liter at the cortical boundary. This salt gradient is maintained by the continual operation of the countercurrent exchange system. Its operation is precisely regulated so that it removes just sufficient salt that the interstitial salt concentration neither increases or decreases.

The salt gradient, thus set up and maintained, is used in conjunction with the hormone ADH (antidiuretic hormone) to reabsorb varying volumes of water from the collecting ducts. The fluid entering the distal portion of the nephron is hypotonic to the surrounding interstitium. Its flow rate of about 16 ml/minute is about 15 percent of the volume filtered at the glomerulus each minute.

After passage through the distal tubule, the fluid enters the collecting ducts. Figure 26-11 shows a portion of a collecting duct. The permeability of the collecting duct is variable depending upon the titer of ADH in the circulating plasma. In *hydropenia* (deficiency of water in the body) the level of ADH is elevated, and this increases the permeability of the walls of the collecting ducts. Therefore, as the fluid descends the collecting duct, water is given up to the salt-laden interstitium and the concentration within the collecting duct can quickly equilibrate with its surroundings. The peak level of urine concentration, therefore, is equal to that of the highest salt concentration in the tips of the renal papillae—about 1200 mOsm/liter. In diuresis (large volume of urine output), the circulating level of ADH is depressed and the collecting ducts are rendered increasingly impermeable. Thus, less water is drawn through the walls of the collecting ducts; the volume of fluid excreted is greater, and its concentration of salts is lower.

Regulation of Sodium and Potassium Excretion

The amount of sodium excreted is determined by the amount filtered through the glomerulus and the amount reabsorbed. The mechanisms that coordinate these two processes to keep a normal level in body fluids are not known.

It is known that if the glomerular filtration rate of sodium increases, the rate of reabsorption of sodium in the proximal tubule increases proportionately. However, the rest of the tubule does not alter its reabsorption rate, so the excretion rate will still increase somewhat.

Adrenocortical steroids, such as aldosterone, deoxycorticosterone, and cortisol increase the rate of tubular reabsorption of sodium. This can be accompanied by the passive reabsorption of chloride ions or it can result from an increase in the rate of exchange of sodium ions for potassium or hydrogen ions.

Potassium ions are the only ions to be both reabsorbed and secreted by the kidney tubules. Reabsorption is confined almost entirely to the proximal convoluted tubule and is so efficient that none or very little potassium is present in the fluid entering the distal convoluted tubule. Secretion of potassium ions occurs in the distal convoluted tubules and collecting ducts. It is here that the exchange transport of sodium and potassium takes place. Sodium is reabsorbed into the capillaries, while potassium is secreted into the tubular fluid.

CONTRIBUTION OF KIDNEYS TO ACID-BASE BALANCE

A normal dietary regimen will produce many acidic by-products. The phosphates released from certain proteins and phospholipids are easily converted to phosphoric acid. Sulfates and sulfur from proteins and polysaccharides are converted to sulfuric acid. Ammonium chloride is converted to urea and a hydrogen ion. The hydrogen ion combines with chloride to form hydrochloric acid. Many of the metabolic intermediates are organic acids, and they contribute to the tendency to lower body pH.

These acidic substances are neutralized by buffer systems that exist in body fluids. The most prevalent buffer is the bicarbonate-carbonic acid one. It is the bicarbonate ion that provides most of the alkali to neutralize acids.

Any acid (for example, HCl) that is released into the

body fluids reacts mostly with bicarbonate to produce carbonic acid and the neutral salt of the acid. The carbonic acid is then lost when it is converted to H_2O and CO_2 and the CO_2 is exhaled via the lungs.

$$\underset{\substack{\text{Hydrochloric}\\\text{acid}}}{HCl} + \underset{\substack{\text{Sodium}\\\text{bicarbonate}}}{NaHCO_3} \rightleftharpoons \underset{\substack{\text{Sodium}\\\text{chloride}\\\text{excreted in}\\\text{urine}}}{NaCl} + \underset{\substack{\text{Carbonic acid}}}{H_2CO_3}$$

$$\underset{\substack{\text{Carbonic acid}}}{H_2CO_3} \longrightarrow \underset{\substack{\text{Water}}}{H_2O} + \underset{\substack{\text{Carbon dioxide}\\\text{exhaled via lungs}}}{CO_2 \uparrow}$$

The consequence of such buffering is the removal of the bicarbonate ion each time an acid molecule is produced. The buffering capacity of the body fluid is progressively depleted as the bicarbonates are used up.

The kidneys recover bicarbonate by reconverting the salt of the acid (NaCl in our example above) to HCl and $NaHCO_3$ by a series of reactions within the tubular cells.

Secretion of Hydrogen Ions

The recovery of bicarbonate ions is linked to a process of hydrogen ion secretion by the tubular cells and cells of the collecting ducts.

Carbon dioxide is made within the kidney cells as a metabolic waste or can be obtained from the circulating plasma (Figure 26-13a). Within the cell, it reacts with water in the presence of a special enzyme named **carbonic anhydrase**. This enzyme favors the formation of carbonic acid, which dissociates slightly to form hydrogen and bicarbonate ions. The presence of an active hydrogen pump secretes the H^+ ions into the lumen of the tubule. This produces an electrochemical gradient so that the inside of the cell is negative and the outside is positive. As a result, sodium ions enter the cell passively in response to the electrical potential. Inside the cell, the sodium combines with the bicarbonate ion to form sodium bicarbonate, which then diffuses out of the cell into the plasma. In this way, not only is H^+ eliminated from the body, but sodium bicarbonate is produced to restore and maintain the buffering capacity of the extracellular fluid.

A second mechanism for raising blood pH is secretion of the ammonium ion (Figure 26-13b). Ammonia, in elevated concentrations, is a poisonous waste product derived from the deamination of amino acids. The liver converts much of the ammonia to a less toxic compound called urea. Urea and ammonia both become part of the glomerular filtrate and are subsequently excreted from the body. Any ammonia produced by the deamination of amino acids in the tubule cells is secreted into the urine. When ammonia (NH_3) forms in the distal and collecting tubule cells, it combines with H^+ to form the ammonium ion (NH_4^+). (The H^+ may come from the dissociation of H_2CO_3.) The cells secrete NH_4^+ into the filtrate, where it takes the place of a positive ion, usually Na^+, in a salt and is eliminated. The displaced Na^+ diffuses into the renal cells and combines with HCO_3^- to form sodium bicarbonate.

As a result of H^+ and NH_4^+ secretion, urine normally has an acidic pH of 6. The relationship of renal

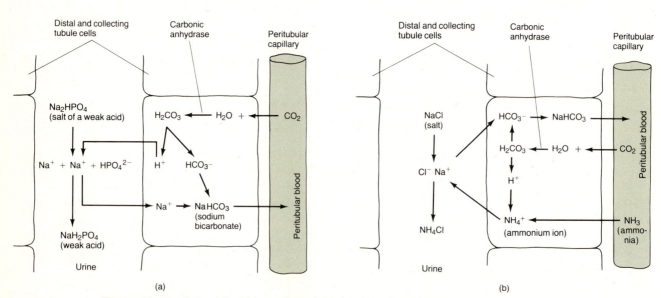

Figure 26-13 Role of the kidneys in maintaining blood pH. (a) Acidification of urine and conservation of sodium bicarbonate by the elimination of H^+ ions. (b) Acidification of urine and conservation of sodium bicarbonate by the elimination of NH_4^+.

RETURN TO HOMEOSTASIS

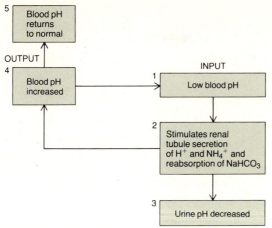

Figure 26-14 Summary of kidney mechanisms that maintain the homeostasis of blood pH.

tubule excretion of H^+ and NH_3 to blood pH level is summarized in Figure 26-14. Exhibit 26-3 summarizes filtration, reabsorption, and secretion in the nephrons.

HOMEOSTASIS

Excretion is one of the primary ways in which the volume, pH, and chemistry of the body fluids are kept in homeostasis. The kidneys assume a good deal of the burden for excretion, but they share this responsibility with several other organ systems.

Other Excretory Organs

The lungs, integument, and alimentary canal all perform special excretory functions (Exhibit 26-4). Primary responsibility for regulating temperature through the excretion of water is assumed by the skin. The lungs maintain blood-gas homeostasis through the maintenance of CO_2. One way in which the kidneys maintain homeostasis is by coordinating their activi-

ties with other excretory organs. When the integument increases its excretion of water, the renal tubules increase their reabsorption of water and blood volume is maintained. When the lungs fail to eliminate enough carbon dioxide, the kidneys attempt to compensate. They change some of the CO_2 into sodium bicarbonate, which becomes part of the blood buffer systems.

Urine

The kidneys perform their homeostatic functions of controlling the concentration and volume of blood and adjusting pH. The by-product of these activities is **urine.** Urine is a fluid that contains a high concentration of solutes. In a healthy person, its volume, pH, and solute concentration vary with the needs of the internal environment. During certain pathological conditions, the characteristics of urine may change drastically. An analysis of the volume and physical and chemical properties of urine tells us much about the state of the body.

Volume

The volume of urine eliminated per day in the normal adult varies between 1,000 and 1,800 ml (2 to 3.5 pt). Urine volume is influenced by a number of factors: blood pressure, blood concentration, diet, temperature, diuretics, mental state, and general health.

Blood Pressure. Normal rapid adjustments to blood pressure are carried out by the pressure receptors of the carotid sinus and aortic arch, but long-term regulation is dependent upon proper maintenance of body electrolyte concentrations and fluid volumes. These are the responsibility of the kidney. The cells of the juxtaglomerular apparatus are particularly sensitive to changes in blood pressure. When renal blood pressure falls below normal, the juxtaglomerular apparatus secretes a hormone called *renin* (Figure 26-15). Renin converts the alpha globulin plasma protein *angiotensinogen* into *angiotensin I*, a decapeptide, which is

Exhibit 26-3 FILTRATION, REABSORPTION, AND SECRETION

Region of Nephron	Activity
Renal corpuscle (endothelial-capsular membrane)	Filtration of glomerular blood under hydrostatic pressure results in formation of filtrate free of plasma proteins and cellular elements of blood.
Proximal convoluted tubule and descending and ascending limbs of Henle	Reabsorption of physiologically important solutes such as Na^+, K^+, Cl^- HCO_3^-, and glucose. Obligatory reabsorption of water by osmosis.
Distal convoluted tubule	Reabsorption of Na^+. Facultative reabsorption of water under control of ADH. Secretion of H^+, NH_3, K^+, creatine, and certain drugs. Conservation of sodium bicarbonate.
Collecting duct	Facultative reabsorption of water under control of ADH.

Exhibit 26-4 EXCRETORY ORGANS AND PRODUCTS ELIMINATED

Excretory Organs	Products Eliminated	
	Primary	Secondary
Kidneys	Water, soluble salts from protein catabolism, inorganic salts	Heat and carbon dioxide
Lungs	Carbon dioxide	Heat and water
Skin	Heat	Carbon dioxide, water, salts
Alimentary tract	Solid wastes and secretions	Carbon dioxide, water, salts, heat

transformed in the lungs to *angiotensin II*, an octapeptide. Angiotensin II raises blood pressure in two ways. It causes constriction of arterioles throughout the body, and it stimulates the adrenal cortex to secrete aldosterone. Aldosterone stimulates the epithelial cells of the distal convoluted tubules, the collecting ducts, and to a lesser extent the proximal tubules to transport Na⁺ back into the blood. As a result, obliga-

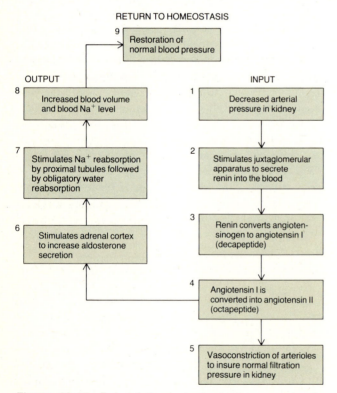

Figure 26-15 Role of the juxtaglomerular apparatus in maintaining normal blood volume.

tory reabsorption of water increases, blood volume increases, and urine volume decreases. By raising the blood pressure, the juxtaglomerular apparatus ensures that the kidney cells receive enough oxygen and that the glomerular hydrostatic pressure is high enough to maintain a normal P_{eff}. The juxtaglomerular apparatus also regulates blood pressure throughout the body.

Blood Concentration. The concentrations of water and solutes in the blood also affect urine volume. If you have gone without water all day and the water concentration of your blood becomes low, osmotic receptors in the hypothalamus stimulate the posterior pituitary to release ADH. This hormone stimulates the cells of the distal convoluted tubule and collecting duct to transport water out of the filtrate and into the blood by facultative reabsorption. Thus, urine volume decreases and water is conserved.

If you have just drunk an excessive amount of liquid, urine volume may be increased through two mechanisms. First, the blood-water concentration increases above normal. This means that the osmoreceptors in the hypothalamus are no longer stimulated to secrete ADH, and facultative water reabsorption stops. Second, the excess water causes the blood pressure to rise. In response, the renal vessels dilate, more blood is brought to the glomeruli, and the filtration rate increases.

The concentration of Na⁺ in the blood also influences urine volume. Sodium concentration affects aldosterone secretion, which in turn affects both sodium reabsorption and the obligatory reabsorption of water.

Temperature. When the internal or external temperature rises above normal, the cutaneous vessels dilate and fluid diffuses from the capillaries to the surface of the skin, where it is lost by evaporation. As water volume decreases, ADH is secreted and facultative reabsorption increases. In addition, the increase in temperature stimulates the abdominal vessels to constrict, so the blood flow to the glomeruli and filtration decrease. Both mechanisms reduce the volume of urine.

If the body is exposed to low temperatures, the cutaneous vessels constrict and the abdominal vessels dilate. More blood is shunted to the glomeruli, glomerular blood hydrostatic pressure increases, and urine volume increases.

Diuretics. Certain chemicals increase urine volume by inhibiting the facultative reabsorption of water. Such chemicals are called *diuretics*, and the abnormal increase in urine flow is called *diuresis*. Some diuretics act directly on the tubular epithelium as they are carried through the kidneys. Others act indirectly by inhibiting the secretion of ADH as they circulate through the brain. Coffee, tea, and alcoholic beverages are diuretics.

Emotions. Some emotional states can affect urine volume. Extreme nervousness, for example, can cause an enormous discharge of urine because impulses from the brain cause dilation of the renal vessels, resulting in an increased glomerular filtration rate.

Physical Characteristics

Normal urine is usually a yellow or amber transparent liquid with a characteristic odor. The color is caused by urochrome, a pigment derived from the destruction of hemoglobin by reticuloendothelial cells. The color varies considerably with the ratio of solutes to water in the urine. The less water there is, the darker the urine. Fever decreases urine volume in the same way that high environmental temperatures do, sometimes making the urine quite concentrated. It is not uncommon for a feverish person to have dark yellow or brown urine. The color of urine may also be affected by diet, such as a reddish color from beets, and by the presence of abnormal constituents, such as certain drugs. A red or brown to black color may indicate the presence of red blood cells or hemoglobin from bleeding in the urinary system.

Fresh urine is usually transparent. Turbid (cloudy) urine does not necessarily indicate a pathological condition since turbidity may result from mucin secreted by the lining of the urinary tract. The presence of mucin above a critical level usually denotes an abnormality, however.

The odor of urine may vary. For example, the digestion of asparagus adds a substance called methyl mercaptan that gives urine a characteristic odor. In cases of diabetes, urine has a "sweetish" odor because of the presence of acetone. Stale urine develops an ammonialike odor due to ammonium carbonate formation as a result of urea decomposition.

Normal urine is slightly acid. It ranges between pH 5.0 and 7.8 and rarely becomes more acid than 4.5 or more alkaline than 8. Variations in urine pH are closely related to diet. Moreover, these variations are due to differences in the end products of metabolism. Whereas a high-protein diet increases acidity, a diet composed largely of vegetables increases alkalinity. Since the ammonium carbonate that forms in standing urine can dissociate into NH_4^+ and form a strong base, the presence of ammonium carbonate tends to make urine more alkaline.

Specific gravity is the ratio of the weight of a volume of a substance to the weight of an equal volume of distilled water. Water has a specific gravity of 1.000. The specific gravity of urine depends on the amount of solid materials in solution and ranges from 1.008 to 1.030 in normal urine. The greater the concentration of solutes, the higher the specific gravity.

Chemical Composition

Water accounts for about 95 percent of the total volume of urine. The remaining 5 percent consists of solutes derived from cellular metabolism and outside sources such as drugs. The solutes are described in Exhibit 26-5.

Abnormal Constituents

If the body's chemical processes are not operating efficiently, traces of substances not normally present may appear in the urine. Normal constituents may appear in abnormal amounts. Analyzing the physical and chemical properties of urine often provides information that aids diagnosis. Such an analysis is called a *urinalysis*.

Albumin. Protein albumin is one of the things a technician looks for when performing a urinalysis. Albumin is a normal constituent of plasma, but it usually does not appear in urine because the molecules are too large to pass through the pores in the capillary walls. The presence of albumin in the urine — *albuminuria* — indicates an increase in the permeability of the glomerular membrane. Conditions that lead to albuminuria include injury to the glomerular membrane as a result of disease, increased blood pressure, and irritation of kidney cells by substances such as bacterial toxins, ether, or heavy metals. Other proteins, such as globulin and fibrinogen, may also appear in the urine under certain conditions.

Glucose. The presence of sugar in the urine is termed *glycosuria*. Normal urine contains such small amounts of glucose that clinically it may be considered absent. The most common cause of glycosuria is a high blood sugar level. Remember that glucose is filtered into the Bowman's capsule. Later, in the proximal convoluted tubules, the tubule cells actively transport the glucose back into the blood. However, the number of glucose carrier molecules is limited. If more carbohydrates are ingested than the body can convert to glycogen or fat, then more sugar is filtered into Bowman's capsule than can be reabsorbed by the cells of the proximal convoluted tubule. This condition, called *temporary* or *alimentary glycosuria*, is not considered pathological. Another nonpathological cause is emo-

Exhibit 26-5   PRINCIPAL SOLUTES IN URINE OF AN ADULT MALE ON A MIXED DIET*

Constituent	Comments
Organic	
Urea: 25	Comprises 60–90 percent of all nitrogenous material. Derived primarily from deamination of proteins (ammonia combines with CO_2 to form urea).
Creatinine: 1.0–2.0	Normal alkaline constituent of blood. Primarily derived from creatine (nitrogenous substance in muscle tissue involved in energy storage).
Uric acid: 0.8	Product of catabolism of nucleic acids derived from food or cellular destruction. Because of insolubility, tends to crystallize and is common component of kidney stones.
Hippuric acid: 0.7	Form in which benzoic acid (toxic substance in fruits and vegetables) is believed to be eliminated from body. High-vegetable diets increase quantity of hippuric acid excreted.
Indican: 0.01	Potassium salt of indole. Indole results from putrefaction of protein in large intestine and is carried by blood to liver where it is probably changed to indican (less poisonous substance).
Acetone bodies: 0.04	Also called ketone bodies. Normally found in urine in small amounts. In cases of diabetes and acute starvation, ketone bodies appear in high concentrations.
Other substances: 2.9	May be present in minute quantities depending on diet and general health. Include carbohydrates, pigments, fatty acids, mucin, enzymes, and hormones.
Inorganic	
NaCl: 15	Principal inorganic salt. About 15 g excreted daily, but concentration varies with intake.
K^+:3.3 Mg^{2+}: 0.1 Ca^{2+}: 0.3	Appear as salts of chlorides, sulfates, and phosphates.
SO_4^{2-}: 2.5	Derived from amino acids.
PO_4^{3-}: 2.5	Occurs in urine as sodium compounds (monosodium and disodium phosphate) that serve as buffers in blood.
NH_4^+: 0.7	Occurs in urine as ammonium salts. Derived from protein catabolism and from glutamine in kidneys. Amount produced by kidney may vary with need of body for conserving Na^+ ions to offset acidity of blood and tissue fluids.

* All values are expressed as the number of grams present in a representative 24-hour collection of urine.

tional stress. Stress can cause excessive amounts of epinephrine to be secreted. Epinephrine stimulates the breakdown of glycogen and the liberation of glucose from the liver into the blood. A pathological glycosuria results from diabetes mellitus. In this case, there is a frequent or continuous elimination of glucose because the pancreas fails to produce sufficient insulin. When glycosuria occurs with a normal blood sugar level, the problem lies in failure of the kidney tubular cells to reabsorb glucose.

Erythrocytes. The appearance of red blood cells in the urine, called *hematuria*, generally indicates a pathological condition. One cause is acute inflammation of the urinary organs as a result of disease or irritation from stones. Whenever blood is found in the urine, additional tests are performed to ascertain the part of the urinary tract that is bleeding. When testing women, care should be taken to make sure the sample was not contaminated with menstrual blood from the vagina.

Leucocytes. The presence of leucocytes and other components of pus in the urine, referred to as *pyuria*, indicates infection in the kidney or other urinary organs. Again the source of the pus must be located and care should be taken that the urine was not contaminated.

Ketone Bodies. Ketone, or acetone, bodies appear in normal urine in small amounts. Their appearance in high quantities, a condition called *ketosis*, or *acetonuria*, may indicate abnormalities. It may be caused by diabetes mellitus, starvation, or simply too little carbohydrate in the diet. Whatever the cause, excessive quantities of fatty acids are oxidized in the liver, and the ketone bodies are filtered from the plasma into the Bowman's capsule.

Casts. Microscopic examination of urine may reveal *casts*—tiny masses of material that have hardened and assumed the shape of the lumens of the tubules. They are flushed out of the tubules by a buildup of filtrate behind them. Casts are named after the substances that compose them or for their appearance. There are white blood cell casts, red blood cell casts, epithelial casts that contain cells from the walls of the tubes, granular casts that contain decomposed cells which form granules, and fatty casts from cells that have become fatty.

Calculi. Occasionally, the salts found in urine may solidify into insoluble crystals called *calculi* or kidney stones. They may be formed in any portion of the urinary tract from the kidney tubules to the external opening. Conditions leading to calculi formation include the ingestion of excessive mineral salts, a decrease in the amount of tubular water, abnormally alkaline or acid urine, and overactivity of the parathyroid glands.

URETERS

Once urine is formed by the nephrons and collecting ducts, it drains through papillary ducts into the calyces surrounding the renal papillae. The minor calyces join to become the major calyces, which in turn unite to become the renal pelvis. From the pelvis, the urine drains into the ureters and is carried by peristalsis to the urinary bladder. From the bladder, the urine is discharged from the body through the single urethra.

Structure

The body has two **ureters**—one for each kidney. Each ureter is an extension of the pelvis of the kidney and runs 25 to 30 cm (10 to 12 inches) to the bladder (see Figure 26-1). As the ureters descend, their thick walls increase in diameter, but at their widest point they measure less than 1.7 cm (0.5 inch) in diameter. Like the kidneys, the ureters are retroperitoneal in placement. The ureters enter the urinary bladder at the superior lateral angle of its base. Since there is no valve or sphincter at the openings of the ureters into the urinary bladder, it is quite possible for cystitis (bladder inflammation) to develop into kidney infection. Three coats of tissue form the walls of the ureters. A lining of mucous membrane, the *mucosa,* composed of transitional epithelium is the inner coat. The solute concentration and pH of urine differ drastically from the internal environment of the cells that form the walls of the ureters. Mucus secreted by the mucosa prevents the cells from coming in contact with urine. Throughout most of the length of the ureters, the second or middle coat, the *muscularis,* is composed of inner longitudinal and outer circular layers of smooth muscle. The muscularis of the proximal third of the ureters also contains a layer of outer longitudinal muscle. Peristalsis is the major function of the muscularis. The third, or external coat of the ureters is a *fibrous coat.* Extensions of the fibrous coat anchor the ureters in place.

Physiology

The principal function of the ureters is to carry urine from the renal pelvis into the urinary bladder. Urine is carried through the ureters primarily by peristaltic contractions of the muscular walls of the ureters, but hydrostatic pressure and gravity also contribute. Peristaltic waves pass from the kidney to the urinary bladder, varying in rate from one to five per minute depending on the amount of urine formation.

URINARY BLADDER

The **urinary bladder** (Figure 26-16) is a hollow muscular organ situated in the pelvic cavity posterior to the symphysis pubis. In the male, it is directly anterior to the rectum. In the female, it is anterior to the vagina and inferior to the uterus. It is a freely movable organ held in position by folds of the peritoneum. The shape of the urinary bladder depends on how much urine it contains. When empty, it looks like a deflated balloon. It becomes spherical when slightly distended. As urine volume increases, it becomes pear-shaped and rises into the abdominal cavity.

Structure

Four coats make up the walls of the bladder. The *mucosa,* the innermost coat, is a mucous membrane containing transitional epithelium. Transitional epithelium is able to stretch—a marked advantage for an organ that must continually inflate and deflate. Stretchability is further enhanced by the rugae (folds in the mucosa) that appear when the bladder is empty. The second coat, the *submucosa,* is a layer of dense connective tissue that connects the mucous and muscular coats. The third coat—a muscular one called the *detrusor muscle*—consists of three layers: inner longitudinal, middle circular, and outer longitudinal muscles. In the area around the opening to the urethra, the circular fibers form an *internal sphincter* muscle. Below the internal sphincter is the *external sphincter,*

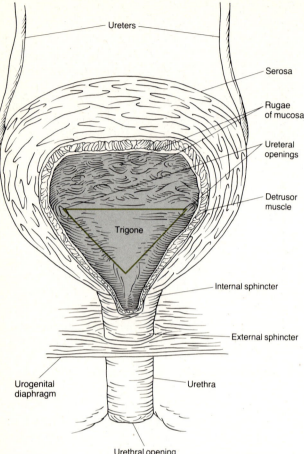

Figure 26-16 Urinary bladder and female urethra.

which is composed of skeletal muscle. The outermost coat, the *serous* coat, is formed by the peritoneum and covers only the superior surface of the organ.

At the base of the bladder is a small triangular area —the *trigone*—that points anteriorly. The opening to the urethra is at the apex of this triangle. At the two points that form the base, the ureters drain into the bladder. The trigone is easily identified because the mucosa is firmly bound to the muscularis so that the trigone is always smooth.

Physiology

Urine is expelled from the bladder by an act called *micturition,* commonly known as urination or voiding. This response is brought about by a combination of involuntary and voluntary nervous impulses. The average capacity of the bladder is 700 to 800 ml. When the amount of urine in the bladder exceeds 200 to 400 ml, **stretch receptors** in the bladder wall transmit impulses to the lower portion of the spinal cord. These impulses initiate a conscious desire to expel urine and

an unconscious reflex referred to as the *micturition reflex.* Parasympathetic impulses transmitted from the spinal cord reach the bladder wall and internal urethral sphincter, bringing about contraction of the detrusor muscle of the bladder and relaxation of the internal sphincter. Then, the conscious portion of the brain sends impulses to the external sphincter, the sphincter relaxes, and urination takes place. Although emptying of the bladder is controlled by reflex, it may be initiated voluntarily and may be started or stopped at will because of cerebral control of the external sphincter.

A lack of voluntary control over micturition is referred to as *incontinence.* In infants about 2 years old and under, incontinence is normal because neurons to the external sphincter muscle are not completely developed. Infants void whenever the bladder is sufficiently distended to arouse a reflex stimulus. Proper training overcomes incontinence if the latter is not caused by emotional stress or irritation of the bladder.

Involuntary micturition in the adult may occur as a result of unconsciousness, injury to the spinal nerves controlling the bladder, irritation due to abnormal constituents in urine, disease of the bladder, and emotional stress due to an inability of the detrusor muscle to relax.

Retention, a failure to void urine, may be due to an obstruction in the urethra or neck of the bladder, nervous contraction of the urethra, or lack of sensation to urinate. Far more serious than rentention is *suppression,* or *anuria*—failure of the kidneys to secrete urine. It usually occurs when blood plasma is prevented from reaching the glomerulus as a result of inflammation of the glomeruli. Anuria also may be caused by low filtration pressure.

URETHRA

The **urethra** is a small tube leading from the floor of the urinary bladder to the exterior of the body (see Figure 26-16). In females, it lies directly behind the symphysis pubis and is embedded in the anterior wall of the vagina. Its undilated diameter is about 6 mm (0.25 inch), and its length is approximately 3.8 cm (1.5 inches). The female urethra is directed obliquely downward and forward. The opening of the urethra to the exterior, the *urinary meatus,* is located between the clitoris and vaginal opening.

In males, the urethra is about 20 cm (8 inches) long, and it follows a different course from that of the female urethra. Immediately below the bladder it runs vertically through the prostate gland, then pierces the urogenital diaphragm, and finally penetrates the penis and takes a curved course through its body.

Structure

The walls of the female urethra consist of three coats: an inner mucous coat that is continuous externally with that of the vulva, an intermediate thin layer of spongy tissue containing a plexus of veins, and an outer muscular coat that is continuous with that of the bladder and consists of circularly arranged fibers of smooth muscle. The male urethra is composed of two membranes. An inner mucous membrane is continuous with the mucous membrane of the bladder. An outer submucous tissue connects the urethra with the structures through which it passes.

Physiology

Since the urethra is the terminal portion of the urinary system, it serves as the passageway for discharging urine from the body. The male urethra also serves as the duct through which reproductive fluid (semen) is discharged from the body.

HOMEOSTATIC IMBALANCES OF THE URINARY SYSTEM

Physical

Floating kidney (ptosis) occurs when the kidney is no longer held in place securely by the peritoneum or its covering of fat and slips from its normal position. Pain occurs if the ureter is twisted or folded. Such an abnormal orientation may also obstruct the flow of urine.

Chemical

If urine becomes too concentrated, chemicals normally dissolved in it may crystallize out, forming **kidney stones (renal calculi).** Common constituents of stones are uric acid, calcium oxalate, and calcium phosphate. The stones usually form in the pelvis of the kidney, where they cause pain, hematuria, and pyuria. A staghorn calculus may fill the entire collecting system; a dendritic stone involves some but not all of the calyces. Severe pain occurs when a stone passes through a ureter and stretches its walls. Ureteral stones are seldom completely obstructive because they are usually needle-shaped and urine can flow around them.

Gout is a hereditary condition associated with an excessively high level of uric acid in the blood. When the purine-type nucleic acids are catabolized, a certain amount of uric acid is produced as a waste. Some people, however, seem to produce excessive amounts of uric acid, and others seem to have trouble excreting normal amounts. In either case, uric acid accumulates in the body and tends to solidify into crystals that are deposited in the joints and kidney tissue. Gout is aggravated by excessive use of diuretics, dehydration, and starvation.

Infectious

Glomerulonephritis or Bright's disease is an inflammation of the kidney that involves the glomeruli. One of the most common causes of glomerulonephritis is an allergic reaction to the toxins given off by streptococcal bacteria that have recently infected another part of the body, especially the throat. The glomeruli become so inflamed, swollen, and engorged with blood that the glomerular membranes become highly permeable and allow blood cells and proteins to enter the filtrate. Thus, the urine contains many erythrocytes and much protein. The glomeruli may be permanently changed, leading to chronic renal disease and renal failure.

Pyelitis is an inflammation of the kidney pelvis and its calyces. **Pyelonephritis** is the interstitial inflammation of one or both kidneys. It usually involves both the parenchyma and the renal pelvis and is due to bacterial invasion from the middle and lower urinary tracts or the bloodstream.

Cystitis is an inflammation of the urinary bladder involving principally the mucosa and submucosa. It may be caused by bacterial infection, chemicals, or mechanical injury.

Hemodialysis Therapy

If the kidneys are so impaired by disease or injury that they are unable to excrete nitrogenous wastes and regulate pH and electrolyte concentration of the plasma, the blood must be filtered by an artificial device. Such filtering of the blood is called **hemodialysis.** *Dialysis* employs a semipermeable membrane to separate large nondiffusible particles from smaller diffusible ones. One of the best-known devices for accomplishing dialysis is the kidney machine (Figure 26-17). A tube connects it with the patient's radial artery. The blood is pumped from the artery and through the tubes to one side of a semipermeable cellophane membrane. The other side of the membrane is continually washed with an artificial solution called the dialyzing solution. The blood that passes through the artificial kidney is treated with an anti-

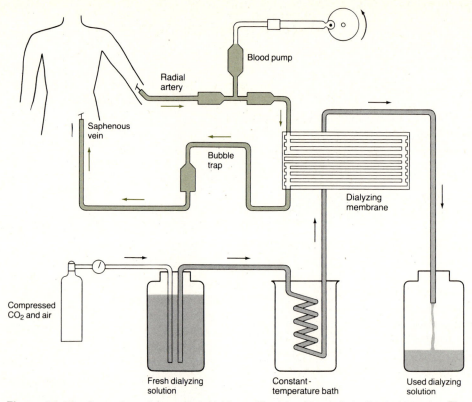

Figure 26-17 Operation of an artificial kidney. The blood route is indicated in color. The route of the dialyzing solution is indicated in gray.

coagulant. Only about 500 ml of the patient's blood is in the machine at a time. This volume is compensated for by vasoconstriction and increased cardiac output.

All substances (including wastes) in the blood except protein molecules and blood cells can diffuse back and forth across the semipermeable membrane. The electrolyte level of the plasma is controlled by keeping the dialyzing solution electrolytes at the same concentration found in normal plasma. Any excess plasma electrolytes move down the concentration gradient and into the dialyzing solution. If the plasma electrolyte level is normal, it is in equilibrium with the dialyzing solution and no electrolytes are gained or lost. The dialyzing solution contains none of the waste products such as urea, therefore they move down the concentration gradient and into the dialyzing solution. Thus wastes are removed and normal electrolyte balance is maintained.

A great advantage of the kidney machine is that the patient's nutrition can be bolstered by the inclusion of large quantities of glucose in the dialyzing solution. While the blood gives up its wastes, the glucose diffuses into the blood. Thus, the kidney machine beautifully accomplishes the principal function of the fundamental unit of the kidney—the nephron.

There are obvious drawbacks to the artificial kidney, however. The blood must be anticoagulated during dialysis, and a large amount of the patient's blood must flow through this apparatus to make it work. To date, no artificial kidney has been developed that can be permanently implanted.

Peritoneal Dialysis

In peritoneal dialysis, the dense capillary network of the peritoneal cavity serves as the dialyzing membrane. By means of a tube, a sterile dialyzing solution containing the proper electrolyte mixture is placed through a small opening into the abdominal cavity. This mixture usually is allowed to remain in the peritoneal cavity for 30 to 45 minutes and is then pumped back out through the tube.

The procedure is repeated until the normal plasma concentrations of the various substances have been achieved. Peritoneal dialysis must be done in a hospital and is generally used during short-term kidney shutdown.

Kidney machines and peritoneal dialysis can be used to quickly and efficiently overcome the possibly fatal effects of drug overdose or poisoning.

CHAPTER SUMMARY OUTLINE

KIDNEYS

1. The kidneys are retroperitoneal and each contains a notch called the hilum through which the renal artery and nerves enter and from which the renal vein, lymph vessels, and ureter exit the kidney.
2. The kidneys use 8 percent of the oxygen utilized by the entire body at rest and receive about 25 percent of the cardiac output.
3. Nephrons are the functional units of the kidneys. They help form urine and regulate blood composition.
4. A nephron is made up of a double-walled globe called Bowman's capsule and a long tubule composed of functionally different regions.
5. The kidneys form urine by glomerular filtration, tubular reabsorption, and tubular secretion.
6. The primary force behind filtration is hydrostatic pressure.
7. If the hydrostatic pressure falls to 50 mm Hg, renal suppression occurs because then the glomerular blood hydrostatic pressure exactly equals the opposing pressures (capsular hydrostatic pressure and blood osmotic pressure).
8. Most substances in plasma are filtered into Bowman's capsule. Normally, blood cells and proteins are not.
9. Tubular reabsorption recovers substances needed by the body, including water, glucose, amino acids, and ions.
10. About 80 percent of the reabsorbed water is returned by obligatory reabsorption in the proximal convoluted tubule, the rest by facultative reabsorption in the distal convoluted tubule and collecting ducts.
11. Some chemicals not needed by the body are discharged into the urine by tubular secretion.
12. The kidneys help maintain blood pH by excreting H^+ and NH_4^+ ions. In exchange, the kidneys conserve sodium bicarbonate.
13. The rate of para-aminohippurate (PAH) filtration increases in direct relation to an increasing PAH concentration in plasma.
14. The ability of the kidneys to produce either hyperosmotic or hyposmotic urine is based on the countercurrent mechanisms.
15. Sodium is reabsorbed into the capillaries, whereas potassium is secreted into the tubular fluid.

CONTRIBUTION OF KIDNEYS TO ACID-BASE BALANCE

1. Acidic substances produced by the diet or by body metabolism are neutralized by buffer systems that exist in body fluids.
2. The consequence of such buffering is the removal of the bicarbonate ion each time an acid molecule is produced.
3. The kidneys recover bicarbonate by reconverting the salt of the acid to HCl and the $NaHCO_3$ by a series of reactions within the tubular cells.

HOMEOSTASIS

1. The primary homeostatic function of the urinary system is to regulate the concentration and volume of blood by removing and restoring selected amounts of water and solutes. It also excretes wastes to help maintain homeostasis.
2. The lungs, integument, and alimentary canal also assume excretory functions.
3. Urine volume is influenced by blood pressure, blood concentration, temperature, diuretics, and emotions.
4. The physical characteristics of urine evaluated in a urinalysis include: color, turbidity, odor, pH, and specific gravity.
5. Chemically, normal urine contains about 95 percent water and 5 percent solutes. The solutes include urea, creatinine, uric acid, indican, ketones, salt, and ions.
6. Abnormal conditions diagnosed through urinalysis include albuminuria, glycosuria, hematuria, pyuria, ketosis, casts, and calculi.

URETERS, URINARY BLADDER, AND URETHRA

1. The paired ureters convey urine from the kidneys to the urinary bladder, mostly by peristaltic contraction.
2. The urinary bladder stores urine. The expulsion of urine from the bladder is called micturition.
3. Lack of control over micturition is called incontinence. Failure to void urine is referred to as retention. Inability of the kidneys to produce urine is called suppression.

4. The urethra extends from the floor of the bladder to the exterior and it discharges urine from the body.

HOMEOSTATIC IMBALANCES OF THE URINARY SYSTEM

1. A floating kidney, or ptosis, occurs when the kidney is no longer held in place securely by the peritoneum, but slips from its normal position.
2. If urine becomes too concentrated, chemicals normally dissolved in it may crystallize out, forming kidney stones or renal calculi.
3. Gout is a hereditary condition associated with an excessively high level of uric acid in the blood.
4. Glomerulonephritis, or Bright's disease, is an inflammation of the kidney that causes the glomerular membranes of the nephrons to become highly permeable, allowing blood cells and proteins to enter the filtrate.
5. Pyelitis, pyelonephritis, and cystitis are other pathological conditions of the urinary system.

Hemodialysis Therapy

1. Hemodialysis is a technique that uses an artificial device to filter the blood, remove wastes, and regulate pH and electrolyte concentration of the plasma.
2. The kidney machine is one of the best-known devices for accomplishing hemodialysis.
3. In addition to removal of wastes and electrolyte replacement, the kidney machine can also bolster the nutrition of the individual. This is accomplished by including large quantities of glucose in the dialyzing solution which will diffuse into the blood.

Peritoneal Dialysis

1. In peritoneal dialysis, the dense capillary network of the peritoneal cavity serves as the dialyzing membrane. This technique also removes wastes, and restores pH and electrolyte balance.
2. Kidney machines and peritoneal dialysis techniques can be helpful in the treatment of drug overdose and poisoning.

REVIEW QUESTIONS

1. What are the functions of the urinary system? What organs compose the system?
2. Describe the location of the kidneys. Why are they said to be retroperitoneal?
3. Prepare a labeled diagram that illustrates the principal external and internal features of the kidney.
4. Describe the blood supply to the kidneys.
5. What is a nephron? List and describe the parts of a nephron starting from Bowman's capsule to the collecting duct.
6. What constitutes the blood supply to the tubules?
7. What is glomerular filtration? Describe the composition of filtrate.
8. Set up an equation to indicate how effective filtration pressure is calculated. What is the cause of renal suppression?
9. What are the major chemical differences among plasma, filtrate, and urine?
10. Define tubular reabsorption. Why is the process physiologically important?
11. What chemical substances are normally reabsorbed by the kidneys?
12. Describe how and where glucose and sodium are reabsorbed by the kidneys.
13. How is chloride reabsorption related to sodium reabsorption?
14. Distinguish obligatory from facultative water reabsorption. How and where is facultative water reabsorption controlled?
15. Define tubular secretion. Why is it important?
16. What is the significance of the secretion of PAH?
17. Define and explain the two countercurrent mechanisms.
18. How is sodium and potassium excretion regulated?
19. Explain the contribution of the kidney to acid-base balance.
20. Contrast the functions of the lungs, integument, and alimentary tract as excretory organs.
21. What is urine? Describe the effects of blood pressure, blood concentration, temperature, diuretics, and emotions on the volume of urine formed.
22. Describe the following physical characteristics of normal urine: color, turbidity, odor, pH, and specific gravity.
23. Describe the chemical composition of normal urine.
24. Define each of the following: albuminuria, glyco-

suria, hematuria, pyuria, ketosis, casts, and calculi.

25. Describe the structure and function of the ureters.
26. How is the urinary bladder adapted to its storage function? What is micturition?
27. Contrast the causes of incontinence, retention, and suppression.
28. Compare the urethra in the male and female.
29. What is a floating kidney?
30. How are renal calculi formed?
31. Explain gout as to causes, effects on the body, and possible treatment.
32. What is Bright's disease? How can it become dangerous?
33. Contrast pyelitis, pyelonephritis, and cystitis.
34. Explain in detail the process of hemodialysis.
35. What are the main advantages and disadvantages of hemodialysis?
36. What is peritoneal dialysis? Correlate this procedure with the treatment for drug overdose and poisoning.

GLOSSARY

A band Anisotropic band, the darker region located in the central part of the sarcomeres of cardiac and skeletal muscle. It is equal in width to the length of the thick myofilaments.

Abdomen That portion of the body that lies between the thorax and the pelvis. It contains viscera.

Abortion The premature removal or expulsion from the uterus of an embryo or nonviable fetus. This can be spontaneous or artificially induced.

Abscess A collection of microbes, leucocytes, and necrotic tissue within a small cavity. A localized collection of pus within a small cavity. Usually named by the location or the causative agent: bacteria, fungi, or protozoa.

Absorption The movement of materials across cell membranes, or epithelia. Can be a mediated process or simple diffusion generally described in nutrient uptake from the digestive tract.

Abstinence Refraining from the use of or indulgence in food, drugs, or sexual intercourse. Often used to describe one method of birth control.

Acceleration An increase in the rate of movement, velocity, growth, or activity. In physics, defines the rate of change of velocity; the derivative of velocity.

Accommodation Adjustments of the eye that take place when objects are viewed at various distances. Mostly accomplished by altering the thickness of the lens.

Acetoacetic acid A four-carbon ketone body formed when the acetyl CoA pool rises to high levels. It is found in the blood and urine of diabetics and during starvation—both are conditions in which fatty acid metabolism is increased.

Acetone A volatile, fragrant ketone body formed in the liver from acetoacetic acid. It is able to diffuse from pulmonary capillaries into alveolar air. Can be detected on the breath of people after prolonged fasting or from uncontrolled diabetes mellitus.

Acetonuria The presence of acetone in urine.

Acetylcholine A chemical transmitter released from nerve endings of cholinergic nerves that diffuses toward the membranes of postsynaptic neurons and muscle fibers. Acetylcholine acts quickly and is destroyed by the enzyme acetylcholinesterase.

Acetylcholinesterase An enzyme located on the postsynaptic membrane that inactivates acetylcholine within 2 msec by splitting it into acetic acid and choline.

Acid Any compound that is able to release hydrogen ions when it is in solution. Those compounds that release H⁺ readily and in large numbers are termed *strong acids*. Those that release H⁺ sparingly and in small numbers are termed *weak acids*.

Acidity A measure of the hydrogen ion concentration, usually expressed in pH units. The higher the H⁺ concentration, the greater the acidity of a solution.

Acidosis A condition in which blood hydrogen ion concentration is higher than normal. A normal value has a pH of 7.4.

Acinus A collection of cells arranged to form a hollow, saclike structure. In glands, acini are often connected so that they resemble a stalk of grapes.

Acromegaly A disease caused by excessive secretion of human growth hormone from the anterior pituitary gland. It is characterized by enlargement of bones and soft tissues, particularly of the hands, feet, and face.

Acrosome   A caplike structure located on the head of a spermatozoan. It contains an enzyme that aids it in penetrating the ovum during fertilization.

ACTH   Adrenocorticotropic hormone.

Action potential   The rapid change in voltage that occurs across the membrane of an excitable cell when given an adequate stimulus. It is due to sudden changes in the membrane permeability—initially to Na^+ and secondly to K^+. The voltage change is an all-or-none response during which the interior voltage changes from a negative to a positive value. Information is carried along nerve fiber membranes by action potentials (nerve impulses).

Activation energy   The energy required to raise a molecule to an excited or *transition state* where the probability is very high that a chemical bond will be formed or broken and a new product made.

Active transport   The movement of molecules or ions across a plasma membrane, usually from a region of low concentration to a region of high concentration. Such a process requires an expenditure of cellular energy (ATP) and the use of a specific carrier molecule.

Acupuncture   The insertion of a needle into a tissue for the purpose of drawing fluid or relieving pain. It is also an ancient Chinese practice employed to cure illnesses by inserting needles into specific locations of the skin.

Adaptation   Occurs when a neuron relays a decreased frequency of action potentials from a receptor even though the strength of the stimulus remains constant.

Addison's disease   A condition resulting from insufficient secretion of hormones from the adrenal cortex. This progressive disease is characterized by a bronzelike pigmentation of the skin, low blood pressure, diarrhea, and digestive disturbances.

Adenine   One of the nitrogenous bases contained in RNA and DNA and nucleotides such as ATP and NAD. It is 6-amino purine, $C_5H_5N_5$.

Adenohypophysis   The anterior lobe of the pituitary gland (hypophysis). It releases several tropic hormones that influence the activity of other endocrine glands.

Adenosine diphosphate (ADP)   The molecule formed when the terminal phosphate is removed from adenosine triphosphate. In biological systems, this reaction provides large amounts of energy to drive energy-requiring reactions.

Adenosine monophosphate (AMP)   The molecule formed when pyrophosphate is removed from adenosine triphosphate.

$$ATP \rightarrow AMP + PPP + energy$$

This molecule is one of the nucleotides used in DNA and RNA.

Adenosine triphosphate (ATP)   The universal energy-carrying molecule manufactured in all living cells as a means of capturing and storing energy. It consists of the purine base *adenine* and the five-carbon sugar *ribose,* to which are added, in linear array, three phosphate molecules. Its formation from ADP + P_i is made possible by energy derived from the breakdown of nutrients: carbohydrates, fats, and protein.

Adenylate cyclase   An enzyme found on the inner surface of cell membranes that catalyzes the conversion of ATP to the cyclic AMP intracellular messenger molecule.

Adequate stimulus   A strength of stimulation just sufficient to depolarize an excitable cell membrane to its threshold voltage. At this point the cell membrane responds with an action potential.

ADH   Antidiuretic hormone.

Adipose   A connective tissue composed of cells that have accumulated large amounts of fat. Adipose tissue derives from loose connective tissue in which the fibroblast cells become fat-storing cells.

Adolescence   The period of life that begins with the appearance of secondary sex characteristics and ends when somatic growth ceases. Encompasses years from about 11 to 19.

ADP   Adenosine diphosphate.

Adrenal cortex   The outer two-thirds of the adrenal gland. The cortex is derived embryologically from mesoderm. It is divided into three zones. The outermost, the zona glomerulosa, secretes mineralocorticoids; the next, the zona fasciculata, secretes glucocorticoids; and the innermost, the zona reticularis, secretes sex hormones.

Adrenal glands   A pair of flattened yellowish masses of tissue located as caps, one on top of each kidney. Each gland consists of an inner core of tissue, termed the adrenal medulla, surrounded by a thicker layer of tissue, termed the adrenal cortex.

Adrenaline   The official name for epinephrine in the British pharmacopoeia. When spelled Adrenaline, it refers to a trademarked commercial preparation of epinephrine.

Adrenal medulla   The inner core of the adrenal gland. It is derived embryologically from ectodermal cells that migrate from the neural crest. It secretes a mixture of the hormones epinephrine (75 percent) and norepinephrine (25 percent).

Adrenergic   Usually employed to describe those nerve fibers that release norepinephrine from their terminal feet. It also describes synthetic or natural chemicals that elicit the effects of either epinephrine or norepinephrine.

Adrenocorticotropic hormone (ACTH) A polypeptide hormone, synthesized, stored, and released from basophilic cells of the anterior pituitary gland. Its primary action is on the cells of the zona fasciculata of the *adrenal cortex* that respond in turn to release glucocorticoids such as hydrocortisone.

Adrenogenital syndrome The result of an overproduction of sex hormones, particularly male androgens, by the adrenal cortex. In females, produces a masculinizing effect.

Adsorption The clinging of a substance to the *surface* of another. The result is to increase the concentration of the clinging substance on that surface.

Aerobic Occurring in the presence of molecular oxygen.

Afferent arteriole The blood vessel that delivers blood into the glomerular capillaries of the kidney. It contains smooth muscle in its walls, and therefore its caliber can be regulated.

Afferent neuron A nerve cell whose cell body is located outside of the CNS. It carries information from peripheral receptors *toward* the CNS.

Affinity A driving force that draws things together. Degrees of affinity are demonstrated. Some components are firmly held together and show high affinity for each other. Others are loosely united and show low affinity for each other.

Afterbirth The placenta and membranes expelled from the uterus after the birth of the fetus.

Agglutination The phenomenon noted when cells in suspension clump or aggregate together. Caused by substances that attach to the cell surfaces and bind neighboring cells to each other.

Agglutinin Any substance that promotes or causes cells in suspension to clump or agglutinate. Some immunoglobulins react with foreign cells to cause agglutination. Certain plant extracts called *lectins* also cause agglutination.

Agglutinogen Any substance, but usually one contained in the outer surface of a cell membrane, that causes the production of an agglutinin type of immunoglobulin.

Albino An organism demonstrating the congenital absence of pigment in the skin, hair, and eyes.

Albumin The most abundant (60 percent) and the smallest (70,000 mol. wt.) of the plasma proteins, which functions primarily to regulate osmotic pressure of plasma.

Albuminuria Albumin present in urine.

Aldosterone One of the mineralocorticoid hormones secreted by the cells of the zona glomerulosa of the adrenal cortex. It acts upon the kidney tubular cells and causes them to reabsorb sodium ions from the tubular fluid while decreasing potassium ion reabsorption.

Alkalinity A measure of the concentration of hydroxide ions (OH^-) in solution. Basic solutions have high alkalinity as opposed to acid solutions, in which alkalinity is very low.

Alkalosis An abnormal condition wherein the concentration of OH^- is elevated in body fluids. Could result from excessive loss of acid (H^+) from the body fluids.

Allantois In embryonic reptiles, birds, and nonhuman mammals, this structure appears as a large sac that functions to store urine. In human embryology, it is not fully formed and is nonfunctional, although its blood vessels develop into those of the umbilical cord.

Allergy Hypersensitivity to a particular foreign substance that in the same dose has no effect on most people. It involves an antibody attack on the foreign material, which is usually adhering to a mucous membrane. This tissue response includes the release of histamine, which is primarily responsible for the allergic symptoms.

All-or-none law Refers specifically to excitable cells. Stimulation causes an action potential to spread over the entire cell or not to occur at all. Inherent in the law is the notion that the response, if it occurs, is a maximal one under existing conditions.

Allosteric enzyme A catalytic protein with two reactive sites. One site catalyzes the specific action of the enzyme. The other site (allosteric) is capable of binding a regulatory molecule that can turn on or off the activity of the catalytic site.

Alpha adrenergic receptor One of two types of receptor sites for norepinephrine and epinephrine. These can be identified and activated or blocked by certain specific drugs. When activated, their effects include an increased peripheral resistance, dilation of pupils, and piloerection. The other type of receptor for norepinephrine and epinephrine is termed β-adrenergic.

Alpha cells Located in the islets of Langerhans, these cells respond to falling levels of blood glucose by releasing *glucagon* into the circulation.

Alpha helix A coiling pattern exhibited by chains of linked amino acids. In this pattern, the bonding constraints imposed compel the backbone of the chain to form a helical coil. In the alpha helix there are 3.6 amino acid residues in one turn of the coil.

Alpha rhythm The EEG of an adult who is awake but resting with eyes closed shows an 8–13 per second wave pattern. This basic pattern is alpha rhythm. The frequency and wave size change when the eyes open or when the subject falls into various levels of sleep.

Alveolus A small saclike pouch. In the lungs, the gaseous exchanges between the blood and air take place through *alveolar* membranes. Also describes the arrangement of cells into a small saclike container in secretory glands.

Amenorrhea The absence or failure of menstrual flow.

Amino acid A relatively simple molecule with molecular weight about 100. Amino acids have the general formula:

$$R-\underset{\underset{NH_2 \text{ (amino group)}}{|}}{\overset{\overset{H}{|}}{C}}-COOH \text{ (carboxyl group)}$$

There are 20 naturally occurring amino acids that appear in biologically formed proteins.

Amino acyl synthetases A group of enzymes, each of which links its own specific amino acid to a corresponding specific tRNA. Considered the first step in protein synthesis.

Ammonia (NH_3) A substance produced during protein destruction. Individual amino acids are eventually deaminated. The ammonia so produced is converted to urea in the liver.

Amnesia A lack or loss of memory.

Amniocentesis A procedure for collecting fetal cells for the purpose of early detection of any possible congenital disorder. Usually done by a surgeon, who inserts a hypodermic needle through the mother's abdominal and uterine walls into the amniotic sac. A sample of amniotic fluid containing cells shed by the fetus is withdrawn. The cells are cultured and later examined for abnormal chromosomes.

Amnion A strong membrane that encloses the fluid-filled *amniotic cavity*, within which the fetus is contained and sustained.

AMP Adenosine monophosphate.

Amphetamine A drug often used (and abused) to stimulate the CNS, increase blood pressure, and reduce appetite. A sympathomimetic drug that produces effects similar to those experienced from stimulation of the sympathetic nervous system.

Amphipathic molecule A molecule that contains within it, at the same time, a polar or ionized region in one location and a nonpolar region at another location. Phospholipids in the cell membrane are amphipathic molecules.

Amplification cascade The sequence of events describing the results of epinephrine on a liver cell. The production of intracellular cyclic AMP triggers production of an active enzyme, which in turn activates another enzyme, and so on. The result is a large, sudden breakdown of glycogen within the liver cell and a rapid outpouring of glucose.

Amylase An enzyme that attacks starch in a random fashion, producing small polysaccharides and disaccharides (maltose). This enzyme is contained in saliva and pancreatic juice.

Anabolism The occurrence of synthetic reactions within cells. The end products are complex compounds useful to the cell.

Anaerobic Refers to a condition in which oxygen is absent. *Anaerobic metabolism* refers to a series of biochemical reactions that occur in the absence of oxygen.

Analgesic Refers to pain-relieving ability without loss of consciousness.

Anaphase A phase in mitosis and meiosis during which the centromeres split and the pairs of resulting chromatids line up on the spindle, then move apart to opposite poles of the spindle and become the daughter chromosomes.

Anaphylaxis An exaggerated antigen-antibody reaction due to prior injection of a foreign material.

Anastomosis A connection between two tubular structures. Blood vessels often deliver blood into common branches to ensure a continual blood flow even if one of the major blood supplies is blocked. Such connections are referred to as anastomotic, and the connecting branches are called collateral channels.

Androgen Any chemical that possesses masculinizing properties, for example, testosterone.

Anemia A decreased value of hemoglobin and/or red blood cells in blood.

Anesthesia A total or partial loss of feeling or sensation. Usually defined with respect to loss of *pain* sensation. Used to permit surgery or any other painful procedure.

Anesthetic A drug or technique that is used to abolish pain sensation.

Aneurysm An outpouching or dilated region in the wall of a blood vessel.

Angina pectoris A sudden chest pain accompanied by a feeling of suffocation and impending death. Due to hypoxia of the myocardium and usually triggered by exertion or excitement.

Angiotensin I A decapeptide (10 amino acids) formed by the action of renin on an α_2-globulin of plasma called angiotensinogen.

Angiotensinogen An α_2-globulin synthesized by the liver and carried in the plasma. When acted upon by renin, secreted from the kidney, it is converted to angiotensin I.

Anion Any negatively charged ion (e.g., Cl^-, HCO_3^-).

Anisotropic Describes the asymmetric behavior of certain crystals or compounds that are doubly refracting or have a double polarizing power. This property is evident in the A band (anisotropic band) of the myofibrillar sarcomere. Basically, it is a property of the myosin molecules.

Antibiotic Literally means antilife. A chemical produced by a microorganism that is able to inhibit the growth of or kill other microorganisms. Specific microorganisms are grown commercially and the antibiotic chemical produced is isolated and used to treat certain infectious diseases of animals and humans.

Antibody A protein molecule produced in lymphoid tissue in response to the introduction of some foreign material into an animal. The antibody molecule reacts specifically with the foreign material (antigen) that caused its production. The newer terminology for antibody is *immunoglobulin*.

Anticoagulant A substance that is able to delay, suppress, or prevent the clotting of blood *in vivo* or *in vitro*.

Anticodon The triplet sequence of nucleotides present in tRNA that is complementary to the triplet sequence of nucleotides (codon) in mRNA.

Antidiuretic hormone (ADH) The hormone stored in the posterior pituitary gland that acts upon the cells of the collecting ducts to permit passive reabsorption of water from urine. In this manner, the volume of the excreted urine is reduced.

Antigen A foreign substance that, when introduced into an animal, causes the production of an immunoglobulin (antibody), which then reacts only with the antigen that promoted its production. Most antigens are protein or conjugated to a protein. A few polysaccharides have been rediscovered that are antigenic.

Antihistamine A drug that counteracts the effects of histamine.

Anuria Cessation of urine production.

Aortic bodies (chemoreceptors) Specialized receptors located in the area of the arch of the aorta. Respond primarily when lack of oxygen becomes severe and life threatening.

Apnea A period of time during which no breathing occurs.

Apneustic center Located in the pons, a collection of neurons that, when permitted to act, produce a maximal and prolonged inspiratory movement.

Apoferritin A protein molecule with a molecular weight of about 460,000. It is present in the mucosal cells of the small intestine and inside liver cells. Functions to bind an iron atom and become ferritin. It serves to store iron.

Aponeurosis A broad, flattened tendon that serves for muscle attachments.

Apothecary A pharmacist licensed to fill prescriptions. Also describes a system of weights and volumes used in compounding and dispensing drugs.

Appendicitis An inflammation of the vermiform appendix.

Appetite A desire or longing for some *specific* food. May not be related to hunger, which is related more to a nutritional or energy requirement for food.

Aqueous humor The watery fluid occupying the anterior and posterior chambers of the eye. Composition is similar to lymph.

Aqueous solution A solution prepared with water as the solvent.

Arachnoid Literally means resembling a spider's web. The name for the delicate, weblike, transparent membrane lining the dura mater. It consists of elastic and collagenous fibers that also stretch across to connect with the pia mater.

Areola A circular area of a different color surrounding a central point (e.g., surrounding the nipple of the breast).

Arrhythmia Irregular beat. In the heart, refers to any variation from the normal pattern of beating.

Arteriole A blood vessel just visible to the unaided eye. Has the highest *percentage* of smooth muscle in its wall. Serves to regulate flow rate and pressure of blood into capillaries.

Arteriosclerosis A thickening and hardening of the walls of the arteries.

Artery Any of the large blood vessels with thick musculoelastic walls that carry blood away from the heart and distribute it to all parts of the body.

Arthritis Inflammation of a joint. Often accompanied by pain, swelling, and redness.

Ascorbic acid Also known as vitamin C. Present in citrus fruits, tomatoes, and green vegetables. Important in protein anabolism.

Aspartic acid One of the 20 naturally occurring amino acids. Included as a possible transmitter substance in CNS excitatory neurons.

Asthma A disease characterized by difficulty in breathing. The smooth muscle of the bronchiolar walls suddenly constricts and narrows the air passageways.

Astigmatism A defect in vision due to the fact that the radius of curvature of the refractive surface in one plane is not the same as that of the radius perpendicular to it. As a result, light rays are not sharply focused on the retina.

Astrocyte Literally a star-shaped cell. Specifically, one type of glial cell found in the CNS that serves to support the function of the neuronal cells. Astrocytes may be involved in myelin formation, transport of materials, and the maintenance of the interstitial fluid.

Atheroma A flattened mass of breakdown material that accumulates in the thickening wall of an artery.

Atherosclerosis The commonest form of arteriosclerosis. An accumulation of cholesterol, and other lipids, within the walls of large and medium-sized arteries. Can lead to dangerous narrowing of the vessel lumen.

ATP Adenosine triphosphate.

Atrioventricular node A collection of specialized conducting cells located in the corner of the right atrium near the interventricular septum. Relays electrical impulses generated from the SA node into the ventricular conducting system.

Atrium A chamber that precedes entrance into another chamber or organ (e.g., the atria of the heart).

Atrophy A wasting away that leads to the decrease in size of a cell, tissue, organ, or part.

Atropine A drug usually derived from plant species such as belladonna or stramonium. Is now produced synthetically. Blocks the receptor sites for acetylcholine (anticholinergic) and produces smooth muscle relaxation in some organs, increases heart rate, and decreases gastric secretions.

Autoimmunity A condition characterized by the production of immunoglobulins by a host against some tissue of its own. Some autoimmune reactions can be very severe and self-destructive.

Autonomic nervous system That division of the nervous system that regulates cardiac and smooth muscle and secretion from glandular structures.

Autophagy Self-eating. Eating of one's own tissues.

Autosome There are 23 pairs of chromosomes in normal human cells. Autosomes are any of the 22 pairs other than the pair of sex chromosomes.

Autotrophic Describing organisms that are able to obtain their nutrition and energy from carbon dioxide and inorganic materials and radiant energy.

AV node Atrioventricular node.

Axon The single long process extending from the cell body of a neuron that carries action potentials on its membrane away from the cell body.

Axoneme The $9+2$ system of microtubules and associated arms and spokes that form the core components of a eucaryotic cilium or flagellum.

B lymphocyte Lymphocyte programed to produce immunoglobulins for circulation in the plasma. When actively producing immunoglobulins, they are called plasma cells.

Bacillus A rod-shaped bacterium; spore-forming.

Bacteria Unicellular microorganisms that have a cell wall around the cell membrane. They are present in every environment and inside all living forms.

Bainbridge reflex A description of the increased heart rate that follows an increased pressure or distension of the right atrium.

Barometer An instrument that measures atmospheric pressure.

Baroreceptor (pressoreceptor) A special receptor that responds to changes in pressure. Baroreceptors are located in the walls of the arch of the aorta and at the point of bifurcation of the common carotid arteries. They constantly monitor blood pressure and generate action potentials that are relayed to centers in the medulla.

Basal body A microtubular structure located within the cell forming the cytoplasmic anchor for the microtubules that extend into the cilia and flagella. They possess a structure identical to *centrioles*—27 microtubules arranged in nine triplet sets forming a circle.

Basal metabolic rate (BMR) A measurement of the minimum energetic cost to just keep an awake person alive. The BMR for an individual is compared to a set of normalized ones for a person of the same sex, age, weight, and height.

Basilar membrane A thin membrane within the inner ear that separates two compartments—the scala tympani and the cochlear duct. On the cochlear duct side of the membrane it supports the organ of Corti—the essential receptor for transducing sound into nerve impulses.

Basophil A cell that takes up and stains with basic dyes. The least numerous of the five types of leucocytes. Granular, with a bilobed nucleus, it may contain heparin in minute amounts.

Bellini ducts The largest tubes within the kidney. These ducts arise from the convergence of many collecting ducts and distal convoluted tubules. They terminate or open up at the tips of the renal pyramids and the urine from these ducts collects in the minor calyces → major calyces → renal pelvis → ureter.

Benign Showing gentleness or kindness. In medicine used to mean not malignant, favorable for recovery.

Beta adrenergic receptors One of two types of receptor sites for norepinephrine and epinephrine. These can be identified and activated or blocked by certain specific drugs. When activated, their effects include an increase in heart rate and strength of contraction, dilation of bronchioles, and glycogenolysis. The other type of receptor for norepinephrine and epinephrine is termed α adrenergic.

Beta-hydroxybutyric acid One of the ketone bodies formed when fat metabolism is accelerated in order to provide energy for the body. Beta-hydroxybutyric acid is formed in the liver, carried in the blood, and excreted in the urine of diabetics and starving people.

Beta oxidation The name given to the sequence of enzyme-catalyzed reactions that break down fatty acids into activated two-carbon molecules called acetyl CoA.

Biconcave Concave on both sides. Consider a round disc that is depressed in the center on both sides. This shape describes the appearance of red blood cells.

Bifurcation The point where a vessel or other tubular structure divides or forks into two continuing branches.

Bile A bitter, alkaline, greenish-yellow liquid secreted by the liver and delivered into the duodenum by the common bile duct. Functions to help solubilize and absorb fats from the small intestine.

Bile canaliculi Small channels that collect the bile components from individual liver cells. Canaliculi run down between double rows (cords) of liver cells into small *bile ductules*, which converge progressively into larger ducts, until the bile leaves the liver within the *hepatic duct.*

Biliflavin One of the end products of hemoglobin breakdown in the liver. It is a yellow pigment that is derived from biliverdin. It is excreted in bile as a waste material.

Bilirubin One of the end products of hemoglobin breakdown by the liver cells. Excreted as a waste material in the bile. It is a red pigment derived from biliverdin by a reduction reaction.

Biliverdin One of the first products of hemoglobin breakdown in the liver cells. It is a green pigment that is converted to bilirubin. Some persists in bile to be excreted as a waste material.

Binary fission Splitting to form two organisms of approximately equal size. For example, bacteria multiply in number by binary fission.

Bipolar cells Nerve cells located within the retina of the eye. These cells are activated by the visual cells (rods and cones) called the first-order cells. The bipolar cells, called second-order, synapse with ganglion cells, the third-order cells, which send their unmyelinated axons into the optic nerve.

Blastocoele The fluid-filled cavity of the *blastula*, the hollow mass of cells, one cell thick, produced by cleavage of a fertilized ovum.

Blastocyst An almost hollow mass of cells formed by the cleavage of a fertilized ovum. Follows the *blastula* stage. The surface layer is more than one cell thick and the interior shows an eccentric inner cell mass of cells.

Bleeding time The length of time required for bleeding to cease from a small puncture wound. Usually an ear lobe is pricked and repeatedly blotted until blood flow ceases. Normal bleeding time is about 2.5 minutes.

Blind spot A portion of the retina insensitive to light. This region of the retina lacks photoreceptor cells and is that part of the retina where the optic nerve leaves.

Blood-brain barrier A combination of anatomical relationships and active transport systems that function to control the rate of formation and composition of the fluid produced from the plasma and bathing the brain.

BMR Basal metabolic rate.

Bohr effect The influence of changing the pCO_2 or pH or both on the combination of hemoglobin and oxygen. The higher the pCO_2, the lower the affinity of Hb for O_2.

Bolus A rounded mass of food ready for swallowing or a mass of food moving through the intestinal tract.

Bomb calorimeter A device constructed to capture all the heat liberated when a substance is burned within it. Essentially, a chamber within a chamber separated by a known volume of water. A bomb calorimeter is used to measure energy available from certain foods.

Bowman's capsule The essentially closed or blind end of the kidney tubule into which the glomerular capillaries intrude.

Bradycardia A slowing of the heart rate, for example, from 80 beats/minute to 60 beats/minute.

Brain stem Expanded portion of the spinal cord as it approaches the cerebrum. It includes the medulla oblongata, the pons varolii, and the midbrain.

Breast Either of two projecting milk-secreting organs on the anterior chest wall of a female. In males they are flat and have no function.

Bright's disease (glomerulonephritis) Inflammation of the glomerular capillaries in the kidney. Can be acute or chronic. Often follows a hemolytic streptococcal infection.

Bronchiole One of the narrower branches of the respiratory tree—about 1 mm or less. As the walls become thinner, bronchiolar subdivisions are recognized; *terminal* bronchioles function only to conduct gases and are then followed by *respiratory* bronchioles with thin-walled outpouchings called *alveoli.*

Bronchitis Inflammation of bronchi.

Bronchogram An x-ray photograph of the bronchial tree of the lung.

Brunner's glands Mucus-secreting glands located in the first part of the duodenum. Compound tubular in structure, they provide extra mucus at the site where pancreatic enzymes are injected into the small intestine.

Buccal glands   Small mucus-secreting glands of the oral cavity located in the mucous membrane covering the cheeks, hard and soft palate, and pharynx.

Buffer solution   A solution containing dissolved chemicals of such nature and in such combination that it will tend to resist any changes in its pH. In plasma, the combination of sodium bicarbonate and carbonic acid functions as a buffer.

Buffy coat   When blood is centrifuged in a tube, the red blood cells are packed in the bottom. Located on top of the packed rbcs is a thin, white, scumlike layer—the buffy coat. It consists of white blood cells and platelets.

Bulbourethral glands   Also called Cowper's glands. A pair of small structures on either side of the base of the penis. During sexual arousal in males, these glands secrete a clear, sticky fluid, that sometimes appears on the tip of the penis.

Bulk flow   The movement or transport of a substance when it is carried suspended or floating in a solution or in a gaseous mixture.

Bundle of His   A small band of conducting tissue located in the interventricular septum adjacent to the AV node. It relays stimuli down its bundle of branches to the apex of the heart to the distributing network of the Purkinje system.

Bursa of Fabricius   An epithelial outgrowth of the cloaca in chick embryos. It contains lymphoid tissue and, before atrophying, is responsible for programing the B-type lymphocyte. The B-cell lymphocyte designation used to describe the immunoglobulin-producing cell was obtained from the *bursa* designation.

Cachexia   A serious condition of general ill health and malnutrition. Term used preceded by some word describing origin or prime site, for example, pituitary cachexia, thyroid cachexia.

Calcitonin (thyrocalcitonin)   A polypeptide hormone synthesized and released from the thyroid gland in response to elevated blood levels of calcium. Causes a lowering of plasma calcium and phosphate levels and inhibits removal of calcium and phosphate from bone.

Calculus   An abnormal solidification of minerals—a stone. Most common sites for calculi are the gallbladder and kidneys.

Calorie (kilocalorie)   The unit of energy used in the study of metabolism. It equals 1000 small calories.

Calorimeter   An apparatus designed to measure the amount of heat given off by an individual.

Calyx   A cup-shaped cavity. In the kidney, the *minor calyces* are the spaces enclosing the pyramids and collecting the urine. Minor calyces deliver the urine into *major calyces* which in turn deliver urine into the renal pelvis.

C-AMP (cyclic AMP)   Cyclic 3',5'-adenosine monophosphate formed from ATP by an enzyme, adenylate cyclase, bound to the inner surface of the cell membrane. C-AMP serves as the intracellular messenger for many hormones.

Cancer   A cellular tumor that if left untreated is fatal. Cancer cells are invasive and can separate from their tissue of origin to take up residence in other regions or tissues of the body (i.e., metastasize).

Candidiasis   An infection with the fungus *Candida*. Usually a superficial infection of wet skin areas or mucous membranes.

Capillary   The smallest of the blood vessels. They are microscopic (7–10 μm) in size. Their walls are the thinnest possible—simple squamous endothelium. They exchange nutrients and oxygen for wastes and carbon dioxide in tissues.

Carbaminohemoglobin   One form by which carbon dioxide is carried in the blood. CO_2 combines with an amino group of the hemoglobin molecules to form the carbamino moiety.

Carbohydrate   A class of compounds, basically *carbon hydrates*—so named because their composition usually follows the formula $(CH_2O)_n$. Included as carbohydrates are the naturally occurring foods such as starches, sugars, glycogen, and celluloses.

Carbonic acid   A weak acid formed when carbon dioxide dissolves in water.

Carbonic anhydrase   An enzyme that catalyzes the reversible reaction of
$$CO_2 + H_2O \rightleftharpoons H_2CO_3$$
Carbon dioxide and water form carbonic acid.

Carbon monoxide (CO)   A colorless, poisonous gas that causes death by asphyxiation. It combines with hemoglobin with great avidity and thus deprives oxygen from combining with hemoglobin.

Carboxypeptidase   An enzyme that attacks proteins by splitting off amino acids one at a time from the C-terminal (carboxy) end.

Cardiac output   The volume of blood pumped from one ventricle of the heart (usually measured from the left ventricle) in one minute. Under normal resting conditions this is about 5 liters/minute.

Cardiopulmonary resuscitation (CPR)   A technique employed to restore life or consciousness to a person apparently dead or dying. It includes artificial respiration (mouth to mouth) and external cardiac massage. When done immediately upon an otherwise dead person, it is often able to revive or keep a person alive until medical measures can be instituted.

Caries (dental) The molecular breakdown (decay) of the calcified parts of teeth. Due to the action of microorganisms on carbohydrates, which is accompanied by decalcification of the tooth. This leads to breakdown of the organic portions.

Carnitine A molecule, $C_7H_{15}NO_3$, used within cells to transfer fatty acids across the inner mitochondrial membrane.

Carotid body Structure located at the bifurcation of the carotid artery. It is a receptor that responds to oxygen lack when it becomes life threatening. The receptor relays action potentials to the respiratory centers in the medulla.

Carotid sinus A dilated region of the internal carotid artery immediately above the bifurcation of the common carotid artery. In the sinus are pressure receptors that monitor blood pressure and relay action potentials to the cardioregulatory and vasomotor centers in the medulla.

Cast A positive copy, such as a mold of a hollow tube, formed from compacted cells, protein, microorganisms, or mineral salts and excreted from the body, for example, a urinary cast formed from precipitated protein.

Castration Removal of gonads, that is, the testes or ovaries, to render the individual incapable of reproduction.

Catabolism Biochemical reactions that break down large or complex substances into simpler compounds.

Catalyst Any substance that increases the rate of a chemical (or biochemical) reaction when it is present and it is not consumed in the net reaction.

Cataract An opacity of the lens of the eye.

Catecholamine An aromatic compound containing the aromatic catechol ring (six-member carbon ring with two neighboring OH groups). Catecholamines function as chemical mediators in the nervous system and include epinephrine, norepinephrine, and dopamine.

Catechol-O-methyl transferase (COMT) One of the enzymes that inactivates catecholamines (epinephrine, norepinephrine, DOPA). It operates by transferring the methyl group from methionine onto the OH group of the catecholamine.

Catheter A long, narrow, flexible tube used for introducing into or removing fluids from a body cavity.

Cation A positively charged ion (e.g., H^+, Na^+, K^+, Ca^{2+}).

Cecum A dilated pouchlike extension at the commencement of the large intestine.

Center A collection of neurons that operate together to regulate or perform one particular function.

Central fovea (of the retina) A shallow pitlike area of the retina composed only of cone cells that has the clearest vision.

Central nervous system (CNS) That portion of the entire nervous system contained within or enclosed by bone; the brain and spinal cord.

Centrifugal A direction away from the center (of rotation). Also used to describe a direction away from or out of the cerebral cortex.

Centrifuge A machine that spins materials at high rates of speed in order to separate components of different densities.

Centrioles Cylindrical organelles located within a cell that migrate to opposite ends of a dividing cell and serve to anchor the mitotic spindle fibers. Centrioles are composed of 27 microtubules arranged in nine parallel sets of triplets arranged to form a circle.

Centripetal A direction toward the center (of rotation). Also used to describe a direction toward or into the cerebral cortex.

Centromere The clear, constricted portion of a chromosome where the two chromatids are joined. It also serves as the point of attachment for the spindle fiber. The centromere location varies in different chromosomes.

Cephalic The head, head region, or toward the head end of the body.

Cephalic phase During the process of digestion, the cephalic phase is the period of time during which salivation and gastric secretion occur *without* any food having been eaten. It occurs because of conditioned reflexes triggered by the smell, sight, or thought of food.

Cerebellum A large portion of the brain divided into two hemispheres located in the posterior cranial fossa behind the brain stem. It functions to refine and coordinate muscular activity.

Cerebral cortex The 2–3 mm thick layer of cells covering the surface of the cerebrum. Referred to as the location for thinking, reasoning, etc. (i.e., the mind).

Cerebrospinal fluid (CSF) A fluid produced by the choroid plexus in the lateral and third ventricles. Similar to lymph, it fills the ventricles of the brain and the subarachnoid spaces surrounding the brain and spinal cord.

Cerebrum (cerebral hemispheres) The cerebrum, composed of two *hemispheres*, is the largest part of the central nervous system. It occupies the superior portion of the cranial cavity.

Cerumen A waxlike secretion located within the auditory canal of the ear.

Cervix A neck or necklike region. The lower, narrowing end of the uterus that connects the lumens of the uterus and vagina.

Chancre The primary sore of syphilis. A papular lesion that forms at the site of infection. Chancres are painless, hardened pimples.

Chemoreceptor A receptor or specialized nerve that is responsive to changes in the concentration of specific chemicals (e.g., taste buds, CO_2 receptors in medulla).

Chemosynthesis The production of carbohydrate molecules from CO_2 and H_2O utilizing energy from chemical reactions as the driving force. This is opposed to photosynthesis, which uses radiant energy to drive similar reactions in different organisms.

Chemotaxis The movement of an organism or cell toward (*positive* chemotaxis) or away from (*negative* chemotaxis) an increasing chemical concentration gradient.

Chief cells (zymogen cells) Cells located in the gastric glands of the stomach. They synthesize and secrete *pepsinogen*, the inactive form of the protein-digesting enzyme *pepsin*.

Cholecystokinin (CCK) A polypeptide hormone synthesized and secreted by mucosal cells in the upper small intestine. It causes contractions of the gallbladder and stimulates enzyme secretion by the pancreas. It is identical to pancreozymin—a name no longer used.

Cholera An infectious disease caused by the microorganism *Vibrio cholerae*. The bacterial toxin blocks sodium reabsorption and stimulates water and salt excretion. The resulting loss of water can cause shock and renal failure. Caused by drinking contaminated water.

Cholesterol Classified as a lipid, it is the most abundant steroid in animal tissues. It is located in cell membranes and is also used for the synthesis of steroid hormones and bile salts.

Cholinergic Describes those nerves that release acetylcholine from their terminations, that is, parasympathetic nerves.

Cholinesterase (acetylcholinesterase) The enzyme that catalyzes the breakdown of acetylcholine to choline and acetic acid. It is located in the postsynaptic membrane.

Chondrocyte A cartilage-maintaining cell.

Chondroitin sulfate A mucopolysaccharide secreted by cartilage-forming cells to form the matrix surrounding the cells.

Chorion The outermost cellular membrane enveloping the developing embryo.

Choroid A thin, dark brown, highly vascularized membrane of the eyeball. It encloses about 80 percent of the posterior portion of the eyeball. It is the layer immediately behind the retina.

Choroid plexus An extensive network of mostly capillary-size blood vessels that projects from the roots of the cerebral ventricles. The choroid plexus produces cerebrospinal fluid.

Chromaffin cells Cells that take up and stain intensely with chromium salt dyes. Such cells are present in adrenal medulla and sympathetic nerves.

Chromatid One of the pair of identical connected nucleoprotein strands. The two are joined at the centromere and separate during cell division. Each goes to an opposite end of the dividing cell and becomes a chromosome of one of the two daughter cells.

Chromatin The dispersed form of the chromosomes during the period of interphase. The nucleoprotein material forms an extensive network throughout the nucleus.

Chromophore Any chemical group that, when attached to a compound, produces a distinct color.

Chromosome In animal cells, located in the nucleus; it contains genetic information that can be utilized by transmission into the cytoplasm via molecules of RNA. Composed of a long strand of DNA and proteins. The number of chromosomes present in the nucleus is characteristic of a species. The human number of chromosomes is 46 per cell nucleus (23 pairs).

Chromotropic Describes an ability to alter the time or rate of an event, for example, changing heart rate.

Chvostek's sign A spasm or twitch of the facial muscles when they, or branches of the facial nerve, are tapped. Occurs in tetany.

Chylomicrons Very small fat droplets found in the intestinal lymphatics and blood following a fatty meal. Absorbed fats are transported in this fashion from the gut to their point of storage or metabolism.

Chyme Semifluid, creamy material produced by the gastric digestion of food.

Chymotrypsin A proteolytic enzyme produced by the pancreas in an inactive form, chymotrypsinogen. Breaks down protein molecules to shorter polypeptide chains.

Chymotrypsinogen An inactive precursor for the enzyme chymotrypsin. Produced in the pancreas and activated by the attack of trypsin.

Cilia Small hairlike processes projecting from the free surface of some cells. Cilia are usually found in large numbers on a cell surface, and they are able to wave or beat in a rhythmic pattern. The interior structure of the cilium consists of 9 pairs + 2 single microtubules (i.e., an axoneme).

Ciliary body Located at the anterior edges of the

choroid of the eyeball. It includes the ciliary smooth muscles, which function in *accommodation*.

Circadian rhythm A repeating cycle of about 24 hours. Applied to describe those events or phenomena that occur with respect to living organisms at about the same time every day.

Circulatory shock A condition that results from a sudden large loss of blood or plasma from circulation. The situation is serious because critical organs (e.g., brain) suffer from O_2 and nutrient lack. Signs include a drop in blood pressure, increase in heart rate, and anxiety.

Circumcision A minor surgical procedure that removes all or part of the prepuce or foreskin from the penis.

Cirrhosis A disease of the liver characterized by loss of normal cellular structure and function and a replacement with fibrous-type tissue. Most likely due to chronic excessive alcohol exposure.

Cisterna A compartment for storing or holding lymph or other body fluids.

Citric acid cycle (Krebs cycle or tricarboxylic acid cycle) A cyclic enzymatic sequence of biochemical reactions that is a final common pathway for the metabolism of carbohydrates, fatty acids, and some amino acids. Fed by acetyl CoA, it produces CO_2, reduced coenzymes, and ATP. The enzymes are dissolved in the matrix of the mitochondria where the resultant reduced coenzymes are oxidized by the eventual utilization of oxygen. ATP production is the ultimate goal.

Clearance The calculated volume of plasma required to account for the amount of substance removed from plasma.

Cleavage The rapid mitotic divisions following the fertilization of an ovum. The total size of the mass of tissue stays about the same, but the size of the resulting cells, *blastomeres*, becomes progressively smaller.

Climacteric A collection of events, physical changes, and feelings that occur to females at the time their reproductive period terminates. Can also be used to describe the changes in males as they experience decreased sexual activity.

Climax The peak period or moments of greatest intensity during sexual excitement.

Clitoris A small elongated body of erectile tissue in the female external genitalia; homologous with the penis in the male.

Clone A strain or a colony of cells or organisms that are derived from a single cell or individual by asexual reproduction.

Clot The end result of a series of biochemical reactions that changes liquid plasma into a gelatinous mass. Specifically, the conversion of fibrinogen into a tangle of polymerized fibrin molecules.

Clotting time The length of time taken for blood to clot after it is withdrawn from the body. The measurement follows standardized procedures involving size of tube, temperature of blood, and amount of tube filling. Normal clotting time by one procedure (Lee-White) is 9–11 minutes. Another method, employing small capillary tubes from which small fragments are broken, produces normal clotting times of between 3 and 5 minutes.

CNS Central nervous system.

Coagulate To form a clot.

Coccus A spherical bacterial cell, usually smaller than 1 μm in diameter.

Cochlea A spirally wound, fluid-filled tube that forms part of the inner ear. It contains the organ of Corti, the essential receptor for transducing sound.

Codon The triplet base sequence in a polynucleotide chain of DNA or mRNA. A codon codes for a specific amino acid to be inserted into a protein.

Coelom During embryonic development, the earliest mesoderm splits into two layers and the cavity so formed is the coelom. In the fully formed or adult individual, the coelom exists in the peritoneal cavity —the space between the viscera and abdominal walls.

Coenzyme An organic, but nonprotein, molecule required to combine with the apoenzyme (protein portion) to produce the active, complete holoenzyme. The coenzymes often serve to transfer atoms (e.g., H^+) or small groups from the enzymatic substrate to some other acceptor. Many coenzymes are derived from vitamins.

Coenzyme A (CoA) Chemically known as *pantothenic acid*. It functions as a carrier of *acyl* groups in enzymatic reactions. CoA is involved in many reactions: oxidation of pyruvate or fatty acids to acetyl CoA; oxidation of α-ketoglutarate to succinyl CoA.

Coenzyme Q A lipid-soluble molecule chemically called *ubiquinone*. A part of the inner mitochondrial membrane, it plays a key role in the carriage of hydrogen atoms in the electron transport system.

Coitus Sexual intercourse. Insertion of penis into the vagina.

Coitus interruptus Sexual intercourse (coitus) in which the penis is withdrawn from the vagina just prior to ejaculation; a widely used but unreliable method of contraception.

Collagen The main supportive protein of the connective tissues. It possesses great strength but no elastic property. When boiled, it is converted to gelatin.

Collaterals Small side branches, as seen in blood vessels and nerve distributions. Important channels that can enlarge to increase blood flow into a region where the main channel has been blocked by a thrombus.

Colloid A distribution of particles in a medium. The particles are larger than ordinary crystalloid molecules, but not large enough to settle out under gravity. They cannot filter through natural membranes (e.g., capillaries).

Colon The first portion of the large intestine. It extends from the cecum to the rectum. Subdivided in ascending, transverse, descending, and sigmoid colon.

Color blindness A colloquial and incorrect term that attempts to describe any deviation in the normal perception of colors. The defect is the result of the lack of one or more of the photopigments of the cone cells. The color discrimination will vary depending on which is missing.

Colostrum A thin, yellow, milky fluid secreted by the mammary glands a few days prior to or after delivery. Contains maternal immunoglobulins important for the newborn's protection.

Coma A deep, prolonged unconscious state from which a patient cannot be aroused.

Complement A group of interacting plasma proteins that is activated by certain antigen-antibody reactions. When triggered, they follow a specific sequence of reactions on the surface of a cell on which the antibody is fixed. The end result is that the membrane is punctured (i.e., lysed).

Compliance The ease with which a tissue stretches. *Surfactant* produced by the alveolar cells increases compliance.

Compression The act of pressing together. A reduction in volume.

Conditioned reflex An action learned by regular association of some physiological function with an unrelated outside event.

Condom A thin sheath, usually of rubber, worn over the penis during sexual intercourse to prevent impregnation or infection.

Conduction The transmission of certain kinds of energy—especially electrical energy as action potentials.

Conductor A material that easily permits the flow of electricity. A substance with very low resistance to current flow. The opposite of a resistor.

Cone A highly specialized light-sensitive cell located in the retina with its outer segment having a conical shape. Cone cells are used in color vision.

Congenital In existence at the time of birth. Innate, inherited.

Connective tissue One of the primary tissues characterized by few cells distributed in a lot of intercellular material. The intercellular material could be a fluid, a gel, a semisolid, or a solid, but it always contains fibers or has a potential for fiber formation.

Contraception The prevention of conception or impregnation. Can be done physically or chemically by preventing spermatozoa from reaching the ovum, by inactivating the spermatozoa, or by delaying the release of the ovum.

Contractility The property of being able to shorten in length when stimulated—as a muscle cell shortens when it is stimulated.

Contraction A shortening in length. A muscle contraction shortens and develops tension by pulling on some fixed part.

Contracture A long-standing muscle contraction that resists passive stretching.

Contralateral Referring to acting upon the opposite side.

Convection The transfer of heat in liquids or gases by the circulation or flow of the component molecules.

Convergence An anatomical relationship where terminations from many neurons synapse upon the same neuron. Also describes the inward movement of the two eyes toward fixation of the same near point.

Cornea The transparent epithelial covering of the anterior portion of the eye. It functions to help focus an image upon the retina.

Cornified An epithelium of the stratified squamous type in which the outer layers of cells have died and become cornified: keratinized.

Coronary Encircling, as a crown. Describes an arrangement of blood vessels, particularly in their distribution through the myocardium.

Coronary sinus The large vein that collects most of the blood that has nourished the heart muscle. It lies on the surface of the heart between the left atrium and ventricle and delivers its blood into the right atrium.

Corpus albicans A white fibrous patch that forms after the corpus luteum in the human ovary regresses.

Corpus luteum A yellow secretory gland in the ovary formed when a follicle has discharged its ovum. It secretes the female hormones, estrogen and progesterone.

Corpus spongiosum A column of erectile tissue that surrounds the urethra in the penis. Its anterior end is expanded into the caplike structure, the *glans penis*.

Cortex The outer layer or portions of an organ as opposed to the internal or more central regions (e.g., adrenal cortex, renal cortex, cerebral cortex).

Corticoids The hormones synthesized by the adrenal cortex. The term includes synthetic drugs with similar effects.

Corticosteroids Any of the steroid molecules synthesized by the adrenal cortex.

Corticotropin Also adrenocorticotropic hormone.

Cortisol One of the steroid hormones produced by the cells of the adrenal cortex. Usually referred to as *hydrocortisone*. A glucocorticoid hormone, it promotes glucose production from noncarbohydrate sources.

Cortisone A hormone produced by the adrenal cortex, but largely ineffective in humans until converted to cortisol. Often used as an anti-inflammatory agent in topical creams.

Countercurrent Current flow in adjacent parallel vessels in opposite directions.

Countercurrent exchange The end result of a countercurrent arrangement of capillary vessels in the medulla of the kidney. These parallel blood vessels remove salts from the interstitium at a rate that just balances their deposition by the countercurrent multiplier system. Therefore, this exchange system maintains the salt gradient originally built up by the countercurrent multiplier system.

Countercurrent multiplication The end result of the action of the long loops of Henle that extend into the medullary pyramid. Although only able to establish a transepithelial gradient of salt of 200 mOsm/liter, a lengthwise gradient of six times that value is set by the countercurrent flow arrangement.

Covalent bond A chemical linkage between two atoms in a molecule. The attractive force results from the sharing of electrons with each other. The more they share, the stronger the bond.

Creatine kinase An enzyme that utilizes ATP to phosphorylate creatine. The end products are ADP + phosphocreatine.

$$ATP + creatine \rightleftharpoons ADP + phosphocreatine$$

Creatine phosphate (phosphocreatine) An important compound used as a reservoir for phosphate-bond energy, particularly in skeletal muscle. Provides an immediate mechanism for ATP re-formation as the ATP is utilized in muscle activity.

$$ADP + PC \rightleftharpoons ATP + creatine$$

Creatinine A waste material derived from the dehydration of creatine. It is excreted in urine.

Cretinism A dwarfing condition accompanied by mental retardation; due to a congenital lack of thyroid hormone.

Crista A projecting, shelflike structure or arrangement of parts. The inner membrane of the mitochondrion is thrown into shelflike projections termed cristae.

Cross bridge One of many projections extending from each end of the thick myofilaments identified in the fibrils within a muscle cell. The cross bridge is the head portion of a myosin molecule. It functions by bonding to the thin myofilament and pulling the myofilaments apart from each other. It releases itself and repeats the pulling activity as long as ATP and Ca^{2+} are available.

Crossing-over The exchanging of chromosomal segments between homologous pairs during the first meiotic division. This produces new combinations of genes in each chromosome.

Cross-matching A test performed prior to blood transfusion to check for compatibility. Red blood cells from donor are mixed with a sample of recipient's serum and then rbc's from recipient are mixed with a sample of donor's serum. If no agglutination occurs, the bloods are deemed compatible.

Cross-reactivity When an immunoglobulin produced in response to the introduction of one antigen also reacts with another antigen. The degree of cross-reactivity varies depending on how good the fit is between the immunoglobulin and the antigen.

Cryptorchidism A developmental defect in which the testes fail to descend into the scrotum.

Cupula A small dome-shaped cap over some structure. The cupula over the semicircular canals of the inner ear.

Curare A toxic extract from a variety of plants. Originally used as arrow poison in South America. Curare attaches to the acetylcholine receptors at neuromuscular junctions, thus blocking the normal effects of acetylcholine. The result is a muscular paralysis.

Current The movement of electrical charges from one point to another through a conductor. The driving force is the potential difference or voltage. (When opposite charges are separated they tend to attract each other; thus the flow potential.)

Cushing's syndrome A collection of symptoms due to hypersecretion of the adrenal cortex. More common to women, the symptoms include: puffiness of the face, neck, and trunk; muscular weakness; dusky complexion with purple striations; amenorrhea; and pain in the abdomen.

Cutaneous Referring to the skin.

Cyanide (sodium cyanide, potassium cyanide) Any compound of cyanogen. Cyanide, used as a very toxic poison to exterminate rodents and other pests, is available as the sodium, potassium, mercuric, etc. salt. It operates by inactivating cytochrome oxidase, the oxygen-utilizing enzyme of the mitochondrion.

Respiration stops and the cells die, and the individual also dies if the dose is sufficient.

Cyanosis A bluish coloration of the skin and mucous membranes caused by an excessive accumulation of reduced hemoglobin in the blood.

Cyclic AMP 3′,5′ See c-AMP.

Cyclic GMP Cyclic 3′,5′-guanosine monophosphate, formed from GTP. It operates as an intracellular messenger to produce effects opposite to c-AMP.

Cyst A closed cavity, lined with epithelium and filled with a liquid or semisolid material. Also a stage in the life cycle of some parasites during which they are enclosed in a tough protective case.

Cystic duct The hollow tube that connects the gallbladder to the hepatic and common bile ducts. It delivers bile into and out of the gallbladder where the bile is stored.

Cystitis An inflammation of the urinary bladder. Can result from many causes, for example, bacterial, allergic.

Cytochromes A group of iron-containing proteins located on the inner mitochondrial membrane. They function to transport electrons originating from various dehydrogenating reactions toward molecular oxygen.

Cytokinesis The term used to describe the cytoplasmic changes that take place during mitosis and meiosis. Distinguished from karyokinesis, which is the term describing the nuclear changes that take place during those processes.

Cytology The discipline that studies the origin, structure, function, and pathology of cells.

Cytoplasm The living material of a cell exclusive of that within the nucleus. The contents of a cell contained between the nuclear and cellular membranes.

Cytosine One of the pyrimidine bases present in DNA and RNA.

Cytotoxin A cellular poison. A chemical that has a destructive effect upon cells of a particular organ. Toxins are named according to the cell type they attack (e.g., neurotoxin).

Dark adaptation The changes that take place within the eye to improve vision in the dark after one moves from a well-lighted region. Changes include dilation of the pupil as well as the regeneration of rhodopsin, the pigment of the rod cells.

Dead air The air that is inhaled but does not participate in the exchange of gases with the blood. It is the air contained in the upper respiratory tree where the conducting vessels are too thick to permit gaseous diffusion. Dead air volume is about 150 ml.

Deamination A chemical reaction in which an amino group ($-NH_2$) is removed from a molecule. Amino acids are deaminated to produce a *keto acid* and ammonia.

Decarboxylation A chemical reaction in which carbon dioxide (CO_2) is removed from a molecule. Keto acids are decarboxylated to yield molecules containing one less carbon atom.

Decerebrate To remove cerebral function by cutting across the brain stem at the level of the anterior colliculi.

Decompression sickness (the "bends") A condition characterized by joint pains and neurologic symptoms. Follows from a too-rapid reduction of environmental pressure or decompression. Occurs in people who have been breathing compressed air at great depths underwater. During decompression, nitrogen that was dissolved easily in body fluid comes out of solution as bubbles. The bubbles can form air emboli and occlude crucial blood vessels.

Defecation Evacuation or elimination of fecal material from the rectum.

Deglutition The act of swallowing.

Dehydrogenation The removal of hydrogen atoms from a molecule. They are usually removed in pairs and collected by a coenzyme.

Dendrite One of the many cytoplasmic processes of a nerve cell that undergoes extensive branching. It serves to *receive* stimuli from many synaptic connections and convey them *toward* the cell body.

Dental plaque A deposit of material onto the surface of teeth that can serve as a medium for bacterial growth.

Deoxycorticosterone A hormone synthesized and secreted by the adrenal cortex. It is a mineralocorticoid that has significant effects on water and electrolyte regulation. Not as powerful as aldosterone but does promote Na^+ reabsorption.

Deoxyhemoglobin (HHb) Reduced hemoglobin.

Deoxyribonucleic acid (DNA) A large double-helical molecule constructed from four different nucleotides. Encoded within the nucleotide sequences is genetic information that is exactly duplicated at the time of cellular division.

Depolarization Used in neurophysiology to describe the reduction of voltage across a cell membrane. Expressed as a movement toward less negative (more positive) voltages on the interior side of the cell membrane.

Dermatome An area of skin that is innervated by all the afferent nerve fibers making up one single posterior spinal root.

Dermis The deeper of the two divisions of skin that consists of loose connective tissue.

Desmosome A junctional complex between adjacent cells. Usually a small patch of dense material where membrane differentiation occurs. The opposing membranes are separated by a narrow space filled with an intercellular material. Fibers anchored in the patches extend out into the cytoplasm of their respective cells.

Dextrins Short polysaccharide chains usually formed from the breakdown of starch.

Diabetes insipidus A metabolic disorder characterized by a large production of dilute urine and great thirst. It is often accompanied by a great hunger, loss of strength, and a wasting away. Caused by a deficient quantity of antidiuretic hormone (ADH), usually caused by injury to the posterior pituitary gland.

Diabetes mellitus A metabolic disorder characterized by large volumes of glucose-containing urine. A greatly elevated blood sugar level is evident due to the lack of insulin release from the beta cells of the islet of Langerhans in the pancreas. This form of diabetes is accompanied by thirst, hunger, wasting away, and weakness. Metabolic disturbances include an increased breakdown of fat, ketosis, acidosis, lipemia, and ketonuria.

Diabetogenic Has the ability to produce symptoms of diabetes mellitus.

Dialysis A method for the separation of the crystalloids from the colloids in the same solution. The solution is pumped through a semipermeable tube surrounded or submerged in water. The crystalloids cross the membrane and enter the water, the colloids cannot and so are retained within the tube.

Diapedesis A method by which easily deformed motile cells squeeze through intact vessel walls. The cells send a narrow protoplasmic arm through an intercellular gap in the vascular wall. The rest of the cell streams through the arm to the other side of the wall until the cell has completely transversed the wall.

Diaphragm The thin sheetlike muscle that separates the thoracic and abdominal cavities. The diaphragm contracts and helps in drawing of air into the lungs.

Diarrhea An abnormal increase in the frequency and water content of the fecal discharge.

Diastole Unless otherwise indicated, refers to the period of the cardiac cycle when the *ventricles* are passive (i.e., not contracting).

Diastolic pressure The minimum blood pressure present in the large arteries. The minimum pressure corresponds to the time of ventricular relaxation. A normal diastolic pressure is about 80 mm Hg.

Dicumarol An anticoagulant originally extracted from spoiled sweet clover. Produces its effect by antagonizing vitamin K. A large number of more potent analogues are now available, but operate in a similar fashion.

Diencephalon (midbrain) That part of the forebrain between the telencephalon and mesencephalon. It contains the thalmus, hypothalmus, and neurohypophysis.

Diethylstilbestrol (DES) A synthetic compound with estrogenic properties. Its molecular structure is very different from the natural steroid hormones. Causes increased contractions of the female genital tract and may cause expulsion of the fertilized ovum or failure to implant. DES has recently been implicated as a cancer-producing drug.

Differential count The relative percentages of the five types of white blood cells. Determined by microscopic identification of 100 white blood cells.

Differentiation The progressive diversification of the cells and tissues of the embryo. The appearance of specialized structural and functional properties.

Diffusion The random movement of molecules or ions that ultimately produces a uniform distribution of all components. Net diffusion is always directed from a region of higher concentration to a region of lower concentration.

Diffusion potential The small voltage that results from the differential diffusion rates of oppositely charged ions. An unequal distribution of the ions results, with more cations in one region and more anions in another—hence the net charge separation and voltage.

Digestion The mechanical and chemical breakdown of food into molecular-sized pieces that allows absorption of the nutrients across the intestinal wall.

1,25-Dihydroxycholecalciferol A hormone that helps regulate metabolism of calcium and phosphate in the human body. This hormone is the most active known derivative of vitamin D. Synthesis begins in skin with aid of UV light from sun, moves into the liver, and is finally completed in the kidney. This hormone promotes absorption of calcium ions from the intestine into circulation.

Dimer A compound consisting of two identical subunits linked by a covalent bond.

Dipeptide Two amino acids covalently bonded to each other.

Diphtheria An acute infectious disease caused by a toxin-producing bacterium, *Corynebacterium diphtheriae*. Symptoms include fever, pain, and inflammation of cardiac and nervous tissues.

Diploid Having two complete complements of chromosomes; designated 2*n*. The normal complement of chromosomes in body cells is diploid. The mature sex cells have only one complete complement of chromosomes, designated *n* and referred to as haploid.

Diplopia Also called *double vision*. A condition where two images of a single object are perceived.

Disaccharide Sugar that is composed of two simpler sugars (monosaccharides), for example, sucrose, maltose, lactose.

Disease A condition of the body when some part or system is not functioning correctly. This can be due to heredity, diet, or environment.

Distal Farther from a predetermined point or plane of reference. In anatomy the plane of reference is a median line. The opposite of distal is *proximal*.

Distal convoluted tubule (DCT) Each of the one million tubules in a kidney is subdivided into five parts. The DCT is located between the loop of Henle and the collecting ducts. This portion of the tubule has some secretory and reabsorbing function.

Diuresis An increased urine production.

Divergence The anatomical arrangement of the axonal terminals as they terminate upon many other neurons. In this way action potentials from the presynaptic cell influence the activity of many others.

DNA Deoxyribonucleic acid.

Dominant Able to override the influence of the complementary gene on the homologous chromosome. The gene that *is* expressed.

DOPA (dihydroxyphenylalanine) The precursor of dopamine and an intermediate product in the biosynthetic pathway of norepinephrine, epinephrine, and melanin. It is used in the treatment of parkinsonism because it can cross the blood-brain barrier and dopamine cannot. DOPA is converted to dopamine in the brain.

Dopamine The immediate precursor of norepinephrine in the biosynthetic pathway to epinephrine. Dopamine is a neurotransmitter in the CNS.

Double bond The stronger linkage resulting between two atoms when *two* electrons are donated by each of the atoms and shared by both. Thus, a total of four electrons are shared between the atoms (e.g., $O{=}O$ represents O_2).

Down's syndrome An inherited defect due to an extra copy of chromosome 21 (i.e., trisomy). Symptoms include mental retardation; a small skull, flattened front to back; a short flat nose; short fingers; and a widened space between first two digits of the hand and foot.

Ductus arteriosus A fetal blood vessel connecting the pulmonary artery to the aorta. This connection diverts blood from flowing into the lungs directly into systematic circulation. Sometimes at birth this vessel does *not* close and must be surgically corrected soon after.

Ductus (vas) deferens The tube that delivers spermatozoa from the epididymis into the ejaculatory duct. It is this tube that is sectioned in a vasectomy.

Duodenum The first 25 cm of the small intestine. Commences immediately beyond the stomach and is continuous with the jejunum.

Dura mater The toughest, most fibrous, and outermost (from the nervous tissue) membrane covering the brain and spinal cord.

Dwarf An abnormally small person. Results from a deficiency of anterior pituitary human growth hormone during the growing years. The result is a miniature person, mentally normal and body parts in proportion. A deficiency of thyroid hormone during the growing years results in a cretin dwarf; mentally retarded and body parts disproportionate.

Dysentery An infectious disease characterized by inflammation and ulceration of the intestine—especially the colon—and accompanied by abdominal pain. Feces contain blood and mucus. Caused by chemical irritants, bacteria, protozoa, or parasitic worms.

Dysmenorrhea Painful menstruation.

ECF Extracellular fluid.

ECG Electrocardiogram.

Ectoderm The outermost of the three primary germ layers which gives rise to the nervous system, the epidermis of skin, and its derivatives.

Ectopic pregnancy Implantation of the fertilized ovum in a locus other than the uterus.

Edema An excessive volume of fluid present in the intercellular tissue spaces of the body. Most likely accumulation is in the subcutaneous tissue.

EEG Electroencephalogram.

Effector The responding component of a reflex. The component that brings about the required response.

Efferent arteriole In the circulation within the kidney, the vessel that carries blood *from* the glomerular capillaries and delivers it to the vasa recta.

Efferent neuron The nerve cell that carries action potentials away from the central nervous system to activate some effector organ or postganglionic neuron.

Ejaculation The ejection or expulsion of semen from the penis. A reflex response effected first by a series of rhythmic contractions of the smooth muscle in the vas deferens, seminal vesicles, ejaculatory duct, and prostate gland followed by vigorous contractions of the skeletal muscles surrounding the root of the penis.

Elasticity The property of being able to stretch and recoil to an original length.

Elastin   A yellow protein that is the essential component of elastic-type connective tissue. It is naturally flexible and elastic.

Electrocardiogram (ECG, EKG)   The graphic recording of the electrical activity occurring in the heart muscle. Standardized procedures for obtaining the recording are used worldwide.

Electrode   A device connected, usually to a wire, that serves to make contact with the object from which or to which electric charges are removed or added.

Electroencephalogram (EEG)   A graphic recording of the electrical activity created by the nerve cells of the brain. Electrodes are secured to the subject's scalp in standardized locations and the voltages are recorded for several minutes from different parts of the scalp.

Electrolyte   Any chemical that dissociates into charged particles (ions) when in solution and thus is able to conduct electricity.

Electron   The smallest unit of negative charge. It weighs 9×10^{-28} g, and is the orbiting component of the atom. When electrons flow they constitute electrical current. When they are ejected from a radioactive substance they constitute the beta rays. The numbers and orbitals of the electrons in an atom determine the atom's chemical and physical activity.

Electron microscope   An instrument that projects on a screen the magnified images of cellular and molecular components. Beams of electrons are focused upon the object. Those electrons passing through the object are collected on a fluorescent screen providing a negative image of the object. The electron microscope is capable of such great resolution and magnification because the electron beams have very short wavelength properties.

Electron transport system   A series of coenzymes and proteins that are integral components of a membrane (inner mitochondrial, thylakoid, or cell) that transfer electrons down a reducing gradient toward a strong oxidizing agent, oxygen.

Embden-Meyerhof pathway   Glycolysis.

Embolism   The sudden blockage of an artery by a clot or air bubble, or foreign material that was carried to the site of blockage by the blood flow.

Embolus   A blood clot carried by the blood flow. The embolus will eventually lodge in and plug a smaller branch of a blood vessel.

Embryo   The developing human during the period from two weeks after fertilization to the end of the seventh or eighth week. After this period it is called a *fetus*.

Emmetropia   When the rays of light entering the eye are focused precisely on the retina. Not nearsighted or farsighted.

Emphysema   An abnormal enlargement usually of the alveoli of lungs either by destruction of the wall separating alveoli or by a great stretching of the wall. The end result is an overall *reduction* in the surface area for gaseous exchange.

Emulsification   A process whereby two completely immiscible fluids are mixed so that one of them is dispersed throughout the other as minute globules. Fats are emulsified in the intestinal secretion with the aid of bile salts.

Endergonic   Describes a requirement for energy input. Certain chemical reactions will take place only if heat (energy) is supplied.

Endocardium   The slick, shiny epithelial (endothelial) lining of the interior chambers of the heart.

Endocrine gland   A ductless-type gland that secretes its product, a hormone, into the circulation. In this way the gland is able to influence the activity of cells or an organ far from the gland.

Endocytosis   The uptake into a cell of substances that are unable to penetrate the cell membrane. The term includes both phagocytosis and pinocytosis, processes in which the membrane first invaginates and then pinches off to enclose some of the surrounding medium.

Endoderm   The innermost of the three primary germ layers of the developing embryo. The endoderm gives rise to the digestive tract, the urinary bladder and urethra, and the respiratory tract.

Endolymph   The fluid within the membranous labyrinth of the inner ear.

Endometriosis   A condition in which tissue identical to the endometrial lining of the uterus occurs in abnormal locations in the pelvic cavity (e.g., ovary, urinary bladder).

Endometrium   The mucous membrane lining the cavity of the uterus. Its thickness and structure change progressively throughout the menstrual cycle.

Endomysium   The connective tissue layer that surrounds each muscle fiber and thus separates each fiber from the adjacent ones.

Endoneurium   The connective tissue layer that surrounds each nerve fiber in a peripheral nerve and thus separates each nerve fiber from the adjacent ones.

Endoplasmic reticulum (ER)   A system of interconnecting membranous tubules, flattened sacs, and vesicles that are dispersed throughout the interior of the cell. *Rough ER* is studded with ribosomes on the cytoplasmic surface. *Smooth ER* lacks the ribosomes.

Endorphins A group of small proteins produced by nerve cells of the brain that are very powerful opiates—painkillers.

Endoscope A thin, tubular instrument that can be inserted into a hollow cavity or tube for the purpose of examining its interior. Some endoscopes are equipped with lights and are able to photograph the interior of organs (e.g., stomach, urinary bladder).

Endothelium The thin, shiny epithelium that lines the interior of all blood and lymphatic vessels. It is continuous into the chambers of the heart where it is called the *endocardium*.

End-plate potential The localized, nonpropagated, slight depolarization that occurs on a muscle cell membrane when its motor neuron terminal releases its acetylcholine onto the end plate of the muscle cell.

Energy The capacity to do work. Energy can be converted from one form into another but cannot be created or destroyed (e.g., mechanical energy into heat).

Enkephalin A protein neurotransmitter found in neurons of the brain. Suspected to be released from those nerve terminals affected by powerful pain-killing drugs—opiates such as morphine.

Enterokinase An enzyme of the upper small intestinal mucosa that activates trypsinogen into trypsin.

Environment The sum total of all surrounding things that act upon or influence in any way the development of an individual.

Enzyme A biological catalyst. A protein (or mostly) molecule that is specific for accelerating the change in some particular substrate without undergoing any net change of its own.

Eosinophil One of the least numerous types of white blood cells. Classed as a polymorphonuclear granulocytic leucocyte, it has a bilobed nucleus and a cytoplasm containing large round granules that stain with acid dyes.

Ependymal cell A cell type that forms the lining membrane of the ventricles of the brain and central canal of the spinal cord.

Epididymis A collection of coiled tubules forming a portion of the male reproductive system. Arises from each *testis* and then becomes continuous with the *vas deferens*. Functions to store spermatozoa.

Epilepsy Chronic but irregular disturbances of brain function that are evidenced as a temporary loss of consciousness up to a generalized attack of violent convulsions.

Epimysium The connective tissue sheath that wraps an entire muscle.

Epinephrine One of the two catecholamines synthe-

sized and secreted by the adrenal medulla. It is a powerful vasopressor, increases blood pressure, stimulates cardiac muscle, causes glycogenolysis. Also called adrenaline (Great Britain).

Epineurium The connective tissue sheath that wraps a peripheral nerve.

EPSP Excitatory postsynaptic potential.

ER Endoplasmic reticulum.

Erection The engorged and stiff state of the penis (or clitoris) resulting from the engorgement of the spongy erectile tissue with blood.

Erogenous zones Those parts of the body which, when excited, produce sexual arousal (e.g., genitals, lips, breasts).

Erotic Arouses or describes sexual desire.

Erythroblastosis fetalis A hemolytic anemia of a new-born child that results from the destruction of the infant's rbc's by antibodies produced by the mother. Usually the antibodies are due to an *Rh blood type incompatibility*. The mother is an Rh negative type, and the fetus is Rh positive. Antigens enter the maternal blood and the mother produces immunoglobulins that are able to cross the placenta and attack the fetal red blood cells.

Erythrocyte A red blood cell. Produced in the bone marrow, their function is to transport oxygen to all body tissues.

Erythropoiesis The production of erythrocytes (red blood cells).

Erythropoietin A glycoprotein that is the blood-borne messenger to the bone marrow that stimulates red blood cell production. Major source is the kidney.

Essential amino acids Those 8–10 amino acids that cannot be synthesized by the human body at an adequate rate to meet its needs. Therefore, they must be obtained from the diet.

Estradiol The most potent of the naturally synthesized estrogens. It is produced by the ovaries.

Estriol A naturally produced estrogen that has only weak estrogenic activity. It is a metabolic product of estradiol and estrone that is excreted in the urine in relatively high concentrations.

Estrogens A group of female steroid sex hormones that are synthesized by the ovary and placenta and to a lesser extent in the adrenal cortex. They are responsible for the development of the female secondary sex characteristics.

Estrone One of the estrogens. Found in the urine of pregnant women. Also prepared synthetically and administered to postmenopausal women.

Euchromatin Those portions of the genetic material (i.e., the chromosomes) that are seen, by electron microscopy, to be dispersed or uncoiled portions of a chromosome. It is believed these are sites of active transcription. The opposite term, heterochromatin, describes the tightly coiled portions.

Eupnea Easy, normal, quiet breathing.

Evagination An outgrowth or outpouching of some part or layer. An outpocketing.

Excitability The readiness to respond to a stimulus. The more excitable the state, the easier or the lesser the strength of stimulation required to obtain a response.

Excitatory postsynaptic potential (EPSP) The slight decrease in negative voltage seen on the postsynaptic membrane when it is stimulated by a presynaptic terminal. The EPSP is a localized event that decreases in strength from the point of excitation.

Excretion The ridding or elimination of wastes by discharging them from the body.

Exergonic Describes a release of energy. When certain chemical reactions take place, they release energy (e.g., heat, light).

Exhaustion A state of inability to respond to stimuli from lack of energy; extremely weak or greatly fatigued. *Heat exhaustion* is an effect of excessive exposure to heat. Signs include *subnormal* temperature, dizziness, headache, nausea. Sometimes called sunstroke.

Exocrine gland Any cell or collection of cells that discharges its secretion through a duct that opens out onto a free surface.

Exocytosis A process of discharging cellular products too big to go through the membrane. Particles for export are enclosed by Golgi membranes when they are synthesized. Vesicles pinch off from the Golgi and carry the enclosed particles to the interior surface of the cell membrane where the vesicle membrane and cell membrane fuse and the contents of the vesicle are discharged.

Exophthalmos A marked protrusion of the eyeballs.

Expiration Exhalation. The breathing out and expelling of air from the lungs into the atmosphere.

Expiratory reserve volume The milliliters of air that can still be exhaled by a maximal effort *after a normal expiration.*

Exponential A mathematical expression (or its graph) that includes numbers or expressions raised to a power.

Exteroceptor A sensory device (receptor) that responds to stimuli that have their origin outside the body (e.g., eyes, ears).

Extracellular Outside of or surrounding cells.

Extracellular fluid (ECF) compartment A term that includes all the body fluids *not* contained within a cell. It includes interstitial fluid, cerebrospinal fluid, blood, lymph, and glandular secretions. It makes up about 40 percent of total body water.

Extrafusal fibers Skeletal muscle cells. However, when describing the muscle spindles that regulate muscle tension, it is necessary to distinguish between the specialized muscle cells of the spindle apparatus as *intrafusal* and the surrounding skeletal muscle cells as *extrafusal fibers.*

Extrinsic clotting path The sequence of biochemical reactions responsible for the production of *prothrombin activator* that is triggered by *thromboplastin* released from injured tissue.

Facilitated diffusion (mediated diffusion) Movement of a molecule across a membrane with the aid of a carrier. The driving force for movement is always down the concentration gradient.

Facilitation The state when a nerve cell membrane is partially depolarized so that an otherwise *less* than normal strength of stimulus can now depolarize the membrane the rest of the way to reach the threshold.

F-actin The major protein component of the thin myofilaments of the muscle fibril. Constructed of G-actin subunits, which polymerize with the aid of Ca^{2+} into the long chains. Two F-actin proteins then twist around each other to form the backbone of the thin myofilament.

Facultative water reabsorption The removal of water from the collecting ducts of the kidney. The volume removed varies due to the amount of ADH in circulation. The volume reabsorbed is in response to the body's need.

FAD Flavin adenine dinucleotide.

False labor Nonproductive pains, resembling the labor pains, which do *not* produce stretching and dilation of the cervix.

Fasciculus A small bundle or cluster. Describes a cluster of muscle or nerve fibers wrapped in a fibrous sheath.

Fat Adipose tissue that is present in the form of soft pads between various organs. Smooths and rounds body contours, and is the most compact form of energy storage. Formed from one molecule of glycerol and three molecules of fatty acids.

Fatty acid An organic acid consisting of CH_2 (methylene) groups bonded together to form chains of 14–22 carbons (16 and 18 are most common), and a carboxyl group (COOH) at one end.

Fe^{2+} Ferrous iron.

Fe^{3+} Ferric iron.

Feces The waste material discharged from the rectum during defecation. It consists of bacteria, undigested food, water, and various secretions (e.g., bile from liver).

Ferritin A complex of the protein apoferritin and iron. In this combined form, iron is stored in the liver and mucosal cells of the upper intestine.

Fertilization The fusion of the spermatozoan and ovum. This usually takes place in the uterine tubes of the female.

Fetus The unborn organism still within the uterus of its mother. The period of development from about the seventh or eighth week until birth.

Fever An elevation of body temperature above its normal value of about 37°C.

Fiber An elongated threadlike strand. It may refer to a whole cell, as a muscle *fiber,* or to a portion of a cell, as in the axon of a nerve cell, which is the nerve *fiber.* Collagen *fibers* are the extracellular aggregations of cellular secretions.

Fibril A subcomponent of a fiber. *Myofibrils* are within the muscle cell (fiber). Collagen fibrils are twisted to make up the collagen fiber.

Fibrillation Small, involuntary, uncoordinated contractions of individual muscle cells. Ventricular fibrillation is characterized by a random, unsynchronized series of muscle cell contractions that are repeated and self-stimulating. A dangerous situation because the effective pumping of the ventricle is greatly reduced.

Fibrin The insoluble protein that results from the action of thrombin on fibrinogen in the clotting process. Fibrin threads tangle and adhere to each other and to roughened surfaces (e.g., cut or torn blood vessels). They entrap cells and "dam up" the leaking blood vessel.

Fibrinogen A soluble high-molecular-weight protein synthesized in the liver. It is converted to a fibrin monomer by the action of thrombin during the clotting of blood.

Fibrinolysis The breakdown or dissolution of fibrin by enzyme action.

Fibroblast A multipotent cell that is present in connective tissue. It is a flat cell with cytoplasmic processes that produces collagen fibers. Fibroblasts can differentiate into chondroblast (cartilage-producing) and osteoblast (bone-forming) cells.

Filtration The passage of fluid and only certain dissolved materials through a membrane due to a hydrostatic pressure gradient. Plasma is filtered through the glomerular capillary walls into Bowman's capsule. Large molecules, such as proteins, and cells cannot pass through the capillary walls.

Filtration fraction The percent of the plasma that enters the kidney (renal plasma flow—RPF) that is filtered in Bowman's capsule (glomerular filtration rate). The value is about 120/700 or 17 percent.

Fimbriae Fingerlike extensions, or a fringe along a free border or edge. At the ovarian end of the uterine tube, fingerlike processes may help guide the released ovum into the opening of the uterine tube.

Final common path The motor neuron that activates a skeletal muscle cell. It is the only route of stimulation for a muscle cell.

Fissure A general term that describes a groove, cleft, or deep fold in a tissue.

Flagellum A long hairlike extension of a cell that is used for propulsion of the cell. It is a major component of spermatozoa.

Flavin adenine dinucleotide (FAD) A hydrogen-carrying coenzyme, important in oxidation-reduction types of reactions inside cells.

Floating kidney (renal ptosis) A kidney dislodged from its normal location. Normally, kidneys are held in place only by fat masses and the peritoneum. If a kidney were to slip greatly out of position, the ureter may fold partially or completely and thus hinder or completely block the passage of urine.

Fluoride The anion of fluorine. Generally, it refers to *stannous fluoride,* SnF_2, which is applied topically to teeth to aid in preventing caries.

Focal length (distance) The distance from a lens on the other side from the incoming parallel rays where the rays come to focus.

Follicle A small sac, pouch, or cavity. Within the ovary, the ova are contained in follicles.

Follicle-stimulating hormone One of the protein hormones produced by the anterior pituitary gland. In the female, it promotes the development and maturation of an ovarian follicle. In the male, it promotes the production of spermatozoa within the seminiferous tubules.

Foreplay A prelude or preparation for sexual intercourse. It consists of a variety of activities intent upon sexual arousal.

Fourth ventricle This is an irregularly shaped cavity located between the medulla oblongata and pons varolii of the brain stem. This cavity is continuous below with the central canal of the spinal cord and above via a canal with the third ventricle. The ventricles and central canal are filled with cerebrospinal fluid.

Fructose A sweet-tasting six-carbon sugar present in honey and many sweet fruits. Fructose and glucose are obtained from the hydrolysis of sucrose (ordinary table sugar).

FSH Follicle-stimulating hormone.

Fundus A general term for the bottom portion of a

hollow organ or that portion farthest from the mouth of the organ.,

Fungiform Shaped like a fungus or mushroom. A wide cap supported by a slender stalk.

Fusimotor fibers Also called the *gamma efferents*. These are nerve fibers that activate the intrafusal fibers of the muscle spindles.

G-actin A muscle protein with a molecular weight of 46,000. In the presence of ATP and Ca^{2+}, it polymerizes to form F-actin.

Galactose A six-carbon sugar usually combined with glucose to form the disaccharide sugar lactose (sugar of milk).

Gallbladder A small pouch that stores bile. It is located under the liver and is filled and emptied of bile via the cystic duct.

Gallstone A solid precipitate that forms and grows by accretion. Usually consists of cholesterol and is formed in the gallbladder or bile duct.

Gamete The mature sex or reproductive cell (e.g., ovum or spermatozoan).

Gammaaminobutyric acid (GABA) A neurotransmitter substance isolated from the brain that has inhibitory effects on certain cells (IPSP).

Gamma efferents The motor nerve fibers that innervate the intrafusal muscle fibers of the muscle spindle.

Gamma globulin A class of protein present in plasma that functions as immunoglobulins (antibodies).

Ganglion A swelling located on a nerve that results from the aggregation of the nerve cell bodies located outside of the central nervous system.

Gangrene Death of a considerable mass of tissue that usually is caused by a loss of blood. It is followed by bacterial invasion and decay or rotting of the tissue.

Gastric Refers to the stomach.

Gastric phase The digestion of food by the enzymes and acid of the stomach that have been triggered by local hormonal reflexes. The presence of food in the stomach causes cells of the stomach to release the hormone *gastrin*, which then stimulates the stomach to secrete.

Gastrin A polypeptide hormone released from the mucosal cells located in the pyloric end of the stomach. The hormone is released by the presence of food in the stomach and acts mainly on the secretory cells of the fundic and body portions.

Gene Classically considered to be the unit of heredity. A nucleotide sequence of DNA that codes for the amino acid sequence of a protein. Other nucleotide sequences (genes) serve to control other regions or genes of the DNA.

Generator potential The graded depolarization that occurs in the region of an afferent neuron adjacent to a receptor cell.

Genetics The study of heredity.

Genitalia The reproductive organs; generally refers to the external structures.

Genotype Describes the entire genetic content of an individual.

Germ cells The primitive cells from which a more complex or specialized cell, tissue, organ, or part develops. Germ cells can give rise to the mature ovum or spermatocyte.

Gestation The time of fetal development from the time of fertilization until delivery.

GFR Glomerular filtration rate.

Giantism Excessive size and stature due to hypersecretion of human growth hormone from the anterior pituitary gland.

Glans penis The cone-shaped cap of erectile tissue at the end of the penis.

Glaucoma An eye disease characterized by increased pressure within the eye. The increased pressure acting on the exciting fibers of the optic nerve causes defective vision.

Glial cells (neuroglia) The *non*neuronal cells of the nervous system that function to support, insulate, nourish, and generally maintain the neurons.

Glomerular filtration rate (GFR) The total volume of fluid that enters all the Bowman's capsules of the kidneys *in one minute*. The value is about 110–120 ml/min.

Glomerulus One of the minute clusters of capillaries in the kidney surrounded and enclosed by Bowman's capsule of the nephron. It is these capillaries that filter plasma fluid into the kidney tubule.

Glucagon A polypeptide hormone secreted by the alpha cells of the islets of Langerhans in response to a falling blood sugar level. Acts primarily on liver cells to cause glycogenolysis.

Glucocorticoids A group of steroid hormones secreted by the adrenal cortex that function to raise blood sugar by drawing from carbohydrate stores or converting fats and proteins into glucose.

Gluconeogenesis The production of glucose from noncarbohydrate sources, for example, from fats or protein.

Glucose A six-carbon sugar, $C_6H_{12}O_6$. It is *the* major energy source for every cell type in the body. Its metabolism is possible by every known living cell for the production of ATP.

Glutamine One of the 20 naturally occurring amino acids. It is deaminated by the kidney cells to yield glutamic acid and ammonia.

Glycerol A simple three-carbon compound with one

OH group on each carbon. It forms the backbone for synthesis of a fat molecule by esterifying one fatty acid to each OH group, thus forming a triglyceride (simple, neutral fat).

Glycine One of the simplest of the 20 naturally occurring amino acids. It may function as an inhibitory neurotransmitter.

Glycocalyx The filamentous or fuzzy coat covering the outside of certain cells. It has adhesive properties, and the major components of the glycocalyx include *glucosphingolipids,* acid *micropolysaccharides,* and *glycoproteins.*

Glycogen A highly branched polymer of glucose containing thousands of subunits. Functions as a compact store of glucose molecules in liver and muscle cells.

Glycogenesis The process that polymerizes glucose to form the large, branching molecule of *glycogen.* Glycogen formation.

Glycogenolysis The process that systematically removes glucose molecules from glycogen, releasing them for further metabolism. Breakdown of glycogen (into glucose monomers).

Glycolysis The sequence of enzymatic reactions that breaks down a molecule of glucose into two molecules of lactic acid. The process yields a net of two molecules.

Glycosuria The presence of an excessive amount of glucose in the urine (more than 1 g/24 hours).

Goiter An enlargement of the thyroid gland visible as a swelling in the front of the neck. It may be a result of hyposecretion or hypersecretion by the gland.

Golgi complex A short stack of flattened sacculated membranes located within cells usually near the nucleus. Some of the newly synthesized proteins are finished in the Golgi by the addition of carbohydrates and packaged in vesicles pinched off from the ends of the membranes.

Golgi tendon organ Stretch receptor located in a muscle tendon; arranged in series with the muscle, it responds when the tendon is pulled either by its own muscle or by passive stretching.

Gonad The organ that produces the sex cells (i.e., the ovary or testes).

Gonorrhea A bacterial infection obtained through sexual intercourse. Symptoms are obvious in the male—pain and inflamed urethra. Caused by the bacterium *Neisseria gonorrhoeae.*

Gout An inherited form of arthritis accompanied by an excess of uric acid in plasma. Peripheral joints are mainly affected (e.g., big toe).

Graafian follicle The aggregation of cells arranged into a hollow sphere containing the ovum and about to shed the ovum from the surface of the ovary.

Gradient The *rate* of rise or fall; the slope or incline; steepness.

Granulocyte A cell containing granules. Usually describes a subgroup of leucocytes (i.e., the neutrophils, eosinophils, basophils).

Gray matter That part of the central nervous system that has a gray color in its natural state. Microscopically, it consists mainly of cell bodies and the unmyelinated parts of nerve cells.

Guanine One of the purine bases present in DNA and RNA.

Gustation The sensation of taste. Tasting.

Gyrus One of the elevations (upliftings) of the surface of the brain caused by infolding of the cortex.

H^+ Hydrogen ion, proton.

H zone One of the transverse bands seen by microscopic examination of muscle cells. Of intermediate intensity, its width corresponds to the distance between the ends of the thin myofilaments.

Hageman factor (factor XII) A plasma protein that initiates the intrinsic clotting pathway. When activated by surface contact, it combines with other trace plasma proteins to produce prothrombin activator.

Hair cells The primary transducing cells for hearing. Located in the organ of Corti, they are activated when tiny hairs protruding from their surface are bent or sheared by the relative movement of the *tectorial membrane.*

Hallucination A sensory experience of something that does not really exist in the world outside of the mind, that is, created from within the brain.

Haploid Containing a single set of chromosomes (*n*) and designated in the number carried in the sex cells. In the human this is 23.

Hapten A substance that by itself is not antigenic, but when bonded to a protein, does elicit production of antibodies directed against it. When the pure substance is later introduced to the antibodies, a reaction between them occurs.

Haustra The sacculations or outpouchings that occur regularly along the length of the colon.

Hay fever A seasonal attack by the pollen of certain plants that triggers an allergic response. Individuals are sensitive to only certain pollens. Symptoms include inflammation of the eyes, tearing, itching, sneezing, and runny nose.

Hb Hemoglobin.

HbO$_2$ (oxygenated hemoglobin) Oxyhemoglobin.

HCG Human chorionic gonadotropin.

Heart sounds Sounds produced by the activity of the

heart. They can be heard with a stethoscope placed on the chest. Generally two distinct sounds are heard; the first is a low-pitched, vibratory sound (lubb) and the second is a higher-pitched, shorter snap (dupp).

Heimlich maneuver   A first aid procedure for choking. Employs a quick upward thrust against the diaphragm that forces air out of the lungs with sufficient force to eject any lodged material.

Hematocrit   An expression of the percentage of the volume occupied by blood cells. Also called the *packed cell volume,* pcv. Usually calculated by centrifuging a blood sample in a graduated tube and then reading off the volumes of cells and total blood.

Hematoma   Blood that has escaped from a blood vessel and collected into a more or less confined area.

Hematuria   Blood in the urine.

Heme   The iron-containing pigmented portion of hemoglobin. One heme molecule is bonded to each of the four polypeptide chains that constitute hemoglobin. Therefore, there are four hemes and thus four iron atoms in every hemoglobin molecule.

Hemocytoblast   The stem cell that, developing along different lines, gives rise to all the different cells of the blood.

Hemodialysis   A technique for the removal of certain materials, usually wastes or poisons, from the blood by delivering the blood into a semipermeable tube through which the materials diffuse down their concentration gradient into a surrounding solution.

Hemodynamics   The study of factors and forces that govern the flow of blood through blood vessels.

Hemoglobin (Hb)   A conjugated protein carried within the red blood cells that is able to combine reversibly with O_2. Consists of four polypeptide chains (α, β, α, β) each of which is bonded to a heme pigment molecule that contains an iron atom.

Hemolysis   The escape of hemoglobin from the interior of the red blood cell into the surrounding medium. Can be a result of toxins or drugs, freezing or thawing, or hypotonic solutions disrupting the integrity of the rbc membranes.

Hemophilia   An hereditary disorder of the intrinsic clotting system that results in spontaneous subcutaneous and intramuscular hemorrhages.

Hemopoiesis (hematopoiesis)   The formation and development of the blood cells.

Hemorrhage   Bleeding. The escape of blood from blood vessels, especially when it is profuse.

Hemorrhoids   The bulging or dilation of the veins of the hemorrhoidal plexus, which is located in the rectum and/or anus. They result from a continuous or prolonged period of increased venous blood pressure.

Hemosiderin   An insoluble form of stored iron in which the aggregates of ferric hydroxide can be seen easily inside cells with a microscope.

Hemostasis   The prevention of blood loss. Can be by normal physiological processes or by surgical means.

Heparin   An anticoagulant found most abundantly in lungs and liver. A mucopolysaccharide extracted from animal tissues that is prepared as sodium heparin and used in prevention and treatment of thrombosis.

Hepatic   Refers to the liver (Greek *hepar* = liver).

Hepatic portal circulation   The distribution of blood flow from the hepatic portal vein. The hepatic portal vein collects the blood flow away from the digestive tract and delivers it, through successively smaller branches, into the sinusoids, which permeate the liver.

Hering-Breuer reflex   The cyclic activity of the nerve fibers activating the muscles of inspiration. Stretch receptors located in the visceral pleura, when activated, send increasing numbers of inhibitory impulses to the inspiratory center in the medulla oblongata.

Hernia   The protrusion of a portion of an organ or tissue into an abnormal location. Commonly seen when a loop of small intestine passes into the inguinal canal.

Herpes   Any inflammatory skin disease demonstrating small clusters of vesicles. Usually restricted to *herpes simplex,* which is an acute viral infection causing vesicles around the edges of the mouth and nose (commonly called *cold sores*).

Hertz (Hz; c.p.s.)   The unit used to measure wave frequency, as in sound. One hertz equals 1 cycle/second.

Heterochromatin   Genetic material when it is visible microscopically as dark staining and tightly coiled. In this condition, the genetic material is inactive or unavailable for transcription.

Heterotrophic   Refers to those organisms that require organic molecules (reduced forms of carbon) for energy and synthesis.

Hexose monophosphate shunt (HMPS)   An alternate pathway for the metabolism of glucose that produces the high-energy coenzyme $NADPH_2$ and a variety of different-sized carbon intermediates.

HGH   Human growth hormone.

HHb   Reduced hemoglobin.

Hilum   A general term that identifies an indented portion of an organ where blood vessels and nerves enter and leave. Also called a hilus.

Hirsutism   An abnormal growth of body and facial hair, especially in women.

Histamine   A normal substance found in most tissues but especially *mast cells* and *platelets*. It is released from injured cells and causes bronchiole constriction and a general vasodilation that can cause a dangerous drop in blood pressure.

Histocompatibility antigens   Those protein cell markers that distinguish one individual from another. Of great importance in tissue and organ transplantations.

Histone   A small basic (positively charged) protein attached to the phosphate group of DNA to form chromatin. Histones may function as gene repressors.

HMPS   Hexose monophosphate shunt.

Homeostasis   The tendency to remain the same, unchanging. The fundamental principle of physiology that describes the mechanisms that keep the internal environment of the body constant so that all cells are able to function optimally.

Homologue   An organ, or part, similar in structure, origin, function, or position to another organ.

Homozygous   Possessing a pair of identical alleles for a particular hereditary characteristic.

Hormone   A chemical substance synthesized by an endocrine gland secreted into and transported by the blood to all parts of the body. However, only certain specific targets (cells, tissues, or organ) are affected and respond.

Human chorionic gonadotropin (HCG)   A hormone synthesized and secreted by the trophoblastic cells of the placenta. It acts on the corpus luteum of the ovary so that it continues to secrete estrogen and progesterone; especially important during the first trimester of pregnancy.

Human growth hormone (HGH)   A protein synthesized by the anterior pituitary gland that promotes growth. It primarily promotes protein synthesis and fat utilization.

Humoral   Refers to fluids in the body (e.g., plasma, lymph) or to reactive molecules produced in the body (antibodies, neurotransmitters). Can also refer to fluids of the eyeball—aqueous and vitreous humors.

Hunger   A craving for food. Correctly used, implies a need based on energy requirements as opposed to *appetite*, which is to satisfy a psychological need.

Huntington's chorea   A rare inherited disease that occurs most frequently in middle age. Patient shows a continual array of highly complex, jerky, dancelike movements, accompanied by mental deterioration. Death usually follows within 15 years.

Hyaline membrane disease (HMD)   A disease affecting only newborn children—especially if premature. A lack of ability to produce *surfactant* in lungs leads to breathing difficulties. Also called *respiratory distress syndrome*.

Hyaluronic acid   An acid mucopolysaccharide containing alternating monosaccharide units of gluconic acid and N-acetyl-D-glucosamine. It is the intercellular matrix material of connective tissue proper. Also present in synovial fluid of joints and vitreous humor of the eye.

Hyaluronidase   An enzyme that breaks down hyaluronic acid by attacking every second linkage in the polymer. Enzyme is present in the venom of certain snakes and spiders and is produced by a variety of bacteria, to enable them to spread through host's tissues.

Hydrocephalus   A condition initiated by a blockage in the normal outflow passages for cerebrospinal fluid from the ventricles of the brain. The overfilling of the ventricles results in an internal pressure enlarging the chambers and pushing the brain outward, causing the crushing of tissue against the skull. The skull itself may enlarge considerably.

Hydrocortisone (cortisol)   A hormone of the adrenal cortex that promotes gluconeogenesis.

Hydrogen bond   A weak interaction between a hydrogen atom that is covalently linked to another atom (i.e., the H is part of a molecule) and another nearby electronegative atom on another molecule.

Hydrogen ion (H^+)   A proton. When the hydrogen atom loses its electron and becomes positively charged, what is left is the proton (H^+).

Hydrolysis   The breaking apart of a compound and adding of water to the two parts—the hydrogen atom is added to one part and the hydroxyl group is added to the other.

Hydrometer   An instrument for measuring the specific gravity of a fluid.

Hydropenia   A severe depletion of body water; a deficiency in total body water.

Hydrophilic   Water loving. Readily absorbs water or interacts with the water molecules so that it can dissolve easily. *Charged* molecules exhibit hydrophilia, whereas *uncharged* molecules exhibit *hydrophobia* (fear of water).

Hydrophobic   Water fearing. Does *not* absorb water and does *not* dissolve in water but comes out of solution. *Uncharged* molecules exhibit hydrophobia, whereas *charged* molecules exhibit hydrophilia (love of water).

Hydrostatic pressure   The force exerted by a fluid whether flowing or stationary.

Hydroxyl ion   The negatively charged OH^- group. When in solution, it increases alkalinity.

Hymen The membrane fold that sometimes partially or completely seals the vaginal opening.

Hyperapnea An abnormal increase in rate and depth of breathing.

Hyperbaric chamber A sealed compartment inside which the gas pressure is greater than atmospheric pressure.

Hypercalcemia An excess of calcium in the blood.

Hypercapnia An excess of carbon dioxide in the blood.

Hyperemia An excess of blood delivered to some region or part.

Hyperglycemia An excess of glucose in the blood. One of the distinguishing symptoms of *diabetes mellitus*.

Hyperinsulinism An excessive secretion of insulin by the beta cells of the pancreas that produces a great drop in blood glucose level.

Hypermetropia (hyperopia) Farsightedness. Distant objects can be seen more clearly than those up close. Condition is due to a refraction error that focuses light rays *behind* the retina; most often because the eyeball is too short front to back.

Hyperphagia Excessive eating. Consuming much more food than nutritional demands require. Gluttony.

Hyperplasia An increase in the *number* of normal cells in their normal arrangement.

Hyperpolarization Increase in the internal negativity across a cell membrane, thus increasing the voltage and moving it farther away from the threshold value. The cell membrane becomes *less* excitable.

Hypersensitivity An exaggerated response to a stimulus. In immunity, an unusual reaction to antigens that involves the tissues where the antigen-antibody reaction occurs.

Hypertension A chronic elevation in arterial blood pressure.

Hyperthermia An elevated body temperature that can be due to bacterial toxins producing a *fever,* or it can be induced for therapeutic purposes.

Hyperthyroidism Excessive secretion of the thyroid hormone, which results in an increased metabolism, goiter, and nervous system disturbances.

Hypertonic Refers to a solution that, when supporting cells, causes a net *outflow* of water from the cells. Thus, the cells tend to shrink.

Hypertrophy An increase in the *size* of otherwise normal cells. Thus, an organ or some part will also enlarge. In response to exercise, muscle cells hypertrophy.

Hyperventilation Breathing in excess of what is required to maintain a normal level of plasma pCO_2.

Leads to alkalosis and apnea. Requires prolonged (3–5 minutes) rapid and deep breathing with emphasis on exhaling. Used clinically as a test for epileptic tendency.

Hypocalcemia A reduction from the normal level of calcium in blood.

Hypocapnia A reduction from the normal value of carbon dioxide in the blood. A reduced plasma pCO_2. Occurs as a result of hyperventilation.

Hypodermic Injected, introduced, or located beneath the skin.

Hypoglycemia An abnormally low concentration of glucose in the blood. Can result from excess insulin (injected or secreted).

Hypophysectomy Removal of the pituitary gland (hypophysis) by surgical means.

Hypophysis The pituitary gland, which is located at the base of the brain and enclosed by a bony pocket, the sella turcica.

Hyposecretion A reduction in the amount of product released by a gland or cell.

Hypotension An abnormally low blood pressure as measured in the large arteries.

Hypothalamic-hypophyseal portal system A collection of veins that deliver blood collected from a *primary* capillary plexus in the hypothalamus to a *secondary* capillary plexus in the anterior pituitary gland.

Hypothalamo-hypophyseal tract A collection of *nerve fibers* that arise from nuclei in the hypothalamus and run down the stalk of the pituitary and terminate in the posterior pituitary gland. The fibers forming these tracts carry chemicals synthesized in the cell bodies located in the hypothalamus down into the posterior pituitary gland where they are stored (e.g., ADH and oxytocin).

Hypothalamus A part of the brain that forms the floor and part of the walls of the third ventricle. It controls and integrates many autonomic mechanisms with the endocrine system. Contains centers that regulate water balance, temperature, sleep, food intake, and secondary sex characteristics.

Hypothermia A low body temperature; often induced as a means of lowering metabolism and thus the need for oxygen. A procedure used in preparation for open heart surgery.

Hypotonic Refers to a solution that, when supporting cells, causes a net *inflow* of water into the cells. Thus, the cells tend to swell.

Hypoxia A low oxygen content insufficient to meet the needs of a tissue.

Hysterectomy The surgical removal of the uterus. Can be done through the vagina or by an incision of the abdominal wall.

I band One of the transverse bands seen by microscopic examination of muscle cells. Of very light intensity, it corresponds to the location of only thin myofilaments.

Ileocecal valve (ileocolic valve) Valve located at the junction of the small and large intestine. It is formed by an insertion of the small intestine into the side of the larger diameter colon. Valve easily permits passage of material from the small intestine into the large, but not in reverse direction.

Ileum The distal, longest portion of the small intestine where most of the nutrients are absorbed from the digestive tract and transported into the blood.

Immunity The condition of being protected or resistant to the invasion of foreign microorganisms or other antigenic substances. Protection comes from the presence of specific immunoglobulins or cells able to function as antibodies.

Immunoglobulins (Ig) Antibodies. Protein molecules synthesized by special lymphocytes (plasma cells) in response to the introduction of antigen. Immunoglobulins are divided into five kinds (IgG, IgM, IgA, IgD, IgE) based primarily on the larger protein component present in the immunoglobulin.

Implantation The early stage of development, when the blastocyst attaches to the uterine wall and begins to burrow its way into the endometrium. Takes place about six days after fertilization.

Impotence Without power. Used primarily in the sexual context to describe the inability of a male to achieve an erection. Thus, sexual intercourse is not possible for that man.

Incontinence The inability to control excretory function (e.g., urinary incontinence, fecal incontinence).

Inducer A molecule with a stronger affinity for the *repressor* molecule than the gene to which the repressor binds. The inducer thus removes the repressor from the DNA molecule, allowing the structural genes to be transcribed.

Inertia The inability to move or react. To stay at rest or continue at constant velocity in a straight line.

Infarct An area of tissue or an organ that is dying or dead because it has been deprived of its blood supply, usually by a thrombus or an embolism.

Infectious mononucleosis An acute infectious disease due to a virus (Epstein-Barr). Symptoms include fever, sore throat, and swollen lymph glands. It is identified by large atypical lymphocytes and immunoglobulins that are able to agglutinate sheep rbc's.

Infertile Not fruitful. Sterile. Lacking the ability to conceive or to cause conception.

Inflammation A localized, protective response to injury. It includes an increased blood flow to the region, which increases its temperature and colors it red. The leakage of plasma and proteins causes tissue swelling and also a coagulation of the tissue fluid, which confines bacteria and toxins.

Infusion The therapeutic introduction of a solution by gravity into a vein.

Ingestion The taking in of food or drugs by mouth.

Inheritance The acquisition of body characteristics and qualities by genetic transmission of information from parents to offspring.

Inhibition The halting or restraining of a process. Poisons have their effect by stopping certain vital biochemical reactions (i.e., inhibiting a process).

Inhibitory postsynaptic potential (IPSP) The slight, confined hyperpolarization created on a nerve cell by the arrival of an action potential on a presynaptic terminal. The chemical released causes a net *loss* of *positive* charge from the cell at the point of contact and the membrane becomes more negative on the inside (i.e., hyperpolarized).

Innervate To supply or distribute nerves to some part. To connect to a nerve.

Inotropic Influencing the *strength* or energy of muscular contraction. Can be positive (increasing strength) or negative (decreasing strength).

Inspiration The act of drawing air into the lungs. This is an active process requiring contraction of muscles. *Expiration* is a passive process that occurs primarily due to the elastic recoil of lungs.

Inspiratory center A collection of inherently active nerve cells located in the medulla oblongata. These cells generate action potentials that are conducted to the inspiratory muscles to cause inspiration and to the expiratory center to initiate expiration.

Inspiratory reserve volume The milliliters of air that can be inhaled beyond a normal inspiration (about 3000 ml in males, 2000 ml in females).

Insulin A protein hormone synthesized and released by the beta cells of the pancreas in response to a climbing blood sugar level. It functions to promote the uptake of glucose by most body cells.

Integral proteins Those protein components of the cell membrane that are removed only with drastic measures, and when they are, cause disruption of the membrane. Many span the entire thickness of the membrane, are amphipathic, and often have transport functions.

Intercalated disc A complex cell junction between two myocardial cells. Consists of several interrupted, steplike portions across a muscle cell that separates two adjacent cells. In longitudinal sections the junction shows a characteristic wavy pattern.

Interferon A protein produced by an animal cell

when it is infected by a virus. The protein then protects other cells from being infected. Research indicates that the interferon prevents the host cell from translating the RNA-type virus particle. It is currently being tested as an anticancer drug.

Internal environment (milieu intérieur) The fluid that bathes every cell in the body and supports its life processes by providing all its nutrients and by removing all the wastes. It is the interstitial fluid compartment—the major part of the extracellular compartment.

Internuncial Refers to those nerves that relay information from one center to another; connecting or communicating neurons.

Interoceptor A sensory device, receptor, that is located within and initiates action potentials from the viscera.

Interphase The time between two successive cell divisions, that is, when the cell is not dividing. During this phase, the individual chromosomes are not evident and the cell is carrying out its special function.

Interstices The small spaces or gaps within a tissue. Often synonymous with the intercellular spaces.

Interstitial Refers to the gaps or spaces within a tissue. The intercellular spaces.

Interstitial cells of Leydig The cells that secrete the male hormones (androgens). Located within the testes interspersed between the seminiferous tubules.

Interstitial cell stimulating hormone (ICSH) The identical hormone in females is LH (luteinizing hormone), so now it is most often called LH and refers to both sexes. In the male it stimulates the testicular Leydig cells to secrete androgenic hormones (e.g., testosterone).

Interstitial fluid That portion of the extracellular fluid that bathes the cells of the body. It is the *internal environment* of the body.

Interstitium Synonymous with the interstitial space—the gaps or openings in a tissue that constitute the intercellular space.

Intestinal phase The digestion of food by enzymes that has been reflexively triggered by the presence of food in the small intestine.

Intracellular fluid (ICF) That portion of the total body water that is contained within the cells (about two-thirds).

Intrafusal fibers (spindle fibers) Muscle cells, modified to respond to changes in tension. They are part of the muscle spindles that function to keep muscle cells under slight tension (tone).

Intrauterine device (IUD) A flexible metal or plastic coil with a nylon thread attached. These are inserted by a physician into the uterus for the purpose of preventing pregnancy.

Intrinsic factor A glycoprotein synthesized and secreted by the gastric mucosa that binds to vitamin B_{12}. In a complex form, intrinsic factors are able to fit into a receptor site on the mucosal cell and thus facilitate the vitamin B_{12} absorption.

Inulin An indigestible polysaccharide composed of fructose subunits. It is used for measuring glomerular filtration rate (GFR).

Invagination An infolding. Seen when a cell takes in *particles* of food. The cell membrane folds inward to form a saclike container that contains the food. The infolding then pinches closed.

Ion Any charged atom or group of atoms. Usually formed when a substance, such as salt, dissolves and dissociates.

Ionic hypothesis Currently held explanation for the changes in membrane voltage during an action potential. Hypothesis is based on selective changes in membrane permeability to sodium and potassium ions.

Ipsilateral On the same side.

IPSP (inhibitory postsynaptic potential) The increase in the internal negativity of the membrane potential so that the voltage moves further from the critical threshold value. The membrane is therefore less likely to be excited.

Irritability The ability to respond to stimuli.

Ischemia Deficiency of blood flow into a part. Can be due to a normal vasoconstriction or to an injury that obstructs the vascular supply.

Islets of Langerhans Clumps of cells and capillaries scattered randomly throughout the pancreas. These cells are endocrine glands that secrete hormones into the capillaries. The cells are divided into three types—the most numerous beta cells that secrete insulin, the much fewer alpha cells that secrete glucagon, and the even fewer sigma cells whose function is unknown.

Isometric contraction An attempted muscle contraction that only develops tension and *no* muscle shortening, when the load is greater than the muscle can move.

Isosmotic Having the same osmotic pressure.

Isotonic contraction A muscle shortening that causes movement of the part; thus does work.

Isotonic solution A solution in which cells can be supported without a net flow of water across the cell membrane.

Isotope A chemical element having the same number of protons as another but having a different number of neutrons (i.e., same atomic number, but different atomic masses).

Isthmus A narrow connection between two larger parts.

IUD Intrauterine device.

Jaundice A yellowish color of the skin and mucous membranes. Bilirubin, normally excreted by the liver in bile, backs up and enters the blood to which it imparts the yellow color.

Jejunum The portion of the small intestine that follows the duodenum and precedes the ileum. The final stages of digestion take place here.

Juxtaglomerular apparatus Consists of the *macula densa* — narrower cells lining the wall in the straight portion of the distal tubule — and the *juxtaglomerular cells* in the wall of the afferent arteriole.

Juxtaglomerular cells Modified cells of the afferent arteriole located almost at the point where the arteriole abruptly branches into a network of capillaries. The cells in the arteriolar wall secrete renin when the blood pressure starts to fall.

Keratin An insoluble, tough protein that is the principal component of skin, nails, and hair.

Keto acid An organic acid with both a carbonyl $(\overset{|}{\underset{|}{C}}=O)$ and a carboxyl $(-\overset{O}{\overset{\|}{C}}-OH)$ group.

Ketone body Substance produced primarily during excessive fat metabolism, such as acetone, acetoacetic acid, and β-hydroxybutyric acid.

Ketonuria Ketone bodies excreted in the urine. Seen in starvation and diabetes mellitus when fat metabolism is the prime source for energy.

Ketosis The condition that describes the excessive increase in concentration of ketone bodies in the body tissues and fluids.

Kilocalorie A large calorie used in metabolic studies. The amount of heat energy required to raise the temperature of 1 *kilogram* of water 1 degree Celsius.

Kinase An enzyme subclass that groups those proteins able to transfer a *high-energy* group, usually from ATP, to some acceptor molecule. Specifically named by the acceptor (e.g., fructokinase, creatine kinase).

Kinesthesia The sensory perception that describes detection of body parts — usually related to the angle changes of joints.

Kinetic energy Energy involved in producing motion or work.

Klinefelter's syndrome (XXY) A condition of males in which three sex chromosomes (instead of the normal number two) are present in all the cells of his body. This trisomy condition is expressed by: infertility, small testes, and some mental retardation.

Krebs cycle TCA cycle. Citric acid cycle.

Kupffer cells Large star-shaped cells that are fixed to the walls of the liver sinusoids and are very active phagocytes.

Kyphosis Hunchback. An exaggerated curvature of the thoracic region of the vertebral column so that a hump appears to be on a person's back.

Labia majora Two folds of skin, extending posteriorly from the mons pubis, which form a cleft into which the urethra and vagina open. The outer surfaces of the labia are pigmented and have hair. The inner surfaces are smooth and possess large sebaceous glands. The labia are analogous to the scrotal sac of the male.

Labia minora Two smaller folds of mucous membrane within the cleft formed by the labia majora. They enclose the space called the *vestibule* into which the openings of the urethra and vagina open. Anteriorly they meet and encircle the *clitoris*.

Labyrinth A complex network of interconnecting canals. The *inner ear cavities* connect to form a complex network of passages.

Lactase An enzyme that hydrolyzes the disaccharide lactose into the two monosaccharide sugars glucose and galactose. Also named β-galactosidase.

Lactate The negatively charged ion formed when lactic acid is in solution. Often used in biochemical terminology to refer to lactic acid.

Lactation The secretion of milk by mammary glands into alveoli.

Lacteal The blind-ended lymphatic vessel located within a villus of the intestinal mucosa.

Lactose The disaccharide sugar consisting of the two monosaccharides glucose and galactose. Also called milk sugar because it is a constituent of milk.

Lamina propria The thin layer of connective tissue underlying the basal lamina of an epithelium.

Larynx A complex of nine cartilages and ligaments that enclose the vocal cords. The anterior thyroid cartilage is identified commonly as the "Adam's apple."

Latent period The period of time between stimulation (production of an action potential) and the onset of response (muscle shortening). Time is in the order of 2–10 milliseconds.

Lateral ventricles Relatively large C-shaped cavities located medially within each cerebral hemisphere and separated from each other by a very thin partition of brain tissue, the *septum pellucidum*.

Lens The transparent, biconvex structure of the eye

that can adjust its curvature to aid in focusing visual images onto the retina.

Lesion A wound or injury that often produces a reduction or loss of function in some part.

Leucocytes The white blood cells (wbc's). Divided into two groups: granular and agranular. Generally function as protective mechanisms aimed at destroying, removing, or inactivating invasive organisms.

Leucocytosis An increase in the total number of leucocytes (wbc's) in circulation (i.e., above 15,000/mm³). Often in response to infection, hemorrhage, or inflammation.

Leucopenia A reduction in the total number of leucocytes (wbc's) in circulation (i.e., below 5000/mm³).

Leukemia A progressive, malignant disease of the bone marrow that results in a rapid and great production of leucocytes and their precursor cells into the circulation.

Leukorrhea A whitish, viscid discharge from the uterus and vagina.

Lever A rigid structure used to move a weight or resistance. A force applied at one point on the lever causes the lever to rotate about a fixed point, the fulcrum.

LH Luteinizing hormone.

Ligament A band or sheet of fibrous tissue that connects bones or cartilages and serves to strengthen the joint.

Ligand A molecule or ion that attaches to another, usually larger, molecule such as a protein. The bonding is weak and the ligand is easily removed.

Limbic system A loosely applied term that generally includes the hippocampus, cingulate gyrus, and septal areas. Associated with autonomic functions, certain emotions, and aspects of learning.

Lipase Any enzyme that catalyzes the breakdown of lipids into glycerol and fatty acids.

Lipids Organic substances that are *insoluble* in water, *soluble* in alcohol, ether, and chloroform. Lipids include neutral fats, fatty acids, waxes, steroids, and phosphatides.

Lipogenesis The synthesis of fat from nonfat food (i.e., from carbohydrate or protein sources).

Lipolysis The breakdown or splitting of fat into smaller units or its component parts.

Lipoprotein A complex molecule containing both lipid and protein parts but having predominantly protein properties. Lipids in plasma are combined, for transport purposes, with proteins.

Local current The resultant flow of charges that occurs in the immediate region or point of stimulation on a nerve or muscle cell membrane.

Loop of Henle That portion of a kidney tubule that emerges from the complex convolutions and forms an outlying loop of the tubule. The arms of the loop are arranged parallel to and in close proximity to each other. About one-eighth of the loops of Henle are much longer than the rest and extend from their origin in the cortex deep into the medullary region.

Lumen The cavity or channel within a tube or hollow organ.

Lumpectomy Surgical removal of a lump. In cystic disease of the breast—which is common in women between 30 and 50 years of age—breast tissue proliferates to form hard masses of tissues (i.e., lumps). When malignancy is a possibility, the lumps may be surgically excised.

Luteinizing hormone (LH) A protein hormone secreted by the anterior pituitary gland that, in the female, promotes the final maturation of the ovarian follicle, ovulation, and the formation of the corpus luteum. In the male, it stimulates the interstitial cells of Leydig to secrete testosterone.

Lymph A transparent, yellowish fluid that is collected into the lymphatic vessels from the interstitial fluid.

Lymphocyte A mononuclear, granulocytic white blood cell. It is the most variable in size of the wbc's (7–20 μm) and is important in the immune response. It produces immunoglobulins.

Lymphoid system A collection of cells and organs that interrelate to produce the antibody machinery. It is divided into three parts: (1) stem cells of the bone marrow, (2) the primary lymphoid organs where lymphocytes are programed (i.e., thymus and bursa equivalent), and (3) the peripheral lymphoid organs (i.e., lymph nodes, spleen, tonsils, and Peyer's patches in the intestinal tract).

Lymphokines A series of soluble products released from the T lymphocytes. They are responsible for an array of reactions produced by the cellular type of immune reaction.

Lysergic acid diethylamide (LSD) A very powerful hallucinogenic compound that produces bizarre behavior, psychosis, and chromosomal damage.

Lysosome A membrane-bound cellular organelle that contains an array of acidic hydrolase-type enzymes. Material brought into cells by endocytic activity is digested when the vessels fuse with the lysosomes.

Macromolecule A very large polymeric molecule (e.g., protein, polysaccharide, nucleic acid).

Macrophage A very large phagocytic cell that is widely distributed throughout the body's connective tissues. Some are sessile, others motile.

Macula densa An area of the distal kidney tubule

where it comes into contact with the juxtaglomerular cells of the afferent arteriole. The two components form the juxtaglomerular apparatus, which is responsible for renin release.

Malignant A situation that will become progressively worse and end in death. Has properties of invasion and rapid spread and multiplication. Usually applied to tumors.

Maltase An enzyme found in the intestinal juice that splits the disaccharide maltose into two molecules of glucose.

Maltose A disaccharide sugar consisting of two glucose subunits. It is formed from the degradation of starches.

Mammal Any vertebrate that feeds its young with milk produced in the mammary glands of the female.

Mammary ducts The tubes that carry the secretion of the mammary glands into the nipple and open onto the surface.

Mammary glands Milk-secreting glandular tissue located on the ventral surface of the female. Consists of about 20 lobes of milk-secreting cells arranged radially around the nipple. The lobes empty their secretions via ducts into the nipple and onto the body surface.

Mammography An x-ray technique specifically for photographing the interior of the mammary glands.

Mast cell A connective tissue cell that contains histamine and heparin granules. The cell releases the chemicals when the tissue is injured.

Mastectomy Surgical removal of a breast (mammary gland).

Masticate To chew. To break up and grind food with teeth in preparation for swallowing and further chemical digestion.

Masturbation The process of reaching orgasm by self-stimulation and manipulation of the genitals.

Matrix That portion of the extracellular substance that forms the continuum between cells and into which the fibers are deposited.

Mediated transport The process that moves nondiffusible molecules through a cell membrane by employing specific carrier molecules. The carriers can function by employing cellular energy (ATP) or by simply moving the substrate down its gradient.

Medulla The innermost portion or the core of an organ or part.

Megakaryocyte A giant cell formed in the bone marrow that fragments to form blood platelets.

Meiosis A special type of cell division restricted to sex-cell production. The end result of two successive cell divisions is cells with half the number of chromosomes found in all other body cells.

Meissner's corpuscle The sensory receptor for the sensation of touch. Found in the skin, especially in palms and soles.

Melanin A dark pigment found normally in the skin and hair and sometimes in tumors.

Melanocyte The cell that synthesizes melanin. It is found in the lower portions of the epidermis of skin, and the number of such cells varies from 5 to 25 percent of all the epidermal cells.

Membrane A thin sheet of tissue that covers or lines a cavity or surface. The outer barrier that surrounds and encloses cells and outlines the internal organelles. Composed primarily of phospholipids and protein molecules.

Membrane potential The voltage present, at any instant, across the cell membrane. Measured with microelectrodes inside and outside the cell, it usually registers "resting" values of about −70 millivolts.

Memory cells Lymphocyte cells that are produced from the repetitive division of an original lymphocyte cell that is activated by a specific antigen. The memory cells do *not* produce immunoglobulins but retain the genetic information for the specific antibody so the organism can respond more rapidly to subsequent challenges of that antigen. These cells are activated to divide rapidly to form clones of immunoglobulin-producing cells and additional memory cells.

Menarche The initial menstrual period or the beginning of menstrual periods.

Meninges The three membranes that cover and protect the brain and spinal cord (i.e., dura mater, arachnoid, and pia mater).

Menopause The cessation of menstrual cycling or that period of time of a woman's life when it occurs. Normally from mid 40s to early 50s.

Menses The cyclic, "monthly" flow of blood and fluid from the uterus of a nonpregnant mature woman.

Menstrual cycle The regularly recurring changes in the levels of sex hormones in the female. The changes repeat approximately every 28 days and are responsible for the changes in the uterine endometrial lining.

Menstruation The cyclic, normal uterine bleeding that occurs about every 28 days in the nonpregnant female.

Mesenchymal cells Multipotential cells that are derived from mesoderm. Present in most tissues, they have the ability to differentiate into various specialized kinds.

Mesentery A double layer of the peritoneal membrane which functions to support or attach various organs to the dorsal body wall.

Mesoderm The middle layer of the three primary germ layers from which are derived connective tissues, blood, and blood vessels, etc.

Messenger RNA (mRNA) The polymer of ribonucleic acids that carries genetic information from the DNA within the cellular nucleus to the ribosomes in the cytoplasm where the proteins are synthesized.

Metabolism The sum of all the biochemical reactions that occur within a live organism. Includes the synthetic (anabolic) reactions and destructive (catabolic) reactions.

Metaphase The stage of mitosis or meiosis during which the chromosomes, present in the form of chromatids, are aligned along the equatorial plane of the dividing cell.

Metarteriole A vessel that is located between arterioles and true capillaries. Appears more as a capillary with a few scattered smooth muscle cells and functions as a precapillary sphincter.

Metastasis The spread of disease from one organ or part to another unconnected, remote part. A characteristic of malignant tumors.

Micelle A supermolecular aggregation usually composed of amphipathic molecules arranged so that the hydrophilic ends protrude into the aqueous phase and the hydrophobic into the nonaqueous.

Microcirculation Generally describes the flow of fluids out of the capillaries at the arteriolar end and back into the capillaries at the venous end. It is this flow of nutrient-filled fluid that supports the life processes of the body cells.

Microelectrode A very fine wire or, usually, a hollow tapered glass tube filled with a conducting solution that is slender enough to be inserted through a cell membrane and cause the least possible damage. Used to study the electrical properties of cells.

Microglia Cells of the central nervous system that support, insulate, protect, and nourish nerve cells.

Microorganism Any small organism that can be seen only with the aid of a microscope. Includes bacteria, molds, yeasts, viruses, and protozoa.

Microtubules Long, hollow, cylindrical structures present within cells. They become most apparent as the spindle fibers during mitosis. Also are involved in cell motility.

Microvillus A tubular extension of the cell membrane. Usually only on one surface of a cell, especially the cells of the proximal convoluted tubule and intestinal epithelium. The microvilli serve to greatly increase the cell's surface area.

Micturition Urination; the elimination of urine from the body.

Milieu intérieur The term introduced by Claude Bernard to describe the internal environment of the body. The fluid that bathes and nourishes all the cells of the body.

Milk letdown A reflex action initiated by the suckling action on a lactating breast. A nervous reflex from the breast to the hypothalamus and posterior pituitary gland causes the hormone oxytocin to be re-released into the blood. The oxytocin stimulates the myoepithelial cells to contract and force the milk from the alveoli into the ducts of the mammary glands.

Mineral An inorganic, homogeneous solid substance. In the body minerals are present as seven major inorganic components: calcium, phosphorus, sodium, potassium, chlorine, magnesium, and sulfur. Other minerals in lesser amounts include iron, copper, iodine, manganese, zinc, cobalt, and fluorine.

Mineralocorticoids Hormones synthesized by cells of the adrenal cortex that are particularly effective in causing the reabsorption of sodium ions in the kidney tubules and the excretion of potassium.

Mitochondrion The double-membraned organelle within cells that contains the enzymes for the Krebs cycle and oxidative phosphorylation. Site of ATP production required for the cellular processes.

Mitosis The process by which the genetic material within the nucleus is duplicated and a complete identical set of chromosomes is passed on to each daughter cell.

Mitral valve The bicuspid valve separating the left atrium and left ventricle of the heart.

Modality Any of the specific sensory entities (e.g., vision, taste).

Molal A unit of concentration. A solution containing 1 mole of a solute dissolved in 1 *kg of pure solvent*.

Molar A unit of concentration. A solution containing 1 mole of a solute *per liter of solution*.

Monoamine oxidase An enzyme that catalyzes the hydrolysis of catecholamine neurotransmitter substances (e.g., norepinephrine, serotonin) into inactive subunits.

Monocyte One of the five types of leucocytes. It is the largest type, 13–25 μm. Contains a large kidney-shaped nucleus. The cells are very active phagocytes.

Monosaccharide Simple sugar. It is a colorless, crystalline substance that has a sweet taste. It *cannot* be broken down to simpler sugars by *hydrolysis*. It has the general formula $(CH_2O)_n$ and is classified by the number of carbons present.

Morphology Describes the form or shape and structure of organisms.

Motor cortex That portion of the cerebral cortex that initiates many of the muscular movements. It is located transversely across the posterior portion of the parietal lobe.

Motor end plate The specialized region of a muscle cell membrane upon which the excitatory nerve cell terminals release their transmitter substance.

Motor neuron An efferent nerve cell that conveys action potentials to muscle cells, glandular cells, and absorptive cells. Motor neurons make something happen.

Motor unit A motor neuron plus all the muscle cells that it innervates. Some motor units contain only 5–10 muscle cells in locations where precision of movement is important. In other locations motor units contain hundreds of muscle cells, as in the large postural muscles.

mRNA Messenger RNA.

Mucin A mucopolysaccharide or glycoprotein which, when dissolved in water, forms the main ingredient of mucus. It is secreted by the salivary glands and cells of the gastric glands.

Mucopolysaccharide A polymer consisting of a *hexosamine* and some form of a modified sugar. Some are modified by the addition of protein. They form the intercellular material of the connective tissues.

Mucus The coating present on certain membranes that serves as a protective layer. It is composed of glandular secretions, inorganic salts, cellular debris, and often white blood cells.

Muscle A tissue that is able to contract or shorten and thereby cause the movement of some part.

Muscle spindle An encapsulated stretch receptor located within skeletal muscles. The receptor is arranged so that the modified muscle cells are parallel to the contracting muscle cells, and they function via nerve fibers to keep the muscle cells under constant tension.

Mutation A genetic change resulting in a permanent alteration in some inheritable characteristic. Explained mechanistically as any change in the sequence of bases in the DNA molecule.

Myasthenia gravis A condition characterized by fatigue and progressive paralysis of skeletal muscles without any sensory losses. Affects primarily muscles of the neck and face. Caused by a decrease in the number of functioning acetylcholine receptors in the motor end plate.

Myelin A wrapping around the axons of some nerve cells. Formed by the concentric coiling of the Schwann cells around the nerve fiber. These cell membranes have high percentages of lipids and thus contribute high electrical resistance or good insulation properties around the fibers. Myelin is interrupted by gaps (nodes of Ranvier) that enable nerve conduction to proceed more rapidly than in unmyelinated fibers.

Myeloid Referring to or derived from bone marrow.

Myocardium Heart muscle.

Myoepithelium Epithelial cells possessing contractile properties. Located in the secretory alveoli of mammary glands, sweat glands, lacrimal glands, and salivary glands.

Myofibrils The contractile structures located within muscle cells. They are composed of thick and thin myofilaments, arranged in parallel array within the fibril. The myofilaments consist mainly of the proteins actin and myosin.

Myofilaments The components of the myofibrils of muscle cells. The thick and thin myofilaments are arranged in parallel interdigitating patterns. When fibrils shorten, the myofilaments slide past each other.

Myoglobin The oxygen-binding, iron-containing conjugated protein complex present in the sarcoplasm of muscle cells. It contributes the red color to muscle.

Myogram The record or tracing produced by the myograph, the apparatus that measures and records the effects of muscular contractions.

Myometrium The smooth muscle layer of the uterine wall.

Myoneural junction The anatomical location of the connection between the nerve terminals and the muscle cells.

Myopia Nearsightedness. An error in refraction that causes images to be focused in front of the retina. The eyeball is too long from front to back.

Myosin The most abundant protein in muscle. It makes up the thick myofilaments and its cross bridges. Myosin has ATPase activity that releases the energy to initiate and cause muscle shortening.

Myotatic reflex A stretch reflex triggered by the abrupt stretching or extending of a muscle.

Myotonia A continuous spasm of muscle; increased muscular irritability and tendency to contract and less ability to relax.

Myxedema An adult condition resulting from insufficient thyroid hormone. Characterized by swollen lips and a thickened nose.

NAD Nicotinamide adenine dinucleotide.

Na⁺-K⁺ ATPase Sodium-potassium ATPase. An enzyme present in most cell membranes that is able to split ATP into ADP and P_i and provide energy for transport processes. Enzyme requires high concentrations of sodium inside the cell and high concentrations of potassium outside the cell.

Narcotic A substance that produces insensibility or stupor.

Near point The closest position at which an object can still be seen distinctly.

Nebulization Conversion into a spray. To treat by using a spray.

Necrosis Death of tissue in a localized area. The cells of the tissue are gradually and sequentially killed or die.

Negative feedback The principle governing control systems. The end result is to reverse the direction of the original stimulus toward the desired or set value. For example, if temperature increases beyond the set or desired point, the control system is *fed back* this information. The system then initiates steps to reverse (turn back; *negative direction*) the increased temperature toward the lower, desired temperature.

Nephron One of the million kidney tubules. The basic functional unit of the kidney which receives the ultrafiltrate of plasma and produces urine.

Neurilemma (sheath of Schwann) The peripheral nucleated cytoplasmic layer of the Schwann cell that encloses the myelin sheath. It is found only around fibers of the peripheral nervous system and it functions to assist in the regeneration of injured axons.

Neuroglia The supporting tissue of the nervous system. The cells of neuroglia aid in myelin formation, transport nutrients and wastes, and help maintain the ionic environment of the neurons.

Neurohypophysis The posterior pituitary gland, which originates during embryonic development from a downgrowth of the base of the brain and remains continuous with the hypothalamus. It stores and releases the hormones ADH and oxytocin, which are synthesized in the hypothalamus.

Neuromuscular junction Myoneural junction.

Neurons The cells of the nervous system capable of conducting action potentials.

Neurosecretion The release from nerve cells of chemicals that serve as neurotransmitters, hormones, or regulating substances.

Neutrophils The most numerous of the white blood cells. They are polymorphonuclear, and possess small cytoplasmic granules that stain with neutral dyes. The cells are active phagocytes and function as first lines of defense against bacterial invasions.

Nicotinamide adenine dinucleotide (NAD) A coenzyme involved in many oxidative reactions. It accepts hydrogen atoms.

Night blindness (nyctalopia) Poor or no vision in dim light or at night. Good vision present during bright illumination.

Nipple The projection located on the surface of the mammary gland. Pigmented, wrinkled, it is the location of the openings of the lactiferous ducts for milk release.

Nissl bodies Granules of ribonucleoprotein seen within the cytoplasm of the nerve cell body.

Nodes of Ranvier The interruptions in the myelin sheath present about every millimeter along the myelinated nerve fibers. The nodes permit the saltatory type of nerve conduction.

Norepinephrine A chemical transmitter synthesized and released by the terminals of postganglionic sympathetic nerve fibers. It is also secreted by the adrenal medulla along with epinephrine.

Novocain Trademark for preparations of *procaine hydrochloride*. Used in solution as a local anesthetic for infiltration, nerve block, and spinal anesthesia.

Noxious Hurtful, harmful to health.

Nucleic acid A long polymer of nucleotides. The phosphate of one nucleotide is bonded to the sugar of the next nucleotide. The organic bases protrude from the sugar-phosphate backbone. DNA and RNA are nucleic acids.

Nucleolus A granular, dark-staining region of the nucleus that includes that portion of DNA coding for rRNA and the multiple copies of rRNA made from the DNA.

Nucleoprotein The protein portion of the chromatin. It probably regulates the activity of the genes encoded in the DNA.

Nucleotide One of the monomeric subunits of nucleic acids. It is a molecule formed by the linking of a purine or pyrimidine base to a sugar and a phosphoric acid.

Nucleus The central portion of an atom made up of protons and neutrons. The large, spherical, membrane-bound organelle present in eucaryotic cells that contains most of the cell's DNA and some RNA. A collection of nerve cell bodies within the CNS.

Nystagmus Involuntary, rapid, repetitive movements of the eyeball in either horizontal, vertical, or rotatory directions.

Obesity Excessive accumulation of fat in the body resulting in an increase in body weight beyond normal limits.

Obligatory water reabsorption The required proportion of water uptake that occurs in the proximal convoluted portions of the nephrons.

Ohm The unit of electrical resistance.

Olfactory Refers to the sense of smell.

Oligodendrocytes One of the cell types that comprises the neuroglia. Nonneuronal cells with long extensions that encircle many of the myelinated nerve fibers of the white matter.

Oocyte One of two stages in the development of the mature ovum. Primary and secondary oocytes are consecutive stages in development of the egg.

Oogenesis The process of formation of the ovum.

Ootid One of the four cells derived from the two consecutive divisions of the primary oocyte.

Operator gene The site on the DNA molecule adjacent to the structural genes which determines whether or not the structural genes will be transcribed. It is the site at which repressor molecules combine.

Operon The genetic unit consisting of an operator gene and the adjacent structural genes which can be repressed or derepressed together.

Opsin A protein of the retinal cells that combines with 11-*cis*-retinal to form the visual pigments. The rod cells have an opsin called scotopsin, and the three types of cone cells have opsins named according to their color of pigment: iodopsin (violet), porphyropsin (red), and rhodopsin (purple).

Opsonization The reaction that renders bacteria and other foreign cells more ready for phagocytosis. Certain antibodies can function as *opsonins* to attack cells and render them more susceptible to phagocytosis.

Optic Refers to the eye or vision or properties of light.

Optic disc The area of the retina where the axons of the ganglion neurons exit the eye as the optic nerve. Since this area is devoid of receptor cells, no image is formed here and thus the optic disc is also referred to as the *blind spot*.

Oral contraceptive ("the pill") A hormonal compound that is swallowed and that prevents ovulation, and thus pregnancy.

Organ A structure within the body made of cells and tissues that operates somewhat independently to perform a specialized function.

Organ of Corti The structure that contains the receptor cells for hearing.

Organelle A membrane-enclosed structure located within a cell. Each type of organelle has a specialized function (e.g., nucleus, mitochondrion, lysosome).

Organic Describes chemical substances containing carbon. Grown or cultivated with the use of animal or plant fertilizers rather than synthetics.

Organism Any individual living thing.

Orgasm The peak and culmination of sexual excitement.

Orifice The opening into or out of any cavity in the body.

Oscilloscope An electronic instrument that amplifies and displays with great accuracy on a fluorescent screen, electrical events and measurements (e.g., action potentials).

Osmoreceptors Cells specialized to respond to very minute changes in the osmotic pressure of their immediate environment.

Osmosis The movement of water across a semipermeable membrane from a region where water is in higher concentration into a region where water is in lower concentration.

Osmotic pressure The pressure generated by osmosis.

Ossicles Small bones, as in the middle ear (malleus, incus, stapes).

Ossification The formation of bone or the conversion of fibrous tissue into a bony substance.

Osteoblast A bone-forming cell.

Osteoclast A large multinuclear cell that removes bone.

Osteocyte The cell of bone. After the osteoblast has laid down the bone and is entrapped in the lacuna (cavity) it becomes the bone maintenance cell, the osteocyte.

Otolith (statolith) One of the minute calciferous granules that is embedded in the gelatinous layer that coats the sensory hairs of the utricle and saccule.

Oval window An oval opening separating the middle and inner ear cavities. It is closed by a membrane and the foot of the stapes.

Ovarian cycle The monthly production and release of a mature ovum from the ovary. The varying hormonal levels stimulate the production of the follicles and the release of an egg each cycle.

Ovary The primary reproductive organ of the female. It produces ova and hormones.

Ovulation The release of an ovum from its follicle on the surface of the ovary.

Ovum The egg or female gamete formed within a follicle of the ovary. It is the female sex cell containing the haploid number of chromosomes. It is fertilized by the male sex cell at the moment of conception.

Oxidation A chemical or biochemical reaction that leaves an atom with an *increased positive* charge. Usually occurs by the atom's loss of electrons. Most biological oxidations result when a molecule loses a pair of hydrogen atoms.

Oxidative phosphorylation The reaction that extracts energy from the electron transport system and uses it to phosphorylate ADP to ATP. This reaction occurs within the inner membrane spheres of the mitochondrion.

Oxygen debt The volume of oxygen required to oxidize the lactic acid produced by muscular exercise. During vigorous exercise, the energy utilized outstrips the ability of the *oxidative reactions* of the muscle cells to replenish the ATP. Consequently anaerobic reactions provide the energy with the resultant production of lactic acid.

Oxyhemoglobin (HbO$_2$) Hemoglobin combined with oxygen.

Oxytocin A peptide molecule containing nine amino acids synthesized in the hypothalamus, stored in and released from the posterior pituitary gland. It stimulates the myoepithelial cells of the alveoli in the mammary glands to contract, promoting milk release. It also causes the pregnant uterus to contract.

Pacemaker That which determines the rate at which a process occurs. In the heart, the sino-atrial (SA) node determines the heart rate. An electronic device that can substitute for the SA node.

Pacinian corpuscle Oval receptor consisting of concentric layers of connective tissue—much like an onion—wrapped around an afferent nerve fiber. The receptors are located in the subcutaneous tissue and are sensitive to pressure applied to the skin.

Packed cell volume In a sample of centrifuged blood, the percent of the total volume occupied by the cells.

PAH Para-aminohippuric acid.

Palpate To examine by feeling and touching.

Pancreas A large elongated gland located behind the stomach. It contains both exocrine and endocrine glands.

Papanicolaou test (Pap test) A cytological staining test for the detection and diagnosis of malignant and premalignant conditions of the female genital tract. Cells scraped from the genital epithelium are smeared, fixed, stained, and examined microscopically.

Papilla A small pimplelike projection or elevation. On the surface of the tongue there are filiform (threadlike), fungiform (knoblike), and circumvallate papillae.

Papillary ducts (ducts of Bellini) The excretory or collecting tubes that empty urine from their openings on the tips of the pyramids into the renal pelvis.

Papillary muscles Located within the ventricles of the heart, these intruding muscle masses anchor the chordae tendineae.

Para-aminohippuric acid (PAH) A chemical used in tests for renal plasma flow. It can be completely cleared from the blood by one passage through the kidney if the concentration is kept low. PAH is *secreted* into the tubular fluid from the blood.

Paraplegic A person who is paralyzed in the lower half of the body.

Parasite A plant or animal that lives upon or within another living organism and obtains some advantage from the relationship.

Parasympathetic Refers to that division of the autonomic nervous system whose preganglionic fibers arise from the brain and sacral portion of the CNS. The postganglionic fibers release acetylcholine from their terminals.

Parasympathomimetic Refers to drugs that imitate the effects of the parasympathetic nervous system.

Parathormone (parathyroid hormone) A hormone synthesized and secreted by the parathyroid glands. It regulates calcium and phosphate levels in plasma. Excess hormone raises plasma calcium concentrations.

Parathyroid glands Usually four small glands situated as two pairs on the posterior surface of the lateral lobes of the thyroid gland. The parathyroids secrete a hormone that regulates calcium and phosphate levels in plasma.

Paraventricular nucleus A sharply defined collection of cells in the wall of the third ventricle that is part of the hypothalamus. Fibers from these cells serve to convey oxytocin from the site of synthesis in the paraventricular nucleus into the cells of the posterior pituitary.

Parietal Refers to the walls of a cavity. Refers to or identifies a proximity to the parietal bones of the skull.

Parkinson's disease (parkinsonism) A slowly progressive disease occurring in late life. It is characterized by a tremor of resting muscle, slowing of voluntary movements, and muscle weakness. It is caused by a decreased level of dopamine release from cells of the substantia nigra of the brain.

Parotid glands The largest of three pairs of salivary glands. They are located below and in front of the ear and secrete a serous type of secretion into the vestibule of the mouth near the second upper molar tooth.

Partial pressure That pressure contributed by a gas in a mixture of gases. If atmospheric pressure is 760 mm of mercury and oxygen is 20 percent of the air, then its partial pressure is 20 percent of the total or about 152 mm Hg.

Parturition Childbirth. Delivery or expulsion of a child from the uterus.

Pathology That branch of medicine that studies the nature, cause, and development of disease as well as the structural and functional changes that it causes.

P-cells Specialized cells located in the nodal tissues of the heart. Electrically unstable, they generate the action potentials responsible for cardiac contractions.

Pedicels Small footlike parts of a structure.

Penicillin An antibiotic substance extracted from cultures of the molds *Penicillium* and *Aspergillus*. These cultures are grown on special media. Penicillin is also manufactured synthetically. It is antibiotic by virtue of its ability to block the production of the cell wall by many bacteria.

Penis The male organ of urinary excretion and copulation.

Pentamer A molecular chain made up of five subunits (monomers).

Pepsinogen The inactive form of the proteolytic enzyme *pepsin*. Pepsinogen is synthesized by the chief cells of the gastric glands and secreted into the lumen of the stomach where it is converted to pepsin by the action of HCl. Pepsin then hydrolyzes proteins into a mixture of proteoses and peptones.

Peptide bond The linkage formed between two amino acids. The amino group (NH_2^+) of one amino acid loses a hydrogen atom and the carboxyl group (COOH) of the second amino acid loses an OH^- (hydroxyl) group. Thus, formation of a peptide bond also produces a molecule of water.

Peptone A short string of about four to eight amino acids (a short polymer). Peptones usually are the result of protein digestion.

Pericardium The fibroserous sac that confines the heart to its location in the thoracic cavity but still permits it freedom to contract.

Perilymph The fluid contained between the bony and membranous labyrinths of the inner ear.

Perimysium The connective tissue layer that wraps each skeletal muscle fascicle and separates it from the others in the muscle.

Perineum The space between the anus and the scrotum or vulva.

Perineurium The connective tissue layer that wraps a bundle of nerve fibers and separates it from the other bundles in the nerve.

Periodontal Refers to the tissues surrounding and supporting the teeth.

Peripheral nervous system (PNS) That portion of the nervous system not enclosed by the skull or vertebral column. It consists of nerves and ganglia.

Peripheral proteins The easily removed proteins of a cell membrane. They are held to the membrane by weak bonding forces and easily removed by solutions that alter the charges on their surface.

Peripheral resistance The hindrance to blood flow encountered within the blood vessels. Due to friction that increases with viscosity, length of blood vessel, and very importantly, the radius of the blood vessel.

Peristalsis The wave of relaxation of circular smooth muscle fibers immediately followed by a wave of contraction of the muscle fibers. In the small intestinal tract this produces a movement that pushes the contents of the tract toward the large intestine.

Peritoneum The serous-type membrane that lines the abdominal and pelvic walls and also wraps and supports the organs in those cavities.

Peritonitis Inflammation of the peritoneum.

Peritubular capillaries (vasa recta) The network of capillaries that envelops the convoluted and loop of Henle portions of the kidney tubules. They take up reabsorbed nutrients from the tubular filtrate and also give up materials that will be secreted into the tubule.

Pernicious anemia A disease in which the number of red blood cells and their content of hemoglobin is much reduced. Primarily due to the malabsorption of vitamin B_{12} in the upper small intestine which fails to secrete sufficient amounts of *intrinsic factor*.

Peroxisome A membrane-bound organelle that contains the enzyme catalase and a variety of other oxidase-type enzymes (e.g., urate oxidase, amino acid oxidase).

Peyer's patches Aggregations of lymphatic tissue in the lower end of the ileum.

pH A measure of the concentration of H^+ in a solution. It is expressed as the negative logarithm of the hydrogen ion concentration: $pH = -\log_{10}(H^+)$. The pH scale extends from 0 to 14 with a value of 7 expressing neutrality, values lower than 7 expressing increasing acidity, and values higher than 7 expressing increased alkalinity.

Phagocytosis The process by which phagocytic cells encircle and take up into their cytoplasm particulate or macromolecular substances. The substances are then digested within the cell. White blood cells are very active phagocytes and are able to remove bacteria and other foreign invaders.

Phantom limb An arm or leg that has been amputated but is perceived to still exist and cause sensations (pain, pressure, etc.).

Pharynx The throat. Located at the back of the mouth and extending up to the nasal cavity and down to the larynx, it is a passageway common to both air and food.

Phasic Appearing or occurring in regularly recurring cycles.

Phenotype The detectable physical, biochemical, and

physiological makeup of an individual as determined by both genetic and environmental influences.

Phenylketonuria (PKU) An inborn error of metabolism that results from lack of the enzyme phenylalanine hydroxylase. The enzyme catalyzes the conversion of phenylalanine to tyrosine and therefore its lack causes phenylalanine to accumulate in body fluids. This in turn leads to mental retardation and other neurologic problems.

Phosphocreatine A high-energy molecule kept as an energy reservoir in muscle cells. Its high energy is released when it breaks down to creatine and phosphoric acid and the energy is used to form ATP from ADP and phosphoric acid.

Phospholipid A subclass of lipid. It consists of a glycerol molecule and two fatty acids bonded by ester linkages to two of the OH groups. The third OH of the glycerol is esterified by phosphoric acid, which in turn is bonded to a nitrogen-containing molecule. The phosphate and nitrogen-containing molecule make up a hydrophilic tail to the otherwise hydrophobic molecule. Such an *amphipathic* molecule forms the bilayer core of most membranes.

Phosphorylation The process of adding a phosphoric acid molecule to another molecule.

Photopigment The complete light-sensitive molecule that undergoes conformational changes when it absorbs energy of specific wavelengths. It consists of a protein *opsin* portion bonded to a *chromophore* pigment portion.

Photoreceptor A sensory organ or cell sensitive to light.

Photosynthesis The process by which green plants convert carbon dioxide and water into carbohydrate using energy of sunlight captured by the photopigment chlorophyll.

Phrenic nerves The motor nerves that arise in the cervical region of the spinal cord and terminate on the muscle cells of the diaphragm. Stimulation of these nerves causes the diaphragm to contract and flatten, which is part of the inspiratory movement.

Physiology The science that attempts to explain how organisms function. It describes the function of the parts and components and the chemical and physical factors involved.

Piloerection The standing upright of hair. Small smooth muscles, attached to the sides of the hair follicles, contract and straighten the follicles so that the hairs stand up.

Pineal gland A small pinecone-shaped gland located on the dorsal surface of the brain protruding partially into the third ventricle. In humans its function is not completely understood.

Pinocytosis The process by which cells take up fluid from their environment. The cells enclose some of the fluid with membrane and draw it into the cell in membrane-bound vesicles.

Piston An exact but movable fitting inserted into a hollow cylindrical chamber. It is designed to compress the volume of gases contained in the cylinder by moving down the cylinder toward the sealed end.

Pitch The quality of sound that describes whether it is high or low. It is dependent on the wave frequency producing it.

Pituicytes The spindle-shaped cells of the posterior pituitary gland that store the hormones ADH and oxytocin.

Pituitary gland (hypophysis) It has been called the "master" gland because its hormones influence the endocrine activity of several other glands. It is located in a bony pocket of the skull called the sella turcica at the base of the brain. It has two parts, the anterior portion, which arises from the buccal epithelium in the embryo, and posterior portion, which arises from an evagination of the base of the brain.

PKU Phenylketonuria.

Placenta The organ formed within the uterus during pregnancy that serves to exchange nutrients and waste materials between the fetus and mother. It is formed by tissue contributions from both fetus and mother.

Plasma The pale yellow fluid in which the cellular parts of unclotted blood are suspended.

Plasma cell The antibody-producing cell derived from divisions of the B-type lymphocyte in response to the introduction of antigen.

Plasmin An enzyme with the specific activity of dissolving blood clots by attacking the fibrin components.

Plasminogen (profibrinolysin) The inactive precursor of plasmin. It is converted to the active enzyme form by urokinase.

Platelets (thrombocytes) Formed by the fragmentation of a large bone marrow cell called a *megakaryocyte*, platelets are about 2–4 μm in size. They are a key component in blood clotting.

Pleura A continuous double layer of a serous-type membrane. The outer layer, the *parietal pleura*, lines the thoracic cavity. The folded back portion forming the inner *visceral pleura* completely encloses the lungs. The enclosed *potential* space is known as the *pleural cavity*.

Plexus A network of interwoven blood vessels, nerves, or lymphatics.

Plicae circulares Circular folds permanently intruding into the lumen of the small intestine. Formed by upfoldings of the mucosa and submucosa of the intestinal wall.

Pneumonia An inflammation of the lungs accompanied by exudation of fluid into the air passages.

Pneumotaxic center A collection of nerve cells in the upper pons that participates in the regulation of breathing. They suppress the activity of the nerve cells of the *apneustic* center located in the lower pons.

Pneumothorax The presence of air or other gas in the pleural space. This causes a separation of the pleural membranes and a collapse of the lung.

Podocytes Specialized epithelial cells of Bowman's capsule that are characterized by many footlike processes. The processes, termed pedicels, wrap around the glomerular capillaries.

Poikilothermic Able to take on a body temperature equivalent to and varying with that of the environment. Characteristic of cold-blooded animals.

Poiseuille's law The relationship between the volume of flow in a tube and the pressure, radius, length, and viscosity.

$$\text{Flow} = \frac{\Delta P \times r^4}{l \times V}$$

Polar bodies The smaller cells resulting from the unequal division of cytoplasm during the meiotic divisions of an oocyte. The polar bodies have no function and are resorbed.

Polarized A condition in which opposite effects or states exist at the same time. In electrical contexts, having one portion negative and another positive. A nerve cell membrane has the outer surface positively charged and the inner surface negatively charged. It is referred to as a *polarized membrane*.

Polar molecule Any molecule having both negatively and positively charged regions. Such molecules are water soluble since they can interact with polar water molecules.

Poliomyelitis A viral disease that in a *minor* illness is characterized by fever, headache, and sore throat. The major form involves the CNS and can produce paralysis and muscular atrophy. Largely controlled by vaccines.

Polycythemia An abnormal increase in the total number of red blood cells in the body.

Polydipsia An abnormal consumption of fluids (frequent drinking) due to a constant excessive thirst. Characteristic of diabetic illnesses with their excess loss of fluids in urine.

Polymer A molecule constructed by the linking together, in a chainlike fashion, of many simpler repeating subunits (monomers).

Polymorphonuclear cells (white blood cells) Cells in which the nucleus is deeply lobed and irregular.

Polyp A protruding growth from a mucous membrane.

Polypeptide A long chain (polymer) of amino acids. Usually identified by the number of linked amino acids (e.g., dipeptide, tripeptide, tetrapeptide). When a polypeptide is very long, it is a protein.

Polyribosome A collection of ribosomes linked together by the threading of a mRNA molecule through each ribosome.

Polysaccharide A long chain of monosaccharides. Starch and glycogen are polysaccharides consisting of the monosaccharide glucose.

Polysome An abbreviation for polyribosome.

Polyuria An excessive production of urine (frequent urination). Characteristic of diabetic illnesses.

Pons varolii (pons) That part of the central nervous system that bridges the medulla oblongata and the mesencephalon.

Positive feedback An unstable and potentially destructive situation that arises when a control system responds to input (sensory information) by *continually increasing* the error rather than by reversing and correcting it.

Posterior pituitary (neurohypophysis) The portion of the pituitary gland that is derived embryologically from neural tissue evaginating from the base of the brain. The cells, pituicytes, of the posterior pituitary gland store the hormones ADH and oxytocin.

Postganglionic neuron The conducting nerve cell that carries action potentials away from a ganglion toward the periphery. Part of the autonomic nervous system, these nerve cells have their cell bodies in a ganglion.

Postsynaptic neuron The nerve cell that is activated by the release of a neurotransmitter substance from another neuron. It carries action potentials away from the synapse.

Potential (difference) A separation of electrical charges such that if they were connected by a conductor, a current would flow.

Potential energy Energy stored or at rest that can be drawn upon to perform work.

Precipitate To bring a substance in solution, out of that solution as visible particles.

Precocious Developed further than the normal range for that age.

Precursor The preceding one. The one that comes before. In biology, it usually refers to an inactive substance that can be changed readily into an active substance.

Preganglionic neuron The nerve cell whose body is

situated in the brain or spinal cord and whose terminals release a neurotransmitter onto the cell body (or dendrites) of a second neuron. The location of the cell terminals of the first neuron and the body of the second cell is termed the ganglion.

Pregnancy The condition of having a developing embryo or fetus within the body.

Prepuce (foreskin) A retractable fold of skin that encloses the glans penis in the uncircumcised male and the clitoris in the female.

Presbyopia The loss of accommodation by the eye with increasing age. The decrease in accommodation results from loss of elasticity of the crystalline lens causing the *near point* to move farther from the eye (i.e., progressive farsightedness).

Pressoreceptors The specialized sensory devices that are located in the walls of arteries designed to measure the lateral pressure on the vascular wall, that is, a type of stretch receptor that signals pressure because the walls give.

Pressure The force per unit area. Also stress or strain.

Presynaptic neuron A nerve cell that carries action potentials toward a synapse.

Primary follicle Within the ovaries, clusters of cells surrounding an ovum.

Primary oocytes The germ cell of the female with $2n$ chromosomes that undergoes meiosis to produce the female gamete with n chromosomes.

Primary spermatocyte The male germ cell derived from the spermatogonial cell with $2n$ chromosomes that undergoes meiosis to produce the male gamete with n chromosomes.

Primary structure of a protein The programed sequence of amino acids in a protein. This information is encoded in DNA and in turn determines the secondary, tertiary, and if applicable, the quaternary structure of a protein.

Primordial The most original or primitive in character. A very simple or early form.

Procarboxypeptidase The inactive form of the enzyme carboxypeptidase. It is secreted by the pancreas into the duodenum where it is activated by another pancreatic enzyme, trypsin.

Procaryote (prokaryote) An organism that does *not* have a true (membrane-bound) nucleus. The genetic material is usually gathered in one region of the cell. Includes bacteria, blue-green algae, and mycoplasmas.

Progestins Certain synthetic and natural compounds that possess progesteronelike activities.

Proglucagon The inactive form of the hormone glucagon. It is synthesized by the alpha cells of the islets of Langerhans. It is activated just prior to secretion by enzymes within the cell.

Proinsulin The molecule synthesized on the ribosomes of the beta cells of the islets of Langerhans. Enzymes within the Golgi apparatus excise a portion of the proinsulin molecule to produce the active form of the hormone, *insulin*.

Prolactin A polypeptide hormone secreted by the anterior pituitary gland of a pregnant female to stimulate milk secretion in the mammary glands.

Pronucleus The material contained within the nuclei of the male and female gametes at the time of fertilization. Each pronucleus possesses n chromosomes so that when the zygote forms, $2n$ chromosomes are present in the nucleus.

Prophase The first stage of cell division in mitosis during which the chromosomes become obvious because of supercoiling of DNA. The first stage of meiosis also.

Proprioception The receipt of information from muscles, tendons, and the labyrinth that enable the brain to determine movements and position of the body and its parts.

Proprioceptors Specialized receptors located in the muscles, tendons, and labyrinth that provide the brain with information about body movements.

Prostaglandins A group of naturally occurring, physiologically active, chemically related, long-chain hydroxy fatty acids. They cause smooth muscle contractions in the uterus and intestine, cause blood vessels to dilate and thus drop blood pressure, and influence the activity of other hormones.

Prostate gland A male gland surrounding the urethra at its origin with the urinary bladder. It contributes a milky fluid to semen.

Prostatitis Inflammation of the prostate gland.

Protein Long-chain molecules made up of amino acids linked by peptide bonds.

Protein-bound iodine (PBI) A diagnostic test used extensively for hyper- and hypothyroidism. Since over 85 percent of the iodine in plasma is bound to protein, variations from the normal range of 4–8 $\mu g/100$ ml of serum indicate the possibility of thyroid malfunction.

Proteolytic Having activity that can break down proteins into shorter-chain-length pieces.

Proteoses A short polypeptide containing 8–12 amino acids. Proteoses are usually the result of partial protein digestion.

Prothrombin An inactive protein synthesized by the liver and released into the blood. It is converted to the active molecule thrombin in the process of blood clotting.

Prothrombin time A measurement of the combined time for the completion of stages II and III of the clotting process. Excess of calcium and tissue thromboplastin is added to citrated or oxalated plasma in a 37°C water bath and the time for clotting recorded. Normal time is between 14 and 16 seconds.

Proton (H^+) The smallest unit of positive charge. It forms the nucleus of the hydrogen atom.

Proximal Nearer the midline of the body or point of attachment.

Proximal convoluted tubule That portion of a kidney nephron that arises immediately from the blind-ended Bowman's capsule. The cuboidal cells that constitute the proximal tubule have microvilli to increase their luminal surface area. The cells are very active reabsorbers.

Psychosomatic Commonly used to refer to those physiological disorders thought to be caused entirely or partly by emotional disturbances.

Ptyalin (α-amylase) An enzyme present in saliva that attacks starches to produce maltose subunits.

Puberty The time of life during which the secondary sex characteristics begin to appear and the capability for sexual reproduction is possible. Usually occurs between the ages of 10 and 15 years.

Pulmonary Refers to the lungs.

Pulmonary circulation The blood flow from the right ventricle, into and from the lungs, and back into the left atrium.

Pulse The rhythmic expansion and recoil of the elastic arteries caused by the ejection of blood from the left ventricle. Pulse rate corresponds to the heart rate.

Pulse pressure The difference between the maximum (systolic) and minimum (diastolic) pressures. Normally a value of about 40 mm Hg.

Pupil The circular, black opening in the center of the iris of the eye. Light enters through the pupil, to focus upon the retina. The amount of light permitted is controlled by dilation or constriction of the pupil.

Purkinje network A distribution of modified muscle cells in the heart that are specialized to conduct action potentials throughout the myocardium.

Pyelitis Inflammation of the renal pelvis.

Pylorus A ring of smooth fibers located at the terminal of the stomach and the entrance of the duodenum. It functions as a sphincter to regulate the movement of food from the stomach into the small intestine.

Pyramid Any pointed or cone-shaped structure or part; for example, the medulla of the kidney has the form of pyramids.

Pyramidal tracts (corticospinal paths) Descending tracts within the spinal cord that travel through the regions of the medulla oblongata termed the pyramids.

Pyruvate The anionic form of pyruvic acid. It is the three-carbon molecule formed by the glycolytic metabolism of glucose. Cells can form lactic acid when oxygen is absent or in short supply or direct the pyruvate into the mitochondria of the cell for oxidative metabolism.

Pyuria The presence of pus in the urine.

QRS wave That portion of the electrocardiogram that records the depolarization of the ventricles.

Quadriplegia Paralysis of all four limbs.

Radioactive Having the property of emitting electromagnetic radiation as a result of nuclear disintegration.

Ramus A smaller branch given off by a larger structure such as a blood vessel, nerve, or bone.

Rapid eye movement (REM) sleep A level of sleep characterized by symmetrical flutter of the eyes and eyelids. Brain wave patterns recorded at this time are similar to those of an *awake* person, therefore often called *paradoxical sleep*.

RAS Reticular activating system.

RBC Red blood cell.

Reabsorption The recovery of dissolved substances otherwise destined for excretion. In the kidney tubule, glucose is actively transported from the filtrate back into the blood, that is, reabsorbed.

Receptive field The area surrounding a nerve terminal, which if stimulated, causes some depolarization of the nerve cell membrane.

Receptor A specialized cell or a nerve cell terminal modified to respond to some specific sensory modality, such as touch, pressure, cold, light, or sound. A specific molecule or arrangement of molecules organized to accept only molecules with a complementary shape.

Recessive Not of controlling influence. In genetics, incapable of expression unless the recessive allele is present on *both* homologous chromosomes.

Reciprocal innervation The activation of those nerve fibers that specifically *inhibit* contractions of the *antagonistic* muscles in any muscle action.

Recombinant DNA Synthetic DNA, formed by joining a fragment of DNA from one source to a portion of DNA from another.

Recruitment The inclusion of additional responding cells as the strength of stimulus increases.

Rectum The terminal portion of the large intestine where fecal material can be temporarily stored. The distension of the walls of the rectum triggers the defecation reflex and the elimination of feces.

Reduced hemoglobin (HHb) Hemoglobin molecules that have given up their oxygen molecules.

Reduction In a chemical reaction, reduction occurs when one of the substances gains electrons (i.e., is reduced in positivity).

Referred pain Pain that is felt at a site remote from the place of origin.

Reflex arc The mechanism by which control systems regulate an activity or process. Usually includes a receptor, afferent path, control center, efferent path, and effector.

Refraction The bending of light as it passes from one medium to another. The process of determining the refractive errors of the eye and then prescribing a correcting lens.

Refractory period A time during which an excitable cell cannot respond to a stimulus usually adequate to evoke an action potential.

REM sleep Rapid eye movement sleep.

Renal corpuscle A tuft of glomerular capillaries surrounded completely by Bowman's capsule. The capillaries filter plasma fluid into the kidney tubule.

Renal plasma flow The milliliters per minute of plasma that flows into the two kidneys in one minute. The value is around 700 ml/min.

Renal sinus The internal space of the kidney into which intrude the calyces, blood vessels, nerves, and fat, and which includes the space of the renal pelvis.

Renal suppression The sudden stoppage of urine production and excretion.

Renin An enzyme released by the kidney into the plasma where it converts *angiotensinogen* into angiotensin I.

Rennin An enzyme secreted by the gastric mucosa of infants which curdles milk. It converts the principal protein of milk, casein, into a more soluble form, paracasein.

Repressor A protein molecule coded for by a *regulator gene*. The repressor is able to bind to the operator gene and prevent the transcription of the adjacent structural genes.

Repressor gene A region of DNA, near an operon, that codes for the synthesis of the protein repressor molecule that inhibits that operon.

Reserpine An alkaloid extracted from the root of *Rauwolfia serpentina*. A drug used as a tranquilizer and an antihypertensive.

Residual volume The volume of air still contained in the lungs after a maximal expiration (about 1100 ml).

Resistance The restriction or hindrance of electrical charge to flow from one point to another. The hindrance encountered by blood as it flows through the vascular system.

Respiration The exchange of oxygen and carbon dioxide between the atmosphere and the cells of the body.

Respiratory cells Collections of nerve cells, located in the medulla oblongata, that initiate inspiration and expiration in an alternating fashion.

Respiratory quotient (RQ) The ratio between the amount of CO_2 produced in a metabolic process to the amount of O_2 consumed.

Respirometer An instrument used to measure the rates, depths, volumes, and patterns of respiratory movements.

Resting potential The voltage that exists between the inside and outside of a cell membrane when the cell is not responding to a stimulus. Voltage is around -70 to -90 millivolts with the inside of the cell negative.

Rete testis A network of tubules emerging from the seminiferous tubules converging and becoming the efferent ducts, which in turn become the epididymis.

Reticular activating system (RAS) An extensive network of branched nerve cells running through the core of the brain stem. When these cells are activated, a generalized alert or arousal behavior results.

Reticular formation A network of neurons running through the core of the brain stem that connects to a region of the thalamus and various ascending and descending pathways. The reticular formation includes many functional centers.

Reticulocyte A new red blood cell that still retains the remnants of an endoplasmic reticulum. Approximately one percent of all rbc's are reticulocytes. Within 1–2 days the remnants of the ER have completely disappeared.

Reticuloendothelial system (RES) A functional system that serves to protect the body by phagocytosing foreign substances. It includes all highly phagocytic cell types: Kupffer cells of the liver, macrophages, leucocytes, etc.

Retina The thin layer of light-sensitive cells, nerve fibers, and pigments that lines the inner surface of the posterior part of the eye. The retina is the primary structure for vision.

Retinene (visual yellow) The pigment portion of the photopigment rhodopsin. When light strikes the rhodopsin molecule in the rod cell, the photopigment breaks into a protein portion and retinene.

Retroperitoneal Behind the peritoneum. The kidneys and ureter are pressed against the abdominal wall behind the peritoneum (i.e., they are retroperitoneal).

Rh factor An inherited antigen that may or may not be present on the surface of red blood cells. About 85 percent of the Caucasians are Rh positive (possess the antigen). *Almost* 100 percent of the Blacks and Orientals possess the antigen.

Rhodopsin A special conjugated protein present in the rod cells of the retina. It is a photosensitive molecule that splits into a visual yellow (retinene) and a protein (scotopsin). In the dark, rhodopsin is re-formed.

Rhythm method A contraceptive technique that requires a couple to refrain from sexual intercourse from shortly before ovulation until after ovulation.

Ribosomal RNA (rRNA) The type of RNA that is transcribed from DNA in the nucleus and enters the cytoplasm to become part of a ribosome.

Ribosome A cytoplasmic particle consisting of two unequal-sized subunits each containing rRNA and proteins. These snowman-shaped particles serve as the site for protein synthesis. Sometimes the ribosomes are free in the cytoplasm and other times they adhere to the endoplasmic reticulum.

Ribonucleic acid (RNA) A polynucleotide, transcribed from DNA, that consists of nucleotides containing the sugar *ribose,* phosphate, and one of the four bases adenine, guanine, cytosine, or uracil. It is present in three types: messenger, transfer, and ribosomal. The three types cooperate to synthesize a protein molecule.

Rickets A disease occurring in infancy and childhood caused by a lack of vitamin D. It is characterized by inadequate calcification of newly formed bone. This causes bones to bend and deform.

Rigor mortis The stiffening of a dead body. As ATP is depleted, the molecular events requiring energy for calcium ion uptake and relaxation of muscle cease. This leaves the cross bridges between thick and thin filaments linked together (i.e., muscle stiffness). The condition begins 3–4 hours after death—sooner if the person had been participating in some very demanding or strenuous activity just prior to death. The condition is maximal in about 12 hours.

RNA Ribonucleic acid.

Rods One of the visual cell types in the retina, cylindrical in shape, containing rhodopsin as the visual pigment.

Round window A circular, membrane-covered opening in the bony wall that separates the inner and middle ear.

RPF Renal plasma flow.

RQ Respiratory quotient.

rRNA Ribosomal RNA.

Rubriblast (pronormoblast) The earliest recognizable cell form of the line developing into rbc's. It has a round nucleus and a uniform distribution of chromatin.

Rugae Ridges or folds. In the stomach, the large lengthwise folds of the mucous membrane seen when the stomach is empty and not distended.

Saccule A little bag or sac. The smaller of the two chambers of the vestibular portion of the membranous labyrinth of the inner ear.

Saliva The clear, alkaline, somewhat viscous secretion present in the mouth. It is produced by the three pairs of salivary glands: parotid, submaxillary, and submandibular. Contains various salts, mucin, and an enzyme.

Salivary glands The three pairs of exocrine glands that secrete a partially viscous but clear, alkaline fluid into the mouth. The fluid contains mucin and the enzyme ptyalin (α-amylase).

Salivation The process of secreting saliva. Can be a result of either a conditioned or an unconditioned reflex.

Saltatory conduction The propagation of an action potential (nerve impulse) along the exposed portions of a myelinated nerve fiber. The action potential appears at successive nodes of Ranvier and therefore seems to jump or leap from node to node.

S-A node Sino-atrial node.

Sarcolemma The cell membrane of the muscle fiber.

Sarcomere One of the repeating subunits along the length of the myofibrils of the muscle cell. It is the contractile unit of muscle delimited by two successive Z lines.

Sarcoplasm The nonfibrillar portion of the muscle cell cytoplasm.

Sarcoplasmic reticulum The endoplasmic reticulum of the muscle cell. It forms a network of saccules and tubes that surround each myofibril. It functions to actively reabsorb calcium ions during relaxation and to release them to cause contraction.

Satiety center A collection of nerve cells located in the medial portion of the hypothalamus, close to the third ventricle. When these cells are activated, the desire for food is abolished.

Saturated fatty acid A long-chain hydrocarbon consisting of repeating CH_2 groups with no double bonds linking any of the carbons.

Scala tympani The lower fluid-filled portion of the divided canal of the cochlea. It receives sound waves from the waving basilar membrane and transmits them to the round window.

Scala vestibuli The upper fluid-filled portion of the

divided canal of the cochlea. It transmits sound waves received from the deflections of the oval window and transmits them onto the basilar membrane.

Scarlet fever (scarlatina) An infection by β-hemolytic streptococci primarily of the pharynx. It produces an elevated body temperature, a generalized flush on the trunk, and pinpoint eruptions on all body parts.

Schizophrenia Any of a group of severe emotional disturbances characterized by misinterpretations, retreat from reality, and hallucinations. Schizophrenics are withdrawn persons who often show bizarre, regressive behavior.

Schwann cell The cell that wraps itself concentrically around certain peripheral nerve fibers. In the process, the cytoplasm thins out and the opposing cell membranes form a double-layered wrapping. A series of Schwann cells, in succession down the length of the nerve fiber, insulate the nerve fiber except for the small gaps called the nodes of Ranvier.

Sclera The tough, white, outer coat of the eyeball. The "white of the eye." It covers about the posterior five-sixths of the eyeball. It is continuous anteriorly with the clear cornea.

Scoliosis A lateral curvature of the vertebral column.

Scotopsin The protein portion of the visual pigment rhodopsin found in the rod cells of the retina. Scotopsin plus the visual yellow pigment retinene form rhodopsin.

Scrotum The pouch of skin and connective tissues that contains the testes and the epididymides.

Sebaceous glands Glands in the skin that secrete an oily substance called *sebum* into the hair follicles.

Sebum A thick, oily substance secreted along with cellular debris from the sebaceous glands of the skin.

Secondary sex characteristics Those features peculiar to males and females that are not directly involved in sexual reproduction. Include distribution of body hair, voice changes, body shape, muscle distribution, etc.

Secondary sex organs Those structures that are *not* responsible for *production* of sex cells but serve in the transport, protection, storage, and delivery of the ova or sperm while they are still in the body.

Secretin A strongly basic polypeptide hormone secreted by the cells of the upper small intestine in response to acid in the duodenum. It causes the pancreas to produce a watery secretion rich in sodium bicarbonate into the small intestine.

Secretion The manufacture and release of a product by a cell or collection of cells forming a gland. The end product that is released from a cell or gland.

Sedimentation rate The number of millimeters, per hour, that red blood cells can settle in a column of blood treated with an anticoagulant.

Semen The thick, whitish fluid produced upon ejaculation by the male. It is comprised of spermatozoa and the secretions of the seminal vesicles, prostate, and bulbourethral glands.

Semicircular canals Three hollow tubes within the temporal bone arranged almost at right angles to each other. They contain the membranous labyrinth with the receptors for equilibrium and circular rotation. The membranes are filled with endolymph and are cushioned in their bony tube by the perilymph outside.

Seminal vesicle A pair of glands in the male, located on either side and slightly posterior to the urinary bladder. They appear as irregular folded sacs and they secrete the bulk of the semen—a fluid rich in fructose—into the beginning of the ejaculatory duct.

Seminiferous tubules Coiled tubules contained within the testes. These tubules contain the germ cells (spermatogonia) that continually divide and thus provide a constant supply of cells that differentiate into the spermatocytes.

Semipermeable membrane A thin layer of tissue—also a cell membrane—that permits the passage of some molecules or ions but not others (a selective process characteristic of the membrane).

Sensory Refers to information detected by specialized receptors. Sensations. A *sensory* neuron relays the information from a receptor to the central nervous system.

Serotonin (5-HT, 5-hydroxytryptamine) A likely neurotransmitter formed in axon terminals from the amino acid tryptophan. It is found in greatest concentration in blood platelets and specialized cells of the intestinal mucosa. It has mood-altering properties that are antagonistic to LSD.

Serous Like serum. Watery, moist, fluid.

Sertoli cells Located within the seminiferous tubule, these cells extend from the basal lamina out into the lumen of the tubule. They are irregularly shaped cells that form a cylindrical channel with adjacent Sertoli cells by forming tight junctions with each other, thus dividing the seminiferous tubule into two concentric compartments extending the entire length of the tubule. They form a "blood-sperm barrier" that can control nutrient and hormone passage to the maturing spermatocytes.

Serum The plasma of blood that has given up its clotting components. The fluid expressed from blood that has clotted.

Sex chromosomes The pair of chromosomes (23rd pair) present in all the cells of all humans which determine the genetic sex of any individual. Designated as the X and Y chromosomes. In genetic males, there is one X and one Y chromosome. In genetic females, both chromosomes are of the X type.

Sexual intercourse Coitus. The insertion of the erect penis of a male into the vagina of a female.

Sickle-cell anemia A genetically inherited condition that results in an abnormal S-type hemoglobin molecule. In regions of low oxygen tension, the S hemoglobin causes the rbc's to fold into sickle-shaped cells. Almost exclusively in Blacks, it causes severe joint and abdominal pain, and ulcerations in the lower extremities.

Sigmoidoscope A tubular instrument filled with a source of illumination that is inserted through the anus into the sigmoid colon for visual examination.

Simmond's disease A condition resulting from a generalized hypofunction of the anterior pituitary gland. Most symptoms are due to loss of pituitary stimulation of the thyroid and adrenal cortex.

Sino-atrial node (S-A node) A collection of modified cardiac cells called P-cells that are located in the posterior wall of the right atrium. The S-A node initiates electrical impulses that cause the heart musculature to contract. Hence, its rate of impulse formation sets the rate for heartbeats.

Sinus A cavity or channel. The dilated blood vessels on the surface of the brain. Also, the air chambers in the cranial bones.

Sinusoid Modified, dilated blood channel. Found in the liver, adrenals, spleen, and pancreas.

Sodium-potassium pump (Na⁺-K⁺ ATPase) An active transport system located in the cell membrane. It transports sodium ions out of the cell and potassium ions into the cell at the expense of cellular ATP. It functions to keep the ionic concentrations of these elements at physiological levels.

Solute The substance dissolved in a solvent.

Solution A homogeneous molecular or ionic dispersion of one or more substances (solutes) in a (usually liquid) dissolving medium (solvent).

Solvent The substance (usually liquid) that dissolves some substance (solute). That component of a solution present in greater amount.

Soma The body (as distinguished from the mind). Body tissue or cells (as distinguished from the sex-cell producing germinal tissue). The cell body (as distinguished from processes, e.g., the nerve cell.)

Somatic Refers to body parts — usually the body wall — as distinguished from the internal viscera.

Somatostatin A hypothalamic hormone (regulating factor) that can inhibit the release of human growth hormone and thyroid stimulating hormone from the anterior pituitary gland. It has also been isolated from the stomach, small intestine, and delta cells of the pancreatic islets.

Somatotropin (STH) Human growth hormone secreted by the anterior pituitary gland.

Somesthetic (somatesthetic) Aware or conscious of body sensations.

Spastic Hypertonic. Muscles are tensed so that movements are awkward and not well coordinated.

Spatial summation When stimuli from different sources arrive *simultaneously* but at *different* sites on the same neuron, their effects can be algebraically summed.

Specific dynamic action (SDA) The increase in metabolic rate caused by eating. Usually about 10–20 percent. It appears to be due to the processing of absorbed nutrients by the liver.

Specific gravity The weight of some substance compared with the weight of an *equal volume* of another standard substance (usually water, which has been given a value of 1.000).

Specific heat The amount of heat required to raise the temperature of 1 g of a substance 1°C.

Spermatic cord The structure extending from the inguinal ring into the scrotum containing the vas deferens, testicular arteries, veins, nerves, lymphatics, and various connective tissue coats.

Spermatids The cells resulting from the division of the secondary spermatocyte. These cells mature to become the spermatozoa (spermiogenesis).

Spermatogenesis The entire process describing the production of spermatozoa — from the initial spermatogonial division to the release of spermatozoa into the rete testes, efferent ducts, and epididymis.

Spermatogonia The undifferentiated male germ cells located in the basal layer of the seminiferous tubules. From this division of these cells come the primary spermatocytes.

Spermatozoan The mature male germ cell produced in the seminiferous tubules of the testes.

Spermiogenesis The maturing process that converts spermatids into mature spermatozoa.

Sphincter A ringlike band of muscle fibers that restricts or closes completely some opening.

Sphygmomanometer An instrument for measuring blood pressure in the arteries.

Spirilla The somewhat rigid spiral-shaped bacteria. Organisms of the genus *Spirillum*.

Spirometer An instrument used for measuring and recording (a spirogram) the breathing patterns of a subject.

Splanchnic Refers to the viscera. Usually any of the large organs in the abdominal cavity (sometimes includes those in the thoracic and pelvic cavities also).

Spore The reproductive structure of fungi, algae, protozoa, etc.

Stapedius A tiny muscle of the middle ear that inserts on the stapes and serves to dampen its movements.

Stapes (stirrup) A small bone of the middle ear that has a stirrup-shaped end that inserts into the oval window.

Starch A polysaccharide obtained from a variety of plants. It is a polymer containing thousands of subunits of glucose.

Starling's law of the heart The stroke volume is directly proportional to the diastolic filling. The energy available for the initial length of the fibers in the muscular walls of the heart.

Statoliths (otoliths) Minute particles of calcium carbonate embedded in the gelatinous mass covering the hair cells of the utricle and saccule.

Stenosis An abnormal narrowing or constriction of a duct or opening.

Sterile Free from any living microorganisms. Unable to conceive or produce offspring.

Sterilization Elimination of all living microorganisms. The rendering of an individual to be incapable of reproduction (e.g., castration, vasectomy, hysterectomy).

Steroid Any of a group of compounds that contain the basic cyclopentophenanthrene ring—four interconnected carbon ring structures. The hormones of the adrenal cortex, sex hormones, cholesterol, and some of the bile salts are steroids.

Stimulus Any irritant, agent, or influence that produces any response or reaction in a receptor, tissue, organ, or organism.

Strabismus A condition in which the visual axes of the two eyes differ. They do not both fix on the same object.

Stress Any environmental influence that causes a physiological response. The concept of stress requires a kind of alert response by the body that can be brief or very prolonged. One component of the response is a release of adrenal steroids.

Stressor Any environmental change that causes stress to an individual.

Stretch reflex (myotatic reflex) A monosynaptic reflex triggered by a sudden stretch of a muscle and ending with a contraction of that same muscle.

Stroke A sudden and severe attack. Usually it refers to sudden vascular lesions in the brain. A hemorrhage, embolism, or thrombosis may deprive some region of the brain of its blood supply. The resulting brain damage can result in paralysis of some body part, vertigo, loss of speech and comprehension, etc.

Stroke volume The milliliters of blood ejected by either ventricle in one contraction (about 70 ml). This volume times the number of beats per minute defines *cardiac output*.

Stroma The supporting connective tissue network in an organ as distinguished by the functional cell characteristic of that organ.

Subatmospheric pressure Any pressure lower than the surrounding environmental (atmospheric) pressure. The pressure in the pleural cavity is subatmospheric.

Subcutaneous Beneath the skin. Hypodermic.

Subliminal Below the threshold. Used most often to describe a strength of stimulus that does not evoke a response or reach a conscious level.

Sublingual glands One of the three pairs of salivary glands. They are located under the tongue. They secrete a mucus-type secretion.

Submandibular gland (submaxillary gland) One of the three pairs of salivary glands. They are located in the mouth just medial to the angle of the mandible. These glands secrete a mixture of both mucus and serous types of secretions.

Substrate A substance upon which an enzyme acts.

Subthreshold A strength of stimulus that causes some depolarization of an excitable cell membrane but not sufficiently to reach the threshold and cause an action potential.

Succus entericus The intestinal juice produced by the glands in the wall of the small intestine.

Sucrase An enzyme that hydrolyzes the disaccharide, sucrose, into the two monosaccharides, glucose and fructose.

Sucrose The disaccharide sugar used as ordinary table sugar. It can be hydrolyzed to produce the monosaccharides fructose and glucose.

Sulcus A groove, valley, furrow. Used to describe the irregular crevices on the surface of the brain.

Summation The algebraic addition of the excitatory and inhibitory effects of many stimuli applied to a nerve cell body. The increased strength of muscle contraction that results when stimuli follow in rapid succession.

Suppuration The formation and discharge of pus.

Supraoptic nucleus A collection of nerve cells in the hypothalamus immediately above the optic chiasma. Its cells secrete a hormone that is transported down the nerve fibers to the posterior pituitary gland.

Surface tension The increased force of attraction between molecules on the surface of a fluid (e.g., water) that causes a nonstretchable "skin" to form on the surface. The same nonstretchable force exists on the moistened surface of the lung alveoli.

Surfactant   Any chemical that is able to reduce the surface tension of fluids. In the lungs, alveolar cells secrete a phospholipid surfactant, that reduces the surface tension of the moistened alveolar surface, allowing greater lung compliance.

Symbiosis   Two dissimilar organisms living together or in close association.

Sympathetic nervous system   The division of the autonomic nervous system whose preganglionic nerve fibers emerge from the central nervous system only from the thoracic and lumbar regions of the spinal cord. The postganglionic fibers release norepinephrine from their terminals.

Sympathomimetic   Producing effects similar to those evoked by stimulating fibers of the sympathetic nervous system.

Synapse   The anatomical location that shows the junction between the processes of two adjacent neurons. The place where the activity of one neuron affects the activity of another.

Synaptic cleft   The narrow gap separating the membrane of an axon terminal of one nerve cell and the membrane of another nerve cell. It is across this gap that the neurotransmitter diffuses to affect the postsynaptic cell.

Synaptic delay   The length of time between the arrival of the action potential at the axon terminal and the membrane potential (IPSP or EPSP) change on the postsynaptic membrane. Usually less than one millisecond.

Syncytium   A collection of identical cells whose cytoplasms have become continuous. The cell membranes between the adjacent cells have been removed and become one continuous membrane.

Syndrome   A set of clinical symptoms that occur together to define an illness.

Syneresis   An internal squeezing process caused by the dispersed component in a gel (e.g., the shrinking of the fibrin threads in a blood clot).

Synovial fluid   A transparent, viscous liquid resembling egg white. It is secreted by the lining layer of synovial joint cavities and bursal and tendon sheaths for protection and lubrication purposes.

Syphilis   A contagious venereal disease caused by the spirochete *Treponema pallidum*. It is transmitted by sexual intercourse. It is divided into primary, secondary, and tertiary stages starting 10 days after infection with swollen lymph glands, and within 2 months fever, skin eruptions, head and joint pains, and eventually, maybe years later, it affects the cardiovascular and nervous systems.

System   Interdependent parts, tissues, organs, and cells that function together to achieve a goal that if acting alone no part could achieve (e.g., the nervous system, digestive system, urinary system).

Systemic circulation   The flow of blood throughout all body parts. It starts from the *left* ventricle of the heart, goes through arteries, arterioles, capillaries, and veins, and finally returns to the *right* atrium.

Systemic lupus erythematosus   An autoimmune disease of connective tissues affecting mainly middle-aged women. It is characterized by skin eruptions, joint pains, fever, and leukopenia. Also shows a very high level of antibodies in circulation.

Systole   The period or time of muscular contraction in the heart. Atrial systole describes contraction of atrial musculature. However, it normally applies to *ventricular contraction* when the term is used alone.

Systolic pressure   The highest pressure measured in the larger arteries. It corresponds to the period of ventricular contraction.

Tachycardia   An excessively rapid heart rate—usually over 100 beats per minute.

Taeniae coli   The three bands of longitudinally arranged smooth muscle in the colon.

Target organ (cell)   The structure responsive to the influence of a hormone. Those cells having the receptor molecules for a hormone.

Tay-Sachs disease   An inherited disorder transmitted as an autosomal recessive strain chiefly affecting children of Jewish ancestry. A defect in lipid metabolism causes an accumulation of sphingolipids in the brain. Symptoms include progressive mental degeneration, loss of vision, paralysis, and death.

T cells   Lymphocyte cells whose precursors were "programed" in the thymus gland during embryonic development prior to their seeding in the spleen, lymph nodes, and other lymphoid tissues. They function in cell-mediated immunity.

Tectorial membrane   A gelatinous membrane in the organ of Corti that overlies and makes contact with the hairs on the receptor cells.

Telodendria   The many small terminal branches of a neuron.

Telophase   The last of the four stages of *mitosis* and the last division of *meiosis*. Defined as the time when the chromosomes arrive at the poles of the cell and cytoplasmic division begins.

Temporal summation   The increased effect on the excitability of a nerve cell membrane caused by stimuli from a particular presynaptic terminal arriving in rapid succession.

Tendon A fibrous bond or cord that attaches a muscle to bone.

Tension The condition of being pulled or being stretched.

Tensor tympani A tiny muscle of the inner ear that tenses the tympanic membrane by pulling on the malleus.

Tertiary structure of proteins The arrangement the long polypeptide assumes by coiling and folding and interacting with different regions or portions of itself. Its final three-dimensional shape.

Testes The paired gonads of the male located outside the body in the scrotum. Testes produce spermatozoa and secrete male sex hormones.

Testosterone The male sex hormone produced by the interstitial cells of the testes. This steroid hormone is responsible for the proper development of the reproductive organs and secondary sex characteristics.

Tetanus An infectious disease caused by the toxin of *Clostridium tetani*. Characterized by tonic muscle spasms and exaggerated reflexes, lockjaw, and arching of the back. A smooth, sustained contraction produced by a series of very rapid stimuli to a muscle.

Tetany A condition due to abnormally low Ca^{2+} levels, parathyroid hypofunction, vitamin D deficiency, and alkalosis. This causes muscle twitches, cramps, convulsions, and a characteristic carpopedal spasm (severe flexion of wrist and ankle joints).

Tetralogy of Fallot A combination of four congenital heart defects: (1) the pulmonary artery is constricted, (2) there is a hole in the interventricular septum, (3) the aorta is moved to the right side so that it receives blood from both ventricles (oxygenated and deoxygenated blood), and (4) the *right* ventricle is enlarged.

Thalamus A large portion of the diencephalon that partly forms the walls of the third ventricle. All sensory input (except smell) makes its last synaptic connection here on its way to the cerebral cortex.

Thebesian vessels Numerous small veins in the muscular walls of the heart that deliver their blood directly into the heart chambers.

Thick myofilaments Part of the myofibrils, they are in the central region of the sarcomere; long, narrow, 12–18-nm diameter structures with angled cross bridges at each end. These thick myofilaments consist of a protein called myosin. Together with the thin myofilaments, they constitute the contractile machinery of the muscle cell.

Thin myofilaments Part of the myofibrils, they extend at right angles from the Z lines. They are 5–8 nm in diameter and contain three main proteins, actin, tropomyosin, and troponin. Together with the thick myofilaments they constitute the contractile machinery of the muscle cell.

Third ventricle A tall, narrow cavity located within the diencephalon between the two thalami. Its floor is the hypothalamus. It communicates posteriorly through the cerebral aqueduct with the fourth ventricle and anteriorly and laterally with the two lateral ventricles via the interventricular foramena.

Thoracic duct The lymphatic vessel that drains the entire lower half and upper left side of the body. It empties into the left subclavian vein just where it joins with the left internal jugular vein.

Threshold potential The membrane voltage that must be reached in order to trigger an action potential.

Thrombin The active enzyme formed from prothrombin which acts to convert fibrinogen to fibrin.

Thrombocytes Blood platelets.

Thromboplastin A factor, or collection of factors, whose appearance initiates the blood clotting process.

Thrombosis The development or presence of a thrombus (blood clot) in a blood vessel. The blood clot often plugs the vessel and prevents any passage of blood past the thrombus.

Thrombus A blood clot within a blood vessel.

Thymine One of the nitrogenous pyrimidine bases in DNA.

Thymus A ductless glandlike body located high in the anterior portion of the thorax. Lymphoid in nature, it is where lymphocytes are "programed" to become T-type cells and mediate cellular immunity.

Thyroglobulin The long protein molecule synthesized within the follicles of the thyroid gland. It is the tyrosine amino acids in the molecule that are iodinated to eventually become the active hormone molecule.

Thyroid gland A bilobed endocrine gland situated in the neck astride the thyroid cartilage. It secretes the hormones thyroxine, triiodothyronine, and thyrocalcitonin.

Thyroiditis Inflammation of the thyroid gland. Can be caused by staphylococcus, streptococcus, or many other bacteria.

Thyroxine (3,5,3',5' tetraiodothyronine) One of the active hormones synthesized and released by the thyroid gland. It is a molecule containing four iodine atoms. Its main effect is to increase the metabolic rate of affected cells.

Tidal volume The volume of air breathed in and out in any one breath. The tidal volume changes with the level of activity. In quiet, resting conditions, the volume is about 500 ml.

Tight junction One of several types of connection between cells. Found in epithelial tissues, the tight junction consists of a sharing of membrane proteins between adjacent cells. It forms a continuous fusion around the circumference of each cell and thus prohibits the passage of molecules between adjacent cells.

Tissue A collection of similarly specialized cells held together by an intercellular material. Four classes of tissue are epithelial, connective, muscular, and nervous.

T lymphocytes T cells.

Tolerance The requirement for increased amounts or strength of a drug to achieve the same initial effect. The state of immunologic acceptance of a foreign molecule, tissue, or organ, for example, in tissue or organ transplants, such as skin grafts and kidney transplants.

Tone A degree of stretch or tension in a muscle. Often regarded as a state of readiness to respond quickly.

Tonic Referring to a condition of *continual* tension or stretch.

Tonofibrils Seen with the electron microscope, fine filamentous structures in the cytoplasm, anchored in the region of the desmosome.

Tonsil A round mound of lymphoid tissue, usually referring to those in the pharynx; palatine, pharyngeal, and lingual.

Toxin A poison. Generally refers to an animal, plant, or bacterial protein product that inhibits some key synthetic process.

Trachea The membranous and cartilaginous tube in the neck which continues the airway from the larynx down to the right and left bronchi.

Tracheostomy An incision of the trachea.

Trachoma A chronic infectious disease of the conjunctiva and cornea of the eye caused by a strain of the bacteria *Chlamydia trachomatis*. It is painful and causes lacrimation and an intolerance to light.

Tract A collection of myelinated nerve fibers within the central nervous system having the same origin, destination, and function.

Transamination The transfer of an *amino* group from an amino acid to an alpha-keto acid. The end result is that the original amino acid becomes an alpha-keto acid and the original alpha-keto acid becomes an amino acid.

Transcription The first transfer of genetic information from DNA. The copying of a DNA gene into a mRNA molecule.

Transferrin A β-globulin in plasma that binds and transports iron.

Transfer RNA (tRNA) A linear ribonucleotide poly-

mer containing about 80 subunits. Structurally it forms internal complementary base-pairing sequences with three regions of outlying loops. One loop contains the *anticodon* to base-pair with the *codons* of the mRNA. There are 64 types of tRNA molecules and each binds to only one amino acid, which it carries to a ribosome for insertion into a newly forming polypeptide.

Translation The construction of a new protein on the ribosome of a cell as dictated by the sequence of codons in the mRNA.

Transmitter substance One of a variety of molecules synthesized within the nerve axon terminals and released from there by the arrival of an action potential. The molecules affect the membrane potential of the next neuron.

Transport maximum (T_{max}) The fastest rate at which cells are able to actively carry molecules through their membranes. In the kidney, the maximum rate of reabsorption or secretion.

Trauma Any physical or psychic injury.

Treppe (staircase phenomenon) The gradual increase in the amount of contraction by a muscle caused by rapid, repeated stimuli of the same strength.

Tricarboxylic acid cycle (TCA cycle) Krebs cycle. Citric acid cycle.

Trichomoniasis An infection with *Trichomonas*, a parasitic flagellate protozoan. Found in the vagina, and in the urinary bladder and urethra of males.

Triglyceride A molecule formed by esterification of three molecules of fatty acid to glycerol. It is the fat of adipose tissue and can be formed from carbohydrate excess.

Trigone A triangular area. In the urinary bladder, a smooth triangular portion of the mucous membrane at the base of the urinary bladder.

Triiodothyronine (3,5,3′) The most active of the thyroid hormones synthesized and released by the thyroid gland. It contains three iodine atoms. Its main effect is to increase the metabolic rate of affected cells.

tRNA Transfer RNA.

Trophic Nourishing. In the endocrine sense, it generally means stimulating.

Trophoblast The outer layer of cells of the blastocyst. It anchors the ovum to the uterine wall and then digests its way deeper to establish connections with the maternal tissues. Initiates the formation of the fetal portions of the placenta.

Tropocollagen The subunit from which collagen forms. Secreted by fibroblast cells, it consists of three polypeptide chains twisted together.

Tropomyosin A protein component of the thin myofilaments in the muscle fibrils. It is a molecule formed from two polypeptides twisted around each other. The length of the molecule corresponds to the length of seven G-actin monomers. It functions to conceal the binding sites on the thin myofilament to which the myosin cross bridges bind.

Troponin A globular protein component of the thin myofilaments in the muscle fibril. It contains three functional sites so that it can bind to the actin chain, to the tropomyosin, and also to calcium ions. It initiates the contractile events when it binds calcium ions.

True labor The process by which the pregnant female expels the fetus from the uterus through the vagina to the outside world. Delivery. Childbirth.

Trypsin One of the proteolytic enzymes secreted by the pancreas. It is an endoproteinase, hydrolyzing linkages containing the carboxyl group of either lysine or arginine. It is secreted in an inactive form called trypsinogen.

Trypsin inhibitor A molecule secreted by the pancreatic acinar cells that prevents the activation of trypsinogen *within* the secretory cells.

Trypsinogen The inactive form of the enzyme trypsin secreted into the duodenum from the pancreas.

T tubules (transverse tubules) Minute, cylindrical invaginations of the muscle cell membrane that carry the surface action potentials deep into the muscle cell. Their close relationship to the sarcoplasmic reticulum causes the release of calcium ions from the interior of the sarcoplasmic reticulum when the electrical impulses arrive.

Tuberculosis An infectious disease caused by the bacterium *Mycobacterium tuberculosis*. It is characterized by the formation of small, grayish, rounded nodules called *tubercles*. Any tissue or organ can be affected, but the most common location is the lungs.

Tumor A new growth of tissue caused by an uncontrolled multiplication of altered cells. It can eventually form an obvious palpable mass.

Turner's syndrome A condition resulting from the *absence* of one of the sex chromosomes with a resulting female phenotype. The afflicted individual is short, possesses undifferentiated gonads, and suffers cardiac defects.

T wave The portion of the electrocardiogram that records the repolarization of the ventricles.

Tympanic antrum The cavity of the middle ear.

Tyrosine An amino acid. It is synthesized from the essential amino acid phenylalanine. It is important as a precursor for the synthesis of the catecholamines, epinephrine, norepinephrine, dopamine, and the thyroid hormones thyroxine and triiodothyronine.

Umbilical cord The long ropelike structure, containing the umbilical arteries and vein, which connects the fetus to the placenta. In the newborn it is about 2 feet long and $\frac{1}{2}$ inch in diameter.

Umbilicus The navel, or "belly button." The site where the umbilical cord was attached to the fetus.

Unilateral One-sided. Affecting or occurring only on one side.

Unmyelinated Refers to nerve fibers lacking myelin.

Unsaturated fatty acid A long-chain hydrocarbon mostly consisting of repeating CH_2 groups with at least one double bond linking two of the carbons in the molecule. A double bond is indicative of the possibility for adding two more hydrogen atoms, that is, it is *not saturated* with respect to the number of hydrogens it could absorb.

Uracil One of the nitrogenous, pyrimidine bases found in RNA but *not* in DNA.

Urea The major nitrogenous end-product of protein metabolism and the major nitrogen-containing component of urine. It is formed in the liver from ammonia obtained from the deamination of amino acids and carbon dioxide.

Urea (ornithine) cycle The sequence of enzyme-catalyzed reactions that describes the synthesis of urea in the liver from ammonia and carbon dioxide.

Uremia A toxic condition produced by the excessive accumulation of urea and other nitrogenous wastes in the blood. Characterized by nausea, headache, vertigo, dim vision, and coma.

Ureter The fibromuscular tube that conveys urine from the renal pelvis to the urinary bladder.

Urethra The tube conveying urine from the urinary bladder to the exterior. In the male it also conveys semen.

Uric acid A nitrogenous waste material derived from the metabolism of purine bases and excreted in the urine.

Urinalysis The physical, chemical, and microscopic analysis or examination of urine.

Urinary bladder A musculomembranous sac located in the anterior portion of the pelvic cavity. It serves to hold urine, which it receives via the ureters and releases through the urethra.

Urine The fluid excreted by the kidney that contains wastes or excess materials and is discharged from the body through the urethra.

Urinometer An instrument for measuring the specific gravity of urine.

Uterine tube Fallopian tube or oviduct. The hollow tube through which the ovum travels from the ovary into the uterus. Fertilization of the ovum normally occurs in the uterine tube.

Uterus The womb. The hollow, smooth muscular walled organ in females in which the developing embryo and growing fetus are nourished. Its cavity opens into the vagina below.

Utricle The larger of the two divisions of the membranous labyrinth. It contains receptors that respond to changes of position and linear acceleration.

Uvula A hanging fleshy mass. Usually refers to the *palatine uvula* that hangs from the soft palate.

Vacuole Any small membrane-bound cavity present in the cytoplasm of the cell.

Vagina The canal in the female extending from the vestibule to the cervix of the uterus. It accepts the penis in sexual intercourse.

Vaginismus Painful spasm of the vagina due to a local hypersensitivity.

Vagotomy Cutting of the vagus nerve or nerves.

Vagus nerve The tenth cranial nerve. It is a mixed nerve, about 90 percent afferent and 10 percent efferent.

Vallate Walled. Having a wall or rim.

Varicose vein An unnatural, permanently distended or stretched vein.

Vasa recta The long U-shaped capillaries that parallel the *long* loops of Henle in the kidney medulla.

Vasa vasorum Small nutrient blood vessels in the walls of the large arteries; "the vessels of the vessels."

Vascular Referring to blood vessels.

Vas deferens The two tubes in the male that carry spermatozoa and the secretion of the seminal vesicles from the epididymis to the ejaculatory duct.

Vasectomy The surgical removal of part or all of the vas deferens. Often done as a birth control method.

Vasoconstriction The narrowing of the lumen of vessels. Causing the diameter of the vessels to decrease.

Vasodilation A greater opening of the lumen of blood vessels. Causing the diameter of vessels to increase.

Vasomotor Changes in the caliber of blood vessels. Causing changes to occur in the diameter of blood vessels—either vasoconstriction or vasodilation.

Vasopressin ADH. Antidiuretic hormone. In addition to its effects on the distal tubules and collecting ducts, it causes the contraction of smooth muscles in blood vessels, digestive tract, and uterus.

Vein A blood vessel that returns blood to or toward the heart.

Venereal disease A contagious disease, acquired by sexual intercourse. Includes *syphilis* and *gonorrhea*.

Ventilation The process of exchanging the air in the lungs with air from the environment.

Ventricle A small cavity. The ventricles of the heart and of the brain.

Venule Any small vessel that receives blood from capillaries. Venules are distinguished from capillaries by the presence of some small amount of connective tissue in their walls.

Vertigo An *illusion* of movement. The sensation that the world is rotating around you, or that you are revolving in space.

Vesicle A small bladder or sac containing liquid. The smallest membrane-enclosed sacs in the cytoplasm of cells (e.g., pinocytic vesicles).

Vestibular apparatus The organ of balance. It includes the membranous sacs of the semicircular canals and vestibule (utricle and saccule).

Vestibule A space or cavity located at the entrance to a canal.

Vestigial Referring to the remnant of some structure that functioned in some previous stage of development of an individual or a species.

Villus A small fingerlike projection from the free surface of a membrane. In the small intestine villi are formed by the upfolding of the submucosal layer. They serve to increase the absorptive surface area.

Virus A complex particle basically consisting of a nucleic acid core and a protein coat. These particles are able to reproduce inside living cells by taking control of their genetic machinery.

Viscera Collectively, all the large internal organs contained in the three great body cavities (thoracic, abdominal, and pelvic). However, in most instances it refers to those in the *abdominal* cavity.

Viscosity A property of matter that describes its inner slipperiness. The friction of the component molecules as they slip over each other.

Visual acuity The minimal distance between two lines that can still be perceived as two lines. It is usually measured with the aid of eye charts.

Vitamins Organic molecules necessary for normal metabolic processes of the body. They are obtained *only* in the diet and are divided into water-soluble and fat-soluble classes.

Vitreous humor The colorless, transparent gel that fills the part of the eye between the lens and the retina.

Vocal cords A pair of horizontal elastic connective tissue folds in the larynx that extend from the left and right sides to meet in the middle of the laryngeal cavity. The space between the two folds is the *glottis*, and all air moving into and out of the lungs passes through it.

Volt   The electrical "driving force." The force necessary to cause one ampere of current to flow against a resistance of one ohm.

Voltage   The electrical potential or electromotive force measured in volts. It measures the potential of separated charges to do work.

Vulva   Collectively, the external female genitalia. It includes the labia majora and minora, mons pubis, clitoris, vestibule of the vagina, vestibular glands, and vaginal orifice.

Wavelength   The distance between the top of one wave and the identical position on a succeeding one.

Weight   Heaviness. The force with which a body is pulled toward the earth by gravity.

White matter   Those parts of the central nervous system that appear white. Made up mostly of myelinated nerve fibers.

Womb   Uterus.

Yin-yang hypothesis   A concept that proposes that equal and opposite forces exist to control processes and behavior. For example, in physiology, sympathetic vs. parasympathetic, cyclic AMP vs. cyclic GMP.

Yolk sac   The extra embryonic membrane that connects with the midgut during early embryonic development.

Z line   A dark, narrow, transverse striation seen in the myofibrils of skeletal muscle. These lines repeat at regular intervals along the length of the fibril. The region between any two Z lines defines a sarcomere. Protruding at right angles to the Z lines are the thin myofilaments.

Zona fasciculata   The middle layer of the adrenal cortex where the glucocorticoid hormones are secreted.

Zona glomerulosa   The outer layer of the adrenal cortex where the mineralocorticoid hormones are secreted.

Zona pellucida   The transparent, noncellular layer of secreted material that surrounds the oocyte in an ovarian follicle.

Zona reticularis   The innermost layer of the adrenal cortex where the steroids with sex hormone activity are secreted.

Zygote   The single cell resulting from the union of a male and female gamete — the sperm and ovum. The fertilized ovum.

INDEX

Page numbers in italics indicate items to be found in artwork. Page numbers followed by the letter *E* indicate items to be found in exhibits.